15th International Congress on Plasma Physics (ICPP 2010) and 13th Latin American Workshop on Plasma Physics (LAWPP 2010)

Journal of Physics: Conference Series Volume 511

Santiago, Chile
8-13 August 2010

ISBN: 978-1-63439-291-4
ISSN: 1742-6588

Printed from e-media with permission by:

Curran Associates, Inc.
57 Morehouse Lane
Red Hook, NY 12571

Some format issues inherent in the e-media version may also appear in this print version.

Printed by Curran Associates, Inc. (2014)

For permission requests, please contact the Institute of Physics
at the address below.

Institute of Physics
Dirac House, Temple Back
Bristol BS1 6BE UK

Phone: 44 1 17 929 7481
Fax: 44 1 17 920 0979

techtracking@iop.org

Additional copies of this publication are available from:

Curran Associates, Inc.
57 Morehouse Lane
Red Hook, NY 12571 USA
Phone: 845-758-0400
Fax: 845-758-2634
Email: curran@proceedings.com
Web: www.proceedings.com

TABLE OF CONTENTS

FUNDAMENTAL PLASMAS

COLD PLASMAS

DENSE PLASMAS, PINCHES, BEAMS AND PULSED POWER

FUSION PLASMAS

ASTROPHYSICS AND SPACE PLASMAS

LASER PLASMA INTERACTION

PLASMA APPLICATIONS, MATERIALS AND PLASMA DIAGNOSTICS

Journal of Physics: Conference Series **511** (2014) 011001 doi:10.1088/1742-6596/511/1/011001

Preface

Written contributions from the participants of the 15th International Congress on Plasma Physics combined with the 13th Latin American Workshop on Plasma Physics

The International Advisory Committee of the 15th International Congress on Plasma Physics (ICPP 2010) and the International Advisory Committee of the 13th Latin American Workshop on Plasma Physics (LAWPP 2010), together agreed to carry out this combined meeting ICPP-LAWPP-2010 in Santiago de Chile, 8–13 August 2010, on occasion of the Bicentennial of Chilean Independence. The ICPP-LAWPP-2010 was organized by the Thermonuclear Plasma Department of the Chilean Nuclear Energy Commission (CCHEN) as part of the official program within the framework of the Chilean Bicentennial. The event was also a scientific and academic activity of the project "Center for Research and Applications in Plasma Physics and Pulsed Power, P4", supported by National Scientific and Technological Commission, CONICYT-Chile, under grant ACT-26.

The International Congress on Plasma Physics was first held in Nagoya, in 1980, and followed by the Congresses: Gothenburg (1982), Lausanne (1984), Kiev (1987), New Delhi (1989), Innsbruck (1992), Foz do Iguacu (1994), Nagoya (1996), Prague (1998), Quebec City (2000), Sydney (2002), Nice (2004), Kiev (2006), and Fukuoka (2008). The purpose of the Congress is to discuss the recent progress and future views in plasma science, including fundamental plasma physics, fusion plasmas, astrophysical plasmas, and plasma applications, and so forth.

The Latin American Workshop on Plasma Physics was first held in 1982 in Cambuquira, Brazil, followed by the Workshops: Medellín (1985), Santiago (1988), Buenos Aires (1990), Mexico City (1992), Foz do Iguacu (1994, also combined with ICPP), Caracas (1997), Tandil (1998), La Serena (2000), Sao Pedro (2003), Mexico City (2005), and Caracas (2007). The Latin American Workshop on Plasma Physics is a communication forum of the achievements of the plasma-physics regional community, fostering collaboration between plasma scientists within the region and elsewhere.

The program of the ICPP-LAWPP-2010 included the topics: Fundamentals of Plasma Physics, Fusion Plasmas, Plasmas in Astrophysics and Space Physics, Plasma Applications and Technologies, Complex Plasmas, High Energy Density Plasmas, Quantum Plasmas, Laser-Plasma Interaction and among others.

A total of 180 delegates from 34 different countries took part in the ICPP-LAWPP-2010. Sixty delegates received economical assistance from the local organized committee, thanks to the support of the International Union for Pure and Applied Physics (IUPAP) and the Chilean Nuclear Energy Commission (CCHEN).

The ICPP-LAWPP-2010 Program was elaborated by the following Program Committee:

Carlos Alejaldre, ITER
Maria Virginia Alves, Brazil
Julio Herrera, Mexico
Günter Mank, IAEA
George Morales, USA
Padma Kant Shukla, Germany
Guido Van Oost, Belgium
Leopoldo Soto, Chile (Chairman)

This Program Committee was formed by selected members from the International Advisory Committee of the ICPP and by selected members from the International Advisory Committee of the LAWPP. In particular, Plenary Lectures and Invited Topical Lectures were selected by the Program Committee from a list of nominated presentations by the International Advisory Committees of both ICPP and LAWPP. Also, the classification of oral and poster presentations was elaborated by the Program Committee. The congress included: 15 invited plenary talks, 33 invited topical talks, 45 oral contributions, and 160 poster contributions.

A major part of the plenary and topical lectures were published in a special issue of the *Plasma Physics and Controlled Fusion*, IOP Publishing (*Plasma Phys. Control Fusion* Volume 53, Number 7, July 2011: http://iopscience.iop.org/0741-3335/53/7). The papers were refereed according to the standards of the journal Plasma Physics and Controlled Fusion.

An large number of the participants sent their contributions articles to this volume of *Journal of Physics: Conference Series*, IOP Publishing. The articles received were reviewed by the local organizing committee and by invited peers. The criteria for review focused on the demand for a consistent research and the clear statement of results. Most of the articles received report the work of research groups where advanced students and young investigators are prominent. Thanks to their enthusiasm, we would like to express our appreciation to the authors.

Previous to the ICPP-LAWPP 2010, an important activity associated to the Latin American Workshop on Plasma Physics took place. This activity was the LAWPP School on Plasma Physics, which was open to participants from over the world, providing basic training to students and young researchers. The School was attended by 44 participants and 6 lecturers from 11 different countries. All participants received economical assistance from the local organizing committee. The topics covered by the school were: general description of plasmas, space and astrophysical plasmas, plasma diagnostic techniques, high temperature and fusion plasmas, and low temperature and industrial plasmas. The organizers of the ICPP-LAWPP-2010 are grateful to the lectures of the LAWPP Plasma Physics School: Luis Felipe Delgado-Aparicio (USA), Homero Maciel (Brazil), and Marina Stepanova, J Alejandro Valdivia, Victor Muñoz, Felipe Veloso, Leopoldo Soto from Chile.

On 27 February, 2010, one of the worst earthquakes in the recorded history of the world struck Chile. Although Santiago was affected little, the region located 200 km South of Santiago was seriously damaged. After this event, the local organizing committee received many messages from members of the plasma physics community around the world expressing their concern. The local organizing committee greatly appreciates the support of the participants from the entire world that decided to come to Chile and attend the Conference. Their solidarity is highly appreciated.

During the celebration of the ICPP-LAWPP in Chile the two pioneers of plasma physics in Chile were affected by grave illness. Albeit that, Dr Hernán Chuaqui, pioneer of experimental plasma physics in Chile participated in the meeting. Alas, Dr Luis Gomberoff, pioneer of the theoretical plasma physics in Chile could not attend. Sadly, Professor Gomberoff died in September 2010 and Professor Chuaqui in July 2012. We would like to remember them with admiration.

Journal of Physics: Conference Series **511** (2014) 011001 doi:10.1088/1742-6596/511/1/011001

The Chairman of the ICPP-LAWPP-2010 is grateful to the members of the Local Organizing Committee of the conference: Karla Cubillos, José Moreno, Cristian Pavez, Felipe Veloso, Marcelo Zambra, Luis Huerta, and Fabian Reyes and to the members of the Program Committee for their work and commitment.

Finally, my personal apology is in order regarding the delay in publishing these proceedings due to an unfortunate sequence of personal and professional circumstances. I would like to thank the *Journal of Physics: Conference Series* for the fast publication of the proceedings, in particular to Ms Sarah Toms for her excellent work and cooperation.

Leopoldo Soto
Chairman of the ICPP-LAWPP-2010
Chilean Nuclear Energy Commission, Chile

Journal of Physics: Conference Series **511** (2014) 011001 doi:10.1088/1742-6596/511/1/011001

Program Committee ICPP-LAWPP-2010

Carlos Alejaldre, ITER
Maria Virginia Alves, Brazil
Julio Herrera, Mexico
Günter Mank, IAEA
George Morales, USA
Padma Kant Shukla, Germany
Guido Van Oost, Belgium
Leopoldo Soto, Chile

ICPP International Advisory Committee

M Virginia Alves, Brazil
Sadruddin Benkadda, France
Robert Bingham, UK
Susana Cappello, Italy
Diego del Castillo-Negrete, USA
Sandra C Chapman, UK
Christine Charles, Australia
C Z Frank Cheng, Taiwan
Robert L Dewar, Australia
Dominique Escande, France
Cary Forest, USA
Vladimir E Fortov, Russia
Manfred Hellberg, South Africa
Osamu Ishihara, Japan
Kimitaka Itoh, Japan
Yasuhito Kiwamoto, Japan
Mark Koepke, USA
Akio Komori, Japan
Siegbert Kuhn, Austria
Jae Koo Lee, South Korea
George J Morales, USA
Volker Naulin, Denmark
Paulo H Sakanaka, Brazil
Kohnosuke Sato, Japan
Abhijit Sen, India
Padma Kant Shukla, Germany
Leopoldo Soto, Chile (Chairman)
Ulrich Stroth, Germany
Richard Sydora, Canada
Shuichi Takamura, Japan
Michael Tendler, Sweden
Milan Tichy, Czech Republic
Chanchal Uberoi, India
Guido Van Oost, Belgium
Frank Verheest, Belgium
Masaaki Yamada, USA
Anatol Zagorodny, Ukraine

Journal of Physics: Conference Series **511** (2014) 011001 doi:10.1088/1742-6596/511/1/011001

LAWPP International Advisory Committee

Carlos Alejaldre, Spain
María Virginia Alves, Brazil
Ibere Caldas, Brazil
Julio Herrera, Mexico
Kohnosuke Sato, Japan
Héctor Kelly, Argentina
Edberto Leal, Puerto Rico
Anthony Peratt, USA
George Morales, USA
Julio Puerta, Venezuela
Ramón Rodríguez, Cuba
Leopoldo Soto, Chile (Chairman)
Michael Tendler, Sweden
Carlos Varandas, Portugal
Malcolm G Haines, UK
Michel de Peretti, France

Local Organizing Committe ICPP-LAWPP-2010

Leopoldo Soto (Chairman), CCHEN
Karla Cubillos (Executive Secretary), CCHEN
José Moreno, CCHEN
Cristian Pavez, CCHEN
Felipe Veloso, CCHEN
Marcelo Zambra, CCHEN
Luis Huerta, Universidad de Talca
Fabian Reyes, CCHEN

Journal of Physics: Conference Series **511** (2014) 011002 doi:10.1088/1742-6596/511/1/011002

Opening Talk
15th International Congress on Plasma Physics (ICPP 2010) 13th Latin American Workshop on Plasma Physics (LAWPP 2010)

- Dr. Gonzalo Gutierrez, President of the Chilean Nuclear Energy Commission.
- Mr. Fernando López, Executive Director of the Chilean Nuclear Energy Commission.
- Members of the International Advisory Committee of the ICPP.
- Members of the International Advisory Committee of the LAWPP.
- Members of the program committee of the combined conferences.
- Prof. Padma Shukla, Chairman of the International Union of Pure and Applied Physics.
- Dr. Günter Mank, International Atomic Energy Agency
- Authorities from Chilean universities.
- Scientists and students from Chile and abroad
- Ladies and gentlemen

Before five hundred and eighteen years ago, the original people of Latin America have their own civilizations and governments, and social values with respect to nature, solidarity and fraternity.

In 1492, the arrival of European people with external economic interests produced a shock on the cultures and civilizations of America. Then, two hundred years ago, the new people product of the consequent mixture of cultures in Latin America, claimed for the European values of liberty, fraternity and equality, achieving independence from the European domain.

Latin America is a continent of contrasts. In particular, Chile is a land of contrasts. Its territory is a long, thin strip of land between the Pacific Ocean and the Andean mountain range, with a wide variety of vegetation and climates. In the North, deserts dominate – Atacama, one of the most arid in the world, rich in minerals, like copper and lithium. The South is known for its cold pine and lush forests. Chile´s central region have a temperate climate, which favors the agriculture.

On behalf of the local organizing committee, we are happy to welcome you.
Last February, our country suffered a great earthquake, one of the strongest history. Although Santiago was little affected, the damage was great in other towns and cities 200 km southern of Santiago. After this event we receive a lot of messages of support from many of you. Thank you very much for your presence here. We appreciate that you decided come from the rest of the world. We highly appreciate your solidarity.

The International Congress on Plasma Physics was first held in Nagoya, in 1980, and is organized every two years since, being the last version in Fukuoka in 2008. The scope of the Congress is to discuss about the recent progress and future view on plasma science, covering a wide range of aspects including fundamental plasma physics, fusion plasmas, astrophysical plasmas, plasma applications, and so forth.

The Latin American Workshop on Plasma Physics was first held in 1982 in Cambuquira, and the last version was in Caracas. The purpose of this event is to provide a forum for display of the achievements of the Latin American plasma physics community, as well as fostering collaboration between plasma scientists within the region and elsewhere.

In Chile, although there were some efforts during the 60´s, sustained research activities in plasma physics begun around 1980. I would like to mention two Chilean pioneers in this regard: Prof. Luis Gomberoff from Universidad de Chile and Prof. Hernán Chuaqui from Pontificia Universidad Católica de Chile. At Universidad de Chile the focus was theoretical and space plasma physics whereas in ~~the~~ Universidad Católica the focus was on experimental plasma physics, especially with transient electrical discharges. Other institutions later joined these start-up efforts: Universidad de Concepción, Universidad de Santiago and Comisión Chilena de Energía Nuclear. Nowadays, after 30 years, we can say that Chile has an active plasma physics community, with around 20 researchers and 20 graduate students.

Both International Committees, ICPP and LAWPP, agreed to carry out this combined meeting in Santiago de Chile, considering the celebration of the Bicentennial of Chilean Independence. We acknowledge this as a recognition of the level achieved by the research activities in plasma physics in Chile. We are grateful and proud for this honor.

This meeting is part of the official program of the Chilean Nuclear Energy Commission (CCHEN) within the framework of the Bicentennial activities. This event is also a scientific and academic activity of the project "Center for Research and Applications in Plasma Physics and Pulsed Power, P^4", of the Comisión Nacional Científica y Tecnológica, CONICYT, Chile, grant ACT-26.

Previous to the ICPP-LAWPP 2010, an important activity associated to the Latin American Workshop on Plasma Physics took place: LAWPP School on Plasma Physics, which was open to worldwide participants with the main objective of providing basic training on plasma physics prior to the Conference. The school had 44 participants from 11 countries. The courses were delivered by lecturers from Brazil, USA and Chile. My special recognition to all of them: Prof. Homero Maciel of Institute Tecnológico Aeronautico, Dr. Luis Felipe Delagado Aparcio of Princeton Plasma Physics Laboratory, USA, Dr. Marina Stepanova, Universidad de Santiago, Dr. J. Alejandro Valdivia and Dr. Victor Muñoz of Universidad de Chile, and Dr. Felipe Veloso of Chilean Nuclear Energy Commission.

The program of these conferences was organized by the Program Committee. In particular, Plenary Talks and Invited Topical Talks were selected from a list of nominated presented by the International Advisory Committees of ICPP and LAWPP. Our acknowledgements to the Program Committee, in particular to Prof. Julio Herrera for his valuable work.

The Book of Abstracts contains the abstracts of invited and contributed papers. Full texts of the invited papers will be published as a special issue of *Plasma Physics and Controlled Fusion*. Contributed papers will be published in an issue of Journal of Physics Conference Series, IOP Publishing.

The main focus of the present conferences is basic and applied plasma physics including the fields of atmospheric and space physics, satellite sciences, medicine, environment, industry, mining, energy, among others.

Journal of Physics: Conference Series **511** (2014) 011002 doi:10.1088/1742-6596/511/1/011002

The challenge of nuclear fusion is also presents in these conferences. Nuclear fusion as clean and endless source of energy. Chile has important reserves of lithium in the desert of Atacama. Lithium is crucial in the nuclear fusion fuel cycle. Therefore, the researches related with nuclear fusion are very exciting for us.

We know that the great challenge of nuclear fusion relies on the international collaboration. In the agenda of the meeting that will take place of the International Advisory Committee of the Latin American Workshop in Plasma Physics, is schedule the discussion of the coordination of a Latin American program on nuclear fusion. The collaboration among Latin American research institutions in the framework of a coordinated program of nuclear fusion surely open to advances and developments that would not be possible to obtain separately.

Finally, I would like to give my gratitude to the Chilean Nuclear Energy Commission as host institutions of these conferences.

Welcome to Chile, enjoy the conference and enjoy our country.

Thank you very much.

Leopoldo Soto
Chairman
ICPP-LAWPP-2010

Santiago, August 9th, 2010.

Journal of Physics: Conference Series **511** (2014) 011003 doi:10.1088/1742-6596/511/1/011003

Speech of the President of the Chilean Nuclear Energy Commission
15th International Congress on Plasma Physics (ICPP 2010) 13th Latin American Workshop on Plasma Physics (LAWPP 2010)

- Dr. Leopoldo Soto, Chairman of the 15th International Congress on Plasma Physics and 13th Latin American Workshop on Plasma Physics.
- Mr. Fernando López, Executive Director of the Chilean Nuclear Energy Commission.
- Professor Padma Shukla, Chairman of the international Union of Pure and Applied Physics.
- Dr. Günter Mank, Head of the Physics Section at International Atomic Energy Agency.
- Authorities from Chilean Universities.
- Scientists and students from Chile and abroad.
- Ladies and gentlemen.

It is my pleasure to give this speech in front of this select audience composed by world class scientists, as well as authorities, students and especial invited that have met in Santiago of Chile on occasion of this 15th International Congress on Plasma Physics and the 13th Latin American Workshop on Plasma Physics.

Our country, as well as other Latin American countries, is about to celebrate its bicentenary next month: that is, two hundred years of independent political life. This has been a long and complex process, with significant progress in some periods, and violent setbacks in others. Indeed, the vast majority most of Latin American countries, after a heroic struggle for self-determination, were no longer colonies and began a period of independent development, inspired by the ideals of liberty, equality and fraternity of the French Revolution and the participatory democracy ideals of the independence of the United States. Besides that, we had our own demands in the region: cooperation with neighbors and focus on building republics for the benefit of its people, looking for an economic development not only based in mining and farming, but also in the industries and manufacturing. In this way, to achieve these goals, our founding fathers established a number of initiatives, such as education open to all because "without it there is no opinion, public spirit nor men constitute a state", as well as other rule to encourage the entrepreneurship and proper markets. Unfortunately, early in the republic this attempt was aborted. Thus, after 200 years, the same powerful ideas still remain to be implemented. In other words, what remains is to obtain independency from the economics and technological point of view. In one word, what remains is to be a developed country.
¿What should be fulfilled in order to reach such a situation?
Of course, certainly education, science and technology are key issues in this regard. We have worked hard in order to reduce the gaps in the access to social welfare, health and culture. However, several problems persist, like in the case of education, where we have tremendous inequalities, according to

Journal of Physics: Conference Series **511** (2014) 011003 doi:10.1088/1742-6596/511/1/011003

the measured by OECD's standards. Chile only spends about 4% of GDP on education, compared to the 7% of GDP recommended by the UN for developed nations, and sadly to say, our educational system maintains a huge segregation across social classes in its schools and colleges. We hope that this situation could be improved in the near future.

In the case of science, we are making efforts to move forward quickly. Our country has institutions, research centres and the proper tools to promote and finance scientific research, both basic and applied, oriented to real applications that will impact in the improvement of the life conditions of our people. There is a vigorous plan of graduate scholarships to train doctors in both Chile and abroad, as well as research grants. However, this is still insufficient. Chile spends only 0.7% of GDP in science and technology, which is clearly not enough to have a real impact in industry.

In Chile, science has developed some areas that have reached an international level. Even though our scientific production could be considered insufficient in quantity, in some cases is considerably good in quality. Therefore, we are proud that these two important events in plasma physics, highly recognized internationally and also in Latin America, are being held in our country. We consider this event as the result of the recognition of the level obtained by our scientists and institutions.

An important goal of our country - for now and in the near future - is to achieve a diversified energy matrix without risks to the environment and natural resources. This is why plasma physics research, which is related to the energy generation by means of nuclear fusion, deserves our great interest, even more due to our important lithium reserves in the Desert of Atacama in the north of Chile, being lithium crucial in the nuclear fusion fuel cycle. That is just one example of the problems we face and where we need to generate adequate public policies: lithium is a strategic resource since we have large reserves and we are the largest producers worldwide. However, incredibly, we have no institution or a systematic research program to address the study of this mineral in a multidisciplinary way. Now, we are repairing this deficiency, with the creation of a Lithium Institute in the Chilean Nuclear Energy Commission, which is part of a strategic plan that restructures the CCHEN and which is detailed in another article (see “Antecedents and perspective on the development of nuclear energy in Chile” in this issue).

We follow with great attention the studies in plasma physics and fusion around the world as well as various international collaborations for its development. One of these is the ITER project. This mega project intends to obtain energy from nuclear fusion at the experimental level. We know that this great challenge of nuclear fusion relies on the international collaboration. But from here, located in Chile, the advances of the international project ITER appear very far. Can we collaborate in this task from Latin America? In several Latin American countries it is possible to find competent scientists and adequate facilities in order to carry out research in plasma physics, material science, computer, and nuclear sciences, among others. All of these are necessary areas to face up the challenge of nuclear fusion. The collaboration from a Latin American program of nuclear fusion will allow us to obtain important advances and developments that would not be possible to obtain alone. Why not try?

Ladies and gentlemen, and dear scientists: I am here as the President of the Chilean Nuclear Energy Commission, the host of this Conference, sure that the hopes in integration of the Latin American people also depends on the integration in science and technology. We are a continent where resources are almost always in a shortage to undertake great projects for scientific development. Nevertheless, integration and common efforts will allow us to succeed in the great task of research and development. We also know that plasma physics offers opportunities of knowledge and applications in areas such as atmospheric and space physics, satellite sciences, medicine, environment, industry, materials, mining, among others fields that are interesting to the development of our country.

Journal of Physics: Conference Series **511** (2014) 011003 doi:10.1088/1742-6596/511/1/011003

Let me finish by acknowledging to all the experts and scientists that are visiting our country for this special event. Receive my best wishes for the success of this congress that starts today.

Thank you very much.

Dr. Gonzalo Gutiérrez
President of Chilean Nuclear Energy Commission

Santiago, August 9th, 2010.

Peer review statement

All papers published in this volume of *Journal of Physics: Conference Series* have been peer reviewed through processes administered by the proceedings Editors. Reviews were conducted by expert referees to the professional and scientific standards expected of a proceedings journal published by IOP Publishing.

2.5D relativistic electromagnetic PIC code for simulation of the beam interaction with plasma in axial-symmetric geometry

Yu M Tolochkevych, T Eu Litoshenko, I O Anisimov
Taras Shevchenko National University of Kyiv, Radio Physics Faculty, Glushkova ave 4g, Kyiv, Ukraine

E-mail: yura.tolochkevych@gmail.com, taras.litoshenko@gmail.com, ioa@univ.kiev.ua

Abstract. 2.5D relativistic electromagnetic PIC code for simulation of the beam interaction with plasma in axial-symmetric geometry was developed. Accurate charge weighting scheme and difference schemes near the system axis were introduced. Simulation tests of electromagnetic wave interaction with inhomogeneous plasma were carried out.

1. Introduction

A lot of problems in plasma electronics (the nonlinear stage of beam-plasma interaction, dynamics of electron bunches in wake fields excited by them in plasma, the virtual cathode formation during the injection of strong electron beams, etc.) can be solved only by means of computer simulation.

The method of particles in cells is used intensively for numerical simulation of plasma [1]. This method is used for various phenomena in space, ionosphere, various discharges, plasma display cells etc.

Electrostatic PIC codes were used widely for the simulation of above mentioned phenomena [2-3]. But using such codes we cannot observe effects of electromagnetic waves propagation or radiation in plasmas. Equations of the particles' motion are not relativistic, but it is necessary to solve the relativistic equations of motion in many cases.

Relativistic PIC codes are also used for plasma simulation (see, e.g., [4]). But usually every type of code is convenient only for some specific problems.

The aim of this work is to present the relativistic electromagnetic code for axial-symmetric beam-plasma systems simulation with a longitudinal magnetic field using the method of particles in cells.

2. Description of the difference scheme

If you don't wish to use the Word template provided, please set the margins of your Word document Two-dimensional cell in cylindrical coordinates (r-z) for the presented axial-symmetric code is shown on Fig. 1. Points are marked where each component of field, potential, charge and current are calculated. Large particle have a shape of the ring which can move along z axis, vary its radius moving in radial direction and rotate azimuthally. It is possible to simulate propagation of electromagnetic waves of E type in two dimensional systems without z component of magnetic field H accordingly.

But in some cases the presence of this component is important (e.g., dynamics of an electron bunch injected along the magnetic field).

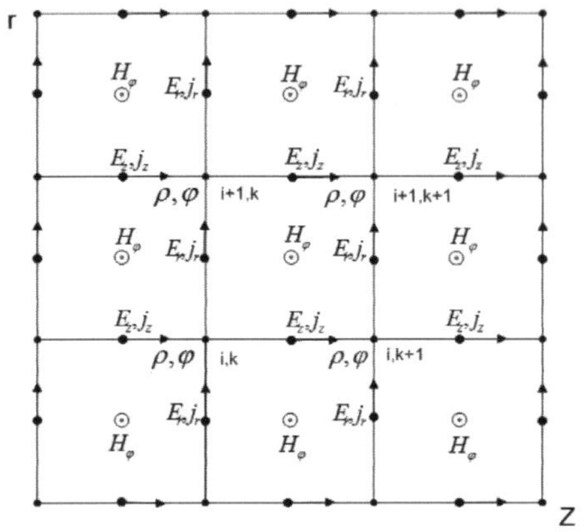

Figure 1. Elementary cell for 2D model

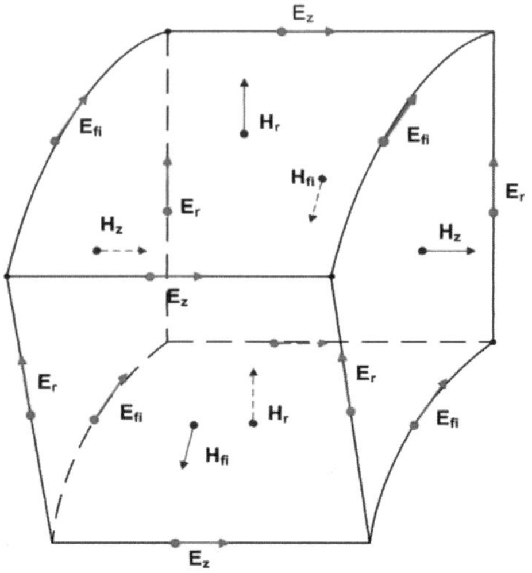

Figure 2. Elementary 3D cell in cylindrical geometry

In the proposed code the space grid remains two-dimensional, but large particles have three components of velocity which results in three components of electric and magnetic field. Therefore electromagnetic waves of H type arise in the system.

For solution of Maxwell equations' set the method of finite differences in time domain (FDTD) was used [5-6]. This method is based on Yee algorithm, which allows finding both electrical and magnetic field in time and space using first pair of the Maxwell equations (the law of Ampere's circuital law with Maxwell's correction and the electromagnetic induction law).

3D elementary cell in cylindrical geometry is represented on Fig.2. Each point where electric field component is calculated is surrounded in a plane perpendicular to its direction by four points, where magnetic field components are calculated. Oppositely, each point where the magnetic field component is calculated is surrounded by four points where electric field components are calculated.

Thus difference schemes for E and H components of the field take a form:

$$E_r\Big|_{i+1/2,k}^{n+1} = E_r\Big|_{i+1/2,k}^{n} - \left[j_r\Big|_{i+1/2,k}^{n+1/2} + \frac{1}{\Delta z}\left(H_\varphi\Big|_{i+1/2,k+1/2}^{n+1/2} - H_\varphi\Big|_{i+1/2,k-1/2}^{n+1/2}\right)\right]\frac{\Delta t}{\varepsilon_0 \varepsilon_{i+1/2,k}},$$

$$E_\varphi\Big|_{i,k}^{n+1} = E_r\Big|_{i,k}^{n} - \left[j_\varphi\Big|_{i,k}^{n+1/2} - \frac{1}{\Delta z}\left(H_r\Big|_{i,k+1/2}^{n+1/2} - H_r\Big|_{i,k-1/2}^{n+1/2}\right) + \frac{1}{\Delta r}\left(H_z\Big|_{i+1/2,k}^{n+1/2} - H_z\Big|_{i-1/2,k}^{n+1/2}\right)\right]\frac{\Delta t}{\varepsilon_0 \varepsilon_{i,k}},$$

$$E_z\Big|_{i,k+1/2}^{n+1} = E_z\Big|_{i,k+1/2}^{n} - . \qquad (1)$$

$$-\left[j_z\Big|_{i,k+1/2}^{n+1/2} - \frac{1}{2i\Delta r}\left(H_\varphi\Big|_{i+1/2,k}^{n+1/2} + H_\varphi\Big|_{i-1/2,k}^{n+1/2}\right) + \frac{1}{\Delta r}\left(H_\varphi\Big|_{i+1/2,k+1/2}^{n+1/2} - H_\varphi\Big|_{i-1/2,k+1/2}^{n+1/2}\right)\right]\frac{\Delta t}{\varepsilon_0 \varepsilon_{i,k}},$$

$$H_r\Big|_{i,k+1/2}^{n+1/2} = H_r\Big|_{i,k+1/2}^{n-1/2} - \frac{\Delta t}{\Delta z \mu_0}\left[E_\varphi\Big|_{i,k+1}^{n} - E_\varphi\Big|_{i,k}^{n}\right]. \qquad (2)$$

$$H_\varphi\Big|_{i+1/2,k+1/2}^{n+1/2} = H_\varphi\Big|_{i+1/2,k+1/2}^{n-1/2} - \left[\frac{1}{\Delta z}(E_r\Big|_{i+1/2,k}^{n} - E_r\Big|_{i,k}^{n}) - \frac{1}{\Delta r}(E_z\Big|_{i+1,k+1/2}^{n} - E_z\Big|_{i,k+1/2}^{n})\right]\frac{\Delta t}{\mu_0},$$

$$H_z\Big|_{i,k+1/2}^{n+1/2} = H_z\Big|_{i+1/2,k}^{n-1/2} - \left[\frac{1}{2(i+1/2)\Delta r}(E_\varphi\Big|_{i+1,k}^{n} + E_\varphi\Big|_{i,k}^{n}) + \frac{1}{\Delta r}(E_\varphi\Big|_{i+1,k}^{n} - E_\varphi\Big|_{i,k}^{n})\right]\frac{\Delta t}{\mu_0}.$$

3. Field calculation near the axis

Solution of equations (1)-(2) has some specific features in the cylindrical coordinate system connected with the calculation of field's components near the system axis. One can see from Fig. 2 that electromagnetic field components E_φ, E_z and H_r are found on the axis, and E_r, H_φ and H_z components are found at the distance $\Delta r/2$ from the axis.

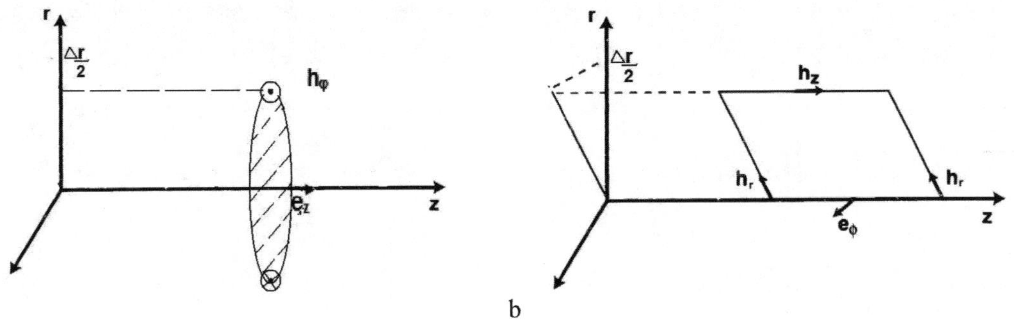

a b

Figure 3. Calculation of electric field components E_z (a) and E_φ (b) near the axis of the system

On the axis H_r=0, but the set of equations (1)-(2) cannot be used for E_z and E_φ calculation near the axis. Therefore, the first Maxwell equation in the integral form should be used in order to obtain the field near the axis:

$$\oint_l H dl = I + \oiint_S \frac{\partial D}{\partial t} dS . \qquad (3)$$

Calculating integral for the cell allocated on the system axis (see Fig.3 a) and taking into account that its radius is equal to $\Delta r/2$, it is possible to obtain E_z:

$$E_z\Big|_{i,k+1/2}^{n+1} = E_z\Big|_{i,k+1/2}^{n} - \left[j_z\Big|_{0,k+1/2}^{n+1/2} - \frac{4}{\Delta r}H_\varphi\Big|_{1/2,k+1/2}^{n+1/2}\right]\frac{\Delta t}{\varepsilon_0\varepsilon_{1/2,k}}, \qquad (4)$$

E_φ is calculated similarly (see Fig. 3 b):

$$E_\varphi\Big|_{0,k}^{n+1} = E_r\Big|_{0,k}^{n} - \left[j_\varphi\Big|_{0,k}^{n+1/2} - \frac{1}{\Delta z}\left[(H_r\Big|_{0,k+1/2}^{n+1/2} - H_r\Big|_{0,k-1/2}^{n+1/2})\right] + \frac{2}{\Delta r}H_z\Big|_{0,k-1/2}^{n+1/2}\right]\frac{\Delta t}{\varepsilon_0\varepsilon_{0,k}}. \qquad (5)$$

On the system axis j_φ=0 and H_r=0, equation (5) can be rewritten in the form:

$$E_\varphi\Big|_{0,k}^{n+1} = E_r\Big|_{0,k}^{n} - \left[\frac{2}{\Delta r}H_z\Big|_{0,k+1/2}^{n+1/2}\right]\frac{\Delta t}{\varepsilon_0\varepsilon_{i,k}}. \qquad (6)$$

4. Weighting procedure in cylindrical coordinates

Procedure of current density weighting also has the specific features in cylindrical geometry. The current density in each node

$$j = \frac{\Delta q}{\Delta t \Delta S}, \tag{7}$$

should be defined so that the continuity equation is fulfilled.

Weighting of the first order was used in the program, where large particles' cross-section has the square shape in the plane (r, z). Therefore, they can distribute the charge to four cells simultaneously. In order to find the current density caused by each large particle it is necessary to trace how the particle passes through each edge of the elementary cell (Fig .4) during each time step of simulation [7].

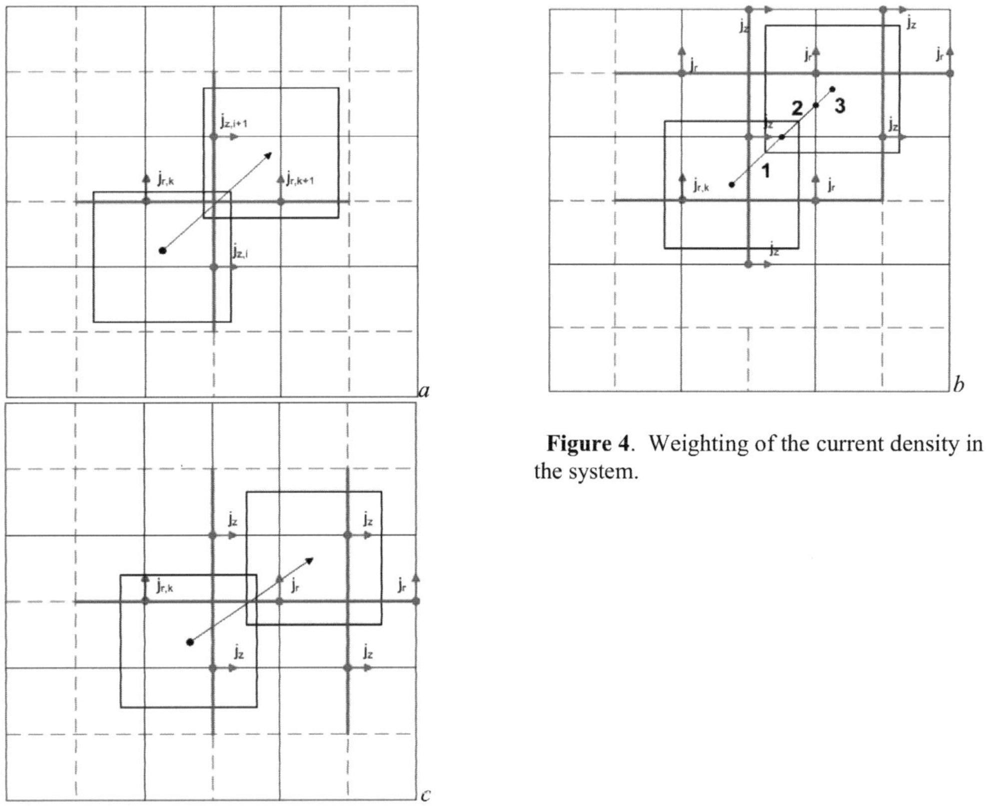

Figure 4. Weighting of the current density in the system.

Maximum distance that particle can pass during one time step should not exceed the cell size (so called Courant condition):

$$v_{\max} \Delta t \le \Delta_x . \tag{8}$$

Let's consider the possible variants of particle current distribution to the neighboring nodes. To reduce the quantity of the possible variants we use Courant condition in more strict form in comparison to (8): limit maximum distance the particle passes per time step to $\Delta_x / 2$.

During its move per time step the particle can give the contribution to 4 sides (Fig.4 a), to 7 sides (Fig.4 b), or to 10 sides (Fig.4 c). It is also necessary to consider that large particle charge density is a function of its radius, in contrast to be constant in the rectangular geometry.

5. Test simulations of electromagnetic wave interacting with inhomogeneous plasma

A simulation volume has a form of cylindrical resonator with radius of 0,4 m and length of 12,8 m. A cylindrical waveguide mode E01 with frequency $f = 400$ MHz is excited by the rod with alternating current. It is located on the axis of the resonator and has a length of 0,37 m. Electromagnetic absorbing layers is imposed on the both sides of resonator.

The system is partly filled with inhomogeneous plasma with linear density profile. The plasma consists of electrons and ions of hydrogen with near zero temperature.

The spatial distribution of z component of electric field for the case of wave interaction with subcritical inhomogeneous plasma is showed on Fig.6. The plasma density changes from $n_p(z = 4.5\text{ m}) = 0.2{\cdot}10^{15}\text{ m}^{-3}$ to $n_p(z = 12.8\text{ m}) = 0.6{\cdot}10^{15}\text{ m}^{-3}$. One can see that in this case the wave remains periodical but changes its wavelength while propagating through the plasma. This happens because subcritical plasma has lower dielectric permeability than vacuum that leads to the decrease of phase velocity of the wave.

The distribution of z component of electric field for the case of interaction with subcritical inhomogeneous plasma with the reflection point on the density profile is showed on Fig.5. The plasma density changes from $n_p(z = 4.5\text{ m}) = 0.6{\cdot}10^{15}\text{ m}^{-3}$ to $n_p(z = 12.8\text{ m}) = 1.5{\cdot}10^{15}\text{ m}^{-3}$. One can see that now the electromagnetic wave reflects from the inhomogeneous plasma. There are also low intensity oscillations of electric field beyond the reflection point which could be explained by generation of second harmonic.

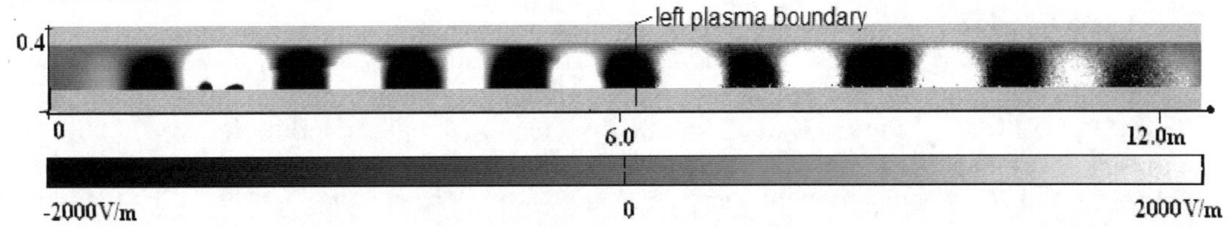

Figure 5. Spatial distribution of z component of electric field for the case of electromagnetic wave interaction with subcritical plasma.

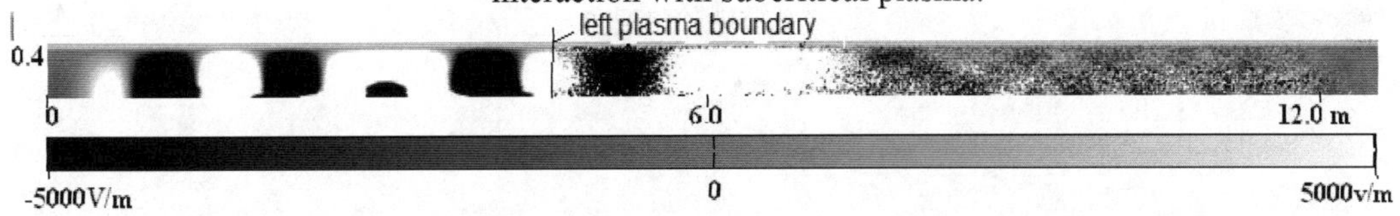

Figure 6. Spatial distribution of z component of electric field for the case of reflection of electromagnetic wave from the inhomogeneous plasma.

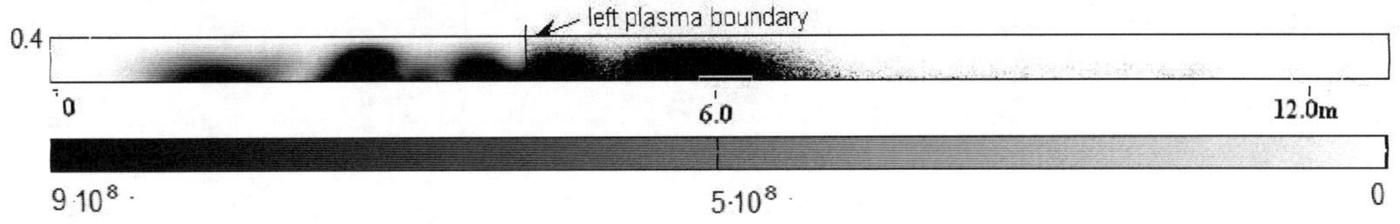

Figure 7. Spatial distribution of intensity of electromagnetic field for the case of reflection of electromagnetic wave from the inhomogeneous plasma.

The distribution of the intensity of the electric field is showed on the Fig.7. One can see that the standing wave is formed in plasma by the incident and the reflected wave. The reflection point is defined by the transversal wave number $k_\perp$. It's well known from the theory [8] that in the case of non-zero angle θ between

Journal of Physics: Conference Series **511** (2014) 012001 doi:10.1088/1742-6596/511/1/012001

wave number and the density gradient the reflection density is

$$n_{ref} = n_{cr} \sin^2 \theta = n_{cr} \frac{k^2 - 2\pi\nu_{01} / R}{k^2}, \tag{9}$$

where $n_{cr} = 1.9 \cdot 10^{15}$ m^{-3} is critical density, R – radius of the resonator, $\nu_{01} = 2.4$– root of the Bessel's function. For parameters of simulation $n_{ref} = 0{,}97 \cdot 10^{15}$ m^{-3}. This value is close to the observed value of $n_{ref,\,exp} = 0{,}8 \cdot 10^{15}$ m^{-3}.

6. Conclusions

The 2.5D electromagnetic relativistic PIC code for simulation of beam- and wave-plasma interaction in cylindrical geometry is presented in the paper. The accurate charge weighting scheme is introduced which allows the charge conservation law to be automatically fulfilled. The correctness of the program is confirmed by the comparison of simulation results with theoretical predictions.

References

[1] Birdsall C, Langdon A 1985 *Plasma Physics via Computer Simulation* (Singapore: McGraw-Hill book company)
[2] Anisimov I, Litoshenko T 2008 *Ukrainian J. of Phys.* **54** 388
[3] Anisimov I, Tolochkevych Yu 2009 *Ukrainian J. of Phys.* **54** 454
[4] Faure J, Clinec Y, Pukhov A, Kiselev S, Gordienko S, Lefebvre E, Rousseau J., Burgy F, Malka V 2004 *Nature* **431** 541
[5] Taflove A, Hagness S 2000 *Computational Electrodynamics: The Finite Difference Time Domain Method, 2nd ed.* (Norwood, MA: Artech House)
[6] Russer P, Siart U 2008 *Finite Domain Methods in Electrodynamics* (Berlin: Springer)
[7] Villasenor J, Buneman O 1992 *Comp. Phys. Comm.* **69** 306
[8] Ginzburg V 1967 *Propagation of electromagnetic waves in plasmas, 2nd edition* (Moscow: Nauka)

On Correction Factor in Scaling Law for Low Pressure DC Gas Breakdown

G Ronchi and M Machida

Instituto de Física "Gleb Wataghin", UNICAMP, Campinas, SP, Brazil.

E-mail: gronchi@ifi.unicamp.br

Abstract. The low pressure gas breakdown described by Paschen's law in Townsend theory, i.e. the breakdown voltage as a function of gas pressure *p* and the electrode distance *d*, provides an accurate description of breakdown in DC discharges when the ratio between inter-electrode gap distance *d* and electrode radii *R* tends to zero. On increasing of the ratio *d/R*, the Paschen's curves are shifted to the region of higher breakdown voltage and higher *pd* values. A modified Paschen's law recently proposed is well satisfied in our measurements. However, the value of constant *b* changes not only due to gas type but also according to electrode gap distance; furthermore, gas breakdown voltages are considerably modified by plasma-wall interactions due to glass tube proximity in the discharge.

1. Introduction

By classical Townsend theory, for a parallel electrodes DC discharge at low pressure, the breakdown voltage is given by [1]

$$V_{br} = \frac{Bpd}{C + \ln pd} \quad \text{with} \quad C = \ln \frac{A}{\ln(1 + 1/\gamma) + 1} \tag{1}$$

where A and B are constants depending on the gas type. This expression arises from only kinetic considerations of gas ionization. In practice, we have a loss of electrons by diffusion in the discharge chamber. When we take into account the hydrodynamic conditions of the system [2], the equation (6) of [3] leads to the same equation (1), but with

$$C = \ln \frac{A}{\ln\left\{1 + \frac{1}{\gamma}\left[1 + \frac{2.4^2}{\alpha V_{br} d}\frac{D}{\mu}\left(\frac{d}{R}\right)^2\right]\right\} + \frac{2.4^2}{V_{br}}\frac{D}{\mu}\left(\frac{d}{R}\right)^2} \tag{2}$$

where μ and D are the mobility and the coefficient of the transverse diffusion of electrons respectively, and α is the first Townsend's coefficient which is a function of pressure, voltage and gap distance. The second term multiplying $1/\gamma$ is usually negligible. So, when we disconsider it in eq. (2), we recover an import result

$$\frac{V_{br(\min)}}{pd_{(\min)}} = B \tag{3}$$

which is known as Stoletov's point and indicate where the ionization capability of electrons is maximum [4]. Here $pd_{(\min)}$ denotes the value of pd when V_{br} is minimum. This ratio is verified experimentally with good agreement [3].

In order to have a simple equation to describe the breakdown voltage behavior, Lisovskiy *et al.* [3] propose the following empirical relationships:

$$V_{br} = V_{br}^{*}\left[1+\left(\tfrac{d}{R}\right)^2\right]^b \tag{4}$$

and

$$pd = pd^{*}\left[1+\left(\tfrac{d}{R}\right)^2\right]^b \tag{5}$$

where $V_{br}^{*} = f(pd)$ and pd^{*} are the ordinary values of Paschen's curves when no diffusion is taken in account, i.e., when the ratio d/R tends to zero in equation (2).

By this way, we get the following Paschen's curves:

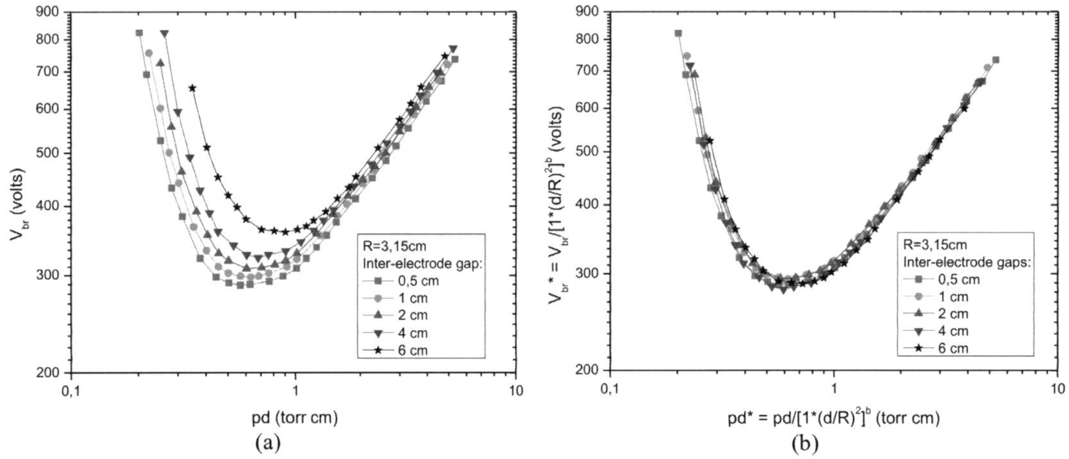

Figure 1. Paschen's curves for nitrogen, (a) without and (b) with the scaling law ($b = 0.12$) [3].

2. Methodology

The discharges were made in chamber called "RFDC device", built for basic low temperature experimental plasma studies. The RFDC device is made of big glass cylinder of 30 cm diameter and 20 cm high, covered by two insulated brass plates. The vacuum is obtained by a mechanical pump and small turbo molecular pump of 70 l/s which can hold working pressure from 10^{-2} to 1 Torr. The DC voltage can reach up to 1000 V.

Before taking measurements, we clean the cathode surface by igniting the dc glow discharge at pressure about 0,1 Torr, with current of 10-30 mA, just to remove the monolayers of gases remaining on the cathode surface in a discharge not yet sufficient to produce the cathode spots.

Using two distinct diameter brass electrodes, 25 mm and 47 mm, with and without a cylindrical glass tube, just enough to cover all the electrodes distance gap inside our chamber, we analyze the dependence of $V_{br(\min)}$ with d/R. From these relationship, we get the $V_{br(\min)}{}^{*}$ and b coefficients.

3. Results and discussion

The behavior of breakdown voltage as a function of d/R ratio is presented in figures 2, 3 and 4. We obtain values of b equal to 0.131, 0.201 and 0.095 for argon, nitrogen and oxygen while Lisovskiy *et al*, using stainless steel electrodes, got 0.16, 0.12, and 0.03, respectively.

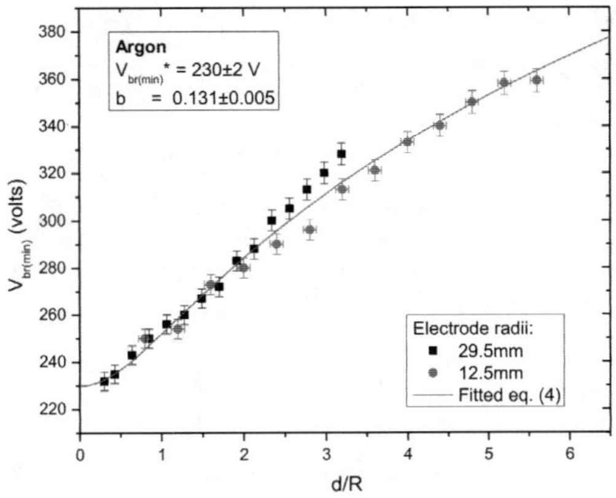

Figure 2. Behaviour of minimum breakdown voltage as a function of *d/R* for argon.

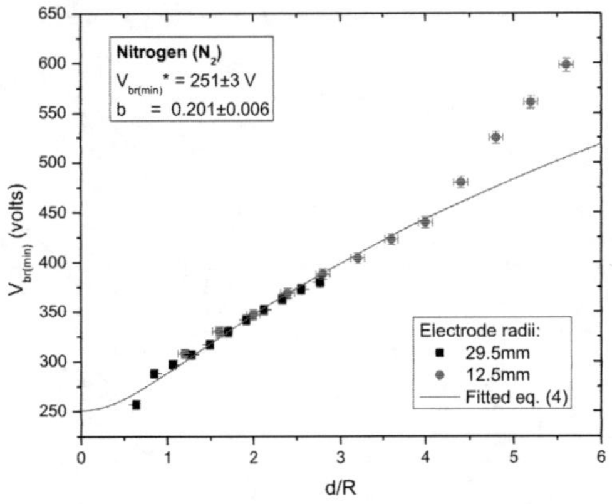

Figure 3. Behaviour of minimum breakdown voltage as a function of *d/R* for nitrogen.

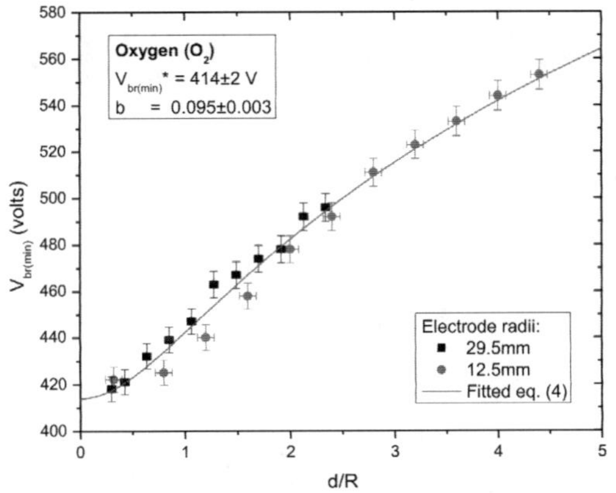

Figure 4. Behaviour of minimum breakdown voltage as a function of *d/R* for oxygen.

These results are in good agreement with the empirical relationship (4), except for nitrogen when *d/R* is greater than 4. For example, for nitrogen at *d/R* = 5.6, *b* should be 0.25 in equation (2) to give the real minimum breakdown voltage $V_{br(\min)}$ with the same ordinary breakdown voltage $V_{br(\min)}^{*}$.

When a glass tube of about same diameter as electrode is added (figures 5, 6 and 7), the proposed scaling law do not provide a good description of minumum breakdown voltage for all range of d/R. Analyzing the breakdown voltage with d/R in the range 0 to 3, the same interval considered by Lisovskiy, we got *b* equal to 0.097 for argon and 0.13 for nitrogen. The oxygen data don't change appreciably on this range.

For the electrode with R = 29.5 mm, the minimum breakdown voltage with glass tube is lower than the normal breakdown voltage, similarly as shown above, but both datas (with glass) do not overlap each other.

Thus we emphasize the large variation in the value of coefficients *b*, particularly for nitrogen, which is similar to value obtained by [3] when the discharge chamber (or glass tube) is large enough just to cover the electrodes; and the non compatibility of the model for high values of *d/R*.

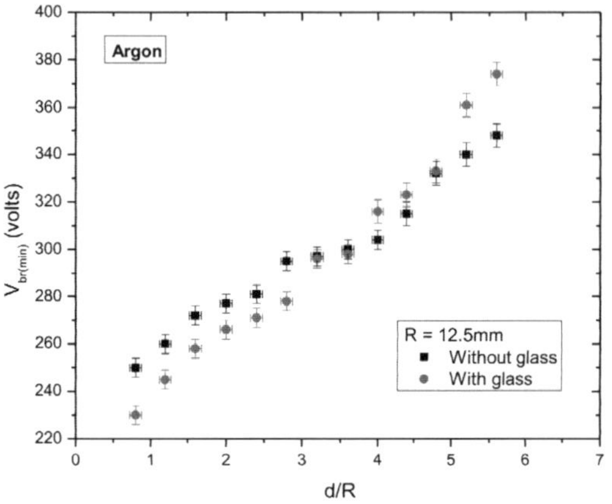

Figure 5. Behaviour of minimum breakdown voltage as a function of *d/R* for argon (with glass tube).

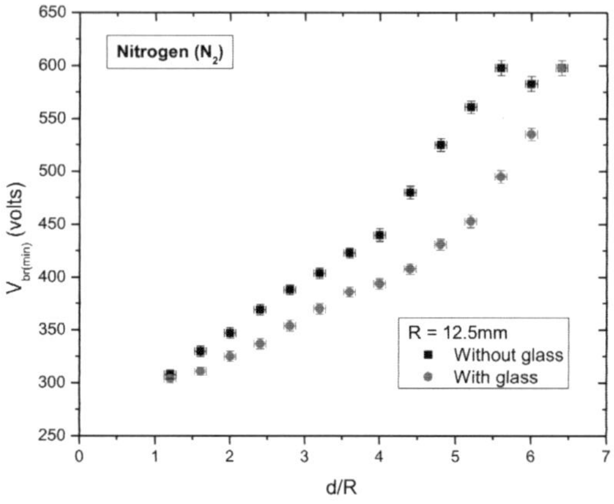

Figure 6. Behaviour of minimum breakdown voltage as a function of *d/R* for nitrogen (with glass tube).

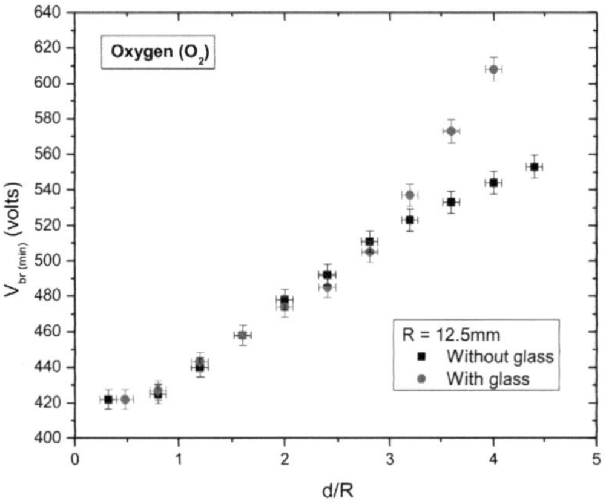

Figure 7. Behaviour of minimum breakdown voltage as a function of *d/R* for oxygen (with glass tube).

4. Conclusion

The simple scaling law proposed by Lisovskiy *et al.* can describe very well the breakdown voltage in a limited range of *d/R* which can differ for each gas. However, the correction factor in the scaling law could depend not only on the gas type but also on others parameters such as D and μ. The diffusion and mobility of electrons and ions have an import role in the discharge ignition, so the presence or absence of wall close to electrodes affects the discharge. They may strongly alter the b coefficient and restrict even more the range of validity of the scaling law for Paschen's curves.

Acknowledgements

This works was supported by CNPq, FAPESP and FAEPEX/UNICAMP.

References

[1] Townsend J S 1915 *Electricity in Gases* (Oxford: Clarendon)
[2] Kolobov V I and Fiala A 1994 *Phys. Rev. E* **50** 3018-32
[3] Lisovskiy V A, Yakovin S D and Yegorenkov V D 2000 *J. Phys. D: Appl. Phys.* **33** 2722-30
[4] Raizer Y P 1991 *Gas Discharge Physics* (Berlin: Springer)

Transversal dynamics of relativistic charged particle beams

R. P. Nunes[1], **and F. B. Rizzato**[2]

[1]Instituto de Física e Matemática, Universidade Federal de Pelotas, Campus Universitário, s/nº. Caixa Postal 354. CEP: 96010-900. Pelotas, RS, Brasil.

[2]Instituto de Física, Universidade Federal do Rio Grande do Sul, Av. Bento Gonçalves, 9500. Caixa Postal 15051. CEP: 91501-970. Porto Alegre, RS, Brasil

E-mail: roger.pizzato@ufpel.edu.br or roger.pizzato@ufrgs.br

Abstract. The idea behind this work is to analyse the transversal dynamics of a relativistic charged particle beam. The beam is azimuthally symmetric, focused by a constant magnetic field and supposed initially cold. While mismatched, nonrelativistic and homogeneous beams oscillate with an invariant cold density profile, it is shown that relativistic homogeneous beams progressively heat and lose an important amount of constituents during its magnetic confinement. This heating process starts with phase-space wave breaking, a mechanism observed before in initially inhomogeneous beams. The results have been obtained with full self-consistent N-particle beam numerical simulations.

1. Introduction

Beams composed by charged particles usually evolve to its equilibrium state with the ejection of some representative amount of its constituents as propagate inside the magnetic focusing channel. From the perspective of the beam phase-space, these ejected particles form a rarefied population around a dense one, which congregate the remaining particles. While the first population is recognized as the beam halo, the second one is denoted as beam core [1][2][3][4].

Halo is a well-known issue in beam physics. It has many undesired implications over the accelerator structure, reducing its lifetime and making hard and increasing the costs of its maintenance [5]. Also, there are many applications in which beam halo has to be, if not suppressed, at least minimized [6].

Although beam halo has been identified and measured during these last years [7], its causes remains not satisfactorily understood yet. A beam is a complex system composed by a very large amount of charged particles submitted to forces generated by the internal and external electromagnetic fields. The internal fields are that generated self-consistently by the beam evolution and the external fields are that produced by the focusing system to control the beam dynamics. The way by which beam particles interact causing halo formation directly depends of the beam initial distribution. For initially cold, quasi-homogeneous and mismatched beams, it has been found that, while the initial spurious inhomogeneity is the forerunner mechanism that allows particles to be progressively excited, in fact the envelope mismatch is the mechanism effectively responsible for halo formation [8]. On the other hand, for initially cold, non-homogeneous but envelope matched beams, it has been found that the forerunner mechanism by which particles are ejected from the core is through phase-space wave-breaking [9]. Once out of the beam core, the ejected particles are permanently excited by a process called charge redistribution [8]. If in this last situation the mismatch is also present, then envelope oscillation acts as other source of energy to excite beam particles [10]. The only difference is that now the coupling is resonant [8].

In the cases mentioned above, nothing has been commented about the relativity of the beam dynamics. In fact, a paraxial approximation has been considered, which implies that, although the beam could be relativistic in the longitudinal direction, its transversal dynamics was purely classic, nonrelativistic. However, notwithstanding the many situations in which this is an adequate approximation, in many others one can assure that this is not. The mass correction introduced by the relativistic effects can be more or less impacting in the dynamics of beam particles and must be understood. For this propose, hereafter an analytical description for the relativistic dynamics of beam particles will be presented.

2. The theory

Suppose an initially homogeneous beam of charged particles evolving inside a linear channel. A conducting pipe with a circular transversal section encapsulates the channel. Solenoids provide the constant magnetic field that permeates the channel. For this system, from Newton 2nd Law, the dynamics of beam particles could be described by the following equation

$$\frac{d}{dt}\mathbf{P} = \mathbf{F}_{\text{lorentz}} = q(\mathbf{E} + \mathbf{v}\times\mathbf{B}), \tag{1}$$

in which $\mathbf{P} = \gamma m\mathbf{v}$ is the relativistic linear momentum and $\mathbf{v}$ is the velocity of the particle, γ is the Lorentz factor, q is the charge and m is the rest mass of the particle, and $\mathbf{E}$ and $\mathbf{B}$ are respectively the electric and magnetic flux density fields inside the accelerator structure.

Since the beam here is azimuthally symmetric, equation (1) could be expressed in cylindrical coordinates. Consider $(\mathbf{e}_r, \mathbf{e}_\theta, \mathbf{e}_z)$ as the unit-vetor base in cylindrical coordinates. In this case, $\mathbf{v} = \dot{r}\mathbf{e}_r + r\dot{\theta}\mathbf{e}_\theta + \dot{z}\mathbf{e}_z$. The fields can be expressed as $\mathbf{E} = E_r\mathbf{e}_r + E_\theta\mathbf{e}_\theta + E_z\mathbf{e}_z$ and $\mathbf{B} = B_r\mathbf{e}_r + B_\theta\mathbf{e}_\theta + B_z\mathbf{e}_z$. Then, in this coordinate system, the equation (1) has the following form

$$\begin{gathered}\frac{d}{dt}(\gamma m\dot{r}) - \gamma m r\dot{\theta}^2 = q\left(E_r + r\dot{\theta}B_0 - \dot{z}B_\theta\right)\\ \frac{1}{r}\frac{d}{dt}\left(\gamma m r^2\dot{\theta}\right) = -q\dot{r}B_0\\ \frac{d}{dt}(\gamma m\dot{z}) = q\dot{r}B_\theta\,.\end{gathered} \tag{2}$$

in which, due to the symmetry of the problem, one has that $E_\theta = E_z = B_r = 0$. $B_z = B_0$, in which B_0 is the density of magnetic flux produced by the solenoids. E_r and B_θ are the fields generated self-consistently by the particle distribution assigned to the beam. The notation $\dot{x}$ means dx/dt with $x = \{r,\theta,z\}$. In the limit $\gamma \to 1$, the system of EDOs in equation (1) reduces to that of a nonrelativistic case.

Solving the ordinary differential equation (ODE) for $\mathbf{e}_\theta$ component, one obtains

$$\gamma m\dot{\theta} = -\frac{qB_0}{2} = \text{constant}. \tag{3}$$

Note that $\gamma m\dot{\theta}$ is now a conserved quantity. The Larmor frequency $\dot{\theta}$ of particles in the external density of magnetic field is a function of the absolute value of velocity $\mathbf{v}$.

With the aid of equation (2) and supposing a changing of variables of t to z one can rewrite the ODE for $\mathbf{e}_r$ component of equation (1) as follows

$$\frac{d^2r}{dz^2} + \frac{\dot{\theta}^2}{\dot{z}^2}r = \frac{q}{\gamma m\dot{z}^2}\left[E_r - \dot{z}B_\theta - \dot{z}B_\theta\left(\frac{dr}{dz}\right)^2\right], \tag{4}$$

in which also has been use the ODE for $\mathbf{e}_z$ component of equation (1).

To turn equation (4) completely soluble, it is necessary to determine both E_r and B_θ. Considering that all charged particles have initially negligible velocity, and being n_b the initial density of charged particles, from the Ampére-Maxwell Law

$$\nabla\times\mathbf{B} = \mu_0\mathbf{J} + \frac{1}{c^2}\frac{\partial\mathbf{E}}{\partial t}. \tag{5}$$

Bearing mind the symmetry of the problem and the stationary characteristic of the fields, the equation above reduces to

$$B_\theta = \frac{\mu_0 q\dot{z}}{2\pi r}Q(r), \tag{6}$$

in which μ_0 is the vacuum magnetic permeability. $Q(r)$ is the charged trapped by a Gauss surface at r

$$Q(r) = \int_0^r n_b(r^*)r^* dr^* d\theta. \tag{7}$$

For the E_r field, solving the Gauss Law

$$\nabla\cdot\mathbf{E} = \frac{\rho}{\epsilon_0} \tag{8}$$

one obtains

$$E_r = \frac{q}{2\pi r\epsilon_0}Q(r), \tag{9}$$

in which ρ is charge density and ϵ_0 is the electric permissivity of the vacuum.

Observe that in fact $B_\theta = B_\theta(Q)$ and $E_r = E_r(Q)$, being possible to express B_θ as a function of E_r. Proceeding in this way

$$B_\theta = \frac{\dot{z}}{c^2}E_r, \tag{10}$$

and inserting this expression in equation (4) with the help provided by equation (3), in view of that $\gamma = \gamma(v)$ by definition, after some extensive algebra, it is possible to achieve the equation

$$\frac{d^2\bar{r}}{d\bar{z}^2} + k_0\frac{1-\left(\frac{d\bar{r}}{d\bar{z}}\right)^2}{1+k_0\bar{r}^2}\bar{r} = K\frac{\left[1-\left(\frac{d\bar{r}}{d\bar{z}}\right)^2\right]^{3/2}}{(1+k_0\bar{r}^2)^{\frac{1}{2}}}\frac{Q(\bar{r})}{\bar{r}}. \tag{11}$$

in which

$$k_0 = \frac{q^2B_0^2}{4m^2\dot{z}^2} \tag{12}$$

is the coefficient of magnetic focusing proportionate by the solenoids and

$$K = \frac{q^2}{2\pi\epsilon_0 m\gamma_z\dot{z}^2} \tag{13}$$

is conventionally denominated the perveance of the beam, with $\gamma_z \equiv 1/\sqrt{1-\beta_z^2}$ and $\beta_z = \dot{z}/c$. To reach at equation (11), also the following change of variables has been also applied

$$\bar{z} = \gamma_z z \text{ e } \bar{r} = \beta_z r. \tag{14}$$

Once in the present case $\dot{z} = \text{constant}$, from the point of view of the equations of motion, the axial length and time has the same meaning. Note that equation (11) is a completely nonlinear equation for the transversal dynamics of beam particles. The second order derivative depends not only of nonlinear functions of $\bar{r}$ but also of $d\bar{r}/d\bar{z}$. The term left to the right of equation (11) can be identified as the source for the particles transversal dynamics, since the fields autoconsistently generated by the evolution of density n_b excite each beam constituent.

3. The results, conclusions, and future works

To explore this new information introduced by relativity, a full self-consistent N-particle beam numerical simulation has been performed. The orbit of each particle that composes the beam has been evaluated with numerical integration of equation (11). As initial condition, the beam particle density has been considered completely homogenous $n_b(\bar{r}, \bar{z} = 0) = N_b/\pi r_b^2$, in which r_b is the beam border, the envelope. For each particle i at the radial coordinate $\bar{r} = \bar{r}_i$, $Q(\bar{r}_i)$ is self-consistently computed. Indeed, $Q(\bar{r}_i)$ represents the influences of the remaining particles that compose the beam over the one at $\bar{r}_i$. The beam particles just interact by the means of the fields inside the accelerator channel. Thus just collective effects are accounted, which is expected in the case of a space-charge dominated beam such is of interest here in this work. The total number of beam particles adopted was $N_b = 10000$. The equation (11) has suffered additional scaling to allow particles at the beam border to be in equilibrium $(d\bar{r}/d\bar{z}|_{\text{border}} = 0)$.

The Figure 1 shows the results obtained with the numerical simulations for two macroscopic beam quantities of interest: beam envelope, in panel (a), and beam emittance, defined as $\epsilon = \sqrt{4(\langle \bar{r}'^2 \rangle \langle \bar{r}^2 \rangle - \langle \bar{r}'\bar{r} \rangle)}$ with $\bar{r}' = d\bar{r}/d\bar{z}$, in panel (b). The angle brackets $\langle \quad \rangle$ represents phase-space averaging. The beam has been supposed initially mismatched in 50%, that is $r_0 \equiv r_b(\bar{z} = 0) = 1.5$, since $r_{\text{eq}} = 1$. As expected, beam envelope for the nonrelativistic case oscillates permanently by an undeterminable axial length of the focusing channel. By the other side, the same does not occur if, for this same beam, the relativistic effects are considered and formally accounted. The beam envelope oscillates by a very short axial distance $\bar{z}$ until it suffers a sharp decay. In panel (b), it is possible to observe that, at this same coordinate, beam emittance experience a suddenly growth. The emittance growth is a macroscopic indicative that microscopically particles are increasing their velocities and improving the amplitude of their orbit in the phase-space. The particles with these attributes are potentially the ones that compose the beam halo and must be investigated.

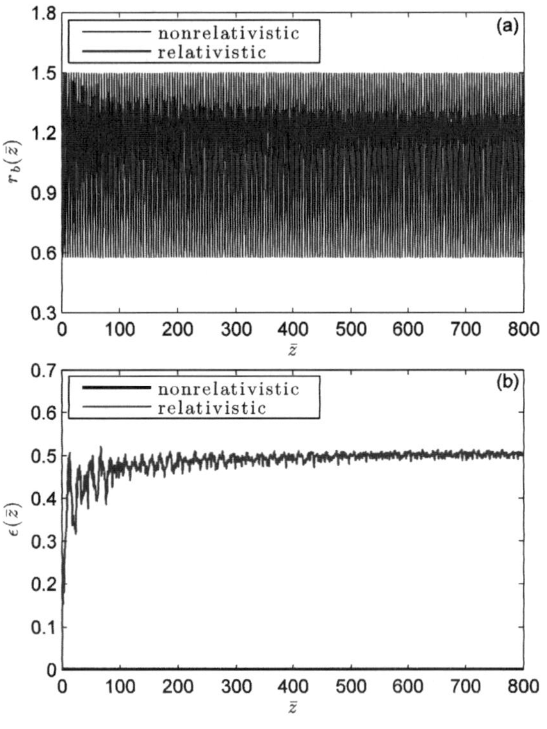

Figure 1. The results of numerical simulations for an initially 50% mismatched, cold and homogeneous beam. In panel (a) the beam envelope r_b and in panel (b) the beam emittance ϵ. While nonrelativistic beams oscillate with fixed amplitude, relativistic beams suffer a strong decay. The emittance growth associated with the relativistic case is pretty perceptible. The beam is composed by $N_b = 10000$.

For this purpose, the beam phase-space dynamics is shown in Figure 2, which has been composed by a set of snapshots at adequate axial lengths $\bar{z}$. In this case, the beam has been considered initially matched for practical purposes. The beam is initially cold, implying that at $\bar{z} = 0$ its appearance at the phase-space is a horizontal line over the $\bar{r}$ axis (panel (a)). However, as the beam propagates inside the focusing channel, a process of redistribution of charge is observed as panel (b) shows. Although initially particles are disposed homogeneously, during its dynamics, particles can accumulate differently along the beam. This gives rise to the propagation of waves inside the beam density. These phase-space waves are self-consistently amplified until a breaking is observed. Some particles are violently ejected from the beam core. Panel (c) and (d) respectively show the axial length right before and right after the first breaking. Panel (e) shows the beam phase-space picture after the second breaking. Many other wave-breakings occur until the beam reaches its equilibrium, which is presented in panel (f).

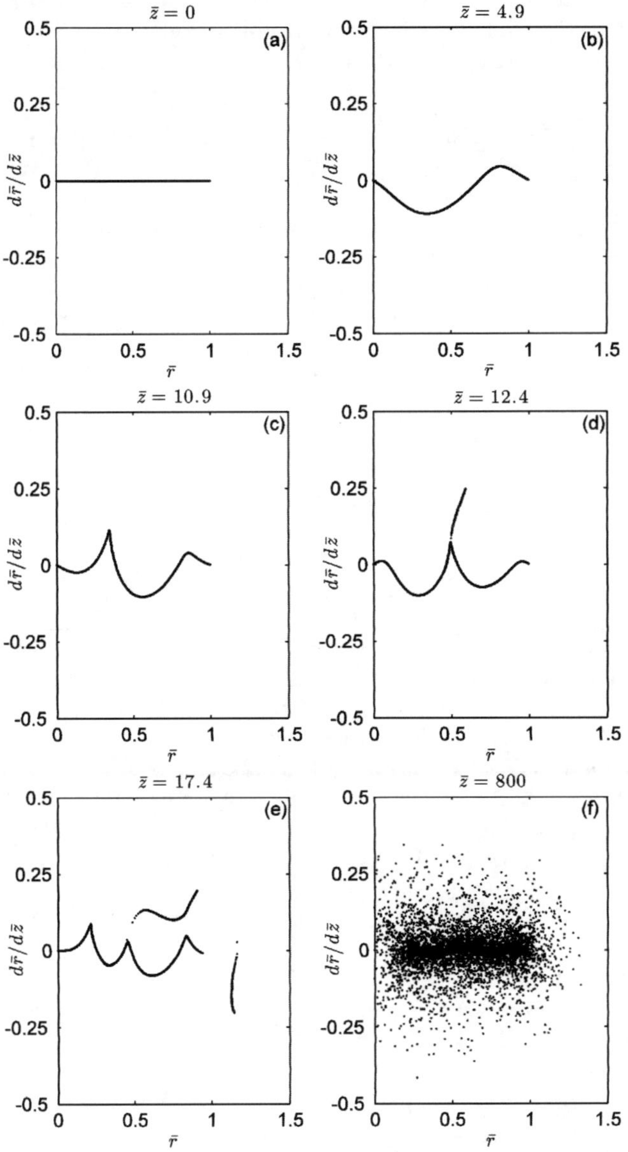

Figure 2. A sequence of snapshots of the beam phase-space dynamics of an initially cold and relativistic beam with matched envelope. In panel (a), the initial beam picture at the phase-space. During its evolution inside the focusing channel, beam charge starts to be redistributed as Panel (b) shows. This process is amplified and the phase–space density wave across the beam breaks. Panel (c) shows the axial coordinate just before and panel (d) just after the first breaking. Many other wave-breakings occur during the beam confinement process (panel (e) shows the second) until the reaching of its equilibrium (panel (f)).

This is pretty interesting since wave-breaking has been observed before in just initially inhomogeneous beams. However, it is shown that also even initially homogeneous beams can undergo wave-breaking if the relativistic effects cannot be neglected. If important, the relativistic effects unconditionally induce the breaking of the density waves inside the beam independently of the character of the initial beam density. And this is of great importance for engineering purposes once particles expelled with the wave-breaking will be the ones that will form the beam halo.

In fact, wave-breaking is associated with the nonlinear behavior of the dynamical equation that describes the motion of each particle that composes the beam. The nonlinearity can be introduced by inhomogeneity or by the inclusion of the relativistic effects to the description of the beam dynamics. Considering that both effects can act together inside the beam, in some particular situation, their superposition can eliminate the breaking of density waves. For some particular kind of initial beam density, the inhomogeneity can compensate the relativistic effects and the wave-breaking phenomenon can be suppressed. This kind of particle density can be readily obtained from equation (11). Imposing that $d\bar{r}/d\bar{z} = 0$ in this equation, solving for the fraction of charge $Q(\bar{r})$, and inserting this result in equation (7), one obtains

$$\bar{n}_b = \frac{1}{2\pi}\frac{\kappa}{K}\frac{2+\kappa\bar{r}^2}{(1+\kappa\bar{r}^2)^{\frac{3}{2}}}. \tag{15}$$

If one employs as initial beam density the function above, no wave-breaking will be observed. The beam spatial inhomogeneity produces forces that, when superposed with the ones introduced by the relativistic correction, cancel each other.

Future works will dedicate time to better understand the effects of the envelope mismatch over the beam relativistic dynamics.

4. Acknowledgements

The authors acknowledge the aid of the National Centre of Supercomputation of the Brazil's South Region (CESUP), located at the Federal University of Rio Grande do Sul (UFRGS). This work has received financial support from the National Council for Scientific and Technological Development (CNPq), and from the Fundação de Amparo à Pesquisa do Estado do Rio Grande do Sul (FAPERGS), both from Brazil.

References

[1] A. Cuchetti, M. Reiser, and T. Wangler, in Proceedings of the Invited Papers, 14th Particle Accelerator Conference, San Francisco, California, 1991, edited by L. Lizama and J. Chew (IEEE, New York, 1991), Vol. 1, p. 251; M. Reiser, J. Appl. Phys. 70, 1919 (1991).

[2] J. S. O'Connell, T. P. Wangler, R. S. Mills, and K. R. Crandall, in Proceedings of the Particle Accelerator Conference, Washington, D.C. (IEEE, New York, 1993), pp. 3657.

[3] R. A. Jameson, in Proceedings of the Particle Accelerator Conference, Washington, D.C. (IEEE, New York, 1993), pp. 3926; R. A. Jameson, Los Alamos Report LA-UR-93-1209, 1993.

[4] R. P. Nunes, R. Pakter, and F. B. Rizzato, Phys. Plasmas 14 (2), (2007); R. P. Nunes, R. Pakter, and F. B. Rizzato, J. Appl. Phys. 104 (1), (2008).

[5] J. M. Lagniel, Nucl. Instrum. Meth. A 345 (1), (1994); J. M. Lagniel, Nucl. Instrum. Meth. A 345 (3), (1994).

[6] T. Wangler, RF Linear Accelerators (Wiley-VCH, Darmstadt, 2008).

[7] C. K. Allen, K. C. D. Chan, P. L. Colestock, K. R. Crandall, R.W. Garnett, J. D. Gilpatrick, W. Lysenko, J. Qiang, J. D. Schneider, M. E. Schulze, R. L. Sheffield, H.V. Smith, and T. P.Wangler, Phys. Rev. Lett. 89 (21), (2002).

[8] R. P. Nunes, and F. B. Rizzato, in Proceedings of the 23rd Particle Accelerator Conference, Vancouver, British Columbia (IEEE, Vancouver, 2009), pp. XXXX.

[9] F. B. Rizzato, R. Pakter, and Y. Levin, Phys. Plasmas 14 (11), (2007); E. G. Souza, A. Endler, R. Pakter, F. B. Rizzato, and R. P. Nunes. App. Phys. Lett. 96 (14), (2010).

[10] R. L. Gluckstern, Phys. Rev. Lett. 73, 1247 (1994).

Nonlinear behavior of double ion beam distributions

P A Muñoz[1] and J A Araneda[2]

Departamento de Física, Facultad de Ciencias Físicas y Matemáticas,
Universidad de Concepción, Concepción, Chile.

E-mail: [1]patricioalmunoz@udec.cl
[2]jaraneda@udec.cl

Abstract. We study the stability and nonlinear evolution of differential streaming between core and beam protons based on the analysis of a nonlinear kinetic dispersion relation and using hybrid simulations describing the self-consistent decay of Alfvén-cyclotron waves. The effects of ion trapping by ion acoustic waves and pitch-angle scattering induced by decay of the Alfvén-cyclotron waves are discussed.

1. Introduction

Circularly polarized Alfvén waves are ubiquitous and frequently observed in space plasmas, in particular in the solar wind [1]. In the last four decades, their decay through parametric instabilities has been the subject of many studies by means of analytical and numerical methods, mostly within fluid MHD theory (see references in [2]). On the other hand, the proton-electron plasma in many of these space environments also shows a proton beam, which changes the behavior of the system contributing with another source of free energy for the instabilities and the evolution of both core and beam proton velocity distribution functions (VDF's). In this sense, the objective of this paper is the analysis of some features related to the presence of the beam not thoroughly understood of these systems within the framework of a hybrid kinetic-fluid dispersion relation, whose results will be compared to simulations which moreover will afford seeing the effects on the evolution of the proton VDF's.

2. Linearized Dispersion relation

2.1. Non-perturbed equilibrium system

Let us consider a finite amplitude circularly polarized Alfvén-cyclotron wave (the *parent* or *pump wave*, denoted by 0) propagating along a background magnetic field $B_s\hat{z}$, in a quasi-neutral plasma composed of two populations of protons: a core (subscript c) and a less dense drifting beam (b), as well as massless isothermal electrons (e). The pump wave has a wave number k_0 and frequency ω_0, and satisfies the exact nonlinear solution of the full set of Vlasov-Maxwell equations [3]:

$$k_0^2 = \frac{\mu_0}{B_s^2}\left[n_{0c}m_p\frac{\omega_{0c}^2}{1-\frac{\omega_{0c}}{\Omega_p}} + n_{0b}m_b\frac{\omega_{0b}^2}{1-\frac{\omega_{0b}}{\Omega_b}}\right], \tag{1}$$

where, if $j = e, c, b$ denotes each specie of the plasma, $\Omega_j = q_jB_s/m_j$, n_{0j} are the densities in equilibrium, and $\omega_j = \omega_0 - k_0U_j$, with U_j the parallel (to the magnetic field B_s) drift speeds with respect to the center of mass frame. The protons will be represented by the distribution function (which satisfies the Vlasov equation)

$$F_j(z,\mathbf{v},t) = n_{0j}\delta(\mathbf{v}_\perp - \mathbf{V}_{\perp j})f_j(z,v_z,t), \tag{2}$$

where

$$\mathbf{V}_{\perp j} = -\frac{\omega_{0j}/k_0}{1-\frac{\omega_{0j}}{\Omega_j}}\frac{B_0}{B_s} \tag{3}$$

are the perpendicular bulk velocities and $f_j(z, v_z, t)$ are spatially uniform drifting Maxwellians. This model [4] has the advantage it keeps the kinetic effects along the longitudinal direction, while the transverse motion of the protons can be modeled by the simpler Hall-MHD equations.

2.2. *Hybrid kinetic dispersion relation*

Now, we consider perturbations about the previously described equilibrium in the form $\delta f \sim \exp(ik_z - i\omega t)$ (and analogously for the rest of the quantities), where $\omega = \omega_r + i\gamma$, with γ the growth rate of the instabilities. Linearizing the Vlasov-MHD-Maxwell set of equations, we obtain a dispersion relation (as appears in [2], with the appropiate changes in the reference system) preserving the same structure as in the pure fluid case, but with a kinetic polytropic coefficient (relating the perturbations of pressure and density) defined as:

$$\gamma_j = 2\left(\xi_j^2 - \frac{1}{Z'(\xi_j)}\right), \tag{4}$$

where $\xi_j = \omega_j/(kv_{Tj})$, $v_{Tj} = \sqrt{2k_BT_{\|j}/m_p}$ are the thermal speeds (m_p: mass of the proton) and $Z'(\xi_j)$ is the derivative of the plasma dispersion function. This dispersion relation, with an infinite number of roots (mostly very damped), has the parameters: amplitude $A = (B_0/B_s)^2$, $\beta_j = 2\mu_0 n_j k_B T_{\|j}/B_s^2$ (=proton thermal pressure/background magnetic pressure), normalized beam density $\eta_b = n_b/n_e$ and relative drift speed $U := (U_b - U_c)/V_A$, with $V_A = B_s^2/(\mu_0 n_e m_p)$ the Alfvén speed (it is useful too the notation $\widetilde{\beta}_j = v_{Tj}^2/V_A^2$, proportional to the temperature). Its ten less damped modes ("daughter" waves) can be identified as:

- $\pm f$: upper/lower sideband forward propagating
- $\pm b$: upper/lower sideband backward propagating
- $\pm beam$: upper/lower beam sideband forward propagating (due to beam)
- $\pm s(b)$: ion (beam)-acoustic high/low phase speed

3. Numcrical solutions and hybrid simulations

3.1. *A typical high drift speed, high beam density case*

We have performed numerical simulations using a hybrid code (i.e.: electrons considered as massless fluid and protons treated kinetically as particles) 1D in space and 3D in velocity [4], in the protons' center of mass frame. The chosen parameters are typical for space plasmas like the fast solar wind or the Earth's bowshock [1].

Thus, in the following figures are shown -by means of the power spectrum of the (longitudinal) density fluctuations- the aforementioned standard modes by comparing with the analytic dispersion relation and their respective growth rates (black lines indicate instabilities) and phase speeds. Colored lines indicate sideband-related waves, and the gray ones are acoustic waves. The parameters used in Figures 1-4 are: $\widetilde{\beta}_e = \widetilde{\beta}_c = \widetilde{\beta}_b = 0.1$, $\eta_b = 0.15$, $k_0V_A/\Omega_p = 0.4$, $U = 1.5$ and $A = 0.3^2$. We have chosen a left-hand polarized forward propagating pump wave ($\omega_0/\Omega_p \approx 0.253$), represented as an open diamond in the diagrams. For the simulation, we choose the system to have a length $L_0 = 2\pi m_0/k_0$, with the mode number of the pump wave $m_0 = 32$. The boundary conditions are periodic with 2048 cells and 400 particles/cell.

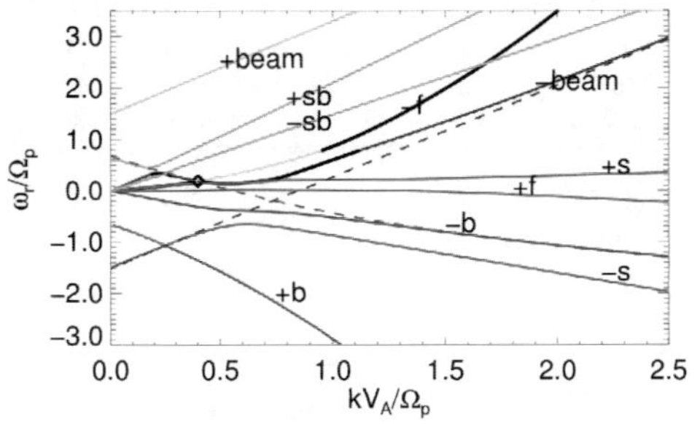

Figure 1. Analytic modes

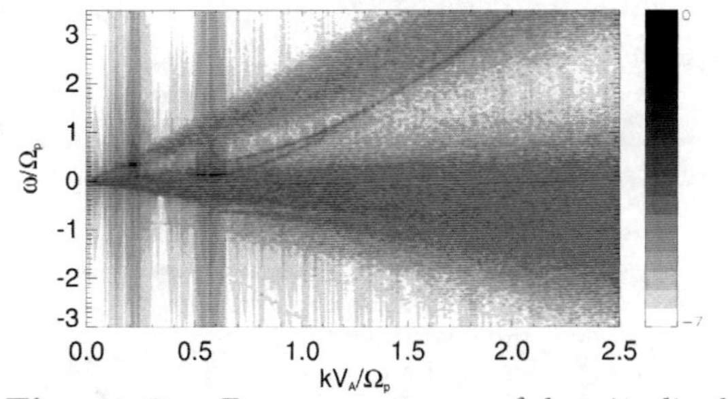

Figure 2. Power spectrum of longitudinal fluctuations δn_e

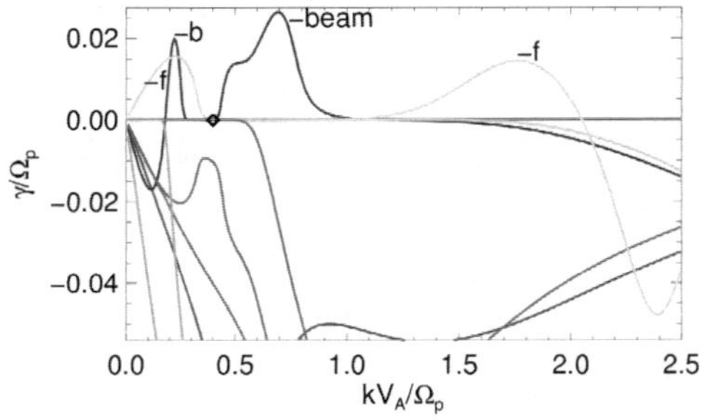

Figure 3. Analytic growth rates

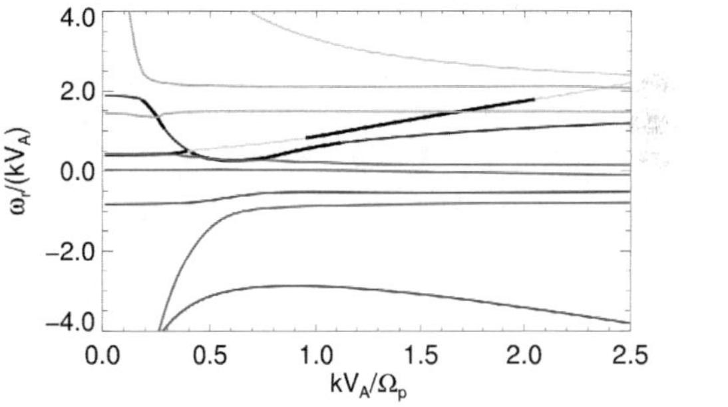

Figure 4. Analytic phase speeds

There are gaps in some branches (whose separation is proportional to η_b) of the dispersion diagram, in particular between the $-b$ and $-beam$. Thus, it is only meaningful to assign these names to the asymptotes showned as dashed lines, but in the gap region we shall only denote a determined branch when the usual electron-proton instabilities [4] are recovered.

Four parametric instabilities are visible and in good agreement with simulations, which can be classified in two groups according to its phase speed $v_\phi = \omega_r/k$:

- Two strong of low phase speed: one for $k < k_0$ in the $-f$ branch ($v_\phi/V_A \approx 0.39$) which extends until $k = 0$ (*modulational*) and other for $k > k_0$ in the $-beam$ branch ($v_\phi/V_A \approx 0.26$), a combined version of the usual beat and decay instabilities.
- Two weak of high phase speed: one for $k < k_0$ in the $-b$ branch (very narrow, $v_\phi/V_A \approx 1.57$) and other for $k > k_0$ in the $-f$ branch (very broad, $v_\phi/V_A \approx 1.54$). These ones are related to the beam-core interaction and its maxima are near to the intersection of the sideband waves with the acoustic $-sb$ branch. They are also present in the fluid approach [5].

We have also verified that the usual conservation relations between the frequency and number wave of the pump wave and of its respective daughter waves are satisfied. This was obtained by means of the time history of the wave power in the magnetic and density fluctuations calculated from the simulations, by analyzing the respective modes with the greater growth rate as time goes by.

In general, the growth rates of all these instabilities are increasing functions of k_0 and A. We can even note that there is another instability of very low phase speed in the -b branch which appears only for a tenous beam and a high k_0 (For example, $\eta_b = 0.025$ and $k_0V_A/\Omega_p = 0.7$). This is also a increasing function of U.

3.2. Effects of the drift speed

To characterize the instabilities and their relationship with the shape of the proton VDF's, we made a parameter survey varying the corresponding parameters related to the core and beam as well as to the pump wave. But in this paper, we analyse principally the behavior related to the drift speed U. A useful way to visualize this (with the analytic dispersion relation) is by means of plots of the maximum growth rates of each instability (and its respective phase speed) vs. U, maintaining the other parameters fixed, in particular k_0. But this last condition, by the relation (1), implies a monotonically decreasing ω_0 with respect to U. The same previous maximum $U = 2.1$ was chosen to be less than the threshold to trigger the linear right-hand polarized instability [6] for this A, because its corresponding growth rate is much higher that the aforementioned parametric instabilities, and therefore it would hide the behavior of these ones. The symbols used in Figures 5-6 are ► and ◄: high phase speed instabilities; ●: modulational; ◇, ○ and □: low phase speed instabilities for $k > k_0$.

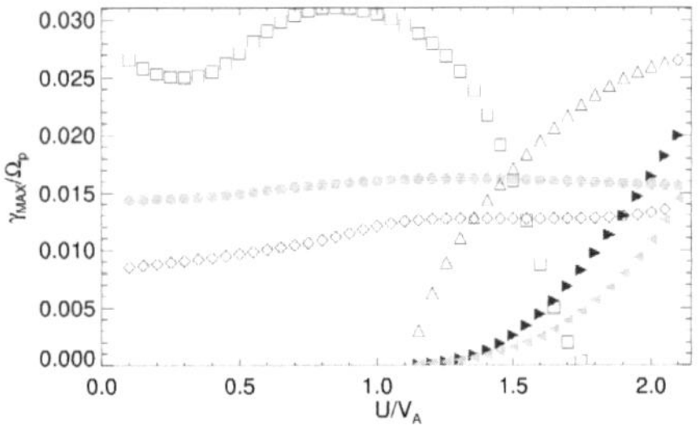

Figure 5. Maximum growth rates for varying U. Same previous parameters (except ω_0) and instabilities colors. Explanation of the symbols in the text.

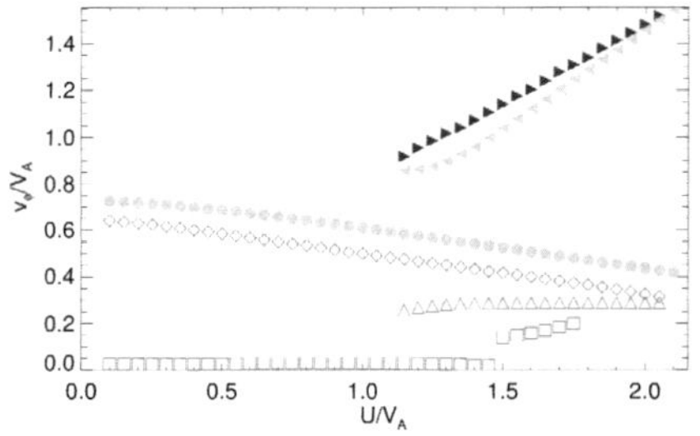

Figure 6. Phase speeds at the maxima growth rates for varying U. See figure 5.

Between $U = 1.2 - 1.7$, the low phase speed instability of the $-beam$ branch has three maxima. In the dispersion diagram ω-k, we can associate these three instabilities with the sectors where the $-beam$ branch is located under or upper the $+f$ branch. In the intersections of both branches, they are stable. Two of these (with the lesser k) are the usual *beat* (○ , over $+f$) and *decay* (□, under $+f$) for an electron-proton plasma [1], and therefore the third one (◇, for the largest k, again over $+f$) is related to the beam. Furthermore, this last one only appears after a threshold $U \approx 1.2$.

The growth rates of the modulational and beat are minimally affected by the beam. On the other hand, the decay instability has a maximum growth rate for $U \approx 0.8$, and for higher U it decreases at the same time that the third low phase-speed instability for $k > k_0$ begins to increase (from $U \approx 1.2$). This processes continues until $U \approx 1.5$, at which both have similar growth rates, and, when U takes higher values, this situation reverses, i.e: decay is growing at the expense of the other instability. For low drift speed $U \lesssim 1.1$, there only exists the well-known core-related instabilities, and are well distinguishable the strong decay and the weaker beat, that together with the modulational, are ubiquitous in a wide range of parameters for $\beta_{c,b} < 1$ and a left-hand polarized pump wave [2].

On the other hand, we can justify that the two high phase speed instabilities are related to the beam because both have maximum growth rates that are monotonically increasing functions of U (and also, but it is not shown here, increasing functions of η_b). However, these instabilities require a relatively high drift speed to be triggered ($U \gtrsim 1.5$). For the high drift speed case $U = 2.1$, the growth rates of the beam-related instabilities are comparable to the corresponding of the electron-proton instabilities. Furthermore, their phase speeds are even higher that for the $U = 1.5$ case, whereas the other core-related instabilites have somehow lower phase speeds. We

Journal of Physics: Conference Series **511** (2014) 012004 doi:10.1088/1742-6596/511/1/012004

can obtain higher growth rates and higher phase speeds for a lesser β_b, i.e.: for a colder beam.

We also analyzed the properties of those instabilities for a right-polarized pump wave. As is shown in past studies within fluid theory [7], for the electron-proton plasma instabilities, it is only the decay instability in the $-b$ branch which has important growth rates for $\beta_{c,b} < 1$. In addition, we also observed several minor instabilities related to the beam in various branches with similar behavior to the high phase speed ones previously mentioned, i.e.: they have a minimum threshold around $U \approx 1.5$ and are also increasing functions of the drift speed.

4. Evolution of the VDF's

The time evolution of the core and beam distributions are shown below in Figures 7-8, for two cases of drift speeds.

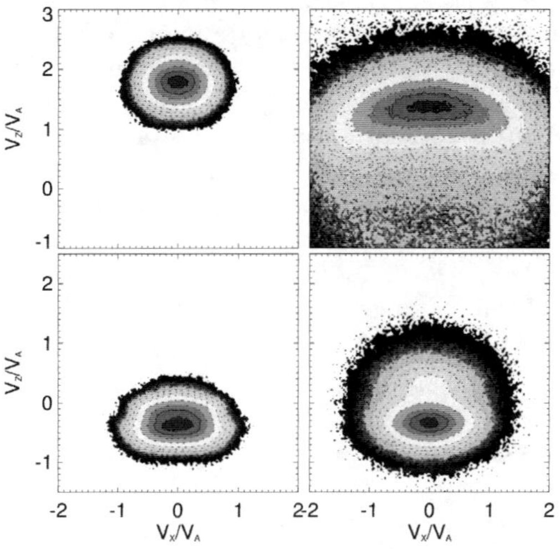

Figure 7. Protons VDF's. Up: Beam, Down: Core. Left: $t = 0\ \Omega_p^{-1}$, Right: $t = 200\ \Omega_p^{-1}$. Same previous parameters. High drift speed case $U = 2.1$.

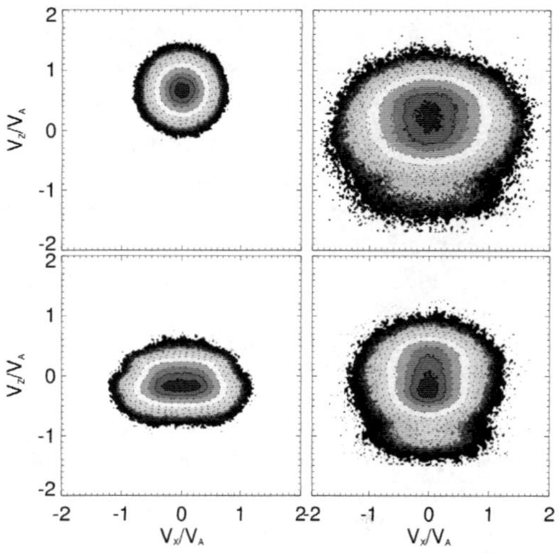

Figure 8. Protons VDF's. Low drift speed case $U = 0.8$. See figure 7

For $U = 2.1$, we note the beam formation in the core population due to trapping by the ion-acoustic waves [8]. In fact, we have two low phase speed instabilities ($v_\phi/V_A < 0.4$) which affect the core VDF, inducing a elongation in the longitudinal direction due to the resonant interactions with the particles. On the other hand, the high-speed parametric instabilities related to the beam affect the center of the beam VDF (because $v_\phi/V_A \approx 1.5$), but its effects are not easily visible. Instead of that, pitch-angle scattering, induced by the parametric instabilities, produces a reshaping of the protons VDF's mainly in the transversal direction. This heating of the beam component is clearly visible by the parabolic shell-like trajectories of its particles. On the other hand, the low phase-speed instabilities also involve some particles in the tail of the beam, elongating this distribution in the core direction.

In the low speed case $U = 0.8$, we can note from the Figure 6 and the previous discussion that the beat ($v_\phi/V_A = 0.53$), decay ($v_\phi/V_A = 0.16$) and modulational ($v_\phi/V_A = 0.64$) affects both the core and the beam, elongating the first in the longitudinal direction towards the beam and the second in opposite direction. Moreover, the pitch-angle scattering in the beam poulation is much less notorious. Thus, both effects produce a more localized (less diffuse) beam VDF in the velocity space, i.e: its heating is lower that for the high drift speed case. It is wortwhile to note that the beam VDF's are closer to the core as the times goes by, while the core goes backwards by momentum conservation in the longitudinal direction [8], but in a lesser proportion due to its larger inertia. On the other hand, the behavior of the core is similar in both cases, although its prolongation is more marked for the low drift-speed case. Therefore, the parametric instabilities plays a crucial role in the morphological evolution of the proton VDF's, producing in general more diffuse distributions in the transversal direction at higher drift speeds.

5. Summary

In the present paper, we have discussed the drift speed dependence of the parametric instabilities of an electron-core-beam proton plasma, and its relation with the proton VDF's, complementing similar studies [2] and [9] (for a beam of α particles instead of protons). The hybrid kinetic dispersion relation was solved numerically, identifying the respective modes. We found a threshold in U for two instabilities of high phase speed related to the beam: a narrow one for $k < k_0$ and a broad one for $k > k_0$. On the other hand, we also have the usual parametric instabilities which appears in an electron-plasma, but they have a lesser phase speed. When we analyzed the effects of these instabilities on the evolution of the proton VDF's, it is possible to note a elongation of the core and beam (in the longitudinal direction) in the places where the excited ion-acoustic waves are resonant with the particles. The main difference when we vary U is for the beam VDF: it is much more diffuse in the velocity space as time goes by for a high drift speed, implicating a greater heating, but mainly due to an induced pitch-angle scattering instead of the effects of ion-acoustic waves resulting from parametric instabilities.

Acknowledgments

One of us (P.A.M.) wants to thank the financial support to attend the ICPP-LAWPP 2010 given by its organizing committee as well by the Dirección de Docencia and Departamento de Física of the Universidad de Concepción.

References

[1] Marsch E 2006 Kinetic physics of the solar corona and solar wind *Living Rev. Solar Phys.* **3** 1
[2] Nariyuki Y, Hada T and Tsubouchi K 2009 Parametric instabilities of circularly polarized Alfvén waves in plasmas with beam protons *J. Geophys. Res.* **114** A07102
[3] Sonnerup B U Ö and Su S Y 1967 Large amplitude whistler waves in a hot collision-free plasma *Phys. Fluids* **10** 462
[4] Araneda J A, Marsch E and Viñas A F 2007 Collisionless damping of parametrically unstable Alfvén waves *J. Geophys. Res.* **112** A04104
[5] Gomberoff L, Gomberoff K and Brinca A L 2002 New parametric decays of proton beam plasma electromagnetic waves *J. Geophys. Res.* **107** A7, 1123
[6] Gomberoff L 2003 Stabilization of linear ion beam right-hand polarized by nonlinear Alfvén/ion-cyclotron waves *J. Geophys. Res.* **108** A6, 1261
[7] Hollweg J V, Esser R and Jayanti V 1993 Modulational and decay instabilities of Alfvén waves: Effects of streaming He^{++} *J. Geophys. Res.* **98** A3, 3491
[8] Araneda J A, Maneva Y and Marsch E 2009 Preferential heating and acceleration of α particles by Alfvén-cyclotron waves *Phys. Rev. Lett.* **102** 175001
[9] Kauffmann K and Araneda J A Parametric instabilities of Alfvén waves in a multispecies plasma: Kinetic effects *Phys. Plasmas* **15** 062106

Collective modes in free plasmas subjected to a radiation field

B V Ribeiro[1], **D D A Santos**[1], **M A Amato**[1,2] **and A L A Fonseca**[1,2]

[1] Instituto de Fsica, Universidade de Brasilia, Campus Darcy Ribeiro, Brasilia, Brasil.

[2] International Centre for Condensed Matter Physics, Universidade de Brasilia, Campus Darcy Ribeiro, Braslia, Brasil.

E-mail: `bruno64bits@yahoo.com.br`

Abstract. In this study we report the effects of an electromagnetic field on the physical properties of free plasmas[3].The calculations are carried out in the semi-classical approximation, i.e., the electromagnetic field is treated classically and the electrons from a quantum mechanical viewpoint. The results show that the collective modes are damped away more smoothly and in a smaller frequency range than those reported by previous studies. An exponential-like decay of the frequencies is readily observable from the plot of the plasmon frequency as a function of the external field amplitude. We successfully recreate the results of previous studies. We also obtain that the single photon processes has a pronounced effect on the decrease of the frequency range of modulation.

1. Introduction

The studies concerning the interaction of electromagnetic fields with electron plasmas using the Schrodinger equation to describe the electron states [2, 3, 4] has been done for a long time. In particular, photon-plasma interactions has been discussed for many authors. See [5] and references therein. Here we extend the work developed in ref. [2] where there is no photon process contributing to the collective modes in the plasma, in what they call a *weak electron-radiation coupling*, and also in references [3, 4] where two limiting cases are discussed, i.e., the strong-field limit such that only multiphoton processes are significant whereas the weak-field regime such that only single-photon processes are significant, respectively.

For this study, the laser beam is treated as a classical plane electromagnetic wave in the dipole approximantion. We consider the laser linearly polarized along the z-axes, with the electric field along the x-axes. The electron states are described by the solution of the Schrodinger wave equation for an electron in the laser field. In this report we take into account a finite number of photons interacting with the electron plasma. This simple extension gives rise to a different dispersion relation of the electron waves. We see that an asymptotic value for the plasmon frequency exists as we increase the radiation wave number, and that this frequency never goes to zero, given a non-zero natural plasma frequency. The collective modes of the plasma are obtained numerically from the zeros of the dielectric function.

[3] The term *free plasma* is used as *unmagnetized plasma*

Journal of Physics: Conference Series **511** (2014) 012005 doi:10.1088/1742-6596/511/1/012005

2. Electron states

For a free plasma under the presence of an electromagnetic wave, the Schrodinger equation for the electrons takes the form:

$$H\psi(\mathbf{r},t) = i\hbar\frac{\partial\psi(\mathbf{r},t)}{\partial t}, \tag{1}$$

where H is the hamiltonian of the system given by

$$H = \frac{1}{2m_e}(\mathbf{p} - e\mathbf{A}(t))^2, \tag{2}$$

$\mathbf{p}$ is the momentum of the electron, $\mathbf{A}$(t) is the vector potential of the radiation

$$\mathbf{A}(t) = \left(\frac{E}{\omega}sin(\omega t)\right)\hat{\mathbf{x}}, \tag{3}$$

and ω is the frequency of the radiation. To solve this equation, given a time-dependent potential, we use a unitary transformation [6, 7] of the form

$$\psi\,(\mathbf{r},t) = U\Phi(\mathbf{r},t), \tag{4}$$

$\Phi(\mathbf{r},t)$ being the solution of the Schroedinger equation for a free electron and U the unitary operator:

$$U = exp\left(\frac{i}{\hbar}\vec{\alpha}(t)\cdot\mathbf{r}\right)exp\left(\frac{i}{\hbar}\vec{\beta}(t)\cdot\mathbf{p}\right)exp\left(\frac{i}{\hbar}\eta(t)\right), \tag{5}$$

where $\vec{\alpha}(t), \vec{\beta}(t)$ and $\eta(t)$ are, respectively, the momentum translation generator, the spatial translation generator and a phase factor. Substituting this into equation (1) we obtain

$$\frac{\partial\psi(\mathbf{r},t)}{\partial t} = \frac{i}{\hbar}\left(\frac{d\vec{\alpha}}{dt}\cdot\mathbf{r} + \frac{d\vec{\beta}}{dt}\cdot\mathbf{p} + \frac{d\eta}{dt}\right)\psi(\mathbf{r},t) + U\frac{\partial\Phi(\mathbf{r},t)}{\partial t}. \tag{6}$$

Multiplying by U^+ and after doing some manipulation:

$$\begin{aligned} H_0 &= \frac{\mathbf{p}^2}{2m_e} + \frac{1}{2m_e}\left(\vec{\alpha}(t) - e\mathbf{A}(t)\right)^2 + \frac{\mathbf{p}}{m_e}\left(\vec{\alpha}(t) - e\mathbf{A}(t)\right) \\ &+ \frac{d\vec{\alpha}}{dt}\cdot\mathbf{r} + \frac{d\vec{\beta}}{dt}\cdot\mathbf{p} - \frac{d\vec{\alpha}}{dt}\cdot\vec{\beta} + \frac{d\vec{\beta}}{dt}\cdot\vec{\alpha} + \frac{d\eta}{dt}, \end{aligned} \tag{7}$$

where H_0 is the hamiltonian for the free electron. We proceed to solve the equation for $\vec{\alpha}, \vec{\beta}$ and η (linear terms in $\mathbf{r}$ and $\mathbf{p}$ and independent terms are set equal to zero) and find:

$$\psi(\mathbf{r},t) = exp\left(\frac{i}{\hbar}F(t)\right)exp\left(i\gamma_0 k_x\left(1 - cos(\omega t)\right)\right)exp(i\mathbf{k}\cdot\mathbf{r})exp\left(-\frac{i}{\hbar}\varepsilon_{\mathbf{k}}t\right), \tag{8}$$

where we have simplified using $F(t) = -2\gamma_1\omega t + \gamma_1 sin(2\omega t)$, $\gamma_0 = \frac{eE}{m_e\omega^2}$, $\gamma_1 = \frac{e^2E^2}{8m_e\omega^3}$ and $\varepsilon_{\mathbf{k}}$ is the energy of the free electron with wave number $\mathbf{k}$.

Equation (8) is the wave function for an electron in an eletromagnetic field given by (3).

We can see that the wave function in (8) forms an orthonormal set. So, it is possible to expand the wave function of an electron in a local potential using (8) as basis:

$$\Psi_k(\mathbf{r},t) = \sum_{\mathbf{k}} a_{\mathbf{k}}(t)\psi(\mathbf{r},t). \tag{9}$$

Assuming a weak local potential of the form

$$\varphi(\mathbf{r},t)=\int d\mathbf{q}\int d\Omega exp(i\mathbf{q}\cdot\mathbf{r})exp(-i\Omega t)\varphi(\mathbf{q},\Omega)+c.c. \tag{10}$$

we determine the coefficients $a_{\mathbf{k}}(t)$ using perturbation theory [8]:

$$\begin{aligned}a_{\mathbf{k+q}}(t) &= e\cdot exp(-i\gamma_0 q_x)\sum_{m,\Omega} i^m J_m(q_x\gamma_0)\varphi(\mathbf{q},\Omega)\\ &\cdot \frac{exp\left(\frac{i}{\hbar}\left(\varepsilon_{\mathbf{k+q}}-\varepsilon_{\mathbf{k}}-\hbar\Omega-m\hbar\omega\right)t\right)}{\varepsilon_{\mathbf{k+q}}-\varepsilon_{\mathbf{k}}-\hbar\Omega-m\hbar\omega-i\zeta}\quad(\zeta\to 0^+),\end{aligned} \tag{11}$$

where we have used the identity

$$exp(i\alpha cos x)=\sum_m i^m J_m(\alpha)exp(imx). \tag{12}$$

Finally, we can write the wave function as

$$\begin{aligned}\Psi_k(\mathbf{r},t) &= exp\left(\frac{i}{\hbar}F(t)\right)exp\left(i\gamma_0 k_x\left(1-cos(\omega t)\right)\right)exp\left(-\frac{i}{\hbar}\varepsilon_{\mathbf{k}}t\right)exp(i\mathbf{k}\cdot\mathbf{r})\\ &\cdot\left\{1+e\sum_{\mathbf{q},\Omega}exp\left(-i\gamma_0 q_x cos(\omega t)\right)\cdot\sum_m i^m J_m(q_x\gamma_0)\cdot\frac{\varphi(\mathbf{q},\Omega)exp\left(-i(\Omega+m\omega)t\right)}{\varepsilon_{\mathbf{k+q}}-\varepsilon_{\mathbf{k}}-\hbar\Omega-m\hbar\omega-i\eta}exp(i\mathbf{q}\cdot\mathbf{r})\right\}\end{aligned} \tag{13}$$

This last wave function corresponds to the electrons of the plasma under the incidence of the radiation given by (3).

3. Dielectric function

Now that we know the states of the electrons via (13), we can obtain the fluctuation of the charge density

$$\rho_k(\mathbf{r},t)=-e\left|\Psi_k(\mathbf{r},t)\right|^2-\rho_k^{(0)}(\mathbf{r},t), \tag{14}$$

$\rho_k^{(0)}(\mathbf{r},t)$ being the charge density in the absence of a weak local potential, *i.e*, the charge density given by the $\psi(\mathbf{r},t)$ distribution. Neglecting terms of higher orders in φ:

$$\begin{aligned}\rho_k(\mathbf{r},t) &= -e^2\sum_{\mathbf{q},\Omega}exp(i\mathbf{q}\cdot\mathbf{r})\varphi(\mathbf{q},\Omega)exp(-i\gamma_0 q_x cos(\omega t))\\ &\cdot\sum_m i^m J_m(q_x\gamma_0)\left\{\frac{exp(-i(\Omega+m\omega)t)}{\varepsilon_{\mathbf{k+q}}-\varepsilon_{\mathbf{k}}-\hbar\Omega-m\hbar\omega-i\eta}+\frac{exp(i(\Omega+m\omega)t)}{\varepsilon_{\mathbf{k+q}}-\varepsilon_{\mathbf{k}}+\hbar\Omega+m\hbar\omega+i\eta}\right\}.\end{aligned} \tag{15}$$

Assuming a maxwellian distribution ($f_{\mathbf{k}}$), we have the total fluctuation as

$$\rho(\mathbf{r},t)=\sum_{\mathbf{k}} f_{\mathbf{k}}\rho_k(\mathbf{r},t). \tag{16}$$

Using, once more, relation (12), we obtain

$$\begin{aligned}\rho(\mathbf{r},t) &= -e^2\sum_{\mathbf{q},\Omega}exp(i\mathbf{q}\cdot\mathbf{r})\varphi(\mathbf{q},\Omega)exp(-i\Omega t)\cdot\\ &\sum_{m,m'} i^{m-m'}J_m(q_x\gamma_0)J'_m(q_x\gamma_0)exp(-i(m-m')\omega t)\cdot\Pi(\mathbf{q},\Omega+m\omega),\end{aligned} \tag{17}$$

Journal of Physics: Conference Series **511** (2014) 012005 doi:10.1088/1742-6596/511/1/012005

where $\Pi(\mathbf{q},\Omega) = \sum_{\mathbf{k}} \frac{f_{\mathbf{k+q}}-f_{\mathbf{k}}}{\varepsilon_{\mathbf{k+q}}-\varepsilon_{\mathbf{k}}-\hbar\Omega-i\eta}$ is the electronic polarizability and m corresponds to the number of photons involved in the process[4]. We take the real part of (17) to calculate the fluctuation. This fluctuation induces a potential in the medium given by Poison equation

$$\nabla^2\varphi_{ind}(\mathbf{r},t) = -4\pi\rho(\mathbf{r},t). \tag{18}$$

Using (17) and the Fourier transform of (18), we obtain

$$\varphi_{ind}(\mathbf{q},\Omega) = \frac{4\pi e^2}{q^2}\varphi(\mathbf{q},\Omega)\sum_m J_m^2(q_x\gamma_0)\Pi(\mathbf{q},\Omega+m\omega), \tag{19}$$

which is the induced part of the *full* local potential:

$$\varphi(\mathbf{q},\Omega) = \varphi_{ext}(\mathbf{q},\Omega) + \varphi_{ind}(\mathbf{q},\Omega) = \frac{\varphi_{ext}(\mathbf{q},\Omega)}{\epsilon(\mathbf{q},\Omega)}, \tag{20}$$

$\epsilon(\mathbf{q},\Omega)$ being the dielectric function of the plasma. The roots of the real part of this dielectric function give us the frequency of the longitudinal waves (collective oscillations, *i.e*, plasmons) in the plasma. We, therefore, separate $\epsilon(\mathbf{q},\Omega)$ into a real and an imaginary part. We proceed to the classical limit by letting $\hbar \to 0$ and, after some algebraic work, we are left with:

$$\epsilon_{real}(\mathbf{q},\Omega) = 1 - \omega_p^2\sum_m J_m^2(q_x\gamma_0)\frac{1}{\lambda^2}\left(1+\frac{2\langle \mathbf{q}\cdot\mathbf{v}\rangle}{\lambda}+\frac{3\langle(\mathbf{q}\cdot\mathbf{v})^2\rangle}{\lambda^2}\right)exp\left(-\frac{\varepsilon_\gamma}{k_BT}\right), \tag{21}$$

where $\lambda = m\omega + \Omega$, ω_p is the plasma natural frequency, $\varepsilon_\gamma = 2\gamma_1\omega$ is the energy of the electromagnetic radiation and the average $\langle g(\mathbf{q},\mathbf{v})\rangle$ is taken with respect to the function f_k.

4. Collective modes

The roots of (21) give us the frequencies of collective oscillations in the plasma as a function of their wave number ($\Omega(\mathbf{q})$). Here we encounter a numerical difficulty, as the sumation over m does not allow us to get an analytical result for this frequencies. However, if we fix a value for **q**, we obtain an expression depending only on Ω, which is easily solvable using a bisection method in FORTRAN language [9]. The sum over m is truncated for a value of m beyond which the terms become neglegible. In this way, we take values of q from zero to about $20000m^{-1}$, and for each q we get a frequency value. So, we can plot the dispersion relation (Ω versus q) of this system for various values of the radiation frequency. In the same manner, we can fix values of E and get the dependece of Ω with E. These plots are shown in figures (1) and (3). We see that, for any given value of ω (the frequency of the radiation), the asymptotic value for the plasmon frequency is the same, the difference is that for larger values of ω the plasmon frequency decays slower. From this we see that, for a radiation with large enough energy, the plasma remains unperturbed.

We can also vary the number of photons involved in the process and see how the curves react for a natural plasma frequency of $6.10^{11}s^{-1}$. This plots are shown in figures (2) and (4).

In figure (2) wee note a clear restriction to the range of frequencies allowed to the plasmons as m increases. So, as more photons participate in the process, the more energetic the plasmons are (given the same value of q). For all plots, the temperature of the plasma is $k_BT = 1,6.10^{-19}J$.

[4] Notice how m appears as the number of $\hbar\omega$ contributing to the polarizability

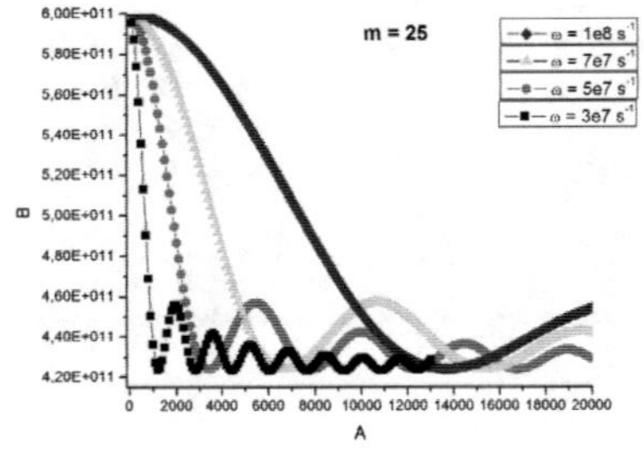

Figure 1. The natural plasma frequency for this case is $6.10^{11}s^{-1}$ and E=10 V/m. Notice that, as we increase the radiation frequency, the dispersion decays slower. The asymptotic value for Ω is about $4,2.10^{11}s^{-1}$.

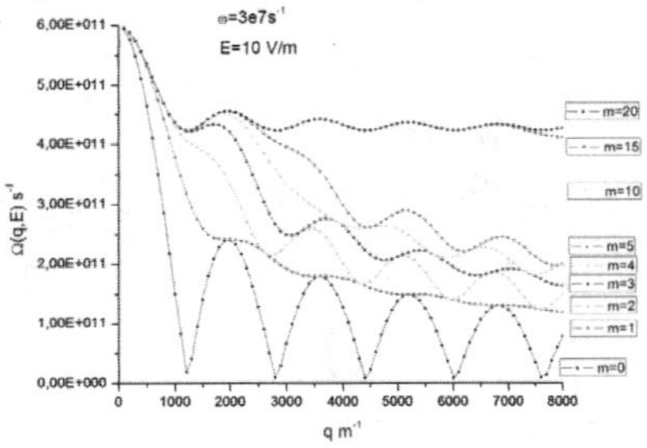

Figure 2. There is a assimptotic value for the plasmon frequency as we increase both m an q. The curve for m=0 reproduces exactly that of ref. [2].

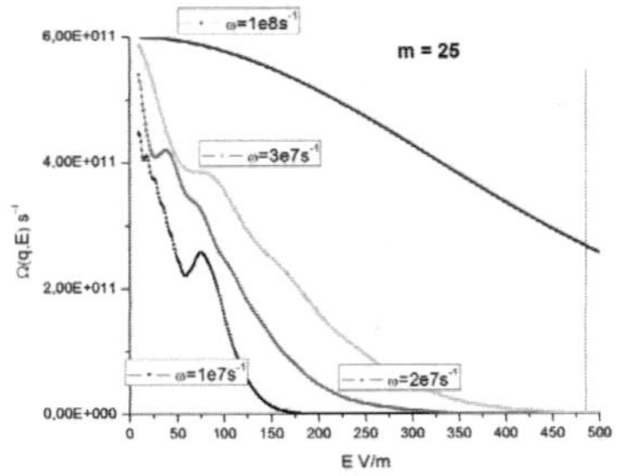

Figure 3. The natural plasma frequency for this case is, also, $6.10^{11}s^{-1}$ and the value of q is generated randomly in the interval of figure(1).For the E dependence we see an almost exponential like decay, which agrees with exponential term that appears in (21).

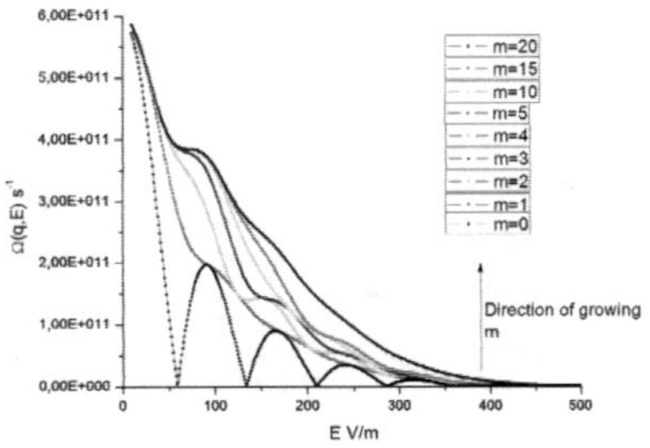

Figure 4. We see the tendency of the curve becoming an exponential-like decay as we increase the number of photons in the process.

5. Conclusion

The number os photons involved in the interaction between the plasma and the electromagnetic radiation is of extreme importance. We see that a single photon "elevates" the range of frequencies of the plasmons (see figure (2)), and, as we increase the number of photons, an assimptotic value appears for these frequencies.

The plasmon frequency decays very rapidly with increasing field amplitude (E) in an exponential-like curve (see figure(3)). As we increase the number of photons in the process, we no longer see an oscillation in the curve of Ω versus E as reported by [2]. We are currently extending these results allowing nonlinearities, and putting forward a full numerical calculation of the dielectric constant.

Acknowledgments
BRV and DDAS aknowledge CAPES for financing their graduate studies. MAA and ALAF aknowledge CNPq for their research grant.

References
[1] Akhiezer A I, Akhiezer I A, Polovin R V, Sitenko A G and Stepanov K N 1967 *Collective Oscillations in a Plasma* (Oxford: Pergamon Press)
[2] Guimaraes A F 2006 *Estudo, utilizando a mecanica quantica, das propriedades dieletricas e do efeito da blindagem dinamica na taxa de aquecimento de plasmas macroscopicos* PhD thesis accepted by the Institute of Physics, University of Braslia (Brasil)
[3] Amato M A and Miranda L C M 1977 *The Phys. of Fluids* **20** 1031-32
[4] Amato M A 1986 *Il Nuovo Cimento* **7** 767-70
[5] Harris E G 1969 in *Advances in Plasma Physics* **3** 157-241 edited by Simon A and Thompson W B (New York: Interscience Publisher)
[6] Lima M B S and Miranda L C M 1978 *J. of Phys.* C **11** L843
[7] Galvao R M O and Miranda L C M 1983 *Am. J. of Phys.* **51** 729
[8] Sakurai J J 1994 *Modern Quantum Mechanics* (New York: Addison-Wesley)
[9] Press W H, Vetterling W T, Teukolsky S A and Flannery B P 1989 *Numerical Recipes in FORTRAN 77: The Art of Scientific Computing 2nd Edition* (London: Cambridge University Press)

Mechanism of nonlinear oscillitons evolved from simple solitary waves

John Z. G. Ma[1] and A. Hirose[2]

[1]Space Science Branch, Canadian Space Agency, 6767, Route de l'Aéroport, Longueuil (Saint-Hubert), Québec, J3Y 8Y9, Canada

[2]Plasma Physics Laboratory, University of Saskatchewan, 116 Science Place, Saskatoon, Saskatchewan, S7N 5E2, Canada

E-mail: [1]johnzg.ma@asc-csa.gc.ca and [2]akira.hirose@usask.ca

Abstract. Besides three conventional, nonlinear simple waves (namely, sinusoidal, sawtooth, and spicky / bipolar) measured in space plasmas, numerous extraordinary nonlinear oscillitons (i.e., amplitude-modulated solitons) have been captured by satellites in last decades. By employing a self-similar, two-fluid (electron-ion) model, we focus on the evolution from single-frequency solitary waves to multi-frequency oscillitons. It is shown that the structures can be triggered by either the charge non-neutrality auto-modulationally, or the electron inertia mutual-modulationally.

1. Introduction

For more than 70 years, a salient nonlinear waveform has been captured in solar-terrestrial space by numerous experiments: a low-frequeny (LF) envelop, while the amplitude of which is modulated violently by higher-frequency (HF) oscillations. In extraterrestrial spaces, detections of such LF-HF "symbiotic" packets began from the 1980s when Voyager 1 & 2 spacecrafts in the upstream of Jupiter's bow shock[1]. In the vicinity of Io, the Galileo spacecraft recorded similar waveforms which were later supposed to be connected with ion-cyclotron (IC) waves[2]. In the Martian upstream region, Mars Global Surveyor confirmed that such structures also exist in the bow shock where the HF signature is exactly at the local proton IC frequency while the LF component is at about seven gyro-periods[3]. In geospace, many high-resolution satellites introduced various nonlinear HF-LF wavefield diagnoses in different plasma regions, such as Freja[4-6], Polar[7, 8], FAST[9, 10], Viking[11, 12], and Cluster[11-13]. Even low-altitude rockets measured this kind of waveforms in auroral regions[14].

Nomenclaturally, this kind of extraordinary nonlinear envelops was described as "oscillitons" (i.e., "oscillations"+"solitons")[15]. In view of nonlinear electromagnetic whistler modes, many authors accepted that oscillitons are excited in multi-ion plasmas in propagations parallel to, and supposedly also, oblique to a static, background magnetic field[16-26]. Impressively, Kourakis & Shukla[27] developed a generic methodological formulation for oscillitons by taking into account either the mutual modulation effect (due to the interaction between HF and LF modes) or the auto-modulation one (due to the self-interaction of a carrier mode). As a matter of fact, in studying large-amplitude, electromagnetic, nonlinear waves in electron-positron plasmas[28], Verheest et al.[29] perceived a sensitive quality, that was unrecognized previously, of the whistler oscillitons through numerical computations: At not too large amplitudes, oscillitons are very similar to single-frequency solitary envelops obtained by solving the Schrödinger equation, either or not the charge neutrality condition is imposed as a priori one.

Intrigued by this work, we hope to make it clear about not only the role played by this condition, but also the presence of electron mass (i.e., the electron inertia), in the excitation of conventional single-frequency solitary envelops and muliti-frequency oscilliton structures. We pay attention to parallel-propagating cases in cylindrical and Cartesian coordinates, respectively, under electrostatic and electromagnetic conditions. More specific descriptions of the modeling are presented in Ma & Hirose [30, 31].

2. Simple nonlinear waves in a nutshell

Since the first soliton (also called a solitary wave) was noticed by John Scott Russell in 1834, and the first prediction was made about the existence of the non-wave structures in plasmas (called the Bernstein-Green-Kruskal, BGK, mode) in 1957, simple-wave (i.e., single-frequency) solitary structures have been observed and modeled in solving different problems relevant to the construction, maintenance, propagation, and effects of the three kinds of solitary structures (sinusoidal, sawtooth, and spiky/bipolar; see a detailed introduction in[32]).

From the FFT spectra of parallel-propagating ion-acoustic (IA) solitary waves in the absence of electron inertia under charge-neutrality conditions[32, 33], it is easy to find that the wave is featured by a dominant peak-frequency, accompanied by its harmonic peaks, due to the fact that the solitary structures do not have pure sinusoidal waveforms. The farther the shape of solitons away from a sine wave, the more the harmonic peaks to be developed (see, for example, two Fig.3 plots in [30, 31]). It deserves to mention here that the dominant peak in the power spectra may not be located at the fundamental frequency. This warns us that, in data analysis to identify in-situ nonlinear waves, a frequency with the largest amplitude may not represent that the wave is excited at that frequency.

3. Auto-modulation process

For a direct comparison to the simple-wave case under charge-neutrality conditions[33], we choose a cylindrical geometry, where the magnetic field $\mathbf{B} = B\hat{\mathbf{e}}_z$ is along the axial z-direction. A self-similar, two-fluid plasma model under charge non-neutrality conditions is employed for a parameterized study to show the modulation of single-frequency solitary structures to multi-frequency oscillitons. By normalizing ion and electron densities (n_i, n_e) with the equilibrium density n_0, coordinates r and z with the ion gyro-radius $\rho_i{=}c_s/\Omega_i$ (where $c_s{=}\sqrt{k_B T_e/m_i}$ is the ion acoustic speed, $\Omega_i = eB/m_i$ is the ion gyrofrequency, k_B is the Boltzmann constant, T_e is the Boltzmann electron temperature, e is the electron charge, and m_i is the ion mass), velocity components (u_r, u_ϕ, u_z) with c_s, time t with $1/\Omega_i$, the perturbed potential φ with $k_B T_e/e$, and writing all dimensionless variables by using $N_i{=}n_i/n_0$, $N_e{=}n_e/n_0$, $R{=}r/\rho_i$, $Z{=}z/\rho_i$, $V_r{=}u_r/c_s$, $V_\phi{=}u_\phi/c_s$, $V_z{=}u_z/c_s$, $\tau{=}\Omega_i t$, $\Phi{=}e\varphi/(k_B T_e)$ (note that φ satisfies $\mathbf{E} = -\nabla\varphi$ where $\mathbf{E}$ denotes the electrostatic electric field) in a self-similar transformation $X = (\alpha_1/M)R + (\alpha_2/M)Z - \tau$ (where M is the Mach number, $\alpha_1{=}\sin\theta$, and $\alpha_2{=}\cos\theta$ in which θ is the inclination angle between the propagation direction and the magnetic field), we obtain a dimensionless set of differential equations as follows:

$$\left.\begin{array}{l}
\left(1-\frac{U_z}{M}\right)\frac{\mathrm{dln}N_i}{\mathrm{d}X}-\frac{1}{M}\frac{\mathrm{d}U_z}{\mathrm{d}X}=\frac{U_r}{R}\,,\ \left(1-\frac{U_z}{M}\right)\frac{\mathrm{d}U_r}{\mathrm{d}X}=-\left(\frac{U_\phi}{R}+1\right)U_\phi \\
\left(1-\frac{U_z}{M}\right)\frac{\mathrm{d}U_\phi}{\mathrm{d}X}=\left(\frac{U_\phi}{R}+1\right)U_r\,,\ \left(1-\frac{U_z}{M}\right)\frac{\mathrm{d}U_z}{\mathrm{d}X}=\frac{1}{M}\frac{\mathrm{d}\left(\frac{N_i}{\xi_T}+\Phi\right)}{\mathrm{d}X} \\
\frac{\mathrm{d}^2\Phi}{\mathrm{d}X^2}=M^2\xi_\omega^2(N_e-N_i)\,,\ N_e=e^\Phi
\end{array}\right\} \qquad (1)$$

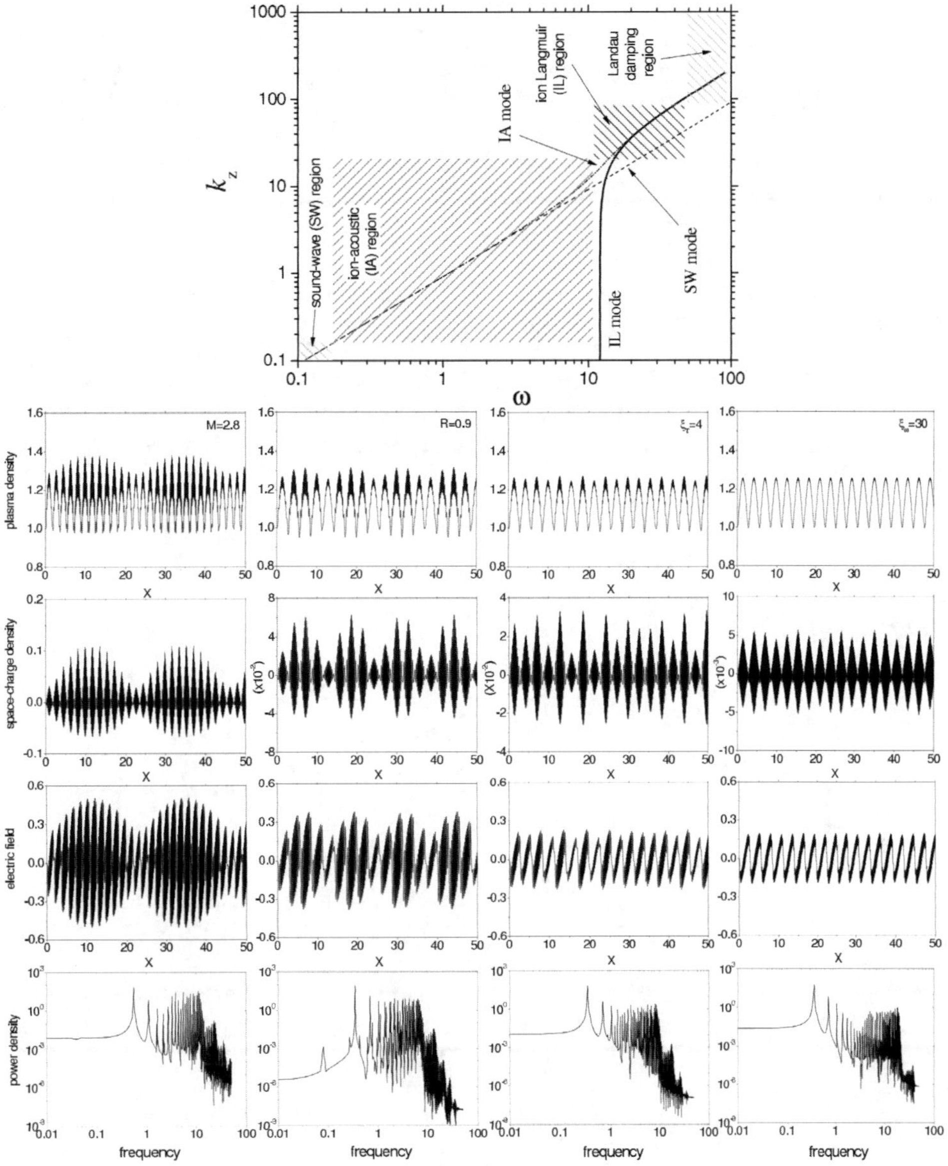

Figure 1. Upper: Background linear oscillation modes in the $\omega-k$ plane with ξ_T=5 and ξ_ω=12. Lower nine panels: Modulated oscillitons by input parameters M (left column), R (middle left column), ξ_T (middle right column), and ξ_ω (right column), respectively, in view of plasma density (top row), space-charge density (second row), electric field (third row), and FFT power density (bottom row), respectively.

in which $\xi_T=c_s^2/v_{Ti}^2=T_e/(2T_{i0})$, and $\xi_\omega=\omega_{pi}/\Omega_i=\rho_i/\lambda_{De}$ where ω_{pi} is the ion plasma frequency, and $\lambda_{De}=c_s/\omega_{pi}=\sqrt{\epsilon_0 k_B T_e/(n_0 e^2)}$ is the electron Debye length. Naturally, $c_s k_{De}=\omega_{pi}$ where $k_{De}=\lambda_{De}^{-1}$ is the electron Debye wave number. Notice that $\theta=0°$ due to the parallel-propagating case. By linearizing Eq.(1), the background linear oscillation modes in the $\omega-k$ plane are shown in the upper panel of Fig.1.

The lower nine panels of the figure exhibit multi-frequency oscillitons via the auto-modulation process under charge non-neutrality conditions. The LF envelop is in the sound-wave (SW) mode, while the HF ingredients include the IA mode and the ion-Langmiur (IL) mode. The amplitudes of the SW wave are violently modulated by the IA oscillations, whileas the upward sides of the IA amplitudes are modulated by the IL oscillations of smaller amplitudes, and the downward sides are modulated by hybrid IA/IL oscillations.

4. Mutual modulation process

In cases where the electron inertia comes into play, there exist electron-ion mutual modulations (see details in [30]). For a convenient comparison to the simple-wave case in the absence of the electron inertia [32], we choose a Cartesian geometry. We use following dimensionless parameters: the electron and ion densities n_e and n_i are normalized by n_0; coordinates $\mathbf{r}=\{x,y,z\}$ by electron Debye length $\lambda_{De}=v_{Te}/\omega_{pe}$; velocity $\mathbf{u}_\alpha=\{u_{\alpha x},u_{\alpha y},u_{\alpha z}\}$ by the acoustic speed $c_s=\sqrt{k_BT_{e0}/m_i}$; pressure p_α by $p_{\alpha 0}=n_0k_BT_{\alpha 0}$; time t by ion plasma period $\tau_{pi}=\omega_{pi}^{-1}$ (notice that $\omega_{pi}\lambda_{De}=c_s$); magnetic field $\mathbf{B}$ by a pseudo-magnetic field $B_0=m_e\omega_{pi}/e$; electric field $\mathbf{E}$ by a pseudo-electric field $E_0=c_sB_0$; and, three coefficients are introduced: $\xi_m=m_i/m_e$, $\xi_T=T_{e0}/T_{i0}$, and $\xi_v=v_{Te}/c$. Using $X=x-Mt$, the dimensionless set of self-similar differential equations of the two-fluid model under electromagnetic conditions in a Cartesian frame is as follows:

$$\left.\begin{aligned}
\frac{\mathrm{d}u_{ex}}{\mathrm{d}X} &= -(E_x+u_{ey}B_z-u_{ez}B_y)/\left(u_{ex}-\frac{\xi_m}{u_{ex}}\right)\\
\frac{\mathrm{d}u_{ey}}{\mathrm{d}X} &= -(S_y-u_{ex}B_z+u_{ez}B_{x0})/u_{ex}\ ,\ \frac{\mathrm{d}u_{ez}}{\mathrm{d}X} = -(S_z+u_{ex}B_y-u_{ey}B_{x0})/u_{ex}\\
\frac{\mathrm{d}u_{ix}}{\mathrm{d}X} &= \xi_m^{-1}(E_x+u_{iy}B_z-u_{iz}B_y)/\left(u_{ix}-\frac{3S_n^2}{\xi_T u_{ix}^3}\right)\\
\frac{\mathrm{d}u_{iy}}{\mathrm{d}X} &= \xi_m^{-1}(S_y-u_{ix}B_z+u_{iz}B_{x0})/u_{ix}\ ,\ \frac{\mathrm{d}u_{iz}}{\mathrm{d}X} = \xi_m^{-1}(S_z+u_{ix}B_y-u_{iy}B_{x0})/u_{ix}\\
\frac{\mathrm{d}E_x}{\mathrm{d}X} &= -\xi_m n_{sc}\\
\frac{\mathrm{d}B_y}{\mathrm{d}X} &= \xi_v^2 S_n\left(\frac{u_{iz}}{u_{ix}}-\frac{u_{ez}}{u_{ex}}\right)/\left(1-\frac{\xi_v^2M^2}{\xi_m}\right)\ ,\ \frac{\mathrm{d}B_z}{\mathrm{d}X} = -\xi_v^2 S_n\left(\frac{u_{iy}}{u_{ix}}-\frac{u_{ey}}{u_{ex}}\right)/\left(1-\frac{\xi_v^2M^2}{\xi_m}\right)
\end{aligned}\right\}\quad (2)$$

in which $S_n=u_{xx0}-M$, $S_y=E_{y0}-MB_{z0}$, $S_z=E_{z0}+MB_{y0}$, $n_e=S_n/u_{ex}$, $n_i=S_n/u_{ix}$, $n_{sc}=n_e-n_i$, $E_y=S_y+MB_z$, $E_z=S_z-MB_y$, and $B_x=B_{x0}$. Following boundary conditions are used: $n_e|_{X=0}=n_0$, $n_i|_{X=0}=n_0$, $u_{xx0}=u_{ex0}|_{X=0}=u_{ix0}|_{X=0}=U_0$-$M$, $E_y|_{X=0}=E_{y0}$, $E_z|_{X=0}=E_{z0}$, $B_x|_{X=0}=B_{x0}$, $B_y|_{X=0}=B_{y0}$, and $B_z|_{X=0}=B_{z0}$. Note that the $u_{\alpha x}$-origin is shifted from "0" to "M", and the density equations and Gauss's law require $S_{ne}=S_{ni}$, written as S_n.

Fig.2 exposes lower-hybrid (LH) oscillitons evolved from single-frequency IA/IC solitary waves under different Mach number, M (top left), and temperature ratio, Te/Ti (top right). To have a qualitative comparison with data, the figure also gives a measurement of "oscilliton" structures in the lower left panel, which was believed to be associated with the LH mode at the plasma sheet boundary (see [34] for details). Under $U_0=(0.2,0,0)$, $E_0=(0.5,0.5,1)$, and $B_0=(0.2,0.5,3)$ with M=1, m_i/m_p=16, T_e/T_i=10, and v_{Te}/c=0.05, the lower right panel depicts a simulation result. Obviously, the electron inertia triggers LH oscillitons characterized by a normal IC-period solitary envelop embedded by LH oscillations which contain higher-frequency but smaller-amplitude IA constituents. This means that the mutual modulation effect between ions and electrons exerts an impact on nonlinear packets via several input parameters like the Mach number and the temperature ratio. The influence of other input parameters (such as ion mass, electron speed) was discussed in [30]. Impressively, Whenever an oscilliton structure is formed, there occurs a density cavity. Its depth can reach up to 20% of the background density, and on both sides of the cavity, there occur density humps. The feature is in accordance with either observations or simulations (see a detailed discussion in [30]).

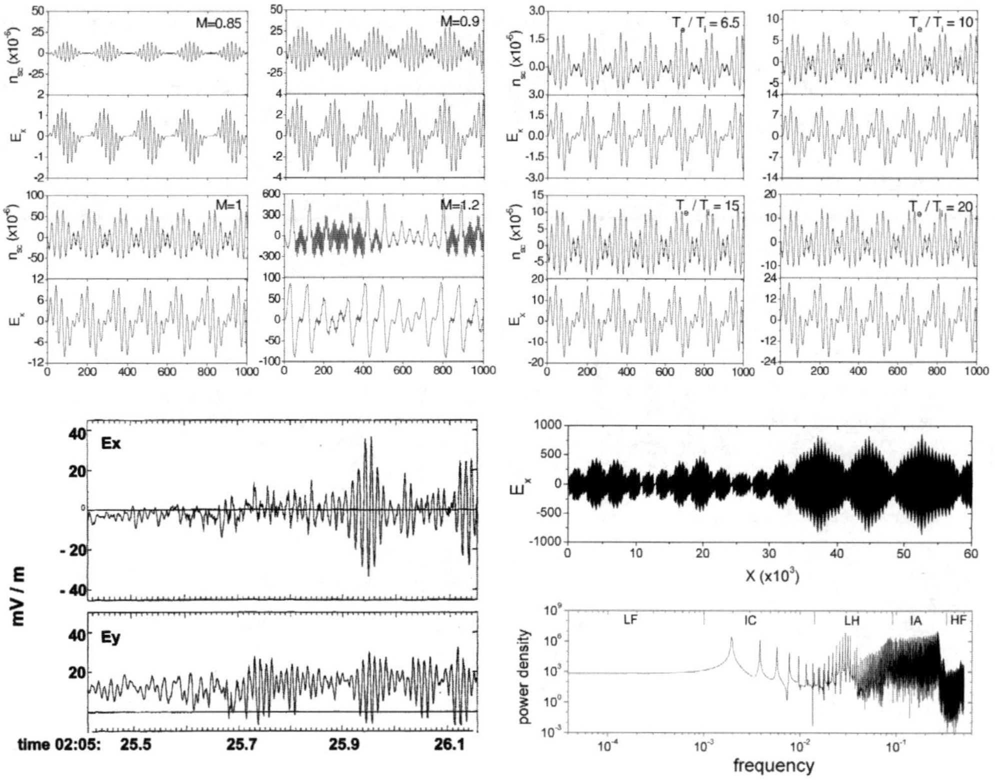

Figure 2. Upper left: Impact of the Mach number M on LH-oscilliton waves. Only space-charge density n_{sc} and electric wavefield E_x-component are shown. Notice that the panels of M=1 is the typical case discussed in the last Section. Upper right: Same as the upper left but the impact of $\xi_T = T_e/T_i$ on LH-oscilliton waves. Lower left: Observed "oscilliton" structures associated with the LH mode by the Polar satellite at the plasma sheet boundary (adapted from Fig.3 of [34]). Lower right: A simulation to observed LH oscillitons in the lower left panel under $U_0 = (0.2, 0, 0)$, $E_0 = (0.5, 0.5, 1)$, and $B_0 = (0.2, 0.5, 3)$ with M=1, m_i/m_p=16, T_e/T_i=10, and v_{Te}/c=0.05. Reproduction of a few figures in [30].

5. Conclusion

Nonlinear oscillitons are evolved from simple solitary waves under certain conditions. If the charge non-neutrality condition is satisfied, an auto-modulation process brings single-frequency solitons into multi-frequency envelops of oscillitons. By contrast, if the electron inertia cannot be neglected, a mutual-modulation process brings solitons into oscillitons.

In the first case, the low-frequency envelop of the oscillitons is in the sound-wave mode, while the high-frequency portion includes the IA mode and the ion-Langmiur mode. The sound-wave amplitudes are violently modulated by the ion-acoustic mode. The upward edges of the ion-acoustic amplitudes are modulated by the ion-Langmiur oscillations of smaller amplitudes, while the downward edges are modulated by hybrid ion-acoustic/ion-Langmiur oscillations.

In the second case, the simple IA/IC waves are developed to lower-hybrid oscillitons where the low-frequency ion-cyclotron envelop is embedded by small-amplitude, high-frequency lower-hybrid oscillations. The packet is superimposed upon by higher-frequency but smaller-amplitude ion-acoustic ingredients. Whenever an oscilliton is triggered, there occurs a density cavity the depth of which can reach up to 20% of the background density, along with density humps on both sides of the cavity.

In either cases, the oscillitons are shown to be sensitive to input parameters (e.g., the Mach number, the electron-ion mass/temperature ratios, and the electron thermal speed).

Acknowledgments
This work is funded by Visiting Fellowships in Canadian Government Laboratories Program, Natural Sciences and Engineering Research Council (NSERC), Canada. AH also acknowledges the supports from NSERC and Canada Research Chairs (CRC) programs.

References

[1] Gurnett D 1985 *Geophys. Monogr.* **35** 207
[2] Kivelson M G, Khurana K K, Walker R J, Warnecke J, Russell C T, Linker J A, Southwood D J, and Polanskey C 1996 *Science* **274** 396
[3] Mazelle C, Winterhalter D, Sauer K, Trotignon J G, Acuña M H, Baumgätel K, Bertucci C, Brain D A, Brecht S H, Delva M, Dubinin E, Øieroset M, and Slavin J 2004 *Space Sci. Rev.* **111** 115
[4] Eriksson A I, Holback B, Dovner P O, Boström R, Holmgren G, André M, Eliasson L, and Kintner P M 1994 *Geophys. Res. Lett.* **21** 1843
[5] Dovner P O, Eriksson A I, Boström R and Holback B 1994 *Geophys. Res. Lett.* **21** 1827
[6] Pécseli H L, Iranpour K, Hotter Ø, Lybekk B, Holtet J, Trulsen J, Eriksson A, and Holback B 1996 *J. Geophys. Res.* **101** 5299
[7] Cattell C, Wygant J, Dombeck J, Mozer F S, Temerin M, and Russell C T 1998 *Geophys. Res. Lett.* **25** 857
[8] Cattell C, Dombeck J, Wygant J R, Hudson M K, Mozer F S, Temerin M A, Peterson W K, Kletzing C A, Russell C T, and Pfaff R F 1999 *Geophys. Res. Lett.* **26** 425
[9] McFadden J P, Carlson C W, Ergun R E, Chaston C C, Mozer F S, Temerin M, Klumpar D M, Shelley E G, Peterson W K, Moebius E, Kistler L, Elphic R, Strangeway R, Cattell C, and Pfaff R 1998 *Geophys. Res. Lett.* **25** 2045
[10] Pottelette R, Ergun R E, Treumann R A, Berthomier M, Carlson C W, McFadden J P, and Roth I 1999 *Geophys. Res. Lett.* **26** 2629
[11] Tjulin A, Eriksson A I, and André M 2003 *Geophys. Res. Lett.* **30** 1364
[12] Tjulin A, André M, Eriksson A I, and Maksimovic M 2004 *Ann. Geophys.* **22** 2961
[13] Santolik O, Gurnett D A, Pickett J S, Parrot M, and Cornilleau-Wehrlin N 2003 *J. Geophys. Res.* **108** 1278
[14] Burchill J K, Knudsen D J, Bock B J J, Pfaff R F, Walis D D, Clemmons J H, Bounds S R, and Stenbaek-Nielsen H 2004 *J. Geophys. Res.* **109** A01219
[15] Sauer K, Dubinin E, and McKenzie J F 2001 *Geophys. Res. Lett.* **28** 3589
[16] Sauer K, Dubinin E, and McKenzie J F 2002 *Geophys. Res. Lett.* **29** 24
[17] Sauer K, Dubinin E, and McKenzie J F 2003 *Nonlin. Proc. Geophys.* **10** 121
[18] Dubinin E, Sauer K, and McKenzie J F 2002 *J. Plasma Phys.* **68** 27
[19] Dubinin E, Sauer K, and McKenzie J F 2003a *J. Plasma Phys.* **69** 305
[20] Dubinin E, Sauer K, and McKenzie J F 2003b *J. Geophys. Res.* **107** 1295
[21] Dubinin E, Sauer K, and McKenzie J F 2003c *J. Geophys. Res.* **107** 1296
[22] Dubinin E M, Maksimovic M, Cornilleau-Wehrlin N, Fontaine D, Travnicek P, Mangeney A, Alexandrova O, Sauer K, Fraenz M, Dandouras I, Lucek E, Fazakerley A, Balogh A, and Andre M. 2007 *Ann. Geophys.* **25** 303
[23] McKenzie J F, Dubinin E, Sauer K, and Doyle T B 2004 *J. Plasma Phys.* **70** 431
[24] Cattaert T, and Verheest F 2005 *Phys. Plasmas* **12** 012307
[25] Sydora R D, Sauer K, and Silin I 2007 *Geophys. Res. Lett.* **34** L22105
[26] Pottelette R, and Berthomier M 2009 *Nonlin. Proc. Geophys.* **16** 373
[27] Kourakis I, and Shukla P K 2005 *Nonlin. Proc. Geophys.* **12** 407
[28] Verheest F, and Cattaert T 2004 *Phys. Plasmas* **11** 3078
[29] Verheest F, Cattaert T, Dubinin E, Sauer K, and McKenzie J F 2004 *Nonlin Proc. Geophys.* **11** 447
[30] Ma J Z G, and Hirose A 2010 *Nonlin Proc. Geophys.* **17** 245
[31] Ma J Z G, and Hirose A 2010 *Ann. Geophys.* (in revision).
[32] Ma J Z G, and Hirose A 2009 *Phys. Scr.* **79** 045502
[33] Ma J Z G 2010 *Astrophys. Space Sci.* 0004-640X (Print) 1572-946X (Online)
[34] Cattell C, Wygant J, Dombeck J, Mozer F S, Temerin M, and Russell C T 1998 *Geophys. Res. Lett.* **25** 857

The Alfvén resonance in pair plasmas

N. F. Cramer
School of Physics, University of Sydney, NSW 2006, Australia
E-mail: cramer@physics.usyd.edu.au

Abstract. Waves propagating obliquely in a magnetized cold pair plasma experience an approximate resonance in the wavevector component perpendicular to the magnetic field, which is the analogue of the Alfvén resonance in normal electron-ion plasmas. Wave absorption at the resonance can take place via mode conversion to the analogue of the short wavelength inertial Alfvén wave. The Alfvén resonance could play a role in wave propagation in the pulsar magnetosphere leading to pulsar radio emission. Ducting of waves in strong plasma gradients may occur in the pulsar magnetosphere, which leads to the consideration of Alfvén surface waves, whose energy is concentrated in the region of strong gradients.

1. Introduction

Because the time and spatial scales associated with nonrelativistic electrons-positron pair plasmas are the same in each species, such pair plasmas have dispersive properties quite different to electron-ion plasmas. A distinctive feature of linear wave propagation in magnetized plasmas with equal numbers of same-mass pair species is that there is no Faraday rotation, so that waves propagating parallel to the magnetic field are linearly polarized, rather than circularly polarized, and that there is no analogue of the ion-acoustic wave [1, 2, 3].

If a cold plasma of equal-mass species is considered, but with different number densities of the species (uncompensated), the wave properties are altered, in that circularly polarized waves propagate along the magnetic field. Such an overall charge neutral plasma can be achieved if some of the charge resides on another species, for example consisting of ions or relatively massive dust particles. If the third species is effectively immobile, it can be thought of a charge sink, and the wave properties are determined by the dynamics of the (relatively) light pair species. Another circumstance of an uncompensated pair plasma occurs in the rotating pulsar magnetosphere, where different overall charges exist in different regions of the magnetospheric plasma in pulsar models.

An important feature of real astrophysical plasmas is that they are in general spatially nonuniform on the large scale and inhomogeneous on shorter scales. Waves in such plasmas experience reflection, transmission and absorption processes related to the local spatial variation of the plasma parameters such as the plasma frequencies and cyclotron frequencies. A particular process of interest in this paper is the Alfvén resonance absorption of wave energy, where, in the context of a normal electron-ion plasma, a fast magnetoacoustic wave propagates obliquely to the magnetic field in a density gradient, and at some point the local value of the Alfvén speed equals the wave phase velocity, or in other words the magnetoacoustic wave couples to the Alfvén wave. Energy absorption, via collisional or mode-conversion processes, occurs at this resonance point where the local value of the wavenumber perpendicular to the magnetic field becomes large [4], and this process in normal plasmas has been invoked as a mechanism for the heating of the solar corona.

The refraction and ducting of waves, launched by emission processes in the nonuniform pulsar magnetosphere, have been invoked as important processes influencing the properties of the emergent pulsar radio waves escaping the magnetosphere, such as pulse profiles [5, 6]. Analyses of refraction of waves in a nonuniform plasma do not usually take into account resonance absorption processes such as the Alfvén resonance, which might play a role in wave propagation in the pulsar magnetosphere.

2. The Dispersion Equation

We consider the case of a cold magnetized plasma, composed of electrons and positrons, plus an effectively immobile species that retains part of the charge, positive or negative. The magnetic field is in the z-direction, and the dielectric tensor components are then:

$$K_{11} = K_{22} = 1 - \frac{\omega_{p+}^2 + \omega_{p-}^2}{\omega^2 - \Omega^2} = 1 - \frac{\omega_p^2}{\omega^2 - \Omega^2}, \tag{1}$$

$$K_{12} = -K_{21} = -i\frac{(\omega_{p+}^2 - \omega_{p-}^2)\Omega}{\omega(\omega^2 - \Omega^2)} = -i\eta\frac{\omega_p^2\Omega}{\omega(\omega^2 - \Omega^2)}, \tag{2}$$

$$K_{33} = 1 - \frac{\omega_p^2}{\omega^2}, \tag{3}$$

where ω_{p+} and ω_{p-} are the positron and electron plasma frequencies, and ω_p is the plasma frequency of the combined positron-electron fluid. Ω is the common cyclotron frequency, and η is a measure of the charge imbalance of the two light species:

$$\eta = (n_+ - n_-)/(n_+ + n_-).$$

All the other dielectric tensor components are zero. We define a function

$$A = K_{11} - \frac{c^2}{\omega^2}k_z^2. \tag{4}$$

The vanishing of this function for normal electron-ion plasmas defines the position of the Alfvén resonance, because for low frequency it corresponds to the point where the phase velocity along the magnetic field equals the Alfvén speed, where a resonance in the perpendicular wavenumber occurs [4]. A plays a similar role, under certain conditions, for a pair plasma.

In uniform plasma, the wave has wavenumber $\mathbf{k}$. If $k_y = 0$, the dispersion equation is:

$$k_x^4(c^2/\omega^2)^2K_{11} + k_x^2(c^2/\omega^2)(-A(K_{33} + K_{11}) - K_{12}^2) + (A^2 + K_{12}^2)K_{33} = 0. \tag{5}$$

3. Alfvén Resonance

We now fix ω and k_z, and vary the plasma density in the x-direction. There are 2 distinct modes. The local k_x^2 may approach zero, i.e. cutoff or strong reflection, or it may become very large, i.e. resonance or strong absorption, as shown in Figure 1.

If $\omega_p \gg \omega$, $|K_{33}| \gg 1$, there are two approximate solutions:

$$c^2k_{x1}^2/\omega^2 = A + K_{12}^2/A, \tag{6}$$

$$c^2k_{x2}^2/\omega^2 = AK_{33}/K_{11}. \tag{7}$$

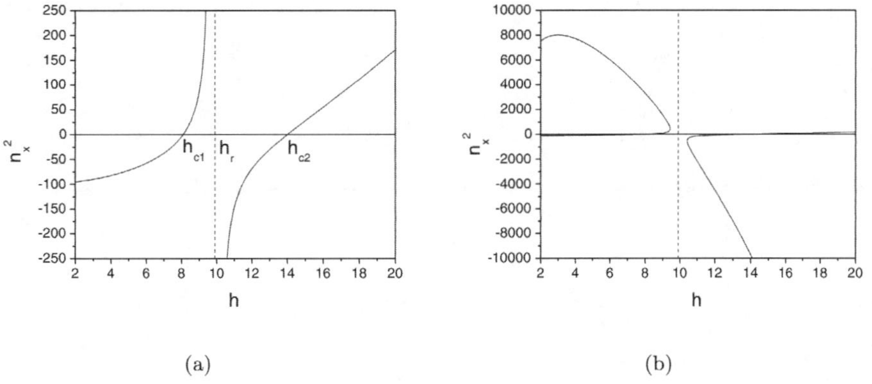

Figure 1. (a) Square of the perpendicular refractive index $n_x^2 = c^2k_x^2/\Omega^2$ plotted against $h = \omega_p/\Omega$, for case $f \ll h$. Here $f = 0.1$, $n_z = 10$ and $\eta = 0.05$. The Alfvén resonance at h_r is shown by the dashed line, and the two cutoffs $h_{c1,2}$ are indicated. (b) As for (a), but the scale of n_x^2 is expanded, to reveal the second high $|n_x|$ mode on each side of the Alfvén resonance

If $A \approx 0$, there is no longer a true Alfvén resonance where $|n_x^2| \to \infty$, but $|n_x^2|$ acquires large values

$$n_{x1,2}^2 \approx \pm K_{12}(-K_{33}/K_{11})^{1/2} \approx \pm i\eta \frac{h^3}{f^2(1+h^2)^{1/2}} \tag{8}$$

If $\omega_p \lesssim \omega$, then $|K_{33}|$ is no longer much greater than unity, and there is no longer a pronounced resonance: the moduli of the refractive indices of the two modes are of comparable size in the vicinity of the resonance point.

The larger the value of $|K_{33}|$, the narrower the stop-band about the resonance point, and the sharper the resonance. Equation (6), representing a (single) mode with a true resonance at $A = 0$ enclosed by two cutoffs, becomes a better approximation for the actual dispersion relation. This mode is equivalent to the fast compressional wave in cold normal ion-electron plasmas with the electron inertia neglected, which also experiences a cutoff-resonance-cutoff triplet [4], with the role of the finite ion-cyclotron frequency due to Hall currents in the case of the normal plasma now being played by the imbalance in the pair plasma densities.

Equation (6) can be written for $f \ll 1$ in terms of the wavenumbers as

$$v_A^2 k_x^2 = \omega^2 - v_A^2 k_z^2 - \frac{\eta^2\Omega^2\omega^2}{\omega^2 - v_A^2 k_z^2}, \tag{9}$$

which shows explicitly the Alfvén resonance ($k_x^2 \to \infty$) occurring where the parallel wave phase velocity equals the Alfvén velocity, with the dispersive term depending on the pair species imbalance factor η. Further details, including use of the wave differential equations to show the existence of the Alfvén resonance, are found in [7].

The large $|n_x|$ mode has the approximate dispersion relation

$$\omega^2 = \frac{v_A^2 k_z^2}{1 + \delta_e^2 k_x^2} \tag{10}$$

where $\delta_e = c/\omega_p$ = electron inertial length. This is the analogue of the Inertial Alfvén wave that occurs in normal plasmas. Surface waves exist on a pair-plasma/vacuum interface for $k_y \neq 0$ (Figure 2). The surface waves may be damped due to Alfvén resonance damping, or coupling into the Inertial Alfvén wave.

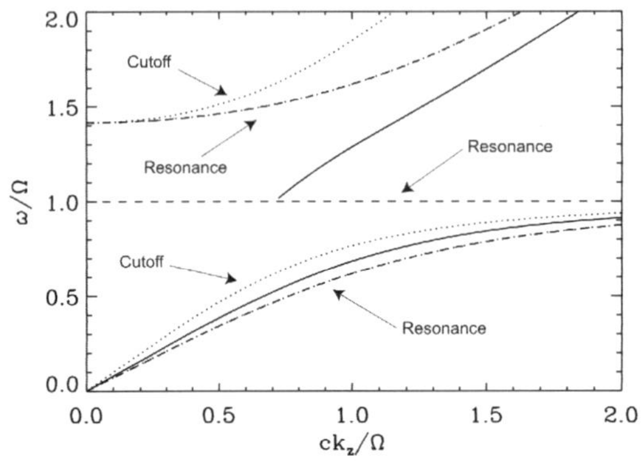

Figure 2. Surface wave dispersion relation (solid curves). Here $h = 1$, $k_y = k_z$ and $\eta = 0$.

4. Conclusion

It has been shown that the Alfvén resonance absorption process occurs in a pair plasma. Analogies are drawn between the resonance-cutoff structures of noncompensated pair plasmas and those of normal electron-ion plasmas with finite ion-cyclotron frequency, with the role of the imbalanced Hall currents in the latter case being played by the imbalanced species densities in the pair plasma.

A true Alfvén resonance can be identified only in the limit of wave frequency much less than the plasma frequency. A long wavelength wave propagating into the "Alfvén resonance" point will partially mode convert into a short wavelength mode which is the pair-plasma version of the "Inertial Alfvén Wave" which occurs in the Alfvén resonance heating of fusion plasmas, and in the form of solitary waves in the Earth's magnetosphere. On a sharp interface between plasmas of differing densities, such as a magnetic flux tube or density duct in the pulsar magnetosphere, a localized surface wave exists. The analogues of Alfvén surface waves on an electron-positron plasma/vacuum interface have been demonstrated. Future work could allow for thermal effects [1] and relativistic effects [8] on the pair plasma dielectric tensor.

References

[1] Stewart G A and Laing E W 1992 *J. Plasma Phys.* **47** 295
[2] Iwamoto N 1993 *Phys. Rev. E*, **47** 604
[3] Zank G P and Greaves R G 1995 *Phys. Rev. E* **51** 6079
[4] Cramer N F 2001 *The Physics of Alfvén Waves* (Berlin: Wiley-VCH)
[5] Arons J and Barnard J J 1986 *Astrophys. J.* **302** 120
[6] Weltevrede P, Stappers B W, van den Horn L J and Edwards R T 2003 *Astron. and Astrophys.* **412** 473
[7] Cramer N F 2010 *Phys. Plasmas* **17** 022103
[8] Luo Q, Melrose D B, and Fussell D 2002 *Phys. Rev. E* **66** 026405

Dense dust structures in cryogenic complex plasma

S N Antipov[1], M M Vasiliev[1], M M Alyapyshev[1,2], O F Petrov[1,2] and V E Fortov[1]

[1] Institution of Russian Academy of Sciences Joint Institute for High Temperatures RAS, Izhorskaya st. 13 bld. 2, 125412 Moscow, Russia

[2] Moscow Institute of Physics and Technology, Institutskii per. 9, 141700 Dolgoprudny, Moscow Region, Russia

E-mail: antipov@ihed.ras.ru

Abstract. In the previous researches of cryogenic complex (dusty) plasma [1] we observed experimentally complex plasma systems formed in cryogenic environments. Particularly it was revealed from the experiments that dust structures with high concentration of macroparticles can be formed. So-called super dense dusty plasma structures in which interparticle distance is comparable with particle size were also observed. Thus concentration of particles was close to concentration of background plasma. Similar formations had unusual properties (sphere-like form, free boundaries, etc.) and represent new object in dusty plasma researches. In the present work new results on experimental investigations of dense dust structures at cryogenic temperatures were presented. The experiments were made by means of recently developed techniques and cryogenic facilities (optical cryostat). Possible nature of the unusual properties of super dense dusty plasma structures was discussed.

1. Introduction

In recent years, plasma with macroparticles, or dusty plasma, has been the subject of extensive investigations. Experiments have revealed various dusty plasma structures in which the observed ordering has been attributed to the Coulomb interaction. This work is devoted to the investigation of a dusty plasma in a dc gas discharge under the conditions where the walls of a gas-discharge tube are cooled to cryogenic temperatures. The experiments performed with a glow-discharge plasma in helium at temperatures of atoms below 100 K have revealed that the specific features of this plasma become more pronounced with decreasing temperature. The first experiments revealed the possibility of forming considerably denser dust structures in the cryogenic plasma as compared to those observed in the discharge at room temperature. Recently, the authors put forward the assumption that the ion Debye length substantially decreases in the cryogenic plasma; furthermore, this phenomenon, in the authors' opinion, is the main factor responsible for the observed approach of dust particles to one another.

This paper reports on the results obtained from the experiments on the generation of dusty plasma structures in a dc glow discharge at cryogenic temperatures in the range from 4.2 to 77 K. The effect of cooling of the neutral component of the gas-discharge plasma on the increase in the density of dusty plasma structures was justified theoretically by analyzing the kinetics of ions in the cryogenic gas discharge. The dependences of the dust particle charge on the gas temperature and the density of the dusty plasma component were calculated using the molecular dynamics method.

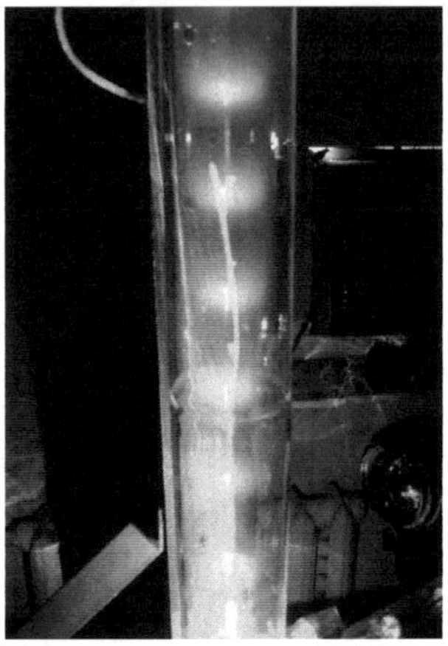

Figure 1. Stratified positive column of the dc glow discharge with dust particles inside striation heads.

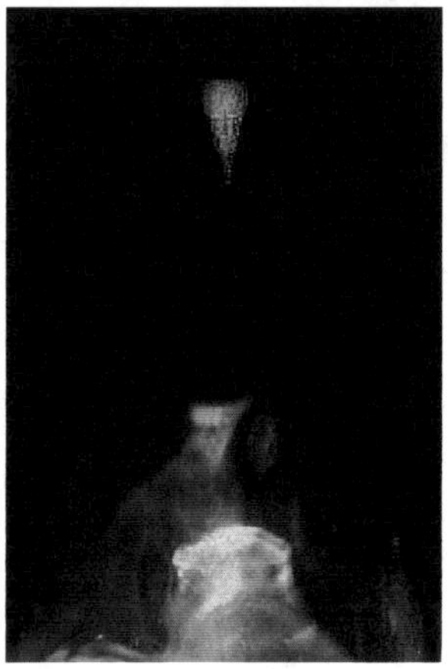

Figure 2. Dusty plasma structure in the dc glow discharge

2. Results of the experimental investigations

In a dc glow discharge, the dust particles acquire a negative charge and are held in standing striations (ionization waves) of the positive column of discharge (Figure 1). An additional glass tube with a narrowing (Figure 2) was placed above the cathode in the lower part of discharge for generation of stable standing striations in a wide range of parameters. Because the function $E/N(NR)$ in a dc glow discharge decreases rapidly in the region of low values of NR (E is the electric field of discharge, and R is the discharge radius), a potential jump arises in the region of narrowing, this jump promotes the emergence of striations in the positive column of discharge. Each striation is a cylindrical potential well in a horizontal plane; in the vertical direction, it is characterized by an abrupt increase in the electric field, which is accompanied by only an insignificant decrease in the electron concentration. The electrostatic force acting on the dust particles in a striation may compensate for the force of gravity and cause the particles to levitate. For the given parameters of discharge, the number of dust particles contained in the striation is limited. In the experiments considered in this paper, the particles were injected until their number in the striation became maximal.

The first cryogenic experiments were performed with a dc glow discharge generated in a vertical glass tube placed inside a cryostat designed in the form of a cylindrical system consisting of two glass Dewar vessels. The outer Dewar vessel served as a heat shield and was filled with liquid nitrogen. The glass discharge tube was located in the inner Dewar vessel under the cover of the cryostat. The upper electrode was a hollow cylindrical anode through which dust particles were injected into the discharge. Dust particles were stored in a container with a perforated bottom and positioned above the anode. The discharge was generated in helium at pressures of the order of 1 Torr (hereafter, the density of the neutral component will be given in terms of a pressure at room temperature) and at discharge currents of the order of 0.1 mA. The scheme of the experimental measurements was described in detail in our previous papers [1, 2], in which we also reported the results of the first experimental observations at temperatures of 77 and 4.2 K. The experimental setup had the possibility of continuously varying and controlling the temperature conditions. The blowing of liquid helium vapors through the inner Dewar vessel made it possible to controllably change the temperature of the walls of the gas-discharge tube in the range from 4.2 to 77 K. The temperature was controlled using a silicon sensor mounted on the outer surface of the wall of the gas-discharge tube near the region of the generation of dust structures.

The temperature sensor was thermally insulated from the outer cooling medium and could measure the temperature of the discharge tube wall with an accuracy of no worse than 1-2 K. In order to illuminate the dust particles, radiation from a diode-pumped solid-state laser with the wavelength 532 nm was introduced into the cryostat through an optical fiber. The observations were performed in the Dewar vessels with a CCD video camera. In order to generate stable striations over a wide range of parameters, the discharge was locally constricted with the use of an additional glass cylinder located in the lower part of the discharge tube. In the upper part, the cylinder narrowed to a capillary with an inner diameter of 0.1 cm. The results of the experiments demonstrated that, both in the discharge at room temperature and in the cryogenic discharge, the dusty plasma structures were formed in the stationary striations generated above the capillary. The observations were carried out in the first striation over the capillary. In the experiments performed at room temperature, we observed the formation of anisotropic crystalline dust structures, which are typical of dc glow discharges and have a preferred direction along the axis of the discharge. The dust particles in this case are ordered in the vertical direction, thus forming threadlike aggregates (dust chains) containing from two to ten particles. The number of chains can vary from several units to several tens, and the total number of particles usually does not exceed 10^2. Figure 2 shows a similar structure formed at the discharge current $I = 0.5$ mA and the gas pressure $p = 2$ Torr. The interparticle distance l_p for these parameters is equal to 500–750 μm. Upon changing over to cryogenic temperatures, both the pattern of the dust structure and the dynamics of dust particles in it become different. At a temperature of 77 K, we observed the formation of large scale dusty plasma structures (with sizes of the order of 10^3 particles) made up of long chains (containing 15-20 particles). The interparticle distance in the chains amounted to 200–250 μm. This circumstance indicates that, compared to the density of dust structures at room temperature, the density of the dust structure at 77 K increases by at least one order of magnitude at the same values of the density of the neutral gas and the discharge current. A further continuous cooling of the discharge is accompanied by a further decrease in the interparticle distances in the dust structure, a transformation of the pattern of the dust structure, and a change in the dynamics of dust particles in it. At temperatures in the range from 30 to 50 K, the interparticle distances decrease to 120–160 μm and the dust structure occupies the entire volume of the striation head. In this case, the dust structure consists of a complex aggregate involving regions of crystalline ordering and convective motions. The chain ordering of dust particles is retained only in the lower part of the structure, and the orientation of the chains is changed in such a way that the larger is the distance between the chain and the axis of the discharge, the greater is the tilt of the chain with respect to the vertical axis (the discharge axis). Upon cooling of the discharge, the interparticle distances decrease until the discharge becomes unstable. The development of discharge instabilities at the aforementioned discharge current prevents the confinement of dust particles in the discharge, and they are deposited on the bottom of the discharge tube. At temperatures below 30 K, the observations of particles were performed after the discharge current was decreased to 0.2 mA, when the discharge was stabilized. The structures formed under these conditions are characterized by a liquid-like behavior. An increase in the density of the dusty plasma structures is observed until the helium is completely condensed in the Dewar vessel, i.e., when the cooling helium vapors transform into the liquid state (4.2 K). Note that, in this case, the conditions of observation of dust particles are substantially deteriorated (the measurement of interparticle distances is strongly complicated). The distances between the dust particles at a temperature of 4.2 K do not exceed 30 μm. The first experimental results obtained provide a qualitatively correct pattern of the variation in the density of the dusty plasma structures during cooling of the discharge to low (cryogenic) temperatures. The data obtained demonstrate a strong nonlinear dependence of the interparticle distance l_p in the dusty plasma structures on the temperature T of neutral atoms. An analysis of the mechanisms of the phenomena under consideration should include a thorough examination of the transformation of the ion velocity distribution function at low temperatures of the gas atoms with the proper allowance made for the ion–atom collisions.

As it was obtained from the first experiments the cooling of the discharge down to very low temperatures (4-5 K) can lead to formation of very dense structures in which interparticle distance is

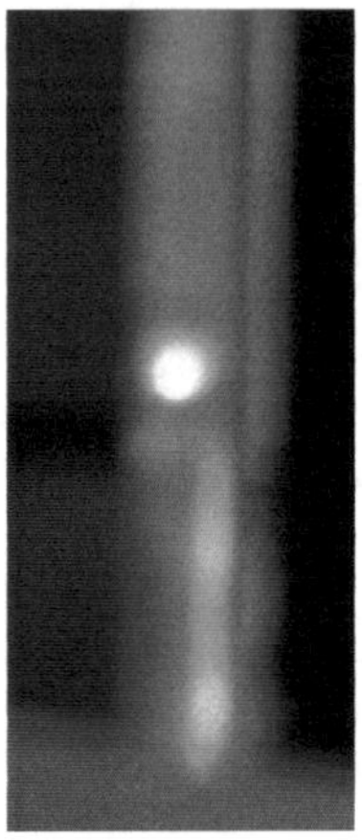

Figure 3. Compact super dense dust sphere of polydisperse Al_2O_3 particles in the dc discharge at 4.2 K.

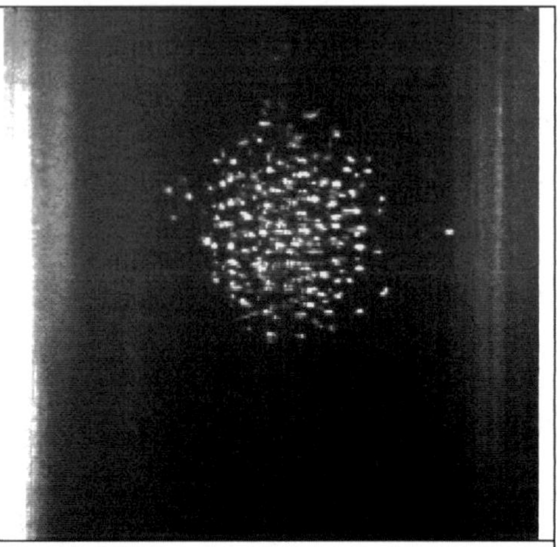

Figure 4. Spheroidizing of dust structures of monodisperse MF 1.1μm-particles in the dc discharge at 77 K.

comparable to the plasma screening length and even to the size of the particles. Figure 3 shows such structure of polydisperse Al_2O_3 particles observed at 4.2 K. The specific feature of this structure is compact spherical form and extremely high concentration of dust particles. Unfortunately, obtained images could not allow as obtaining any structural and dynamic structure characteristics because of the poor conditions for observation. For the purpose of observing individual particles in super dense dust structures at cryogenic temperatures the new optical cryostat with flat round windows was developed. By means of the new cryostat we observed the phenomenon of spheroidizing of dust structures at the discharge cooling (Figure 4). Such a sphere-like structure was obtained as a result of decreasing of pressure and current of the discharge at 77 K. Here the interparticle distance is about 100 μm. Thus, one can expect the formation of superdense compact dusty plasma structures (spheres) of monodisperse particles at 4.2 K.

For the estimation of the dust charge in super dense structure at 4.2 K we can use the quasineutrality condition:

$$n_d Z_d + n_e = n_i,$$

where $n_i \sim 2\times10^9$ cm^{-3} for the experimental conditions [1]. Assuming that the conditions $n_e << n_i$ is valid (because of the energetic (~1 eV) electron loss on the dust particles) we can achieve

$$Z_d \sim 20.$$

This result seems to be rather realistic since such a small dust charge does not lead to sufficient Coulomb repulsing of dust particles. Since plasma parameters in the dc discharge striations are anisotropic, such sphere-like form can be result of surface tension of the structure caused by forces of non electric nature (ion drag force, thermophoretic force, etc.).

This study was supported by the Program "Thermal Physics and Mechanics of Intense Pulse Actions" of the Presidium of the Russian Academy of Sciences, by the NWO Project No. 047.016.020 and by the CRDF Project No. RUP2-2891-MO-07

References

[1] Antipov S.N., Asinovskii E.I., Fortov V.E., Kirillin A.V., Markovets V.V., Petrov O.F., and Platonov V.I., *Phys. Plasmas*, **14**, 090701 (2007)

[2] Antipov S.N., Asinovskii E.I., Kirillin A.V., Maiorov S.A., Markovets V.V., Petrov O.F., and Fortov V.E., *JETP*, **106**, 830 (2008)

Soliton, double layer and shock formation in pair-ion plasmas

S. Mahmood[1,2], H.Ur Rehman[1,2], W. Masood[1] and N. Imtiaz[1]

[1]Theoretical Plasma Physics Division, P.O. Nilore Islamabad, Islamabad

[2]Department of Physics and Applied Mathematics, PIEAS P.O. Nilore Islamaabad Pakistan

shahzadm100@gmail.com

Abstract. Nonlinear electrostatic structures are studied in unmagnetized pair-ion plasmas. The low amplitude solitons and double layer structures are obtained using reductive perturbation method in non-dissipative and ideal plasmas. It is found that both electrostatic potential hump (compressive) and dip (rarefactive) solitons and double layers structures are obtained depending on the temperature ratio between and positive and negative ion species. The Kadomtsev-Petviashvili-Burger (KPB) is also derived by taking into account the weak transverse perturbations and kinematic viscosity of both positive and negative ions plasmas. Both two-dimensional rarefactive and compressive shocks and solitons solutions are obtained using Tan hyperbolic method. The structure dependence on temperature ratios between two ion species is also discussed. The present study may have some relevance for understanding of the formation of electrostatic structure in laboratory produced pair-ion plasmas.

1. Introduction

The collective modes in relativistic and nonrelativistic electron-positron (EP) pair plasmas have been studied with its application to astrophysical plasmas [1-5] and references therein. However, it is not an easy task to generate and maintain EP pair plasma in laboratory experiments for longer times due to their short annihilation time in comparison with plasma period to study collective effects. Therefore, in order to study the collective modes for same mass but oppositely charged particle plasma, a pair-ion (PI) fullerene plasma was produced in the laboratory [6-7]. The PI plasma is quite different from the usual electron-ion (EI) plasmas, where difference in masses of ions and electrons induces different time and length scales. The space time symmetry is maintained in pure pair-ion (PI) and electron-positron (EP) plasmas, because mobility of charge particles is the same in electromagnetic fields in this case. Three electrostatic modes have been observed parallel to the magnetic field in pure fullerene plasma experiment. A lot of theoretical research work has already been published on electrostatic waves in PI plasmas under the condition that temperatures of the species are not equal [8-12].

The manuscript has been organized in the following way: The nonlinear dynamic equations for electrostatic waves in PI plasmas are presented in Sec.II. The Kortewge-de Vries (KdV) equation is obtained using reductive perturbation method in Sec. III. The extended KdV (EKdV) equation is also obtained for low amplitude double layer (DL) structures in PI plasmas in Sec. IV. The Kadomtsev-Petviashvili-Burgers (KPB) equation for two dimensional soliton and dissipative shock solutions are derived for PI plasmas in Sec. V. Finally, conclusion on nonlinear electrostatic structures has been presented in Sec.VI.

Journal of Physics: Conference Series **511** (2014) 012009 doi:10.1088/1742-6596/511/1/012009

II. Nonlinear set of equations for pair-ion plasma

We are considering pure pair-ion (PI) unmagnetized plasma. Two fluid model has been used to study the electrostatic waves for isothermally heated positive and negative ions of same mass. The normalized set of continuity and momentum equations for positive and negative ions in one dimension can be written as follows,

$$\frac{\partial n_+}{\partial t} + \frac{\partial}{\partial x}(n_+ v_+) = 0, \tag{1}$$

$$\frac{\partial n_-}{\partial t} + \frac{\partial}{\partial x}(n_- v_-) = 0, \tag{2}$$

$$\frac{\partial v_+}{\partial t} + (v_+ \frac{\partial}{\partial x}) v_+ = -\frac{\partial \Phi}{\partial x} - \frac{1}{n_+}\frac{\partial n_+}{\partial x}, \tag{3}$$

$$\frac{\partial v_-}{\partial t} + (v_- \frac{\partial}{\partial x}) v_- = \frac{\partial \Phi}{\partial x} - \frac{\beta}{n_-}\frac{\partial n_-}{\partial x}, \tag{4}$$

The Poisson equation is given by

$$\frac{\partial^2 \Phi}{\partial x^2} = (n_- - n_+), \tag{5}$$

where the normalized electrostatic potential $\Phi=(e\varphi/T_+)$ has been defined. The perturbed velocities for positive and negative ions are normalized by ion thermal velocity due to positive ions $v_{T+}=(T_+/m)^{1/2}$ and perturbed number density by unperturbed densities n_0. The temperature ratio of negative and positive ion species $\beta=T_-/T_+$ has been defined. The space and time co-ordinates, x and t are normalized by Debye length $\lambda_{D+}=(T_+/4\pi n_0 e^2)^{1/2}$ and ion plasma frequency by $\omega_p=(4\pi n_0 e^2/m)^{1/2}$ respectively. In equilibrium, we have $n_{0+}=n_{0-}$ (say n_0).The isothermal ion pressure is defined as $p_\pm=\gamma n_\pm T_\pm$ and $\gamma=1$.

III. Derivation of Korteweg-de Vries equation in pair-ion plasmas

In this section, we apply the reductive perturbation method to solve nonlinear Eqs.(1-5) and obtain Korteweg-de Vries (KdV) equation for electrostatic soliton in unmagnetized pure PI plasmas. The perturbed quantities can be expanded in the power series of ε as follows:

$$n_j = 1 + \epsilon n_j^{(1)} + \epsilon^2 n_j^{(2)} + \ldots \tag{6}$$

$$v_j = \epsilon v_j^{(1)} + \epsilon^2 v_j^{(2)} + \ldots \tag{7}$$

$$\Phi = \epsilon \Phi^{(1)} + \epsilon^2 \Phi^{(2)} + \ldots \tag{8}$$

Here all perturbed quantities are functions of x and t, while ε is a small $0<\varepsilon<1$ expansion parameter characterizing the strength of the nonlinearity, such that the stretched variables are introduced in the standard fashion as follows:

$$\xi = \epsilon^{1/2}(x - \lambda t), \quad \tau = \epsilon^{3/2} t \tag{9}$$

Using Eqs. (6-7) and (9) in Eqs (1-5) and collecting the lowest order terms, we obtain

$$\lambda = \sqrt{\frac{1+\beta}{2}}. \tag{10}$$

which is the linear phase velocity of acoustic wave in PI plasmas. Now collecting, the next higher order and after some simplification we obtain the following Korteweg-de Vries (KdV) equation for the pure PI plasmas as follows [11],

$$\partial_\tau \Phi^{(1)} + A\Phi^{(1)}\partial_\xi \Phi^{(1)} + B\partial_\xi^3 \Phi^{(1)} = 0, \tag{11}$$

where nonlinear and dispersive coefficients are defined by A and B respectively,

$$A = \frac{[(3\lambda^2 - 1)(\lambda^2 - \beta)^3 - (3\lambda^2 - \beta)(\lambda^2 - 1)^3]}{2\lambda(\lambda^2 - 1)(\lambda^2 - \beta)[(\lambda^2 - \beta)^2 + (\lambda^2 - 1)^2]},$$

$$B = \frac{(\lambda^2 - 1)^2(\lambda^2 - \beta)^2}{2\lambda[(\lambda^2 - \beta)^2 + (\lambda^2 - 1)^2]},$$

The soliton solution of Eq. (11) is given by,

$$\Phi^{(1)} = \phi_m \operatorname{sech}^2\left(\frac{\eta}{W}\right), \tag{12}$$

Where $\varphi_m=3U/A$ and $W=(4B/U)^{1/2}$ are amplitude and the width of electrostatic soliton in pure PI plasmas, and η is the transformed coordinate with respect to a wave frame moving with velocity U $\eta=\xi-U\tau$. The weakly nonlinear electrostatic excitation in pure PI plasmas parallel to magnetic field is possible only when a slight difference of temperature between the positive and negative ions (i.e., $T_+ \neq T_-$) exist. The numerical values of PI plasma density $1-2\times10^8 cm^{-3}$ at $E_e = 100eV$ while temperatures of C_{60}^+ and C_{60}^- ions i.e., T_+ and T_- respectively lies in the range of $0.3-0.5\ eV$ have been reported in the laboratory experiments [7].

IV. Derivation of extended KdV equation for double layers in pair-ion plasmas

An electrostatic double layer is a structure in plasma which consists of two parallel layers with opposite electrical charge. These sheets of charge cause a strong electric field and correspondingly a sharp change in voltage (electrical potential) across the double layer. The positive and negative ions, which enter the double layer are accelerated, decelerated or reflected by the electric field. In order to find DL we use the following stretching for independent variables described as follows,

$$\xi = \varepsilon(x - \lambda t),. \qquad \tau = \varepsilon^3 t \tag{13}$$

The perturbed quantities are expanded in the powers of ε as defined earlier. Now collecting the higher order terms of continuity and momentum equations of ions $\sim\varepsilon^3$ and Poisson equation $\sim\varepsilon^4$ and using relation of parameter given in Eq.(10) after some simplification, we obtain the EKdV equation given by :

$$A\,\partial_\tau\varphi_1 + Q\,\partial_\xi\varphi_1^2 + B\,\partial_\xi\varphi_1^3 + \partial_\xi^3\varphi_1 = 0 \tag{14}$$

where the coefficients are defined as

$$A = 2\lambda\left[\frac{1}{(\lambda^2 - \beta)^2} + \frac{1}{(\lambda^2 - 1)^2}\right]$$

$$B = \left[\frac{(3\lambda^2 - 1)^2}{2(\lambda^2 - 1)^5} - \frac{\left(2\lambda^2 - \frac{1}{3}\right)}{(\lambda^2 - 1)^4} + \frac{(3\lambda^2 - 1)^2}{2(\lambda^2 - \beta)^5} - \frac{\left(2\lambda^2 - \frac{\beta}{3}\right)^2}{(\lambda^2 - \beta)^4}\right]$$

$$Q = \frac{(3\lambda^2 - \beta)}{2(\lambda^2 - \beta)^3} - \frac{(3\lambda^2 - 1)}{2(\lambda^2 - 1)^3}$$

Now using the transformation η=ξ-Uτ, where U is the speed of the co-moving frame with the speed of the soliton. The above EKdV equation can be described in the form of energy integral equation as follows,

$$\frac{1}{2}\left(\frac{\partial \varphi_1}{\partial \eta}\right)^2 + V(\varphi_1) = 0 \tag{15}$$

where the Sagdeev potential is defined as

$$V(\varphi_1) = -\frac{UA}{2}\varphi_1^2 + \frac{Q}{3}\varphi_1^3 + \frac{B}{4}\varphi_1^4 \tag{16}$$

For the formation of DLs, the Sagdeev potential must satisfy the following conditions,

$$V(\varphi_1) = 0, \quad \frac{dV}{d\varphi_1} = 0 \text{ at } \varphi_1 = 0 \text{ and } \varphi_1 = \varphi_m.$$

$$\frac{d^2V}{d\varphi_1^2} < 0 \text{ at } \varphi_1 = 0 \text{ and } \varphi_1 = \varphi_m \tag{17}$$

The above conditions are satisfied only if, coefficients Q and U described as

$$Q = -\frac{3}{2}B\varphi_m \quad , \quad U = -\frac{1}{2}\frac{B}{A}\varphi_m^2$$

The Sagdeev potential satisfying the DLs boundary conditions is given as follows,

$$V(\varphi_1) = \frac{B}{4}\varphi_1^2(\varphi_m - \varphi_m)^2 \tag{18}$$

Using Eq.(e38), the solution of energy integral equation (e35) can be written as,

$$\varphi_1 = \frac{\varphi_m}{2}\left[1 - \tanh\left\{\sqrt{\frac{-B}{8}}\varphi_m(\eta - U\tau)\right\}\right] \tag{19}$$

where amplitude and width of the DLs are defined as

$$\varphi_m = -\frac{2}{3}\frac{Q}{B} \quad \text{width} = \frac{\sqrt{\frac{-B}{8}}}{\varphi_m}$$

In order to find the real solution of Eq.(19) B<0 must hold. In case, Q<0 compressive and for Q>0 rarefactive DLs are formed, where $\beta=T_-/T_+$ has been defined. It is clear from Eq.(19) that DL structures will be formed only when there occurs a small difference of thermal energies between pair ion species.

V. Derivation of Kadomtsev-Petviashvili-Burgers (KPB) equation for pair-ion plasmas

In order to derive the KPB equation in PI fullerene plasmas the dissipation in the system is taken into account through kinematic viscosity of the ions in the system. The normalized form of continuity, momentum, and Poisson's equations in two dimensions are [12]:

$$\frac{\partial n_+}{\partial t} + \frac{\partial}{\partial x}(n_+u_+) + \frac{\partial}{\partial y}(n_+v_+) = 0,$$

$$\frac{\partial n_-}{\partial t} + \frac{\partial}{\partial x}(n_- u_-) + \frac{\partial}{\partial y}(n_- v_-) = 0,$$

$$\frac{\partial u_+}{\partial t} + u_+\frac{\partial u_+}{\partial x} + v_+\frac{\partial u_+}{\partial y} = -\frac{\partial \phi}{\partial x} - \frac{1}{n_+}\frac{\partial n_+}{\partial x} + \frac{\eta_+}{n_+}\left(\frac{\partial^2}{\partial x^2} + \frac{\partial^2}{\partial y^2}\right)u_+,$$

$$\frac{\partial v_+}{\partial t} + u_+\frac{\partial v_+}{\partial x} + v_+\frac{\partial v_+}{\partial y} = -\frac{\partial \phi}{\partial y} - \frac{1}{n_+}\frac{\partial n_+}{\partial y} + \frac{\eta_+}{n_+}\left(\frac{\partial^2}{\partial x^2} + \frac{\partial^2}{\partial y^2}\right)v_+,$$

$$\frac{\partial u_-}{\partial t} + u_-\frac{\partial u_-}{\partial x} + v_-\frac{\partial u_-}{\partial y} = \frac{\partial \phi}{\partial x} - \frac{\beta}{n_-}\frac{\partial n_-}{\partial x} + \frac{\eta_-}{n_-}\left(\frac{\partial^2}{\partial x^2} + \frac{\partial^2}{\partial y^2}\right)u_-,$$

$$\frac{\partial v_-}{\partial t} + u_-\frac{\partial v_-}{\partial x} + v_-\frac{\partial v_-}{\partial y} = \frac{\partial \phi}{\partial y} - \frac{\beta}{n_-}\frac{\partial n_-}{\partial y} + \frac{\eta_-}{n_-}\left(\frac{\partial^2}{\partial x^2} + \frac{\partial^2}{\partial y^2}\right)v_-,$$

$$\frac{\partial^2 \phi}{\partial x^2} + \frac{\partial^2 \phi}{\partial y^2} = n_- - n_+.$$

Here, the ion fluid velocities along x-axis and y-axis i.e., u_α and v_α respectively, are normalized by the ion-thermal speed $v_{T+}=(T_+/m)^{1/2}$, ion density n_α is normalized by equilibrium density n_0 and electrostatic wave potential φ is normalized by T_+/e . The space and time variables are normalized in the units of positively charged ion Debye length $\lambda_{D+}= (T_+/4\pi n_0 e^2)^{1/2}$ and its plasma frequency $\omega_p= (4\pi n_0 e^2/m)^{1/2}$ respectively. The temperature ratios of the ion species are defined as $\beta=T_-/T_+$ whereas the normalized viscosity coefficients are $\eta_+ = \nu_+/(\lambda_+ V_{T_+})$ and $\eta_- = \nu_-/(\lambda_+ V_{T_+})$ for positively and negatively charged ion fluid, respectively and $\nu_\pm = \mu_\pm/mn_0$ have been defined. The viscosity coefficients of the same mass fluids are different since their temperatures are assumed to be different. The kinematic viscosity is defined as $\nu_\alpha = \mu_\alpha/mn_\alpha$ (where μ_α is defined as the dynamic viscosity of the $\alpha-$ species in PI plasmas). The dynamic viscosity of fully ionized gas has been analyzed by Braginskii [13] and in the absence of magnetic field, the dynamic viscosity of ions is given by

$$\mu_{i\alpha} = \frac{0.406\, m_i^{\frac{1}{2}}(k_B T_\alpha)^{\frac{5}{2}}}{Z^4 e^4 \ln\Lambda} = 2.21\times 10^{-15}\,\frac{T_\alpha^{\frac{5}{2}} A_i^{\frac{1}{2}}}{Z^4 \ln\Lambda}\,\frac{gm}{cm\sec}$$

where A_i is the atomic weight of ions, Z is the charge number and $\ln\Lambda$ is the Coulomb logarithm. The above defined dynamic viscosity of ions by Braginskii will be used for numerical studies in our model for PI species.

In order to study the two dimensional electrostatic waves in unmagnetized pair-ion plasmas, we define the stretching of the independent variables as :

$$\zeta = \epsilon^{1/2}(x - \lambda t),\ \chi = \epsilon y,\ \tau = \epsilon^{3/2} t,$$

where ε is a small expansion parameter and λ is wave phase velocity .Using reductive perturbative technique, we expand the perturbed quantities about their equilibrium values in powers of $\sim\varepsilon$ such that

$$
\begin{aligned}
n_\alpha &= 1 + \epsilon n_\alpha^{(1)} + \epsilon^2 n_\alpha^{(2)} + \epsilon^3 n_\alpha^{(3)} + \cdots, \\
u_\alpha &= \epsilon u_\alpha^{(1)} + \epsilon^2 u_\alpha^{(2)} + \epsilon^3 u_\alpha^{(3)} + \cdots, \\
v_\alpha &= \epsilon^{\frac{3}{2}} v_\alpha^{(1)} + \epsilon^{\frac{5}{2}} v_\alpha^{(2)} + \epsilon^{\frac{7}{2}} v_\alpha^{(3)} + \cdots, \\
\phi &= \epsilon \phi^{(1)} + \epsilon^2 \phi^{(2)} + \epsilon^3 \phi^{(3)} + \cdots.
\end{aligned}
$$

The value of η is assumed to be small, i.e., $\eta_\pm = \epsilon^{1/2}\eta_{0\pm}$ where $\eta_{\pm 0}$ is $O(1)$. Substituting above relations in the dynamics equations of PI plasmas and collecting different powers of ε, the lowest order equation yields the dispersion relation of acoustic wave in PI plasmas is given in Eq.(10). From next higher order equations in ε we obtain the KPB equation for pair-ion plasmas as follows

$$
\frac{\partial}{\partial\zeta}\left(\frac{\partial\Phi}{\partial\tau} + A\Phi\frac{\partial\Phi}{\partial\zeta} + B\frac{\partial^3\Phi}{\partial\zeta^3} - C\frac{\partial^2\Phi}{\partial\zeta^2}\right) + D\frac{\partial^2\Phi}{\partial\chi^2} = 0, \tag{20}
$$

$$
A = \frac{[(3\lambda^2-1)(\lambda^2-\beta)^3 - (3\lambda^2-\beta)(\lambda^2-1)^3]}{2\lambda(\lambda^2-1)(\lambda^2-\beta)[(\lambda^2-1)^2+(\lambda^2-\beta)^2]},
$$

$$
B = \frac{(\lambda^2-1)^2(\lambda^2-\beta)^2}{2\lambda[(\lambda^2-1)^2+(\lambda^2-\beta)^2]},
$$

$$
C = \frac{\eta_{0+}(\lambda^2-\beta)^2 - \eta_{0-}(\lambda^2-1)^2}{2[(\lambda^2-1)^2+(\lambda^2-\beta)^2]},
$$

$$
D = \frac{\lambda}{2},
$$

where $\Phi \equiv \phi^{(1)}$, A and C are the coefficients of non-linearity and dissipation whereas B and D are the coefficients of predominant and weak dispersion respectively. Using the transformation $\xi = k(\zeta + \chi - U\tau)$, where k is the dimensionless nonlinear wave number. There are number of methods to solve the nonlinear partial differential equations (NLPDE's). However, when the partial differential equation in a system is formed by the combined effect of dispersion and dissipation, the most convenient and efficient method to solve the NLPDE is tanh method [13], which gives the solution of KPB equation as follows,

$$
\begin{aligned}
\Phi(\zeta,\chi,\tau) = {} & \frac{6}{25}\frac{C^2}{AB}\left[1 - \tanh\frac{C}{10B}\left\{\zeta + \chi - \left(\frac{6C^2}{25B} + D\right)\tau\right\}\right] + \\
& \frac{3}{25}\frac{C^2}{AB}\left[\sec h^2\frac{C}{10B}\left\{\zeta + \chi - \left(\frac{6C^2}{25B} + D\right)\tau\right\}\right].
\end{aligned} \tag{21}
$$

Where k=C/10B . The above solution will exist only if $T_+ \neq T_-$ holds. The coefficients A , B and C become undetermined at $T_+ = T_-$ and therefore no shock and soliton structures can be obtained. If we assume the dissipative coefficient C equal to zero, then the KPB equation reduces to KP equation which admits soliton solution given as

$$
\Phi(\zeta,\chi,\tau) = \frac{12B}{A}\sec h^2[\zeta + \chi - U\tau]. \tag{22}
$$

Where U=-(4B+D) .The parametric dependence of the electrostatic shocks and solitons for above equations has been shown and discussed in detail in [12]. Both compressive and rarefactive two dimensional solitons can be obtained from Eq.(22) depending on the value of β i.e., the value of the temperature ratio between between positive and negatively charged ions in PI plasmas (i.e., $T_+ < T_-$ and $T_+ > T_-$). Similarly the shock structures from Eq.(21)can be obtained for the cases $T_+ < T_-$ and $T_+ > T_-$ respectively. It can be seen from the figures that for $T_- > T_+$ case, only compressive shocks are obtained while for $T_+ > T_-$ case, rarefactive shocks are formed. This happens due to the fact that for $T_+ > T_-$, the nonlinearity changes sign (it becomes negative in this case contrary to the positive nonlinearity sign in the $T_- > T_+$ case) and hence the formation of the rarefactive shocks. The situation reverses for $T_+ > T_-$ case owing the change of the sign of the nonlinearity.

VI. Conclusion

To conclude, linear and nonlinear electrostatic waves have been studied in unmagnetized PI plasmas. The nonlinear partial differential equations such as KdV, EKdV and KPB equations are obtained using reductive perturbation method. It is found that KdV equation gives both electrostatic hump and dip soliton solutions in unmagnetized pure PI plasmas. The EKdV equation gives both the compressive and rarefactive electrostatic DLs structures in PI plasma depending on the value of temperature ratio of both PI species. The study of DLs in PI plasma is important due to its application in accelerating or decelarating the charged particles in plasmas. The two dimensional soliton and shock structures are also obtained from KPB equation by taking into account the dissipation through kinematic viscosity of the PI species. It is found that electrostatic structures are formed only when there is a slight difference of temperature between positive and negative fullerene ions remains in PI plasma. The results are general and may be applicable to explain some salient features of the formation of nonlinear electrostatic structures in laboratory produced PI plasmas.

One of the author SM is thankful for the Executive Director COMSATS Hq. Islamabad for approving the travel grant (No. 12(26)DGIA/Gen/STG/2581 dated July 01, 2010) to attend the Latin American Workshop on Plasma Physics (LAWPP) and International Congress on Plasma Physics (ICPP) from 01-13 August 2010 at Santiago, Chile. The author is also thankful to the local Organizing Committee for their financial support of living during the conference.

References

[1] H. R. Miller and P. J. Witta, Active Galactive Nuclei (Springer, Berlin, 1987) p-202
[2] M. C. Begelman, R. D. Blandford and M. D. Rees, Rev. Mod. Phys. **56**, 255 (1984)
[3] M. E. Gedalin, J. G. Lominadze, L. Stenflo and V. N. Tsytovich, Astrophys. Space Sci. **108**, 393 (1985)
[4] Y. M. Yu, P.K. Shukla and L. Stenflo, Astrophys. J. **309**, L63 (1986)
[5] G. P. Zank and R.G. Greaves, Phys. Rev. E **51**, 6079 (1995
[6] W. Oohara and R.Hatakeyama, Phys. Rev. Lett. 91, 205005 (2003)
[7] W. Oohara, Y. Kuwabara and R. Hatakeyama, Phys. Rev. E **75**, 056403(2007)
[8] F. Verheest, Phys. Plasmas **13**, 082301(2006)
[9] P.K. Shukla and M. Khan, Phys. Plasmas **12**, 014504 (2005)
[10] A. E. Dubinov ,I. D. Dubinova and V. A. Gordienko , Phys. Plasmas **13**, 082111(2002)
[11] S. Mahmood, H. Ur. Rehman and H. Saleem , Phys. Scr. **80**, 035502 (2009)
[12] W. Masood, S. Mahmood and N. Imtiaz, Phys. Plasmas **16**, 122306 (2009)
[13] L. Spitzer, *Physics of Fully Ionized Gases*, (John Wiley & Sons, USA, 1962) p-146
[14] W. Malfliet , Am. J. Phys. **60,** 650 (1992)

Ion acoustic mode in permeating plasmas

J. Vranjes and S. Poedts
Center for Plasma Astrophysics,
and Leuven Mathematical Modeling and Computational Science Research Centre (LMCC),
Celestijnenlaan 200B, 3001 Leuven, Belgium.
E-mail: jvranjes@yahoo.com; Stefaan.Poedts@wis.kuleuven.be

Abstract. In the presence of an electron drift relative to ions, the ion acoustic wave is subject to the kinetic instability which takes place if the directed electron speed exceeds the ion acoustic speed. The instability threshold is very different in the case of one quasi-neutral electron-ion plasma propagating through another quasi-neutral (target) plasma. The threshold velocity of the propagating plasma may be well below the ion acoustic speed of the static plasma.

1. Introduction

An electron drift (current) relative to static ions can make the ion acoustic wave growing due to the kinetic instability which takes place if the directed electron speed exceeds the ion acoustic speed. The growth rate of the instability is given by

$$\omega_i = \left(\frac{\pi}{8}\right)^{1/2} k c_s \left[\left(\frac{m_e}{m_i}\right)^{1/2}\left(\frac{v_{e0}}{c_s} - 1\right) - \tau^{3/2}\exp\left(-\frac{\tau}{2}\right)\right]. \quad (1)$$

Here, $c_s = (\kappa T_e/m_i)^{1/2}$ is the ion sound speed, and $\tau = T_e/T_i$. The necessary condition $v_{e0} > c_s[1 + (m_i/m_e)^{1/2}\tau^{3/2}\exp(-\tau/2)]$ may imply a rather high electron current magnitude. The instability described by (1) is a purely kinetic, collision-less, electron inertia effect. Its two-fluid counterpart [1, 2], however, is a strictly electron-collision effect.

However, a much lower instability threshold may be obtained in the case of two interpenetrating (permeating) plasmas. This is then a *current-less* instability, where now the Doppler shift of ions appears to play the main role. Such permeating plasmas can be created in lab conditions, while in space this happens to be a rather frequent situation, for example in the case of colliding astrophysical clouds, in the propagation of plasmas originating from the explosions of novae and supernovae and moving through the surrounding plasmas, and also in the case of solar and stellar winds. In the case of the solar wind, one electron-ion quasi-neutral component is generated mainly in the polar regions (fast wind), while the other electron-ion quasi-neutral component (slow wind) originates from lower latitudes, yet at large distances from the origin there may be overlapping between the two. A most obvious example of such permeating plasmas is seen also in the solar atmosphere where plasma streams from lower layers are seen continuously propagating along magnetic field structures towards and through the solar corona. In all of these examples we have one fast flowing plasma whose parameters we shall denote by the subscript f, and one slow moving (target) plasma with the parameters denoted by the subscript s. In fact, the target plasma can be non-moving (or 'static', hence the same index 's'), or if it does move with some constant velocity $\vec{v}_{s0} = v_{s0}\vec{e}_z$, the corresponding equations for the whole combined plasma system can conveniently be written in the coordinate frame moving with this same velocity. Therefore, in the forthcoming text the corresponding velocity of the other component $\vec{v}_{f0} = v_{f0}\vec{e}_z$ will be used to describe just the difference between the two.

2. Derivations and results

In the case of singly-charged ions, the quasi-neutrality conditions for the two plasmas in the equilibrium read:

$$n_{fi0} = n_{fe0} = n_{f0}, \quad n_{si0} = n_{se0} = n_{s0}. \tag{2}$$

In general $n_{s0} \neq n_{f0}$, and the same holds for the temperatures of the two separate quasi-neutral plasmas $T_{fe} \neq T_{se}$, $T_{fi} \neq T_{si}$. We use the plasma distribution function for the species j,

$$f_{j0} = \frac{n_{j0}}{(2\pi)^{3/2} v_{Tj}^3} \exp\left\{ -\frac{1}{2v_{Tj}^2}\left[v_x^2 + v_y^2 + (v_z - v_{j0})^2\right]\right\}, \tag{3}$$

where $n_{j0} = const$, and $v_{Tj}^2 = \kappa T_j/m_j$. From the linearized Boltzmann kinetic equation and for longitudinal electrostatic perturbations $\sim \exp(-i\omega + ikz)$, we obtain

$$f_{j1} = -\frac{iq_j}{m_j}\frac{\vec{E}_1}{\omega - \vec{k}\vec{v}}\frac{\partial f_{j0}}{\partial \vec{v}}.$$

Further, using the Ampère law $0 = \mu_0\vec{j} + (1/c^2)\partial\vec{E}/\partial t$, and the expression for the macroscopic current $\vec{j} = \sum_j q_j \int \vec{v} f_j d^3\vec{v}$, we arrive at

$$1 + \sum_j \frac{1}{k^2\lambda_{dj}^2}[1 - Z(b_j)] = 0. \tag{4}$$

Here, $b_j = (\omega - kv_{j0})/(kv_{Tj})$, $\lambda_{dj} = v_{Tj}/\omega_{pj}$, $\omega_{pj}^2 = q_j^2 n_{j0}/(\varepsilon_0 m_j)$, and

$$Z(b_j) = \frac{b_j}{(2\pi)^{1/2}}\int_c \frac{\exp(-\zeta^2/2)}{b_j - \zeta}d\zeta$$

is the plasma dispersion function, with the integration over the Landau contour c. The dispersion equation (4) describes the plasma (Langmuir) and the ion acoustic oscillations.

In the case $v_{sj0} = 0$, $v_{fj0} = v_{f0}$, expanding Eq. (4) in the ion acoustic frequency range on conditions

$$kv_{Tsi} \ll |\omega| \ll kv_{Tse}, \quad |\omega - kv_{f0}| \ll kv_{Tfe},\ kv_{Tfi}, \tag{5}$$

where $v_{Tsj}^2 = \kappa T_{sj}/m_j$, $v_{Tfj}^2 = \kappa T_{fj}/m_j$, we obtain the dispersion equation for the ion acoustic wave

$$\Delta(k,\omega) \equiv 1 + \frac{1}{k^2\lambda_d^2} - \frac{\omega_{psi}^2}{\omega^2} - \frac{3k^2 v_{Tsi}^2\omega_{psi}^2}{\omega^4}$$

$$+i\left(\frac{\pi}{2}\right)^{1/2}\left[\underbrace{\frac{\omega\omega_{pse}^2}{k^3 v_{Tse}^3}}_{c} + (\omega - kv_{f0})\left(\underbrace{\frac{\omega_{pfe}^2}{k^3 v_{Tfe}^3}}_{a} + \underbrace{\frac{\omega_{pfi}^2}{k^3 v_{Tfi}^3}}_{b}\right) + \frac{\omega\omega_{psi}^2}{k^3 v_{Tsi}^3}\exp\left(-\frac{\omega^2}{2k^2 v_{Tsi}^2}\right)\right] = 0. \tag{6}$$

Here, $1/\lambda_d^2 = 1/\lambda_{dse}^2 + 1/\lambda_{dfe}^2 + 1/\lambda_{dfi}^2$, and $\lambda_{dse} = v_{Tse}/\omega_{pse}$ etc. In view of the conditions (5) the inertia of the ion acoustic (IA) mode is provided mainly by the ions of the static (target) plasma.

In standard electron-ion plasma with streaming electrons and for $T_i \ll T_e$ it yields

$$1+\frac{1}{\lambda_{de}^2 k^2}-\frac{\omega_{pi}^2}{\omega^2}+i(\pi/2)^{1/2}\left\{\frac{\omega_{pe}^2(\omega-kv_{e0})}{k^3 v_{Te}^3}+\frac{\omega_{pi}^2\omega}{k^3 v_{Ti}^3}\exp\left[-\omega^2/(2k^2 v_{Ti}^2)\right]\right\}=0. \quad (7)$$

From this one can further obtain the growth rate (1) and the frequency $\omega^2 \simeq k^2 c_s^2$.

The only terms in Eq. (6) that may produce a growing mode are those with the Doppler-shift, the two other terms in the imaginary part always give the Landau damping. The electron part in the Doppler-shifted term is usually negligible because $a/b = (T_{fi}/T_{fe})^{3/2}(m_e/m_i)^{1/2} \ll 1$. Assuming also that

$$c/[b(\omega-kv_{f0})] = |\omega/(\omega-kv_{f0})|(n_{s0}/n_{f0})(T_{fi}/T_{se})^{3/2}(m_e/m_i)^{1/2} \ll 1,$$

the dispersion equation becomes

$$1+\frac{1}{\lambda_d^2 k^2}-\frac{\omega_{psi}^2}{\omega^2}-\frac{3k^2 v_{Tsi}^2\omega_{psi}^2}{\omega^4}$$

$$+i(\pi/2)^{1/2}\left\{\frac{\omega_{pfi}^2(\omega-kv_{f0})}{k^3 v_{Tfi}^3}+\frac{\omega_{psi}^2\omega}{k^3 v_{Tsi}^3}\exp\left[-\omega^2/(2k^2 v_{Tsi}^2)\right]\right\}=0. \quad (8)$$

The frequency of the ion acoustic wave which follows from this is

$$\omega_r^2 = (k^2\lambda_d^2\omega_{psi}^2/2)[1+(1+12\lambda_{si}^2/\lambda_d^2)^{1/2}]. \quad (9)$$

For an ordinary electron-ion plasma with s-species only, it takes the shape $\omega_r^2 = (kc_s^2/2)[1+(1+12T_{si}/T_{se})^{1/2}]$.

The growth rate from (8), $\omega_i \simeq -Im\Delta(k,\omega_r)/[\partial(Re\Delta)/\partial\omega]_{\omega\simeq\omega_r}$, becomes

$$\omega_i = -\left(\frac{\pi}{8}\right)^{1/2}\frac{\omega_r^3}{k^2 v_{Tfi}^2(1+6k^2 v_{Tsi}^2/\omega_r^2)}\left\{\frac{\omega_r-kv_{f0}}{kv_{Tfi}}\frac{n_{f0}}{n_{s0}}+\frac{\omega_r}{kv_{Tsi}}\frac{T_{fi}}{T_{si}}\exp\left[-\omega_r^2/(2k^2 v_{Tsi}^2)\right]\right\}. \quad (10)$$

In the limit $T_{fe} \gg T_{fi}$, $T_{se} \gg T_{si}$ and setting $\omega_r \simeq kc_{se}$, the growth rate becomes

$$\omega_i = \left(\frac{\pi}{8}\right)^{1/2} kc_{se}\left(\frac{T_{se}}{T_{fi}}\right)^{3/2}\left\{\left(\frac{v_{f0}}{c_{se}}-1\right)\frac{n_{f0}}{n_{s0}}-\left(\frac{T_{fi}}{T_{si}}\right)^{3/2}\exp\left[-T_{se}/(2T_{si})\right]\right\}. \quad (11)$$

From this the instability threshold is

$$v_{f0} > c_{se}\left[1+\frac{n_{s0}}{n_{f0}}\left(\frac{T_{fi}}{T_{si}}\right)^{3/2}\exp\left(-\frac{T_{se}}{2T_{si}}\right)\right]. \quad (12)$$

Here, $c_{se}^2 = \kappa T_{se}/m_i$. Equation (12) has a remarkable properties. It is seen that here the first temperature ratio contains only the two *ion* temperatures, and in addition the electron-ion mass ratio vanishes. In fact, the mass ratio can appear, yet in the present case it would contain the *two ion masses* (that are taken as equal). For the singly charged two ion species, instead of n_{s0}/n_{f0} in Eq. (12) we would have $(n_{s0}/n_{f0})(m_{fi}/m_{si})$. Consequently, in the present case the instability threshold may become considerably lower. This is particularly valid for a flowing plasma being accelerated into a less dense but hotter static target plasma. One example of that kind may be the plasma from the lower solar atmosphere propagating through hot upper layers (e.g., spicules, and flows along magnetic loops in general).

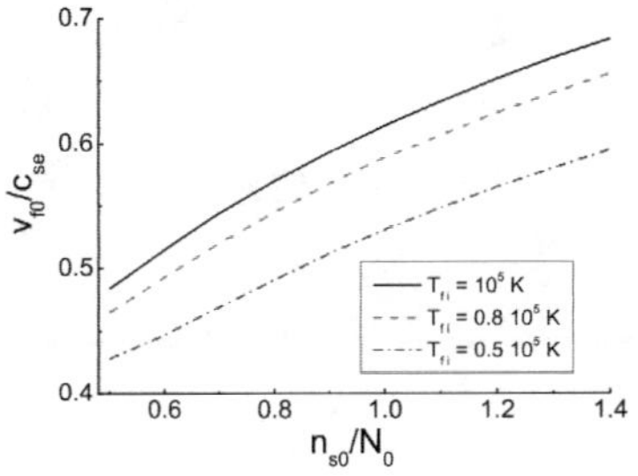

Figure 1. The critical (threshold) values of the flowing plasma velocity for the ion-acoustic wave instability in terms of the number density of the static (target) plasma. Here, $c_{se}^2 = \kappa T_{se}/m_i$, $N_0 = 10^{17}$ m^{-3}.

However, in the general case the threshold velocity v_{f0} for the currentless instability may in fact be *below* the sound velocity of the target plasma. As an example, the full Eq. (6) is solved numerically by taking the wavelength $\lambda = 0.3$ m, and $T_{se} = 10^5$ K, $T_{si} = 10^3$ K, $n_{f0} = 10^{17}$ m^{-3}, and n_{s0} is varied in the range shown in Fig. 1. The critical value of the flowing plasma velocity is shown for several values of the temperature of the flowing plasma. For the velocities above the given lines the IA mode becomes growing. For these parameters, in Eq. (9) the term $\lambda_{si}^2/\lambda_d^2 = 0.07$ and also $\lambda_d \omega_{psi} = 0.38 c_{se}$. Therefore, contrary to Eq. (1) or (12), the instability threshold (c.f. Fig. 1) is in fact well *below* the IA speed of the static (target) plasma. For lower T_{fi} the instability threshold is reduced because of a lower Landau damping by the flowing ions.

The cases of dominant ion (or electron) terms discussed above are equivalent to studying also electron-ion plasmas containing an additional ion (or electron species) [3]-[5]. As special cases, these include also pair-ion [6]-[9], negative-ion [10], electron-positron-ion plasmas, and also dusty plasmas with opposite charges on grains [11]. Such plasmas have been extensively studied in the recent past. The general streaming kinetic instability presented above is directly applicable to all these plasmas as special cases.

Acknowledgments

The results presented here are obtained in the framework of the projects G.0304.07 (FWO-Vlaanderen), C 90347 (Prodex), GOA/2009-009 (K.U.Leuven). Financial support by the European Commission through the SOLAIRE Network (MTRN-CT-2006-035484) is gratefully acknowledged.

[1] Vranjes J, Tanaka M Y and Poedts S 2006 *Phys. Plasmas* **13** 122103
[2] Vranjes J, Pandey B P and Poedts S 2007 *Phys. Plasmas* **14** 032106
[3] Vu H X, Wallace J M and Bezzerides B 1996 *Phys. Plasmas* **1** 3542
[4] Vranjes J, Petrovic D, Pandey B P and Poedts S 2008 *Phys. Plasmas* **15** 072104
[5] Jones W D, Lee A, Gleman S M and Doucet H J 1975 *Phys. Rev. Lett.* **35** 1349
[6] Oohara W and Hatakeyama R 2003 *Phys. Rev. Lett.* **91** 205005
[7] Vranjes J and Poedts S 2005 *Plasma Sources Sci. Technol.* **14** 485
[8] Saleem H 2007 *Phys. Plasmas* **14** 014505
[9] Kourakis I, Esfandyari-Kalejahi A, Medhipoor M and Shukla P K 2006 *Phys. Plasmas* **13** 052117
[10] Song B, D'Angelo N and Merlino R L 1991 *Phys. Fluids B* **3** 284
[11] D'Angelo N 2001 *Planet. Space. Sci.* **49** 1251

Kinetic dust acoustic mode in inhomogeneous partially magnetized plasma

J. Vranjes and S. Poedts

Center for Plasma Astrophysics,
and Leuven Mathematical Modeling and Computational Science Research Centre (LMCC),
Celestijnenlaan 200B, 3001 Leuven, Belgium.

E-mail: jvranjes@yahoo.com; Stefaan.Poedts@wis.kuleuven.be

Abstract. The dust acoustic mode in inhomogeneous plasmas is discussed in the regime when the electrons and ions are magnetized while at the same time the dust grains remain un-magnetized. Although the dynamics of the light species is strongly affected by the magnetic field, the dust acoustic mode may still propagate in practically any direction. The inhomogeneity implies a source of free energy for an instability that develops through the diamagnetic drift effects of the magnetized species. It is shown that this may be a powerful mechanism for the excitation of dust acoustic waves. The analysis presented in this work is also applicable to plasmas containing both positive and negative ions and electrons, provided that at least one of the two ion species is un-magnetized.

1. Introduction

Inhomogeneous plasmas imply the presence of a source of free energy that can make some plasma modes growing. In an electron-ion plasma, when both electrons and ions are magnetized, the corresponding growing low-frequency mode is the drift mode. However, in some situations, in the perturbed state the heavier species (ions) may in fact behave as un-magnetized regardless of the presence of a magnetic field. In view of the mass difference, for electrons we may have at the same time $\omega \ll \Omega_e = eB_0/m_e$. Electrostatic modes propagating in a plasma with such properties will have the basic features of the ion acoustic (IA) mode, in spite of the magnetized electrons. Within the two-fluid theory such a mode in an inhomogeneous plasma may become growing [1] provided the simultaneous presence of collisions and the mentioned equilibrium density gradient perpendicular to $\vec{B}_0$. Within the kinetic theory the mode is also growing due to purely kinetic effects. The growth rate is similar to the standard drift wave instability [2], with the necessary condition that the wave frequency is below the electron diamagnetic drift frequency $\omega_{*e} = v_{*e}k_\perp$.

In a plasma containing also some dust grains, and in the presence of some small perturbations, the time and space scales at which the grain dynamics develops is orders of magnitude different as compared to the above mentioned drift and IA wave scales. Hence, both the ions and electrons may be well magnetized, while the dust grains may not feel the effects of the magnetic field. One may then expect that the grain dynamics remains within the dust acoustic (DA) wave description although the lighter species are magnetized and thus have very different dynamics in the directions parallel and perpendicular to the magnetic field vector. In an inhomogeneous plasma, the dynamics of the electrons and ions will be described by a set of equations that follow from the standard drift wave theory, with possibly the only difference regarding the finite ion inertia effects that at the DA wave scale can simply be ignored. The resulting dispersion equation may then yield a growing DA wave, with the instability driven by the density gradient of the magnetized light particles. This type of the dust acoustic, gradient-driven instability will be studied in the present work. In particular, we shall use a model in which the equilibrium dusty plasma is confined by the magnetic field, implying that all species are magnetized, while the perturbations will be studied. Two specific cases will be studied, viz. i) an electron-ion plasma with negatively charged grains, and ii) a plasma containing positively charged grains and negative ions and electrons.

2. Derivations and results

We shall assume a plasma placed in an external magnetic field $\vec{B}_0 = B_0\vec{e}_z$, with a density gradient of the species j in the x-direction ($j = e, i, d$). The charge number Z_d of the dust grains is assumed to be constant (which is appropriate for frequencies far below the mean attachment frequencies for electrons and ions so that the equilibrium quasi-neutrality condition reads $n_{i0}(x) = n_{e0}(x) + Z_d n_{d0}(x)$. A varying grain charge is known to introduce an extra mode of oscillations [3]. In such a geometry, the species will have the above described equilibrium diamagnetic drift velocities $\vec{v}_{*j} = \vec{e}_z \times \nabla p_j/(q_j n_j B_0)$. The perturbed number density of the species j is $n_{j1} = \int_\infty^\infty f_{j1} d^3\vec{v}$, where f_{j1} is the corresponding distribution function. In the perturbed state, we shall continue to treat the light species as magnetized. However, the dust grains are taken as un-magnetized, implying wave-lengths that are above the ion and electron gyro-radii, yet below the grain gyro-radius, or/and the wave frequencies that are below the electron and ion gyro-frequency, but above the grain gyro-frequency. Hence, the derivation will be different for magnetized and un-magnetized species. The former yields for the perturbed number density

$$n_{j1} = -\frac{q_j n_{j0}\phi_1}{\kappa T_j}\left\{1 + (\omega - \omega_{*j})\sum_m \frac{\Lambda_m(\beta_j)}{\omega - m\Omega_j}\left[W\left(\frac{\omega - m\Omega_j}{|k_z|v_{Tj}}\right) - 1\right]\right\}. \quad (1)$$

Here, $\beta_j = k_\perp^2\rho_j^2$, $\Lambda_m(\beta_j) = I_m(\beta_j)\exp(-\beta_j)$, $W(\chi) = (2\pi)^{1/2}\int_{-\infty}^{+\infty}\eta\exp(-\eta^2/2)d\eta/(\eta - \chi)$, I_m is the modified Bessel function of the first kind and order m. We use the local approximation with small perturbations of the form $\sim \hat{f}(x)\exp(-i\omega t + ik_y y + ik_z z)$; $\hat{f}(x)$ is the x-dependent amplitude and $|d/dx| \ll |k_y|$. For frequencies much below Ω_j, we keep only the term $m = 0$ and, being interested in the DA spatial and time scales, it is good enough to use $\Lambda_0(\beta_j) \simeq 1$ for both electrons and ions. For similar reasons ($|\chi| \ll 1$) we shall use the approximate expression $W(\chi) \simeq 1 - \chi^2 + \ldots + i(\pi/2)^{1/2}\chi\exp(-\chi^2/2)$. This yields for electrons and ions

$$\frac{n_{j1}}{n_{j0}} = -\frac{q_j\phi_1}{\kappa T_j}\left[1 + i\left(\frac{\pi}{2}\right)^{1/2}\frac{\omega - \omega_{*j}}{|k_z|v_{Tj}}\exp\left(\frac{\omega^2}{2k_z^2 v_{Tj}^2}\right)\right]. \quad (2)$$

Regarding the perturbed dust dynamics, we shall assume frequencies satisfying the conditions $\omega \gg \Omega_d$, and $\omega \gg \omega_{*d}$. The latter is particularly easily satisfied for grains with a high charge number. Further in the text, these assumptions will be supported by the parameters which we use. In such a frequency limit for dust grains we have

$$\frac{n_{d1}}{n_{d0}} = \frac{eZ_d\phi_1}{\kappa T_d}\left[1 - \mathcal{Z}\left(\frac{\omega}{kv_{Td}}\right)\right]. \quad (3)$$

Here, $\mathcal{Z}(\alpha) = [\alpha/(2\pi)^{1/2}]\int_{-\infty}^{+\infty} d\xi\exp(-\xi^2/2)/(\alpha - \xi)$ is the plasma dispersion function. Note that the argument α here comprises the total k, which describes the essential fact that the grain perturbations are independent on the direction of the magnetic field vector $\vec{B}_0 = B_0\vec{e}_z$.

After expanding $\mathcal{Z}(\alpha) \simeq 1 + 1/\alpha^2 + \ldots - i(\pi/2)^{1/2}\alpha\exp(-\alpha^2/2)$, what is valid if $|\alpha| \gg 1$ and $|Re(\alpha)| \gg |Im(\alpha)|$, and using Eqs. (2) and (3), the quasi-neutrality condition, which in the present case reads $n_{i1} = n_{e1} + Z_d n_{d1}$, yields the dispersion equation:

$$\Delta(\omega, \vec{k}) \equiv \frac{k^2 Z_d^2 n_{d0}}{m_d \omega^2} - \frac{n_{i0}}{\kappa T_i} - \frac{n_{e0}}{\kappa T_e}$$

$$-i\left(\frac{\pi}{2}\right)^{1/2}\left[\frac{\omega}{k v_{Td}}\frac{Z_d^2 n_{d0}}{\kappa T_d}\exp\left(-\frac{\omega^2}{2k^2 v_{Td}^2}\right) + \frac{(\omega - \omega_{*i})}{k_z v_{Ti}}\frac{n_{i0}}{\kappa T_i}\exp\left(-\frac{\omega^2}{2k_z^2 v_{Ti}^2}\right)\right.$$

$$\left. + \frac{(\omega - \omega_{*e})}{k_z v_{Te}}\frac{n_{e0}}{\kappa T_e}\exp\left(-\frac{\omega^2}{2k_z^2 v_{Te}^2}\right)\right] = 0. \tag{4}$$

The real part of Eq. (4) yields the DA wave frequency $\omega_r^2 = k^2 c_d^2$, $c_d^2 = \kappa T_e T_i n_{d0} Z_d^2/[m_d(n_{i0}T_e + n_{e0}T_i)]$. The growth rate (and damping) γ is given approximately by $\gamma \simeq -Im(\Delta)/[\partial Re(\Delta)/\partial\omega]_{\omega=\omega_r}$, and this yields

$$\gamma = -\left(\frac{\pi}{2}\right)^{1/2}\frac{m_d c_d^3 k}{2Z_d^2 n_{d0}}\left[\frac{\omega_r n_{d0} Z_d^2}{k v_{Td}\kappa T_d}\exp\left(-\frac{\omega_r^2}{2k^2 v_{Td}^2}\right)\right.$$

$$\left. + \frac{(\omega_r - \omega_{*i})}{k_z v_{Ti}}\frac{n_{i0}}{\kappa T_i}\exp\left(-\frac{\omega_r^2}{2k_z^2 v_{Ti}^2}\right) + \frac{(\omega_r - \omega_{*e})}{k_z v_{Te}}\frac{n_{e0}}{\kappa T_e}\exp\left(-\frac{\omega_r^2}{2k_z^2 v_{Te}^2}\right)\right]. \tag{5}$$

The diamagnetic drift separates charges so that in the present case ω_{*i} has the sign opposite to ω_{*e}. The shape of Eq. (5) reveals that, for a positive ω_{*e}, the only term capable of changing the sign of γ and yielding the growth is in principle the electron term, provided that $\omega_r < \omega_{*e}$. The dust term in Eq. (5) is typically always negligible in comparison to the others. To check if the instability is really possible, we make the ratio of the ion and electron terms γ_i/γ_e that are within the brackets [ ] in Eq. (5). As long as $\omega_r^2 \ll k_z^2 v_{Te}^2, k_z^2 v_{Ti}^2$, the exponential terms are practically equal to unity and play no role. Also, having $\omega_r \ll \omega_{*e}, \omega_{*i}$, it is seen that the instability may take place only provided that

$$|\gamma_i/\gamma_e| \simeq \left|\frac{n'_{i0}}{n'_{e0}}\right|\left(\frac{T_e m_i}{T_i m_e}\right)^{1/2} = \left|1 + \frac{Z_d n'_{d0}}{n'_{e0}}\right|\left(\frac{T_e m_i}{T_i m_e}\right)^{1/2} < 1. \tag{6}$$

Hence, here the instability implies unrealistically high values of the ion temperature and the excitation of DA waves by electron terms in such dusty plasmas containing negatively charged grains is unlikely. An exception could be a specific case in which the electron and dust density gradients have opposite signs, and, in addition, nearly balance each other so that the term $|1 + Z_d n'_{d0}/n'_{e0}| \simeq |1 - Z_d n_{d0} L_e/(n_{e0} L_d)|$ becomes very small, implying a nearly flat ion density profile, i.e., $L_e/L_i \ll 1$. Note that in standard electron-ion plasma (e.g., in the case of the drift wave) exactly the same γ_e term appears, resulting in the instability mentioned in the introductory lines, while in the same time it is usually $\omega_r^2 > k_z^2 v_{Ti}^2$, or $\omega_r^2 \gg k_z^2 v_{Ti}^2$. The total ion contribution is thus considerably reduced and the electron terms are capable of making the mode growing. The reason for the absence of the electron-driven instability in the present dusty plasma case is clearly only due to the fact that the frequencies are low and the exponential term in the ion part plays no role, i.e., $\omega_r^2 \ll k_z^2 v_{Ti}^2$.

Under typical laboratory conditions in the past, grains are negatively charged by electron attachment and this corresponds to the case discussed above. However, more recent experiments dealing with the photoelectric charging [4], and the presence of negative ions in a dusty plasma yield positively charged dust grains. In space and astrophysical plasmas, grains can have both positive and negative charges, sometimes even at the same time, or in the form of neighboring layers containing opposite charges on the grains. This peculiar behavior of a dusty plasma is due to several competing charging mechanisms, like the attachment of plasma particles due to inelastic collisions, secondary electron emission and photoemission. Dusty plasmas containing negative ions imply a reduced charging of the grains due to electron attachment and dust grains can become positively charged, this especially in the case when positive ions are much lighter than negative ones. In such cases the diamagnetic velocities of electrons and any other negatively charged magnetized species have the same sign, and therefore their destabilizing effects will add up.

In view of that, performing a similar procedure as above, it can easily be shown that such plasmas are additionally destabilized, with the effects of negatively charged magnetized species (ions or grains) being similar to those of electrons. In the equilibrium we have $Z_d n_{d0}(x) = n_{i0}(x) + n_{e0}(x)$. Here, the density gradients of electrons and negative ions are assumed to be of the same sign, and they are balanced by the variation of the density of positively charged grains. In the perturbed state the dispersion equation is obtained from $Z_d n_{d1} = n_{i1} + n_{e1}$. The real part of the dispersion equation will again yield the same frequency, while the growth rate will be similar as Eq. (5), except for the ion contribution that now contains the negative sign in the critical term $\omega - \omega_{*i}$. If the dust term in Eq. (5) is negligible, it is seen that as long as $T_i \leq T_e$ the dust-acoustic mode will always be growing whenever $\omega_r < \omega_{*i}$. In case of plasmas with hot negatively charged magnetized ions $T_i \geq T_e$, the instability becomes determined by the electron term. The instability in the present case follows from the fact that both electrons and singly charged negative ions together change the sign of the growth rate γ.

In Fig. 1, we present the growth rate for positively charged grains and negative ions for an arbitrary set of plasma parameters, and in terms of the angle of propagation $\theta = \arctan(k_z/k_y)$ between the wave-vector $\vec{k}$ and the y-axis. The electron and negative-ion number densities are taken equal to $n_{i0} = n_{e0} = n_0 = 10^{18}\ \mathrm{m}^{-3}$, and we take $Z_d = 10^3$ yielding $n_{d0} = 2 \cdot 10^{15}\ \mathrm{m}^{-3}$. We assume m_i to be equal to proton mass, for grains we take $m_d = 10^{-16}$ kg, and we assume a plasma with $L \equiv L_e = L_i = (d \log n_0/dx)^{-1} = 1$ m. For the wave-number we take $k = 10^3\ \mathrm{m}^{-1}$ so that $L/\lambda = 160$, $\lambda = 2\pi/k$, and $\rho_d/\lambda = 3$, where $\rho_d = v_{Td}/\Omega_d$ and the magnetic field is taken as $B_0 = 0.2$ T.

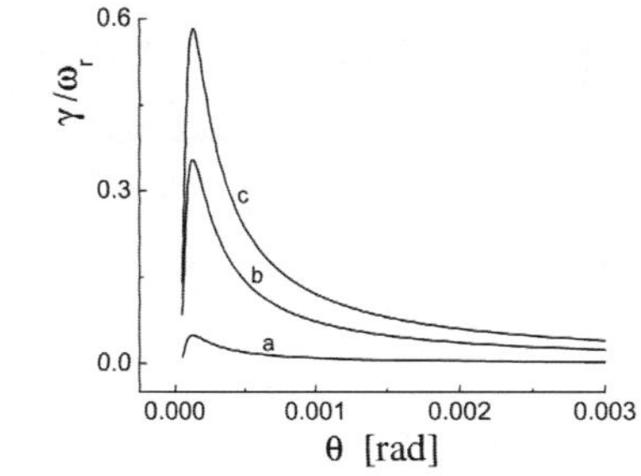

Figure 1. The normalized growth rate γ/ω_r in terms of the angle between $\vec{k}$ and the y-axis, for $T_e = T_i = 0.1, 0.5, 1$ eV (lines a, b, c respectively).

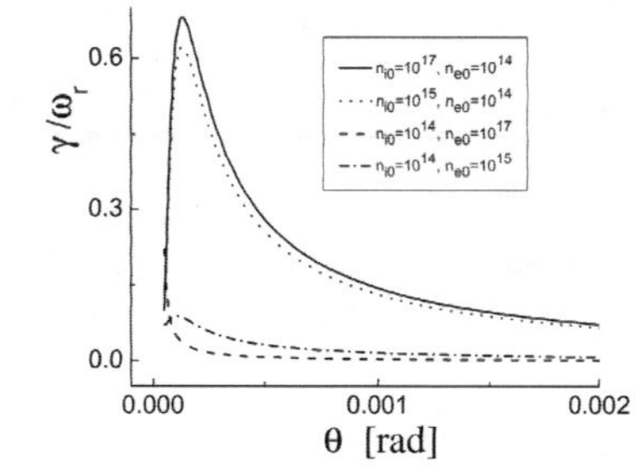

Figure 2. The normalized growth rate γ/ω_r in terms of the angle between $\vec{k}$ and the y-axis, for several values n_{i0} and n_{e0}, and for $n_{d0} = (n_{i0} + n_{e0})/Z_d$.

The lines a, b, c in Fig. 1 correspond to three different values of the electron and ion temperature $T_e = T_i = T = 0.1, 0.5, 1$ eV, respectively, where in the same time $T_d = T/50$.

The growth rate is strongly dependent on both the angle of propagation and the temperature of the light plasma species that are responsible for the growth of the mode. Note that ω_r here takes values between 400 and 1260 Hz while $\Omega_d = 0.3$ Hz, and ω_{*d} changes between 0.01 Hz and 0.1 Hz, so that indeed we have properly assumed limits for the local approximation, un-magnetized grains, and negligible dust diamagnetic drift effects.

We stress that taking any other value for the dust temperature, even going to the limit $T_d \sim T$, yields exactly the same shape for the growth rate. This is because for these parameters the dust contribution in Eq. (5) is completely negligible. Similarly, the growth rate does not change at all by varying n_0, and this holds as long as n_{i0}, n_{e0} are kept equal. This may be expected from Eq. (5) because the density n_0 cancels out by the dust term $Z_d n_{d0}$ that is in the denominator in front of the brackets. However, for any other values $n_{i0} \neq n_{e0}$ the growth rate changes. This is presented in Fig. 2 for several number densities n_{i0}, n_{e0} (per cubic meter) and for the case $T = 0.5$ eV, $T_d = T/50$. Here, the corresponding dust density is given by $Z_d n_{d0} = n_{i0} + n_{e0}$.

The described mechanism of excitation of dust acoustic waves in inhomogeneous plasmas with magnetized light species is very likely to occur in any inhomogeneous environment. In such situations, the DA wave will practically always be growing and the growth-rate can easily be far above the wave frequency. The mode is most easily excited in a very narrow angle around a direction that is almost perpendicular to the magnetic field vector and the density gradient. In astrophysical clouds the proposed instability may become an initial triggering mechanism for the fragmentation, especially if self-gravity effects are also taken into account. The kinetic analysis presented here is in fact the most appropriate for such astrophysical plasmas with a small amount of collisions. Artificially produced grains under laboratory conditions in the recent past are also positively charged [4, 5, 6], and the proposed instability discussed here will undoubtedly work there too. However, the plasmas in these recent experiments include positively charged grains, electrons and negative ions, but, in addition to this, also a variable amount of positive ions [5], or negative grains [4]. The simplified model with three species only, discussed in the text, can easily be generalized to include the effects of these additional species. In both cases presented here it is in fact the ion density gradient that crucially determines the mode behavior.

Acknowledgments
The results presented here are obtained in the framework of the projects G.0304.07 (FWO-Vlaanderen), C 90347 (Prodex), GOA/2009-009 (K.U.Leuven). Financial support by the European Commission through the SOLAIRE Network (MTRN-CT-2006-035484) is gratefully acknowledged.

[1] Vranjes J and Poedts S 2009 *Phys. Plasmas* **16** 022101
[2] Krall N. A. and Book D 1969 *Phys. Rev. Lett.* **23** 574
[3] Vranjes J, Pandey B P and Poedts S 2001 *Phys. Rev. E* **64** 066404
[4] Sickafoose A. A., Coldwell J. E., Horányi M and Robertson S 2000 *Phys. Rev. Lett.* **84** 6034
[5] Merlino R. L. and Sun-Hyun Kim 2006 *Appl. Phys. Lett.* **89** 091501
[6] Rosenberg M and Merlino R. L. 2007 *Planet. Space Sci.* **55** 1464

Dissipation mechanism in 3D collisionless magnetic reconnection

Keizo Fujimoto

Computational Astrophysics Laboratory, RIKEN, 2-1 Hirosawa, Wako, Saitama 351-0198, Japan

E-mail: fkeizo@riken.jp

Abstract. Dissipation processes responsible for fast magnetic reconnection in collisionless plasma are investigated using 3D electromagnetic particle-in-cell simulations. The present study compares two simulation runs; one with small system size in the current density direction, and the other with larger system size. In the case with small system size, the reconnection processes are almost the same as those in 2D reconnection, while in the other case the drift kink mode evolves along the current density and deforms the current sheet structure drastically. Although fast reconnection is achieved in both the cases, it is found that the dissipation mechanism is very different between them. In the case without kink mode, the electrons transit the electron diffusion region without thermalization, so that the magnetic dissipation is supported by the inertia resistivity alone. On the other hand, in the kinked current sheet, the electrons are not only accelerated in bulk, but they are also partly scattered and thermalized by the kink mode, which results in the anomalous resistivity in addition to the inertia resistivity. It is discussed that in 3D reconnection the thickness of the electron current sheet becomes larger than the local electron inertia length.

1. Introduction

Magnetic reconnection is a common process in space and laboratory plasmas, releasing the energy stored in a compressed magnetic field into plasma kinetic energy. The process can change the global field line topology and cause strong acceleration of plasma particles, so that it is considered to play a key role in explosive phenomena such as geo-magnetospheric substorms and solar flares. However, these phenomena commonly occur in collisionless plasmas where the Coulomb collision frequency is so small that the magnetic dissipation is hardly generated in the classical sense. In fact, in the resistive magnetohydrodynamic (MHD) framework based on the Spitzer conductivity, the rate of reconnection is unrealistically small to explain the explosive energy release in space [1, 2]. In order to enhance the reconnection rate, the usual fluid simulations that cannot treat explicitly the dissipation processes employ ad-hoc anomalous resistivity or/and numerical resistivity [3, 4, 5]. However, it has been suggested that the rate of reconnection depends on the resistivity model [3, 4] and, in the global simulations, the global response of substorms and flares is sensitive to the parameterization of the resistivity [5, 6]. These results of the fluid simulations indicate that the dissipation processes, which take place in very small region called the diffusion region formed around the magnetic X-line, can have an impact on the large-scale dynamics of reconnection and the relevant phenomena. In this context, the present study focuses on the dissipation mechanism in collisionless reconnection, using the

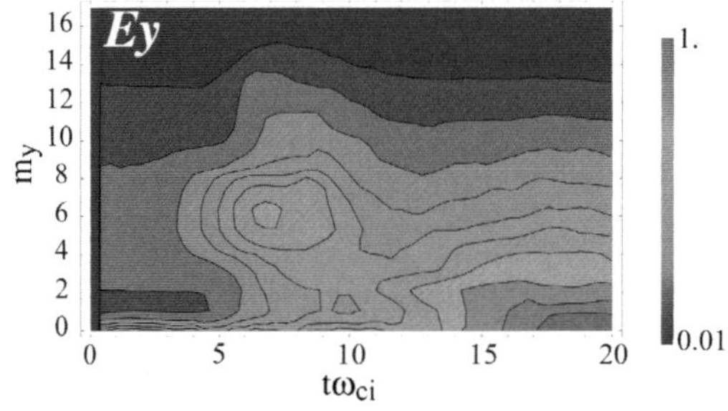

Figure 1. Time history of the wave mode amplitude for Run2 propagating in the y direction in the yz plane through the X-line.

particle-in-cell (PIC) simulations. It is found that the process in 3D reconnection is significantly different from that in 2D reconnection. In 2D case, the dissipation is totally supported by the electron momentum transport away from the X-line due to the Speiser-type motion, while in 3D case a kink mode excited along the current density causes the electron thermalization which gives rise to the electric resistivity.

2. Simulation model

The simulations are performed using an electromagnetic PIC code with the adaptive mesh refinement (AMR) and particle splitting-coalescence method [7]. The AMR technique subdivides and removes computational cells in accordance with refinement criteria so as to efficiently enhance the local spatial resolution dynamically in time. The particle splitting-coalescence method is implemented in order to control the number of particles per cell in the refinement regions. The present study examines two 3D simulations with different system size in the current density direction. The system boundaries are periodic in the x and y directions, and conducting wall in the z direction. The simulations are carried out using a Harris-type current sheet with the magnetic field $B_x(z) = -B_0 \tanh(z/L)$ and the number density $n(z) = n_{ps}\,\mathrm{sech}^2(z/L) + n_b \tanh^2(z/L)$, where L is the half width of the current sheet. We choose $L = 0.5\lambda_i$ and $n_b = 0.044 n_{ps}$ with λ_i the ion inertia length. The system sizes are $l_x \times l_y \times l_z = 31\lambda_i \times 1.9\lambda_i \times 31\lambda_i$ (Run1) and $31\lambda_i \times 7.7\lambda_i \times 31\lambda_i$ (Run2). The mass ratio, velocity of light, and temperature ratio are $m_i/m_e = 25$, $c/V_A = 10$, and $T_i/T_e = 5.0$, respectively. The grid spacing is set as $\Delta_{l_B} = 0.24\lambda_i$ for the base level cells and $\Delta_{l_D} = 0.03\lambda_i$ for the dynamic range level cells.

3. Simulation results

Magnetic reconnection is initiated with a small perturbation to the magnetic field, which produces the X-line in the current density direction at the center of the xz plane in the simulation domain. In both Run1 and Run2, the first growing mode has relatively short wavelength with the mode number $m_y \approx 6$ for Run2 and propagates in the direction almost perpendicular to the initial magnetic field (see figure 1). This mode is the lower hybrid drift instability (LHDI). The LHDI is driven by the diamagnetic current that is supported by the density gradient, and grows quickly around the edge of the current sheet (figure 2a). The LHDI activity modifies the current sheet structure so as to compress the current sheet, which facilitates the onset of fast reconnection. However, once the fast reconnection has been triggered, the LHDI activity is drastically suppressed around the diffusion region, because the density gradient almost

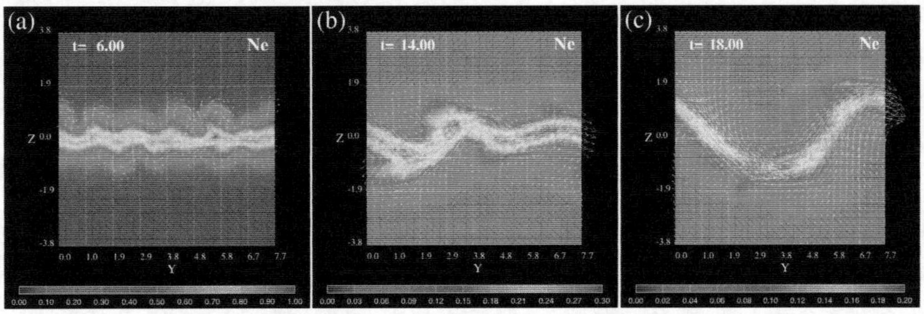

Figure 2. Electron number density for Run2 in the yz plane through the X-line at (a) $t\omega_{ci} = 6.0$, (b) $t\omega_{ci} = 14.0$, and (c) $t\omega_{ci} = 18.0$. Arrows show the electron flow vector in the plane.

disappears from the vicinity of the X-line. In Run1, because of the small system size in the y direction, the current sheet behavior is quite similar to that in 2D simulation after the onset of the fast reconnection, On the other hand, in Run2, the drift kink mode with $m_y \approx 2$ evolves along the current density (see figures 1 and 2b). This mode eventually coalesces into a large wavelength mode with $m_y = 1$ (figure 2c) as the tearing mode grows and the ion inertia scale expands in the current sheet. As shown in figure 1, the kink mode survives under reconnection (with the dc electric field), so that the magnetic dissipation proceeds in the kinked current sheet.

The reconnection electric field averaged over the X-line reaches around $E_R \approx 0.1V_{A0}B_0$ in the both the runs, where $V_{A0} = B_0/\sqrt{\mu_0 n_b m_i}$ is the Alfvén velocity in the lobe region. Nevertheless, it is found that the dissipation mechanism in the electron diffusion region is significantly different between the two cases. Figure 3 shows the evolution of the energy density in the v_y-v_z space of the electrons which were sampled in the upstream region of the electron diffusion region. In Run1 (figure 3a), once the electrons enter the electron diffusion region at $t\omega_{ci} \approx 18.0$, they are quickly accelerated in bulk due to the reconnection electric field. However, the thermal energy (dashed curve) is hardly enhanced during the acceleration. This indicates that in 2D reconnection the electrons are coherently accelerated in the electron diffusion region and the momentum gained by the reconnection electric field is transported away from the region due to the Speiser-type motion. Thus, the magnetic dissipation in Run1 is provided by the inertia resistivity alone [8, 9, 10]. On the other hand, in Run2, most electrons enter the diffusion region at $t\omega_{ci} \approx 16.6$,

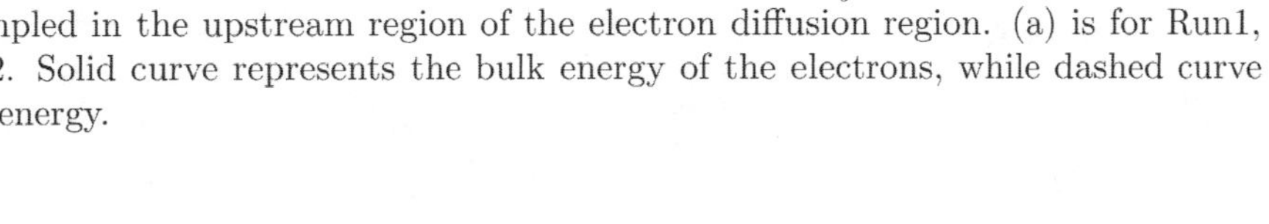

Figure 3. Time evolution of the energy density in the v_y-v_z space for the electrons which were sampled in the upstream region of the electron diffusion region. (a) is for Run1, and (b) for Run2. Solid curve represents the bulk energy of the electrons, while dashed curve denotes thermal energy.

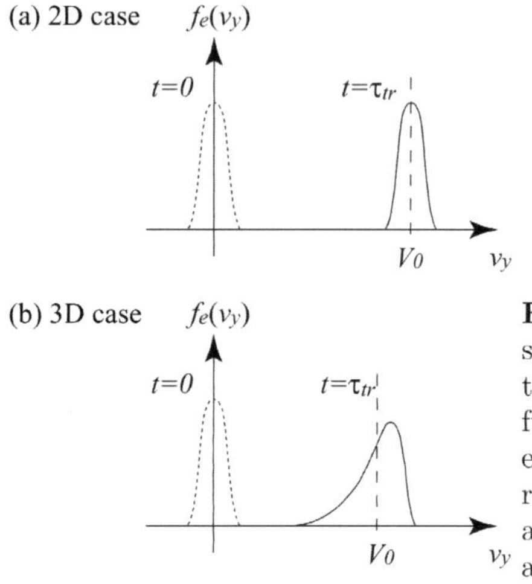

Figure 4. Schematic diagram describing the change in the electron distribution function in the electron diffusion region from $t = 0$ (when the electrons enter the electron diffusion region) till $t = \tau_{tr}$ (where τ_{tr} is the averaged transit time) for (a) 2D case and (b) 3D case.

and about 70% of them are ejected before $t\omega_{ci} \approx 17.4$ without significant thermalization, giving rise to the inertia resistivity (see figure 3b). However, a fraction of the electrons stay longer and are accelerated more around the X-line. Figure 3b shows that these electrons are not only accelerated in bulk, but they are also strongly thermalized after $t\omega_{ci} \approx 17.4$. The thermalization occurs efficiently when the high-energy particles turn the curve in the kinked current sheet. Therefore, we can conclude that the high-energy electrons are thermalized as a result of particle scattering due to the kink mode. Since the electron thermalization results in the reduction of the averaged momentum along the X-line, it causes the electric resistivity which may be called the anomalous resistivity.

4. Discussion and conclusions

The present study has investigated the dissipation mechanism in collisionless magnetic reconnection using the 3D electromagnetic PIC simulations. We have examined two simulation runs with different system size in the current density direction. In the run with larger system size, the drift kink mode evolves along the current density and reconnection proceeds in the kinked current sheet. On the other hand, in the run with smaller system size, the kink mode is not allowed to grow, so that the current sheet behavior is quite similar to the 2D case. In both the cases, a fast reconnection with $E_R \approx 0.1V_{A0}B_0$ is achieved. However, it is found that the dissipation mechanism is quite different between the two cases. In the case without the kink mode, that is, the 2D reconnection case, the electrons transit the electron diffusion region without thermalization, so that the magnetic dissipation is supported by the inertia resistivity alone. On the other hand, in the case with the kink mode, the electrons are not only accelerated in bulk, but they are also partly scattered and thermalized by the kink mode. Since the electron thermalization results in the reduction of the averaged momentum, that is, the current density, it causes the electric resistivity in addition to the inertia resistivity.

The electron behavior in the electron diffusion region is described schematically in figure 4a for the 2D case and in figure 4b for the 3D case. In the 2D case, since the electron momentum gained by the reconnection field becomes the bulk momentum totally, the bulk velocity along the X-line is provided by $V_0 = -(eE_y/m_e)\tau_{tr}$, where τ_{tr} is the averaged transit time through the electron diffusion region and is estimated as $\tau_{tr} \approx -\delta_e/V_{in}$ with δ_e the half width of the electron current sheet and V_{in} the electron inflow velocity. If we use $V_0 \approx -(1/\mu_0 en)(B_{in}/\delta_e)$ from the

Ampére's law with B_{in} the magnetic field at the inflow region and $E_y \approx -V_{in}B_{in}$, we can obtain $\delta_e \approx \lambda_e$, where λ_e is the electron inertia length. However, in the 3D case, the electrons which are accelerated by the reconnection field are partly scattered in the kinked current sheet. In this case, the bulk velocity (in absolute value) must be less than that in the 2D case under the same reconnection rate, that is, $V_0 < (e|E_y|/m_e)\tau_{tr}$, which leads to $\delta_e > \lambda_e$. Thus, in 3D reconnection where the electron scattering occurs in the electron diffusion region, the current sheet thickness should be larger than the local electron inertia length.

Acknowledgments
This work has been partly supported by a collaborative research project at STE laboratory, Nagoya University, and by the JSPS Grant-in-Aid for Research Activity Start-up (21840061).

References
[1] Sweet P A 1958 *Electromagnetic Phenomena in Cosmical Physics* ed B Lehnert (London: Cambridge University Press) pp 123–134
[2] Parker E N 1963 *Astrophys. J., Suppl. Ser.* **8** 177
[3] Biskamp D 1986 *Phys. Fluids* **29** 1520
[4] Ugai M 1995 *Phys. Plasmas* **2** 388
[5] Raeder J, McPherron R L, Frank L A, Kokubun S, Lu G, Mukai T, Paterson W R, Sigwarth J B, Singer J H and Slavin J A 2001 *J. Geophys. Res.* **106** 381
[6] Kuznetsova M M, Hesse M, Rastätter L, Taktakishvili A, Toth G, De Zeeuw D L, Ridley A and Gombosi T I 2007 *J. Geophys. Res.* **112** A10210
[7] Fujimoto K and Sydora R D 2008 *Comput. Phys. Commun.* **178** 915
[8] Speiser T *Planet. Space Sci.* **18** 613
[9] Tanaka M 1995 *Phys. Plasmas* **2** 2920
[10] Fujimoto K and Sydora R D 2009 *Phys. Plasmas* **16** 112309

On the influence of metastable states and the behavior of the EEDF in the characterization of the negative glow of a N_2-Ar discharge by OES

Isola L.[1], **López M.**[1], **Gómez B. J.**[1] **and Guerra V.**[2]
[1]Instituto de Física Rosario (CONICET-UNR) 27 Febrero 210 Bis. (2000) Rosario, Argentina.
[2]Instituto de Plasmas e Fusão Nuclear, Instituto Superior Técnico, 1049-001 Lisboa, Portugal
E-mail: isola@ifir-conicet.gov.ar

Abstract.
Optical emission spectroscopy (OES) is an essential diagnostic technique in many plasma systems, such as those used for surface treatments or fabrication of thin films. Despite the simplicity of application of OES, its interpretation is not straightforward. In particular, it requires the use of models, which due to the complexity and variety of discharge conditions, have not yet been fully understood [1]–[3].

In addition, Langmuir probes have been widely used to characterize plasmas. They allow the measurement of several parameters of interest, such as the electron density and temperature, as well as the determination of the electron energy distribution function (EEDF) by numerical derivation of the characteristic $V - I$ [4] or by probe-current modulation [5].

In this work, some second positive system bands in the negative glow of an Ar-N_2 plasma at a pressure of 2.5 Torr were investigated both by OES and Langmuir probes, for different mixture concentrations. The main purpose of this study was to verify how metastable states and the behavior of the EEDF may influence the interpretation of OES data.

1. Introduction

Optical emission spectroscopy (OES) methods based on spectral band head ratios have been used to determine electron properties of N_2 plasmas, such as electron temperature and density [1, 2, 6]. However, these methods cannot be directly applied to $Ar - N_2$ gas mixtures because of the high excitation transfer of N_2 from $Ar(^3P_{2,0})$ metastable atoms [6]. In fact, this process produces a characteristic vibrational and rotational distribution in the *second positive system* (SPS) [7, 8], with an overpopulation of high rotational levels ($K > 30$), producing an overlap between the vibrational bands belonging to the same vibrational family. Hence it is not possible to identify the contribution of each vibrational band in the measured spectrum.

Therefore, the simulation of the bands, taking into account the possible excitation processes, becomes essential.

On the other hand, the methods based on the band head ratio use a Maxwellian EEDF, but this assumption is often inaccurate.

In this work, we studied the discharge kinetics, through the analysis and fitting of the band structure at 380.49 nm ($N_2(C^3\Pi_u^+, v' = 0 \rightarrow B^3\Pi_g^+, v'' = 2)$) and 375.54 nm ($N_2(C^3\Pi_u^+, v' = 1 \rightarrow B^3\Pi_g^+, v'' = 3)$), belonging to the SPS, and the EEDF behavior with the variation of the Ar concentration in the gas mixture.

2. Experimental set-up

The plasma reactor used consists of a stainless steel chamber of 254 nm in diameter and 360 nm in height, with two side glass windows (Figure 1). The work pressure and the gas flow rate were kept constant at 2.5 Torr and 100 ml/min respectively. The discharge was generated between a central disc, which works as a cathode, and the walls of the chamber. A variable pulsed source with a frequency of 100 Hz and duty cycle of 0.7 was used, keeping the voltage constant at −500 V.

The optical emission spectra from the discharge were recorded with a Jarrell-Ash monochromator with Czerney-Turner mounting. The slit entry was set at 25 μm, whereas the grating used was 3600 grooves/mm. A lens focused the light to the monochromator, and a linear arrangement of 1024 photodiodes was used as detector.

Concerning the probe measurements, a cylindrical Langmuir probe consisting of a tungsten wire of radius $R = 15\,\mu m$ and $L = 60$ mm in length was employed. It was fed by a variable (from 0 to −40 V) source. A small AC sine voltage with frequency 1 Khz was superimposed on the DC probe bias, then the second derivative of the probe characteristic was obtained directly from measuring the current component with frequency 2 Khz [9]. This measurement was performed with a Lock-in amplifier. It is worth mentioning that the measurements were performed in a DC plasma due to the impossibility of measuring with the Lock-in in a pulsed plasma.

The electron energy distribution function $g(\varepsilon)$ can then be obtained from the second derivative of the probe characteristic curve,

$$g_e(\varepsilon) = \frac{2m}{e^2 A}\left(\frac{2\varepsilon V}{m}\right)^{\frac{1}{2}} \frac{d^2 I_e}{dV^2} \tag{1}$$

where ε denotes the electron energy, m the electron mass, V is the probe voltage with respect to the plasma potential, A is the area of the probe, and the maximum of the first derivative has been taken as the plasma potential.

Both the optical emission spectroscopy and the Langmuir probe measurements were made at the same position in the discharge chamber, marked in Figure 1 by h, being $h = 10.25$ mm.

3. Measurements

3.1. Spectroscopy

In this work, bands of the SPS were studied by spectroscopy for different mixture concentrations. Typical spectra are shown in Figure 2.

3.2. Fitted Bands

The bands were fitted using the least squares method. The band intensities $v' = 0 \rightarrow v'' = 2$ and $v' = 1 \rightarrow v'' = 3$ of the SPS $N_2(C^3\Pi_u) \rightarrow N_2(B^3\Pi_g)$ (with band heads in 380.49 nm and 375.54 nm) are given by :

$$I_{CB}(\lambda, T_{rot}, N_o, \delta\lambda, \Delta\lambda, X) = N_o(\lambda, v', T_{rot}) A_{CB}(v', v'') \sum_{i=P,R,Q} \sum_{\Omega=0,1,2} \sum_{J'} \exp\left(\frac{-BhcJ'(J'+1)}{k_B T_{rot}}\right) \exp\left(\frac{-4\ln(2)\left[\lambda + \Delta\lambda - \lambda^i_\Omega(J')\right]^2}{(\delta\lambda)}\right) S^i_\Omega(J')X(J') \tag{2}$$

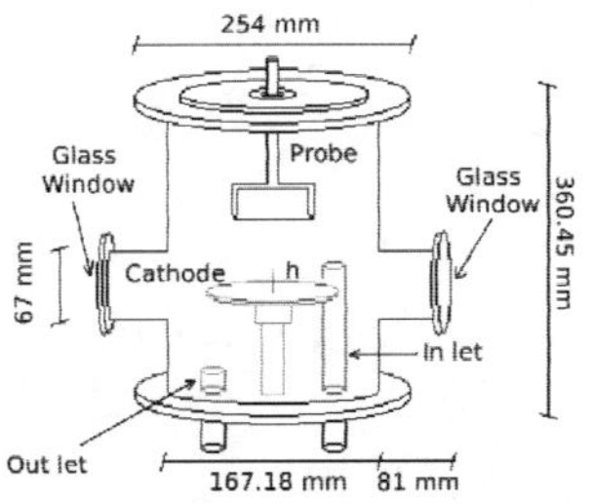

Figure 1. *Experimental set-up*

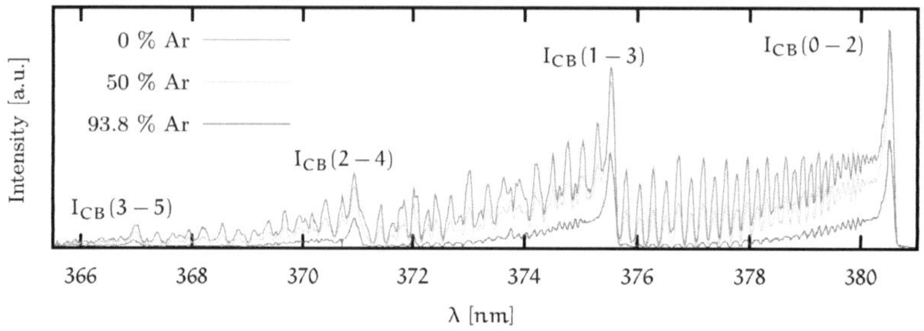

Figure 2. *Vibrational bands with* $\Delta v = v' - v'' = -2$ *belonging to the SPS, for different* Ar *concentrations.*

where N_o includes the spectral response, the density in the upper level and partition function, $\delta\lambda$ the full width at half maximum of the rotational line, $\Delta\lambda$ a small shift in the wavelength, $A_{CB}(v', v'')$ is the Einstein coefficient [10], B the rotational constant, $S^i_\Omega(J'_i)$ is the Hönl-London factor [11] and $X(J')$ is a fitting parameter that allows the alternation between K even and odd that occurs because of the collision with $Ar(^3P_2)$, when we increase the Ar concentration (see Figure 3). Finally, in the bands area of 375.54 nm and 380.49 nm there are some lines of Ar

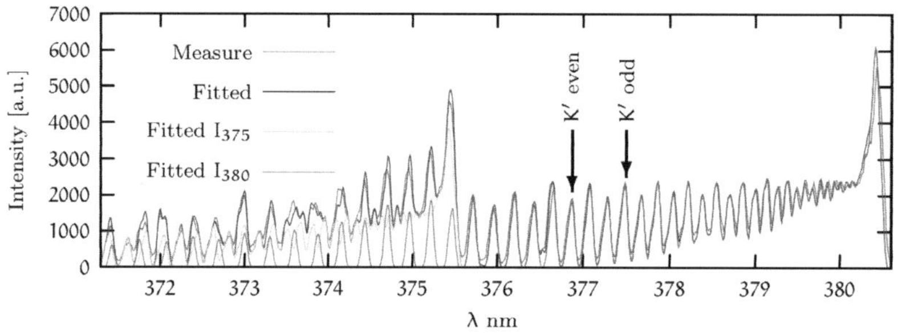

Figure 3. *Overlapping bands* 380.49 nm *and* 375.54 nm *in a discharge with high* Ar *concentration (93.75% Ar). Measured and fitted bands.*

and Fe, which were taken into account to fit these bands, using the line intensities as fitting parameters.

3.3. Kinetic model

To analyze these spectra we used a simple model for the negative glow kinetics, described by the following assumptions:

(i) The main excitation reactions of state $N_2(C, v')$ are

$$e + N_2(X, 0) \rightarrow e + N_2(C, v') \tag{3}$$

$$N_2(A) + N_2(A) \rightarrow N_2(X) + N_2(C, v') \tag{4}$$

$$Ar(^3P_{2,0}) + N_2(X, 0) \rightarrow Ar(^1S_0) + N_2(C, v') \tag{5}$$

(ii) Levels $N_2(C, v')$ and $N_2^+(B, v')$ are primary lost by radiative decay.

3.4. Model Discussion

3.4.1. Excitation by e When $N_2(C^3\Pi_u)$ is populated by electron impact from the ground state, the vibration distribution can be approximated by the Franck-Condon coefficient [10]. The rotational distribution is taken to be unchanged in the excitation by electron impact, then the $N_2(C^3\Pi_u)$ rotational distribution will correspond to the $N_2(X^1\Sigma_g^+)$ rotational distribution. Note that the rotational population is distributed between the P, Q and R branches, belonging to $C^3\Pi_u$ level, following the Hönl-London factor [11].

3.4.2. Excitation by $N_2(A)$ The pooling reaction (4) produces an overpopulation of the level $N_2(C^3\Pi_u, \nu = 1)$ (see Table 1 [12]). This overpopulation in $\nu = 1$ is not observed in our N_2 discharge, giving an *a posteriori* confirmation that the pooling reaction does not contribute significantly to the production of the $N_2(C^3\Pi_u)$ state.

3.4.3. Metaestable Ar The $Ar(^3P_0)$ is less efficient than $Ar(^3P_2)$ in excitation transfer to $N_2(C^3\Pi_u)$. In addition, its concentration is smaller, then all the excitation transfer is in fact produced by $Ar(^3P_2)$.

The collision (5) is an exothermic process (Table 2), then it produces excited molecules $N_2(C)$ with large rotational energies, particularly for $\nu' = 0$. This process also produces an intensity alternation between even and odd K' rotational levels, which is not expected in a Π state [13]. The population of different $N_2(C)$ vibrational states cannot be explained by the Franck-Condon factor and the rate coefficients shown in Table 1 (from $\nu = 0$) [14] should be used.

Table 1. Franck-Condon coefficients and rate coefficients $[10^{-11}cm^{-3}s^{-1}]$

	$q_{0\nu'}$	$k^{\nu'}_{N_2(A,0)-N_2(A,0)}$	$k^{\nu'}_{Ar-N_2}$
$N_2(C,0)$	0.55	2.6 ± 0.1	2.9 ± 0.3
$N_2(C,1)$	0.31	4.1 ± 0.2	0.60 ± 0.07
$\frac{N_2(C,1)}{N_2(C,0)}$	0.56	1.58	0.21

Table 2. Exothermic $\Delta E(^3P_j, \nu')$ (meV) for the excitation transfer process

Initial State	$\nu' = 0$	$\nu' = 1$	$\nu' = 2$	$\nu' = 3$
$^3P_0$	701	444	205	-28
$^3P_2$	526	269	30	-203

The addition of vibrational energy to $N_2(X)$ gives increased vibrational excitation in $N_2(C)$, up to at least $\nu' = 4$ [8].

4. Results

4.1. Spectroscopy

Figure 4 shows how the intensity of the bands at 380.49 nm and 375.54 nm increase with the Ar concentration. Notice that the raise in the intensity of the band at 375.54 nm is slower. This can be seen very clearly in Figure 5. It should be note that the band intensities were calculated from the area under the fitted bands.

To perform a preliminary analysis of the results, it can be assumed that most of the $N_2(X)$ molecules have a vibrational state $\nu = 0$. Thus, for a N_2 discharge the ratio between the $N_2(C, \nu' = 1)$ and $N_2(C, \nu' = 0)$ densities should be close to the Franck-Condon coefficient ratio, while for a high Ar concentration discharge, this ratio should be close to the relation between the rate coefficients $K^{\nu'}_{N_2-Ar}$.

Figure 5 shows how this ratio changes from 0.65 to 0.39, which agrees qualitatively with the ratio between Franck Condon factors (0.56) and the ratio between the rate coefficient (0.21). Obviously, this difference is obtained because there is a vibrational distribution in the ground state $N_2(X)$.

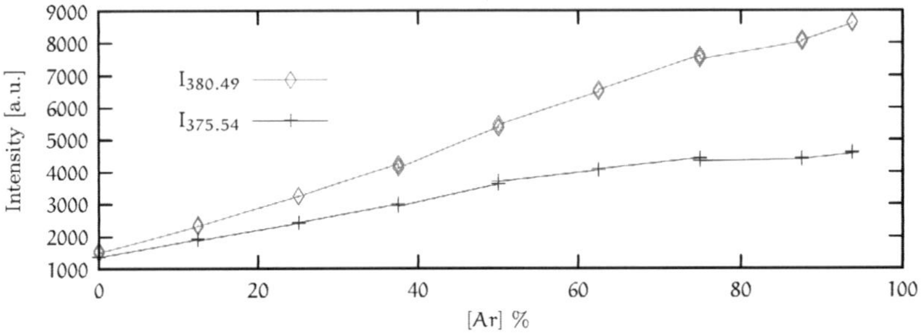

Figure 4. *Emission intensity of bands* 380.49 nm *and* 375.54 nm *as a function of the* Ar *concentration*

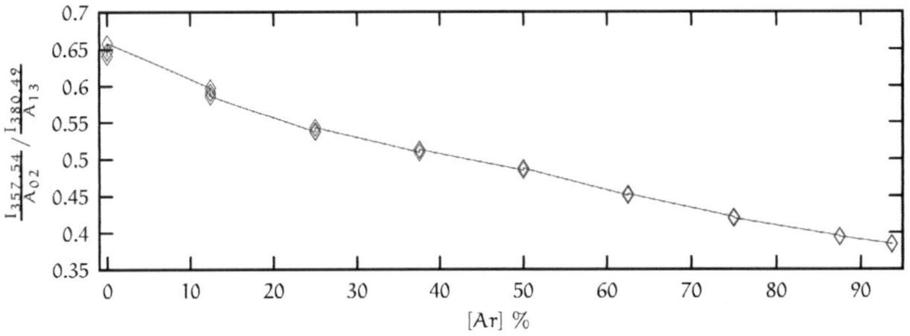

Figure 5. *Ratio between* $\frac{I_{375.54}}{A_{CB}(1,3)}$ *and* $\frac{I_{380.49}}{A_{CB}(0,2)}$ *as a function of the* Ar *concentration*

Another important result is observed with rotational temperatures of the SPS bands (Figure 6). Both bands have the same temperature (approximately 740 K) at 100 % of N_2, which agrees with the fact that the main mechanism of excitation is the collision with electrons from the ground state (reaction (3)), as proposed by the model. Then, when the Ar concentration is increased, the rotational temperatures reach 1960 K for 375.54 nm and 2510 K for 380.49 nm, in 93.75 % of Ar where the main mechanism is the reaction (3). This higher temperature of 380.49 nm can be understood because the energy excess to excite the level $N_2(C, v' = 0)$ is greater than the energy excess to excite the level $N_2(C, v' = 1)$ (See Table 2).

4.2. Lagmuir Probe

Figure 7 shows the EEDFs as a function of the Ar concentration. As can be seen, the electron density and electron temperature increase with the Ar content. Moreover, the EEDF deviates from a Maxwellian distribution in all cases. Work is in progress to include this effects in a model allowing the determination of the electron energy and temperature in $N_2 - Ar$ mixtures.

5. Conclusions

The rotational and vibrational structures of the SPS of N_2 are strongly influenced by the collisions between N_2 molecules and metastable Ar atoms. Therefore, any scheme designed to characterize the plasma by OES must take into account this excitation mechanism.

The relation between the populations of levels $N_2(C, v' = 1)$ and $N_2(C, v' = 0)$ gives information on the excitation mechanisms of $N_2(C)$. However, to perform quantitative analysis

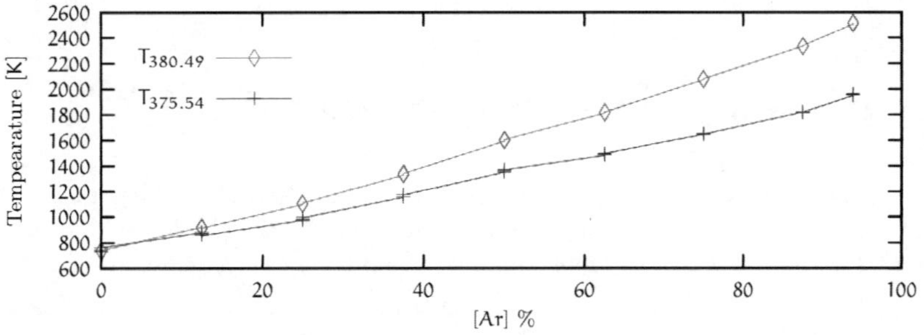

Figure 6. *Rotational Temperature as a function of the* Ar *concentration*

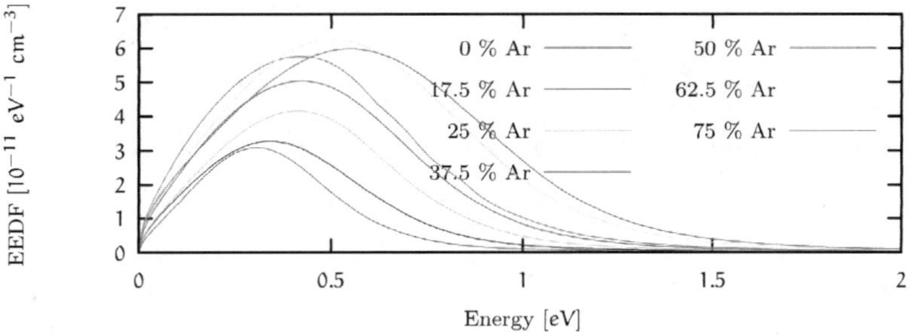

Figure 7. *EEDF as a function of the* Ar *concentration*

it is necessary to know the vibrational distribution function of $N_2(X)$, which is not easy to measure.

The electron density and temperature increases with the Ar content of the mixtures. Furthermore, the EEDFs differ quite significantly from Maxwellian distributions. This reveals the need of using EEDFs with an accurate information on the electron kinetics in the discharge in order to realistically characterize the plasma by OES.

Acknowledgments
The authors would like to thank Ing. Aldo Marenzana for this valuable collaboration in the development of the instrumentation. We award a grateful acknowledgment to Conicet (Argentina) and to the "Agencia Nacional de Promoción Científica y Tecnológica" (National Agency for Scientific and Technologic Promotion), Argentina, PICT 12-01981/2008-0374 and Josefina Prats Fundation.

References
[1] Isola L M, Gómez B J and Guerra V 2010 Journal of Physics D: Applied Physics **43** 202
[2] Zhu X and Pu Y 2008 Plasma Sources Science and Technology **17** 1–6
[3] Debal F, Bretagne J, Jumet M, Wautelet M, Dauchot J P and Hecq M 1998 Plasma Sources Science and Technology **7** 219–229
[4] Magnus F and Gudmundsson J T 2008 Review of Scientific Instuments **79**
[5] Dietrich S, Christ-Koch S, Fantz U and Crowley B 2007 28 ICPIG, July 15-20, 2007, Prague, Czech Republic
[6] Britun N, Gaillard M, Ricard A, Kim Y M, Kim K S and Han J G 2007 J. Phys. D: Appl. Phys. **40** 1022–1029
[7] Nguyen T D and Sadeghi N 1983 Chemical Physics **79** 41–55
[8] Setser D W and Stedman D H 1970 J. Chem. Phys. **53** 1004–1020
[9] Banner G R, Friar E M and Medicus G 1963 The review of scientific instruments **34** 231–237
[10] Lofthus A and Krupenie P H 1977 J. Chem. Phys. Ref. Data **6** 113
[11] Hartmann G and Johnson P C 1978 J. Phys. B: Atom Molec. Phys. **9** 1597–1612
[12] Piper L G 1988 J. Chem. Phys. **216** 231–239
[13] Derouard J, Nguyen T D and Sadeghi N 1980 J. Chem. Phys. **12** 6698–6705
[14] Sadeghi N, Cheaib M and Setsers D W 1989 J. Chem. Phys. **90** 219–231

Experimental study of a glow discharge of CH_2-Cl_2

A. Gomez[1], L. Salazar-Flores[2], H. Martinez[3], P.G. Reyes[1] and F. Castillo[4].
[1]Facultad de Ciencias, Universidad Autónoma del Estado de México, Instituto Literario 100, Centro, 50000, Toluca, Estado de México, México.
[2]Posgrado en Ciencias Físicas, Universidad Nacional Autónoma de México, Avenida Universidad 1000, 05000, México D.F., México.
[3]Instituto de Ciencias Físicas, Universidad Nacional Autónoma de México, Apartado postal 48-3, 62251, Cuernavaca, Morelos, México.
[4]Instituto de Ciencias Nucleares, Universidad Nacional Autónoma de México, Apartado Postal 70-543, 04510, México DF., México.

E-mail: hm@fis.unam.mx

Abstract. Low-pressure DC plasma discharges sustained in a glow discharge of CH_2-Cl_2 are studied. This plasma was generated at a pressure of 1.0 Torr, power of 20 W and a flow of 12 l/min. The diagnostic has been made by optical emission spectroscopy (OES) using a spectrometer. The principal species observed were: were at 339.61, 358.60 and 377.96 nm for $C_2(c'^1\Pi_g$-$b^1\Pi_u)$; at 392.50 nm for $C_3('\Pi_u$-$'\Sigma^+_g)$; at 431.42 nm for $CH(A^2\Delta$-$X^2\Pi)$; at 778.28 nm for Cl; at 657.80 nm for C^+; at 471.90 and 487.30 nm for H_2; at 380.61 nm for $CH^+(A'\Pi$-$X'\Sigma)$ and at 317.73 nm for $HCL^+(A^2\Sigma$-$X^2\Pi)$. In the study, special attention was given to the behavior of material deposition in the electrodes. The time dependence of material deposition in the electrodes was also investigated. When the discharge time of CH_2CL_2 plasma was increased, the mass deposition was increased. It can observe that the mass deposition display an increased behavior until 30 min, and then shows an saturation behavior. Fourier Transform Infrared (FTIR) spectroscopy was applied for the surface analysis. The variations of the size of the thin film may change the subsequent chemical reactions in the gas phase significantly. As the reaction time was increased signal from structural fragment containing chlorine atom were loosed, which can explain the behavior observed in the chlorine signal using emission spectroscopy measurements.

1. Introduction

There is great interest in the testing and development of effective abatement technologies for chlorinated volatile organic compounds (CVOCs). Thermal catalysis of dichloromethane, CH_2Cl_2(DCM), has been investigated by many groups using a variety of catalysts in oxidative environments. It is now well-known that particle production and trapping are ubiquitous in all radio frequency discharge. The great interest for dust plasmas results from the various domains concerned by dust formation in the gas discharge. Particles are the main source of contamination in the manufacture of electronic devices [1]. The formation of dusts immersed in plasmas is useful in understanding the numerous universe plasmas such as nebulae or comet tail [2]. Also, surface modification of material is a growing research field having applicability in many areas. In order to improve adhesive properties of polymeric films, chemical etching, termal/flame treatment, and acid/alkali swelling. References 3, 4 and 5 are some surface modification techniques available, but the plasma glow discharge treatment occupies a special importance, due to plasma modifies the surface without affecting the bulk properties.

In this study, we investigated the glow discharge of dichloromethane (CH_2Cl_2) in DC plasma. Optical emission spectroscopy was used to study a glow discharge of CH_2Cl_2 at total pressure of 1.0 Torr, a power of 20 W and a flow of 12 l/min. One of the most interesting questions about this kind of plasma is the influence of thin solid film formation on the electrodes. So, the goal of this work, it is to explore the optical emission spectroscopy (OES) in a low-pressure plasma discharge of CH_2-CL_2. The material deposition will be analyze with the aid of a scanning electron microscope (SEM) equipped with an energy dispersive X-ray (EDX) and in order to analyse the material deposition at different discharge times, Fourier Transform Infrared (FTIR) spectroscopy will be applied for the surface analysis.

2. Experimental setup

The experimental apparatus and technique to measure the DC plasma emission spectrum is identical to that described in previous work [6, 7]. A brief description is reported here. The system consists of two stainless steel circular plane electrodes, 3 mm thick and 30 mm in diameter. The electrodes are positioned at the center of the reaction chamber with 5 mm gap spacing. The CH_2-Cl_2 was injected into the reaction chamber through the bottom flange. The same connection was used for the pressure sensor (MKS, Type 947 convection enhanced Pirani). The DC plasma was produced at a pressure of 1.0 Torr and a flow pf 12 l/min. An ultra pure CH2-Cl_2 was used in the course of the measurements. The discharge power supply (Bertran serie 105, 1 KW) was maintained at an output of 800 Volts and a current of 25 mA (20 W). The base pressure of the plasma chamber of $3.39 \text{ X } 10^{-3}$ m^3 volume was $2.0\text{X}10^{-2}$ Torr, achieved using a mechanical pump (SD-301Varian Pump). The vacuum chamber was purged with Ar gas several times at a pressure of 1.0 Torr in order to remove the background gas. In order to monitor the plasma discharge, in front of the lateral plane of the electrodes was located an optical fiber (Solarization-resistant UV and fiber diameter size of 400 μm) that was connected to the entrance aperture of a high-resolution Ocean Optics Inc. Spectrometer Model HR2000CG-UV-NIR The inlet and outlet slits were 5 μm wide. The data were obtained in a single accumulation of 9 s integration times. The wavelength scan interval was of 0.47 nm. The fiber optics assembly was positioned to collect the emission light appearing on the CH_4-N_2-Ar plasma.

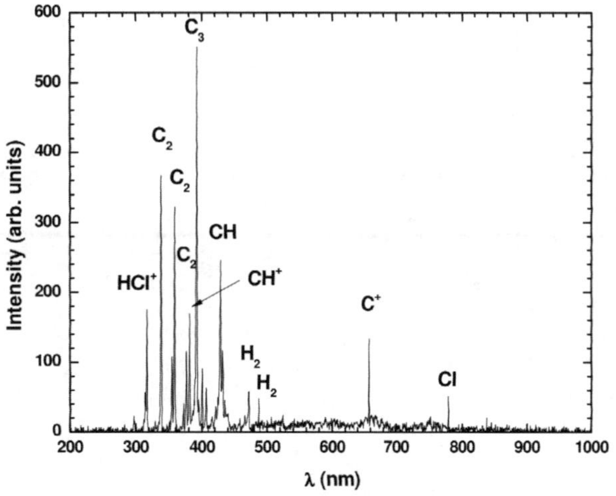

Figure 1. Emission spectra of CH_2-Cl_2 plasma at 1.0 Torr.

3. Results and discussion

The chemical and optical properties depend on the deposition pressure. However, the variation in deposition pressure may change not only the collision frequencies of molecules/atoms but also the plasma properties, such as the intensity of UV emission and the charged particles flux. To investigate the plasma properties, the plasma emission was measured with a CCD-array spectrometer (Ocean Optics) with an optical fiber though the Pyrex glass chamber. The plasma-ions flux reaching the cathode can be evaluated from $J=I_D/S$, where I_D is the current discharge (25 mA) and S is the electrode area (7.07 cm^2). So, the plasma- ions flux is ~3.41 x 10^{18} cm^{-2} min^{-1}. A typical OES measurement of CH_2-Cl_2 glow discharge plasma at a pressure of 1.0 Torr is displayed in Figure 1, at 20 W of power. This allowed analysis of the most luminous area, which corresponds to the negative glow near the cathode dark space. Only the most intense spectral lines and bands of the plasma within the 200-1000 range are quoted [8]. The principal species observed were: at 339.61, 358.60 and 377.96 nm for $C_2(c'^1\Pi_g$-$b^1\Pi_u$); at 392.50 nm for C_3('Π_u-'Σ^+_g); at 431.42 nm for CH($A^2\Delta$-$X^2\Pi$); at 778.28 nm for Cl; at 657.80 nm for C^+; at 471.90 and 487.30 nm for H_2; at 380.61 nm for CH^+(A'Π-X'Σ) and at 317.73 nm for HCL^+($A^2\Sigma$-$X^2\Pi$).

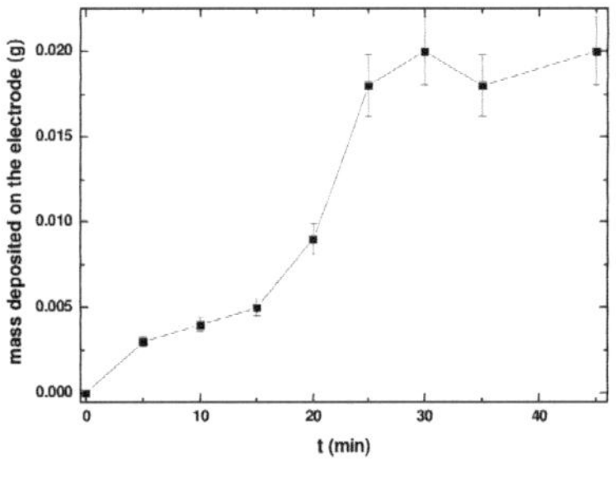

Figure 2. Mass deposition on the electrode

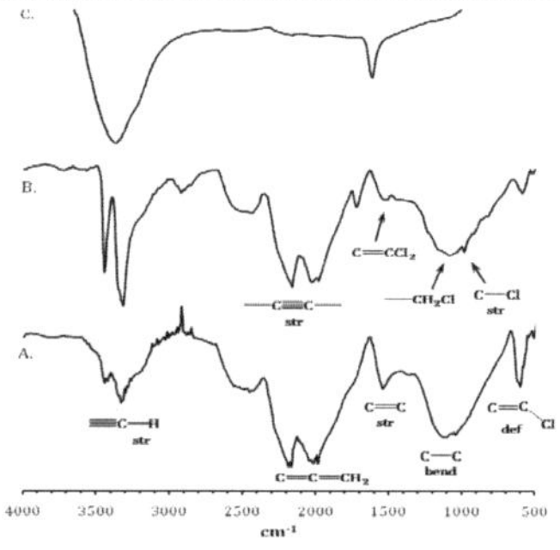

Figure 3. Infrared spectra of MD in an electrode using a CH_2Cl_2 plasma. A. 5 min; B. 15 min; C. 30 min.

At the present plasma condition (V=280 Volts, d=5 mm, P= 1-10 Torr) and assuming T_g= 400 K, the At the present plasma condition (V=800 Volts, d=10 mm, P= 1 Torr) and assuming T_g= 400 K, the ion gas density (CH_2-Cl_2) is $2.41x10^{16}$ cm^{-3}, which with n_e □ 10^{10} cm^{-3} (assumed), we obtained an ionization degree of □10^{-6}. As the ionization degree is < 10^{-4}, it may neglect electron-electron interaction [9]. Also, in this case the reduced electric field E/N is 331.95 Td (E is the field strength). The measurement conditions, such as the integration time of the spectrometer and the position of the optical fiber, were fixed for every measurement. The limitation factor on long-time measurements was the deposition of yellow-orange layer on the electrodes causing sparks in the discharge. The influence and elimination of this kind of tholins requires further investigations.

Figure 2 depicts the mass deposited on the electrode as a function of the discharge time. The mass deposited shows a increasing between 0 and 30 min, with a approximated growth rate of 87.93 mg/min. And then display saturation behaviour.

In order to analyse the material deposited (MD) in the electrode at different discharge times, FTIR spectroscopy was applied for the surface analysis. Attenuated total reflectance (ATR) measurements were performed on MD using a Fourier Transform Infrared (FTIR) spectrometer from PerkinElmer model GX in the mid-IR region of 500-4000 cm^{-1}. The ATR crystal used was diamond that has absorbance bands in the region 2,500-1,650 cm^{-1}. The sample (50 mg of MD) was placed on the diamond crystal without sample preparation. FTIR spectra of the untrated (pristine) sample of MD prodeuced in CH_2CL_2 plasma during 5, 15 and 30 minutes, are shown in Figure 3. As can be seen from Figure 3, the IR spectra are structure less as the residence time of CH_2Cl_2 in the plasma was increased. It behavior is explained on basis the growing of material particles as it's were deposited in the electrode. In the plasma neutral radical reactions occurs as a consequence of the high density of neutral particles with high energies, all of them from the dissociation of the CH_2Cl_2. Radicals can react together give more complexes structures by a step by step mechanism, increasing in size the particles as the reaction time was increased. As can be seen from the IR spectra, the structure less is the result of a lowering of the chemical complexity, increasing the multiple carbon-carbon bonds and eliminating chlorine and hydrogen atoms from the substrate. The main bands in IR spectra are well assigned in the A spectra, signals at 3400 to 3300 cm^{-1}, are attributed to a C-H stretching joined to *sp* carbon, comparatively with C-H with it as sp^3, which are observed at 3500 cm^{-1}, as the reaction time was increased its signals rest structure up to observe a total coalescence at maximum reaction time, spectra C. The signal attributed to *sp-sp* carbon stretching was observed at 2180 cm^{-1}, signal at 2010 cm^{-1} was attributed to C-H stretching when carbon atom is sp^2 type joined to a second sp^2 carbon atom. Its signals were totally loosed at maximum reaction time, spectrum C. Signals from sp^2-sp^2 carbons stretching were observed at 1580 cm^{-1}, its can be correctly signed to sp^2-sp^2 carbon stretching joined to chlorine atoms as and structure of type $C{=}CCl_2$. Single sp^3-sp^3 carbon bonds bending were observed at 1120 cm^{-1}, a signal also signed to $-CH_2Cl$ fragment. Other signals from C-Cl stretching were observed at 1030 cm^{-1}, additional to 620 cm^{-1} signal which is attributed to a deformation of the fragment C=C-Cl. As the reaction time was increased signal from structural fragment containing chlorine atom were loosed, which can explain the behavior observed in the chlorine signal using emission spectroscopy measurements.

4. Conclusions

An experimental study of optical emission spectroscopy of a low-pressure DC plasma discharge sustained in a CH_2-Cl_2 glow discharge was studied. This plasma wase generated at a pressure of 1.0 Torr, power of 20 W and a flow of 12 l/min. The principal species observed were from the molecular bands of $C_2(c'^1\Pi_g\text{-}b^1\Pi_u)$, $C_3(`\Pi_u\text{-}`\Sigma^+_g)$, $CH(A^2\Delta\text{-}X^2\Pi)$, H_2, $CH^+(A'\Pi\text{-}X'\Sigma)$ and $HCL^+(A^2\Sigma\text{-}X^2\Pi)$, and atomic lines of Cl and C^+.

The mass deposited shows a increasing between 0 and 30 min, with a approximated growth rate of 87.93 mg/min. And then display saturation behaviour.

Fourier Transform Infrared (FTIR) spectroscopy was applied for the surface analysis. The variations of the size of the thin film may change the subsequent chemical reactions in the gas phase significantly. As the reaction time was increased signal from structural fragment containing chlorine atom were loosed, which can explain the behavior observed in the chlorine signal using emission spectroscopy measurements.

ACKNOWLEDGMENTS

The authors are grateful to A. González, O. Flores, and José Rangel for technical assistance. This research was partially sponsored by DGAPA IN-105010, CONACyT 28714.

References

[1] G.M. Smadi, G.Y. Kong, R. N. Carlile, and S.E. Beck, J. Vac. Sci. Technol. B10, 30 1992.
[2] B. Draine and B. Sutin, Astrophys. J. 320, 803 1987.
[3] B. Westerlind, A. Larson, and M. Righdal, Int. J. Adhes. Adhesives 148, 115 (2002).
[4] L. Gerenser, J. Elma, M. Mason, and J.M. Pochan, Polymer 26,1162 (1985).
[5] D. Briggs and C.R. Kendall, Int. J. Adhes. Adhesives 2, 13 (1982)
[6] M. Calixto-Rodríguez, H. Martinez, Y. Peña, O. Flores, H.E. Esparza-Ponce, A. Sánchez-Juárez, J. Campos-Alvarez, P.G. Reyes, Applied Surface Science 256, 2428-2433 (2010).
[7] Villa M, Calixto-Rodriguez M, Martínez H, Poveda J C, Reyes P G, Altuzar P 2010, Plasma Science and Technology 12
[8] Pearse R W B, Gaydon A G, *in The identification of molecular spectra,* University Printing House Cambridge, 1976.
[9] Janda M., Martišovitš V., Morová M., Machala Z., Hensel K. 2007 *Eur. Phys. J. D* 45 309

Electron inertia effect on floating plasma potential

V N Duarte and R A Clemente
Instituto de Física Gleb Wataghin
Universidade Estadual de Campinas
Cidade Universitária Zeferino Vaz, Campinas-SP 13083-859, Brazil

E-mail: vnduarte@ifi.unicamp.br

Abstract. The steady one-dimensional planar plasma sheath problem, originally considered by Tonks and Langmuir, is revisited. Two-fluid equations for cold ions and isothermal electrons, including terms for particle generation and electron inertia, have been numerically integrated together with Poisson equation. The inclusion of electron inertia in the model allows us to obtain the value of the plasma floating potential as resulting from an electron density discontinuity at the walls, where the electrons attain sound velocity and the electric potential is continuous. Results from numerical computation are presented in terms of plots for densities, electric potential and particles velocities. Comparison with results from literature, corresponding to electron Maxwell-Boltzmann distribution (neglecting electron inertia), are also shown.

The steady one-dimensional planar plasma sheath problem, originally considered by Tonks and Langmuir [1], is a classic problem in plasma physics which is still object of study and discussions [2-22]. One of the universal assumptions in approaching the problem is to neglect electron motion and consider their distribution to be simply Maxwellian. By one side this simplify the equations, but when the floating plasma electric potential has to be evaluated, an unsatisfactory hypotheses for the electron flux to the walls has to be introduced, estimating it with the one side randomic flow associated to the electron Maxwellian distribution itself. Here, we want to present a model of the problem which allows properly taking into account electron inertia and estimating the plasma floating potential in a consistent way.

Let us consider an infinite partially ionized plasma, with isothermal electrons and cold single ionized ions, embedded between two perfectly absorbing parallel walls. Since electrons have a greater mobility than ions, the walls should become negatively charged, in such a way that ions are attracted to them. The system should tend to a steady state with equal electron and ion fluxes (vanishing electric current density) with the walls at a characteristic value of electric potential, named floating potential. In order to maintain the steady state, some mechanism for electron and ion production is necessary, this will be ascribed to electron collisions with neutrals (assumed uniformly distributed and infinitely massive) and will be taken into account through a particle generation term proportional to some power γ of the electron density (typically $\gamma = 0$, 1 or 2) like $G(x) = \nu\, n_0\, [n_e(x)/n_0]^{\gamma}$, where $n_e(x)$ represents the electron density, $n_0 = n_e(0)$ and ν may be regarded as the ionization rate divided by the electron density at the center of the configuration ($x=0$), ion-electron collisions will be neglected. The problem can be reduced to consider continuity and momentum conservation equations for electrons and ions together with Poisson equation.

In the literature, following Tonks and Langmuir approach, the plasma is usually divided in two regions, one in which the plasma is almost neutral and ion electron pairs are produced, commonly called pre-sheath, and the sheath, close to the walls, where ion motion is supersonic (according to Bohm criterion [2]) and electron and ion densities differ appreciably. This approach generate serious troubles when a matching between the two regions has to be done [3], but actually it is not necessary when the equations are solved numerically, as shown by Self [4], who numerically solved the whole problem with Maxwellian electron and free falling ions produced locally by electron collisions with neutrals and starting from rest. Here also a numerical treatment will be adopted and in order to compare the present approach to that of Self, we adopt a similar normalization for the different variables entering the problem.

Using Gaussian units, n_0 for the electron density at the middle of the configuration, T_e and $-e$ for the electronic constant temperature and charge, $m_{e,i}$ for the electron/ion mass and k for the Boltzmann constant it is possible to define the following dimensionless dependent variables: densities $\tilde{n}_{e,i}(y) = \frac{n_{e,i}(y)}{n_0}$; velocities $\tilde{u}_{e,i}(y) = \sqrt{\frac{m_e}{kT_e}}u_{e,i}(y)$; particles flux densities (assumed to be equal for both species) $\Gamma(y) = \tilde{n}_{e,i}(y)\tilde{u}_{e,i}(y)$ and electric potential $\phi(y) = -\frac{e\,V(y)}{k\,T_e}$ (positive defined), where the dimensionless independent variable $y = \frac{x}{L}$ has been defined in terms of $L = \frac{1}{\nu}\sqrt{\frac{kT_e}{m_i}}$ a typical discharge width. With such notation the corresponding two-fluid one-dimensional equations for continuity, momentum conservation and electric potential can be conveniently written [5, 6]:

$$\frac{d\Gamma}{dy} = \sqrt{\frac{2m_e}{m_i}}\tilde{n}_e^{\gamma}\,, \tag{1}$$

$$\frac{m_i}{m_e}\frac{d}{dy}\left(\frac{\Gamma^2}{\tilde{n}_i}\right) = \tilde{n}_i\frac{d\phi}{dy}\,, \tag{2}$$

$$\frac{d}{dy}\left(\frac{\Gamma^2}{\tilde{n}_e}\right) + \frac{d\tilde{n}_e}{dy} = -\tilde{n}_e\frac{d\phi}{dy}\,, \tag{3}$$

$$\frac{d^2\phi}{dy^2} = -\frac{2}{\alpha^2}(\tilde{n}_e - \tilde{n}_i)\,. \tag{4}$$

Ion temperature has been neglected and a new constant $\alpha = \sqrt{\frac{\nu^2 m_i}{4\pi n_0 e^2}} = \frac{\sqrt{2}\lambda_D}{L}$, where $\lambda_D = \sqrt{\frac{kT_e}{4\pi n_0 e^2}}$ is the usual Debye length, has been introduced. In order to start the set of equations it is necessary a charge density unbalance at the middle of the configuration (y=0). Assuming $\tilde{n}_i(0) = 1 + \varepsilon$, it is possible to relate α to ε by imposing the vanishing of the first derivative of $\tilde{n}_i$ at y=0 in equation (2), which leads to $\varepsilon = 2\alpha^2$.

Equations (1-4) have been numerically integrated using the subroutine NDSolve of *Wolfram Mathematica* software [23], for a wide range of α^2, γ=0, 1 and 2, and m_i corresponding to the representative cases of hydrogen (H) or mercury (Hg). The integration is performed with boundary conditions imposed at the center and assuming symmetry about it. In all cases the integration has to be aborted when electrons reach the sonic isothermal velocity, i.e., $\tilde{u}_e \to 1$, since in that limit $\frac{d\tilde{n}_e}{dy}$ in equation (3) diverges. We ascribe to such singular behavior the attainment of plasma floating potential at the wall, which depends on the chosen values of α and $\frac{m_i}{m_e}$. In the following, numerical results are presented for several cases and compared with Self results.

In figure 1, dimensionless electric potential profiles are plotted as functions of y for H, three representative values for α (0.1, 0.01 and 0.001) and γ=1. In figure 2, dimensionless electric potential profiles are plotted as functions of y for H, α=0.01 and γ=0, 1 and 2. Filled curves represent results from the present model, while dashed ones indicate the corresponding solutions given by Self (for free falling ions and which do not suffer from any singularity in the range of integration). Our curves have to be stopped at the corresponding plasma floating potential owing to the attainment of sonic velocity by the electrons. The little displacement of the curves is due to the fact that in the present model ions lack any spread in their velocities.

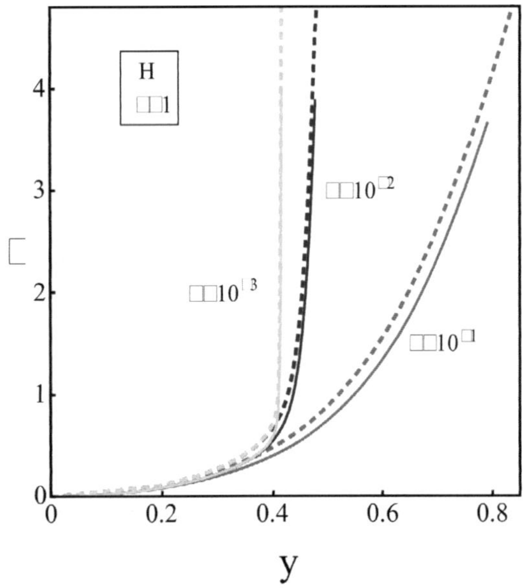

Figure 1: Potential profiles with characteristic values for α (0.1, 0.01 and 0.001) with γ=1 for atomic hydrogen (H). Filled curves correspond to the model described by the present work, while dashed curves indicate solutions given by Self.

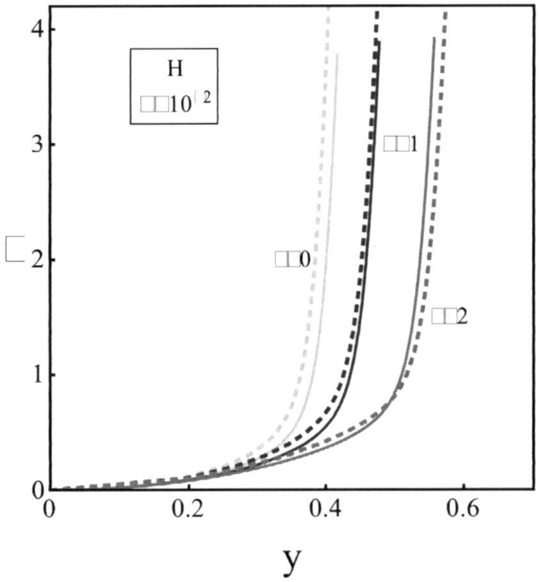

Figure 2: Potential profiles with characteristic values for γ (0, 1 and 2) with α =0.01 for atomic hydrogen (H). Filled curves correspond to the model described by the present work, while dashed curves indicate solutions given by Self.

It is interesting to note that Self solutions are independent of the ionic to electronic mass ratio for a given α, since he neglected electron inertia and used a convenient normalization of the spatial coordinate, so that the ionic mass disappeared explicitly from the characteristic equation of his model and the wall position (which does depend on the ionic to electronic mass ratio) is deduced *a posteriori*, equating the ion flux to the one side randomic flux of the Maxwellian electrons.

In order to better show the differences between Self results and ours, in table 1 we compare values of the electric floating potential and wall position corresponding to H and Hg for some typical values for α and different γ. η_w and s_w correspond to Self values for wall dimensionless potential and dimensionless wall position, while ϕ_w and y_w correspond to ours. As it can be seen, differences are not really significant, but the inclusion of electron inertia allows us to define the floating wall potential in consistency with the model equations.

Journal of Physics: Conference Series **511** (2014) 012015 doi:10.1088/1742-6596/511/1/012015

Table 1: Comparison for dimensionless plasma wall potentials and dimensionless wall position obtained by Self (s_w, η_w) and by the present work (y_w, ϕ_w) for H and Hg and different values of γ and α.

γ=0

	H				Hg			
α	η_w	s_w	ϕ_w	y_w	η_w	s_w	ϕ_w	y_w
0.1	2.98	0.618	3.29	0.665	5.47	0.723	5.81	0.769
0.01	3.41	0.398	3.78	0.417	6.03	0.414	6.39	0.432
0.001	3.53	0.353	3.91	0.364	6.18	0.355	6.56	0.366

γ=1

	H				Hg			
α	η_w	s_w	ϕ_w	y_w	η_w	s_w	ϕ_w	y_w
0.1	3.29	0.748	3.65	0.791	5.94	0.893	6.30	0.925
0.01	3.50	0.469	3.87	0.478	6.16	0.487	6.53	0.494
0.001	3.55	0.415	3.93	0.416	6.20	0.417	6.58	0.418

γ=2

	H				Hg			
α	η_w	s_w	ϕ_w	y_w	η_w	s_w	ϕ_w	y_w
0.1	3.43	0.910	3.79	0.932	6.08	1.075	6.42	1.082
0.01	3.54	0.568	3.91	0.557	6.19	0.586	6.56	0.574
0.001	3.56	0.504	3.93	0.485	6.21	0.505	6.58	0.487

In figure 3, a typical result for ion and electron densities as function of position, corresponding to H, α=0.01 and γ=1, is shown. As it can be seen, ion and electron densities differ considerably at the wall position. Inside the figure, an enlargement of the detail for electron density close to the wall is shown. For comparison, when the electron density of our model is plotted together with that corresponding to a Maxwell-Boltzmann distribution, they practically overlap up to very close to the wall. The difference increases for larger values of α and diminishes for smaller α. The abrupt fall of the electron density of our model is due to the fact that electron velocity is reaching the sonic one at the wall.

In figure 4 we show, for H and the same parameters as in figure 3, the ion and electron dimensionless velocity profiles. The ion velocity is multiplied by a factor $\sqrt{m_i/m_e}$ so that unity corresponds to ion sound velocity as defined in literature for cold ions. Comparing with figure 3, it is possible to appreciate that when the ionic velocity is close to unity, ion and electron density are still very close, while they are well separated when the ions reach the wall and their dimensionless velocity is about 2.7 their sonic one. The electron velocity profile clearly shows how the electrons reach their thermal velocity at the wall.

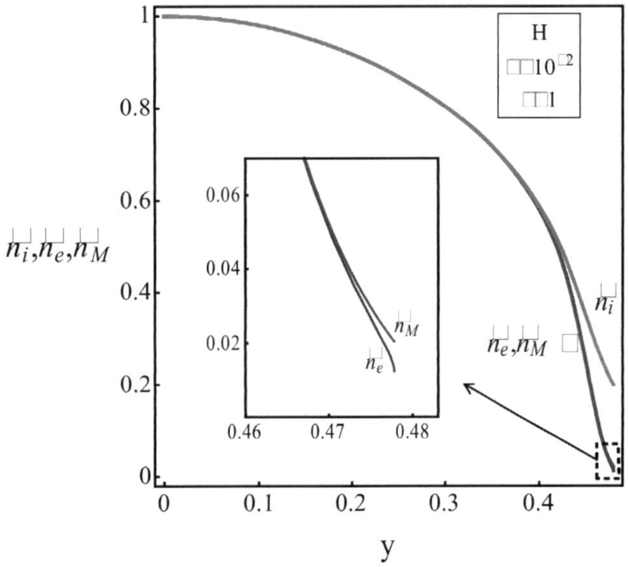

Figure 3: Dimensionless density profiles for electrons ($\tilde{n}_e$) and ions ($\tilde{n}_i$) together with electron density due to Maxwellian distribution ($\tilde{n}_M$). In order to show precisely the difference between our electron density and the commonly assumed Maxwell-Boltzmann distribution, we enlarged in the same figure a small region near the wall, where $\frac{d\tilde{n}_e}{dy} \to -\infty$.

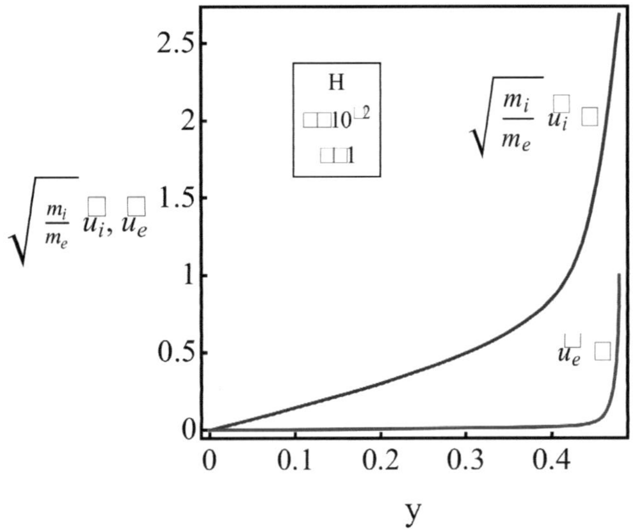

Figure 4: Ion (upper curve) and electron (lower curve) velocities, normalized to their own sound velocities, from the center to the wall for H and the same parameters as in figure 3. As $\tilde{u}_e$ approaches 1, the integration has to be aborted as electron density derivative, in the electron momentum equation, diverges.

As it can be appreciated, the inclusion of electron inertia in the treatment of the classic plasma sheath problem introduces little changes to the floating plasma wall potential when compared with results from models that simply assume Maxwell-Boltzmann distribution for electrons and neglect their velocity. The differences are essentially due to an abrupt fall in electron density close to the end of the discharge, while the velocity with which electrons impinge on the wall is the same (insofar as electron temperature is maintained constant). In our model, electrons attain their sonic speed at the wall, and then a shock should build up there, where electron density and velocity approach discontinuities while the electric potential is continuous. Such description may be a physically attractive alternative to the usual assumption of equating the ion flux to the one side randomic electron flux.

Acknowledgments

The authors would like to thank Fundação de Amparo à Pesquisa do Estado de São Paulo (FAPESP) for financial support under Grants 2008/55778-8 and 2009/11835-0.

References

[1] Tonks L and Langmuir I 1929 *Phys. Rev.* **34** 876
[2] Bohm D 1949 *The Characteristics of Electrical Discharges in Magnetic Fields*, ed A Guthrie and R K Wakerling (New York, McGraw-Hill) pp 77-86
[3] Caruso A and Cavaliere A 1962 *Nuovo Cimento* **26** 1389
[4] Self S A 1963 *Phys. Fluids* **6** 1762
[5] Bittencourt J A 2004 *Fundamentals of Plasma Physics* (New York, Springer) pp 193-212, 279-91
[6] Kino G S and Shaw E K 1966 *Phys. Fluids* **9** 587
[7] Allen J E 2009 *Plasma Sources Sci.. Technol.* **18** 014004
[8] Friedman H W and Levi E 1967 *Phys. Fluids* **10** 1499
[9] Riemann K-U 2009 *Plasma Sources Sci. Technol.* **18** 014007
[10] Stangeby P C and Allen J E 1970 *J. Phys.* A: *Gen. Phys.* **3** 304
[11] Ingold J H 1972 *Phys. Fluids* **15** 75
[12] Riemann K-U 1991 *J. Phys.* D: *Appl. Phys.* **24** 493
[13] Harrison E R and Thompson W B 1959 *Proc. Phys. Soc.* **74** 145
[14] Bissell R C, Johnson P C and Stangeby P C 1989 *Phys. Fluids* B **1** 1133
[15] Godyak V A and Sternberg N 1990 *IEEE Trans. Plasma Sci.* **18** 159
[16] Alterkop B 2004 *J. Appl. Phys.* **95** 1650
[17] Deka U and Dwivedi C B 2010 *Braz. J. Phys.* **40** 333
[18] Morales Crespo R, Fernández Palop J I, Ballesteros J, Hernández M A, Lucena-Polonio M V and Díaz-Cabrera J M 2010 *Plasma Sources Sci. Technol.* **19** 025012
[19] Robertson S 2009 *Phys. Plasmas* **16** 103503
[20] Lam S H 1964 *Phys. Fluids* **8** 73
[21] Senda I 1995 *Phys. Plasmas* **2** 6
[22] Hu P N and Ziering S 1966 *Phys. Fluids* **9** 2168
[23] Mathematica 7.0, Wolfram Research Inc.

Optical Characterization of Glow and Afterglow Regions of Ar/O_2 Microwave Plasma: Effect of Applied Power and Gas Flow

J V da Maia, R S Pessoa, A S da Silva Sobrinho, M Massi and H S Maciel
Plasmas and Processes Laboratory – Department of Physics, Technological Institute of Aeronautics – ITA, São José dos Campos/SP, 12228-900, Brazil.

E-mail: argemiro@ita.br

Abstract. In this work, the effect of microwave power and gas flow on formation of atomic oxygen (O) was investigated by actinometry method in the two different regions of Ar/O_2 microwave plasma: the glow and afterglow regions. The experiments were performed keeping constant the Ar/O_2 gas flow ratio at proportion of 3/2 and varying the microwave power from 200 to 1100 W and the total gas flow from 50 to 300 sccm. The results showed that the production of O depends on both discharge parameters investigated and that the microwave power has the greater influence in the production of O. At afterglow region the production of O tends to be uniform and not dependent on the gas flow and microwave power. This may be due to homogenization of the processes of recombination and loss in the gas phase and to the walls along the afterglow region.

1. Introduction

The afterglow discharge region of oxygen microwave plasma has been used with great interest in surface modification of materials. In the case of polymeric material, the major interest is to modify the surface energy in order to improve the adhesion properties [1]. In the case of inorganic and metallic materials the major interest is the surface cleaning (oil removing, for example) [2] and sterilization of medical instruments [3].

In this scope, the densities of neutral and metastable atomic oxygen in the reactor are of great concern since the processing speed depends on the flux of these particles onto substrate, which is greatly influenced by process conditions such as gas pressure and applied power. Moreover, it is known that the addition of argon in the oxygen plasma produces a significant enhancement of the dissociation rate of oxygen gas [4]. At the same time, it creates a condition for a simple and low cost measurement of atomic oxygen (O) concentrations by comparing the emission from argon and oxygen atoms [5].

Nowadays, there are several experimental techniques to measure the amount of O formed during an electrical discharge in a gas, and one of the more commonly used is optical emission spectroscopy (OES) [6]. Through the use of this technique, in conjunction with the actinometry method [7, 8], the relative amount of O formed in gas phase can be estimated through the intensities of the spectral lines of oxygen OI (844.6 nm) and argon ArI (750.4 nm), i.e. the I_{OI}/I_{ArI} ratio.

In this work, the effect of microwave power and total gas flow on formation of O was investigated by actinometry method in the glow and afterglow regions of Ar/O_2 microwave plasma. The experiments were performed varying the microwave power and the total flow, while keeping constant the Ar/O_2 ratio. In order to characterize the glow and afterglow regions, the OES measurements were realized at three different points along the microwave plasma.

2. Experimental setup

A schematic of microwave plasma system used in this investigation is presented in Fig. 1. The gas is injected in a 10 mm outer diameter quartz tube that is connected to a 200 mm diameter and 1300 mm length Pyrex tube (pos-discharge chamber). The system is pumped by mechanical and root pumps. The total Ar/O_2 gas flow was varied from 50 to 300 sccm. The microwave plasma was generated inside the 10 mm quartz tube coupled to a 2,45 GHz rectangular waveguide and resonant cavity. The microwave power was varied from 200 to 1100 W. The waveguide is of adjustable size and has three manual stubs used as matching network system.

The pressures were varied from 12 to 32 mTorr in the lower and higher flow condition, respectively.

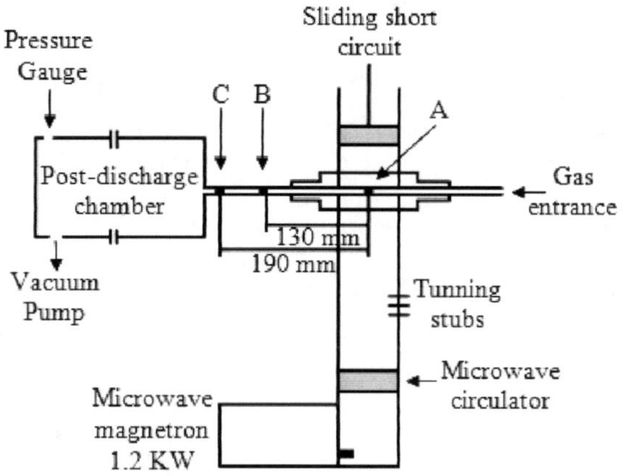

Figure 1: Experimental setup used for generation of microwave plasma. The optical spectra were obtained in the position A, B and C

For the accomplishment of the optical characterization of glow and afterglow regions of Ar/O_2 microwave plasma an optic fiber was installed at three points (A, B and C, Figure 1) of quartz tube in order to collect the light emitted from the plasma and the afterglow regions. These fibers are connected externally to an optical emission spectrometer (Ocean Optical USB4000) with resolution of 1.5 nm, operating in the range of 200-850 nm.

The actinometry method consists in correlate the optical emission spectrum, more specifically the intensity ratio of the gas of interest (in this case, the oxygen) with the intensity of the gas used as actinometer. For oxygen plasmas, the reactive specie to be studied is the atomic oxygen O(3p) with emission at 844.6 nm and the actinometer gas was used the argon Ar(4p) with emission at 750.4 nm. The actinometry is a qualitative method to estimate the concentration of certain active specie formed in the plasma or to observe its behavior with the variation of plasma parameters. The qualitative concentration of oxygen [O] is obtained using equation 1. The [Ar] is controlled inserted in the plasma, and the I_{OI} and I_{ArI} are measured using an optical spectrometer. The weak point of this method is to obtain the reaction constant k(O), that take into account all kinetic reactions on the plasma. This constant also strongly depends of the plasma parameters (microwave power, pressure, gas flow, etc.) and oxygen argon gas ratio.

$$I_{OI}/I_{ArI} = k(O)[O]/[N] \qquad (1)$$

In order to certify the results obtained by actinometry, quantitative method as gas titration has been used by several researchers [9-11]. The results obtained by these author shows that the actinometric method produces results that are in good agreement with the titration method an Ar/O_2 gas ratio up to about 2. Based on the literature, the behavior of the atomic oxygen concentration for different plasma conditions was studied using an Ar/O_2 ratio of 1.5.

3. Results and discussions

Surface plots of optical intensities ratio, I_{OI}/I_{ArI}, as function of total gas flow and microwave power, obtained at positions A, B and C are presented at Figure 3a, 3b and 3c, respectively. The surfaces representing the trends of experimental data (black points) were created using the *spline* method.

As expected, the more important variation on the values of I_{OI}/I_{ArI} with variation of microwave power and with total gas flow occurs in point A. This point is the center of microwave applicator where the highest microwave field intensity is concentrated. For a given microwave power the value of I_{OI}/I_{ArI} decrease with the increase of the total gas flow. On the other hand, the I_{OI}/I_{ArI} for a given total gas flow increase with the increase of microwave power. However, the increase is less pronounced for high gas flow.

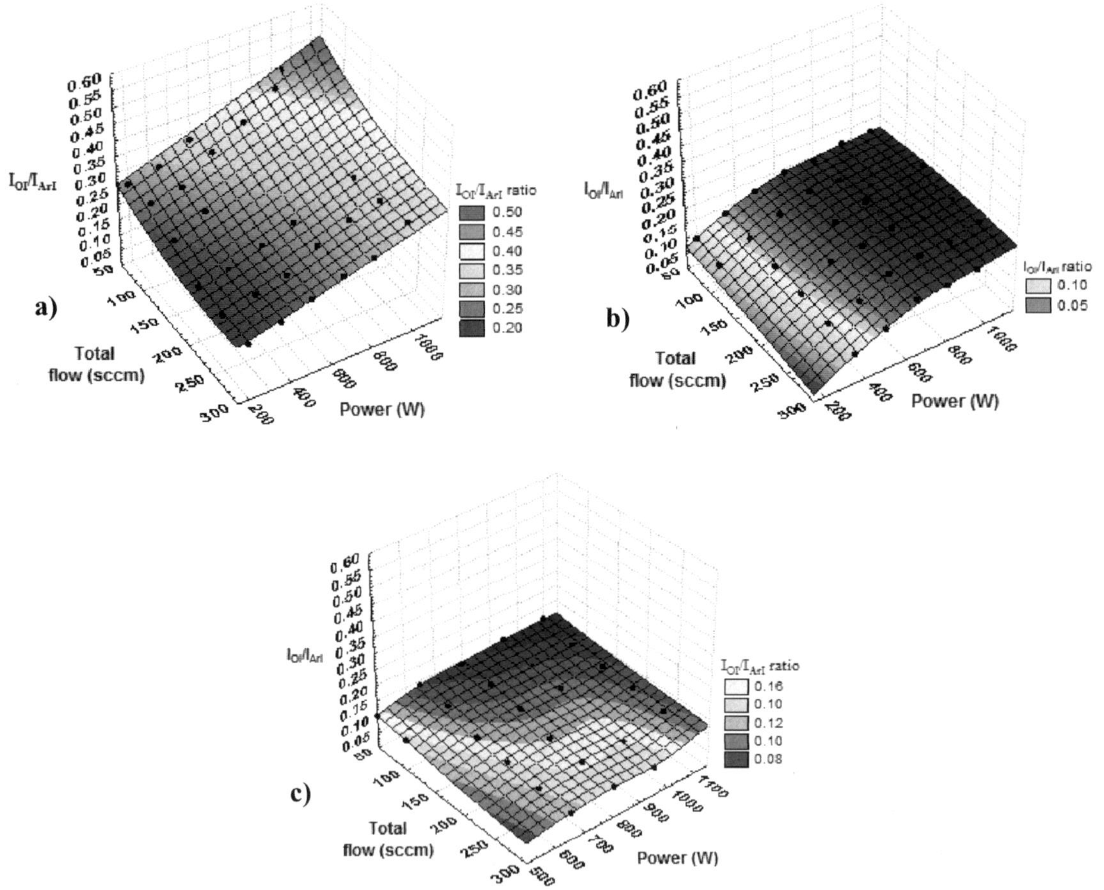

Figure 3: I_{OI}/I_{ArI} as a function of microwave power and total gas flow rate along the microwave discharge axis at: a) point A; b) point B; and c) point C.

The lower rate of atomic oxygen production with the increase of total gas flow may be related with the increase of the pressure inside the quartz tube. The pressure varied from 14 mTorr to 34 mTorr for the total gas flow variation from 50 sccm to 300 sccm, respectively. The increase of pressure decreases the mean free path and the residence time of the gas molecules in the microwave field area, which increase the number of molecules that will not acquire enough energy to ionize the gas. The decrease of the mean free path also contributes to the recombination process. All these processes contribute to the decrease of the atomic oxygen with the increase of gas flow.

The higher values of atomic oxygen obtained for low gas flow are mainly related with the lower pressure inside the quartz tube, which increase the molecules mean free path, and the increase of residence time of the molecules inside the microwave field region. These effects contribute to the increase of the dissociation process [12, 13].

Figure 4 shows the dependence of I_{OI}/I_{ArI} with microwave power on the three measuring points for a total gas flow of 100 sccm. Here, we can better observe the variation of I_{OI}/I_{ArI} along plasma region. In the afterglow regions (positions B and C) the I_{OI}/I_{ArI} values to not change for microwave power higher than about 650 W, indicating that the creation and recombination processes in the volume and in the tube walls are in equilibrium in the afterglow region.

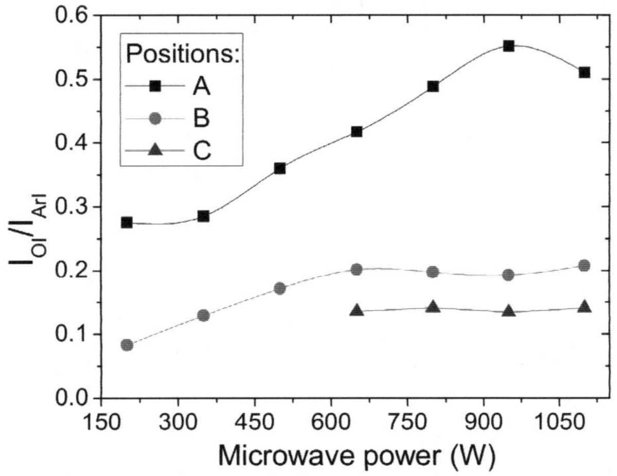

Figure 4: I_{OI}/I_{ArI} as function of microwave power for total gas flow rate of 100 sccm measured at the positions A, B and C.

4. Conclusions

Using the OES technique associated with the actinometry method was possible to qualitatively study the behavior of atomic oxygen production in glow and afterglow regions of microwave plasma. The behavior of I_{OI}/I_{ArI} was investigated as function of discharge microwave power (200 - 1100 W) and total gas flow (50 - 300 sccm) for an Ar/O_2 gas flow ratio of 3/2.

The I_{OI}/I_{ArI} and consequently, the relative concentration of atomic oxygen showed different behaviors with microwave power and total gas flow along the discharge axis. Within the range of parameters analyzed the higher production of atomic oxygen can be associated to the elevated microwave power and low gas flow.

Concerning the variation of concentration of O along the plasma axis, it was observed that in the afterglow regions the atomic oxygen concentration decreases due to the recombination processes in the volume and in the tube walls. In the afterglow regions the creation and recombination processes are equivalent and the atomic oxygen concentration are kept constant for microwave power higher than about 650 W.

References

[1] Moraes J H, Da Silva Sobrinho A S, Maciel H S, Dutra J C N, Massi M, Mello S A C and Schreiner W H 2007 *Journal of Physics D: Applied Physics* **40** 7747.
[2] Gaillard M, Raynaud P and Ricard A 1999 *Plasma Polym.* **4** 241.
[3] Ricard A and Monna V 2002 *Plasma Sources Sci. Techn.* **11** 150-153.
[4] Piper L G and Rawlins W T 1986 *J. Phys. Chem.* **90** 320-325.
[5] Ershov A and Borysow J 2007 *Plasma Sources Sci. Techn.* **16** 798-802.
[6] Fuller N C M, Malyshev M V, Donnelly V M and Herman I P 2000 *Plasma Sources Sci. Technol.* **9** 116-127
[7] Coburn J W and Chen M 1980 *J. of Applied Physics* **51** 3134-3136.
[8] d'Agostino R, Cramarossa F, de Benedictis S and Ferraro G 1981 *J. Appl. Phys.* **52** 1259.
[9] Henriques J, Villeger S, Levaton J, Nagai J, Santana S, Amorim J and Ricard A 2005 *Surf. And Coat. Techn.* **200** 814-817.
[10] Granier A, Chérea D, Henda K, Safari R and Leprince P 1994 *J. Appl. Phys.***75** 104-114.
[11] Ricard A, Henriques J, Cousty S, Villeger S, Amorim J 2007 *Plasma Process and Polym.***4** 965-968.
[12] Vesel A, Mozetic M, Balat-Pchelin M 2007 *Vacuum* **81** 1088-1093.
[13] Balat-Pchelin M, Badie J M, Berjoan R, Boubert P 2003 *Chem. Phys.* **291** 181-194.

Spectroscopic Characterization Of RF Hydrocarbon Plasmas For DLC Coatings

F Guzmán, M Ruiz, E Valderrama[1], M Favre, H Bhuyan, E S Wynham, and H Chuaqui

Departamento de Física, Pontificia Universidad Católica de Chile, Casilla 306, Santiago 22, Chile

E-mail: faguzman@uc.cl

Abstract. In previous experiments, using a RF plasma generator at low power, we have grown diamond like carbon films (DLC), using a hydrocarbon plasma, without the use of a catalyzer precursor[1]. In order to better understand the relationship between the properties of the resulting DLC films and the RF plasma parameters, we have performed detailed optical emission spectroscopy (OES) observations of RF hydrocarbon plasmas, under identical conditions to those used for DLC films growth. The experiments were performed in a low power capacitively coupled RF (13.6 MHz) plasma generator, operating at 30 W. Working gases were acetylene and hydrogen. The optical emission of the RF plasma was recorded, under different operational conditions, with a spectrometer, and an f-f light collection arrangement was used to achieve spatial resolution. The relative emission of the different molecular species was found to depend on the C_2H_2:H_2 flow ratio. Based on the OES observations, we present a description of plasma conditions in a low power RF hydrocarbon plasma reactor.

1. Introduction

Radio frequency (RF) plasmas are widely used to grow different kinds of carbon coatings with sub-micron size characteristic features and varied structural properties [2], which depend on the structural order and the micro/nanostructure of the coatings [3]. Carbon spheres can be produced with control in crystallinity and diameter [4], allowing a wide spectrum of applications. Carbon spheres have been synthesized with different methods including chemical vapour deposition (CVD), pyrolisis of hydrocarbons [5], and plasma enhanced chemical vapour deposition (PECVD). In most of the cases the studies aim to produce large amounts of carbon spheres, with low spread in sizes. As opposed to other methods [4, 5, 6], growing of carbon spheres using PECVD in RF plasmas often requires the use of a catalystic agent. Moreover, when produced in RF plasmas, carbon spheres are collected upon deposition over the chamber floor, or over a heated Si substrate. In order to overcome the use of a catalyst and to improve on the collection capability, we have investigated the formation of carbon spheres using a RF hydrocarbon plasma, produced with mixtures of acetylene and hydrogen, at different mixing ratios. In Plasma-Based Ion Implantation and Deposition (PBII&D) [7] a target sample is surrounded by a plasma and pulse biased to a high negative potential relative to the chamber

[1] Present address: Alameda Applied Sciences Corporation, 3077 Teagarden Street, San Leandro, CA 94577, U.S.A.

wall. Ions generated in the surrounding plasma are accelerated across the sheath formed around the sample and implanted normally into the surface of the targets. We have used the PBII&D technique to study the formation of micro/nanostructured a-C:H layers of sub-micron size carbon spheres on a silicon (100) substrate, without substrate heating, and in the absence of any template or catalyzer. The films have been characterized using a variety of different diagnostics, including atomic force microscopy (AFM) and scanning electron microscopy (SEM). In order to understand better the relationship between the properties of the resulting DLC films and the RF plasma parameters, we have performed detailed optical emission spectroscopy (OES) observations of RF hydrocarbon plasmas, under identical conditions to those used for DLC films growth. The identification of different molecular species in the C_2H_2 discharge plasma spectra, and their evolution as a function of the C_2H_2/H_2 mixing ratio is analyzed in the context of the structural and morphological properties of the DLC films.

2. Experimental details

The DLC deposition experiments were performed in a low power, capacitively coupled RF (13.6 MHz) plasma generator, operating at 30 W. The electrode assembly consists of stainless steel plates separated 7 cm. The anode, top plate, is 10 cm in diameter, and the water cooled cathode, bottom plate, is 8 cm in diameter. Working gases are acetylene and hydrogen. The flow rate of acetylene is fixed at 3 sccm, and the flow of hydrogen is varied to obtain 1:0, 1:2, 1:4, 1:8 and 1:10 C_2H_2:H_2 flow ratios. A schematic of the experimental set–up is shown in figure 1. Due to the high asymmetry of the electrodes, a negative dc bias is induced in the power electrode, of the same magnitude as the applied RF voltage, which results in a plasma potential of a few volts, which favours application of the PBII&D technique. These conditions reduce sputtering and surface etching, which are known to be strong at high values of the plasma potential [8]. Plasma potential, electron density and electron temperature, were measured, from the $I - V$ characteristics of a RF-compensated Langmuir, in inert gases like argon and helium, at equivalent operational conditions. Further details on the experimental arrangement have been published elsewhere [1].

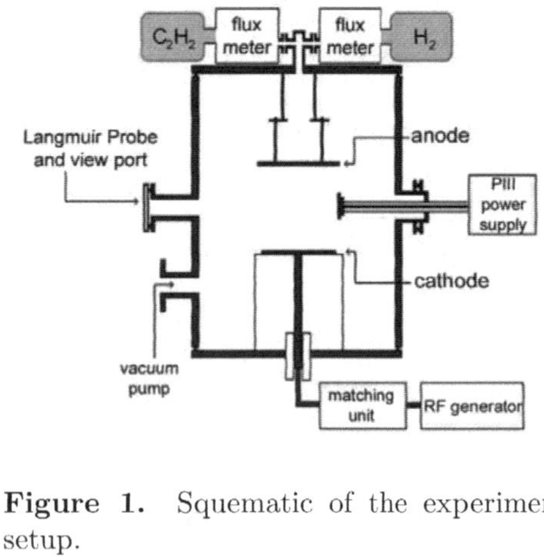

Figure 1. Squematic of the experimental setup.

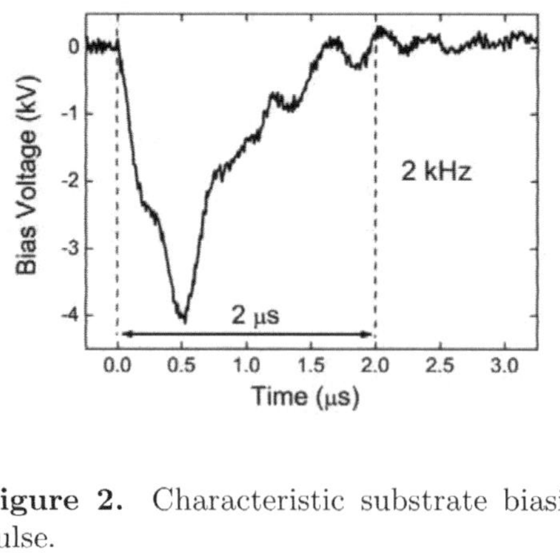

Figure 2. Characteristic substrate biasing pulse.

The plasma potential was measured to be less than 30 V, the electron temperature was between 1 and 3 eV, and the electron densities between 10^9 and 10^{10} cm^{-3}, for RF powers below 100 W [9]. Direct measurements in the acetylene RF plasma were not reliable due to fast contamination of the probe surface. The optical emission of the RF plasma was recorded, under different operational conditions, with a SpectraPro 275 spectrometer, using a 1200 grove/mm grating, using a 30 ms integration time. The PBII&D power supply is based on two HV IGBT's driving two transformers, which are discharged in series, to produce a negative short pulse of 30 kV, 2 s max width, at a 2 kHz rate, 400 ns rise time. These short pulses are known to improve the conformal implantation by decreasing the width of the plasma sheath. The voltage was measured with a high voltage probe and the current was measured with an optically coupled isolator phototransistor. A characteristic 4 kV voltage signal, corresponding to substrate biasing used throughout this experiments, is shown in Fig.2. The depositions were carried out on mirror-like polished silicon substrate of (100) orientation and 500 μm thickness. The substrates were positioned in an off-axis configuration, away from the main discharge plasma to reduce irradiation damage from the plasma. The deposition time was 30 min.

3. Results and discussion

Figure 3 shows SEM images of the DLC films under different C_2H_2:H_2 mixing ratios. The surface morphology is dominated by carbon spheres, with characteristic diameter below 1 μm. The size and size distribution is seen to depend on the C_2H_2:H_2 ratio. Measurements of rms roughness, made with an AFM microscope in tapping mode, showed characteristic values ranging from 170 nm up to a deep minimum of $\sim$3 nm, at a mixing ratio of C_2H_2:H_2=1:2 [1].

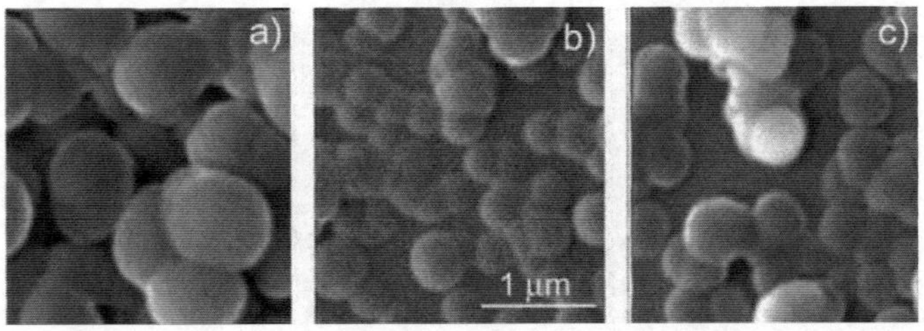

Figure 3. SEM images of DLC films obtained with, a) pure acetylene, b) 1:4, and c) 1:8 C_2H_2:H_2 mixing ratios.

Figure 4 shows OES spectra of the hydrocarbon plasma for different C_2H_2:H_2 mixing ratios. In addition to atomic hydrogen, prominent molecular emission from H_2, CH, C_2 and $C_4H_2^+$, among other active radicals, is clearly identified. This identification is noteworth, as during the processes of DLC deposition, neutral hydrocarbon species are considered to be important precursors for film growth [10]. Additionally, C_2 dimer is known to play a significant role in the formation of nanodiamond films [11]. On inspecting the hydrocarbon spectra, a general and expected trend to be noticed is the increase in the intensity of atomic hydrogen emission as the hydrogen content of the hydrocarbon plasma increases. It is also observed that no molecular hydrogen emission band are observed when pure acetylene is used. In order to visualize the effect of the hydrogen content, figures 5 and 6 show the dependence of the molecular emission intensity, as a function of the C_2H_2:H_2 mixing ratio, for different radicals, as identified in figure 4. The rms roughness of the DLC films increases sharply as the mixing ratio becomes bigger that 1:4. On comparing with figures 5 and 6, this increase coincides with a decrease in the relative intensity of molecular emission lines. It can then be inferred that a low roughness is favored by a higher content of active hydrocarbon radicals in the RF plasma. Carbon films can be characterized in terms of the dominant C-bondings, ranging from a material with 100% sp^2 sites (graphite) to a material with $\sim$100% of sp^3 sites (defected diamond). Raman analysis of the resulting carbon films indicate that the increase in the flow of hydrogen move the sp^3

content to a lower values [9], which is in agreement with the literature, where the use of ion implantation cause the conversion of sp^3 bonds to sp^2 bonding [12]. Here, a combination of high content of hydrogen ions in the RF plasma, together with the negative high voltage biasing of the substrate, can explain the observed tendency to a higher graphite–like structure at higher hydrogen mixing ratios.

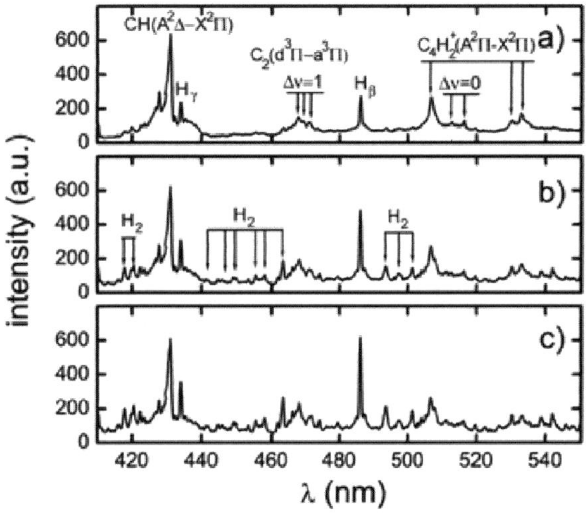

Figure 4. Characteristic emission spectra of the RF plasma obtained with, a) pure acetylene, b) 1:2, and c) 1:4 C_2H_2:H_2 mixing ratios.

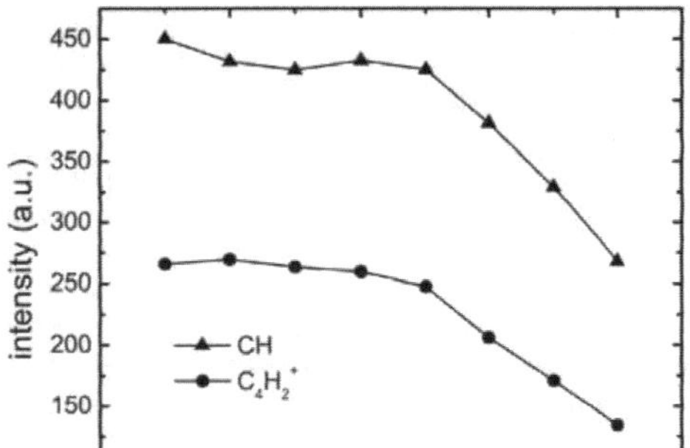

Figure 5. Dependece of CH and $C_4H_2^+$ molecular emission intensity, as a function of the C_2H_2:H_2 mixing ratio.

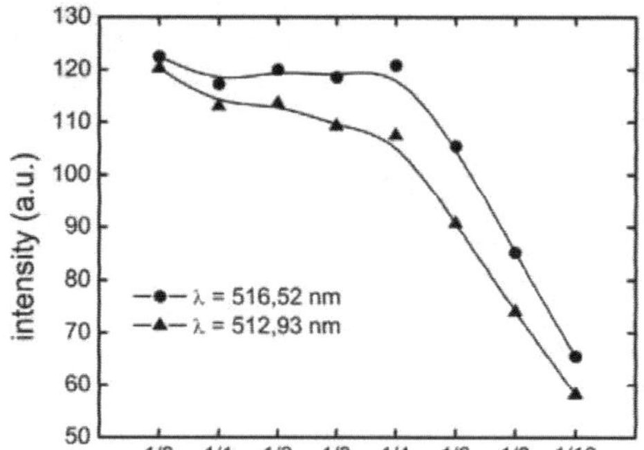

Figure 6. Dependece of C_2 molecular emission intensity, at two different wavelenghts, as a function of the C_2H_2:H_2 mixing ratio.

4. Conclusions

OES has been used to characterize a low power C_2H_2:H_2 RF plasma under different conditions of the flux ratio, in relation with nanostructured DLC film deposition, using PBII&D technique.

Detailed emission spectra have been obtained, which allowed several active species in the RF plasma to be identified, including neutrals, ions and active radicals. A qualitative correlation between DLC film composition and roughness with plasma composition has been established. Further analysis of the spectral information is being performed in order to establish better quantitative correlations with the resulting DLC films.

Acknowledgments
This work has been funded partially by bridge grant 5/2010 from VRI-PUC and the Physics Department. F Guzmán acknowledges financial support from CONICYT. M Ruiz acknowledges a financial support from MECESUP.

5. References

[1] Valderrama E, Favre M, Bhuyan H, Ruiz H M, Wyndham E, Valenzuela J, Chuaqui H 2010 *Surf. Coat. Technol* **204** 2940
[2] Lacerda R G, Marques F C 1998 *Appl. Phys. Lett* **73** 617.
[3] Caruso F 2001, *Adv. Mat.* **13** 11.
[4] Qian H S, Han F M, Zhang B, Guo Y, Yue J, Peng B 2004, *Carbon* **42** 761.
[5] Jin Y Z, Gao C, Hsu W K, Zhu Y, Huczko A, Bystrzejewski M, Roe M, Lee C Y, Acquah S, Kroto H, Walton D R M 2005, *Carbon* **43** 1944.
[6] Cai P, Feng L 2008, *Mat. Chem. Phys.* **108** 1.
[7] Conrad J R 1987 *J. Appl. Phys.* **62** 777.
[8] Ueda M, Gomes G F, Berni L A, Rossi J O, Barroso J J, Beloto A F, Abramof E, Reuther H 2000 *Nucl. Instr. and Meth. in Phys. Res. B* **161** 1064.
[9] Valderrama E 2009 Ph.D thesis, Pontificia Universidad Católica de Chile.
[10] Zhou J, Fisher E R 2006 *J. Phys. Chem. B* **110**, 21911.
[11] Gordillo–Vzquez F J, Albella J M 2004 *Plasma Sources Sci. Technol.* **13** 50.
[12] McCulloch D G, Gerstner E G, McKenzie D R, Prawer S, Kalish R 1995 *Phys. Rev. B* **52** *850.*

Analysis Of The Different Zones Of Glow Discharge Of Ethyl Alcohol (C_2H_6O)

C. Torres[1*], P. G. Reyes[1], J. Mulia[1], F. Castillo[2] and H. Martínez[3]

[1*]Estudiante de Posgrado, Facultad de Ciencias, Universidad Autónoma del Estado de México, Instituto Literario 100, Centro, 50000, Toluca, Estado de México, México.
[1]Facultad de Ciencias, Universidad Autónoma del Estado de México, Instituto Literario 100, Centro, 50000, Toluca, Estado de México, México.
[2]Instituto de Ciencias Nucleares, Universidad Nacional Autónoma de México, Apartado Postal 70-543, 04510, México DF., México.
[3]Instituto de Ciencias Físicas, Universidad Nacional Autónoma de México, Apartado Postal 48-3, 62251, Cuernavaca, Morelos, México.

E-mail: pgrr@uaemex.mx

Abstract. The aim of this work is to explore the emission spectroscopy of ethyl alcohol in some regions, also is determine the result elements of the glow discharge, the spectrums were observed in a range of 200 at 1100 nm in the different zones inside of the tube at different distances of 20 and 30 cm. The elements are: in anode region C_6H_5 (483.02nm), CHO (519.56nm) and H_2 (560.47nm), in the positive column CO_2^+ (315.52 y 337.00nm), O^+ (357.48nm), CH^+ (380.61nm) and CO^+ (399.73nm); in the cathode region we observed O^+ (391.19nm), CHOCHO (428.00nm), CO^+ (471.12nm) and H_2 (656.52nm). C_6H_5, CHO y H_2 species occurring in all regions analyzed varying the glow discharge emission intensity.

1. Introduction

The study of low-pressure discharges in alcohol steam has been the subject of many investigations due to their importance in some medicals applications; that is why, it is very important in the development of new cleaning and polishing techniques from biomedicine. Also, the study of electronic and vibrational processes as well as electron impact ionization and recombination processes that can occur in such plasmas [1].

At the present moment of enormous development in material science, organic compounds are frequency used as reactive components in various plasma chemical or technological applications. Organic compounds are also used in the course of various analytical procedures and or are present as constituents in the analyses of natural samples. Thus, these species reach the gas plasma (either a chemical reactor or analytical plasma source), taking part in different elementary processes. Therefore, the understanding of the influence of such components on the physical and analytical characteristics of plasma is of essential importance in analytical atomic spectrometry, as well as in plasma chemistry and plasma processing. This effect has been investigated mainly in flames and inductively coupled plasma and seldom in DC plasmas [2, 3].

The main feature of plasma created in glow discharges is that the electrons and the other chemical species within the discharges are not in thermal equilibrium. In fact, for such low pressure glow discharges, electrons may acquire enough energy to cause molecular fragmentation through inelastic collisions [4]. Therefore, cold plasmas are extremely reactive and contain chemical species that would not be produced in thermal equilibrium. Thus the study of chemical kinetics in such discharges is important for the understanding of the possible pathways that occur in the synthesis of complex compounds produced during glow discharges.

The purpose of this work was reported the results of an optical emission spectroscopic study of an ethyl alcohol glow discharge.

2. Experimental setup

DC glow discharge plasma in C_2H_6O was generated in a plasma apparatus [5, 6] of the Advanced Physics Laboratory of de Science Faculty UAEMex, which is shown in figure 1. The system consisted of two copper circular plane electrodes, 3 mm thick and 30 mm in diameter. The electrodes are positioned at the center of the reaction chamber with 38 cm gap spacing. The ethyl alcohol steam is introduced by a flowmeter (MATHESON TRI-GAS) with 3 psig of constant flux; the pressure of the atmosphere (1.0 Torr) of ethyl alcohol in the tube was measured with a vacuum SenTorr (VARIAN VACUUM GAUGE Type 0536). The plasma chamber was initially evacuated to a base pressure of 1.0 x 10^{-3} Torr with a mechanic pump (ALCATEL Type 2002BB). The discharge power (SPELLMAN SA4) was maintained at an output of 338 V and a current of 8 mA (2.7 W), which was measured using a digital Fluke multimeter model 70 III. The active species generated in the glow discharge was studied by plasma emission spectroscopy. The spectrums were observed in a wavelength range of 200 at 1100 nm in different zones between the electrodes by a spectrometer (StellarNet EPP2000) equipped with a 300 lines-mm^{-1} holographic grating. The wavelength scan interval was 0.1 nm and the dwell time was 500 ms. The data were obtained in a single accumulation with a 0.5 s integration time. Also, to obtain the spectrum in cathode, we used an ACTON scanning monochromator (SpectraPro 2150i) equipped with a 1200 lines/mm holographic grating. The wavelength scan interval was 0.1 nm and the dwell time, 500 ms. The data were obtained in a single accumulation with a 0.5 s integration time. The spectrum (200 to 1100 nm) of the emission cell was recorded with a photo multiplier tube (HAMAMATSU model PD471).

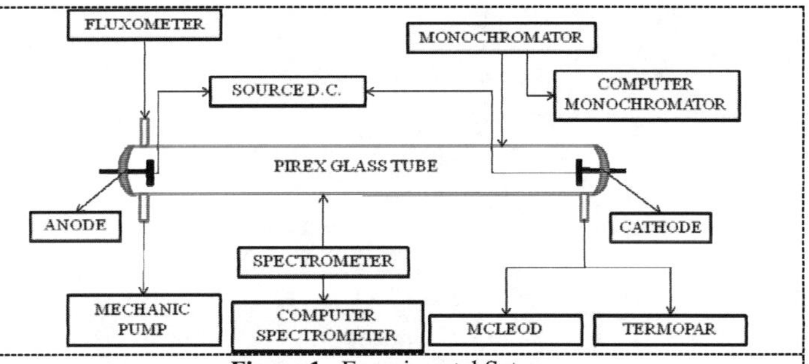

Figure 1. Experimental Setup.

3. Results and discussion

The analysis of the area between the cathode and anode was done by optical emission spectroscopic measurement of C_2H_6O glow discharge plasma at a pressure of 1.0 Torr and it is shown in Fig. 2. Identified species in the spectra in the cathode region were O^+ (391.19nm), CHOCHO (428.00nm), CO^+ (471.12nm), C_6H_5 (483.02nm), CHO (519.56nm) and H_2 (560.47 and 656.52nm), while at 20 and 30 cm from the cathode appear several species added to the observed in the cathode, that are: CO_2^+ (315.52 y 337.00nm), O^+ (357.48nm), CH^+ (380.61nm) and CO^+ (399.73nm), also we observed that the band of H_2 (656.52nm) vanished at these positions. Finally, on the anode region, there are only present the following species C_6H_5 (483.02nm), CHO (519.56nm) and H_2 (560.47nm).

Fig. 3 displays the intensity of the species observed between anode and cathode as a function of the distance from the cathode. The most important species observe at 20 and 30 cm from the cathode are O^+, CH^+ and CO_2^+, which vanished in the anode and cathode region. That may be due to they represent the luminous part of the discharge between the anode dark space and the edge of the Faraday dark space. It appears to exhibit a steady and uniform luminosity. The gas in this zone is in an ionized state and the species observed can be produced by electron impact and recombination processes that they are two of the basic processes in the discharge.

While, we observed less species in the anode zone due to the anode in the glow discharge does not emit any particle; it is merely collects the electrons. The voltage drop across this zone, the ions that are generated in this zone enter to the positive column. This is accomplished by the electrons emerging from the positive column and gaining energy in the anode fall, thus causing a light glow at the anode surface [7].

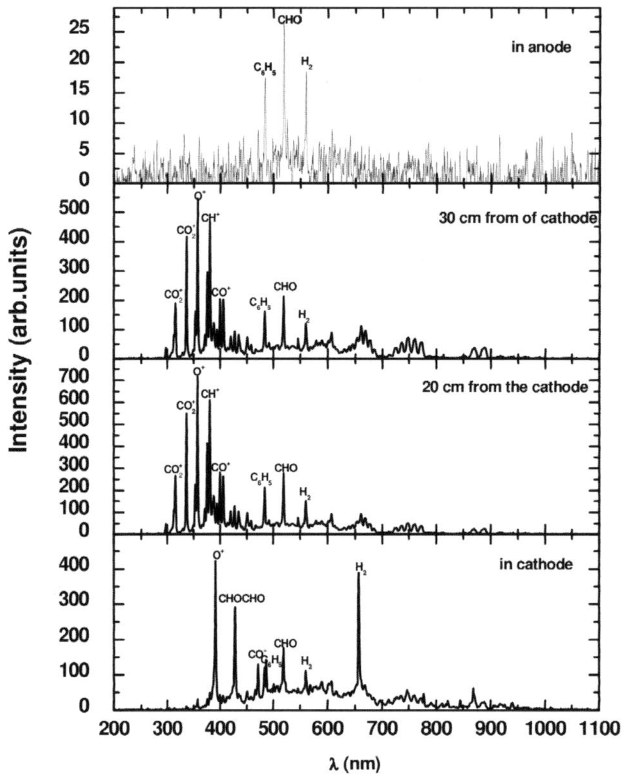

Figure 2. Spectroscopic emission C_2H_6O

Also, in the cathode zone present less species due to this zone of the discharge carries most of the voltage. The physical processes in this zone can be described as follows. Electrons are emitted from the cathode by two mechanisms: the bombardment of positive ions and incident radiation. Because their initial energy is low, the electrons from a thin sheath of electron space charge changes in which the current is carried mainly by the positive ions moving toward the cathode. As the electrons leave this thin sheath, they are accelerated in the high field in a region of net positive charge (positive column).

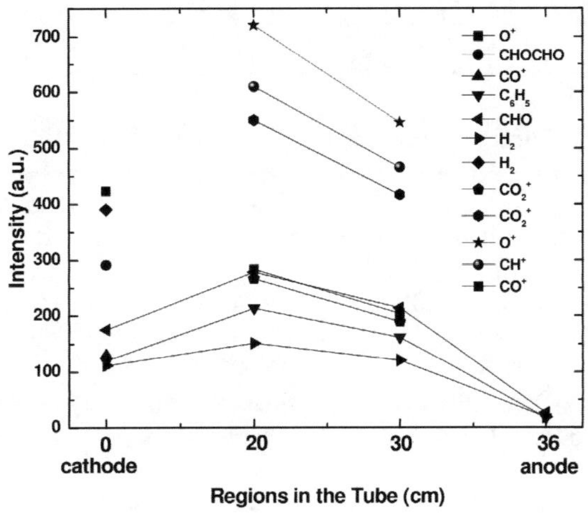

Figure 3. Intensity of species as a function of the distance from the cathode.

4. Conclusions

The experimental study mainly focused on measurements of optical emission spectroscopy of an ethyl alcohol glow discharge at a pressure of 1.0 Torr, a power of 2.7 W and a flow constant of 3.0 psig was studied. The main regions between electrodes were analyzed. It was observed several species that they are produced in different regions within the discharge tube, being the most intense: in anode region C_6H_5 (483.02nm), CHO (519.56nm) and H_2 (560.47nm), in the positive column CO_2^+ (315.52 y 337.00nm), O^+ (357.48nm), CH^+ (380.61nm) and CO^+ (399.73nm) and in the cathode region O^+ (391.19nm), CHOCHO (428.00nm), CO^+ (471.12nm) and H_2 (656.52nm). C_6H_5, CHO and H_2 species are generated in all analyzed regions, but with different intensity.

5. Acknowledgments

The authors are grateful to Nayeli Rodrìguez and José Rangel for technical assistance. This research was partially sponsored by UAEM 2648/2008U, student grant ENLACE-UAEM.

6. References

[1] G Cartry, L Magne and G Cernogora, J. Phys. D: Appl. Phys. 32, 1894–1907 (1999).
[2] L. F. Nascimento, et al, Planet. Space Sci. Vol. 46, N^0 8, 969-74, 1998.
[3] A. De Visscher, et al, Plasma Sources Sci. Technol. **17**, 015004 (11), 2008.
[4] M. Jimenez, C. Yubero and M. D. Calzada, J. Phys D: Appl. Phys. **41**, 175201 (6), 2008.
[5] P.G. Reyes, E. Mendez-Martinez, D. Osorio-González, F. Castillo, H. Martínez, Journal of Physics: Conference Series, 207, 012020 (2010).
[6] E. F. Méndez-Martínez, P. G. Reyes, D. Osorio-González, F. Castillo and H. Martínez, Plasma Science and Technology 12(3), 314319 (2010).
[7] F.F. Chen and J.P. Chang, in Lecture Notes on Principles of Plasma Processing, Plenum/Kluwer Publishers, 2002.

CHARACTERIZATION A LOW PRESSURE OF PLASMA OF METHANOL (CH_4O) ALCOHOL

M. Villa[1*], C. Torres[1], P. G. Reyes[1], D. Osorio[1], F. Castillo[2], H. Martínez[3]

[1*]Estudiante de Posgrado, Facultad de Ciencias, Universidad Autónoma del Estado de México, Instituto Literario 100, Centro, 50000, Toluca, Estado de México, México.
[1]Facultad de Ciencias, Universidad Autónoma del Estado de México, Instituto Literario 100, Centro, 50000, Toluca, Estado de México, México.
[2]Instituto de Ciencias Nucleares, Universidad Nacional Autónoma de México, Apartado Postal 70-543, 04510, México DF., México.
[3]Instituto de Ciencias Físicas, Universidad Nacional Autónoma de México, Apartado Postal 48-3, 62251, Cuernavaca, Morelos, México.

E-mail: pgrr@uaemex.mx

Abstract. The aim of this work is to explore the emission spectroscopy of Methanol alcohol plasma in some regions, also is determine the result elements of the glow discharge, the spectrums were observed in a range of 200 at 1100 nm in the different zones inside of the tube at different distances of 20 and 30 cm. The elements are: in anode region: C_7H_7 (451.06nm), C_6H_5 (483.02nm), CHO (519.56nm), H_2 (560.46nm), C (607.02nm), H_2 (661.46nm); cathode region: O_2 (391.04nm), CHOCHO (428.00nm), H_2 (656.52nm); to 20 and 30 cm region: O_2^+ (297.00nm), CO_2^+ (315.52nm), O^+ (357.48nm), C^+ (388.00nm).

1. Introduction

Conversion of hydrocarbon fuels to methanol will promote their efficient utilization, since methanol can easily be converted to hydrogen using low-temperature heat energy [1]. The conventional chemical process for methanol synthesis from CO and H requires high pressure and high temperature. A plasma chemical process is a possible alternative to the conventional methanol synthesis, since chemical reactions can be promoted at lower pressure and temperature conditions. In flue gas cleaning or decomposition of volatile organic compounds, nonthermal plasma process has been widely investigated, and promising results have been obtained [2]. There are, however, a few reports on direct methanol synthesis using nonthermal discharge plasma. It has reported that methanol synthesis can be made from CH and O by pulsed discharge plasma. Other used glow discharge or spark discharge in reduced pressure condition and produced methanol from CH and HO. This experimental result indicated an important advantage of nonthermal plasma process that can promote endothermic chemical reactions. Energy efficiency of methanol production in these previous works, however, was still insufficient to substitute the conventional method of methanol synthesis. Further studies are necessary to optimize the type of plasma and gas condition and to improve electrical power conversion efficiently and gas conditions. A combination of nonthermal plasma and catalyst might improve methanol production efficiency. Also, the methanol decomposition can be carried out more effectively under the conditions of cold plasma [3]. In this paper, the aim of this work is to study alcohols plasma, in order to know their characteristics, properties and their applications and try to improve or obtain new applications; first we have to design experiment, and then produce Methanol alcohol gas to realize glow discharge alcohol plasma and finally study atomic and molecular reaction in the process of plasma.

2. Experimental setup

DC glow discharge plasma in CH_4O was generated in a plasma apparatus [4, 5] of the Advanced Physics Laboratory of de Science Faculty UAEMex, which is shown in figure 1. The system consisted of a Pyrex glass tube of 48 cmm length and 4.7 cm diameter with two copper circular plane electrodes,

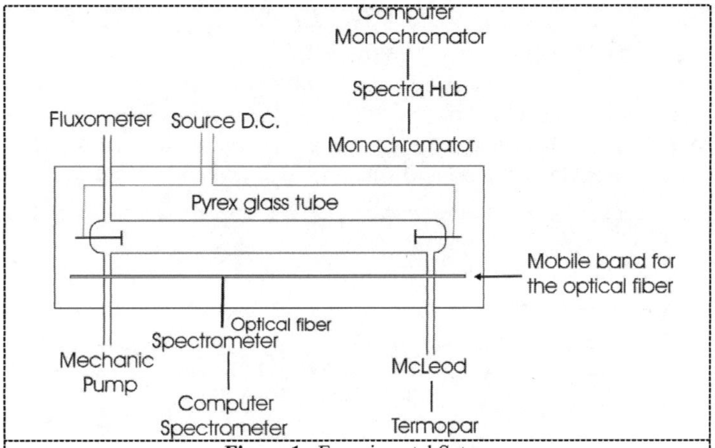

Figure 1. Experimental Setup.

3 mm thick and 30 mm in diameter. The electrodes are positioned at the center of the reaction chamber with 38 cm gap spacing. The methanol alcohol steam is introduced by a flowmeter (MATHESON TRI-GAS) with 3 psig of constant flux; the pressure of the atmosphere (1.3 Torr) of methanol alcohol in the tube was measured with a vacuum SenTorr (VARIAN VACUUM GAUGE Type 0536). The plasma chamber was initially evacuated to a base pressure of 1.0 x 10^{-3} Torr with a mechanic pump (ALCATEL Type 2002BB). The discharge power (SPELLMAN SA4) was maintained at an output of 338 V and a current of 8 mA (2.7 W), which was measured using a digital Fluke multimeter model 70 III. The active species generated in the glow discharge was studied by plasma emission spectroscopy. The spectrums were observed in a wavelength range of 200 at 1100 nm in different zones between the electrodes by a spectrometer (StellarNet EPP2000) equipped with a 300 lines-mm^{-1} holographic grating. The wavelength scan interval was 0.1 nm and the dwell time was 500 ms. The data were obtained in a single accumulation with a 0.5 s integration time. Also, to obtain the spectrum in cathode, we used an ACTON scanning monochromator (SpectraPro 2150i) equipped with a 1200 lines/mm holographic grating. The wavelength scan interval was 0.1 nm and the dwell time, 500 ms. The data were obtained in a single accumulation with a 0.5 s integration time. The spectrum (200 to 1100 nm) of the emission cell was recorded with a photo multiplier tube (HAMAMATSU model PD471).

3. Results and discussion

The analysis of the area between the cathode and anode was done by optical emission spectroscopic measurement of CH_4O glow discharge plasma at a pressure of 1.0 Torr and it is shown in Fig. 2. Identified species in the spectra in the cathode region were: O_2 (391.40nm), CHOCHO (428.00nm), C_7H_7 (451.06nm), C_6H_5 (483.02nm), CHO (519.56nm), C (607.02nm), H_2 (656.52nm), while at 20 and 30 cm from the cathode appear, it observed the same species but with more intensity, the specie O_2

(391.4nm) and less intensity the specie H_2 (656.52nm), with respect cathode region. Finally, on the anode region, there are present the same species without the O_2 (391.4nm) specie.

Fig. 3 displays the intensity of the species observed between anode and cathode as a function of the distance from the cathode. The most important species observe at 20 and 30 cm from the cathode are CHO, C_6H_5, H_2 and C^+, which decreases their intensity in the anode and cathode region. That may be due to they represent the luminous part of the discharge between the anode dark space and the edge of the Faraday dark space. It appears to exhibit a steady and uniform luminosity. The gas in this zone is in an ionized state and the species observed can be produced by electron impact and recombination processes that they are two of the basic processes in the discharge.

While, we observed less species in the anode zone due to the anode in the glow discharge does not emit any particle; it is merely collects the electrons. The voltage drop across this zone, the ions that are generated in this zone enter to the positive column. This is accomplished by the electrons emerging from the positive column and gaining energy in the anode fall, thus causing a light glow at the anode surface [6].

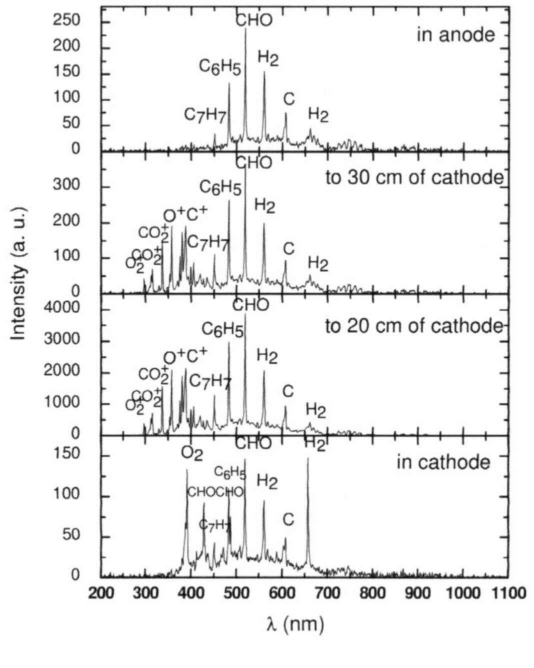

Figure 2. Emission spectra.

Also, in the cathode zone present less species due to this zone of the discharge carries most of the voltage. The physical processes in this zone can be described as follows. Electrons are emitted from the cathode by two mechanisms: the bombardment of positive ions and incident radiation. Because their initial energy is low, the electrons from a thin sheath of electron space charge changes in which the current is carried mainly by the positive ions moving toward the cathode. As the electrons leave this thin sheath, they are accelerated in the high field in a region of net positive charge (positive column).

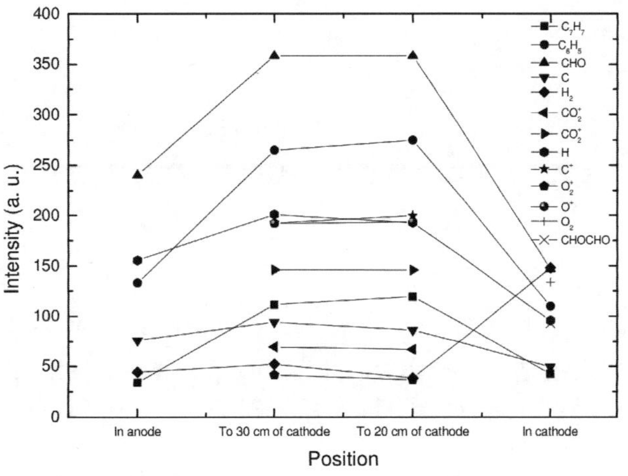

Figure 3. Intensity of species as a function of the distance from the cathode.

4. Conclusions

The experimental study mainly focused on measurements of optical characterization of a methanol alcohol glow discharge at total pressure of 1.0 Torr, a power of 42.25 W and a flow constant of 3.0 psig was studied. The experiment was realized observing the emission from the DC plasma was scans in the range of 200-1100 nm; the main regions between electrodes was been analyzed. It was observed that there are lines and bands that correspond to identical species in different regions within the discharge tube, being the most intense are: O_2 (391.40nm), CHO (519.56nm), C_6H_5 (483.02nm), H_2 (656.52nm), CHOCHO (428.00nm), C_7H_7 (451.06nm), C (607.02nm), H_2 (560.46nm).

5. Acknowledgments

The authors are grateful to Nayeli Rodríguez and José Rangel for technical assistance. This research was partially sponsored by UAEM 2648/2008U, student grant ENLACE-UAEM.

6. References

[1] Tetsuya Akitsu, et al, Surface & Coating Technology, **193**, 29-34, 2005.
[2] Mamoru Okumoto, et al, IEEE Trans. on Industry Appl. Vol. **34**, N^0 5, 940-44, 1998.
[3] Mamoru Okumoto and Akira Mizuno, Catalysis Today, **71**, 211-17, 2001.
[4] P.G. Reyes, E. Mendez-Martinez, D. Osorio-González, F. Castillo, H. Martínez, Journal of Physics: Conference Series, 207, 012020 (2010).
[5] E. F. Méndez-Martínez, P. G. Reyes, D. Osorio-González, F. Castillo and H. Martínez, Plasma Science and Technology 12(3), 314319 (2010).
[6] F.F. Chen and J.P. Chang, in Lecture Notes on Principles of Plasma Processing, Plenum/Kluwer Publishers, 2002.

Investigation of a klystron with a pseudospark-sourced electron beam

D Bowes[1], H Yin[1], A W Cross[1], W He[1], A D R Phelps[1], K Ronald[1], X Chen[2], D Li[2]

[1] SUPA Department of Physics, University of Strathclyde, Glasgow, G4 0NG, UK
[2] Department of Electronic Engineering, Queen Mary University of London, London, E1 4NS, UK

E-mail: david.bowes@strath.ac.uk

Abstract. Investigation has been undertaken at the University of Strathclyde to simulate, design and construct a klystron amplifier with an operating frequency of 94 GHz. A novel beam source shall be utilized in the form of a pseudospark discharge, a form of low-pressure, high-current plasma discharge which produces an electron beam possessing high brightness as well as self-focusing properties. The seed signal for the klystron will be fed in using a tapered dielectric-lined iris coupling system.

1. Introduction

Extremely high frequency (EHF) radiation (30-300GHz) has wide potential applications in both scientific research and commercial use, including plasma diagnosis, medical imaging and advanced communications, etc. All these applications require compact and inexpensive radiation sources. The klystron is an excellent choice for EHF generation, due to its high gain, high efficiency and robustness as well as the fact that it may be scaled in size in order to achieve higher frequency operation [1]. Due to the decrease in size as the frequency is increased, there is a need for the electron beam current density to increase in order to achieve reasonable output powers. A pseudospark discharge is a viable possibility in this respect, due to the emitted linear beam's characteristic properties such as high current density and high brightness as well as being self-focusing during its propagation. A pseudospark driven microklystron will be a compact and simple device as no external magnetic guiding field is required.

2. Pseudospark beam experiments

A pseudospark is an axially symmetric, self-sustained, transient, low pressure (typically 50-500 mTorr) gas discharge in a hollow cathode / planar anode configuration with the potentially useful property of the formation of an electron beam during the breakdown process [2-3]. During a pseudospark discharge, low temperature plasma is formed as an abundant electron and ion source and may be classed as a low work function surface that facilitates electron/ion extraction by voltage application at different polarities. A 10kV voltage was applied to a pseudospark of configuration: 1mm cathode aperture, 1mm anode aperture and 6mm gap separation as shown in figure 1. The anode was grounded and there was no external applied guide magnetic field.

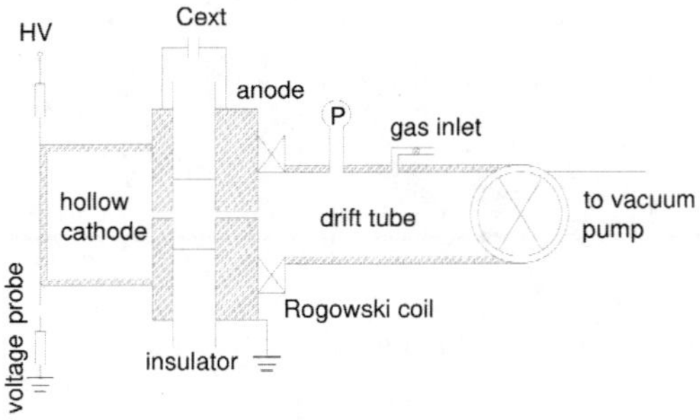

Figure 1: Experimental setup of pseudospark beam experiment.

At a pressure of 100mTorr, an output current of 4 A was measured as shown in figure 2, which demonstrates an output current of a similar scale to that observed in previous 3 mm aperture experiments [4-10]. This also shows the possibility to further scale down the pseudospark to become a micron sized discharge to drive a microklystron.

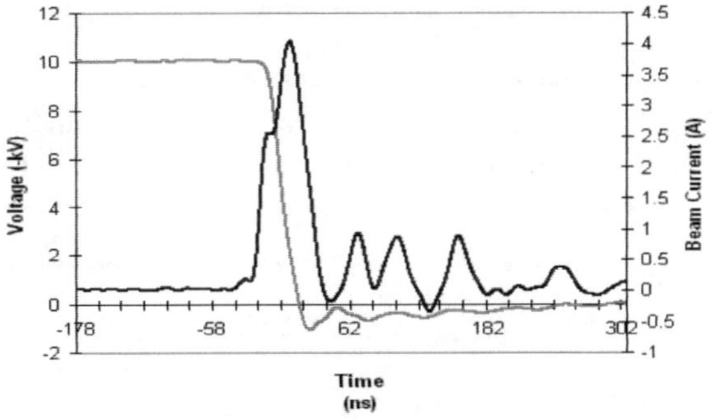

Figure 2: Current and voltage traces of single-gap pseudospark discharge.

3. Klystron simulations

The klystron is a linear-beam vacuum device which operates via using an input RF signal in order to velocity modulate a beam progressing along a length of drift tube. This seed signal induces signal amplification at the cavity resonant frequency coupled to the beam which may then be extracted via an output cavity further along the drift tube.

A multi-cavity klystron with an operating frequency of 94 GHz was designed and simulated using the particle-in-cell (PiC) code MAGIC. An 8 kV electron beam of 15 mA dc current modulated with an input RF signal of 200 mW was focused by a 0.6 T magnetic field. MAGIC-2D results revealed a strong amplification signal with a peak power output of 19.3 W at a millimetre wave input power of 200 mW, corresponding to a device gain of 19 dB and efficiency of 16%.

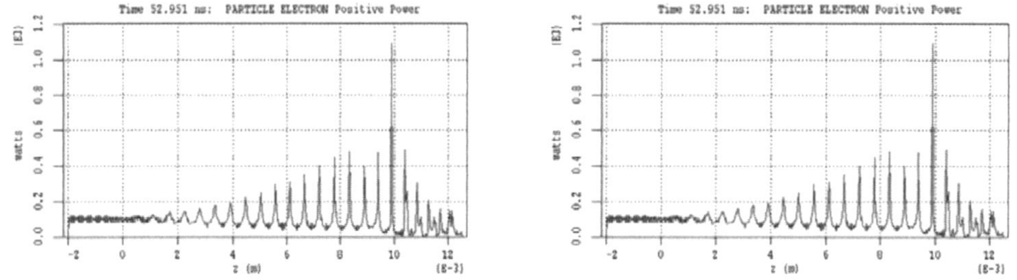

Figure 3: Electron positive power along length of beam.

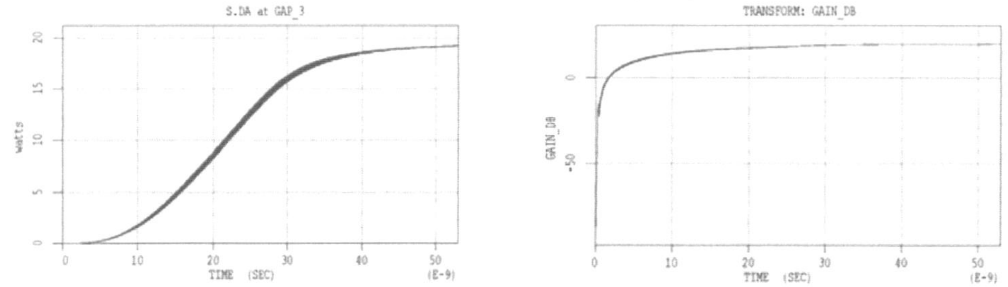

Figure 4: MAGIC-2D simulated output power, left, and gain.

Simulations were also performed using a modified current and voltage pulse in an attempt to better model device performance with a realistic pseudospark pulse. Voltage was stepped down to an average of 8 kV for 15ns before dropping to zero, whilst a current pulse of peak current 15 mA was timed to coincide with the voltage step, as shown in figure 5.

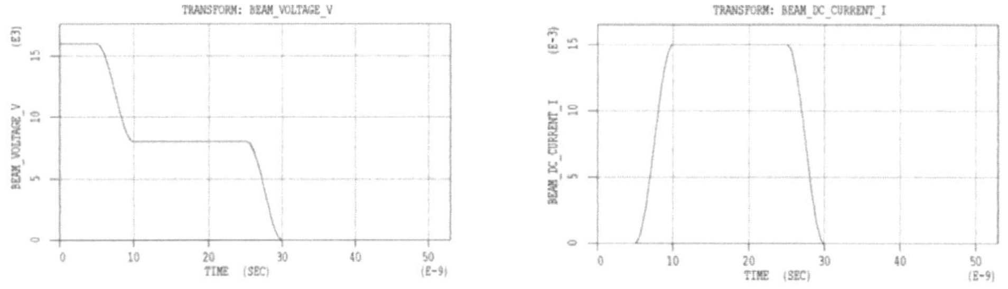

Figure 5: Pulsed current and voltage traces, 15 ns duration.

By varying the device interaction length a peak output power of 8.86 W was observed, coinciding with a device efficiency and gain of 7.4% and 16.5 dB respectively.

4. Coupler design

Owing to the precise dimensions involved with high-frequency klystron design and manufacture, coupling was an issue due to the cavities possessing dimensions significantly smaller than waveguide of corresponding frequencies. A solution has been proposed via coupling the input and output cavities to a tapered length of waveguide, gradually increasing to standard waveguide dimensions. In order to compensate for the difference in cutoff frequency, the taper is lined with a dielectric material, thus altering the effective cutoff frequency along the length of the coupler. The proposed structure was then modeled in CST Microwave Studio.

5. Summary and ongoing work

A three-cavity klystron which operates at a frequency of 94 GHz, using a pseudospark discharge as a beam source, is currently under investigation. Through the use of the EM modeling packages MAGIC-2D and CST Microwave Studio, it has been possible to construct a realistic model of device performance. Manufacturing and experimental measurements shall begin soon on the 94 GHz klystron, as well as testing of the coupling system and the use of a puffed-gas input system for pseudospark discharges.

References

[1] Siegel P H 2002 *IEEE T Microw. Theory* **50** 910
[2] Gunderson M A and Schaefer G 1990 NATO ASI Ser. B, New York: Plenum
[3] Frank K and Christiansen J 1989 *IEEE Trans. Plasma Sci.* **17** 748
[4] Yin H, He W, Cross A W, Phelps A D R and Ronald K 2001 *J. Appl. Phys.* **90** 3212
[5] Yin H, Cross A W, Phelps A D R, Zhu D, He W and Ronald K 2002 *J. Appl. Phys.* **91** 5419
[6] Yin H, Cross A W, He W, Phelps A D R, Ronald K, Bowes D and Robertson C W 2009 *Phys. Plasmas* **16** 063105
[7] Cross A W, Yin H, He W, Ronald K, Phelps A D R and Pitchford L C 2007 *J. Phys. D Appl. Phys.* **40** 1953
[8] H. Yin, A.W. Cross, W. He, A.D.R. Phelps, and K. Ronald, 2004 *IEEE Transactions on Plasma Science,* **32***,Special Issue on Pseudospark Physics and Applications,* 233-239.
[9] H. Yin, A. W. Cross, A. D. R. Phelps, W. He and K. Ronald, 2004 *Nucl. Instr. and Meth. in Phys. Res. A*, **528** 378
[10] A.W. Cross, H. Yin, D Bowes, W. He, K. Ronald and A.D.R. Phelps, 2009 *NewDevelopments in Nonlinear Plasma Physics, Proceedings of the 2009 ICTP Summer College on Plasma Physics and International Syposium on Cutting Edge Plasma Physics, pp380*

Neutron emission characterisation at the FN-II Dense Plasma Focus

F. Castillo-Mejía, I. Gamboa-de Buen, J.J.E. Herrera-Velázquez and José Rangel-Gutiérrez

Instituto de Ciencias Nucleares, Universidad Nacional Autónoma de México
A. P. 70-543, Ciudad Universitaria, 04511 Coyoacán, D.F., México

ciro@nucleares.unam.mx, herrera@nucleares.unam.mx

Abstract. Plasma foci are efficient plasma based neutron sources, when deuterium is used as the filling gas. The dense plasma focus FN-II is a small device (4.7 kJ), in which the emission of deuterium fusion neutrons (2.45 MeV) are studied. The system produces an average neutron yield of (5.3 ± 0.5) x 10^8 neutron/shot in 4π sr at ~ 350 kA peak discharge current and 2.75 torr deuterium operation. Three methods are currently used; silver activation counters and CR-39 nuclear track detectors, for time integrated and angular distribution studies, and BC 400 scintillators coupled to photomultiplier tubes for spectra studies. In the latter case, we compare signals due to neutron reflections in the laboratory with those obtained with a collimated beam in a paraffin shielded detector. Regarding the angular distribution of the neutron emission, it has been found to have isotropic and anisotropic components, the former giving the largest contribution. Also, the neutron spectrum, measured at 90° of the axis device, is broadened, peaking at energies slightly larger than 2.45 MeV. These can be interpreted as the consequence of coexisting neutron generation mechanisms, which will be discussed in this work. The correlation between neutron and hard X-ray yields is also discussed, and a possible interpretation in terms of different neutron generation mechanisms is attempted. There has also been a dosimetric study of the laboratory with TLD dosimeters, which will be presented in this paper.

1. Introduction

The mechanisms of neutron production in a plasma focus device are usually interpreted by both thermonuclear and non-thermonuclear (beam-target) mechanisms [1-6]. Several methods have been developed for the determination of the neutron generation mechanisms, including measurements of the neutron yield and its anisotropy [7-11] and the energy spectra of neutrons emitted from the plasma focus device [12-14]. Among these methods, the latter has advantages compared to the others. It provides information on the production mechanisms of neutrons and the energy distribution of reaction ions. The standard time of flight (TOF) spectroscopic method allows the determination of the emitted neutron energy and the number of the particles emitted from the source with high accuracy. Since the neutrons in question are not relativistic, in a pulsed neutron source their energy can be directly determined from $E = 1/2mv^2$, where the velocity v is obtained from the distance between the source and the detector, and time interval between the shot and the time at which they are recorded at the detector. This method can be used when the emitting source consists in a short pulse, even if there is a broad energy distribution, so long as the average energy of the neutrons is not too large.

2. Experimental setup

This work was performed using the FN-II small plasma focus device, operated at the Instituto de Ciencias Nucleares, UNAM, whose details have been described in Ref. 5. It was operated at 37 kV with 4.8 kJ stored energy. The anode is made of oxygen-free copper, 40 mm long, with a 50 mm diameter. The co-axial cathode is formed by twelve copper rods, 8 mm diameter each, arranged on a squirrel cage configuration around a circle of 100 mm diameter. The insulator is an annular *Pyrex®* tube, 12 mm long, located at the base of the anode. As reported in Ref. 9, the neutron yield and its uniformity increased considerably when the diameters of the insulator and the inner electrode matched, probably due to an improvement of the current sheath breakdown, aided by the edge effect of the electrode close to the insulator. The energy storage is provided by four 1.863 μF capacitors in parallel, and the discharge is triggered by a simple mushroom electrode spark gap. The device was operated in its neutron optimized regime, corresponding to a 2.7 torr pure deuterium gas pressure [9]. In this regime a peak focus current of 350 kA and an average yield of 1.6×10^8 neutrons per shot have been routinely obtained.

The neutron absolute yield for each shot is measured by two Geiger-Muller (GM) activation counters. These are calibrated by counting in the saturation method [15-16] using an Am-Be neutron source with 2×10^{-6} ns^{-1} intensity. Five detectors are used for neutron time of flight measurements and hard X-ray monitoring. They are placed at different distances from the device. They consist in cylindrical *Bicron©* BC-400 scintillators coupled to *Hamamatsu©* R1250, No. RA2457 photomultipliers, powered by two C4840 and three HTV C762-01 *Hamamatsu©* power supplies, through *Ortec©* 269 photomultiplier bases. The electronics, and the characteristics of each detector are as similar as possible. They are shielded from light and electromagnetic noise by custom made aluminum casing. Their time resolution is in the order of 5 ns. The signals of the detectors are displayed on two 300 MHz and one 500 MHz Tektronix oscilloscopes, along with the current derivative signal of the plasma focus circuit, obtained with a Rogowski coil.

3. Results and discussion

For the set of shots analysed in this work, the neutron average yield, as measured by GM-activation counters is about $(5.3 \pm 0.5) \times 10^8$ neutron/shot into 4π sr and the anisotropy (the neutron yield at 90° with respect to the neutron yield at 20° with respect to the device axis) is found to be 1.15.

The time difference between the x-ray peak and the neutron signals, at a given distance is defined as the reference time for signal processing, taking into account a correction due to the time of flight of the hard X-rays. Also, the peak current derivative signal (dI/dt) can be selected as the time reference for the neutron signal time analysis.

All detectors show the separation of the neutron and x-ray signals properly, as shown in Figure 1. At short distances, the x-ray and neutron signals may overlap each other. The environmental effects will change the detectors' signals, especially in the case of the detectors placed farther away. Neutron signals are recognized by the TOF between the anode and each detector.

3.1 Shielding

A problem that is always present, and is actually exacerbated in cases when the neutron yield is particularly high, is the effect of X-ray and neutron scattering by the laboratory. In fact, when the yield is particularly high, the X-ray peak widens and overlaps the neutron peak in such a way that the correspondence between the yield obtained from the GM-counters and that obtained from the integrated neutron signals is lost. It would therefore be strongly desirable to be able to shield the detectors from this effect. In Fig. 2 we show typical examples of some attempts in this direction. The three signals belong to the same shot (9726). The first signal on top left is for an unshielded detector. It can be seen how the X-ray signal slightly overlaps the neutrons one. The signal on top right, comes from a detector fully surrounded by blocks of paraffin, 8 cm. thick, with a 5 cm. hole, in order to collimate the neutron beam. As seen, this attempt is unsuccessful; rather than obtaining a collimate beam, the neutrons are thermalised by the paraffin, and the neutron signal is smeared. In the third signal, bellow, the detector is shielded by blocks of lead, 5 cm. thick. This shows to be a good means of reducing the hard X-rays, and allows for good time-of-flight measurements, since the difference between the first X-ray peak and the arrival of the first neutrons is clearly seen.

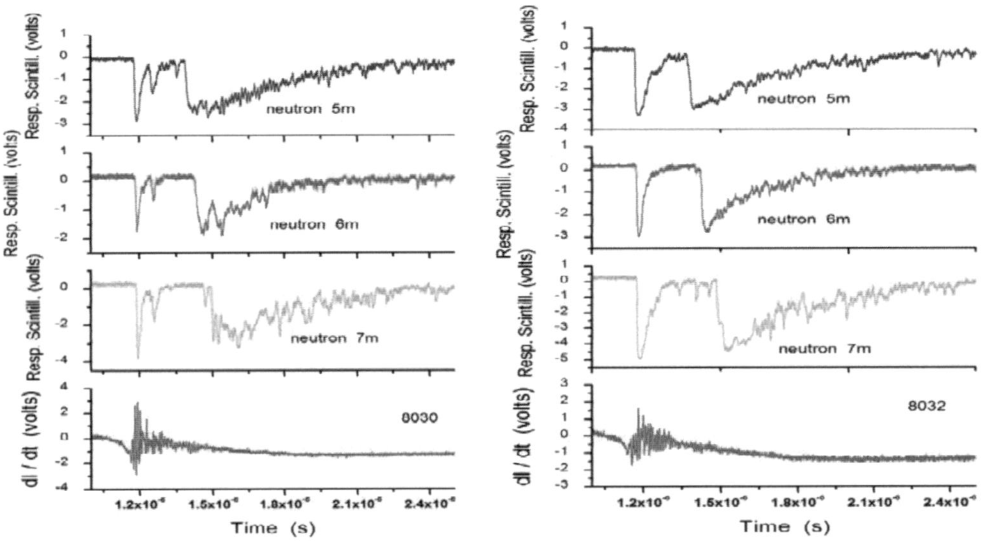

Figure 1. Scintillator-photomultiplier signals at 5, 6 and 7 m from the source, along with the Rogowski coil (current derivative) signal. The shot on the left shows the case of a double pinch. While it remains well resolved for the hard X-rays with increasing distance, the double neutron peak is smeared at 7 m.

3.2 Neutron spectra

Since deuterium is being used as filling gas, the fusion neutrons obtained should have 2.45 MeV energies, coming from a thermonuclear source. By taking the time difference between the first X-rays and the first neutrons to reach the detector, it is possible to have a good estimate of the neutrons energy. Fig 3 shows histograms of the energies obtained from a series of shots, for detectors placed at 5, 6 and 7 m. from the source. The peaks are clearly placed at 2.45 MeV, which probably indicates a strong thermonuclear component. However, the energies are widely spread form 1.7 up to 3.2, and in some case, up to 4 MeV. These deviations need to be explained by means of the existence of a strong beam-target effect. Since the detectors are placed at 90° from the axis of the plasma column, in this case we are not talking about the beam-target effect due to axial acceleration of deuterons, which gives rise to the well known anisotropy in these devices. This means there must be an azimuthal acceleration of deutrons, so the spread in energies is due to Doppler shifts from collisions from both approaching and receding accelerated ions, with background ions. The fact that the peak still remains close to the expected 2.45 MeV energy is consistent with the earlier finding that the neuron yield has both isotropic and anisotropic components. At 90° the isotropic component, which might be due to thermonuclear effects, would be expected to prevail.

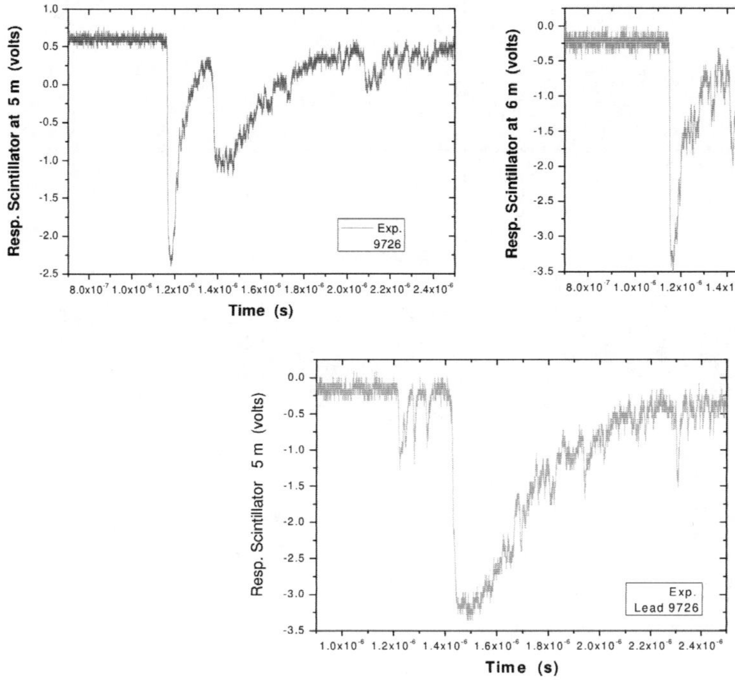

Figure 2. Signals from three different detectors with different shielding, for shot 9726. The signal on top left is from an unshielded detector. The one on top right is from a detector shielded with 8 cm. thick paraffin with a 5 cm. hole. The idea was to have a collimated neutron beam, but the neuron signal is actually smeared by thermalisation. The signal in the bottom is from a detector shielded by 5 cm. thick lead blocks. The hard X-ray peak is successfully reduced, allowing better conditions for time-of-flight measurements.

3.3 Dosimetric measurements

For safety reasons, dosimetric measurements have been made, placing thermoluminiscent dosimeters (TLD) at five different locations in the laboratory, both close to the device, and away from it, behind a 80 cm. thick concrete brick wall. Both TLD-600 and TLD-700 dosimeters have been used. The results are shown in Table 1. The doses are found to be rather small, but above the background radiation. Further study of this should be made.

	Gamma rays	Neutrons	
Package (distance m)	Dose (mGy)	Equivalent dose (mSv)	Equivalent dose per shot (mSv/shot)
1 (2.54)	0.23	5.65	0.019
2 (2.40)	0.21	4.95	0.017
3 (4.81)	0.20	3.06	0.010
4 (7.62)	0.21	3.80	0.013
5 (7.10)	0,20	3.24	0.011

Table 1. Dosimetric measurements using TLD-600 and TLD-700 dosimeters placed at different locations in the laboratory and distances from the source.

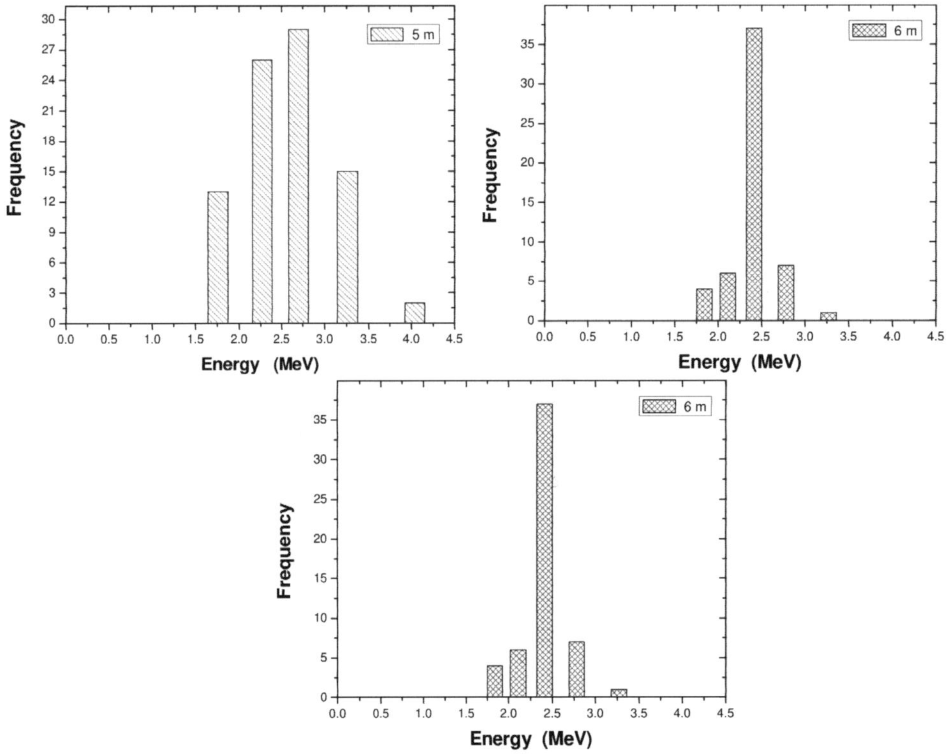

Figure 3. Neutron energies measured in detectors at 5, 6 and 7 m from the source. While the peak of the 2.45 MeV neutrons is clearly seen in the 6 and 7 m cases, the energies are widely spread between 1.7 and 3.2 MeV for some of the shots.

4. Conclusions

The use of the time-of-flight technique is very useful for a better understanding of the neutron spectra, and therefore the neutron generation mechanisms in plasma focus devices. However, some technical problems arise, which need to be addressed. The main one is that the neutron peak comes both from neutrons which travel directly from the source to the detector, as well as from neutrons scattered by the laboratory. Therefore, in this work only the first neutrons (the beginning of the neutron peak) are used. Even in of only the early neutrons are used, the X-ray peak tends to overlap the neutron one, specially in the cases of high performance shots. This requires the use of shielding. Shielding with paraffin in order to collimate the incoming neutrons has turned out to be unsuccessful, since the paraffin tends to thermalise the scattered neutrons, rather than stopping them. Probably boron should be added to obtain the necessary effect. On the other hand, lead is useful to reduce the X-ray peaks, and to allow good time-of-flight measurements.

It was found that at 90° the early neutrons are mostly around the 2.45 MeV energy, although in a few cases lower and higher energies are obtained, revealing azimuthal beam-target effect.

From dosimetric measurements, it was found that the radiation levels in the laboratory, while relatively small, are above the background radiation. It may be important to conduct similar studies in other plasma foci laboratories, in order to asses their safety.

Acknowledgement
This work is supported by DGAPA-UNAM grant IN120409.

References

[1] Castillo Mejia F, Milanese M, Moroso R and Pouzo J, *J. Phys. D: Appl. Phys.* **33**, 14 (2000).
[2] Hübner K, Bruhns H and Steinmetz K, *Phys. Lett. A* **69**, 269 (1978).
[3] Bernard A, et al., *Phys. Fluids* **18**, 180 (1975).
[4] Jager U and Herold H, *Nucl. Fusion* **27**, 407-23 (1987).
[5] Vikhrev V V, *Sov. J. Plasma Phys.* **12,** 262 (1986).
[6] Trubnikov B A, *Sov. J. Plasma Phys.* **12**, 271 (1986).
[7] Castillo Mejia F, Milanese M, Moroso R and Pouzo J, *J. Phys. D: Appl. Phys.* **30**, 1499 (1997).
[8] Czaus K, et al., *Nukleonika* (Suppl.) **46**, 77, (2001).
[9] Castillo F, et al., *Brazilian J. Phys.* **32**, 3, (2002).
[10] Aliaga-Rossel R and Choi P, *IEEE Trans. Plasma Sci.* **26**, 4, 1998.
[11] Castillo F, Herrera J J E, Rangel J, Milanese M, Moroso R, Pouzo J Golzarri J I ad Espiniosa, G, *Plasma Phys. Control. Fusion* **45**, 289 (2003).
[12] Brzosko J S, Robouch B V, Klobukowska J, *Fusion Technol.* **12** 71 (1987).
[13] Bostick W H, Kilic H, Nardi V, Powell C W, Nuclear Fusion, **33** 413 (1993).
[14] Johnson M G, et al., *Nuclear Instrum. and Meth.* **591**, 417 (2008).
[15] Slaughter D R and Pickles W L, *Nucl. Instrum. Methods* **160**, 87 (1979).
[16] Gentilini A et al., *Nucl. Instrum. Methods* **172**, 541 (1980).

Journal of Physics: Conference Series **511** (2014) 012022 doi:10.1088/1742-6596/511/1/012022

Soft X-Ray Emission Analysis Of A Pulsed Capillary Discharge Operated In Nitrogen

MP Valdivia, JC Valenzuela, ES Wyndham, M Favre, H Chuaqui, and H Bhuyan

Facultad de Física, Pontificia Universidad Católica de Chile, Ave. Vicuña Mackenna 4860, Santiago, Chile.

E-mail: mpvaldi1@uc.cl

Abstract. We present results from a pulsed capillary ns discharge source, operated in Nitrogen and N/He mixtures, in an alumina capillary 2.1mm long with outer diameter of 6.3mm and inner diameter of 1.6mm. The electrical energy stored is 0.5J with peak current of 6kA. Fast charging from an IGBT based pulsed power circuit allows operation at 35-600 Hz with voltages in the range of 18-24kV. Characteristic time-integrated N/He spectra were recorded and analyzed for values of 20-200 Å, with clear evidence of He-like Nitrogen emission at 28.8Å, which represents a possible source for water window soft x-ray microscopy. Filtered diode measurements reveal the influence of axial electron beams, generated by hollow cathode dynamics, on the x-ray emission in the range of 300-450 eV. We discuss optimal voltage applied and pressure conditions for soft x-ray generation. Time-integrated MCP images of a filtered slit-wire system delivered clear evidence of full wall detachment with $\sim 500\mu m$ in radial size for the entire emission range and $\sim 200\mu m$ for the emission in the 300-450 eV range.

1. Introduction

Over the last decade capillary discharges have been studied as bright soft x-ray sources [1] [2] [3]. This type of plasma discharge, emitting in the water window range (280 - 540 eV), as a source of He-like Nitrogen emission at 2.88 nm, is a candidate for "table-top" soft x-ray microscopy. In the present work, x-ray emission spectra from the plasma are analyzed to find the optimal conditions for Nitrogen He-like emission. In order to optimize the discharge, parameters such as voltage, frequency, and pressure, are varied. The axial production of intense electron beams, generated prior to breakdown by the hollow cathode effect, has been studied previously [4]. The soft x-ray output is analyzed considering different electrode configurations, which strongly influence the parameters of the x-ray emitting plasma. To distinguish between electron beams, produced by hollow cathode effect and fluorescence, and x-ray emission, we use a set of magnets and collimators, which avoids confusion in some of the diagnostics techniques employed, as it has been noticed before [3]. Here the use of filtered MCP slit-wire imagery allows an approximation of the maximum energy of our source, as well as an estimate in size for different radiation ranges.

2. Experimental details

The Pulsed Capillary Discharge system uses as energy storage a parallel plates capacitor. This system is composed of two circular plates, filled with flowing water, which serves both as coolant and dielectric. The total capacitance of the system is approximately 1.6nF. The discharge takes place inside an alumina capillary, 2.1mm long with outer diameter of 6.3mm and inner diameter of 1.6mm. The charging circuit delivers approximately 0.5J directly to one capacitor plate in less than 1μs, delivering voltages of 18-24kV. This charging circuit has been discussed in detail previously [5]. The upper half of the capacitor plates and alumina capillary are shown in Figure 1. In addition, Figure 2 shows the general diagram of the capillary system, including the placement for diagnostics used in the present study.

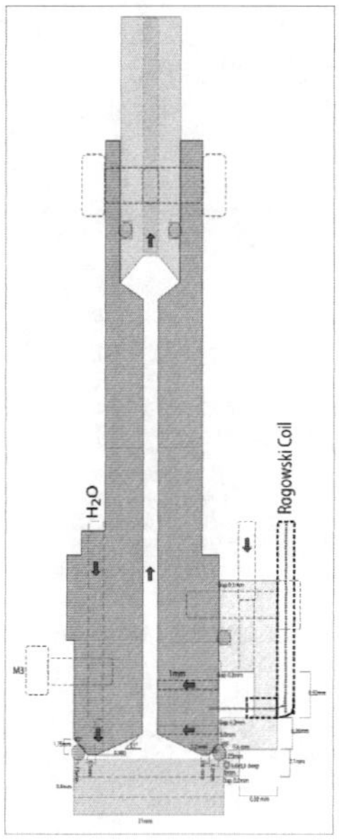

Figure 1. Capacitor and capillary

The gas mixture is fed from cathode to anode maintaining a constant flow of gas by pumping through the anode side. Pressure is measured both at the anode (ground) and cathode side. The system is mounted on standard KF fittings to allow mounting of a wide variety of diagnostics. Voltage is measured by an external resistive compensated monitor and current is measured by a single groove Rogowski coil. Diagnostics are placed on axis at the anode side. The diagnostics consist of a negatively biased Faraday cup, which doubles as a bare XRD detecting electrons and x-rays, filtered wide-band planar diodes, and an EUV Rowland circle spectrometer with CCD interface, which records spectra in the range of 20-400Å.

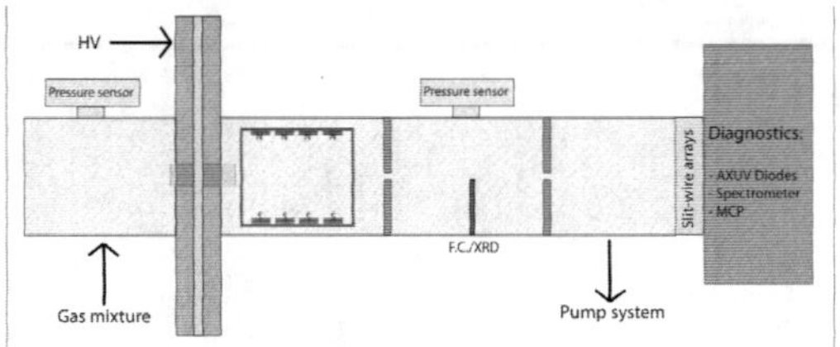

Figure 2. Capillary system diagram

3. Results and discussion

3.1. Slit-wire MCP images

Using the intensity obtained from MCP slit-wire images (100, 120, 150 and 200μm wires) and measuring intensity profiles [5], we obtain a value of under 1mm in diameter size for the case of an unfiltered image, which clearly shows full wall detachment. When filtering the system with 1μm of Ti, a diameter of under 400μm is found. We conclude the hotter plasma emitting in the desired energy range is located on axis, as well as the cooler plasma. The cooler plasma may surround the hotter plasma or take place at different time ranges, either before and/or after the more energetic emissions. This remains to be determined by further analysis.

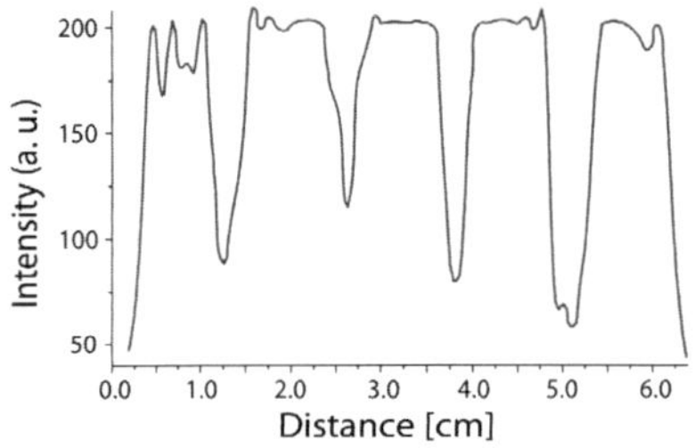

Figure 3. Intensity profile for unfiltered image

3.2. Time-integrated spectra

We present comparative spectra for a N/He mixture. Total intensity of emission increases with voltage applied, but as voltage increases, so do impurity lines from wall ablation. The Oxygen lines begin to rise while Nitrogen lines decrease, compared to those of Oxygen, as voltage reaches 24kV. This determines the maximum value of voltage to be used if impurities are to be avoided.

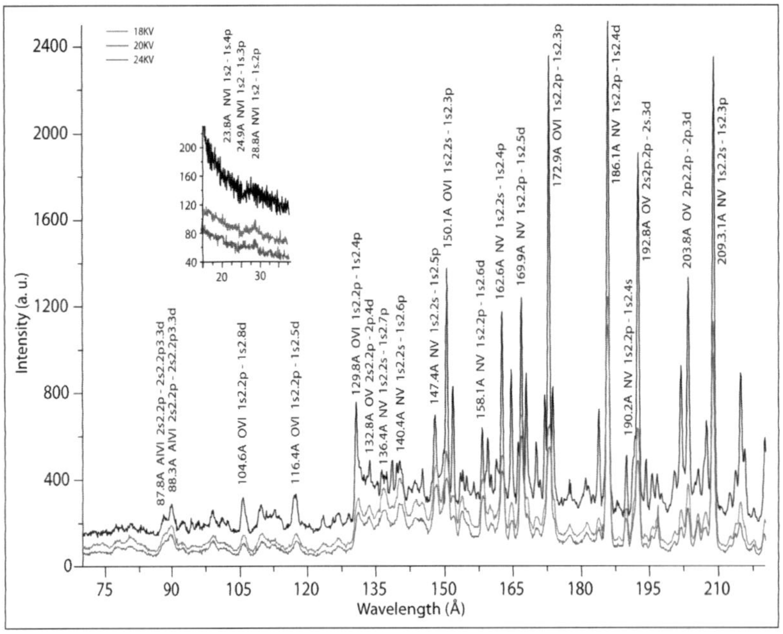

Figure 4. Spectra for N/He discharge at two different values of voltage

3.3. Electronic diagnostics

A 20kV discharge is shown in Figure 5. The Faraday Cup and Bare XRD signal shows a slight negative peak after voltage breakdown, which we attribute to electron beams originated at the hollow cathode region. The positive signal indicates x-ray emission, which extends well after 40ns. The signal from the AXUV 1μm Ti filtered diode indicates emission at 300-450eV range. Therefore, the emission line at 28.8Å, shown in Figure 4, can be attributed to the x-ray signal observed at this particular time range, whereas total x-ray emission continues well after this time at lower energies.

For lower values of anode pressure we observe a slight increase in x-ray emission in the F.C. signal for constant cathode pressure. Also, the AXUV signal shows time delay in x-ray emission as pressure decreases at the anode. For varying cathode pressure, we observe an extended emission in the F.C. signal for lower values, coinciding with an increase in impurities, as seen in spectral data.

The influence of pressure parameters in e-beam and x-ray production must be further investigated to fully describe the dynamics of the discharge.

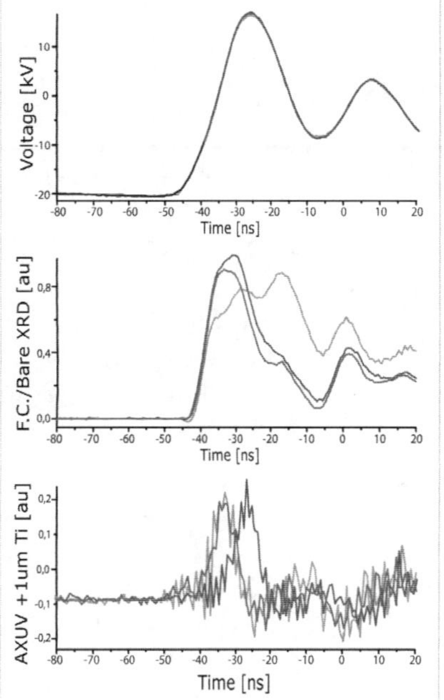

Figure 5. Voltage, FC/XRD and AXUV

4. Summary

We have presented a possible source for soft x-ray microscopy. We have discussed the optimal parameters for soft x-ray production and found the time range at which the He-like Nitrogen emission at 28.8Å takes place. We also report the source size indicating full wall detachment and found that the desired emission comes from a much smaller plasma volume, given by a 200μm radius, on axis, radiating at 300-450eV.

Acknowledgments

The research project, FONDECYT 110005, has been financed by Chilean government. We thank EPPRA sas. for an important contribution to purchasing the spectrometer used.

References

[1] Tamas M, Jancarek A, Nevrkla M, Vrbova M, Vrba P, 2007, Proc. 28th ICPIG, Prague, Czech Republic, 3P15-16

[2] Vrba P, Jancarek A, Vrbova M, Scholozova L, Fojtik A, Tamas M, Havlikova S, 2004, J. Plasma Fusion Res. SERIES, Vol. 6

[3] Wyndham ES, Favre M, Valdivia MP, Valenzuela JC, 2009, Proc. IEEE Pulsed Power conference, PPC .09 pp. 688-692

[4] Avaria G, Guzmán F, Ruiz M, Favre M, Wyndham E, Bhuyan H, Chuaqui H, 2009, Plasma Sources Sci. Technol. 18 045014

[5] Wyndham ES, Favre M, Valdivia MP, Valenzuela JC, Chuaqui H, and Bhuyan H, 2010, Rev. of Sci. Instrum. 81, 093502

[6] Choi P, Dumitrescu C, Wyndham E, Favre M, Chuaqui H, 2002, Rev. Sci. Instrum. 73, 2276

A Compact Ultrafast Capillary Plasma Discharge As an Intense XUV Source

J. C. Valenzuela, M. P. Valdivia, E. S. Wyndham, M. Favre, H. Chuaqui and H. Bhuyan

Facultad de Fsica, Pontificia Universidad Catlica de Chile, Casilla 306, Santiago, Chile

E-mail: jcvalen1@uc.cl

Abstract. A fast nanosecond capillary discharge for fast repetition rate operation has been designed and tested. Operated in Ar and Ar/He mixtures X-ray emission from ionization stages from Ar VIII to XII have been observed according to capillary dimensions and operating voltages. Intense electron beams are formed as a result of the transient hollow cathode effect. The emission characteristics are presented for capillaries of two lengths, 21 and 26 mm, and at three diameters, 0.8, 1.6 and 3.2 mm. In addition, time resolved observations allow comparison of theory and experiment of the emission in relation to the current and applied voltage. We find from the spectra a strong dependence on the capillary dimensions, capillary operating conditions such as voltage, gas mix and pressure, diameter and length with clear evidence of wall ablation under some conditions. This has important consequences on the long term reliability of the discharge. The energy output at 4.9 nm (Ar IX emission) and the in-band (13.5 nm 1%) is also presented.

1. Introduction

Pulsed capillary discharges have been studied as a possible source of EUV and soft X-ray radiation. More recent work has concentrated on the potential applications of the capillary dicharges as a 135 Å source for EUV lithography [1] and as a water-window x-ray microscope source at 28.9 Å. Also Laser operation have been shown to be possible in the extreme ultraviolet [2]. The range of time scales of the discharges, plasma volume, energy delivered to the discharge is very varied. Most of the discharges store tens, or more, of Joules and operate at currents which allow pinching of the plasma columns. I contrast our discharge operates at energies less than one joule and time scales of 10 ns [3]. The skin depth scale of our discharge is comparable with its diameter, so it does not show the characteristic pinching as the normal capillary discharges. Here, rather than pinching and forming a small hot plasma column, which is the mechanism commonly used in bigger diameter, the transient hollow cathode effect plays an important role in forming a low temperature precursor, through the generation of intense axially colimated high energy electron beam just prior to breakdown. This favor conduction on axis with minimal wall damage. The early non Maxwellian electron beam favors the ionization of higher states and bright EUV source is obtained.

We present typical spectra in argon discharges where ArVIII to ArX show the most prominent lines. Wall ablation is clearly observed from oxigen lines present in the spectra. This results combined with filtered diodes, XRD's and Faraday cups show the emission characteristics of intense electron beams and the plasma X-ray emssion.

2. Experimental Description

Very fast discharges are possibles when the energy stored is an integral part of the experiment's electrodes [3], [4]. This configuration of low inductance, allows an approximately 20 ns half period current waveform and a maximum current of order 10 kA.

Two diferent capillary lengh have been used, 21 mm and 36 mm. A schematic diagram of the capillary discharge for the 21 mm case is shown in Fig. 1. We used three different internal diameter for each capillary length: 0.8, 1.6 and 3.2 mm, in all the cases the external diameter was 6.35 mm and made of alumina.

In the Fig. 1 we can see a parallel plate capacitor configuration of 1.6nF, which acts as the energy storage medium. Water is used both as dielectric and as coalant. A cooling agent is needed becuase of the high repetition rate posible with this cofiguration (up to 1 kHz). The capacitor was charged up in 1 μs with a fast IGBT based charger [8].

A differential pressure scheme is used in order to enhance spontaneous plasma formation in the hollow cathode region. This is achieved by pumping the cathode region through the capillary, allowing a pressure gradient to be established between the cathode and the anode.

The observations shown here were taken in Ar-He mixture. The helium presence prolongs the life-time of the capillary and improves the the spectral characteristics changing by the Ar/He ratio.

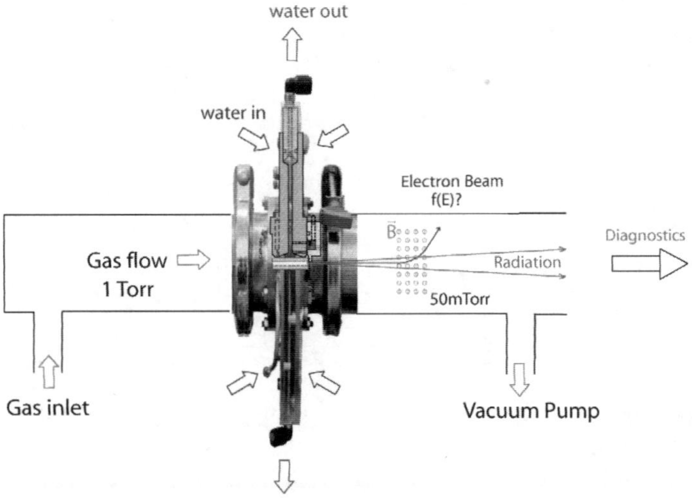

Figure 1. Experimental Set-Up. A schematic was put in front of a real picture of the capillary discharge.

A range of diagnostics have been used to meassure the electrical properties of the capillary and the emission characteristics. A groove at the anode region works as a Rowoski coil, voltage is measured with a resistive voltage monitor. Immediatly downstream the anode four pairs of permanent magnets bend the electron beam away from the detectors. Beyond the magnetic field we used the AXUV HS-5 diodes series [5] and a photoconducting diamond [6], with an appropriate filter, to see the temporal evolution of certain lines and to calculate the in band energy. At the same position that the diodes the GIMS (Grazing Incident Monochromator Spectrometer) spectrometer was the principal diagnostics for the species in the plasma [9]. It is based on Rowland circle system with a CCD interface. Time integrated emission spectrum in the ranges 20-300Å was possible. Emission intensity were estimated using the responsivities 0.25A/W for the diodes [5] and $4x10^{-4}$ A/W for the PCD [7].

3. Results and Discussion

In Fig. 2 we show important features of the capillary phenomenology for a discharge at 24 kV and 150 Hz in Ar/He using a 21 mm long and 1.6 mm inner diameter alumina tube. In the upper diagram, it is the current close to sinusoid after the first 5 ns with a half period of 20 ns. Maximum current is 4.8 kA for these conditions. The Faraday cup pin is slightly displaced off-axis so as not to obstruct the view of the filtered diodes downstream. The lower set of traces show the filtered diode response. As the aluminum filtered diode extends its response from 171 Å well into the longer EUV wavelengths, its signal is the longer lasting, starting with the bare XRD signal and extending well into the period of damped oscillating current. As will be seen below, from the spectrum integrated in time, the aluminum filter is sensitive to the lower ionization states of argon as well as oxygen impurities, while the titanium filter permits observation of the highest argon ionization stages, Ar X to XII, while the polycarbonate responds principally to Ar IX emission in the vicinity of 49 Å. If we consider the temporal evolution, the rising edges of filter signals show a clear tendency from longer to shorter wavelength as the plasma attains higher temperature. The width of the Ti-filtered diode and the polycarbonate filtered diamond detector is noticeably less, consistent with the duration of a hot plasma of approximately 4 ns duration.

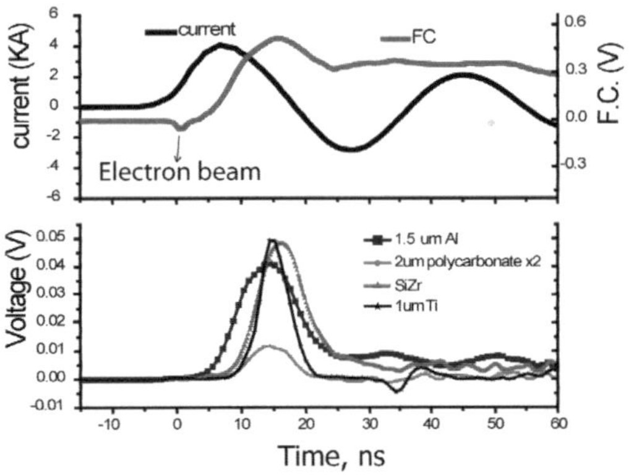

Figure 2. Current and the temporal evolution of the filtered diodes.

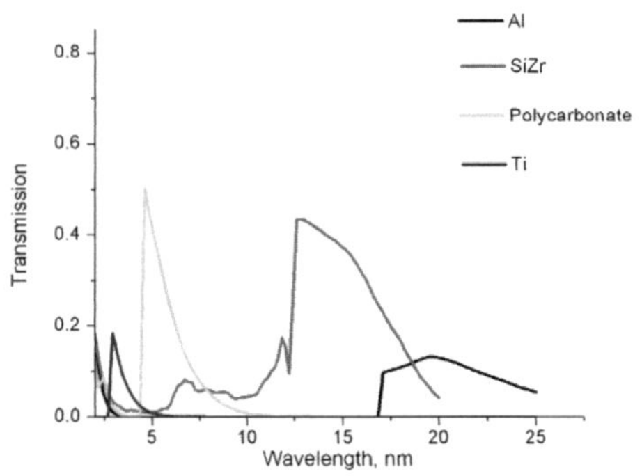

Figure 3. The diferents filters used in the experiment.

The Fig. 4 and 5 were taken at the same conditions than the electrical signal shown before. In Fig. 4 the dominant feature is the Ne-like argon 2p6-2p5.3s transition at 49 Å. Higher ionization states are seen when the voltage is increased, up to ArXII have been seen when the capillary is charged at 27kV. When the voltage is increased, O and Al lines appear in this region, showing erosion of the capillary wall. On observing Fig. 5, OVI lines are clearly visible. The most prominent argon line is 3s-4p transition from ArVIII at 159 Å.

The spectrometer covers 30-400 Å, but the emssion beyond 250 Å falls off markedly and is not discussed here.

Intensity calculations on Fig. 6 show the intensity per pulse of some of the most interesting windows. For example, the He-like Ar lines at 49 Å are possible to see when a PCD is used with 2 μm of polycarbonate filter. Similarly the in band energy flux at 135 Å is posible to estimate usig a AXUV diode filtered with 100/200 Å of Si/Zr, which wavelength is of great interest to the EUV lithography comunity. In Fig. 7 the importance of the length and the diameter is establish. We can see that the smaller the diameter the more intense is the emission. Even though the life-time is reduced by at least one order magnitude. The best diameter to maximize

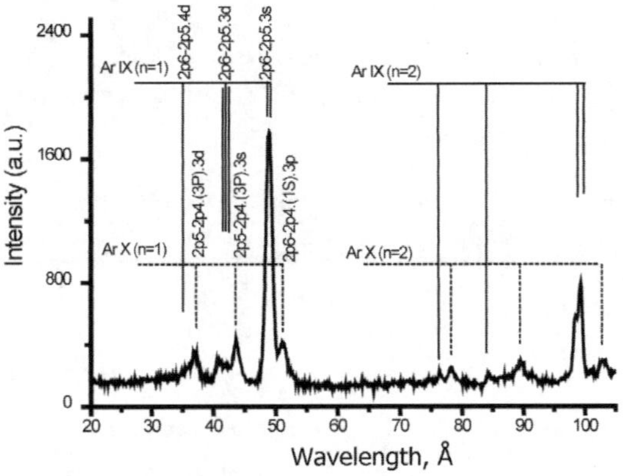

Figure 4. Ne-like argon lines are the most prominent.

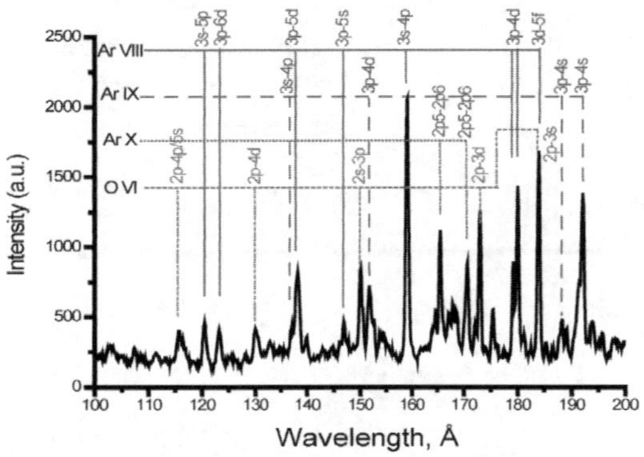

Figure 5. O VI lines show that erosion of the capillary wall is present.

intensity/life-time is 1.6 mm. If we consider the 1.6 mm diameter and 21 mm length capillary, the total energy per pulse in the window 30-200Å integrated over 2π steradians is 11mJ. This result shows that the yield is 1/100 of the total initial stored energy.

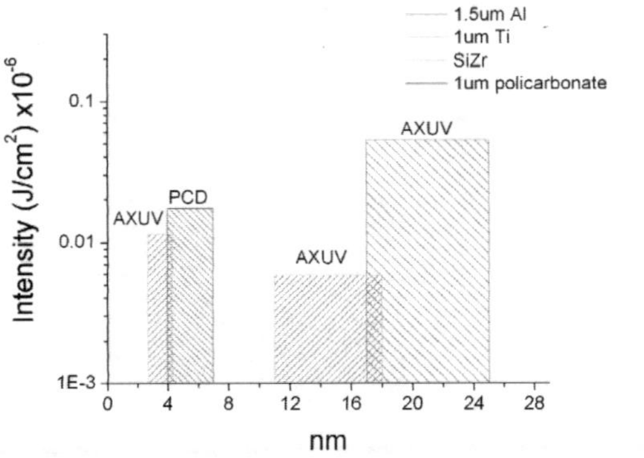

Figure 6. Intensity calculated from the signal of the diodes.

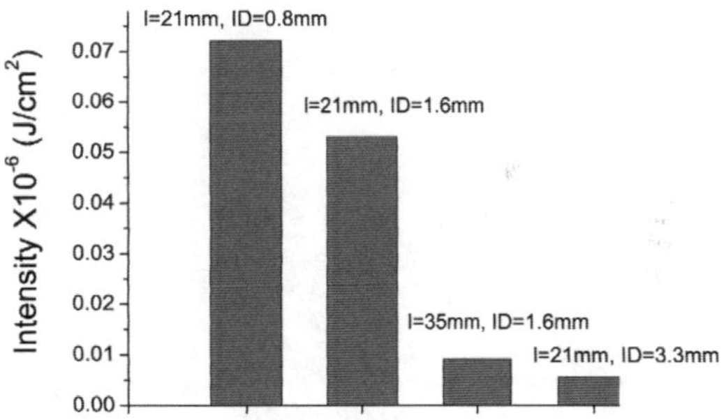

Figure 7. Comparison of the diferents lengths and diameters.

4. Conclusion

We have shown some preliminary results showing that this experiment may be used as a EUV radiation source. We have measured the soft X-ray output for two capillary lengths and three capillary diameters. The 1/100 efficiency, together with the high repetition rate could yield an output power as high as 10W. Wall damage remains a very important problem when the capillary is operated a high frequency and high voltage. Without doubt this will be a critical parameter to evaluate in the future. Shorter capillary, different gases and higher voltage are parameteres that in the future will be studied thoroughly to improve the spectral composition and the emitted intensity.

5. Acknowledgments
The research project has been financed by the Chilean government under project FONDECYT 1100065. We thank EPPRA SAS of France for an important contribution to purchasing the soft x-ray spectrometer used here.

References
[1] Zakharov S. V., Zakharov V. S., Novikov V. G., Mond M. and Choi P. 2008 *Plasma Sources Sci. Technol.* **17** 024017

[2] Rocca J. J., Shlyaptsev V., Tomasel F. G., Cortázar O. D., Hartshorn D. and Chilla J. L. A. 1994 *Phys. Rev. Lett.* **71** 2192

[3] M. Favre, P. Choi, H. Chuaqui, I. Mitchell, E. Wyndham and A.M. Leñero 2003 *Plasma Sources. Sci. and Tech.* **12** 78-84

[4] M.Favre, E.Wyndham, A.M.Leero, F.Suzuki, J.Valenzuela, G.Avaria, M.Ruiz, H.Bhuyan P.Choi 2008 *Plasma Sources Sci. and Technol.* **17** 024011 2008

[5] www.ird-inc.com

[6] www.aasc.net

[7] Rick B.Spielman 1995 *Rev. Sci. Instrum.* **66** No. 1

[8] E. S. Wyndham, M. Favre, M. P. Valdivia, J. C. Valenzuela, H. Chuaqui and H. Bhuyan 2010 *Rev. Sci. Instrum.* **81** 093502

[9] A. Shevelko 2005 A EUVL application specific wide range spectrometer design *Sematech Inc. EUVL Symposium* Poster Sessions `www.sematech.org/meetings/archives/litho/euvl/7470/Poster/FinalS2/2-ME-11\%20Shevelko_Setpoint\%20Poster.pdf`

Dense plasma focus PACO as a hard X-ray emitter: a study on the radiation source

L Supán[1], S Guichón[1], M Milanese[1,3], J Niedbalski[1], R Moroso[1], H Acuña[2] and F Malamud[2]
[1]IFAS, Universidad del Centro and CONICET, Pinto 399 (7000) Tandil, Argentina.
[2]IFIMAR, Universidad Nacional de Mar del Plata and CONICET, Funes 3350 (7600) Mar del Plata, Argentina.

Abstract. The radiation in the X-ray range detected outside the vacuum chamber of the dense plasma focus (DPF) PACO, are produced on the anode zone. The zone of emission is studied in a shot-to-shot analysis, using pure deuterium as filling gas. We present a diagnostic method to determine the place and size of the hard X-ray source by image analysis of high density radiography plates.

1. Introduction

The dense plasma focus device (DPF) [1,2] produces a hot dense plasma by the rapid magnetic compression of a plasma cylinder formed at the end of a coaxial electrode system (see figure 1). When pure deuterium is used as filling gas, fusion reactions are produced in about 100 ns of the hot plasma lifetime. These reactions could be produced by two mechanisms: thermonuclear and beam-target. The thermal mechanism produces a soft X-ray emission (of some keV) [3,4] originated by bremsstrahlung of electrons on deuterons of the thermal plasma. The non-thermal mechanism produces high-energy ion beams and, in opposite direction, electron beams that collide with the anode zone originating a short pulse of hard X-ray emission [5]; both mechanisms coexist [3]. High definition radiographs have been obtained and related studies have been made in the DPF of our Lab in previous works ([7], [8]).

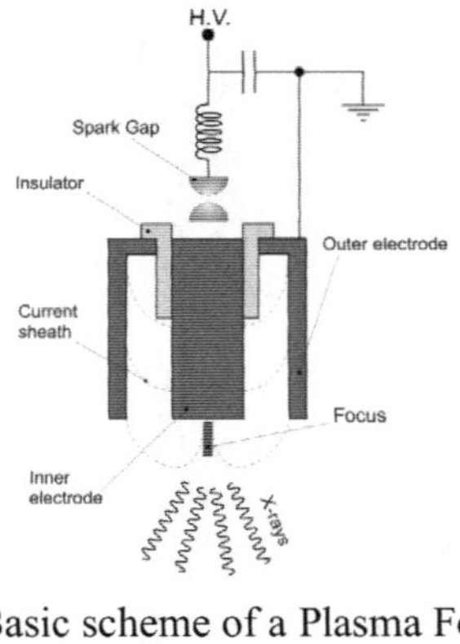

Figure 1. Basic scheme of a Plasma Focus device.

[3] Author to whom any correspondence should be addressed. E-mail: milanese@exa.unicen.edu.ar

2. The Experiment

The Dense Plasma Focus PACO (Spanish "Plasma Auto COnfinado" = auto confined plasma) is a small one: 2 kJ of energy in the capacitor bank, 4 μF, 31 kV of charging voltage, 250 kA of maximum short-circuit current. It has been designed through a bi-dimensional code [10], optimised in the sense of the neutron yield. The filling gas is deuterium (1.7 mb), in order to use neutron yield as a diagnostics. The current derivative was registered with a Rogowski coil, the voltage between anode and cathode (top side) is measured with a resistive voltage divider. The neutron yield (time integrated) was measured with a silver activation counter (SAC) and the hard X-ray and neutron pulses (time resolved) with a Bicron® scintillator-photomultiplier tube (SPMT).

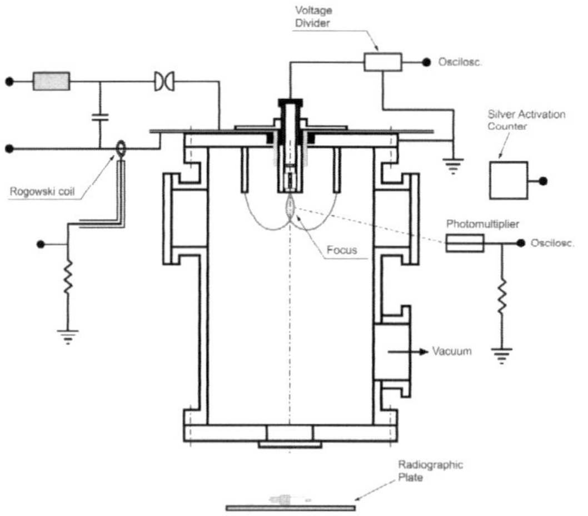

Figure 2. Diagram of the experimental setup for radiographies.

In each discharge a radiographic plate for mammography (Kodak®) was placed in the axial end-on direction to register an image of the obstacle interposed between the source and the plate. Mammographic plates were used since they have demonstrated to be more suitable to obtain a better spatial resolution than the conventional plates for thorax radiographies[6].

Due to the location of the PACO's chamber, the maximum focus–radiographic plate distance was 1.20 m. The anode of PACO, made in OFHC copper, has a tungsten disk (20 mm in diameter) in its free end.

3. Procedure and Results

The idea of the experiment was simply to put an obstacle between the source and the radiographic plate and analyze the image on the plate to infer the size of the source of radiation. The obstacle consisted in a copper plate with a thickness

$$c = (0.50 \pm 0.05)\ \text{mm}$$

located at a distance a below the anode of

$$a = (380 \pm 5)\ \text{mm}$$

Radiographic plates where located at a distance

$$b = (800 \pm 5)\ \text{mm}$$

below the copper plate. On the other hand, it was used a pair of wires placed vertically in order to determine the position of the focus in each shot. Taking into account that the shadow projected by them points towards the source, drawing lines along these shadows and intersecting them it was possible to determine the focus position, from which distances were measured.

The system was carefully aligned and each radiography was irradiated with only one shot. Radiographic plates with sufficient exposition to radiation were used in our analysis.

4. Analysis

The obtained radiographic plate was digitalized by means of digital scanner (Canon® CanoScan LIDE 25). From the digital image (in jpg format) the curves needed for the analysis were extracted, without modifying the image.

The obtained curves are the ones corresponding to profiles in grayscale along specific lines of interest over the image. Such curves result to be proportional to the received radiation on every zone of the plate.

The following figure shows one of the obtained images after digitalization:

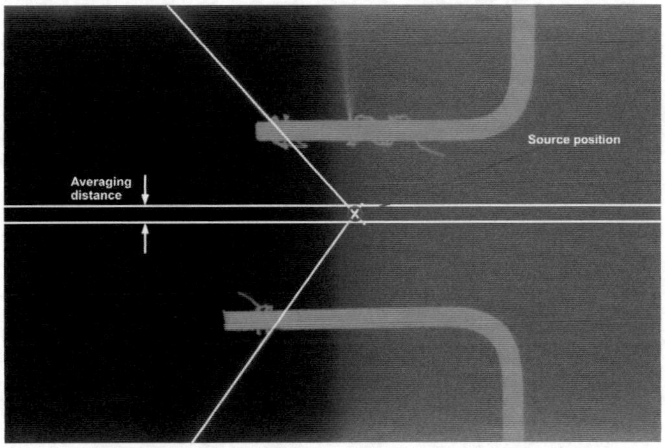

Figure 3. Digitalized image of the radiographic plate.

In order to obtain the curve corresponding to the received radiation intensity (grayscale) on every pixel on the plate along the horizontal line passing through the position of the focus (see figure 3) the values of intensity were averaged taking into account some values above and below that line. These curves were taken as a data, performing an analysis by means of a numerical procedure, described below.

4.1. Procedure

The radiation from the source arrives at the radiographic plate passing through the copper plate. Different zones will have distinct intensities which are related to the area of the source that it not obstructed by the obstacle. Below the copper plate edge the intensity reaches its mean value, which is in agreement with the fact that at this point half of the radiation coming from the source is not attenuated. See figure 4 and 5.

Therefore, the intensity of the radiation on every point of the radiographic plate is related to the integral of the source intensity profile. By modifying the limits of that integral, the different intensity values registered at every point of the radiographic plate can be obtained.

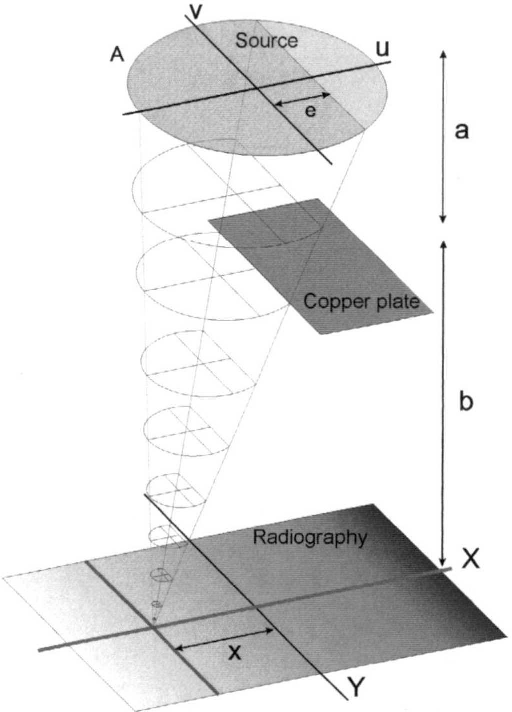

Figure 4. Scheme of the experimental configuration used to obtain the intensity profile curves.

The obtained profiles represent the intensity of radiation $I[x]$ which imprints the radiographic plate. A characteristic curve is shown in figure 5.

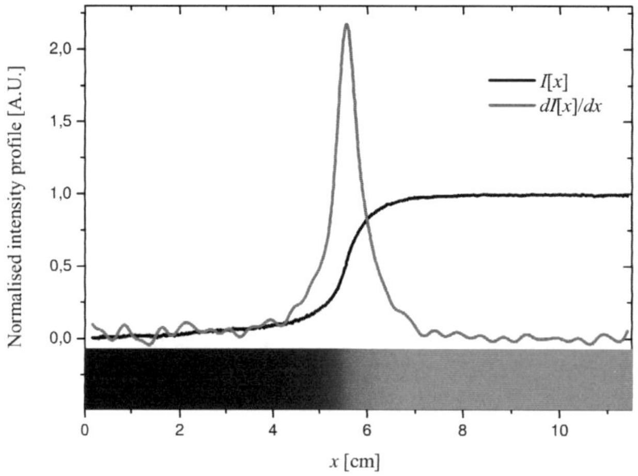

Figure 5. Intensity profile obtained $I[x]$. The derivative of $I[x]$ is also shown.

The image registered on the radiographic plate is a magnified image of the copper plate edge. The distance x, measured in the (x,y) coordinate system, and e, measured in the (u,v) system, are related by a simple expression, i.e.,

$$e = \frac{ax}{b} \tag{1}$$

The obtained curves $I[x]$ are proportional to the integral of the intensity profile of the source $S[u]$.

$$I[x] \propto \int_s^{\frac{ax}{b}} S[u]du = T[\frac{ax}{b}] - T[s] \tag{2}$$

The integration limits are given by a arbitrary minimum value s (which in principle, at most, can be equal to the anode radius) of null intensity and the position e. A function proportional to $S[x]$ can be obtained by deriving equation 2:

$$\frac{dI[x]}{dx} \propto S[\frac{ax}{b}] \tag{3}$$

The derivative of the intensity profile $I[x]$ is shown in figure 5. The FWHM of this curve was calculated in a series of seven radiographs and the obtained result, which is representative of the characteristic size d of the emitting zone. The obtained value is

$$d = (0.25 \pm 0.02) \text{ cm}$$

5. Remarks

In this preliminary work, by means of a relatively simple procedure it could be obtained an estimation of the value of a characteristic size of the hard X-ray emitter source in the PACO Plasma Focus device. The method gave a value of a few millimeters.

So, the method has proved to be suitable for measuring a characteristic size of the source and to have quite convenient properties such as their simplicity and easiness to implement, or the advantage to be a low-cost method. But there are still some questions concerning to the reliability of the method that we expect to explore and improve in a future.

One way to explore has to do with the experimental set up, and it might consist of varying the geometry to change the obtained data in order to get, for instance, curves where some points or zones become clearer to identify, etc.

Another refinement, experimental as well as theoretical, it could be to change that configuration and implement a new one in which the approximation of parallel rays is no longer valid, so the model has to take into account even some possible effects of anisotropy of the emitted radiation which imprint the radiographic plates.

Or maybe to design a different configuration, perhaps a variation of this, with which it can be possible to explore the possibility that the source is actually slightly bigger than we could expect. If that were the case, it might mean that, although the main contribution to the X-ray hardness is due to the bremsstrahlung of electrons against the anode, the source might have another contribution that makes it bigger, as it might be the bremsstrahlung of electrons against vaporized W in this case, in general the metal of the anode.

Another fact to take into account is that the range of energy over which the mammographic plates have an appreciable response is of the order of some tens of keV. So it might be possible that in fact the zone of the source responsible for the hardest components of the radiation can be smaller than the obtained value, eventually of the order of the electron imprints on the anode surface. We expect to explore that possibility in the near future.

6. References

[1] Mather J W 1871 *Dense Plasma Focus Methods of Experimental Physics* ed R Lovberg and H Griem (New York: Academic) 9B p 187

[2] Imshennik V, Filippov N and Filippova T 1973 *Nuclear Fusion* **30** 929

[3] Castillo-Mejía F, Milanese M, Moroso R and Pouzo J 2000 *Nuclear Fusion* **30** 929

[4] Castillo-Mejía, M. Milanese, R. Moroso and J. Pouzo, *Journal of Physics* D: *Applied Physics* **33** 141-7

[5] Serban A and Lee S 1997 *Plasma Sources Sci. Technol.* **6** 78

[6] Acuña H, Milanese M, Moroso R and Pouzo J 2002 *European Physical Journal* D **21** 97

[7] Milanese M, Niedbalski J, Castillo-Mejía F, Moroso R, Supán J, Guichón S, Acuña H, Malamud F, Barbaglia M and Marcazzó J 2009 *XXIX Int. Conf. on Phenomena in Ionized Gases* (Cancún, México: July 12-17 2009)

[8] Pouzo J, Milanese M and Moroso R 2002 *AIP Conf. Proc.* 669 p 277

[9] Castillo-Mejía F, Gamboa I, Herrera J, Rangel J and Villalobos L 2008 *Applied Phys. Letters* **92**, Issue 5, Web version: 051502-1/-2.

[10] Vargas J, Gratton F, Gratton J, Bruzzone H and Kelly H 1976 *Proc. of the 6th. Int. Conf. on Plas. Phys. and Contr. Fusion Res.* (Vienna: IAEA publications) vol 3 p 289.

[11] Castillo-Mejía F, Milanese M, Moroso R, Pouzo J and Santiago M 1998 *25th EPS Conf. on Contr. Fusion and Plasma Phys.* vol. 22C p 2686-9.

Journal of Physics: Conference Series **511** (2014) 012025 doi:10.1088/1742-6596/511/1/012025

Measurements of ions emission using ToF method and CR39 SSNTD in a Small Plasma Focus device of Hundreds of Joules

José Moreno[1,2], José Pedreros[3] and Leopoldo Soto[1,2]
[1] *Comisión Chilena de Energía Nuclear, Chile*
[2] *Center for Research on Plasma Physics and Pulsed Power – P^4, Chile*
[3] *Universidad de Santiago de Chile, Chile*

E-mail: jmoreno@cchen.cl

Abstract. Ion beam emission in plasma focus (PF) discharges was originally investigated to explain the strong forward anisotropy observed in the neutron emission when D_2 is used as filling gas in the PF devices. Several properties of emitted deuteron beam, including its angular distribution and energy spectra in PF devices operating at energies from 1kJ to 1MJ, have been measured. At present there is a growing interest in the development of very small PF devices operating under 1kJ. As part of the very low energy (< 1kJ) PF devices physics characterization program carried out at the Chilean Nuclear Energy Commission, the charged particle emission is being studied when hydrogen (H_2) and mixture (H_2+ %Ar) is used as filling gas in the PF device. The experiments have been performed in a plasma focus device of 400 joules (PF-400J). In order to estimate the ion beam energy spectrum and its ionization degree, by means of the time of flight method, a graphite collector system operating in the bias ion collector mode was constructed. On the other hand, measurements of the energy and flux of the ions have been corroborated using CR39 SSNTD. Preliminary results mainly show the presence of hydrogen ions with an average energy of 40keV. Also, in some cases, copper (from anode) and aluminum (from insulator) ions with energies lower than 1keV have been observed.

Introduction

It is an interesting phenomenon that high energy ions with an energy of more than 1 MeV are produced in plasma focus devices [1,2]. The energy of these ions is sometimes larger than hundred times the charging voltage of the capacitor bank, (table 1). The mechanism of the production of these ions has been investigated for several decades [3,4]; however, the physical mechanism of the production of these ions remains unclear. These energetic ions are considered to play an important role in the production of the intense neutron flux in the plasma focus device when using deuterium gas [5,6]. The determination of ion beam characteristics is very important not only in understanding the mechanism of production of high-energy ion or neutron, but also for their potential applications.

In general, the studies of ions beam have been performed in plasma focus devices that operate with energies from 1kJ to 1MJ. In this work, preliminary results of some characteristics of the emitted ions from a plasma focus devices of 400 joules [5] are presented. In particular, the ion energy and ion current density is measured with a graphite, biased ion collector [6]. Furthermore, a magnetic spectrometer is used to obtain preliminary images of ion beams on a film of SSNTD, CR39, with a pinhole that enables differential vacuum to be performed between the PF and spectrometer chambers. The CR39 is used in the focal plane of the spectrometer to detect the ions passing through the magnet. The special feature of this method is that the ions produce tracks in two particular regions on the CR39: neutral band and the ions spectrum. The neutral position is used as reference to obtain the ions energy spectrum, since they travel through the magnet without deflection. The CR39 detector is etched to enlarge and develop ion tracks, so that they can be viewed under and optical microscope.

Table 1. Measurements of ion beams at different PF devices.

Device	Energy of the Bank	Gas	Pressure	Ionic State	Ion Energy	Technique
PF_UC	**1.8kJ**	methane hydrogen	0.2-0.45Torr 0.5-1.3Torr	C+4 C+5 H+	**0.4-0.6MeV** **0.9-1.1MeV** **0.2-0.4MeV**	ToF
	4.75kJ	nitrogen		N+1,N+2,N+3	**0.17-4.0MeV**	ToF
Bostick et al.	**5.4kJ**	deuterium			**0.3 to 0.5MeV** **1-9MeV**	ToF
Gullickson & Sahlin	**76kJ**	deuterium			**8MeV**	
Krompholz et al	**1kJ**	deuterium			**0.1 to 0.4MeV**	ToF & Mspec
Takao et al	**19kJ**	hydrogen			**0.15 to 2MeV**	
PGN (Kucinski et al.)	**3.6kJ**	deuterium	2 - 6 torr		**> 0.05MeV**	SSNTD CN y CR-39
PFIPS (Heo & Park)	**0.6kJ**	argon	0.1-0.7Torr		**<E>=0.2MeV**	
PF-20 IPJ, Poland	**12kJ**	deuterium			**0.22-1.0MeV** **1.7 to 2.5MeV**	NTD
IPF (Stuttgart)	**56kJ**	deuterium			**0.3 to 3.0MeV**	
PGN	**4kJ**	deuterium			**0.08 to 0.5MeV**	CN ó CR39
PF-II	**4.75kJ**	nitrogen			**0.05 to 1MeV**	

The Device and Diagnostics

The experiments presented here were conducted in a plasma focus device operating at the energy level of hundreds of joules (PF-400J, 880nF, 20-35kV, 176-539J, ~300ns time to peak current, dI/dt ~ $4x10^{11}$ A/s)[5]. This device was designed and constructed at the Chilean Nuclear Energy Commission, CCHEN. The dimensions for copper anode and its alumina insulator, in PF-400J, were: radio of the anode r_a = 6mm; inner radio of the copper cathode r_c = 15.5mm; overall length of the anode z = 28mm; length of the alumina insulator l = 21mm. The discharge chamber was filled with hydrogen at pressures of 5 to 12mbar, and mixture (hydrogen + 5% of argon) at pressure of 1 to 9mbar. A charging voltage of 27 kV was used. Diagnostics include voltage and current derivative usual probes, a BPX-65 PIN diode masked with 10μm thick aluminium filter radially located at 100mm from upper end of the anode to record the x-ray signal from the side-on plasma emission, and a Faraday Cup used as an ion collector. The ion collector is a conical probe designed to maintain an impedance of 50Ω[6]. It is composed by a conical inner electrode of graphite of 15mm in length with base diameter 7mm to reduce the secondary electron emission (SEE) and a brass external electrode separated by Teflon insulator. The ion detector chamber, which acts as a drift tube, is separated from the PF chamber by a 0.5mm diameter hole. In order to minimize the ion beam attenuation in the background gas medium the ion detection chamber was kept at a lower pressure (~10^{-2} mbar) than the main chamber, using additional pumping on that particular zone. The graphite collector inside the drift tube was used to detect the axial ion beams emitted from the PF and it was placed 222mm away from the anode tip. The collector is negatively biased to a dc -200V. Due to the negative biasing, the ion probes also act as XRD. Thus, the x-ray emission recorded by the PIN diode and the x-ray signal in the ion probe can be used to establish a time correlation between plasma evolution and ion beam emission during the formation process.

The magnetic spectrometry was performed mainly in hydrogen discharges. In these experiments, it was used an arrangement consisting in two rounds bar ceramic magnets (NdFeB) of 2.22 cm diameter and 2.54 cm long, which are clamped by two disks maintaining a certain pole gap as shown in figure 1a. The magnets are also separately connected to a high voltage power supply to produce a DC electric field parallel to the magnetic field in the gap. In the figure 1b is shown a radial distribution of magnetic field.

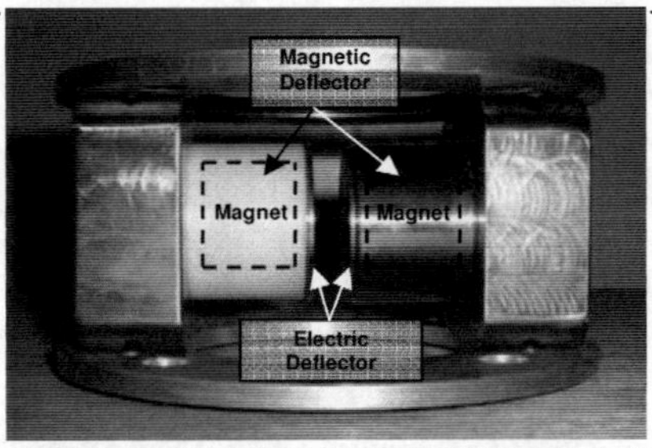

Figure 1a. Electro-Magnetic deflector.

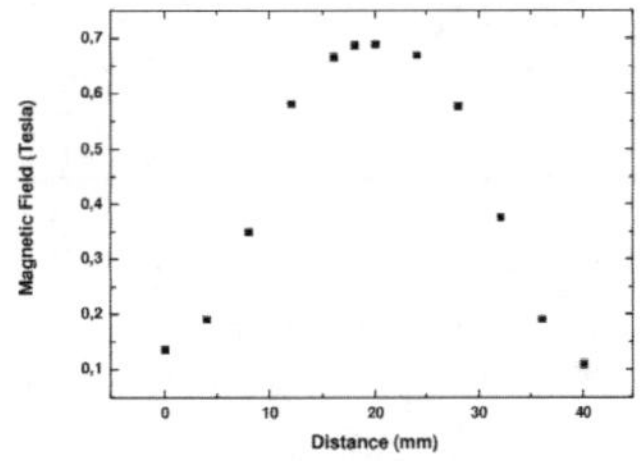

Figure 1b. Radial distribution of magnetic field.

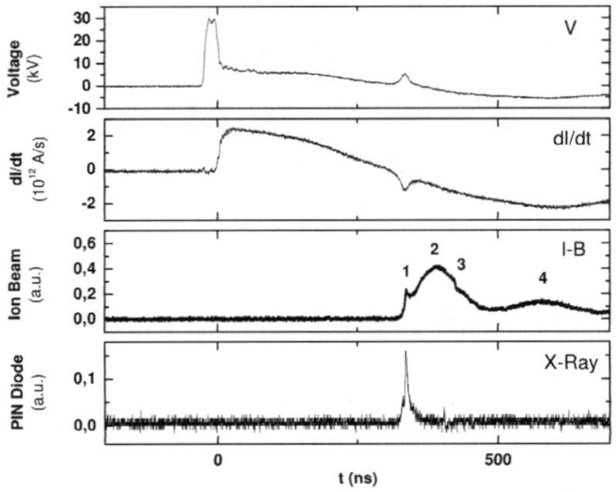

Figure 2. Characteristics signals at 7 mbar of mixture, H_2 + Ar(5%). From top to bottom: voltage, time derivative of the discharge current, ion probe signal (FC) and PIN diode signal. Label 1 represents x-ray signal in FC, and labels 2, 3 y 4 identify three distinctive ion peaks.

Results and Discussion

Figure 2 shows simultaneous characteristic oscilloscope signal of voltage, current derivative, ion probe and PIN diode corresponding to operation in mixture at 7 mbar (H2+Ar(5%)). Even though the electrode geometry is not the optimal to obtain a good pinch, it is possible to observe the dip in the voltage and current derivative signals, which is seen to coincide with both the x-ray pulse in the FC and the PIN diode signals. This kind of peak on these detectors has been reported earlier as x-rays emitted during pinching [6,7]. Time of flight (TOF) measurements performed using two ion probes indicate that the time delay between probes for the different beam components with simultaneous emission at a time which, within the time resolution, coincides with the emission of the x-ray pulse. These facts allow single ion probe signals to be used for the ion energy spectrum measurements [7].

The time correlation between x-ray and ion beam emission indicates that in PF discharges ion acceleration takes place at some time close to the maximum radial compression phase. Therefore, it is reasonable to assume that ion acceleration takes place in existing plasma, with a multi-charged ion composition. Under these conditions, all ions are accelerated by the same electric field and the energy gained in the acceleration process is proportional to the ion charge. This assumption allows using the following relation between the TOF of hydrogen ions, t_H, and other species ions (X) of charge state n, t_X:

$$t_X = t_H \sqrt{M_X / nM_H} \tag{1}$$

where M_H and M_X are the mass of the hydrogen and unknown ions, respectively.
Using equation (1), and assuming that the first ion peak, labelled 2, in the ion probe signals corresponds to hydrogen, the charge state of the different subsequent ion peak was determined from the average values of the different measured TOF.
The velocity, energy and density of ions were estimated using TOF technique [2,8]. The ion velocity is estimated by taking the ratio of the distance to the flight time of ions from source to detector. The time of flight of ions is deduced by correlating the FC signal with voltage or current derivative probe. A statistical procedure was used to assign a characteristic TOF to the different peaks in the ion probe signals, prior to charge state identification [7]. First, a peak deconvolution routine was applied to the ion probe signal, assuming a gaussian shape for the different peaks. Second, characteristic TOFs were assigned to the onset and maximum value of the deconvolved peaks. Third, using equation (1) and assuming that the first ion peak, after x-ray peak, in the ion probe signals corresponds to hydrogen, the charge state of the different subsequent ion peak was determined from the average values of the different measured TOF. A detailed explanation on the analysis procedure appears in the reference 7. The ion velocity (v) thus obtained can be used to calculate the energy associated with the ions, which are reaching the FC at different instant, by the simple formula:

$$E = \tfrac{1}{2} M_X v^2 \tag{2}$$

where M_X is the atomic mass of unknown ion. The number density of ion having velocity v and charge q is given by

$$n_i = \frac{V}{RqAv} \tag{3}$$

where V is the maximum voltage of the ion pulse, R is the resistance across which the output of the ion pulse is taken (50Ohm) and A is the area of the aperture.
The figure 3 shows the distribution of the ions energy for all data obtained in hydrogen (H_2) and mixture (H_2 + 5% of Ar). In hydrogen, it is clearly observed that the ions energy is distributed in three narrow regions with an energy maximum around of 50keV. The region of higher energy, R3, should correspond to hydrogen ions, the middle region (R2), in relation to the equation (1), should be Cu^{+8} ions, and the first region (R1), should be Cu^{+} ions. These copper ions appear form the electrode erosion made by the plasma. On the other hand, when hydrogen-argon mixture is used, the energy range is larger and an energy maximum of 270keV is obtained. In this case, the determination of the ion species is more complicated, due to the greater difficulty in measuring the dip time in the voltage or current derivative signal. However, it is possible to distinguish a first region, lower than 1keV, that could correspond to Cu^{+} or Cu^{+2} ions. It is also possible to observe a second wide region of greater energy. On this wide region, it is possible to find ions of hydrogen and/or Al^{+3} (aluminium are eroded from the alumina insulator). However, additional data and/or diagnostics are needed to discriminate these ions. .

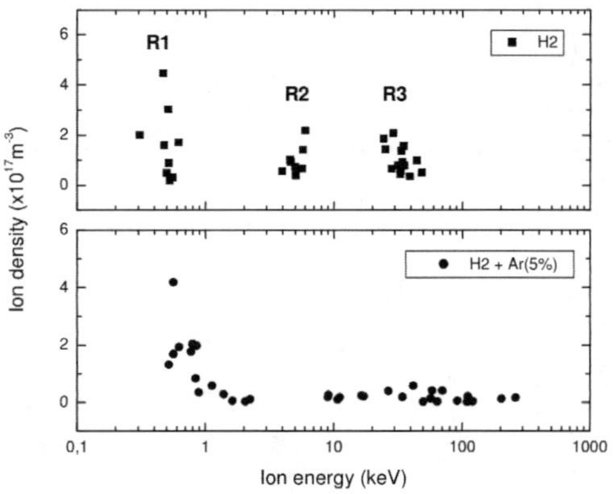

Figure 3. Ion density distribution in function of the ion energy to measurements in hydrogen (H_2) y mixture (H_2 + 5% de Ar).

The track signals obtained after a series of 500 discharges in hydrogen gas on CR39 detector through magnetic spectrometer (figure 4) show different distinct areas of the spectrum. These areas can be defined as: neutral band (1), non-hydrogen charged particles (2) and hydrogen spectrum (3) [9].

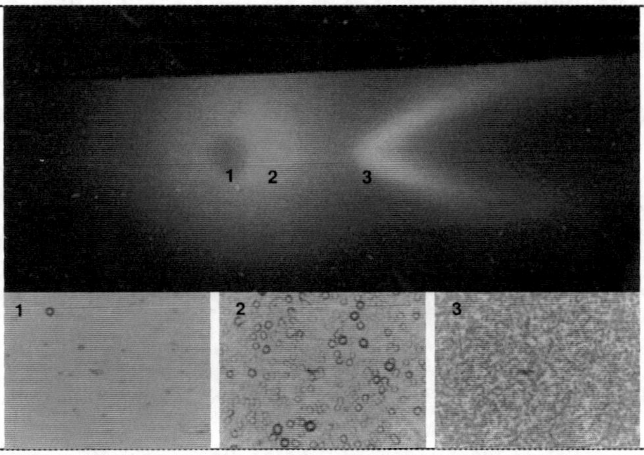

Figure 4. Typical image of ion spectrum on the CR39 detector obtained with a series of 500 discharges in hydrogen operating to 27kV. 1) Neutral band, 2) non-hydrogen charged particles and 3) hydrogen spectrum.

The result of spectral analysis of these zones is shown in the figure 5, where it is possible to observe that the highest distribution of ions is lower that 250keV and it corresponds to the hydrogen spectrum (zone 3 on figure 4). This is in agreement with the results obtained with the ion energy distribution using Time of Flight method in hydrogen discharges previously discussed. In this case, the energies of hydrogen ions are over the 40keV.

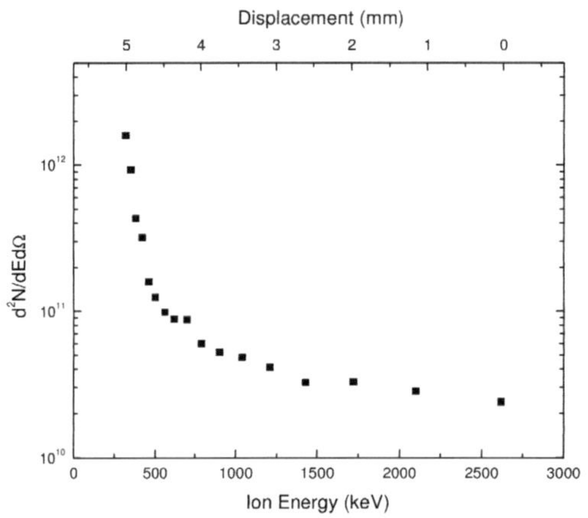

Figure 5. A plot of typical individual ion spectrum emitted from the plasma focus pinch, for a hydrogen working gas pressure of 12 mbar. The displacement is considered from zone 1 to zone 3 in figure 4

Two complementary diagnostic techniques were used to study high energy ion beams form a very small PF device (PF-400J). The obtained ranges of energies vary according to the operating gas of the discharge. In the case of hydrogen discharges, the ion energy range is from 0.5keV to 50keV; and in discharges using gas mixtures (H_2 + Ar) the ion energy range increases from 0.5keV to 270keV. When pure hydrogen is used, narrow energy beams of copper ions coming from electrode erosion are measured with the Faraday cups; which are not identifiable with the CR39 detector. Even though the charging voltage is lower than 30kV, it was possible to measure ion beams with energies larger than 200keV, indicating the presence of an acceleration mechanism inside of the pinched plasma as observed in larger machines. However, it is necessary to increase the statistics, mainly in the case of mixture, and the analysis method of the CR39 detector in order to obtain further conclusions.

References

[1] K. Takao, Y. Doi, S. Hirata, M. Shiotani, I. Kitamura, T. Takahashi, and K. Masugata, Jpn. J. Appl. Phys. **40**, 1013-1015 (2001).
[2] W. H. Bostik, H. Kilic, V. Nardi, and C. W. Powell, Nucl. Fusion. **33**, 413-420 (1993).
[3] M. Sadowski, J. Zebrowski, E. Rydygier, and J. Kucinski, Plasma Phys. Control. Fusion **30**, 763-769 (1988).
[4] M. Zakaullah, I. Akhtar, G. Murtaza, and A. Waheed, Phys. Plasma **6**, 3188-3193 (1999).
[5] P. Silva, J. Moreno, L. Soto, L. Birstein, R. E. Mayer, and W. Kies, Appl. Phys. Lett. **83**, 3269-3271 (2003).
[6] S. R. Mohanty, H. Bhuyan, N. K. Neog, R. K. Rout, y E. Hotta, Jpn J. of Appl. Phys. **44**, 5199-5205 (2005).
[7] H. Bhuyan, H. Chuaqui, M. Favre, I. Mitchell, and E. Wyndham, J. Phys. D: Appl. Phys. **38**, 1164-1169 (2005).
[8] G. Gerdin, W. Stygar, and F. Venneri, J. Appl. Phys. **52**, 3269-3275 (1981).
[9] M. V. Roshan, S. V. Springham, A. Talebitaher, R. S. Rawat and P. Lee, Plasma Phys. Control. Fusion **52**, 085007 (2010).

Journal of Physics: Conference Series **511** (2014) 012026 doi:10.1088/1742-6596/511/1/012026

Current measurement by Faraday effect on GEPOPU

Correa N[1], Chuaqui H[1], Wyndham E[1], Veloso F[2], Valenzuela J[1], Favre M[1] and Bhuyan H[1].

[1]Departamento de Física, Pontifica Universidad Católica de Chile, Av. Vicuña Mackenna 4860, casilla 306, Santiago 22, Chile.

[2]Present address: Departamento de plasmas termonucleares, CCHEN, casilla 188-D, Santiago, Chile.

E-mail: nacorrea@puc.cl, hchuaqui@fis.puc.cl

Abstract. The design and calibration of an optical current sensor using BK7 glass is presented. The current sensor is based on the polarization rotation by Faraday effect. GEPOPU is a pulsed power generator, double transit time 120ns, 1.5 Ohm impedance, coaxial geometry, where Z pinch experiment are performed. The measurements were performed at the Optics and Plasma Physics Laboratory of Pontificia Universidad Católica de Chile. The verdet constant for two different optical materials was obtained using He-Ne laser. The values obtained are within the experimental error bars of measurements published in the literature (less than 15% difference). Two different sensor geometries were tried. We present the preliminary results for one of the geometries. The values obtained for the current agree within the measurement error with those obtained by means of a Spice simulation of the generator. Signal traces obtained are completely noise free.

1. Introduction

The Faraday effect offers the potential for making devices that can measure electrical currents leaving the flow of the current unperturbed. This is due to the fact that the Faraday effect results from the interaction between the magnetic field that is produced by the current and a beam of light that passes through a material subjected to that field. When a linearly polarized beam propagates inside a dielectric material, its polarization is rotated under the influence of the parallel component of the magnetic field generated by the current I. The Faraday rotation Φ_F in the presence of a magnetic field $\vec{B}$, is given by the following relationship:

$$\Phi_F \equiv \int_L \nu \vec{B} \cdot d\vec{l} \tag{1}$$

Where ν, is the verdet constant of the dielectric material and L is the interaction lenght. The total rotation angle will be given by the integration of the equation above, along the path of the laser, i.e., along a straight line. The length of the integration is equal to the length of our BK7 glass rod (Edmund Optics Hexagonal light pipe, uncoated), which is 125mm. With this consideration in mind, we can rewrite equation (1) for our geometry which gives:

$$\Phi_F = \frac{\mu_0 I \nu}{\pi} \left[\left| \ln(B + \sqrt{B^2 - x^2}) - \ln(R + \sqrt{R^2 - x^2}) \right| \right] \tag{2}$$

Where A and B are the respective ends of the rod ($\overline{AB} = 125mm$), R is the distance between the centre of the electrode and one end of the rod, x is the distance between the centre of the electrode and the centre of the glass bar.

1.1. The apparatus: GEPOPU

The apparatus used to carry out our current measurement was GEPOPU, a pulsed power generator situated in our laboratory which consists of different stages. The first one is a Marx capacitor bank, used to multiply the input charging voltage. The Marx bank has 8 capacitors charged by a DC high voltage power supply in parallel and discharged in series using a number of gaseous switches filled with nitrogen. The switch voltage is adjusted by varying the gas pressure at a fixed electrode separation. Triggering the switches causes the erection of the Marx, giving rise to an output voltage of $8V_0$, where V_0 is the voltage used to charge each capacitor. The design of this capacitor bank uses a plus-minus charge configuration reducing the number of switches by a factor of two. This configuration has a lower jitter and faster rise time of the output voltage, compared to a single polarity configuration. The entire marx generator is inmersed in a tank filled with transformer oil, to prevent flashover.

The second stage of the generator, corresponds to a transmission line. In GEPOPU we have one line with two sections of matched impedance. The first part stores the energy being supplied by the Marx bank, which is transfered through a line spark gap to the transfer section. The duration of the current pulse is twice the transit time of the line, 120 ns. The line gap is filled with nitrogen. In order to improve the risetime of the current on the discharge camera (and last stage), a peaking gap separates the transfer section from the discharge chamber. For detailed information, see [1].

2. Experimental Section

Two experimental setups were used in this research.

The first one, was used to perform measurements of the verdet constant for two different materials: BK7 glass and acrylic, by means of the application of an AC current. The power supply used was a transformer 220V-10V, whose current was limited by the use of two resistances in series. The first one, R_1 composed of four ceramic resistors of 3.3Ω each in parallel and R_2 in series with this combination, 0.1Ω. The voltage across R_2 is proportional to the applied current. The value of the current was changed by means of the values of R_1.

A linear polarizer is used to adjust the laser polarization to 45°, to obtain the maximun signal difference between both polarization states. The bar of material was placed in the interior of the coil, the laser beam passes through it. The entire experimental setup used to measure the verdet constant is shown in figure 1.

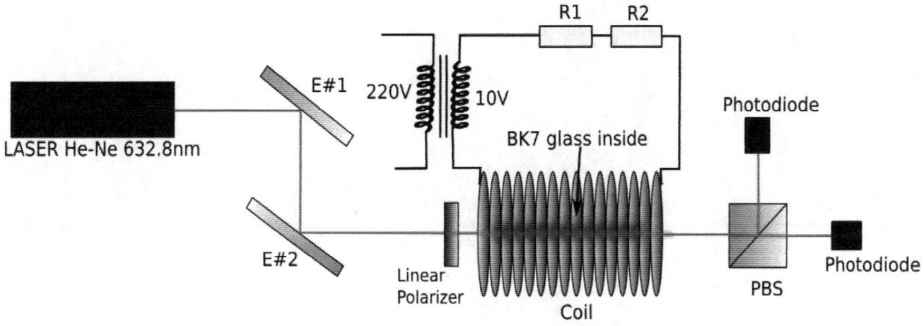

Figure 1. setup used to measure the verdet constant.

The second experiment, was used to perform the current measurement at the GEPOPU generator under short-circuit. A brewster angle polarizer was used to produce a linearly polarized beam, before the analizer a $\lambda/2$ was used to adjust the photodiode (FDS100-Thorlabs) signals to achieve equal values. The PBS is the analizer to provide both polarization states. Two lenses are used to feed the laser signals to a fibre optics, each of them going to a fast photodiode. The material used to measure the current, in this case a BK7 glass bar is situated on a groove on the solid brass short circuit.

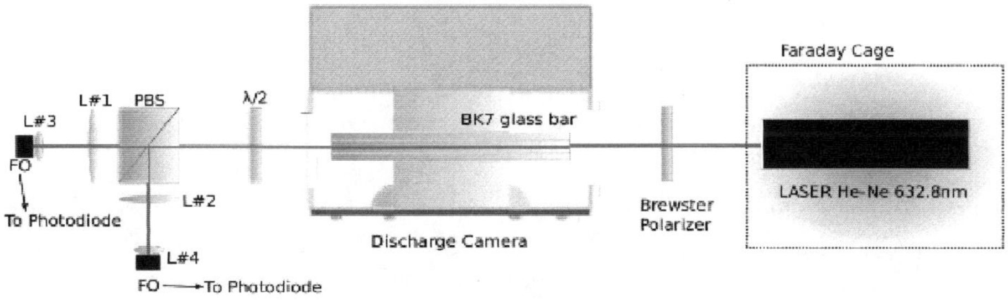

Figure 2. setup used to measure the current on the generator.

3. Results and Discussion

The verdet constant of the two different materials obtained, gives us a difference of less than 15%, compared to the values found in the literature[2],[3].

$$\nu_{BK7} = 2.1501 \pm 0.0036 \left(\frac{\circ}{Tcm}\right) \qquad \nu_{acrylic} = 1.4863 \pm 0.0081 \left(\frac{\circ}{Tcm}\right) \tag{3}$$

With these values, we proceed to measure the current using the setup in figure 2, using BK7 glass bar. The signals obtained using BK7 glass for both polarization states is shown in the graph below, together with the trace of the rogowski coil, the traditional method to measure the current in this type of generator.

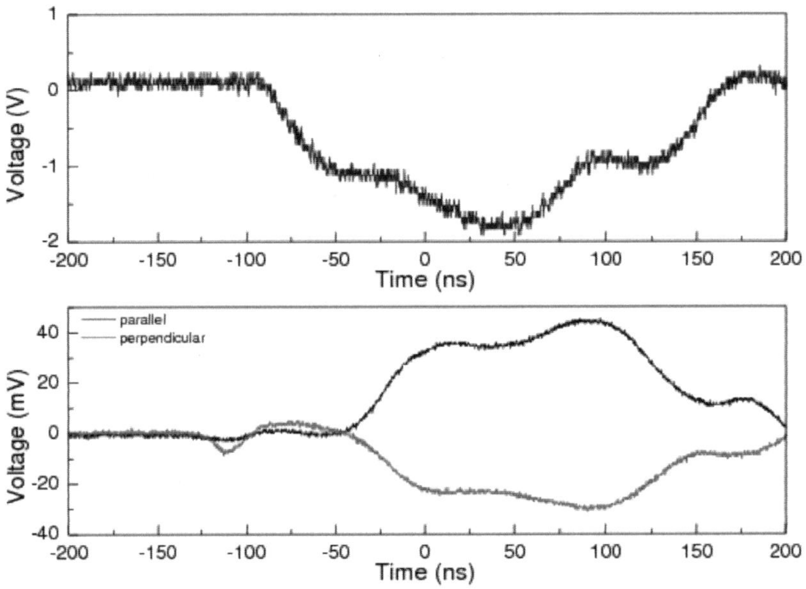

Figure 3. The top trace is the integrated rogowski coil, while the lower oscilloscope traces are the parallel and perpendicular polarizations.

It is apparent that the traces shown on figure 3 have similar forms, the difference in starting time between the rogowski coil signal (grey trace) and the optical signals is due to different lenghts of cable.

The graph shown in figure 4 corresponds to the current obtained by the application of the Faraday rotation formula (2). The difference between both signals is mainly due to the calibration factor. The apparent current prepulse on the photodiode signals at $\sim 25ns$ on the left trace and $\sim 125ns$ on the right trace, is due to electrical noise generated by the Marx bank.

On the next traces we can see that our results agree with the simulations, the traces have similar form and rise times (figure 5). In the image, the perpendicular trace has not been plotted.

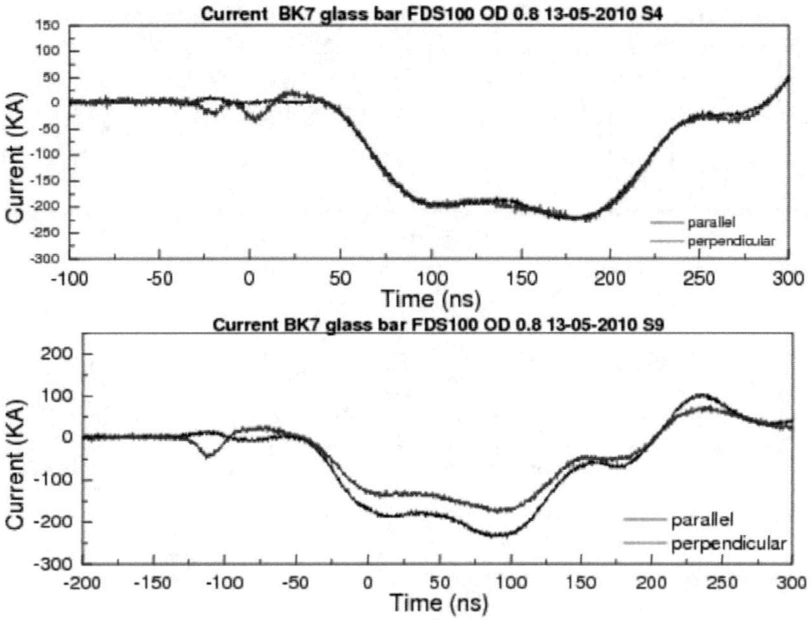

Figure 4. Traces of current obtained by the application of the Faraday rotation formula. In the lower graph, the difference between both signals is due to the calibration factor.

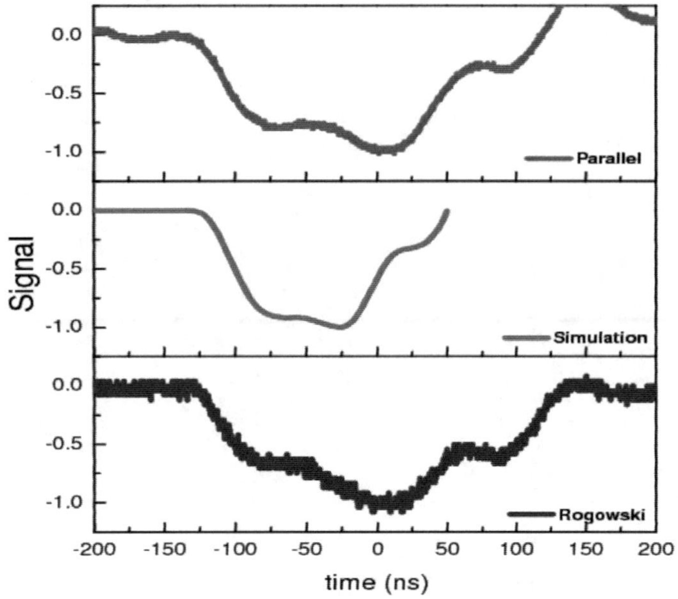

Figure 5. Normalized traces of current obtained from an optical signal (top), simulation (centre) and rogowski coil (bottom).

4. Conclusion

A non-invasive way to perform current measurement has been presented. A method of caracterization of different materials has been developed which gives an error of less than 15%.

Journal of Physics: Conference Series **511** (2014) 012026 doi:10.1088/1742-6596/511/1/012026

The traces obtained using Faraday rotation show a good responde in time and have similar shape to those obtained by simulation or by a rogowski coil. Work is in progress to achieve a consistent photodiodo calibration and the development of a new geometry, which considers a closed optical path around the current, in order to avoid errors due to asymmetry of the current distribution.

Acknowledgments

The present work has been funded by Proyecto PUENTE $N°$ 9/2010, VRAID, and Departamento de Física, Pontificia Universidad Católica de Chile.

References

[1] M. Favre, "Pulse Technology", Proc. Symposium on Small Plasma Physics Experiments (ICTP, Trieste, Italy) 1989, p. 170.
[2] E. Munin et al., J. Phys. D: Appl. Phys. 1992, 25 1653.
[3] Eul Ha Hwang and Byoung Yoon Kim, Meas. Sci. Technol. 2006, 17 2015.

Initial stages in hundreds of Joules plasma focus operating in deuterium – argon mixtures: Preliminary results

Felipe Veloso, Leopoldo Soto, Cristian Pavez, José Moreno

Comisión Chilena de Energía Nuclear – Chile

Center for Research and Applications in Plasma Physics and Pulsed Power – P4 Chile

fveloso@cchen.cl

Abstract. In plasma focus research, the mixture of deuterium with noble gases has been one of the attempts for both improving the radiation yield and for understanding of the origin of the fusion reactions. However, differences on the neutron emission from the plasma focus operating with gas mixtures according to the device energy have been reported. In an attempt to improve the understanding of such experimental observations, experiments in a very low energy plasma focus device (400J) working with gas mixtures are presented. Different stages prior to the final compression of the plasma focus discharges were studied; such as the breakdown phase and the plasma sheath movement. The electrical signals analysis shows differences on the breakdown voltage but similar axial sheath velocities. Nonetheless, these preliminary observations must be corroborated with additional experimental data.

1. Introduction

The plasma focus (PF) [1,2] is a z-pinch type discharge, where an axial current flowing through a cylindrical plasma provides both energy input and magnetic field for its confinement. In these discharges, the plasma is generated from the ionization of the background gas using two coaxial electrodes separated by an insulator. This ionization takes place over the insulator surface and the pulsed current flows over the resulting plasma. As the current generate its own magnetic field, there is a resulting Lorentz force that produces the detachment of the plasma from the insulator, forming a dynamic plasma sheath. When this plasma sheath arrives to the top of the coaxial electrodes, the magnetic force drives a radial implosion forming z-pinch over the central electrode. This pulsed plasma device has been widely studied in basic plasma physics and applied research related to x-ray, ions and neutrons sources (the latter, from D-D fusion reactions). Most of the research in PF devices has been performed in large machines (in terms of size, weight and stored energy), but during the last years there has been an effort on the miniaturization of these devices in order to use them for field applications [3]

The neutron yield obtained from PF devices vary according to electrode geometry, deuterium filling pressure [4,5] and stored energy [6]. Variations on the discharge characteristics have been made in order to enhance the neutron yield further from the scaling laws. Some of these are the modification of the anode geometry, inclusion of deuterated targets and noble gas doping to the discharge. The latter has shown different variations on the total neutron yield according to the device energy changes.

Experiments performed in a 28kJ PF device using neon, argon and krypton as doping gas, show a decrease in the total neutron yield for in comparison to pure deuterium discharges. They also observed an increase in the neutron emission anisotropy and changes in the pinching phase according to the atomic number of the doping [7]. Other experiments in a 3.3kJ PF device evidenced variations from 0.8- to 1.8-fold the neutron yield in pure deuterium, depending of the percentage of argon doping [8]. For energies below 1kJ, the total yield is improved drastically compared to pure deuterium operation. For a 500J PF device, it was found an increase of 5-fold and 8-fold for neon and argon respectively [9] and 3.6-fold on argon and krypton doping [10]. Another experiment in a 200J PF device found an increase up to 30-fold using krypton doping [11]. These studies (below 1kJ) [9,11] also observed variations on the implosion time of the discharge. A summary of these results is shown in table 1

Table 1. Summary of previous results on deuterium + noble gas on PF devices.

Device	Device characteristics	Mixture	Total neutron yield changes
DPF – 78 [7]	28kJ T / 4 ~ 1,5 μs	D_2 + Ne D_2 + Ar D_2 + Kr	0.3x to 0.6x 0.4x to 0.8x 0.7x to 0.9x
UNU / ICTP PFF [8]	3,3 kJ T / 4 ~ 2,5 μs	D_2 + Ar	0.8x to 1.8x
AASC DPF [9,10]	500 J T / 4 ~ 600ns	D_2 + He D_2 + Ne D_2 + Ar D_2 + Kr	1x (No change) 1x to 5x 1x to 8x Up to 3x
FMPF-1 [11]	200 J T / 4 ~ 400ns	D_2 + Kr	0.2x to 30x

The present work is motivated by these reported differences on neutron yield variations when noble gas doping is used. In contrast to the existing published studies, this work does not concentrate in the total neutron yield from the gas mixtures, but in the previous phases; i.e. the breakdown over the dielectric and the plasma sheath movement. It is expected that a proper understanding of these previous phases using gas mixtures will give some keys for the physical processes that produce the observed differences in the final neutron emission.

2. Experimental setup

The experiments were carried out in the very low energy plasma focus device available at our lab, named PF-400J [12]. This PF device operates at energies of 400J its capability of producing fusion neutrons have been previously demonstrated [12,13]. The ratio of the effective length of the anode to its radius is nearly the unity, which means they are classified as hybrid-type PF devices. A schematic diagram of the PF electrode configuration, together its dimensions are shown in figure 1. In order to diagnose the discharge, the voltage and current derivative were measured using resistive divider and Rogowski coil respectively.

In this set of experiments, the final working pressure of the gas mixture was achieved by partially filling the vacuum chamber with each gas at the desired concentration on each case. The main gas on these discharges is deuterium and the argon concentration was varied from 0%Ar (i.e. pure deuterium) to 10%Ar. In order to avoid contaminations on the gas mixture from the material erosion, both gases were continuously renewed using a stationary gas flux, being frequently monitored using a capacitive vacuum gauge.

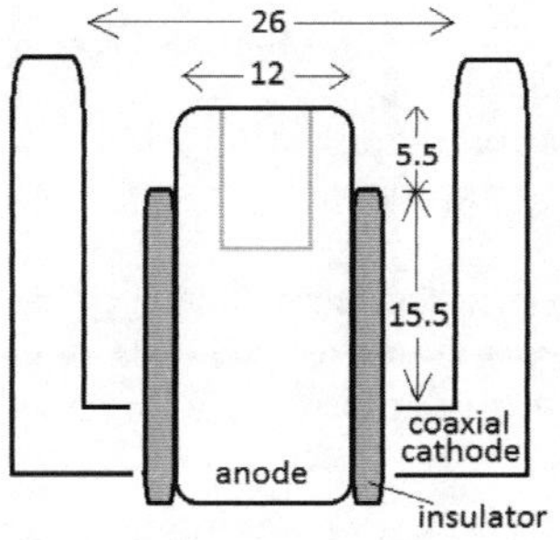

Figure 1: Schematic diagram of electrode configuration. Dimensions are in mm.

3. Results

Even if the breakdown voltage in a pulsed device depends on the voltage rise, triggering system and other generator characteristics, the trends and differences among different pressure or gases should still be observed. For studying the breakdown in our discharge, the filling pressure of the chamber was varied from 5mbar to 16mbar with different mixture percentages. In order to ensure the breakdown over the dielectric surface, the electrode configuration (particularly, the cathode) was slightly modified. The cathode can be separated in two parts, the 'floor' of the PF discharge and the outer coaxial electrode which consist on eight uniformly spaced copper rods. For this part of the experiment, the cathode rods were removed. Therefore, the breakdown can only occur to the remaining cathode (discharge 'floor') having no other path to it than the dielectric surface. Although the plasma sheath evolution will be different in this modified electrode configuration, the breakdown phase will be the same than usual PF configuration.

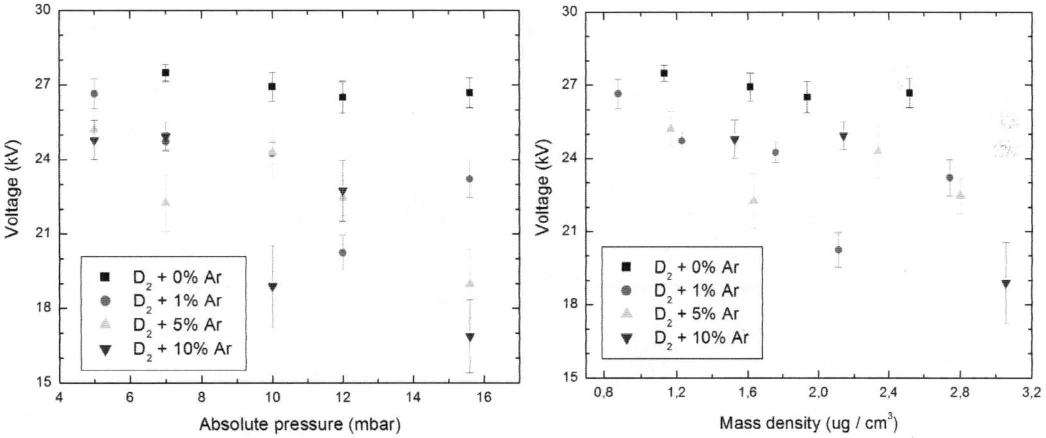

Figure 2: Measured breakdown voltages for different deuterium-argon mixtures

Figure 2a shows the measured breakdown voltages as a function of the absolute total pressure. The vertical error bars are given from shot to shot variation. From this figure, it can be seen than the breakdown voltages for pure deuterium discharges are always higher than any percentage of argon doping. This behavior can be explained from the Pashen's curve for breakdown in gases. For these pressure values, the argon Pashen curve has lower breakdown voltages than hydrogen [14]. Even when the Pashen breakdown has been defined for static voltages, the differences among gases should be maintained for pulsed voltages. On the other hand, previous studies on PF gas mixtures claimed that one of the parameters than should be kept constant for optimizing performance is the mass density of the background gas (in comparison to pure gas discharges) [7,15]. Figure 2b shows the breakdown voltage as function of mass density of the background gas for different argon percentages. It can be seen that regardless of the argon concentration in the gas mixture, the breakdown voltage for pure deuterium is always higher. This observation is valid for every mass density studied here.

In order to study the evolution of the discharge, the time varying inductance has been obtained. This information can be extracted using circuit analysis, where both generator and plasma are included. To compute the time varying inductance, the following equation is used [16]

$$L_p(t) = \frac{\int_{t0}^{t} V(t)dt + \left(L_0' + L_p(t_0)\right)I(t_0)}{I(t)} - L_0' \qquad (1)$$

where *V(t)* and *I(t)* are the measured voltage and current, respectively; t_0 is the first peak on the dI/dt signal and corresponds to the time when the initial gas breakdown is complete and equation (1) can be applied. L_o' is the inductance from the plasma to the voltage probe connection. In order to get the inductance information, details on the processing method of the electrical signals can be found elsewhere [16,17]. Even though the inductance calculations do not consider any specific load geometry, the temporal trends can be associated to the different phases of the plasma evolution; i.e., the axial and the radial phase. The axial phase has a relatively constant slope, which lasts up to a sudden increase in the inductance value. This sudden increase is associated to the radial phase. After some few nanoseconds the inductance slope diminishes, which is associated to the later expansion of the plasma after focusing. Calculated inductances for single shots of different gas mixtures can be seen on Figure 3. For these shots, the mass density was kept constant on 1.4μg/cm^3, which is the equivalent mass density of the optimal deuterium pressure for neutron emission on this device (9 mbar)

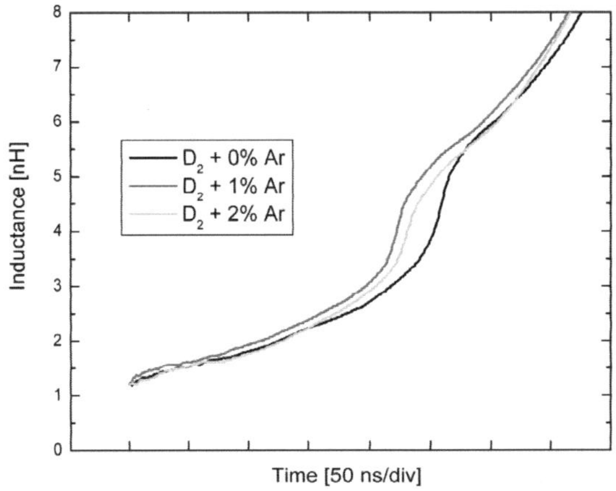

Figure 3: Calculated time varying inductance

Since the geometrical calculation of the inductance in a coaxial geometry has a linear dependence with the electrodes height; it is possible to relate the slope of the inductance on the axial phase with the average axial speed of the plasma sheath. The slopes of the axial phases of the inductance traces in Figure 3 do not show significant variation according to the argon doping on the discharge. From this observation, it is possible to state that the average axial speeds are nearly the same for the pure and doped deuterium discharges, i.e., the global dynamics of the plasma sheath do not change as far as the mass density remains constant. Even though this observation can be rather obvious given that the Lorentz force does not depend on the working gas, it states that plasma components have larger kinetic energy (due to the argon ion mass), which can be later transfer to the pinched plasma in the radial phase. On the other hand, the inductance values are also nearly equal for the studied gas mixtures. Since the inductance is a geometrical measurement of the plasma sheath, it is possible to observe that the shape and curvature for the plasma sheath have not significant variation for these argon doping levels. However, additional data are needed in order to either corroborate or discard such observations.

4. Conclusions

Preliminary results on very low energy plasma focus using deuterium-argon mixtures were presented. Using a 400 joules plasma focus, it was found that the argon doping on deuterium discharges reduces the breakdown voltage required to produce the initial plasma sheath over the insulator surface. This observation is independent on the mixed gas pressure and the load mass density. On the other hand, the time varying inductance calculations shown that the average axial speeds of the plasma sheath and its geometrical shape and curvature remain nearly constants. These observations indicate that the global dynamics of the discharge remains unchanged with the argon doping; but the energy of the plasma components differ. This difference can be possibly related to the differences on the final neutron emission of such configurations. Nonetheless, further experimental data which includes neutron yield measurements are required in order to obtain a deeper understanding of the phenomena.

The authors would like to acknowledge financial support from the programs PBCT-ACT26 and PSD-01, Center for Research and Applications in Plasma Physics and Pulsed Power – P^4 Chile

References

[1] J W Mather, Phys Fluids Suppl. **7**, 5 (1964)
[2] N V Filippov, T I Filippova, and V P Vinogradov, Nucl. Fusion Suppl Pt. **2**, 577 (1962)
[3] L Soto et. al., "Plasma focus of 2J for field applications. Pinch evidence", in this proceeding.
[4] F N Beg, M Zakaullah, M Nizar and G Murtaza, Modern Phys. Lett. **B6**, 593 (1992)
[5] C Moreno, H Bruzzone, J Martinez and A Clausse, IEEE Trans. Plasma Sci. **28**, 1735 (2000)
[6] L Soto, Plasma Phys. Control. Fusion **47**, A361 (2005)
[7] R Aliaga-Rossel and P Choi, IEEE Trans. Plasma Sci. **26**, 1138 (1998)
[8] S L Yap, C S Wong, P Choi, C Domitrescu and S P Moo, Jap. J. Appl. Phys. **44**, 8125 (2005)
[9] B L Bures, M Krishnan and Y Eshaq, AIP Conf. Proc. **1088**, 195 (2009)
[10] B L Bures, M Krishnan, R E Madden and F Blobner, IEEE Trans. Plasma Sci. **38**, 667 (2010)
[11] R Verma, P Lee, S Lee, S V Springham, T L Tan, R S Rawat and M Krishnan, Appl. Phys. Lett. **93**, 101501 (2008)
[12] P Silva, J Moreno, L Soto, L Birstein, R E Mayer and W Kies, Appl Phys Lett **83**, 3269 (2003)
[13] L Soto, P Silva, J Moreno, M Zambra, W Kies, R E Mayer, A Clausse, L Altamirano, C Pavez and L Huerta, J. Phys. D: Appl. Phys **41** 205215 (2008)
[14] A M Howatson, Introduction to Gas Discharges, Pergamon Press, London (1965)
[15] M Favre, P Silva, P Choi, H Chuaqui, C Domitrescu and E Wyndham, IEEE Trans. Plasma Sci. **26** 1154 (1998)
[16] H Bruzzone, H Acuña, M Barbaglia and A Clausse Plasma Phys. Control. Fusion **48** 609 (2006)
[17] J W Mather and P J Bottoms, Phys. Fluids **11** 611 (1967)

Potentiality of a table top plasma focus as X-ray source: Radiographic applications

Cristian Pavez[1,2], Marcelo Zambra[1,2], Felipe Veloso[1,2], José Moreno[1,2], Leopoldo Soto[1,2]

[1]Comisión Chilena de Energía Nuclear – Chile

[2]Center for Research and Applications in Plasma Physics and Pulsed Power – P4 Chile

cpavez@cchen.cl

Abstract. Images radiographic testing on different elements both in biological and inorganic samples are obtained using the X-ray radiation coming from a small plasma focus of at least 350J (880 nF, 40 nH, 28kV, T/4 ~ 300ns). The experimental device is operated using hydrogen as filling gas in a discharge region limited by a volume of around 70 cm^3. The X-ray radiation is monitored shot by shot by means of a scintillator-photomultiplier system located outside of the vacuum chamber at 2.3m far away from the emission region. Angular distribution measurements of the accumulated X-ray dose are carried out using TLD-100 dosimeters which are radially distributed around the emission region center. There are two different arrays for the dosimeter which are placed in two different radial positions outside the discharge chamber. In each array, the TLDs dosimeters are uniformly located and distributed respect to the symmetry axis of the plasma column. An estimation of the energy spectrum of X-ray by means of the filters techniques is presented. The potential of this table top plasma focus is discussed according to its size, the quality of the radiographies, the effective equivalent energy and the dosimetric characteristics.

1. Introduction

The plasma focus configuration corresponds to a dynamical pinch pulsed discharge that is developed between two concentric electrodes. The electrodes are connected to a pulsed power generator, basically composed by a capacitor bank and a high voltage switch [1, 2]. An insulator covers partially the anode bottom surface where the discharge starts, connecting the anode with the cathode when the energy is transferred from the capacitor bank to the electrodes configuration at the moment that the switch is closed. The Lorentz's force detaches and accelerates axially the current sheath. After the current sheath runs over the upper end of the central electrode, the plasma sheath is radially compressed to a small region forming a dense and hot plasma column (pinch). In most of PF devices, the evolution of the plasma sheath up to forming the pinch, takes a few microseconds, and less than 500 ns in a new generation of fast PFs. The maximum pinch compression should be close to the peak current in order to achieve the best efficiency. Depending on the energy of the pulse power generator, the current in the pinch varies from tens of kA to about MA. The pinch generates beams of ions and electrons, and ultra-short X-ray pulses. The duration of these pulses is of the order of tens to hundreds of nanoseconds. Using deuterium gas, PF devices produce fusion D–D reactions, generating fast-neutron pulses (~2.5 MeV) [3]. From the beginning in the research in plasma focus, a growing interest has been placed in the potential use of its powerful pulsed radiation, as alternative source for application both in medicine and industry. If well, alternative sources of pulsed radiations as the X-pinch, presents advantageous characteristics in radiographic applications that require intense sources of micrometer size [4]. However, this kind of sources has the technical disadvantage of the repetition regime, unlike to the performance that could have a plasma focus device, especially those of low energy.

Radiographic applications using dense transient PF devices operating at some kJ or greater energy have been obtained in several laboratories in both biological radiography [5,6] and introspective radiography of non- biological objects [7,8]. Many authors call the devices in the range of the kJ as table top plasma focus.

A new generation of table-top PF devices, working at energies lower than 1 kJ, has been recently developed. These devices have portable sizes and they are reliable to operate in repetitive mode. Several of these compact table top small PF devices have been recently developed, most of them with energies ranged between 400J to 0.1J [3,9-16]. Another advantage that favors the development of small plasma focus devices is the scaling observed in the dimensions of the pinch [17]. It has been shown that the plasma pinch column size linearly scales with the anode radius a, while the pinch radius is $\approx 0.12a$ and the pinch length is $\approx 0.8a$. So, the source size of hard X-rays will not be greater than $2a$

In this paper, we present experimental results at X-ray applications, both in biological samples and inorganic objects, using a small table-top plasma focus of only 350J (400J nominal) [11]. Measurements of the angular distribution of the mean dose X-rays emitted for hydrogen discharges were made. From the high quality of radiographic images, the angular distribution of the mean dose and the effective equivalent energy measured, the potential of a small plasma focus as a source of high X-ray intensity pulsed is discussed.

2. Experimental setup

The experiments were carried out in a small and compact plasma focus device of only 400J (nominal energy), total weight of 64.5kg (including generator, electrode configuration and discharge chamber) and a discharge region limited by a volume of around 70 cm^3. The PF-400J is powered by an 880 nF capacitor bank operating up to 30 kV, and an external inductance of 38 nH, reaching current peak ~ 127 kA with a quarter period ~300 ns [11]. For the results reported here the device was operated in hydrogen flow, in a pressure range of 7.5 to 10.5 mbar. The voltage in the capacitor bank ranged between 27 and 29kV, with the spark gap operating in self breakdown regimen. Typical voltage and current were obtained using a resistive divider and a classical Rogowski coil, respectively.

The efficiency and level of radiation emitted by the device was monitored in every shot using a Photonis XP2262b photomultiplier (5 cm diameter) combined with a 5 cm long BC400 plastic scintillator; this diagnostic was placed, perpendicular to the axial axis, at 2.3 m far away from the chamber and directed to the anode top level. To control exposure of radiographs, the number of shots with voltage level on the photomultiplier ≤ -1.5V was used as reference.

The angular distribution measurements of the accumulated X-ray dose are carried out using TLD-100 dosimeters which are radially distributed around the emission region center. There are two different arrays for the dosimeter which are placed in two different radial positions outside of the discharge chamber, one at 23 cm and the other at 35cm. In each array, the TLDs dosimeters are uniformly located and distributed respect to the symmetry axis of the plasma column, those that were exposed to 600 shots with signal level in the photomultiplier ≤ -1.5V. Figure 1 shows a detailed cut PF400J device where design features are observed as well as the distribution of thermoluminescent detectors TLD100. The arcs that contain the TLD detectors are in vertical planes and their intersection forms an angle of 20°. This ensures that the arcs do not interfere each other and they could see portions of the discharge chamber with the same geometric characteristics (thickness and without extensions of window for observation).

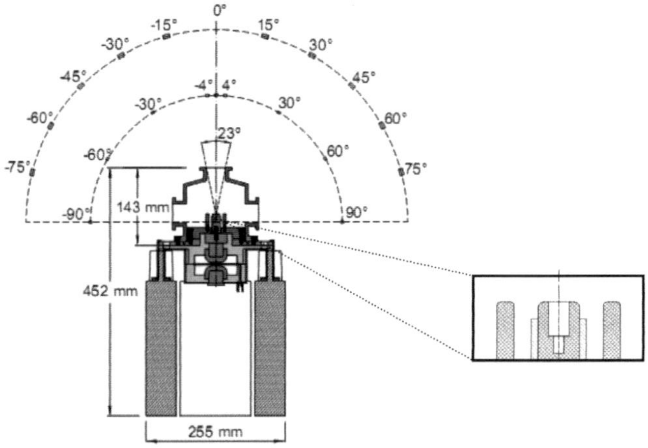

Figure 1: Transversal cut of the PF-400J generator and detector angular distribution TLD100.

For the radiographic record two systems were used. One of these corresponds to a commercial radiographic film, Agfa Curix ST-G2, in an AGFA CURIX Ortho Regular chassis, was placed in the axial axis, outside of the discharge chamber at distances ranging from 17 cm to 45.5 cm from the top of the anode until the radiographic chassis. In all cases, the object was attached to the radiographic chassis. The X-rays travel towards this diagnostic after crossing a 41 mm diameter aluminium flat window of 1.5 mm thickness. The second method was used to radiograph a tooth placed 20.5 cm faraway from the source, for which we used a commercial D-speed film from KODAK, for application in odontology (here the tooth was attached to the film). To determine the X-ray effective mean energy, measurement of the attenuation coefficients of metallic slabs of different materials and thicknesses were made [18]. For radiographic registration of these samples the radiographic chassis was used

3. Results

Figure 2 shows the typical digitized images of radiographic obtained during the experiments when a biological samples element is exposed to the radiation shots of the PF-400J plasma focus device.

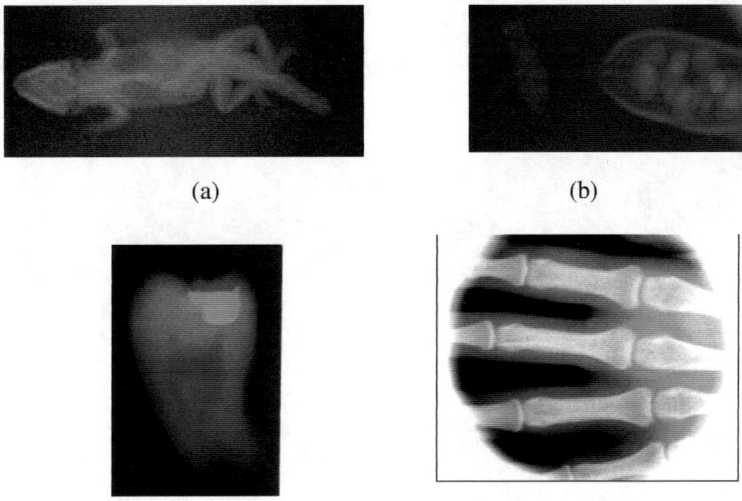

Figure 2: Radiography of biological samples, (a) Tiny lizard; (b) A seed and an insect (c) a tooth and (d) human hand, which were exposed to 5, 10, 15 and 16 shots respectively, with signal level in the photomultiplier $\leq$ -1.5V.

Figure 3 shows two radiographies taken to inorganic samples. The BNC connector was taken to distance of 17cm from the source and this was exposed to 5 shots, whereas the spark plug was taken to distance of 23cm from the source and using 12 shots.

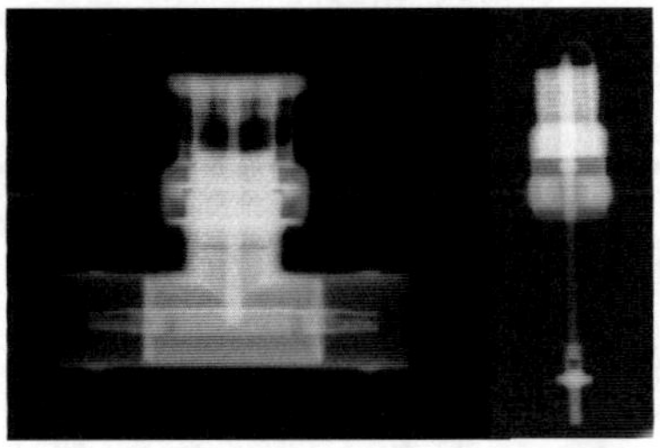

Figure 3: Radiographic images of a typical T-BNC connector and a spark plug.

The results of the X-ray effective mean energy are summarized in the graph showed in Figure 4 [18]. From the information of the graphic two aspects can be remarked: on the one hand, the spectral range ranging from effective energies around to 40keV up to ~120keV, which can be controlled for the material of the target inside of the anode cavity. On the other hand, exposure to X-ray radiation for a given object can be controlled selecting an appropriate number of shots.

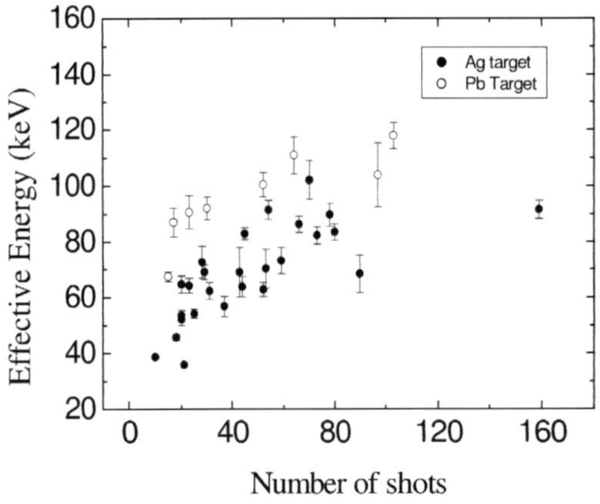

Figure 4: Calculated hard x-ray effective mean energy against the number of shots over a radiographic film when two different anode targets are used.

The graphic of Figure 5 shows dose values measured in the TLD detector array exposed to 600 shots, with a reference level in the signal from the photomultiplier ≤ -1.5V. In the information displayed there, there is clearly a dose concentration in a cone that not exceeds 10° respect to the symmetry axis of the discharge that is determined mainly by the depth of the cavity at the top of the anode. The X-ray shielding in excess of 15 degree angle is determined by the anode wall of the cavity in addition to the walls of the discharge chamber. The mean dose at the axis is 0.017mS.

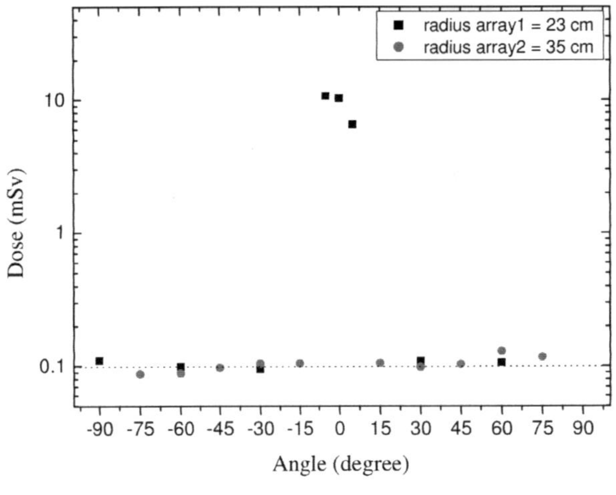

Figure 5: Angular distribution measurements of the accumulated X-ray dose.

4. Discussion and conclusion

From the results shown in this paper, the performance of the PF-400J as a X-ray source, operated at energies in the capacitor bank ranging from 320J up to 370J, can be summarized as follow: Radiographies of high quality of both biological and non-biological objects are achievable using a small table-top plasma focus of only hundred of joules. The spectral range of the effective mean energy can be controlled; either changing the target material inside of the anode or controlling the number of shots. In future applications and considering the duration time of the X-ray pulse (32ns FWHM), radiographies with a single shot and moving objects will be developed. For small plasma focus devices with a cavity in the anode top, the shielding generated by the anode walls are favorable to the concentration of radiation in a cone of emission in the axial direction, and thus operating conditions are safer for the handler of the device.
The device PF400J presents favorable characteristics in X-ray emission, both in the spectral feature and source size, which motivate the development of a pulsed radiation device-oriented to medical and industrial applications, with a radiation dose distribution well directed, ensuring a more secure system to any potential operator.

The authors would like to acknowledge financial support from the PSD01 PBCT-Chile, ACT26 PBCT-Chile grants, the partial support of FONDECYT 11090377 grant. The authors also are very grateful for the given technical assistance by the personnel of the Dosimetry Department of CCHEN.

References

[1] J W Mather, Phys Fluids Suppl. **7**, 5 (1964)
[2] N V Filippov, T I Filippova, and V P Vinogradov, Nucl. Fusion Suppl Pt. **2**, 577 (1962)
[3] L. Soto, "New Trends and Future Perspectives on Plasma Focus Research", *Plasma Phys. Control Fusion* **47**, A361 (2005).
[4] F N Beg, K krushelnick, P Lichtsteiner, A Meakins, A Kennedy, N. Kajumba, G. Burt, and A E Dangor, Appl Phys Lett **82**, 4602 (2003).
[5] S Hussain, S Ahmad, M Z Kanh, M Zakaullah, and A Waheed, Jounal of Fusion Energy, **22,** 3 (2004).
[6] F Castillo, I Gamboa-deBuen, J J Herrera, J Rangel, and S Villalobos, Appl Phys Lett **92**, 051502 (2008)
[7] C Moreno, V Raspa, L Sigaut, and R Vieytes, Appl Phys Lett **89**, 091502 (2006)
[8] S Hussain, M Shafiq, R Ahmad, A Waheed, and M Zakaullah, Plasma Sources Sci. and Technol. **14**, 61-69 (2005).
[9] P Silva, J Moreno, L Soto, L Birstein, R E Mayer and W Kies, Appl Phys Lett **83**, 3269 (2003)
[10] L Soto, P Silva, J Moreno, M Zambra, W Kies, R E Mayer, A Clausse, L Altamirano, C Pavez and L Huerta, J. Phys. D: Appl. Phys **41** 205215 (2008)
[11] P Silva, J Moreno, L Soto, L Birstein, R.E Mayer, and W Kies, Appl. Phys. Lett. **83**, 3269 (2003).
[12] L Soto, C Pavez, J Moreno, M Barbaglia, and A Clausse, Plasma Sources Sci. and Technol. **18**, 015007 (2009)
[13] L Soto et. al., "Plasma focus of 2J for field applications. Pinch evidence", in this proceeding.
[14] C Pavez and L Soto,. IEEE Transaction on Plasma science, **38**, N°5,(2010).
[15] R Verma, R S Rawt, P Lee, M Krishnan, S V Springham, and T L Tan, Plasma Sources Sci. and Technol. **51**, 075008 (2009)
[16] R Verma, R S Rawt, P Lee, M Krishnan, S V Springham, and T L Tan, IEEE Transaction on Plasma science, **38**, N°4,(2010).
[17] Lee S and Serban A 1996 IEEE Trans. Plasma Sci. 24 1101.
[18] M Zambra, P Silva, C Pavez, D Pasten, J Moreno, and L Soto, Plasma Phys. Control. Fusio*n* **51**, 125003 (10pp) (2009).

Experimental Study on the Optimization for Neutron Emission in a Small Fast Plasma Focus Operated at Tens of Joules

A Tarifeño-Saldivia[3,2] and L Soto[1,2]

[1] Comisión Chilena de Energía Nuclear (CCHEN), Casilla 188-D, Santiago, Chile
[2] Center for Research and Applications in Plasma Physics and Pulsed Power, P4, Chile
[3] Universidad de Concepción, Chile

atarisal@gmail.com

Abstract. This work reports results of a systematic experimental study dealing with the optimization for neutron emission of the PF-50J plasma focus. The device was operated in a repetitive mode at repetition rates of 0.1-0.5 Hz. Optimal configurations, neutron emission rates, observed anisotropy, analysis and contrast of "good" and "bad" shots are currently presented. Additionally, engineering aspects on the neutron emission were also studied, such as contaminants removal circuit and chamber design.

1. Introduction.

PF-50J is a small fast plasma focus device operated at CCHEN in Chile. The device has characteristic parameters given by: 160*nF* equivalent capacitance, ~ 40*nH* total inductance in short circuit, 150*ns* first quarter of period, 25 - 35*kV* charge voltage, energy E ~ 50 - 100*J*, 50-70*kA* peak current in short circuit. In the past, this device has been reported to be the first plasma focus which emits fusion neutrons when operated with deuterium at energies of only tens of joules [1]. It has been also demonstrated that this device reproduces the dynamics and high density scenarios observed in machines operated at energies several orders of magnitude higher [2]. Notwithstanding the systematic research already done in PF-50J, the optimization of the device as a neutron has not been studied in detail. In comparison with other devices, PF-50J is a extremely fast and low energy plasma focus. While in conventional devices the complete evolution of the discharge to form the pinch last typically 0.5-10μs, in PF-50J all the process including sheet formation and lift-off, axial and radial transit, pinch and rupture of the column last around 150*ns*. Consequently, the electrodes results in a hybrid configuration ($L_{eff} /2r_a < 1$, L_{eff} effective anode length and r_a anode radius), where the effective insulator length (L_{ins}) is several times the effective anode length as well. With this electrode geometry, the lift-off stage has similar duration than the axial and radial stages. In the most plasma focus devices the quotient L_{eff}/L_{ins} has a value in the range 2-10, while r_{cat}/r_a is around 2.5 (r_{cat} cathode radii), in PF-50J L_{eff}/L_{ins} ~ 0.17 and r_{cat}/r_a ~ 4.5 [3].

Journal of Physics: Conference Series **511** (2014) 012029 doi:10.1088/1742-6596/511/1/012029

The drive parameter is defined as $S = I_{peak}/r_a \cdot P^{1/2}$ where I_{peak} is the peak of current and P the filling gas pressure. It has been found that S in the range 70-84 $kA/cm \cdot mbar^{1/2}$ for neutron optimized devices [4, 5]. Nevertheless, setting up device operation in this range for S does not ensure optimal operation. Optimization of plasma focus devices is usually done by an experimental survey in parameters as charge voltage (V_c), pressure, effective anode length and anode radii. In this work is presented an study on the optimization of PF-50J device as neutron source. Optimization is done at

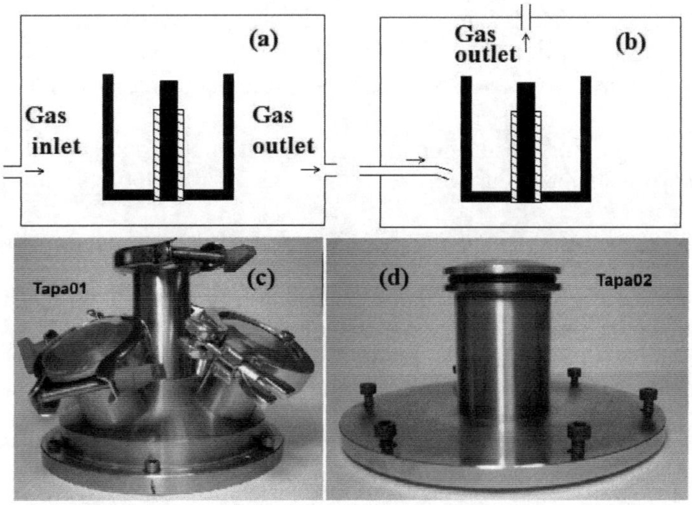

Figura 1. (a) Normal gas flow circuit, (b) Modified gas flow circuit, (c) Tapa01 chamber cup and (d) Tapa02 chamber cup.

fixed charge voltage and anode radii. One of the motivations is to study experiments with $r_{cat}/r_a \sim 2.5$, as well as to study influence of gas flow and chamber geometry in neutron emission.

2. Experimental setup and diagnostics.

The electrode configuration consist of an central stainless steel anode of 3 *mm* radii and a hollow of 4.5 *mm* diameter. The hollow deep is 18 *mm*. The insulator is made from alumina, 1/4" inner diameter and 3/8" outer diameter. The insulator effective length is 24.5 *mm*. The cathode is made from six stainless steel 5 *mm* diameter rods, 34 *mm* effective length. The inner cathode radius is 6.5 *mm* and the rod axis radius 8.75 *mm*. At the base of the bottom cathode plate and next to the insulator base a field distortion element was placed. Discharge is operated in autorupture at 28.5 *kV* charge voltage. Deuterium in flow is used as filling gas.

Neutron diagnostics consist in two detection systems based on ^{3}He proportional counters tubes for neutron yield measurements [6, 7] and arranges of plastic scintillators coupled to photomultipliers neutron measurements with temporal resolution. The ^{3}He detection systems were calibrated in efficiency only by using a standard isotopic fast neutron source (^{252}Cf). Neutron production is obtained by the use of a recently developed methodology to count events from the net output detector signal area and detector's efficiency [7].

3. Description of experimental studies.

The first study is related with optimization of PF-50J as neutron source. For this purpose effective anode lengths of 3.7 *mm*, 4.6 *mm*, 5.5 *mm*, 6.3 *mm* and 6.8 *mm* were used. Filling gas pressure was varied in the range 2-12 *mbar*. This study was done at normal flow circuit (see fig. 1.a). Neutron emission rate per pulse was measured from series of experiments at different operative conditions (*P* and L_{eff}). Besides the emission rate, some pieces of the neutron yield distribution are obtained. In particular,

- *Non null emission rate (NNE)*. It is the emission rate calculated only from shots registering neutron emission.
- *10% highest emission rate (10%S)*. It is the emission rate calculated from the 10% of shots with highest emission.
- *Best shot (BS)*. It is the neutron emission of the best shot in each series of experiment.

Each series of experiments has a sample size in the range of 150-300 observations.

The second study is focused on some engineering aspects of PF-50J device as a neutron source. The interest is to study the influence of the level of gas flow, gas flow circuit and size of the discharge chamber in the pulsed emission rate. Two setups for the level of gas flow were tested: medium and high. For the gas flow circuit two configurations were studied, in the first namely "normal", gas inlet and outlet are placed side in the chamber and far from the electrode configuration (fig 1.a). In the second namely "modified", gas inlet is placed side-on close to the electrode configuration, while gas outlet is placed end-on (fig. 1.b). Size chamber is modified by using two kinds of chamber cups, namely "Tapa01" and "Tapa02", as could be seen in figs. 1.c and 1.d. Four experiments were done with different setups as shown in table 1.

Table 1. Experimental setup for study on engineering issues.

Experiment	Chamber		Gas Flow Level		Flow Circuit	
	Tapa01	***Tapa02***	***High***	***Medium***	***Normal***	***Modified***
Serie1		•		•	•	
Serie2	•			•	•	
Serie3		•		•		•
Serie4		•	•			•

4. Results and analysis.

Plots for neutron emission rate as a function of pressure for different effective anode length are shown in figure 2. In this plot the second axis correspond to the emission rate estimated over a solid angle of 4π if source would be isotropic. Optimal configuration is obtained from the experiments to be

$$L_{ef} = 6.3\ mm \text{ and } P = 5.3\ mbar,$$

when device is operated at $r_a = 3$ *mm*, $r_{cat}/r_a = 2.17$, $V_c = 28.5$, autorupture, 0.1-0.5 Hz. With this setup neutron emission rate end-on is 240,5 ± 28 *n/pulso·sr* and 198,4 ± 22 *n/pulso·sr* , the estimated total neutron emission is 214,2 ± 17,4 *n/pulso*. In the plasma focus literature is usually reported the device neutron yield. This quantity is related to the neutron production observed in the so called "good" shots. In our case, we define "good" shots as the experiments inside the 10%S piece of the neutron yield distribution. For PF-50J in the optimal configuration reported in this work, the neutron

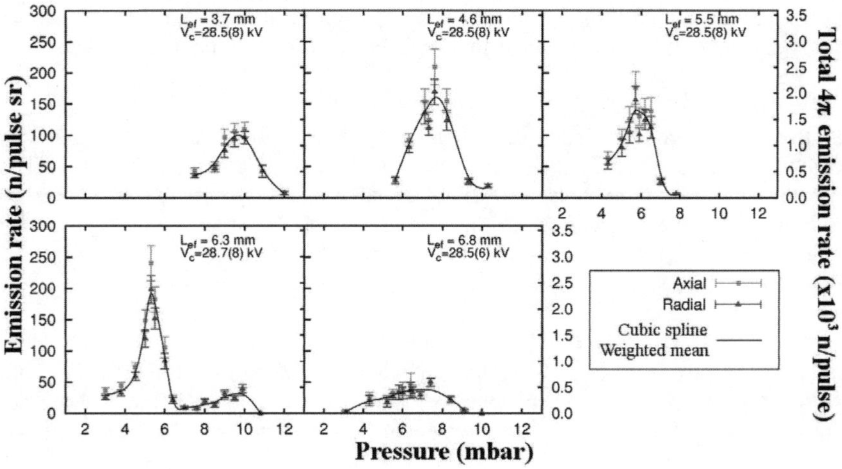

Figure 2. Neutron emission rates for different pressure and effective anode length. Operational conditions are D_2 in flow, $r_a = 3$ *mm*, $r_{cat}/r_a = 2.17$, autorupture.

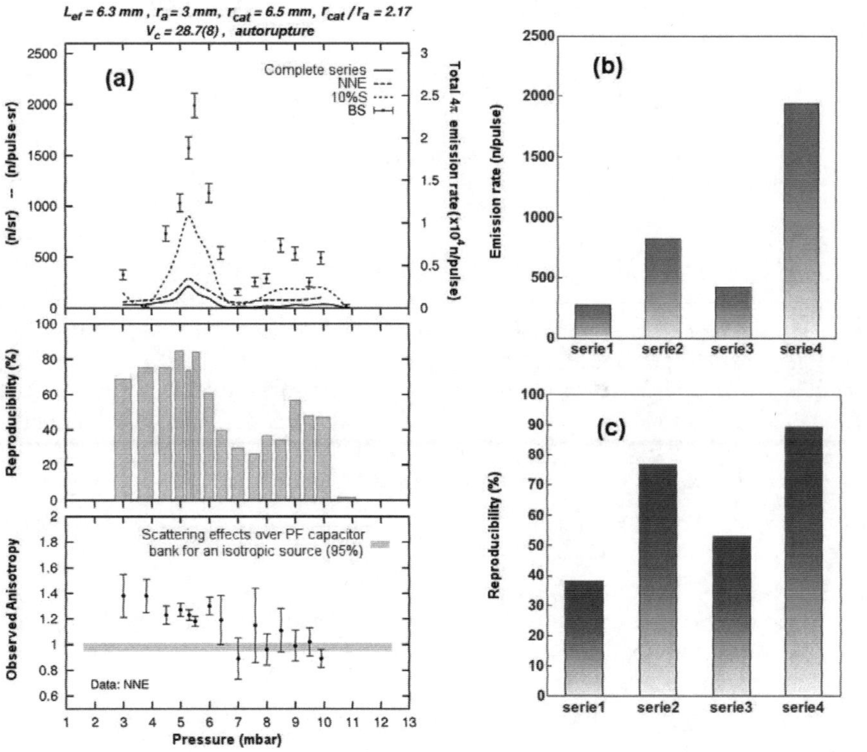

Figure 3. (a) Neutron emission, reproducibility and observed anisotropy for optimal configuration. (b) and (c) Emission rate and reproducibility for experiments on engineering issues.

yield is 1.3×10^4 *n/pulse,* which is in agreement with neutron scaling laws for devices operated with energies up to 1 *MJ* [7, 8]. The most intense shot registered during the experiments was $2.6 \pm 0.2\times10^4$ *n/pulse*.

Pieces of the neutron yield distribution (NNE and 10%S) follow the same trend than neutron emission rate, as could be seen in fig. 3.a . Consequently, the found device neutron optimization is still valid if only non null emission shots or "good" shots were been considered instead of neutron emission rate. From the same figure is concluded that reproducibility does not change in the same way than neutron emission rate, it means that configurations with low emission rate could still a significant reproducibility. Nevertheless, high reproducibility (R > 80%) is only obtained close to the optimal configuration.

Observed anisotropy is compared in figure 3.a with the anisotropy, for 95% confidence band, generated from an isotropic source when neutrons scatter in the capacitor bank of the device. This isotropic source anisotropy was obtained experimentally considering that detected scattered neutrons contribute in first order to the observed emission rate. From this results it is concluded that device emission is not isotropic and is more intense in the end-on direction. At this point it is necessary to point out that absolute anisotropy (without scattering contributions) is inaccessible to be measured in our device. The observed anisotropy for the optimal configuration is $A = Y_{end\text{-}on} / Y_{side\text{-}on} = 1.36$.

In figure 3.b and 3.c are shown histogram charts for emission rate and reproducibility of experiments in table 1. It is usually claim that contaminants removing by change the gas or working with gas flow is of importance for device performance. Our results indicate that gas flow circuit and chamber design are also of importance. The most efficient setup for impurities remotion (serie4) shows an improvement of almost nine fold over the worst setup (serie1). The effects of chamber design should be related with dispersion of remaining plasma which is ejected axially after pinch disruption.

5. Conclusions and final remarks.

The optimization of a small fast plasma focus device operated at 67 *J* was experimentally studied. It was found that the highest emission rate is obtained when device is operated with D2 in flow, 5.3 *mbar* 28.5 *kV* nominal charge voltage, autorupture, 3 *mm* anode radii, $r_{cat}/r_a = 2.17$ and $L_{ef} = 6.3$ *mm.* Neutron yield is 1.3×10^4 *n/pulse.* It was also observed that remotion of contaminants generated by the discharge are of supreme importance if a high emission rate per pulse is expected. A pulsed neutron source based on repetitive device similar to PF-50J should considering as a part of the engineering the gas flow level, gas flow circuit and a suitable chamber design in order to maximize neutron emission rate.

6. Acknowledgments.

Authors would like to acknowledge financial support from the programs PBCT-ACT26 and PSD-01, Center for Research and Applications in Plasma Physics and Pulsed Power – P^4 Chile.

References

[1] Soto L, Silva P, Moreno J, Zambra M, Kies W, Mayer R E, Clausse A, Altamirano L, Pavez C and Huerta L 2008 *J. Phys. D: Appl. Phys.* **41** 205215.

[2] Tarifeño-Saldivia A, Pavez C, Moreno J and Soto L "Dynamics and Density Measurements in a Small Plasma Focus of Tens of Joules Emitting Neutrons" *IEEE Trans. Plasma Science*, in

press.

[3] Soto L 2010 Private comunication.

[4] Lee S and Serban A 1996 *IEEE Trans. Plasma Science* **24** 1101.

[5] Soto L, Pavez C, Tarifeño A, Moreno J and Veloso F 2010 *Plasma Sources Sci. Technol.* **19** 055017.

[6] Moreno J, Birstein L, Mayer R E, Silva P and Soto L 2008 *Meas. Sci. Technol.* 19 087002.

[7] Tarifeño-Saldivia A 2011 *Experimental Study of a Tens of Joules Small Fast Plasma Focus Discharge Emitting Neutrons*. Phd thesis, Universidad de Concepción, Chile (In Spanish).

[8] Milanese M and Pouzo J 1988 Neutron-yield scaling laws for plasma focus devices. Small Plasma Physics Experiments p66.

Modeling of the pinch electron-density distribution in a small Plasma Focus

A. Márquez[1], J. González[2], A. Tarifeo[4] , C. Pávez[4], A. Clausse[3] and L. Soto[4]
[1]CNEA and Instituto Balseiro, 8402 Bariloche, Argentina
[2]CNEA-CONICET and Instituto Balseiro, 8402 Bariloche, Argentina
[3]CNEA-CONICET and Universidad Nacional del Centro, 7000 Tandil, Argentina
[4]CCHEN, Comisin Chilena de Energa Nuclear, Casilla 188-D, Santiago, Chile.

E-mail: arielmqz@gmail.com

Abstract. In this work, a model of the plasma dynamics was developed for a Plasma Focus device. Using MHD theory, coupled differential equations were generated to describe the macroscopic behaviour of pinch compression phase. In order to provide these equations with appropriate initial conditions, the modelling of previous phases was included by describing the plasma sheath as planar shockwaves. The equations were solved numerically and the results were contrasted against experimental measurements performed on the device PF-50, in order to evaluate the model's capability of predicting focus conditions at the time of maximum compression. Pinch dimensions and radial electron density profile were satisfactorily adjusted.

1. Introduction

Dense plasma pinching is a phenomenon occurring when an intense current is driven in a plasma column generating an azimuthal magnetic field that produces autoconfinement [1]. Plasma pinches are among the most efficient ways to achieve very high temperatures and particle densities, and have wide applications in radiation production and pulse power [2]. In plasma focus devices, the plasma pinch is the result of the time evolution of the electrical discharge that travels along coaxial electrodes. A charged capacitor bank is connected to the closed end of the electrodes through a gaseous switch (spark gap), and the discharges normally start over the insulator separating the electrodes. Then, the plasma sheath takeoff axially and radially accelerated by the magnetic field generated by the current itself. After the current sheath runs over the open end of the central electrode, the plasma becomes rapidly compressed into a small column resulting in very hot and dense plasma pinch. If the filling gas is Deuterium, fusion reactions are achieved, and a neutron pulse is observed. In this work, an extension of González et al. [3] model is presented, introducing a lumped parameter model of the pinch based in the Von Karman integral method (Von Karman and Biot [4]), to estimate pinch radius and electron density. The model predictions are compared with available experimental data of PF-50 plasma focus device, operating at CCHEN [5].

2. Model Description

The model uses Von Karman approximations, assuming the following plasma velocity $-v-$ and density $-\rho-$ profiles in the plasma cylinder (pinch):

$$v_r(r,z,t) = V(t)\Big(\frac{r}{R(t)}\Big)^a \tag{1}$$

$$v_z(r,z,t) = \frac{C(t)z}{H(t)} \tag{2}$$

$$\rho(r,z,t) = \rho_0(t) - \Big(\rho_0(t) - \rho_R(t)\Big)\Big(\frac{r}{R(t)}\Big)^b \tag{3}$$

where R is the radius of the plasma cylinder (which extends from $-H$ to H in z), $V = dR/dt$, $C = dH/dt$, $\rho_0 = \rho|_{r=0}$ and $\rho_R = \rho|_{r=R}$. Constants a and b are adjunting Von Karman parameters. This aproximations are used for solving MHD equations; for this purpose, aditional hypothesis are introduced. The first is assuming the ideal gas relation as equation of state for plasma. This relation is then used to convert internal energy contained in the pinch to pressure, assuming also equallity between electron and ion numerical densities and temperatures. Internal energy is calculated using energy balance equations for the whole PF system. The second aproximation assumes an axial and superficial pinch current density, which can be expressed as follows:

$$\bar{J} = \frac{I}{2\pi r}\delta\Big(r - R(t)\Big)\hat{z} \tag{4}$$

where $\bar{J}$ is the current density and I is the total pinch current. The pinch expands freely against the chamber gas at the filling pressure in the axial direction. The mentioned simplifications are introduced into the radial and axial momentum conservation equations, obtaining the following final expressions:

$$\frac{dV}{dt} = \frac{1}{\Big(\tilde{\rho}R^2H\Big)}\frac{d\Big(\tilde{\rho}\,VR^2H\Big)}{dt} - V\left[\frac{b\frac{d\rho_0}{dt} + (a+2)\frac{d\rho_R}{dt}}{b\rho_0 + (a+2)\rho_R} + \frac{C}{H} + \frac{2V}{R}\right] \tag{5}$$

$$\frac{d}{dt}\Big(\tilde{\rho}\,VR^2H\Big) = -\frac{I^2\mu_0 H}{8\pi^2 R} + (\gamma-1)RH\left[\Big(\frac{E}{M_p} - \frac{C^2}{6}\Big)\frac{\rho_R + b\rho_0}{b+1} - \frac{V^2}{2}\,\frac{(2a+1)\rho_R + b\rho_0}{(2a+b+1)(2a+1)}\right] \tag{6}$$

$$\frac{dC}{dt} = \frac{2(\gamma-1)}{H}\left[\frac{E}{M_p} - \frac{(b+2)}{(a+1)(2a+b+2)}\,\frac{\Big(b\rho_0 + 2(a+1)\rho_R\Big)}{\Big(b\rho_0 + 2\rho_R\Big)}\frac{V^2}{2}\right] \tag{7}$$

where E is the time-dependant total energy of the pinch, M_P is the mass of plasma trapped inside the plasma column −considered as constant once the pinch is formed−, γ is the heat capacity ratio for the ion gas, μ_0 is the vacuum permeability and $\tilde{\rho}$ is an effective mass density defined by

$$\tilde{\rho} = \frac{\rho_0}{a+2} - \frac{\rho_0 - \rho_R}{a+b+2} \tag{8}$$

The initial value problem is solved for the coupled variables R, H, V, C, ρ_0, ρ_R, E and two PF electrical circuit variables: LI (product of total inductance of the PF system and the current) and Q (charge in the capacitor bank). The initial conditions are given by the final value of the variables describing the stage prior to compression, as described in Reference [3].

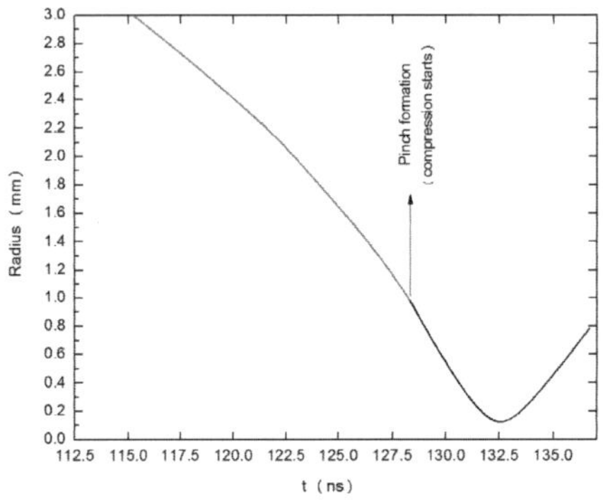

Figure 1. Radial position $R(t)$ of the cylindrical planar piston followed by the pinch compression.

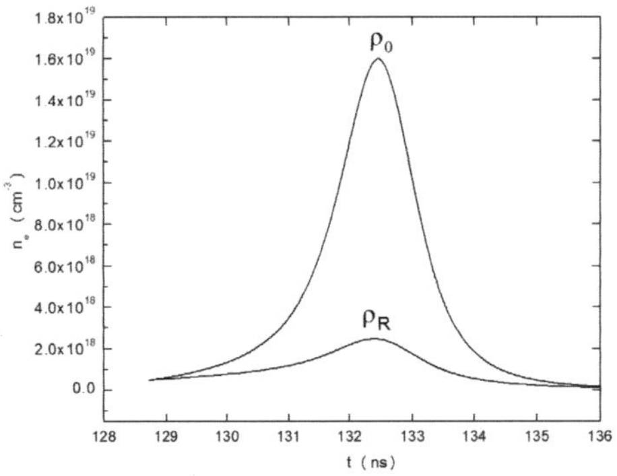

Figure 2. Density evolution during the compression phase at $r = 0$ and $r = R(t)$.

3. Results

PF-50 was modelled, operating at 29 kV with a deuterium filling pressure of 250 Pa. Calculated pinch radius and electron density evolution using $b = 2$ and $a = 0.3$ are shown in Figures 1 and 2 respectively. Using relation (3), the density profile at the time of maximum compression was reconstructed. This result is exhibited in Figure 3, where the experimental points are also shown. It can be seen that the results show good agreement with the experiments using a Von Karman exponent $b = 2$. The minimum pinch radius predicted is 0.12 mm, which should be compared with estimations from the interferograms $\approx$ 0.28 mm. Nevertheless, nearly 80% of the total mass trapped inside the measured pinch is contained within 0.12 mm radius (calculated value). The selected values for a and b are those that best fit the experimental data. In particular, if b is incremented to fit data curvature, the pinch radius obtained is smaller. However, the magnitude of the calculated maximum density is relatively unsensitive to a and b changes when compared to initial-condition variations [3], so the model is successful in predicting this value.

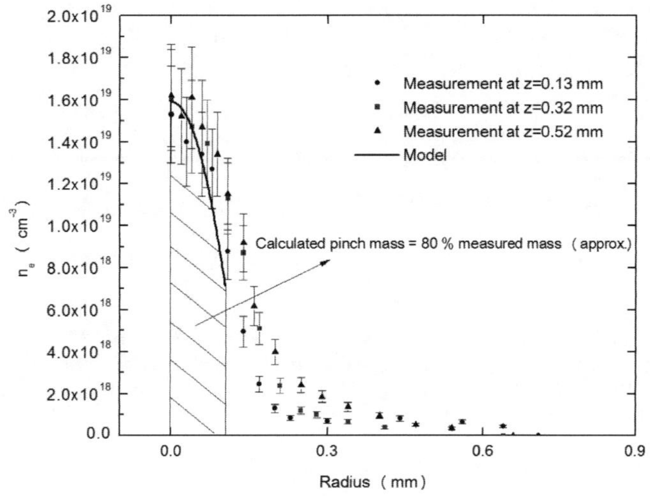

Figure 3. Radial profile of the electron-density inside the pinch at the time of maximum compression. Model (curve) and experimental data measured at different axial positions (points).

4. Conclusions

A lumped parameter model based on Von Karman approximations for calculating pinch compression in Plasma Focus discharges has been presented. Numerical calculations showed good agreement with experimental data obtained in PF-50, particularly in predicting the electron density profile and values inside the pinch. Also a good representation of total mass trapped is observed, although external pinch radius is underestimated. These results are very useful to study neutron production mechanisms, oriented to develop a conceptual design tool, to optimize neutron yield as a function of the device parameters.

References

[1] Liberman M 1999 *Physics of High-Density Z-Pinch Plasmas* (Springer-Verlag)
[2] C Moreno e a 2002 *Braz. J. Phys.* **32** 20–25
[3] J Gonzlez e a 2009 *IEEE Transactions on Plasma Science* **37** 2178–2185
[4] VonKarman and Biot 1940 *Modeling of the Dynamic Plasma Pinch in Plasma Focus Discharges Based in Von Karman Approximations* (McGraw-Hill)
[5] A Tarifeno e a 2009 *AIP Conf. Proc.* **1088** 227–230

Numerical simulation of current sheet instabilities in a small plasma focus

F Casanova[1], A Tarifeño[2], F Veloso[2], L Soto[2] and A Clausse[1,3]
[1]CNEA-CONICET and Universidad Nacional del Centro, Pinto 399, (7000) Tandil, Argentina.
[2]CCHEN, Comisión Chilena de Energía Nuclear, Casilla 188-D, Santiago, Chile.
[3] Corresponding author.

E-mail: clausse@exa.unicen.edu.ar

Abstract. A comparison between numerical simulations of discharges in the Plasma Focus device PF-50J using the code Cshock is presented. The code was able to reproduce kinetics and circuit parameter experimental measurements as well as the particular current sheath instability observed in this specific device.

1. Introduction

The computer aided optimization of plasma-focus (PF) devices requires the calculation of the sheath movement and the sensibility of the process to geometrical and operational parameters. This calculation involves the tracking of the ionizing shock accelerated to hypersonic velocities. During the displacement, the shape of the current sheath changes, inducing, in turn, variations on the boundary conditions, the forces and on the dissipation rates. CShock [1,4] is a numerical code of PF discharges based in a two-dimensional description of the current sheath (CS), which is represented by means of a set of dynamic conical elements moving and reshaping themselves according to the conservation equations of mass, momentum and energy [2]. The code showed good performance in tracking the current sheath in various PF and recently it was used to calculate the neutron production in a pinch [3]. In the present paper, Cshock is applied to simulate the CS evolution in a small plasma focus where the rundown and runover stages overlap.

2. The model

A PF consists basically of two coaxial electrodes, inner anode and external cathode, separated by an insulator material at one end. The space between electrodes is filled with low pressure gas and, when pulsed high voltage is applied between the electrodes, a discharge starts over the insulator from which the plasma sheath take-off axially accelerated by the magnetic field generated by the current itself. In cases where the cathode is just a planar plate, the current sheath expands freely radially away from the axis. The evolution of the mass (m) and momentum (mv) of a CS element of area A (Fig. 1), length l, velocity v, located at radius r, follows:

$$\frac{dm}{dt} = \rho A v \tag{1}$$

$$\frac{d(mv)}{dt} = \frac{\mu_o}{4\pi}\frac{l}{r}I^2 \tag{2}$$

Where ρ is the stagnant density and I is the electrical current, which is calculated from the equation of an electrical circuit with variable inductance:

$$\frac{d}{dt}\left[\left(L_{ext} + L_g\right)I\right] + R_{ext}I + \frac{Q}{C} = 0 \tag{3}$$

$$\frac{dQ}{dt} = I \tag{4}$$

In Eqs. (3) and (4), L_{ext} and R_{ext} are the external inductance and resistance, L_g is the variable inductance associated to the coaxial gun, and Q and C are the bank charge and capacity.

Figure 2 shows a 3D visualization of the current sheath representation. The sheet is re-structured every step in order to maintain a uniform number of nodes per unit length (in the present case 3 nodes per mm).

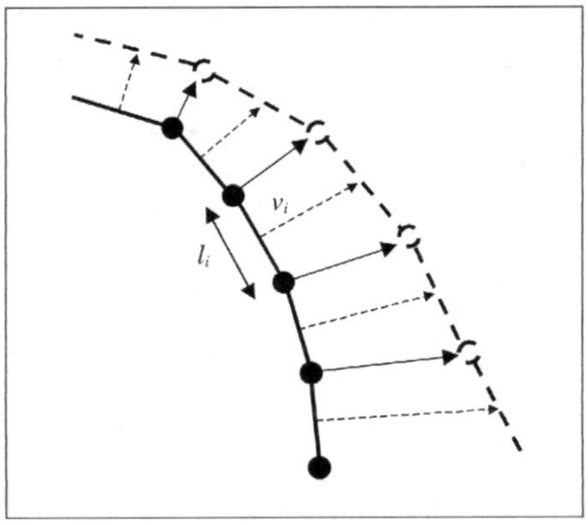

Figure 1. Diagram of the numerical calculation of the movement and shaping of the current sheath.

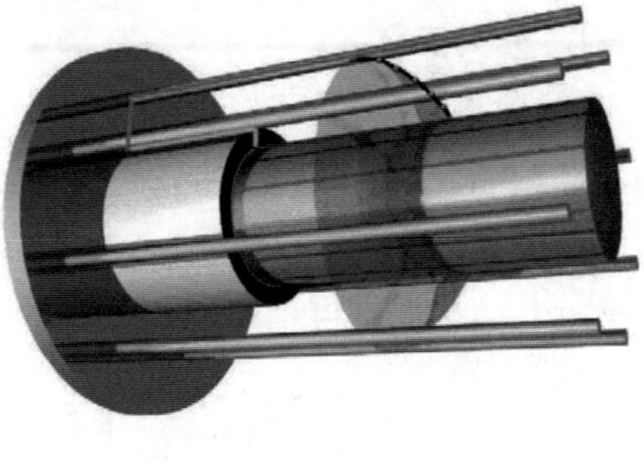

Figure 2. Visual representation of the discretization of the current sheath.

3. Results

Cshock was applied to calculate the shape evolution of the CS in the PF device PF-50J [1], which was compared with experimental visualizations and electrical diagnostics. The electrical parameters used in the simulation were: capacity 158 *nF*, external inductance 41.6 *nH*. The geometrical parameters were: anode radius and length 3 and 27.9 *mm* respectively, insulator radius and length 4.7625 and 23.9 *mm* respectively. The operation parameters were: D_2 filling pressure 6 *mbar* and charging voltage 29 *kV*. It should be stressed that Cshock does not require the introduction of sweeping parameters since, unlike the planar pistons models [5-6], the shape of the current sheet is solved along with the evolution of the electrical and thermodynamic variables.

Figure 3 shows the evolution of the electrical current and the minimum radial position of the sheath. It can be seen that the numerical calculations reproduces well the experimental trend. Figure 4 compares the calculated shape of the current sheath at different times with the shape deduced from interferometrical measurements. It is interesting to observe that the code was able to reproduce a singular elbow observed immediately after the lift-off of the plasma sheet. The singularity is produced in the code by an instability brought about by a local accumulation of mass which start at the border of the insulator. Figure 4 shows the mass distribution along the CS, where a peak density appears where the CS presents a singular elbow.

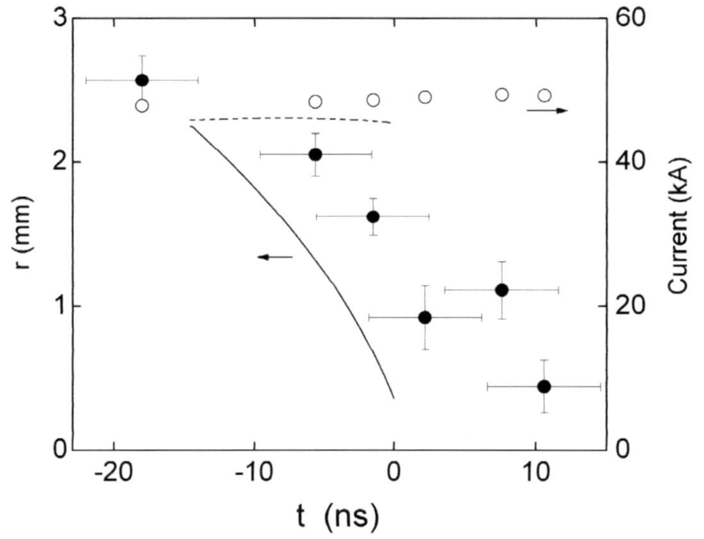

Figure 3. Evolution of the radial position of the current sheet (full and solid) and the electrical current (void and dashed). Experiment (points) and numerical results (curve). The time *t* is measured relative to the pinch.

4. Conclusions

A two-dimensional model of the current sheath dynamic shaping in discharges on a small 50 kJ plasma focus was presented. The code was able to reproduce kinetics and circuit parameter experimental measurements as well as the particular current sheath instability observed in this specific device. The singular trend of the sheath observed in the initial stages of the discharge is reproduced by the numerical model, where it is caused by an instability brought about by a local accumulation of mass which start at the border of the insulator.

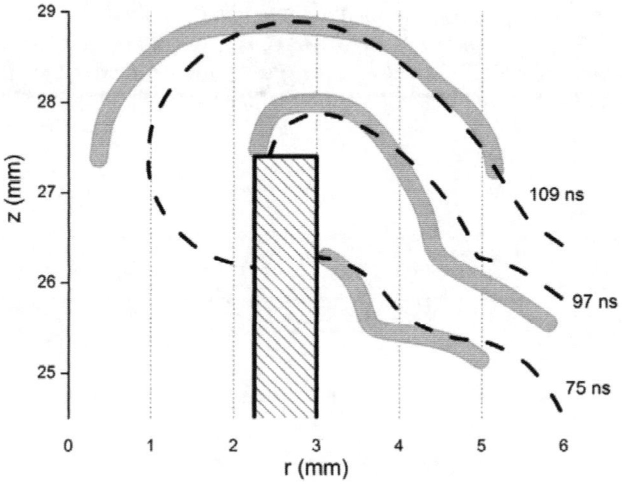

Figure 4. Shape of the current sheet at different times. Numerical (dash), experimental (solid grey). The numbers at the right indicate the corresponding time after the closure of the spark gap.

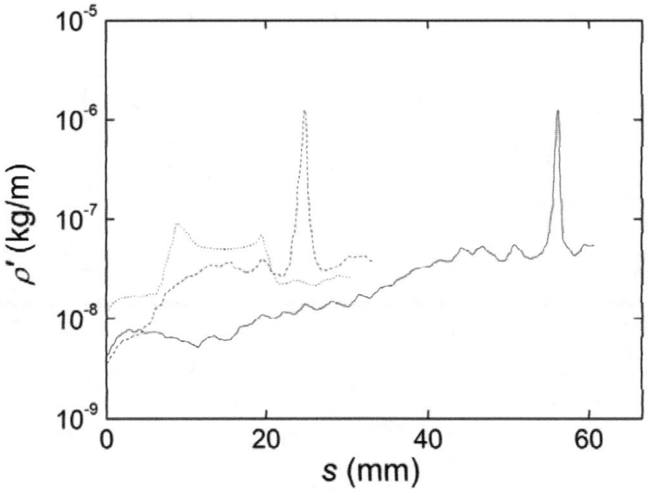

Figure 5. Linear density of the nodes of the current sheet at 75 ns (dots), 97 ns (dashes) and 109 ns (solid). The coordinate *s* represents the position along the CS starting at the anode.

References

[1] L. Soto, P. Silva, J. Moreno, M. Zambra, W. Kies, R. Mayer, A. Clausse, L. Altamirano, C. Pavez, L. Huerta. *J. Physics D: Applied Physics* **41** (2008) 205215 (7pp).

[2] F. Casanova, C. Moreno, A. Clausse, *Plasma Physics and Controlled Fusion Journal* **47** (2005) 1239-1250.
[3] F. Casanova, A. Clausse, *Mecánica Computacional* **28** (2009) 2385-2391.
[4] A. Tarifeño, C. Pavez, J. Moreno, L. Soto, *AIP Conf. Proc.* **1088** (2009) 227-230.
[5] J. González, M. Barbaglia, F. Casanova, *Brazilian Journal of Physics* **39** (2009) 633-637.
[6] S. Lee, *Radiations in Plasmas*, ed B McNamara, Singapore: World Scientific (1984) 978–87.

Non-radioactive Source for Field Applications Based in a Plasma Focus of 2J: Pinch evidence

Leopoldo Soto[1, 2], Cristian Pavéz[1, 2], José Moreno[1, 2], José Pedreros[3], and Luis Altamirano[4]

[1] Comisión Chilena de Energía Nuclear (CCHEN), Casilla 188-D, Santiago, Chile
[2] Center for Research and Applications in Plasma Physics and Pulsed Power, P4, Chile
[3] Universidad de Santiago, Chile
[4] Dicontek Ltda., Chile

E-mail: lsoto@cchen.cl

Abstract. A portable plasma focus for field applications operating at two joules have been designed and constructed. Pinch evidence in the current derivative signal have been obtained. Also the dependence of the time to pinch , t_p, over the filling pressure, p_0, ($t_p \alpha \sqrt{p_0}$) was observed. Therefore, the device is working as a plasma focus.

1. Introduction.

Small and compact plasma focus devices are not only interesting for pure plasma research. Also are especially suitable for field applications. More remarkable is that it constitutes a safe radiation on-off source. The plasma focus is especially suited for pulsed neutron applications with a reduced danger of contamination compared to conventional isotopic radioactive sources. A passive radioactive source of neutrons with similar energy (~ 2.5 MeV, for instance 252Cf or Am/Be) emits continuously, carrying inconveniences in handling and storing. In turn, plasma-focus sources would not have such activation problems. The pulses radiation (neutrons and X-rays) has a duration of a few nanoseconds, in particular X-rays have a wavelength of nanometers or less, thus it is reasonable to call them "nanoflashes".

2. The device.

Using the experience obtained in the scalability of plasma focus devices [1-5], a compact and portable device was designed and constructed. This device was designed for operation with few kilovolts (10kV or less) with a stored energy of 2J and a repetition rate of 10Hz without external cooling. The electrical parameters are: capacitance 180nF, inductance 40nH, voltage operation from 5 to 8kV, energy stored from 2-5J, peak current from 10 to 17kA achieved in ~ 100ns. The size of the device is of the order of $25x25x20cm^3$. The weight is less than 3kg. Figure 1 shows a photograph of the device.

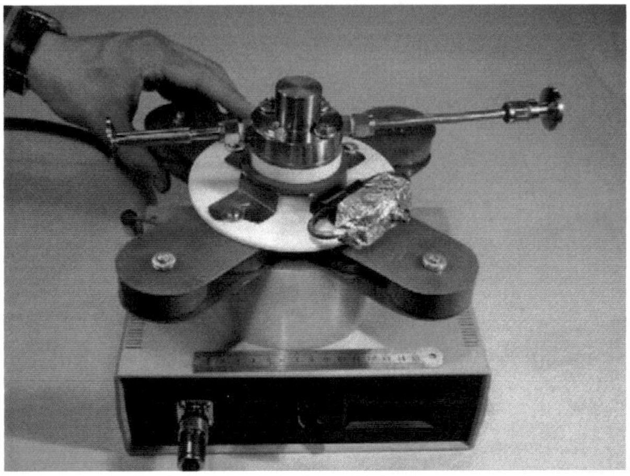

Figure 1. PF-2J: a portable plasma focus device.

3. Results.

Electrical characterization for discharges in hydrogen was developed. Evidence of pinch was obtained in the current derivative dI/dt and current I signals (figure 2). Also the dependence of the time to pinch, t_p, over the filling pressure, p_0, ($t_p \alpha \sqrt{p_0}$) was observed (figure 3).

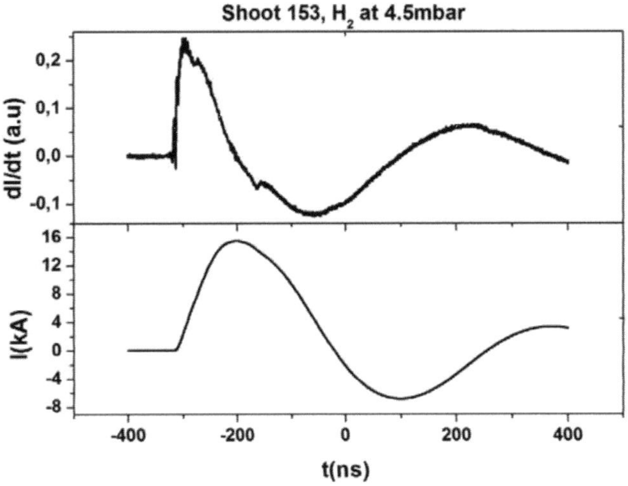

Figure 2. Electrical signals of the current derivative dI/dt and current I. The typical dip in the dI/dt and the drop in I, that correspond to the pinch effect is observed.

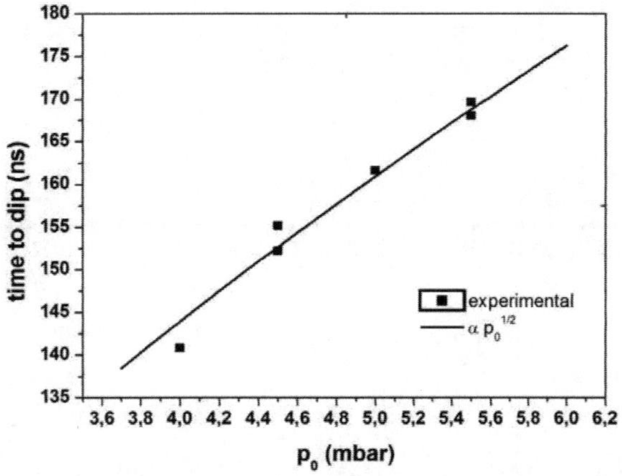

Figure 3. Time to dip vs filling pressure for discharges in hydrogen. The time to dip dependence on pressure, $\alpha \sqrt{p_0}$, shows that the device is working as plasma focus.

4. Conclusions.

A compact and transportable plasma focus device was designed and constructed, PF-2J as a step to design a non-radioactive source for field applications. The device is operated with a charging voltage less than 10kV and with a stored energy less than 5J. Evidence of pinch was obtained from signals of current derivative. In addition, the dependence of the time to pinch, t_p, over the filling pressure, p_0, ($t_p \alpha \sqrt{p_0}$) was observed. Future works include discharges in deuterium and neutron emission characterization. A neutron flux of the order of 10^4-10^5 n/s is expected for operation rates of 10 to 100Hz. In addition a discharge camera with optical windows will be constructed to study de plasma dynamics.

5. Acknowledgments.

Authors would like to acknowledge financial support from the grants PBCT-ACT26 and PSD-01 from CONICYT.

References.

[1] P. Silva, J. Moreno, L. Soto, L. Birstein, R. Mayer, and W. Kies, App. Phys. Lett. 83, 3269 (2003).
[2] L. Soto, P. Silva, J. Moreno, M. Zambra, W. Kies, R. E. Mayer, A. Clausse, L. Altamirano, C. Pavez, and L. Huerta, J. Phys. D: App. Phys. 41, 205215 (2008).
[3] L. Soto, C. Pavez and J. Moreno, M. Barbaglia and A. Clausse , Plasma Sources Sci. and Technol. 18, 015007 (2009).
[4] C. Pavez and L.Soto, IEEE Trans. Plasma Sci. 38, 1132 (2010).
[5] L. Soto, C. Pavez, A. Tarifeño, J. Moreno and F. Veloso, Plasma Sources Sci. and Technol. 19, 055017 (2010).

Journal of Physics: Conference Series **511** (2014) 012033 doi:10.1088/1742-6596/511/1/012033

Pinch density measurements in compact plasma foci of 400J and 50J

Ariel Tarifeño-Saldivia[2,3], Cristian Pavez[1,2] and Leopoldo Soto[1,2]

[1] Comisión Chilena de Energía Nuclear (CCHEN), Casilla 188-D, Santiago, Chile
[2] Center for Research and Applications in Plasma Physics and Pulsed Power, P4, Chile
[3] Universidad de Concepción, Chile

E-mail: lsoto@cchen.cl

Abstract. A Mach-Zehnder interferometer using a pulsed Nd-YAG laser (600mJ, 532nm, 8ns) was implemented to measure the electron density and the dimensions of the pinch column in two sub-kJ compact plasma focus devices operating at hundred joules (PF-400J) and tens of joules (PF-50J).

1. Introduction.

Interesting features observed in plasma focus devices operating from 1MJ to 1kJ are that the pinch radius, r_p, and pinch length, z_p, scale with the anode radius, a, as $r_p \sim (0.1\text{-}0.2)\ a$, $z_p \sim (0.8\text{-}1)\ a$, and the density of the pinch is of the order of 10^{25} m^{-3} on the axis. To study if these features appears in plasma focus devices operating in the sub-kJ range, pulsed interferometry diagnostics were applied to a plasma focus operating at hundred joules, PF-400J, and to a plasma focus operating at tens of joules, PF-50J.

2. Experimental details and results.

A refractive optical system was implemented in order to measure the electron density and the dimensions of the pinch column in two sub-kJ compact plasma focus devices, (PF-400J, 400J, 130kA achieved in 300ns [1] and PF-50J, 70J, 60kA achieved in 150ns [2]). A Mach-Zehnder interferometer using a pulsed Nd-YAG laser was implemented (600mJ, 532nm, 8ns). Figure 1 shows interferograms close to the pinch time for the PF-400J and figure 2 shows the respective electron density profile obtained from the Abel inversion. Figure 3 shows an interferogram close to the pinch time for the device PF-50J and figure 4 shows the respective electron density profile.

An average electron density of $(8.4 \pm 1.3) \times 10^{24}$ m^{-3} was measured on the axis in the PF-400J operated in H_2 at 5mbar, and of $(1.5\pm0.2) \times 10^{25}$ m^{-3} in the PF-50J operated in D_2 at 6mbar. The pinch radius observed with this method in the PF-400J was of the order of $0.12a$ and between $0.1a$ to $0.2a$ in the PF-50J, where a is the anode radius. The corresponding line density was measured in $(8.6\pm1) \times 10^{18}$m^{-1} for the PF-400J operated in H_2 and in $(2.2\pm0.3) \times 10^{18}$ m^{-1} for the PF-50J operated in D_2. With these values the Bennett temperature can be calculated, being ~300eV for the PF-400J and ~250eV for the PF-50J.

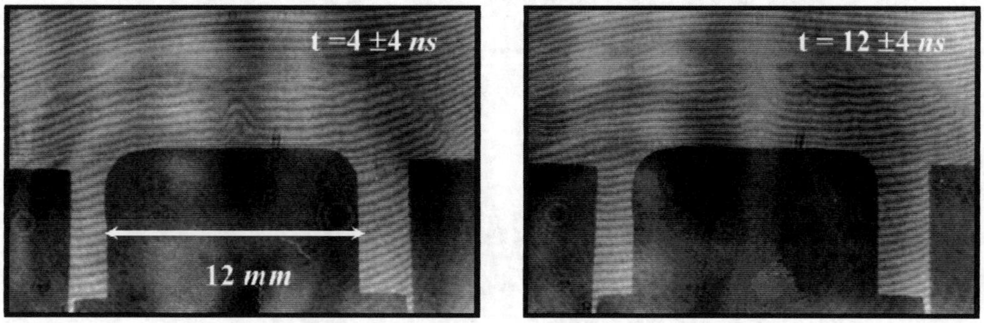

Figure 1. Interferograms close to the pinch time for a discharge in hydrogen at 5 mbar in the PF-400J [3]. Time zero corresponds to the dip time in the current derivative signal. Anode radius is 6mm. Pinch radius ~0.85mm, pinch length ~4mm.

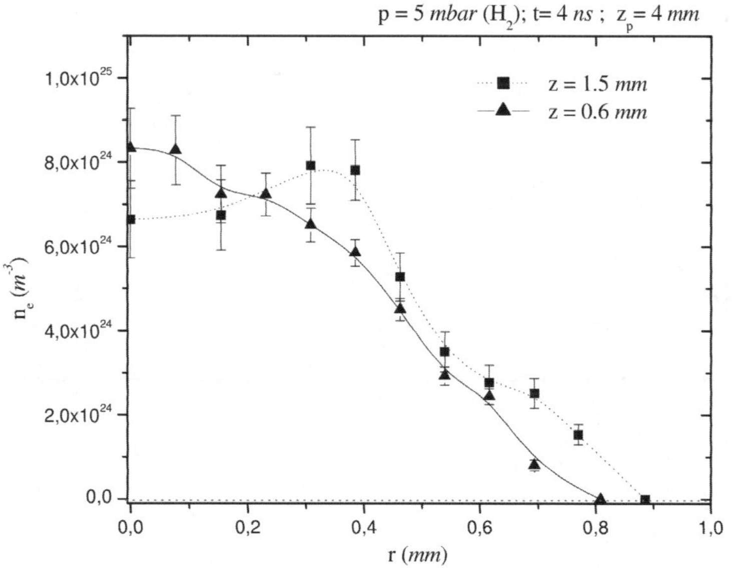

Figure 2. Electron density profile obtained from the Abel inversion of the interferogram of figure 1 at 4ns after the dip in the current derivative [3]. The profiles are at two different position z from the anode top. The line density, N, was measured in (8.6±1) $\times 10^{18}$ m^{-1}. Pinch radius has a value of (0.85±0.04)mm.

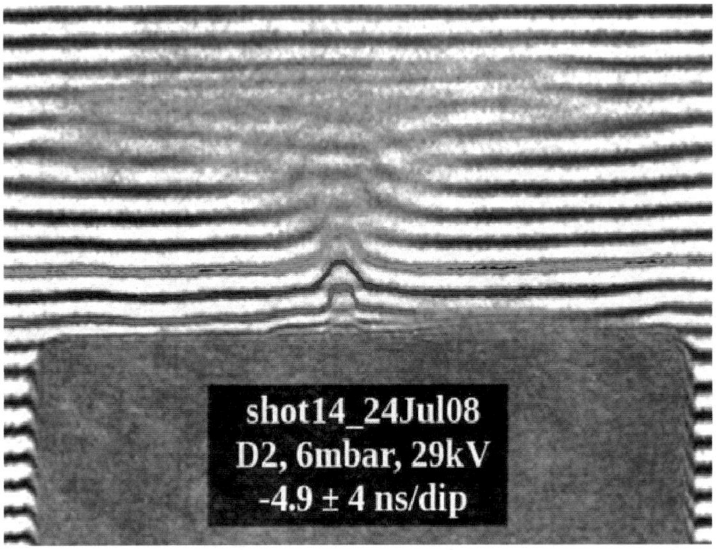

Figure 3. Interferogram close to the pinch time for a discharge in deuterium at 6 mbar in the PF-50J. Time zero corresponds to the dip time in the current derivative signal. Anode radius is 3mm. Pinch radius ~0.3mm, pinch length ~2.5mm.

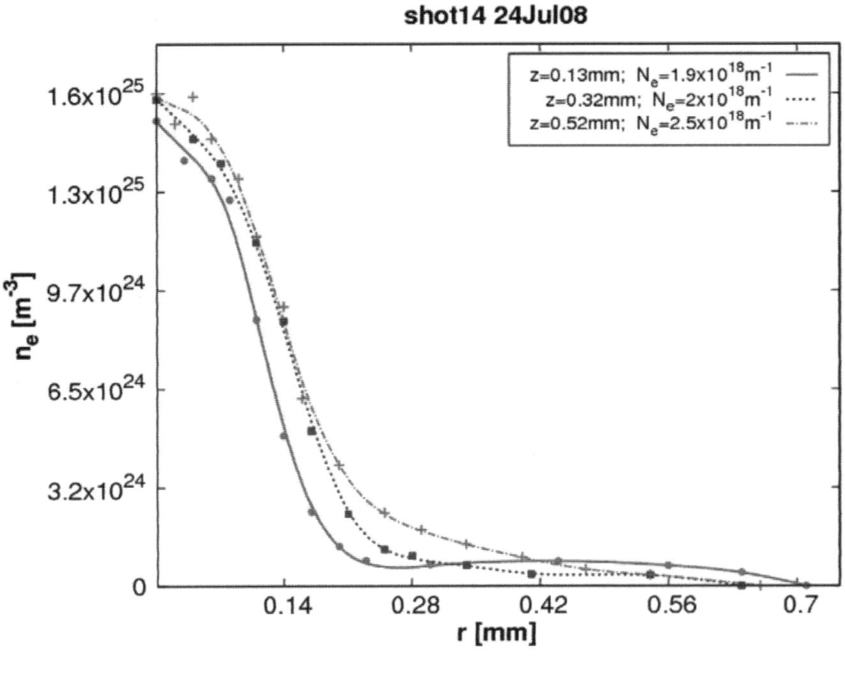

Figure 4. Electron density profile obtained from the Abel inversion of the interferogram of figure 3 at 4.9ns before the dip in the current derivative. The profiles are at three different position z from the anode top. The line density, N, was measured in (2.2±0.3) $\times 10^{18}$ m^{-1}. Pinch radius can be measured in ~0.3mm.

Journal of Physics: Conference Series **511** (2014) 012033 doi:10.1088/1742-6596/511/1/012033

3. Conclusions.

The results show that the the pinch radius, r_p, and pinch length, z_p, scale with the anode radius, *a*, as $r_p \sim (0.1\text{-}0.2)\ a$, $z_p \sim (0.8\text{-}1)\ a$, and the density of the pinch is of the order of 10^{25} m^{-3} on the axis in the PF-400J and in the PF-50J. Using the Therefore, the pinch conditions, and electron density observed in larger machines are also present in these experiments operated at only hundred and tens of joules.

4. Acknowledgments.

Authors would like to acknowledge financial support from the grants PBCT-ACT26 and PSD-01 from CONICYT.

References.

[1] P. Silva, J. Moreno, L. Soto, L. Birstein, R. Mayer, and W. Kies, App. Phys. Lett. **83**, 3269 (2003).
[2] L. Soto, P. Silva, J. Moreno, M. Zambra, W. Kies, R. E. Mayer, A. Clausse, L. Altamirano, C. Pavez, and L. Huerta, J. Phys. D: App. Phys. 41, 205215 (2008).
[3] C. Pavez and L. Soto, Physica Scripta T131, 014030 (2008).

Optical Emission Spectroscopy Observations Of Fast Pulsed Capillary Discharge Plasmas

G Avaria[1], M Ruiz, F Guzmán, M Favre, E S Wyndham, H Chuaqui and H Bhuyan

Departamento de Física, Pontificia Universidad Católica de Chile, Casilla 306, Santiago 22, Chile

E-mail: `mfavre@fis.puc.cl`

Abstract. We present time resolved optical emission spectroscopic (OES) observations of a low energy, pulsed capillary discharage (PCD). The optical emission from the capillary plasma and plasma jets emitted from the capillary volume was recorded with with a SpectraPro 275 spectrograph, fitted with a MCP gated OMA system, with 15 ns time resolution. The discharge was operated with different gases, including argon, nitrogen, hydrogen and methane, in a repetitive pulsed discharge mode at 10-50 Hz, with, 10-12 kV pulses applied at the cathode side. The time evolution of the electron density was measured using Stark broadening of the H_β line. Several features of the capillary plasma dynamics, such as ionization growth, wall effects and plasma jet evolution, are inferred from the time evolution of the optical emission.

1. Introduction

Pulsed Capillary Discharges (PCD) are versatile transient plasma devices which are currently being used in different applications, including lasing in the extreme ultraviolet (EUV) [1], EUV and soft x-ray sources for lithography and microscopy [2, 3], and plasma jets production [13], among others. The transient behaviour and characteristic geometry of a PCD, a narrow open ends tube with a large aspect ratio, establishes natural conditions for the occurrence of the hollow cathode effect (HCE) [5], and the emission of plasma jets [6]. Electron beams emitted from the hollow cathode region (HCR) favor on-axis initiation of the PCD. Important issues in PCD investigations are the study of the plasma dynamics and the measurement of characteristic plasma parameters with time and space resolution. Time resolved optical emission spectroscopy is a suitable diagnostic to measure both, electron and density and temperature. The first is achieved by Stark broadening measurements of suitable transition lines and the second, by line intensity ratios or Boltzmann plots. OES also allows the dominant plasma species to be identified. This is important in order to evaluate potential applications of PCD as EUV and soft x-ray sources. We have performed OES observations of the capillary plasma, at different stages of the capillary discharge evolution, operating with different gases. Stark broadening of the H_β line was used to measure the time evolution of the electron density. Identification of Oxygen and Aluminium line emission allowed to identify the effect of capillary wall plasma in the electric breakdown phase of the discharge.

[1] Present address: NSF ERC for Extreme Ultraviolet Science and Technology, Department of Electrical and Computer Engineering, Colorado State University, Fort Collins, CO 80523, USA

2. Experimental Details

The experimental observations have been performed in a low inductance high voltage PCD device. The capillary is an alumina tube, 55 mm long, 1.6 mm internal diameter. The discharge was operated with different gases, including argon, nitrogen, hydrogen and methane, at pressures between 0.6 and 1.4 Torr in the HCR, in a repetitive pulsed discharge mode at 10-50 Hz, with, 10-12 kV pulses applied at the cathode side. The stored energy is less around 30 mJ per pulse. Characteristic current is of the order of 1 kA, in a 10 ns FWHM pulse. A schematic of the capillary assembly is shown in figure 1. Spectroscopic measurements were performed with a SpectraPro 275 spectrograph with a MCP gated OMA system. The system was operated in pulsed mode synchronized with the discharge voltage, allowing registration of spectra at different times during the plasma evolution. Time resolution was determined by a 15 ns integration time. Optical emission was observed axially, to investigate the time evolution of the capillary plasma, and at the capillary exit, transversal to the axis, to investigate plasma jets emitted from the capillary discharge, as shown schematically in figure 2. Electron beams were observed with a scintillator-photomultiplier arrangement, coaxial with the capillary, at the anode side [7]. Further details of the experimental set-up and OES arrangement have been published elsewhere [5].

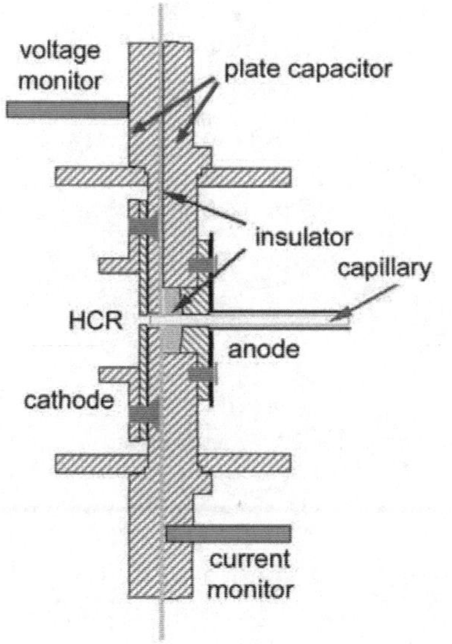

Figure 1. Capillary discharge assembly.

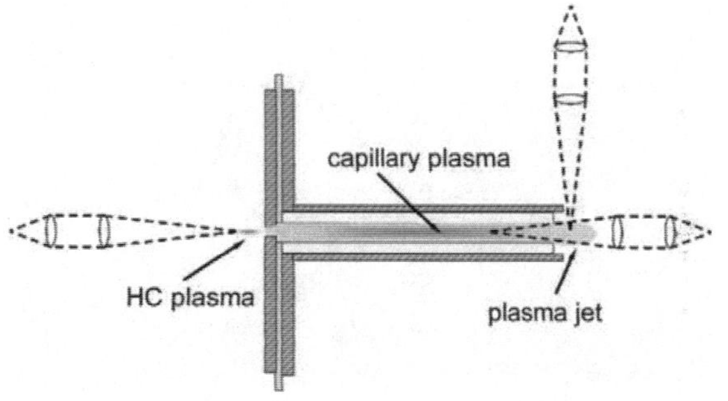

Figure 2. Plasma emission observation arrangement.

3. Results and Discussion

Figure 3 shows the voltage, HC emission and electron beam signals for an Argon discharge. Four distinctive phases are identified in the time evolution of the different signals, as defined in the figure. 1) the hollow cathode phase, 2) the ionization wave phase, 3) the electric breakdown and hot capillary plasma phase, and 4) the plasma expansion and jet emission phase. Phase 1 takes place before electric breakdown in the capillary, and correspond to the formation of a HCR plasma, due to the hollow cathode effect [7]. In phase 2, still before full electric breakdown in the capillary, a fast ionization wave propagates from the cathode towards the anode, along the capillary, giving rise to plasma formation. This has been establised by experimental observations using capacitive probes along the capillary [7] and by a theoretical model of ionization wave generation [8]. This process is associated with a sharp increase in HC plasma emission and strong electron beam activity. Phase 3 corresponds to electric breakdown inside the capillary, with hot plasma formation, in association with a very fast, ~10 ns, current pulse of kA level [9]. In phase 4 capillary plasma is expelled from the capillary, at the anode side, in the shape of a plasma jet [6]. Figure 4 shows the time evolution of Argon emission from the HCR, in the 460-500 nm band. $t = 0$ ns corresponds to electric breakdown. The intensity of the spectral emission follows clearly the light emission curve from the HC, seen in figure 3. The spectral lines correspond to ArII emission. Detailed analysis of the HCR plasma indicate that the characteristic electron temperatures and electron densities are in the ranges 1.4-1.8 eV, $10^{15} - 10^{16}$ cm^{-3}, respectively, which is produced by collisional processes involving electrons accelerated by the electric field leaking inside the HCR, with no additional heating [5].

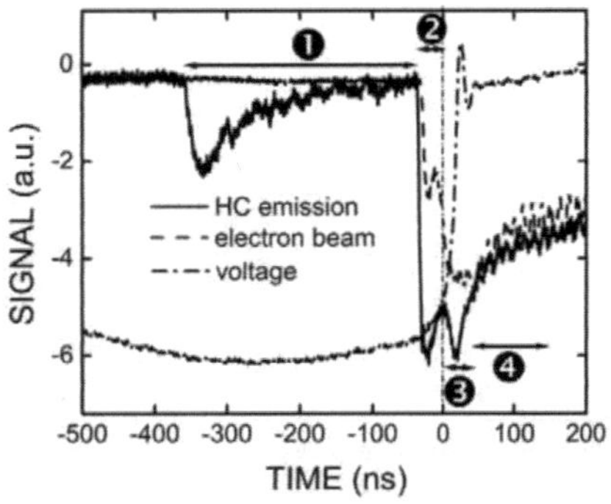

Figure 3. Time evolution of voltage, HC emission and electron beam signals, for an Argon discharge. $t = 0$ corresponds to electric breakdown in the capillary.

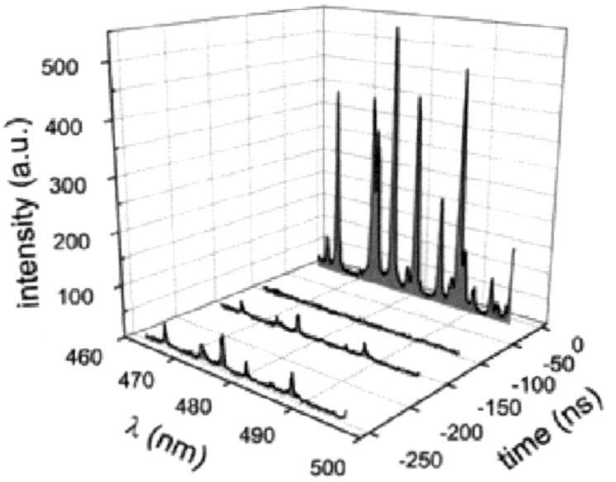

Figure 4. Time resolved Argon plasma emission from the HCR, during the hollow cathode and ionization wave phases.

Figure 5 shows axial observations of Argon emission, in the 430-470 nm band, during the ionization wave and hot capillary plasma phases. The emission intensity is orders of magnitude higher than that in the HC phase, as seen in figure 3. Figure 6 shows a time resolved spectrum, at a characteristic time which corresponds to peak current. Several ArII lines are identified. OII d AlII lines are also observed, which can be ascribed to emission from wall species ablated during the electric breakdown process. A point to note is that in this spectral band no emission lines corresponding to higher Argon ionization states exist, so the presence of ArII lines can not be taken by itself as an indication of a low temperature plasma at this phase.

Figure 7 shows the discharge current, together with the time evolution of the electron density, obtained from the Stark broadening of the H_β line [10], in Hydrogen discharges, at 0.75 Torr. It is seen that the electron density rises together with the current pulse, reaching a peak value close to $3 \cdot 10^{16}$ cm^{-3}, and later decays in a characteristic time scale of ~80 ns, which corresponds to half of the peak value. The peak value is consistent with a fully ionized Hydrogen plasma, as the average initial neutral Hydrogen atom density is $\sim 2.5 \cdot 10^{16}$ cm^{-3}. Figure 8–a shows a synthetic spectrum of a carbon plasma, calculated using a Saha-LTE model [11], assuming a 4

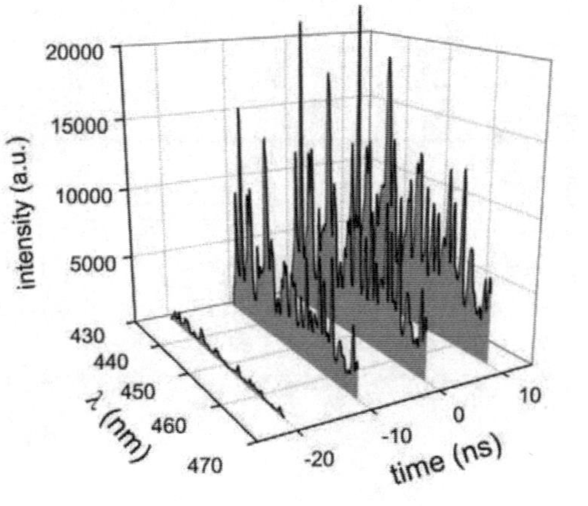

Figure 5. Time resolved Argon plasma emission from the capillary plasma, during the ionization wave and hot capillary plasma phases.

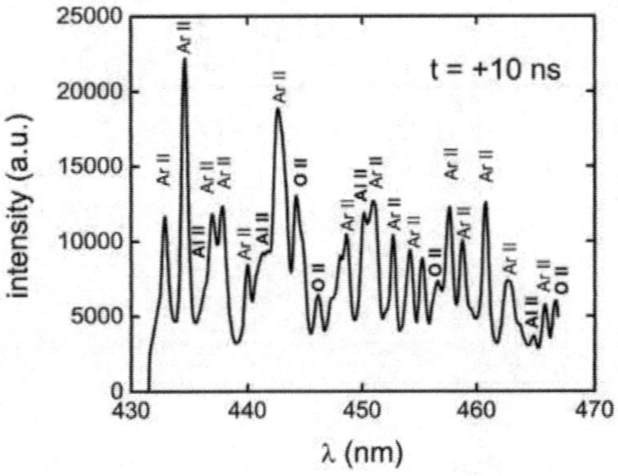

Figure 6. Time resolved Argon spectrum, corresponding to axial emission, 10 ns after electric breakdown.

eV, $3 \cdot 10^{16}$ cm^{-3} plasma. Under these conditions, the dominant lines in the spectral band correspond to CIII and CIV transitions. The relative line intensity matches reasonably well an actual spectrum corresponding to the axial emission of a methane capillary plasma, 10 ns after electric breakdown, shown in 8-b. At later times, greater than 80 ns after electric breakdown, the CIV lines almost disappear, giving place to CII lines, and very strong AlIII lines, from wall ablated species. An important point here is the fact that synthetic carbon emission spectra calculated under Saha-LTE equilibrium, assuming electron temperatures 0.5 eV either above or below 4 eV, result in relative line intensities which do not match the observed spectrum at 10 ns. Further estimations of the electron temperature in the case of an Argon plasma, using the Boltzmann plot technique, and assuming a partial local equilibrium model (pLTE), resulted in electron temperatures around 1.5 eV. The temperature was found to remain nearly constant for up to 500 ns after electric breakdown [12]. This is probably due to the fact that the temperature estimation relaies on ArII transitions, which might not satisfy adequately the conditions for pLTE.

Figure 9 shows the time evolution of the spectral emission from an Argon capillary plasma, recorded transversal to the capillary axis, at the capillary output, during the plasma expansion and jet emission phase. Maximum emission is achieved around 100 ns after electric breakdown, and later decays in a time scale of the order of 300 ns. A similar trend is observed in figure 10, which shows the time resolved H_β emission from a Hydrogen capillary plasma, recorded at different axial positions, beyond the capillary output. Again, maximum emission is achieved around 100 ns after electric breakdown and last for around 300 ns. The axial decay in plasma emission can be be attributed to plasma expansion in the low pressure region at the anode side. The Hydrogen spectra show also the presence of emission lines associated with OIII transitions arising from wall material [12]. Preliminary experiments reported elsewhere show that the capillary plasma jets can be used for thin film deposition, taking advantage of the wall ablation by using different materials at the anode end [13].

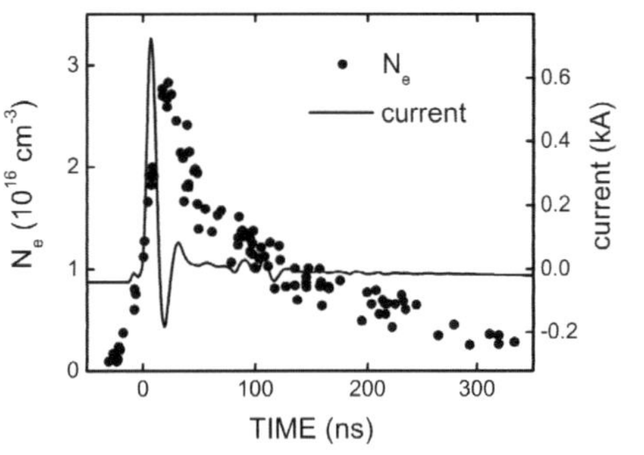

Figure 7. Time evolution of the electron density and discharge current.

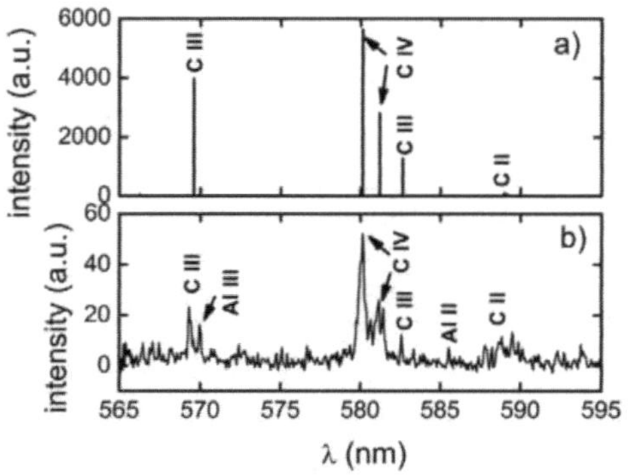

Figure 8. a) Synthetic spectrum of a carbon plasma, calculated using a Saha-LTE model, b) Methane emission spectrum, 10 ns after electric breakdown.

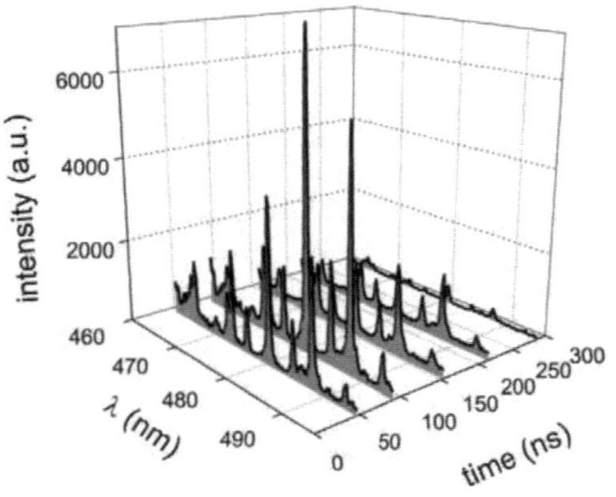

Figure 9. Time resolved Argon plasma emission at the capillary output, observed transverse to the capillary axis.

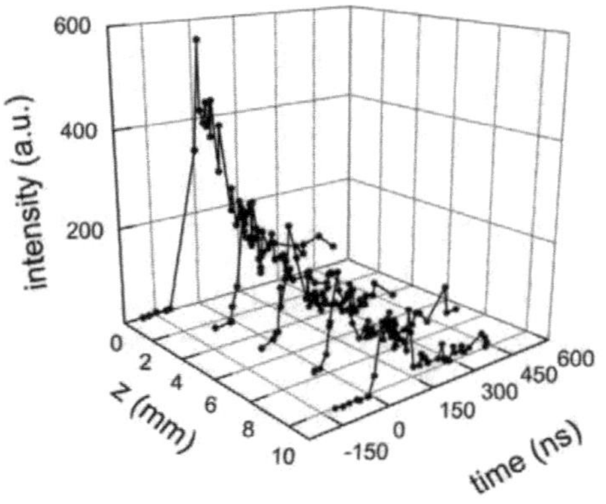

Figure 10. Time resolved evolution of the H_β line observed transverse to the capillary axis, at different axial positions.

4. Conlusions

We have performed detailed OES observations of the plasma emission in low energy, fast PCD. The overall features of the time evolution of the plasma emission are in agreement with the characteristic phases of a PCD, as established earlier using electron beam and capacitive probe measurements. We have measured the electron density, with time resolution, and inferred a characteristic electron temperature. Even though previous experiments confirm the effect of the HC electron beams in favouring on–axis discharge initiation, the presence of emission lines associated with capillary wall species, Aluminum and Oxygen, in the ionization wave, electric breakdown and hot capillary plasma, and plasma jet phases, indicates that the electron beam activity first, and the eV temperature capillary plasma later, ablate wall species, which contribute to the overall plasma emission. These OES measurements confirm the unique features of the PCD, by which a low inductance design allows transient plasmas of a few eV temperature to be generated with a stored energy of less than 50 mJ per pulse. Operation with different gases has been demonstrated, which allows the PCD to be suitable as an EUV radiation source or a pulsed plasma device for thin film deposition. Further analysis of the spectral data is in progress to characterize better the PCD plasma.

Acknowledgments
This work has been funded partially by bridge grant 5/2010 from VRI–PUC and the Physics Department. F Guzmán acknowledges financial support from CONICYT. M Ruiz acknowledges financial support from MECESUP.

5. References

[1] Sakai Y,Takahashi S, Hosokai T, Watanabe M,Kim G H, and Hotta E 2010 *J. Appl. Phys.* **107** 083303
[2] Zakharov S V, Zakharov V S ,Novikov V G, Mond M and Choi P 2008 *Plasma Sources Sci. Technol.* **17** 024017
[3] Wyndham E S, Favre M, Valdivia M P, Valenzuela J C, Chuaqui H and Bhuyan H, 2010 *Rev. Sci. Instrum*
[4] Bhuyan H, Favre M, Valderrama E, Avaria G, Wyndham E, Chuaqui H, Baier J Kelly H, Grondona D and Marquez A 2009 *Appl. Surf. Sci.* **255** 3558
[5] Avaria G, Guzman, Ruiz M, Favre M, Wyndham E S, Bhuyan H and Chuaqui H 2009 *Plasma Sources Sci. Technol.* **18** 045014
[6] Caballero L S, Chuaqui H, Favre M, Mitchell I and Wyndham E S 2005 *J. Appl. Phys.* **98** 023305
[7] Favre M, Choi P, Chuaqui H, Mitchel I, Wyndham E S and Leñero A M 2003 *Plasma Sources Sci. Technol.* **12** 78
[8] Rutkevich I, Mond M, Kaufman Y, Choi P and Favre M 2000 *Phys. Rev. E* **62** 5603
[9] Choi P and Favre M 1998 *Rev. Sci. Instrum.* **69** 3118
[10] Griem H R. 1964 *Plasma Spectroscopy* (McGraw-Hill Book Company)
[11] Ralchenko Y, Kramida A E, Reader J and NIST ASD Team (2010). NIST Atomic Spectra Database (ver. 4.0.0), [Online]. Available: *http://www.nist.gov/physlab/data/asd.cfm* [2010]. National Institute of Standards and Technology, Gaithersburg, MD.
[12] Avaria G 2008 Ph.D. thesis, Pontificia Universidad Católica de Chile
[13] Bhuyan H, Favre M, Valderrama E, Avaria G, Wyndham E S, Chuaqui H, Baier J, Kelly H, Grondona D and Marquez A 2009 *App. Surf. Sci.* **255** 3558

Generation and application of the soft X-ray laser beam based on capillary discharge

Oleksandr Frolov[1], Karel Kolacek[1], Jaroslav Straus[1], Jiri Schmidt[1], Vaclav Prukner[1] and Andrey Shukurov[2]

[1]Institute of Plasma Physics AS CR, v.v.i., Za Slovankou 1782/3, 182 00, Prague, Czech Republic

[2] Faculty of Mathematics and Physics, Charles University in Prague, Ke Karlovu 3, 121 16 Prague , Czech Republic

frolov@ipp.cas.cz

Abstract. In this work we report on the generation and characterization of a focused soft X-ray laser beam with intensity and energy density that exceed the threshold for the ablation of PMMA. We demonstrate a feasibility of direct ablation of holes using a focused soft X-ray laser beam. Ablated craters in PMMA/gold-covered-PMMA samples were obtained by focusing the soft X-ray Ar^{8+} laser pulses generated by a 46.9 nm tabletop capillary-discharge-pumped driver with a spherical Si/Sc multilayer mirror. It was found that the focused beam is capable by one shot to ablate PMMA, even if the focus is significantly influenced by astigmatism. Analysis of the laser beam footprints by atomic force microscope shows that ablated holes have periodic surface structure (similarly as Laser-Induced Periodic Surface Structure) with period ~2,8 μm and with peak-to-peak depth ~5-10 nm.

1. Introduction

Table top soft X-ray lasers based on capillary discharge produce hot and dense plasma which is a brilliant source of the EUV and soft X-ray radiation. For many years, researchers are looking for compact tabletop soft X-ray lasers that could be routinely used in application in numerous disciplines including

- physics of thin layers and physics of surfaces/interfaces,
- biological applications including imaging of cellular and sub-cellular structures, studies of membranes, radiation damage studies,
- materials science for real-time lattice dynamics studies,
- phase transition studies,
- element-state-selective radiation chemistry (so called molecular surgery),
- radiography, interferometry, shadowgraphy of dense plasmas with micrometer resolution,
- real-time studies of fast chemical reactions and liquid-phase chemistry,
- non-linear optics at ultrashort wavelengths,
- precise studies of atomic physics phenomena such as photoexcitation, photoionisation, inner-shell processes and threshold Raman processes,

- development and testing of new kind x-ray detectors for various applications e.g. in biomedicine and astrophysics.

In future, when the wavelengths of soft X-ray radiation drop below 1 nm, the following applications will also be realized:

- a time integrated difractometry of biological and inorganic materials,
- a medical diagnostic by coherent soft X-rays.

Moreover, certainly some of important applications still remain unforeseeable.

With demonstrated multi-hertz repetition rate operation and average powers of several mW and mJ-level pulse energy at 46.9 nm, capillary discharge-pumped lasers are the first compact lasers that reached this goal [1].

It is necessary to realize that besides laser-pumped and discharge-pumped soft X-ray lasers there are several alternative possibilities how to achieve coherent radiation in soft X-ray region: synchrotron sources [2]-[3], free electron lasers (FEL's) [4]-[5] and harmonic up-conversion of high-power visible/IR lasers [6]-[7]. However, due to their size and cost - these sources are unsuitable for massive application.

In this work, we report on the ablation of PMMA/gold-covered-PMMA samples irradiated with an intense focused 46.9 nm laser beam. Firstly, the soft X-ray laser beam was focused on PMMA sample covered by a thin (transparent) layer of gold. In this case obtained footprints show that energy of the soft X-ray laser beam is sufficient for ablation of this layer (unfortunately, the focus is influenced by astigmatism). Secondly, PMMA sample was ablated by soft X-ray radiation emitted from the capillary. It was demonstrated that our Ar^{8+} laser with wavelength of $\lambda = 46.9$ nm (the energy of which has not yet been reliably measured, but is estimated to be tens or hundreds of μJ) is capable to ablate PMMA/gold-covered-PMMA samples by a single shot.

2. Experimental apparatus

The high current flowing through the capillary results in pinching of the Ar plasma to form the hot thin plasma column along the axis. The experimental setup is shown schematically in Figure 1. The high current pulses used for excitation of the capillary plasma are produced by a Marx generator, coupling section and pulse forming line. This water filled fast capacitor is discharged through a self-breakdown spark gap pressurized with SF_6 gas to alumina capillary. The ~23 cm long capillary is directly joined to the spark gap (having one common electrode). The gas filling and pumping part is

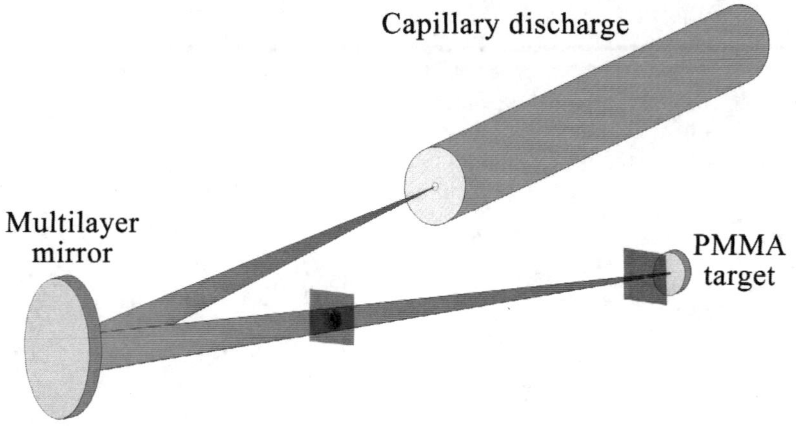

Figure 1 – Schematic presentation of experimental setup used to focus soft X-ray laser beam

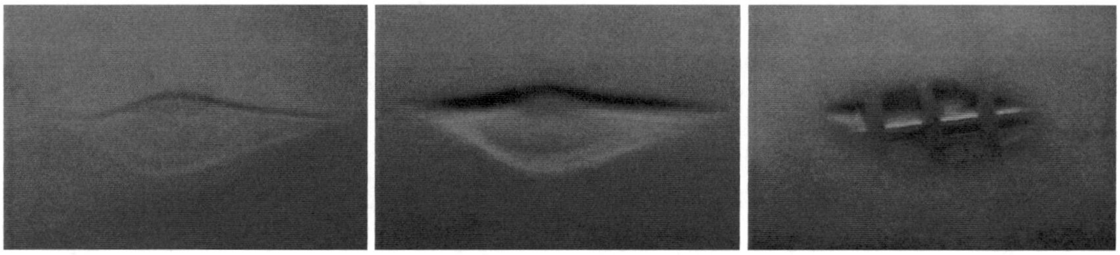

Figure 2 – PMMA ablated by one shot of Ar^{8+} laser (Left), PMMA ablated by five shots (Middle), PMMA ablated through Ni grid (step 100x100 μm, free windows 70x70 μm, traverses 30 μm) by five shots (Right); all these three pictures are in the same scale and with false colours.

attached to the outer end of the capillary (through the orifice in the outer electrode). A pre-ionization current precedes the main discharge (typically it is initiated 10 μs prior to the spark gap breakdown and its current has usually amplitude ~10 A) and it is produced by an independent circuit. A detailed description of our capillary discharge driver CAPEX can be found in the recently published paper [8].

The laser beam is focused with a spherical (R=2100 mm) Si/Sc multilayer-coated mirror placed in a vacuum chamber. In our first experiments footprint of the soft X-ray laser beam was on-line detected by a "phosphor" screen (ZnS, density 8 mg/cm^2)/photographic camera assembly. Nowadays, the soft X-ray laser beam is visualized by ablation of easily ablative material (PMMA/gold-covered-PMMA samples).

3. Experimental results

At first, the focused laser beam was registered in a visible region by the "phosphor" screen, which was covered by 0.4 μm aluminium filter to avoid the detection of visible plasma radiation [9]. It was found that the far field laser beam profile with a diameter ~5 mm has an annular shape (similarly as in [10]). In case of near field the output laser beam profile with a diameter ~ 0.6 mm has an intense single peak. Furthermore, the beam profile in the region around laser beam waist was influenced by astigmatism. In consequence of possible ambiguous interpretation of these results in visible region we are obliged to look for new diagnostic tools. During the later experiments PMMA/gold-covered-PMMA samples were selected as a target for focused soft X-ray laser beam.

First experiments with ablation of PMMA sample show that the focused beam of our laser is capable to ablate PMMA by one shot, even if the focus is significantly influenced by astigmatism (see Figure 2 – visible photo-pictures taken through optical microscope with geometrically reflected illumination) [11]. Analysis of these footprints by optical microscope between crossed polarisers shows that the sample has no residual stress caused by local heating. It seems that the ablation is not a result of thermal effects of radiation, but of quantum ones only. Unfortunately, following analysis of ablated PMMA samples by scanning electron microscope was not successful. Finally, the laser beam footprints were analysed by atomic force microscope (AFM). It was found that ablated part has on its

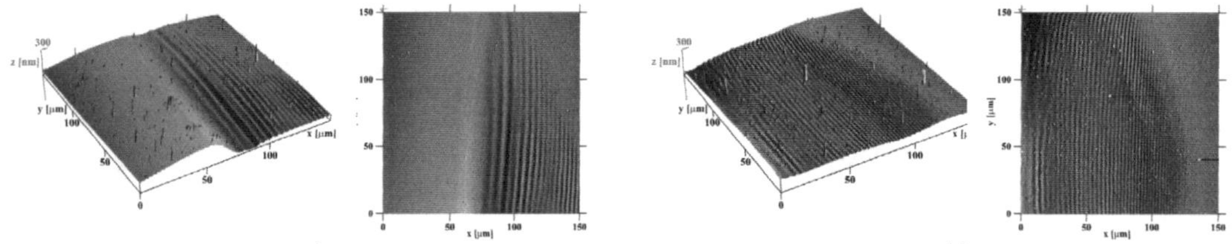

a) b)

Figure 3 – Analyzed parts (150x150 μm) of the spot shown in the middle of the Figure 2: a) 3D (Left) and 2D (Right) region in the right corner. b) 3D (Left) and 2D (Right) region in the middle-bottom part of the spot. Both parts contain the ablated and the not-ablated regions.

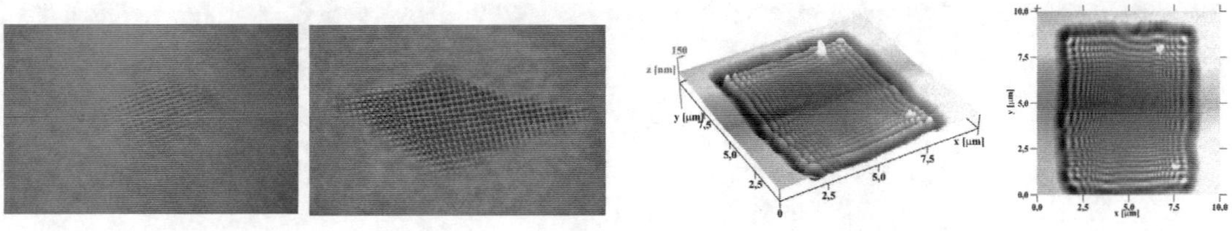

a) b)

Figure 4 – a) PMMA samples ablated by one shot of Ar^{8+} laser (Left) and five shots (Right) through Au grid (step 12,5x12,5 μm, windows 7,5x7,5 μm) – microphotograph, geometrically reflected illumination ; b) ablated single window in 3D (Left) and 2D (Right) mapped by AFM.

surface more or less contrast periodic structure with period ~2.8 μm and with peak-to-peak depth ~5-10 nm (see Figure 3). It offers to attribute these structures to LIPSS-II ("incoherent" Laser-Induced Periodic Surface Structure) [12], which depend mainly on fluency and interaction time of the laser beam, in other words on "melting effects" of the radiation. Contrary, LIPSS-I ("coherent") depend on the wavelength, coherence, polarization and incidence angle of the laser beam [13].

The last experiment should have answer, how grid with smaller characteristic pattern influences LIPSS. Therefore, PMMA sample was ablated through Au grid (step 12,5x12,5 μm, windows 7,5x7,5 μm) (see Figure 4 a)). Analysis of these laser beam footprints exposed through Au grid was realized by atomic force microscope. The period of the diffraction pattern (see Figure 4 b)) changes from ~800 nm (at the edge of window) down to ~125 nm (in the middle of the window). A clearly visible diffraction pattern (esp. at the edges) suggests that hardly any melting takes place.

As already mentioned, analysis of our first ablated PMMA samples by scanning electron microscope was not successful, but for this analysis the PMMA sample was covered by a thin (~40 nm) layer of gold. As it turned out, already the first test shots at this layer showed that the energy of our laser is sufficient for ablation of this layer. Gold-covered-PMMA samples were analyzed by AFM microscope. Typical ablated laser beam footprint (~100 μm) at focus (~1427 mm from the mirror) is shown in Figure 5 a) (Left). Laser beam footprint (~250 μm) taken far from the focus (~1450 mm from the mirror) has an annular shape similarly as in earlier experiments in visible region (see Figure 5 a) Right). Phase contrast on the right-top image in Figure 5 b) indicates that physical properties of underlying substrate are significantly different from those of the gold film. Analysis of the ablated spot in AFM working in extended TUNA mode (in which local conductance was

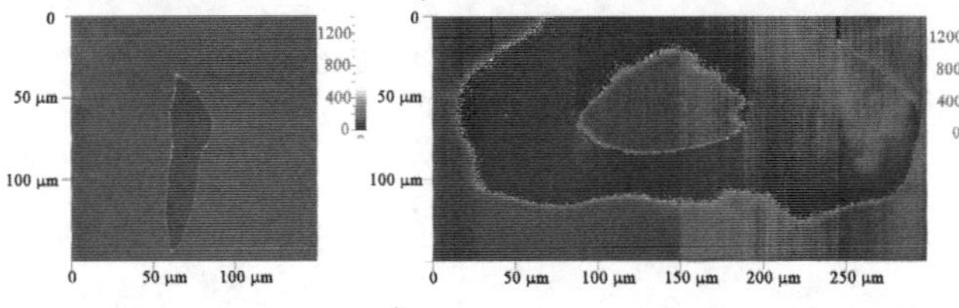

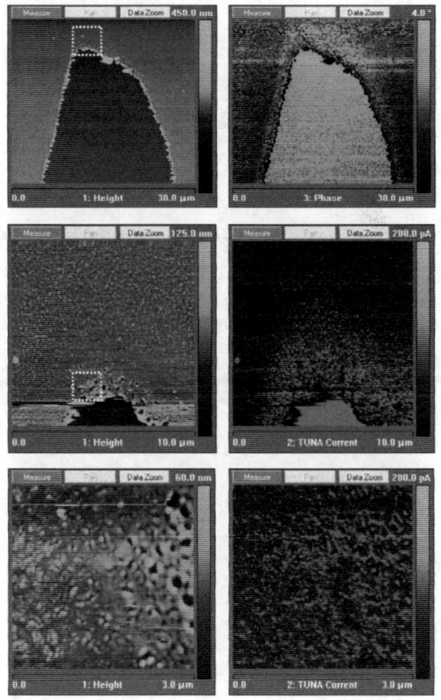

a) b)

Figure 5 – Analysis of the gold-covered-PMMA by AFM microscope: a) whole ablated laser beam footprint at focus (Left) and out of focus (Right); b) scaled-up edge of footprint in AFM tapping mode, height (Left) and phase (Right) images (Top), scaled-up edge of footprint in AFM extended TUNA mode, height (Left) and TUNA current (Right) images (Middle and Bottom).

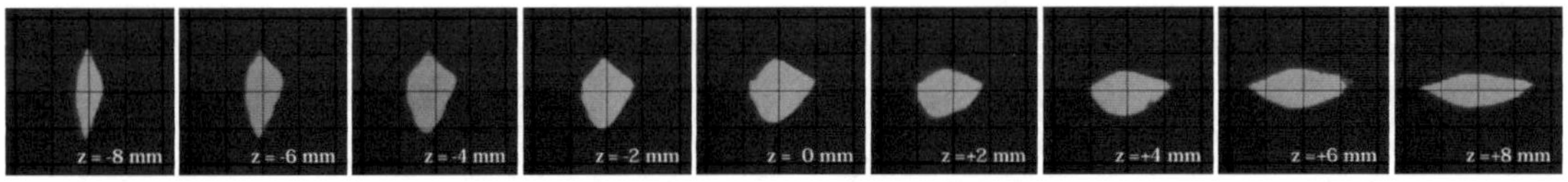

Figure 6 – Laser beam footprints on gold-covered-PMMA, the images were taken through optical microscope (grid 125x125 μm/div); false colours.

measured) (Middle and Bottom right images in Figure 5 b)) show that the composite area, where gold is intermixed with PMMA, spreads several microns from the border of the laser spot. Any melting effects on the edge of ablated laser beam footprint are not observed. We scanned the laser beam footprint in Z-direction trying to find the best focus. It was found that astigmatism of our soft X-ray radiation is very strong, astigmatic difference being ~16 mm (see Figure 6). Later it turned out that the distortion of focal spot is caused by distorted mirror shape. Analysis of the last results with ablation of gold-covered-PMMA confirmed our assumption that in our case the quantum ablation plays a dominant role and the thermal effects are negligible.

4. Conclusion

Our laboratory is one of several successful laboratories in the world, where Ar^{8+} ion laser is in routine operation and where discharge based soft X-ray lasers are further studied. It was demonstrated that focused beam of our soft X-ray laser ($\lambda = 46.9$ nm) is capable by one shot to ablate PMMA/gold-covered-PMMA samples, even if the focus is significantly influenced by astigmatism and mirror distortion. It was found that in case of large ablated pattern the PMMA sample has on its surface a periodic structure with period ~2,8 μm (like LIPSS-II structure). Contrary, at small ablated pattern a diffracted pattern (with periodicity down to ~125 nm) was observed. Analysis of PMMA/gold-covered-PMMA samples by atomic force microscope working in different modes shows that the ablation is not a result of thermal effects of radiation, but of quantum ones only.

References

[1] C. D. Macchietto, B. R. Benware, J. J. Rocca, Opt. Lett. 24, 1115 (1999)
[2] D. T.Attwood, K.Halbach, and N. J.Kim, Science 228, 1264 (1985)
[3] R. Coisson, Appl. Opt. 34, 904 (1995)
[4] J. Rossback, in X-ray Lasers, edited by S.Svanberg and C.G.Wahlstrom (Institute of Physics, University of Reading, Berkshire 1996), Vol. 1511, p.436 (1996)
[5] Free-Electron Lasers Challenges, edited by P. G. O'Shea and H. E.Bennett, Proc. SPIE, Bellingham, WA, Vol. 2988 (1997)
[6] Z. Chang, A. Rundquist, H. Wang, M. M. Murnane, and H. C. Kapteyn, Phys. Rev. Lett. 79, 2967 (1997)
[7] A. Rundquist, C. G. Durfee III, Z. Chang, C. Herne, S. Backus, M. Murnane, and H. C. Kapteyn, Science 280, 1412 (1998)
[8] K. Kolacek, J. Schmidt, V. Bohacek, M. Ripa, O. Frolov, P. Vrba, J. Straus, A. A. Rupasov, A. S. Shikanov, Plasma Physics Reports 34, 162-168 (2008)
[9] O. Frolov, K. Kolacek, J. Schmidt, J. Straus, V. Prukner, Conference Proceedings of ICPIG 2009, Vol.2, p.733 (2009)
[10] B. N.Benware., A. Ozols, J. J. Rocca, I. A. Artioukov, V. V. Kondratenko, A. V. Vinogradov, Optics Letters 24, 1714-1716 (1999)
[11] K. Kolacek, J. Straus, J. Schmidt, V. Prukner, O. Frolov, J. Sobota, T. Fort , H. Biederman, A. Shukurov, V. Holy, V. Hajkova, A. A. Rupasov, A. S. Shikanov, Proceedings of workshop IRNANO-2009, in print
[12] J. F. Young, J. E. Sipe, H. M. van Driel, Phys. Rev. B 30, 2001-2015 (1984)
[13] J. E. Sipe, J. F. Young, J. S. Preston, H. M. van Driel, Phys. Rev. B 27, 1141-1154 (1983)

Journal of Physics: Conference Series **511** (2014) 012036 doi:10.1088/1742-6596/511/1/012036

A model of hard X-rays emission from free expanding Plasma-Focus discharges

E. Fogliatto[1], **J. González**[1,2,3], **M. Barbaglia**[3] **and A. Clausse**[3]

[1] Instituto Balseiro, R8402AGP - Bariloche, Argentina

[2] Comisión Nacional de Energía Atómica, R8402AGP - Bariloche, Argentina

[3] CNEA-CONICET and Universidad Nacional del Centro 7000 - Tandil, Argentina

E-mail: fogliate@ib.cnea.gov.ar

Abstract. A planar-piston model of the hard x-ray production in Plasma-Focus devices is presented. The model applies Von Karman approximations to represent the inner structure of the pinch. The hard x-ray emission is calculated assuming Bremsstrahlung from the collision on the anode base of an electron current running away from the pinch.. The model was applied to analyse the experimental data of a small Plasma Focus without surrounding cathode, founding good agreement.

1. Introduction

Plasma-Focus (PF) devices produces hot ($\sim 0.3 - 1\,keV$) and dense ($\sim 10^{25}\,m^{-3}$) dynamic pinches inducing ionization by means of high voltage discharges in coaxial geometries. Recently, small devices were developed aiming to applications of the pulsed radiation emissions of the pinch. Miniature devices were constructed following a special configuration derived from capillary discharges, where the surrounding cathode is replaced by a circular plate located at the base of the gun. This feature allows the current sheath to expand freely during the run down.

During the compression x-rays are produced by two mechanisms. A soft x-ray source inside the pinch is produced by line radiation, Coulomb interaction between charged particles and recombination. A hard x-ray source of Bremsstrahlung photons are produced by electrons that escape from the pinch and hit the anode base material. This work presents an extension of a lumped parameter code of free expanding PF [1], introducing a model of electron runaway and hard x-ray production. The model is validated with experimental data of a small PF device.

2. Model Description

A PF are basically two coaxial electrodes separated by an insulator material at one end (the anode at the interior). The interelectrode space is filled with low pressure gas and pulsed high voltage is applied between the electrodes. The discharge starts over the insulator from which the plasma sheath takes-off axially accelerated by the magnetic field generated by the current itself. After the current sheath runs over the open end of the central electrode, the plasma becomes rapidly compressed into a small column resulting in very hot and dense plasma (pinch). In cases where the cathode is just a planar plate, the current sheath expands freely radially away from the axis. In the lumped parameter model, the current sheet is represented by planar pistons following a snow-plough approximation. Initially the sheet takes the shape of the insulator and then expands in axial and radial direction, keeping the cylindrical symmetry and capturing gas with certain efficiency [2]. A constant linear mass density is assumed so that gas particles are re-distributed uniformly all over the sheet surface as they are captured, and this approximation is used to estimate the sheet thickness. For details of the evolution equations of the pistons see [1].

The pinch starts when the sheet collapses at the open end of the anode. The pinch is modeled as a plasma cylinder compressed by the Lorentz force and obeying the integral MHD equations. The inner structure of the plasma column is represented by means of Von Karman approximations of velocity and density profiles. The detailed set of equations of the pinch dynamics can be found in [3]. In the present version a current density profile depending on the radial coordinate r is included according to:

$$J(r,t) = K(t)\left[\frac{r}{R(t)}\right]^{\alpha} \tag{1}$$

where R is the pinch radius, K is the current density at the external boundary, and α is a shape parameter. This current profile brings about axial and radial components of electric field inside the pinch, which can be calculated using the Ohm's law:

$$\mathbf{E} + (\mathbf{V}\times\mathbf{B}) = \eta\,\mathbf{J} \tag{2}$$

Where $\mathbf{E}$, $\mathbf{V}$, $\mathbf{B}$, η and $\mathbf{J}$ represent the electric field, velocity, magnetic field, Spitzer's resistivity and current density respectively. Considering a magnetic field with only an azimuthal component

$$B_\theta(r,t) = \frac{1}{2}\frac{\mu_0 i(t)}{2\pi r} \tag{3}$$

where μ_0 is the vacuum permeability and $i(t)$ is the pinch current, then the electric field is given by:

$$\mathbf{E} = E_r(r,z,t)\,\hat{\mathbf{e}}_\mathbf{r} + E_z(r,z,t)\hat{\mathbf{e}}_\mathbf{z} \tag{4}$$

$$E_z(r,z,t) = \frac{i(t)}{2\pi R^{\alpha+2}}\,r^{\alpha}\left[\eta\,(\alpha+2) - \frac{r^{a+1}}{R^{a}}V_{rad}\mu_0\right] \tag{5}$$

$$E_r(r,z,t) = \frac{2Z}{H}\frac{\mu_0}{2\pi R^{\alpha+2}}\,i(t)V_{ax}\,r^{\alpha+1} \tag{6}$$

where V_{ax} and V_{rad} are the axial and radial velocities of the cylindrical surfaces respectively. According to Dreicer theory, a fraction of the pinch electrons escape from the pinch accelerated by the axial field if the magnitude of the axial field is higher than the critical value:

$$E_d = 3,9.10^{-10}\frac{n\left[\frac{1}{cm^3}\right]}{KT[eV]} \tag{7}$$

n being the particle pinch density.

Assuming that the energy of a single runaway electron, e, is proportional to an effective voltage, gives

$$e \sim q\,H\,(E_z - E_d) \tag{8}$$

where H is the pinch length and q is the electron charge. Hence, using the appropriate conversion factors, the total energy produced by Bremsstrahlung in the anode base can be estimated as proportional to the square of the incident electron energy.

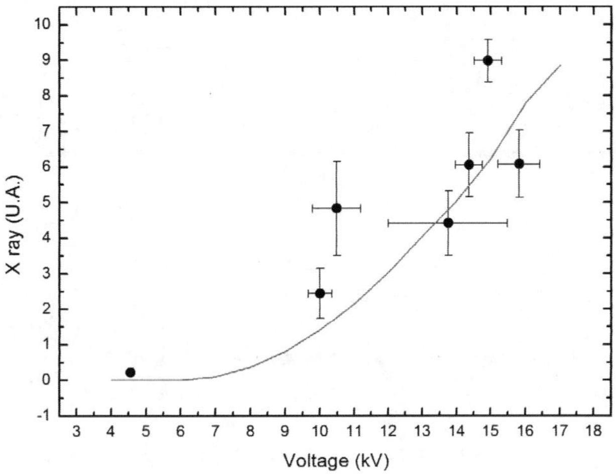

Figure 1. Measured vs. calculated hard x-ray intensity

3. Results

In order to validate the model, the numerical results were compared with experimental measurements from a Plasma Focus with open cathode. Table 1 details the geometric and electrical parameters of the device [4]. Fig. 1 shows the measured dependence of the hard X-ray intensity on the charging voltage. The detectors were positioned at 90 degrees from the PF axis, and the Deuterium filling pressure was 16.4 mbar. The curve in Fig. 1 shows the numerical results obtained with the present model. The constant parameters used in the calculations are listed in Table 1. It can be seen that the model is able to follow the experimental trend. In particular, note that below 9 kV very low intensities were observed, which is in agreement with the absence of runaway due to the Dreicer condition. Another remarkable result is that a high exponent of the current density profile was needed to fit the experiments ($\alpha = 50$), indicating that the current is concentrated at the pinch border. The latter is in agreement with previous neutron production models where the current was assumed concentrated at the external boundary [5].

4. Conclusions

A lumped parameter model for estimating hard x-ray production in Plasma Focus discharges has been presented. Numerical calculations show good agreement with experimental data, particularly in predicting the dependence of the radiation intensity with the charging voltage. The model set out here offers a useful tool to calculate and design of open-cathode PF devices applied to production of pulsed beams of charged particles and hard x-rays.

References

[1] González J, Barbaglia M, Casanova F and Clausse A 2009 *Brazilian Journal of Physics* **39** 633–637
[2] Fogliatto E 2010 Investigación en generación de pulsos de radiación x en equipos plasma focus de baja energía Instituto Balseiro, Bariloche, Argentina (in Spanish)
[3] Márquez A 2010 Caracterización y modelado de fuentes pulsadas de neutrones del tipo plasma focus, Instituto Balseiro, Bariloche, Argentina (in Spanish)

Journal of Physics: Conference Series **511** (2014) 012036 doi:10.1088/1742-6596/511/1/012036

[4] Barbaglia M, Bruzzone H, Acuña H, Soto L and Clausse A 2009 *Plasma Physics and Controlled Fusion* **51** 045001 (9pp)
[5] González J, Brollo F and Clausse A 2009 *IEEE Transactions on Plasma Science* **37** 2178–2185

Symbol	Parameter	Value
C	Bank capacity	0.8 μF
V	Charging voltage	18-28 kV
L_e	External inductance	65 nHy
z	Anode length	0.191 cm
R_a	Anode radius	0.31 cm
p_0	Filling pressure	16.4 $mbar$
α_{sg}	Characteristic frequency of the spark gap	350 ns^{-1}
t_0	Characteristic delay of the spark gap	400 ns
ξ	Efficiency parameter	0.2
α	Current shape parameter	50

Table 1. Parameters of the experiment and the model

SCR-1: Design and Construction of a Small Modular Stellarator for Magnetic Confinement of Plasma.

L Barillas[1*], V I Vargas[1], A Alpizar[1], J Asenjo[1], J M Carranza[1], F Cerdas[1], R Gutiérrez[1], J I Monge[1], J Mora[1], J Morera[1], H Peraza[1], V Queral[2], C Rojas[1], D Rozen[1], F Saenz[1], G Sánchez[1], M Sandoval[1], H Trimiño[1], J Umaña[1], L F Villegas[1]**

[1]PlasmaTEC, Instituto Tecnológico de Costa Rica, Central Campus, Cartago, Costa Rica; P.O. Box 159-7050 Cartago, Costa Rica

[2]Laboratorio Nacional de Fusión, CIEMAT, Madrid, Spain

[1*] laura.plasma.tec@gmail.com, barillas-moral2@asme.org
[1**] ivargas@itcr.ac.cr, ivanvargasblanco@gmail.com

Abstract. This paper describes briefly the design and construction of a small modular stellarator for magnetic confinement of plasma, called Stellarator of Costa Rica 1, or SCR-1; developed by the Plasma Physics Group of the *Instituto Tecnológico de Costa Rica*, PlasmaTEC. The SCR-1 is based on the small Spanish stellarator UST_1, created by the engineer Vicente Queral.
The SCR-1 will employ stainless steel torus-shaped vacuum vessel with a major radius of 460.33 mm and a cross section radius of 110.25mm. A typical SCR-1 plasma will have an average radius 42.2 mm and a volume of 8 liters (0.01 m^3), and an aspect ratio of 5.7.
The magnetic resonant field will be 0.0878 T, and a period of 2 (m=2) with a rotational transform of 0.3. The magnetic field will be provided by 12 modular coils, with 8 turns each, with an electrical current of 8704 A per coil (1088 A per turn of each coil). This current will be fed by a bank of cell batteries. The plasma will be heated by ECRH with magnetrons of a total power of 5kW, in the first harmonic at 2.45GHz. The expected electron temperature and density are 15 eV and 10^{17} m^{-3} respectively with an estimated confinement time of 7.30 x 10^{-4} ms. The initial diagnostics on the SCR-1 will consist of a Langmuir probe, a heterodyne microwave interferometer, and a field mapping system.
The first plasma of the SCR-1 is expected at the end of 2011.

1. Introduction and motivation

The world as we know it depends on energy, in particular electrical power. The electrical power that we consume in commerce and industry is largely generated using resources such as coal and gas that pollute the environment, are non-renewable and are being depleted [1]. To reduce the negative impact of electricity generation and ensure the long term supply of energy, there is a strong motivation to develop a safe, clean, efficient and renewable energy source: nuclear fusion is a potential candidate for achieving these aims may fulfil these requirements [1].

Some of the main obstacles in fusion research are the large budgets and facilities required to research the typical plasmas used to achieve fusion. Despite this, between 2005 and 2007, a small

Stellerator (UST_1) was designed, built and operated by just one person: the Spanish engineer Vicente M. Queral, on a budget of under $4000; proving that low-cost techniques to build accurate-acceptable quality stellarators are possible [2].

The low volume of research on small size magnetic confinement devices and the need to contribute to the world's energy solution research motivated the Plasma Physics Group of the *Instituto Tecnológico de Costa Rica* (ITCR), PlasmaTEC, to build on the achievement of Queral and design a new upgraded and improved model based on the UST_1: the Stellarator of Costa Rica 1 (SCR-1). This model will be, so far, one the few small stellarators ready to diagnose some of the main parameters of the plasma.

2. The SCR-1 Project

2.1. Structure and areas of the SCR-1 Project

The SCR-1 Project is run mainly by undergraduate students from a range of engineering degrees under the supervision of experts and professionals. The structure of the project involves about 25 engineering and physics undergraduate students, mainly from the ITCR, 1 graduated physic and 4 supervisor engineers, all of them under the leadership of two plasma physics and nuclear fusion doctors. This kind research structure had never been presented before in Costa Rica.

The project is divided into the following areas: Coil Systems and Layer Materials, Vacuum Systems, Magnetic Fields, Power Supply Systems, Heating, Diagnostics, Simulation, Safety, Data Acquisition and Control Systems, and finally, the Administrative, Technical, and General Supervising areas. Each area represents a part or system of the SCR-1 and it is constituted by at least 3 students, except for the Supervising areas, which are logistical areas managed by the graduated engineers and the doctors. A diagram of the SCR-1 Project structure is show in Figure 1.

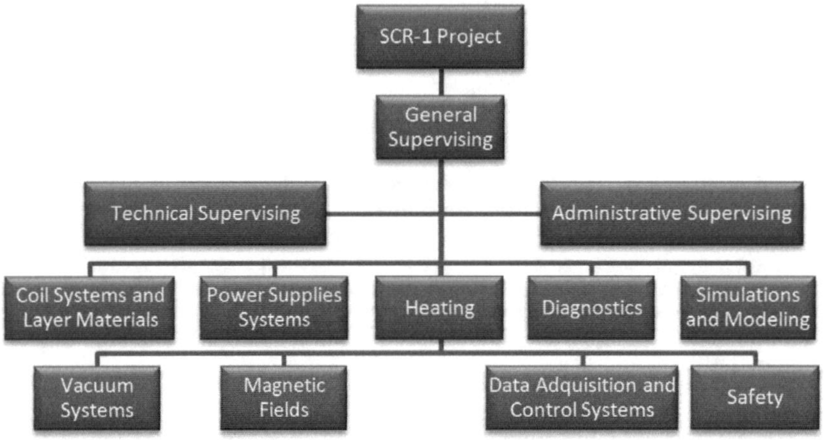

Figure 1. Diagram of the SCR-1 Project structure

A description of each part or area of the SCR-1 will be described, including its characteristics and parameters. At the end, the expected plasma parameters will be mentioned too. Before starting, it is important to mention that the SCR-1 Project is on design and construction phase at this time, so some parameters may change in the future.

2.1.1. Vacuum Systems. This area is responsible for designing the vacuum vessel, as well as selecting the equipment for reaching the desired vacuum. As vessel, is has been designed and selected a torus-shaped vacuum chamber, with major radius of 238 mm and a cross section radius of 80 mm (this is shown on Figure 2). The vessel will have 14 ports available for different applications, as seen in Figure 3. The chamber design has more ports than the actual diagnostics and probes require, which is good because that will let PlasmaTEC incorporate more components if needed.

The vessel material will be austenitic 304L grade stainless steel, with an approximated thickness of 3 mm; these parameters depend on the commercial availability of the material and its price.

The minimum pressure that the vacuum pump can reach is 10^{-9} Torr, and the expected pressure in the vacuum vessel will be 10^{-6} Torr.

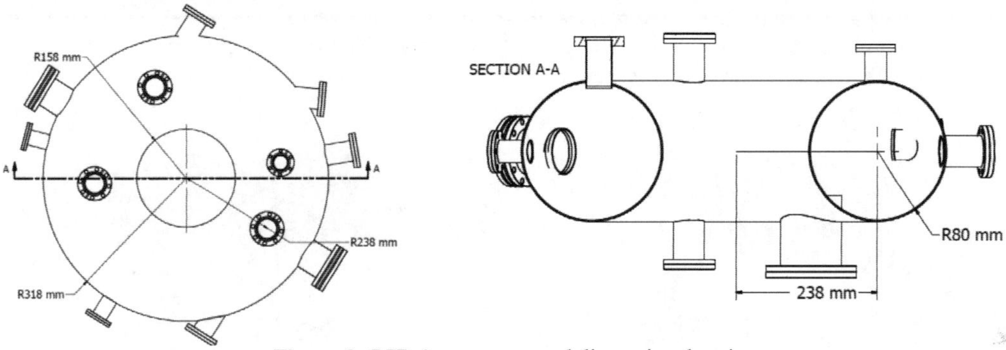

Figure 2. SCR-1 vacuum vessel dimension drawing.

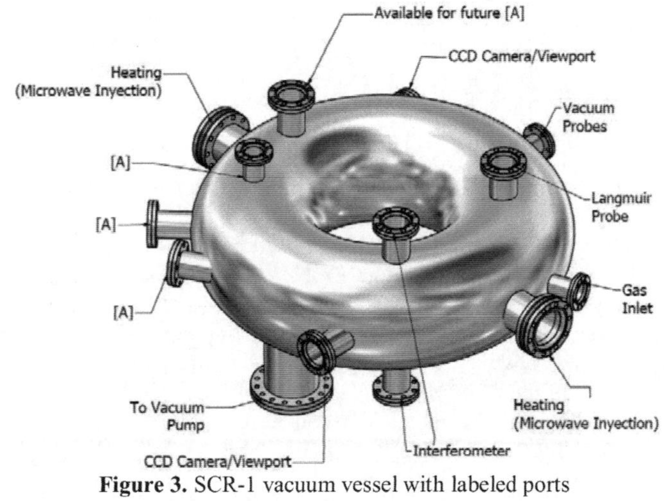

Figure 3. SCR-1 vacuum vessel with labeled ports

2.1.2. Coil Systems and Layer Materials. The SCR-1 will have 12 modular copper coils, with 8 turns each one. The electrical current passing through each coil will be 8704 A approximately, this means, 1088 A per turn. This area is in charge of positioning the coils, that is why they are currently investigating on materials that can be set over the vacuum vessel and mechanized the grooves to place the coils. The first option, as shown in Figure 4, is to set a layer of plaster over the vacuum chamber and mechanized the grooves with a special machining device that is being designed and built, in order to obtain the complex coil geometries required. This special machining device will have the flexibility to allow for alternative coil shapes to be implemented in future magnetic confinement devices.

The Coils Systems and Layer Materials area is also in charge of assemble and test the field mapping system.

2.1.3. Power Supplies Systems. The electrical supply system will consist of an array of 82 lead-acid electrochemical cells (battery bank); each cell will have a nominal voltage of 2V and an electrical storage capacity of 200 Ah. Due to the electrical requirements of the SCR-1, these cell characteristics are necessary to ensure the functioning of stellarator during a plasma discharge. The battery bank will have a final voltage of 164V and will be able to deliver a current of 1088A at least during one duty cycle of the stellarator. It is estimated that each cycle will last for about 4s; however, the battery bank capacity will allow us to increase the length of the cycle.

The system will also use a variable load connected between the bank and coils system in order to maintain a constant current flow during each work cycle, dissipate excess power, counteract the changes in wire resistance due to temperature and automate the delivery of current pulse.

A drawing of the battery bank is presented in Figure 5.

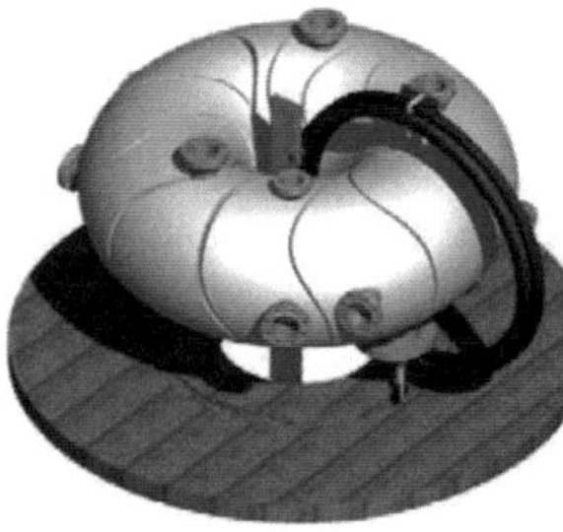

Figure 4. Device mechanizing the coil grooves into the layer material (plaster) of the vacuum vessel.

Figure 5. Representation of the battery bank: an array of 82 lead-acid cells.

2.1.4. Heating. The plasma in the SCR-1 will be heated with electron cyclotron resonance heating (ECRH), through 2 magnetrons of 2.5kW each (5 kW total), at 2.45 GHz in the first harmonic. These were chosen due to commercial availability of magnetrons with this frequency, plus other calculation made in order to have the best heating possible, with the simplest systems (this reduces costs).

2.1.5. Diagnostics Systems. As mention, this stellarator will be able to diagnose plasma when it will be ready. The diagnostics consist of a Langmuir Probe (Figure 6) and a Heterodyne Microwave Interferometer, especially design for the current application: a small stellarator. It is important to mention that the interferometer was developed along with a computer program to receive the data of the lineal density of the plasma (Figure 7). Both diagnostics were design by 2 PlasmaTEC engineering students with the collaboration of the *Laboratorio Nacional de Fusión*, in Madrid, Spain.

Also a field mapping system will be installed using an e-gun and a CCD camera, in order to superposition the frames taken by the camera and compare the experimental field lines with the ones simulated; this will validate that the calculations made for the SCR-1 are correct.

2.1.6. Magnetic Fields. The magnetic resonant field for the Stellarator of Costa Rica 1 will be 0.0878 T, and a period of 2 ($m = 2$) with a rotational transform $\iota = 0.3$.

2.1.7. Simulations and Modeling. A JAVA code named SimPIMF was developed by V. Queral to calculate three-dimensional (3D) magnetic fields. Later, the code evolved, and was able to calculate/simulate by field line tracing: Poincaré plots, rotational transform and magnetic well profile, plasma size, orbit simulation with drifts, particle losses, other 'plasma' parameters, minimum distance between coils, and optimization of such parameters by iterative generation of parametric 3D coils [2]. In Figure 8 are presented some vacuum magnetic surfaces on 3 different toroidal angles obtained on Poincaré simulations for UST_1, which also work for SCR-1.

This area work together with Magnetic Fields area and they both are currently developing new codes for simulations, in order to predict some plasma behaviour and implement modifications if needed into the current design of the SCR-1.

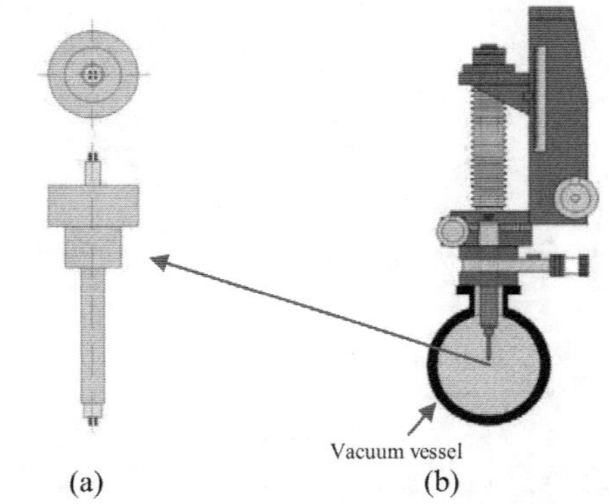

Figure 6. Diagrams of: (a) Close-up of the Langmuir Probe tip. (b) Langmuir Probe system especially designed for the SCR-1.

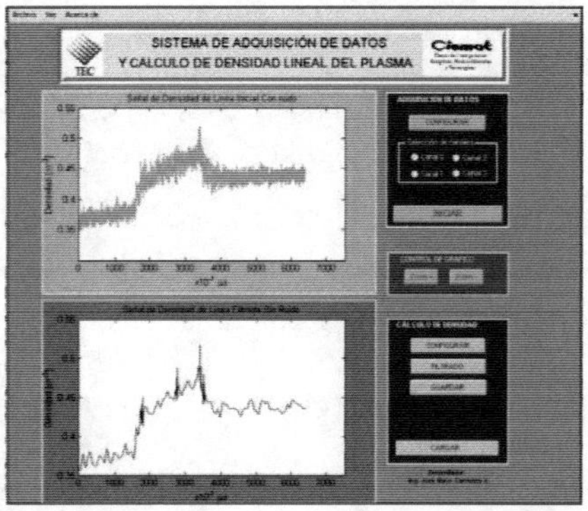

Figure 7. Computer program to receive the data of the lineal electron density of the plasma from the SCR-1's microwave interferometer.

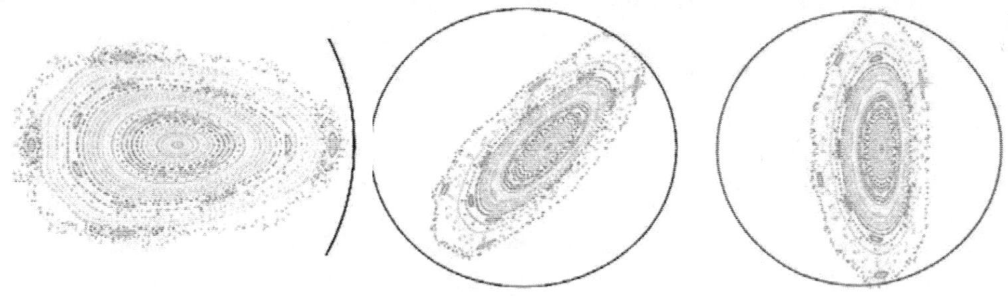

(a) (b) (c)

Figure 8. Simulations on SIMPIF in Poincaré plots to obtain vacuum magnetic surfaces at: (a) $\Phi = 0^o$, (b) $\Phi = 45^o$, (c) $\Phi = 90^o$

2.1.8. Data Acquisition and Control Systems. The responsibility of this group is to automate all possible systems of the SCR-1, in order to accurately launch the plasma shooting sequence. They are also in charge of any data acquisition in the project. The actual work of this group will start just after the SCR-1 is built, but right now the area also supervises the rest of the areas before buying any equipment, with the intention of recommend products that contribute with the automation of the SCR-1.

2.1.9. Safety. This is consider the most important and respected area of the project, because its main purpose is aim to achieve the goals of the project according to any necessary aspect that guarantees the safety of the people and the equipment, as well as the correct development of the operations. They must establish a protection and risk reduction system that relate from the development and operation of each stage of the SCR-1; and also, they must have knowledge on technical aspects of the SCR-1 and human issues.

2.1.10. Plasma parameters: The following are the expected plasma parameters and other characteristics.

- Minor plasma radius: 42.2 mm
- Electron temperature: 15 eV
- Electron density: 10^{17} m^{-3}
- Estimated confinement time: 7.30×10^{-4} ms
- Volume: 8 liters (0.01 m3)

- Aspect ratio: 5.7

The first SCR-1 plasma is expected at the end of 2011.
Finally, Figure 9 describes a computer image of the SCR-1 shown in layers.

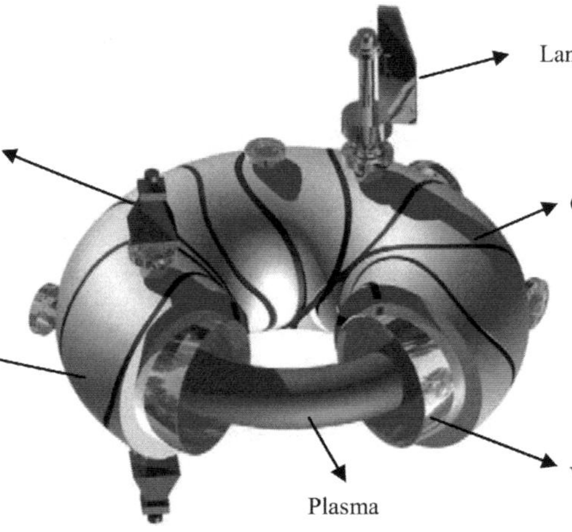

Figure 9. Computer image of the SCR-1shown in layers

3. Final comments

It is important to emphasize that the experience of designing and building of this small stellarator provides important opportunities for students, especially for undergraduates, to develop the skills required for future research by working on real research; engaging with the real engineering problems involved and finding their solutions; and contributes to the hands-on experience that is required before graduating. It is also very significant for PlasmaTEC to work on the design and construction of the first stellarator of Latin American; we hope that this can bring more research opportunities to our countries.

We also hope that the SCR-1 inspires more universities to develop similar devices, like this one, that can work well for didactical purposes.

We appreciate any kind of help, comment or suggestion. Please contact us if any question. Further information about the SCR-1 and other PlasmaTEC projects are available at http://www.tec.cr/sitios/Vicerrectoria/vie/investigacion/plasma/Paginas/default.aspx

Acknowledgments

The authors would like to thank CIEMAT and the *Laboratorio Nacional de Fusión*, in Madrid, Spain for the support provided throughout this project.
Thanks to Professor Sebastian Tallens, Ph.D.[1], for providing technical advice to the project.
Special thanks to Professor Lisandro Araya, MSc. Eng.[1], for helping us taking this work to Chile to the ICPP-LAWPP 2010.

References

[1] Eliezer S and Eliezer Y 2001 *The Fourth State of matter: An Introduction to Plasma Physics* 2nd ed (London: Institute of Physics Publishing) chapter 5 pp 103-120
[2] Queral V 2008 UST_1, a small, low-cost stellarator. *Stellarator News.* (2008) 118

Alfvén and Helicon Wave Resonances Measured with Demodulator Circuit in TCABR

[1]**P. G. P. P. Puglia, A. G. Elfimov, L. F. Ruchko**
Laboratrio de Fsica de Plasmas, Instituto de Fsica da Universidade de So Paulo
E-mail: [1]paulogiovane@gmail.com

Abstract.
The measurements of the RF driven magnetic field oscillations with the analog demodulator circuitry in the TCABR tokamak are presented. This circuitry works as the synchronous detector that is used to multiply the signal from magnetic probes by the reference signal, and then to filter out the low frequency signal. The output signal carries the information about the amplitude and the phase of the RF driven plasma oscillations, and can be acquired by the TCABR data acquisition system during all RF pulse.

The experiments were carried out in the regimes with the low-density plasma $n = 1 - 9 \times 10^{16}/m^3$ and magnetic field $0.04T$ when helicon wave resonances can be excited. The antennae were fed by the RF generators whose frequency was varied in the range $2 - 5MHz$ with variable time $\Delta t = 1 - 5ms$. The signals acquired with the analog demodulator circuitry were compared with the results obtained after numerical simulations of the data acquired directly with the high frequency oscilloscope. The very good agreement was demonstrated that makes possible to use the demodulator circuit measurements with long RF pulse duration in future experiments.

1. Introduction

The measurements of the externally induced Alfvén like waves in tokamak plasmas are very important for fusion plasma diagnostics. They can permit to measure the isotope composition of the fusion plasmas and q(r) profile in tokamaks [*see Elfmov et. al. [1]*]. Now this diagnostic technique is intensively developing at the TCABR tokamak. The experimental equipment consists of the RF generator with variable frequency, Alfvén wave antennae, magnetic probes [2] for wave detection, and a demodulator circuit [3].

The frequency of the excited Alfvén waves is in the range of approximately 4-5 MHz. The frequency of the data acquisition system of the TCABR is limited up to 3 MHz. That is why an analogous demodulator circuit has been developed in order to ensure the possibility to measure higher frequencies directly. Only one antenna module was used in these experiments but we are going to operate with two antenna module in order to separate toroidal mode numbers $N = 1$ and 2 in the future.

Usually, in order to increase the efficiency the RF power generators, they work with the resonant load, and the Alfvén antenna constitutes part of the resonant circuit that operate at fixed frequency. In the case of the sweeping frequency, it is difficult to ensure the impedance matching of the generator output and the resonant load in a large frequency range. In order to overcome this difficulty, the RF amplifiers with wide frequency band were developed using fast RF power MOSFETs in the switching regime. They can supply RF power up to $1kW$ in the frequency bandwidth of $f = 0.05 - 5MHz$. In this case, the frequency and the phase of the current in the RF antennae is controlled by the functional generator.

2. Demodulator Circuit

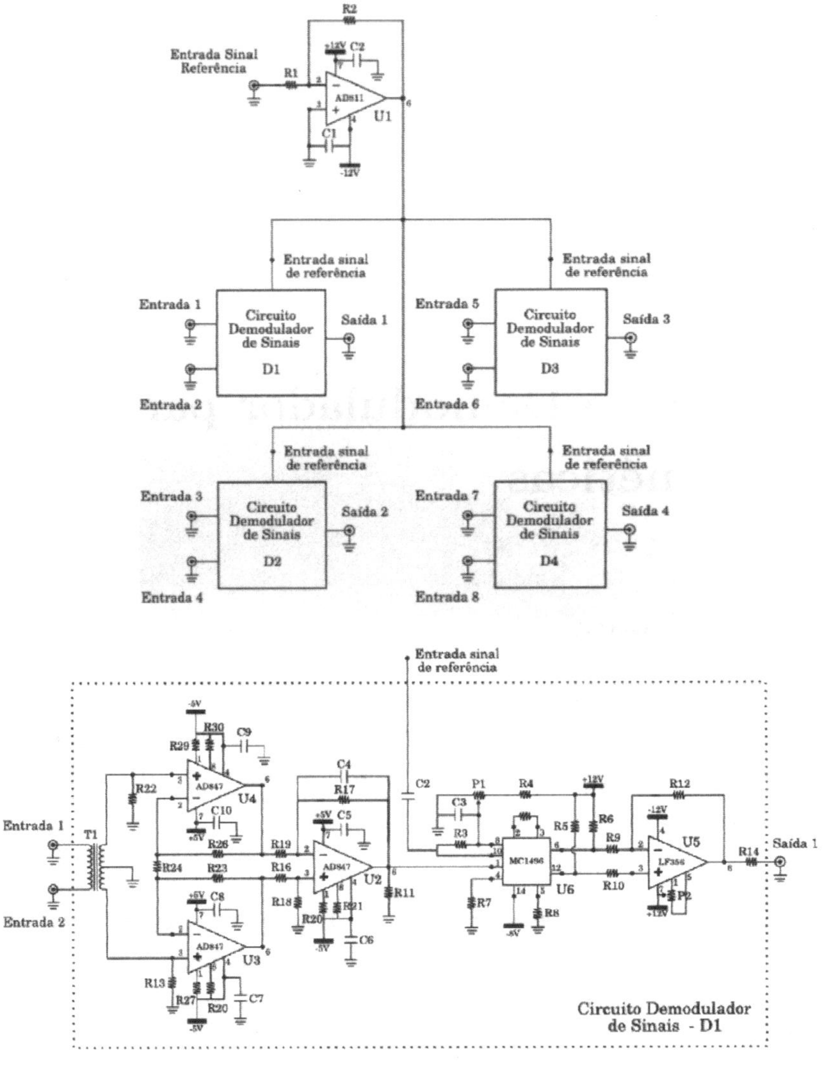

Figure 1. Schematics of the demodulator circuit.

A demodulator circuit shown in Fig.1 was developed [3] to solve the problem to measure the high frequency magnetic field oscillations. This circuit multiplies the signals from the magnetic censors [2] and the RF antenna current measured by a Rogowski coil, which is used as the reference signal. This multiplication gives rise to components of sum and subtraction of the frequencies from the antenna current and from the magnetic censors. As a result, a low frequency component ¡2kHz appears, which can be filtered out and acquired by the data acquisition system. This signal carry the information about amplitude and the phase of the magnetic field oscillations.

Filters and differential amplifiers were used to diminish noise from the input signal and a low-pass filter was used in the output stage because we have interest in the low frequency signal. There are four input channels, to use in conjunction with the four magnetic coils installed in the TCABR, of which two are positioned to measure the toroidal and the other two the poloidal magnetic field.

Characteristic curves of the demodulator circuit are presented in Fig.2. In Fig.2(a), we show the selectivity of the demodulator to the signals at the frequency of the reference signal and workable linear part of the characteristic curve is demonstrated in Fig.2(b).

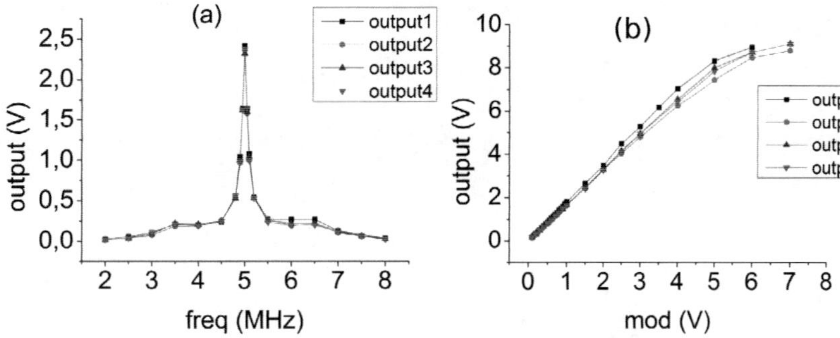

Figure 2. Characteristic curves of the demodulator circuit.

3. Data Analysis

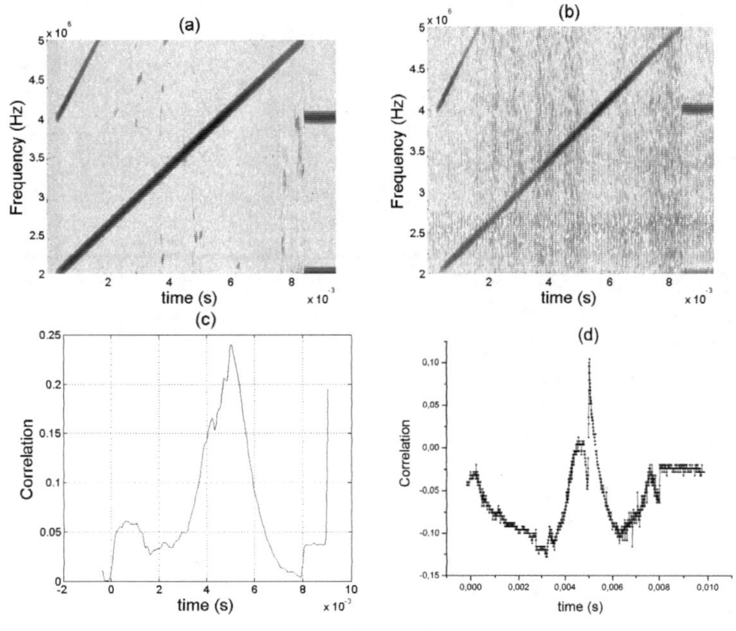

Figure 3. Spectrogram of (a) antenna and of (b) magnetic coil, (c) digital multiplication of (a) and (b). (d) Signal from demodulator circuit, for Shot 1.

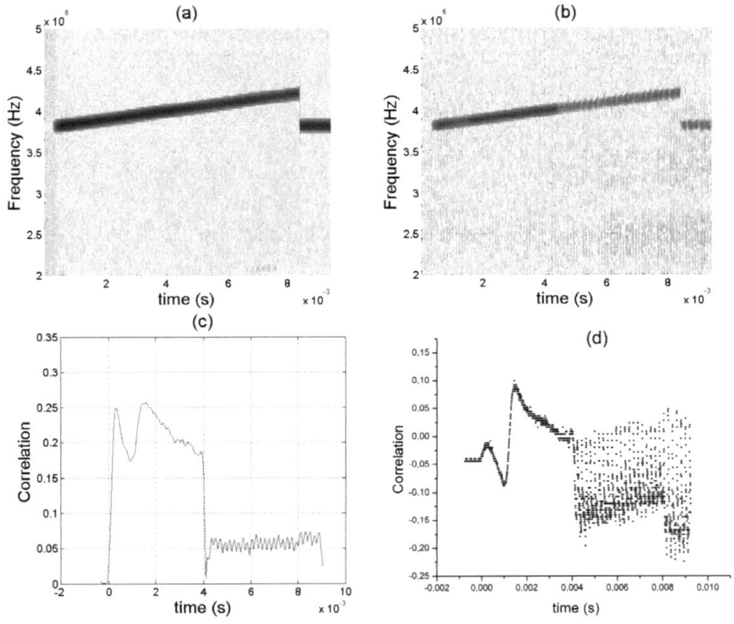

Figure 4. The same values as in Figure 3 for Shot 2.

The equipment was tested in the TCABR tokamak in regimes of the vacuum chamber cleaning by RF discharges with the frequency of $5kHz$. The plasma density was $n = 1 - 9 \times 10^{16}/m^3$ and the magnetic field of $0,04T$, both much lower than the normal regime of the TCABR. Temperature was estimated as $25eV$. In the plasma with these parameters, we could not excite Alfvén waves but it is possible to excite helicon waves. To find helicon resonances, the frequency scan of the generator was fulfilled during $8ms$ in the desired frequency range $2 - 5MHz$.

As an alternative to the circuit, a high sampling frequency oscilloscope is used to acquire data directly from the magnetic coils, as well as the signal of the antennae current. Then, both signals are digitally multiplied and the result is compared with the output of the demodulator circuit. We note that the multiplication of the signals permits to find the cross-correlation between them. When the correlation is high, it can indicate the excitation of the plasma eigenmodes. The cross-correlation signal and the analogical product at the output of demodulator circuit both depend on the phase shift between the input signals. We adjust the length of the signal cable that comes from the antenna to the demodulator circuit in order to compensate the instrumental phase errors and to maximize the signal. As the output signal from the circuit presents the cross-correlation, it is higher if the two input signals are in phase. To simulate this experiment digitally, we have introduced variable phase differences, and thus we could get good coincidence with the experimental data.

Here we present the results from two different shots in Figs.3 and 4 where Fourier analysis of the antenna current (a) and of the data from the magnetic coil (b) are shown. We can observe that both stay in a good agreement, it means that the magnetic field oscillation propagates from the antenna through the plasma with time delay due to the distance about $1m$ from the magnetic coil to the antenna. The signal is very small in the plasma absence. The other figure represent the signal from the demodulator circuit acquired from the magnetic coil (d) and the corresponding digital simulations, so that we can compare both signals. We note that the resonance frequency $f \sim 4MHz$ is found in the shot №1 (Fig.3) at $t \approx 0,005s$. This resonance is identified as the helicon wave after comparison with numerical calculations.

4. Numerical Calculations

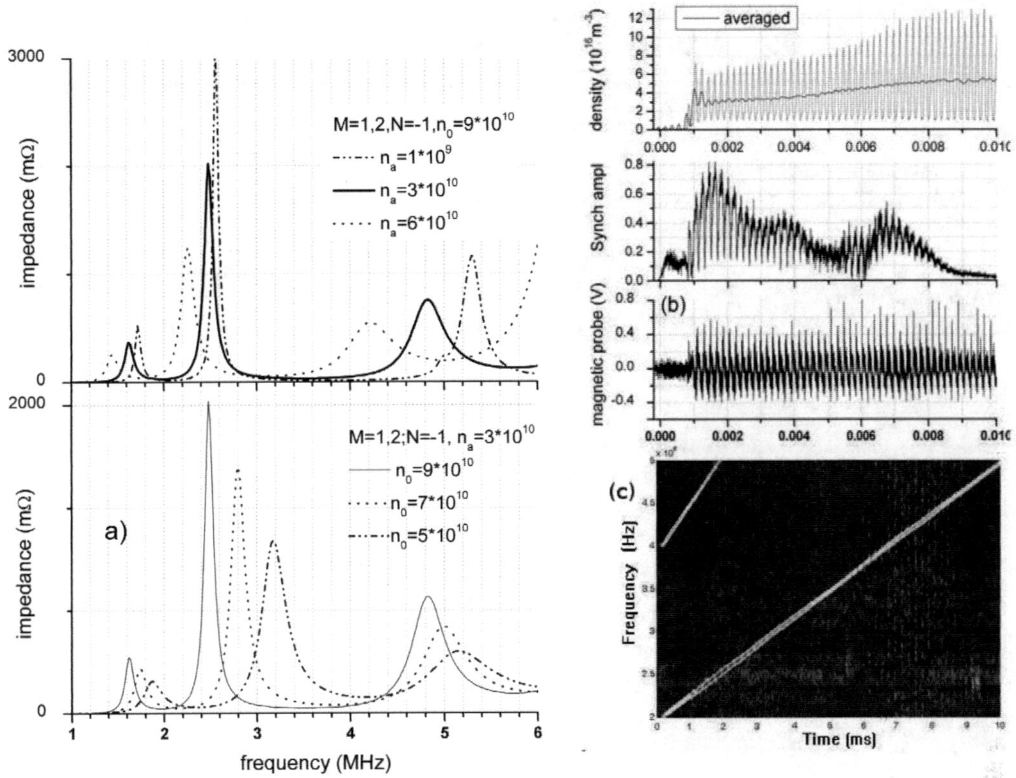

Figure 5. Simulation of helicon waves (a) for parabolic density profile $2 - 9 \times 10^{10}/cm^3$ in the center, and temperature $25eV$. (b) shot that show better agreement with helicon wave simulation, (c) spectrogram of same shot. Synch ampl is similar to the cross-correlation before mentioned, but in this case is calculated between the magnetic probe signal and density fluctuations, measured with a Langmuir probe.

To simulate the helicon waves we used a numerical code which was previously developed for Alfvén waves studies, where parameters were changed for the cleaning discharge mode in TCABR. The simulation was done using k-spectrum related to one antenna module in the frequency range from $1-6MHz$. We choose the lowest toroidal and poloidal modes ($M = \pm1, \pm2$ and $N = \pm1$) for simulations and the results are presented in the Fig.5(a). The density variation measured with a Langmuir probe, which is used in calculations, is presented in upper frame of Fig.5(b), the cross-correlation signal synch ampl is shown in the medium. Finally, the magnetic probe signal is presented in the low frame. As we can see there are resonances for some frequencies, one of them, in particular, is near the frequency $4MHz$ that is close to the frequency shown in Fig.3. The simulation is in agreement with analytical expression for helicon waves $\omega = c^2 k_{\|} k \omega_{ce}/\omega_{pe}^2$. The good agreement with the simulation can be seen in the spectrogram of the magnetic probe signal in Fig.5(c) for the same shot. Unfortunately the demodulator circuit did not capture this shot.

5. Conclusion

The experiments have shown a good agreement between the measurements made by the demodulator circuit and by the calculations of the cross correlation between the magnetic probe and the reference signals, which simulate the circuit behavior. The demodulator circuit is successfully tested, and ready for measurements of Alfvén wave resonances in tokamak plasmas, as soon as TCABR is operational. It is especially suitable for fast findings of the plasma resonances, which can be calculated more accurately later by the digital technique with some high sampling frequency equipment.

It is demonstrated that magnetic probe measurements of sweeping frequency launched by antenna are confirmed by numerical calculation that gives possibility to identify the plasma density. In the future we will make measurements with two antenna modules in more dense plasmas.

Acknoledgements

We are thankful to G. Paganini Canal for density measuraments and FAPESP.

References

[1] Elfimov A G, Ruchko L F, Galvão R M O, *Nucl. Fusion* **46**, 2006
[2] Lerche E A 1998 (Master Thesis, IFUSP, São Paulo)
[3] Guidolin L C B M 2007 (Master Thesis, IFUSP, São Paulo)

Analysis of the electron temperature measurement in TCABR tokamak by Electron Cyclotron Emission and Infrared Thomson scattering diagnostics

O C Usuriaga[1], F O Borges[1], A G Elfimov[1], R P da Silva[1], M H Ono[1], P G P P Puglia[1], M P Alonso[2], J H F Severo[1], I C Nascimento[1], E K Sanada[1], W P de Sá[1], R M O Galvão[1,3], J I Elizondo[1].

[1]Institute of Physics, University of São Paulo, SP, Brazil.
[2]Associação EURATOM/IST, Instituto de Plasmas e Fusão Nuclear - Laboratório Associado, Instituto Superior Técnico, 1049-001 Lisboa, Portugal.
[3]Brazilian Center for Physics Research, Rio de Janeiro, RJ, Brazil.

E-mail: usuriaga@if.usp.br

Abstract. This work presents the experimental analysis of the central electron temperature measured by the electron cyclotron emission (ECE) radiometer and the infrared Thomson Scattering (ITS) diagnostic. The detection of the ECE radiation is done by a heterodyne scanning radiometer that works at the second harmonic extraordinary mode, in frequency range from *50* to *85GHz*, which allows measurement of the radial profile of electron temperature with good spatial and temporal resolutions. The ITS diagnostic uses a Neodymium Glass laser (wavelength $1.054\ \mu m$). This ITS diagnostic measures the electron temperature in the center of plasma column one time during plasma shot. Results also show a discrepancy between the two diagnostics in the electron temperature measurement in the presence of Magnetohydrodynamics activity that gives an explanation for this apparent inconsistency.

1. Introduction

The TCABR tokamak in operation at the Institute of Physics of University of São Paulo (main parameters: Major radius $R_0 = 61.5\ cm$, Minor radius $a = 18\ cm$, Plasma current $I_P < 120\ kA$, Plasma current duration $t_P < 150\ ms$ and Toroidal magnetic field $B_0 = 1.14\ T$) has a broad scientific research program that includes interaction of RF waves [1 and 2], magnetic islands and MHD oscillations [3 and 4], transport barriers by electrically polarized electrode [5] and plasma rotation [6, 7 and 8]. TCABR tokamak is equipped with a wide variety of diagnostics, being the Electron Cyclotron Emission (ECE) and infrared Thomson scattering (ITS) diagnostics used to measure of the plasma electron temperature.

The ECE radiometer has long been the only diagnostic for providing the electron temperature of plasma in tokamak TCABR. This radiometer has the major drawback of the impossibility of measuring the electron temperature beyond the cutoff density due to the conditions of accessibility. This restriction of access due to the low toroidal field, *1.14T*, imposes a limit on the measurement of electron temperature. So we can only use this diagnostics for electron densities less than $2.5{\times}10^{19}\ m^{-3}$ in the central position of the plasma column. This value is lower than required for most research in TCABR, such as interaction with biased electrode where the electron density in the center of the

plasma to reach a value of up to 4.0×10^{19} m^{-3}. This limitation causes, for practical purposes, this diagnostic is of limited use.
Recently, in collaboration with the Instituto Superior Técnico in Lisbon, has made it possible to install a new system for the diagnostic of electron temperature using Thomson scattering. This system allows the measurement of electron temperature in the center of the plasma column in a given time regardless of the value of the electron density. Preliminary results on the measurement of electron temperature in the central column of plasma, have been obtained using both diagnostics [9]. These initial results showed an acceptable agreement for discharges during ohmic heating and a discrepancy with Alfvén waves. Continuing to collect data for different situations TCABR plasma, we found some systematic differences in the value of electron temperature provided by both diagnostics. It was noted that this discrepancy is directly related to the presence of strong MHD activity. In this paper we present the situations in which there is agreement and disagreement between measurements made by both diagnostics and give a possible explanation for the apparent dual behavior.

2. Measurement of electron temperature from ECE emission

The detection of ECE radiation is done by means of a heterodyne scanning radiometer that works at the second harmonic extraordinary mode (*X-mode*) with frequency range from *50* to *85GHz*, which allows measurement of the radial profile of electron temperature with good spatial and temporal resolutions. The system can operate in a fixed frequency mode (time resolution *10 μs*), or in a sweeping mode that allows the radial temperature profiles to be obtained. Since the radial resolution at the center of the plasma column of *1.1 cm* for toroidal field $B = 1.14T$, and circular cross section beam $\sim 3.8\ cm^2$. The plasma millimeter-wave emission is received by a Gaussian Antenna located in the equatorial plane at the TCABR of the vacuum vessel.
In tokamak machines with low toroidal magnetic fields and high plasma densities, such as the TCABR tokamak, the accessibility conditions impose restrictions to the detection of the ECE. The accessibility conditions are examined based on the model of *cold plasma approximation* to describe the propagation of microwaves through the plasma. From numerical calculations it was found for the TCABR, due to the harmonic overlapping (Fig. 1a), the detection of the second harmonic, in the lower field side, is limited to the *52-74 GHz* band (*~70%* of the plasma column, which correspond to the radial range between approximately *12 cm* and *-8 cm*).

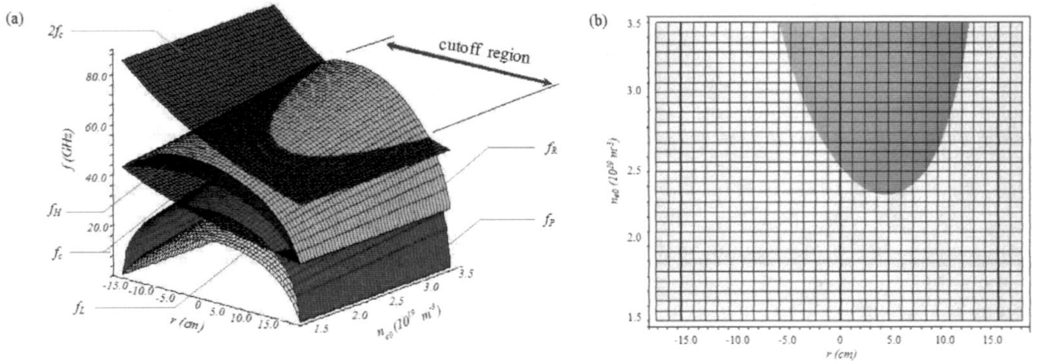

Figure 1. (a) The harmonic overlapping and cutoff frequencies for the TCABR plasma. Frequencies involved are f_c, cyclotron frequency, f_P, plasma frequency, f_H, upper hybrid frequency, f_L, left hand frequency, and f_R, right hand frequency. (b) The area in gray color represents the region not accessible by the second harmonic plasma frequency.

The accessibility conditions for the second harmonic (X mode) restrict the maximum electron density to values bellow $n_{e0}=2.5\times10^{19}$ m^{-3}, as can be seen in Fig. 1b. Due to this cutoff, the electron temperature cannot be measured for densities greater than this.The ECE radiometer has been calibrated absolutely; the method used basically consists in determining the output voltage of the radiometer as a function of frequency for two blackbodies with different temperatures [10]. From the results it appears that the accessibility conditions determined by the electron density and magnetic field in plasma TCABR, impose limitations on the use of the ECE diagnostic. However, the measure of electron temperature for densities below the cutoff region has been very useful in experiments in TCABR, even as the opposite of ITS where the temperature measurement is made in a single instant of time, the ECE diagnostics provides the electron temperature throughout the plasma discharge.

3. Measurement of electron temperature from Thomson scattering

The infrared Thomson scattering diagnostic is based upon a Neodymium Glass Laser with up to *4.5 Joules* per laser pulse which has a duration of approximately *50 ns*. The laser emission has central wavelength in the infrared part of the spectrum, $\lambda = 1.054\ \mu m$, and *2 nm* spectral width at half height.
In Thomson scattering experiments, the radiation from a high-energy pulsed laser is vertically directed into the Tokamak plasma. The spectrum of scattered radiation is analyzed at 90° by simple collecting objective composed by one plan-convex and one plan-aspherical lens. The electron temperature can be determined from shape of the measured Thomson spectrum. A full description of infrared Thomson scattering diagnostic is given in ref. [9]. The infrared Thomson scattering system is shown in Fig. 2. A collection fiber transport the collected light to a set of 3-channel filtered polychromator that use avalanche photodiode detectors (APDs), ref. [11]. The scattered radiation is measured by polychromator and digitized by oscilloscope. Analyzing the signal-ratio of the detector signals (Fig. 3) we can obtain the electron temperature in center of plasma column during one plasma shot with a spatial resolution of $\approx 18\ mm$ height and $\approx 3\ mm$ diameter.

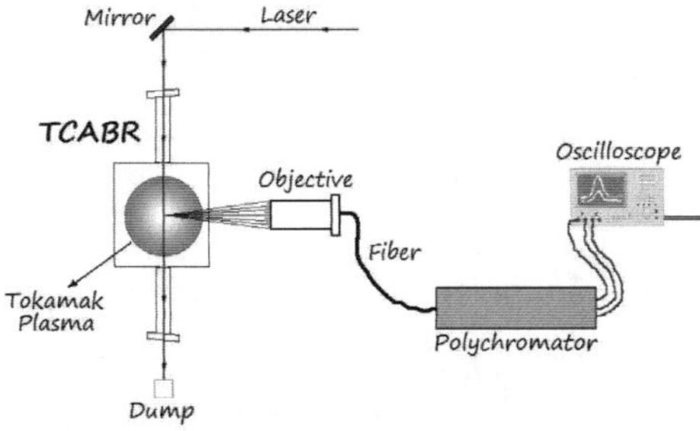

Figure 2. Schematic layout of TCABR infrared Thomson scattering system.

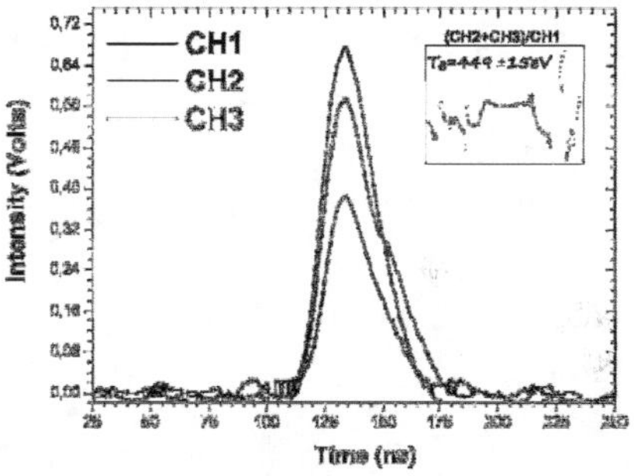

Figure 3. Thomson scattering signals from 3-channels polychromator. The lower graph is the signals-ratio.

4. Results

Figure 4 shows the results of measurements on the line-averaged electron density in the center of the plasma column and electron temperature in the same region measured by the ECE diagnostic. The lines indicate fits through the data. The top graphic shows the line-averaged density versus current plasma, and the bottom graphic shows the electron temperature versus the current plasma. The results were divided into two groups, the first group corresponds to a density of $(1\text{-}1.2) \times 10^{19}\ m^{-3}$ (solid circles), and the second group between $(1.25\text{-}1.45) \times 10^{19}\ m^{-3}$ (open circles). What is clear is the behavior of the electron temperature to the density, in this region of plasma, shown by line fit, the electron temperature decreases as we increase the density. From this we can deduce that the results obtained by the ECE diagnostic are in full agreement with what was expected in the plasma of TCABR. After careful reviews of the electron temperature, measured simultaneously by the ECE and ITS diagnostics of a relatively large number of discharges, discrepancies were found in many of them. The question is: what is happening so that there is an apparent contradiction in the temperature measurement made by both diagnostics? To provide an answer to this question, more detailed reviews are carried out. Then we use the experimental results obtained for ohmic discharges, such as the ECE radiometer, ITS, and Mirnov coils to thereby obtain information on electron temperature in the center of the plasma column. For comparison only are selected plasma discharges doing not overstep the limits of cutting the ECE diagnostic, imposed by the electron density, i.e. line-averaged electron density range $\overline{n}_e=(1 - 1.5) \times 10^{19}\ m^{-3}$. For line density, $\overline{n}_e>1.5\times 10\ m^{-3}$ (the central density $n_{e0}>2.5\times 10^{19}\ m^{-3}$) the cutoff in ECE radiation was observed. In Figure 5, showing two typical discharges; the first with no MHD activity (shot 24472) and the second with strong MHD activity (shot 24231). The dashed line indicates the instant it is connected to the Thomson scattering diagnostic ($t=73\ ms$ for 24472 and $t=90ms$ for 24231 discharge); this instant of time coincides with the measurement of electron temperature for both diagnostics. The evolution of temporal profiles in shot 24231, we can see clearly that the MHD oscillation, Mirnov coil signal, results in a decreased electron temperature, the radiometer ECE signal.

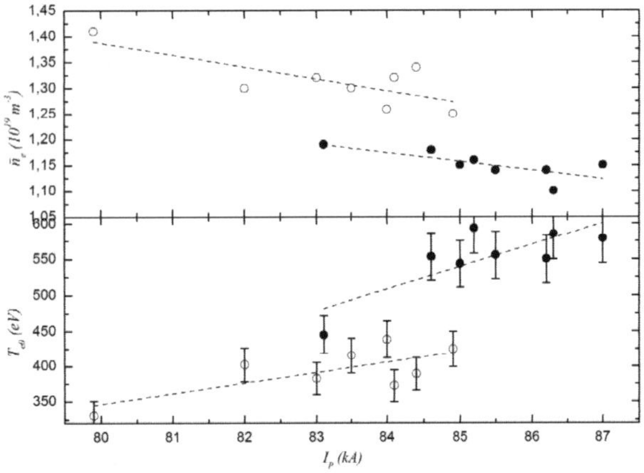

Figure 4. Overall result of the electron density and temperature in the center of the plasma column for ohmic discharges in TCABR tokamak. Top: the line-averaged density versus current plasma. Bottom: the central electron temperature, measured with ECE diagnostic, versus the current plasma. The lines indicate fits through the data.

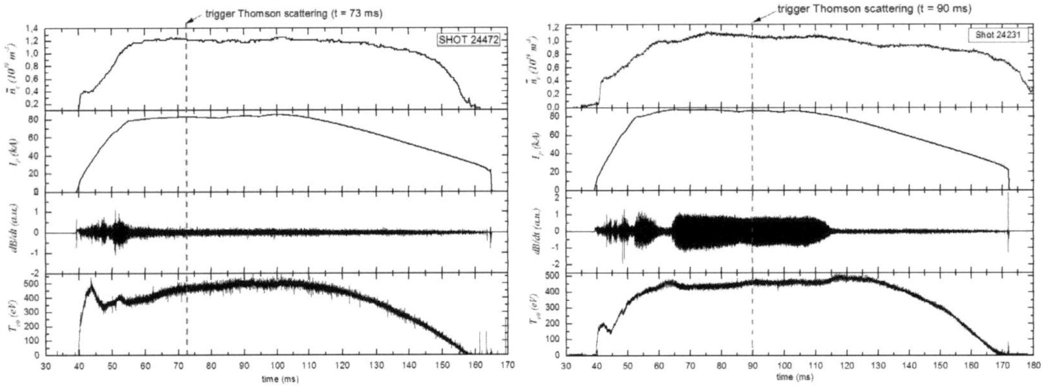

Figure 5. Typical discharge in TCABR tokamak, shot 24472 with very low MHD activity and the shot 24231 with strong MHD activity. Shows the temporal profiles obtained for of averaged-line density ($\overline{n}_e$), current plasma (I_P), Mirnov coil and electron temperature (signal ECE). The dashed line indicate trigger Infrared Thomson scattering system.

Making up the Fourier analysis of time signals of ECE in both discharges (Fig. 6a), we observe the presence of dominant frequency of $\sim 2.4\ kHz$ (and subsequent harmonics), which corresponds to the oscillation frequency of the sawtooth (safety factor $q<1$). The sawtooth oscillations appear in shot 24472, during ohmic discharge phase (~60ms to ~140ms), and shot 24472 appear early ($t\sim 60\ ms$), corresponding to initial phase of the discharge with duration approximately of $10ms$, and final phase ($t\sim 110\ ms$) (Fig. 6b). Then performed the Fourier spectral analysis of signals measured by Mirnov coils in both discharges (Fig. 6c), proved the existence of a dominant mode with a frequencies of 12.5

kHz to 14 kHz. This frequency is not present in the spectrogram ECE for shot 24472. The same is not true for shot 24231, where the ECE signals and Mirnov coils present similar frequency ~12.5 kHz, in spectral graphs, constant during the time intervals (~*70 ms* to ~*110 ms*). The position feedback system of TCABR guarantees that substantial horizontal and vertical displacements of the plasma column do not occur in this time interval.

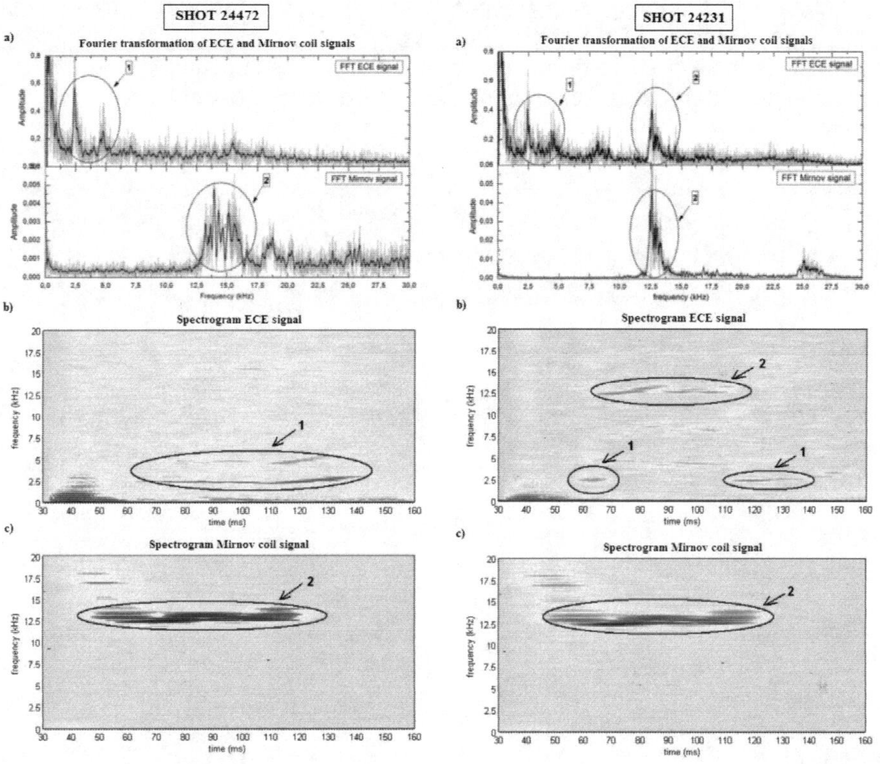

Figure 6. a) Fourier transformations of ECE and Mirnov coil signals, b) power spectrum of ECE radiometer data located central plasma column and c) power spectrum of Mirnov coil located on the outboard midplane, for shots 24472 and 24231 respectively.

The Figures 7 shows the central electron temperature, measured simultaneously by Thomson scattering and ECE diagnostic, in the ohmic phase as a function of density, In Figure 7a, the Thomson scattering data confirm the measured values by ECE when there is no MHD activity. The opposite occurs when there is a strong MHD activity; this is evident in Figure 7b.

5. Discussions and Conclusion

The electron temperature is measured by Infrared Thomson scattering and ECE diagnostics in TCABR experiments. Analyzing the temperature of a set discharges, we can confirm that the temperature obtained by ECE diagnostics decreases with increasing of the electron density due to imposing a cutoff layer. Some discrepancies appear in temperature measurements with the ECE and ITS diagnostics when they are carried out simultaneously. From a detailed analysis of the results, we conclude that the signals in discordant measurements have a strong MHD activity, which appears in the temperature profile, this result is observed in the analyzed frequency ECE spectrum.

An important factor is that ECE and ITS diagnostics are positioned along toroidal direction with a angle gap ≈*120* degrees, that does not give possibility to observe the same plasma region. Others important differences in this analysis are; the temporal resolution of the diagnostics (ECE ≈*10μs* and

ITS ≈*50ns*) and observed plasma volume by ECE ≈*6.46cm*^{3} and by ITS ≈*0.13cm*^{3}. The parameters of one can conclude that the measured temperature by the ITS is a instantaneous measurement, and the electron temperature measured by ECE has temporal and spatial averaging.

In summary, we conclude that both diagnostics appear as different during MHD activity. The ITS be measured at a point of radius (with higher spatial resolution), and in a very short time interval likes a few nanoseconds, and ECE observe the emission of radiation from a larger region of the plasma and during the time interval approximately 200 time greater. In the sense, this discrepancy in the measurement of electron temperature during MHD activity happens in reality because the temperature is acquired by both diagnostics in the different manner, and is inherently influenced by parameters of a technical nature.

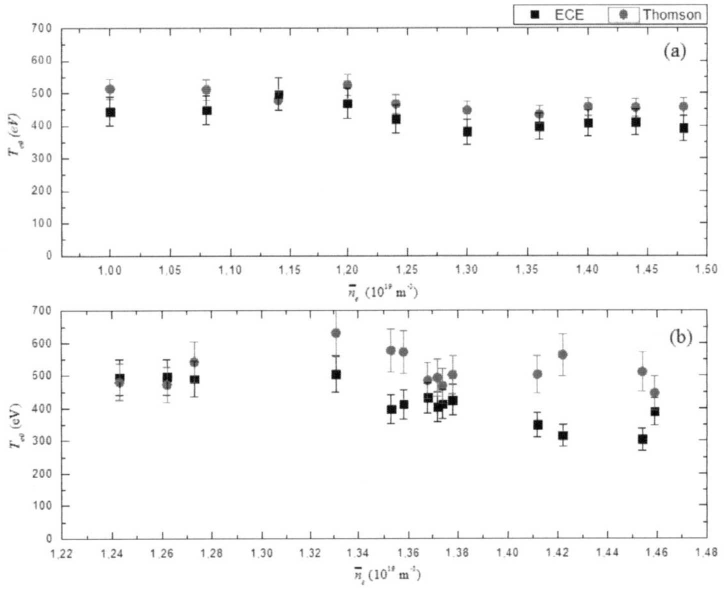

Figure 7. (a) Electron temperature without MHD activity e (b) temperature perturbed by large MHD mode activity. ECE radiometer temperature (solid squares), Infrared Thomson scattering temperature (solid circles). The errors bars for T_e in both diagnostic are of the order of 10%.

Acknowledgments

This work was partially supported by CNPq (National Council for Research and Development – Brazil) and by FAPESP (Foundation for Supporting Research of the State of São Paulo). The authors wish to thank to the "Instituto Superior Técnico – IST" (Portugal) for collaboration in the implementation of the Thomson scattering diagnostic and tokamak TCABR team.

References

[1] Galvão R M O, Kuznetsov Yu K, Nascimento I C, et al. 2001 *Plasma Physics and Controlled Fusion*. v. **43**, pp. 1181
[2] Ruchko L F, Lerche E A, Galvão R M O, et al. 2002 *Brazilian Journal of Physics*. V. **32** pp 57
[3] Severo J H F, Nascimento I C, Tsypin V S, et al. 2004 *Physics of Plasma* **11**(2) 846
[4] Heller M V A P, Caldas I L, Ferreira A A, et al. 2005 *Czechoslovak Journal of Physics* **55** 265
[5] Nascimento I C, Kuznetov Yu K, Guimarães-Filho Z O, et al. 2007 *Nuclear Fusion* **47**, 1570
[6] Severo J H F, Tsypin V S, Galvão R M O, et al. 2002 *Brazilian Journal of Physics* v.**32**, pp. 13
[7] Severo J H F, Nascimento I C, Kuznetov Yu K, et al. 2007 *Review of Scientific Instruments* **78** 043509
[8] Severo J H F, Nascimento I C, Kuznetov Yu K, et al. 2009 *Nuclear Fusion* **49**, 115026
[9] Alonso M P, Figueiredo A C A, et al. 2010 *Journal of Physics: Conference Series* **227** 012027
[10] da Silva R P, Fonseca A M M, Vuolo J H, et al. 2004 *Brazilian Journal of Physics* v. **34**, 1771
[11] Alonso M P, Bemi L A, Severo J H F, et. al., 2008 *Plasma and Fusion Science* **996**, 192

Comparison of Plasma Visible Spectral Emissions Between Nova-UNICAMP and TCABR Tokamaks

F do Nascimento [1], M Machida [1], G Ronchi [1], L M F Schmutzler [1], J H F Severo [2], I C Nascimento [2] and E K Sanada [2]

[1] Instituto de Física "Gleb Wataghin", UNICAMP, Campinas, S. P., Brazil.
[2] Instituto de Física, USP, São Paulo, S. P., Brazil.

E-mail: fellypen@ifi.unicamp.br

Abstract. A comparison between the visible spectrum emissions observed on the Nova-UNICAMP and TCABR tokamak plasmas has been made in this work using a hand top HR4000 Ocean Optics spectrometer equipped with a CCD detector. A number of 58 emission lines, in which 22 are common to both machines, have been observed and identified. The differences in the observed spectrum can be explained by the difference in the time integration used in these measurements and by the materials compositions of the limiter and electrode in the TCABR tokamak. Nearby peak emissions have been separated using multi-peak Gaussian fit curves obtaining separation between peak centres with the same order of the spectrometer resolution. The HR4000 spectrometer can be routinely used to monitor the impurity species in the visible spectrum from the different tokamak windows.

1. Introduction

In this work we present a comparison between visible spectrum emissions, from 350 to 800 nm, observed on Nova-UNICAMP and TCABR tokamak plasmas. One of the objectives of this work is to complement the information presented in the reference [1], where a similar comparison was made but just in the VUV region.

The Nova-UNICAMP [2] is a very small iron core tokamak where the main scientific program is the development of new diagnostics for plasma fusion [3-5]. The Nova-UNICAMP tokamak has the major and minor plasma radius equal to $R = 0.30$ m, $a = 0.06$ m, characteristic plasma current $I_P = 10$ kA, toroidal field $B_T = 0.8$ T, and plasma discharge duration of 10 ms. On the other hand, TCABR is bit bigger tokamak and has the following plasma parameters, major radius $R = 0.61$ m, minor radius $a = 0.18$ m, discharge current $I_P = 100$ kA, toroidal magnetic field $B_T = 1.1$ T, maximum average density n_e $(1 - 4.5) \times 10^{19}$ m^{-3}, $T_e(0) \approx 500$ eV, $T_i(0) \approx 200$ eV, duration of the stationary phase of the discharge 60 ms. This machine has a broad scientific program that includes, interaction of RF waves [6, 7], magnetic islands and MHD oscillations [8, 9], transport barriers by electrically polarized electrode [10] and plasma rotation [11-13].

The measurements have been made using a hand top HR4000 Ocean Optics spectrometer equipped with a Toshiba CCD detector, model TCD1304AP, which is a linear array with 3648 pixel elements and external trigger. The HR4000 is a very small spectrometer that uses a Czerny-Turner mount with 10.16 cm focal length, a diffraction grating of 600 grooves/mm and an entrance slit with fixed size (25 μm). This spectrometer has a spectral resolution (FWHM) of 0.53 nm.

The measurements were made from limiter window, without focusing on it, for the Nova-UNICAMP, and from tangential window where Alfvén antenna is located for the TCABR tokamak.

2. Observed spectra

The spectra obtained in the two tokamaks are shown in Figure 1, and the emission lines identified are listed in Table 1. We have observed and identified a number of 58 emission lines, in which 22 are common to both machines, 32 were detected only on TCABR and 4 only on Nova-UNICAMP.

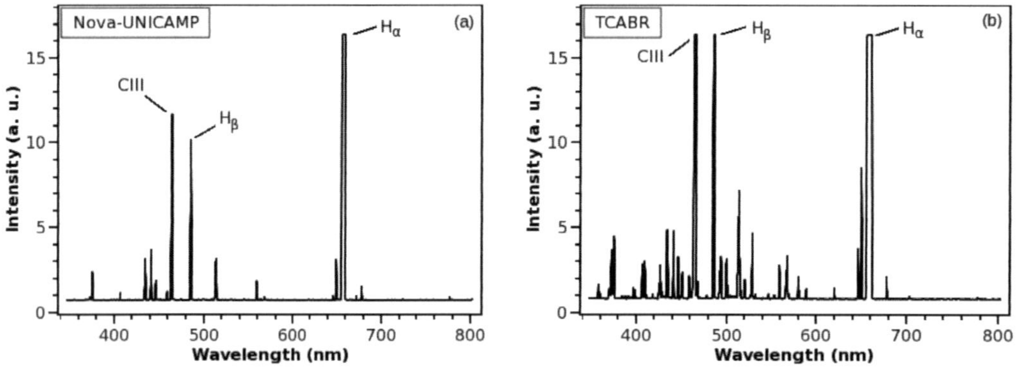

Figure 1. Spectral lines from 350 to 800 nm obtained in (a) Nova-UNICAMP and (b) TCABR tokamaks. This TCABR spectra corresponds to shot #24128.

3. Analysis and results

The three most intense lines, H_α at 656.28 nm, H_β at 486.13 nm and carbon CIII at 464.74 nm, are commonly observed in tokamaks and other fusion devices that use H_2 as primary fuel.
As we can see in Table 1, the number of emission lines observed in TCABR is larger than the observed in the Nova-UNICAMP tokamak. The carbon emission lines appear more frequently in the TCABR tokamak due to the graphite limiter used in this machine.

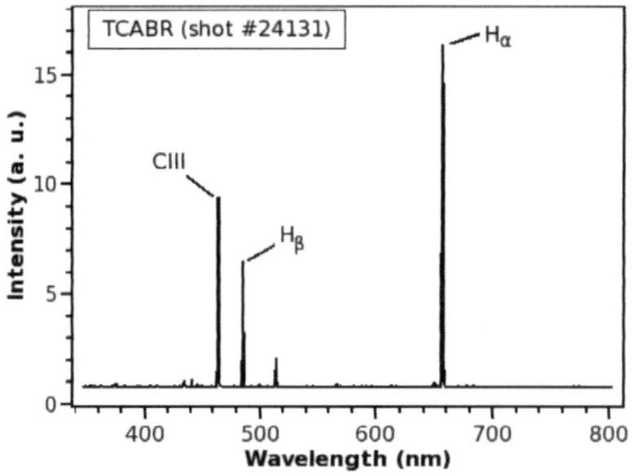

Figure 2. Visible spectra obtained for shot #24131 in TCABR tokamak.

We also note that spectral lines from elements like *Fe* and *Cu* appears only on TCABR, probably due to the presence of an Alfvén antenna and some diagnostic tools installed on it, like multi position Langmuir probes. The absence of these lines on the Nova-UNICAMP tokamak also can be due to a better wall treatment done in this machine.

Moreover, the different integration times used in these acquisitions (5 ms for Nova-UNICAMP, due to shorter plasma discharge time, and 25 ms for TCABR), can result in much intense emission lines. Data acquisition using 5 ms in TCABR was made for shot #24131 (Figure 2) and we verified that, in this case, the spectrum of both machines are more similar.

Table 1. Spectral lines observed on the Nova-UNICAMP and TCABR tokamaks with their respective wavelengths in nanometers. Theoretical wavelength values were taken from the reference database NIST [14].

Line #	Element	λ Nova	λ TCABR	λ NIST	Line #	Element	λ Nova	λ TCABR	λ NIST
1	Fe I		357.90	358.12	30	C III	466.04		466.59
2	Fe I		359.36	359.46	31	O II		467.33	467.62
3	Fe I		360.55	360.67	32	Fe I		468.24	468.36
4	O III		370.22	370.34	33	O IV		478.44	478.64
5	O IV	372.79	372.73	372.90	34	N III	485.36		485.38
6	Fe I		373.66	373.71	35	H_β	486.04	485.91	486.13
7	O I	375.23	375.05	375.21	36	O V		492.85	493.03
8	O III	375.84	375.82	375.99	37	O II		494.14	494.30
9	Fe I		381.85	382.04	38	N II	500.12	500.04	500.15
10	O II		384.75	384.79	39	C II	513.34	513.09	513.29
11	Fe I		388.44	388.63	40	C II	514.27	514.24	514.35
12	C II		391.87	392.07	41	O II		520.49	520.67
13	O II		397.13	397.33	42	C VI		528.77	529.05
14	O IV		399.37	399.51	43	Fe I		532.45	532.80
15	C III	407.15	406.86	407.03	44	O II	558.52	558.62	558.32
16	O II		409.61	409.82	45	O III	559.08	559.03	559.27
17	O II		411.84	412.05	46	N II		566.31	566.66
18	C III		418.53	418.69	47	N II	567.63	567.69	567.96
19	Cu II		425.20	425.56	48	C IV		579.85	580.13
20	C II		426.51	426.73	49	C II		588.84	588.98
21	Fe I		428.60	429.14	50	C III		620.35	620.56
22	H_γ	434.31	433.89	434.05	51	N III	646.56	646.49	646.70
23	O II	434.85	434.75	434.74	52	N II		648.07	648.21
24	O II	441.31	441.25	441.39	53	O V	649.98	649.90	650.02
25	O II	446.46	446.45	446.45	54	H_α	657.10	656.24	656.28
26	N III		451.00	451.09	55	O II	672.12		672.14
27	O II	459.03	458.91	459.10	56	C II	678.34	678.28	678.39
28	C III	464.15		463.91	57	O III		702.87	703.14
29	C III	464.86	464.61	464.74	58	O I	777.26	777.15	777.19

Some peak emissions were separated using multi-peak Gaussian fit curves, obtaining separation between peak centers with the same order of the spectrometer resolution. Examples of this analysis are shown in Figure 3.

Excepting line number 55, the peak emission that appears only in the Nova-UNICAMP, do not appear in the TCABR because they are jeopardized by the nearest peak in its spectra and can not be separated using the multi-peak fit procedure.

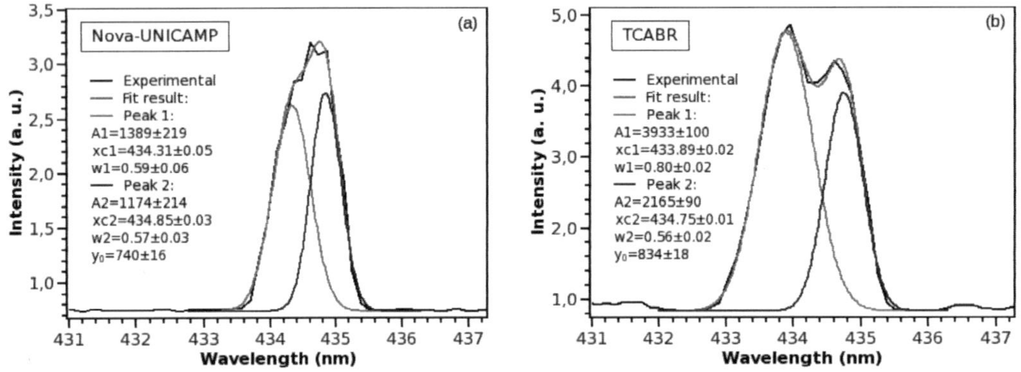

Figure 3. Examples of multi-peak fits used to separate emission lines in (a) Nova-UNICAMP and (b) TCABR spectra. The Gauss model ($y = y_0 + (2/\pi)^{1/2} \cdot (A/w) \cdot exp[-2 \cdot (x - x_c)^2/w^2]$ was used for the fits. Both graphs show the H_γ (434.05 nm) and OII (434.74 nm) emissions.

4. Conclusions and considerations

The HR4000 spectrometer can be routinely used to monitor the impurity species in the visible spectrum from the different window positions of tokamaks. More interesting results can be obtained if a better diffraction grating is used. Combining it with multi-peak fitting methods or deconvolution analysis more spectral lines could be identified. Therefore, this kind of spectrometer shows a good versatility due to very low cost and mobility.

Both machines should present the same visible spectral emissions if the observations are made in plasma regions with similar behavior. This is expected because the same occurs in VUV region [1].

Moreover, we noted that some emission lines are isolated in the spectra, and they can be used for ion temperature measurements.

Acknowledgements

This work was supported by CAPES, RNF, IAEA, CNPq and FAPESP.

References

[1] Machida M, Arsioli B S, Nascimento F *et al* 2009 *J. Plasma Fusion Res. SERIES* **8** 636-639
[2] Daltrini, A M, Machida M *et al* 2002 *Brazilian J. of Physics* **32** 26-29
[3] Daltrini A M and Machida M 2005 *Review of Scientific Instruments* **76** 053508
[4] Monteiro M J R, Machida M, Daltrini A M *et al* 2002 *Brazilian J. of Physics* **32** 54-56
[5] Machida M, Daltrini A M, Severo J H F *et al* 2008 *Plasma and Fusion Science Book Series: AIP Conference Proceedings* **996** 235
[6] Galvão R M O, Kuznetsov Yu K *et al* 2001 *Plasma Phys. and Controlled Fusion* **43** 1181-1190
[7] Ruchko L F, Lerche E A, Galvão R M O *et al* 2002 *Brazilian J. of Physics* **32** 57-64
[8] Severo J H F, Nascimento I C, Tsypin V S *et al* 2004 *Physics of Plasma* **11** 846-848
[9] Heller M V A P, Caldas I L, Ferreira A A *et al* 2005 *Czechoslovak J. of Physics* **55** 265-270
[10] Nascimento I C, Kuznetsov Yu K *et al* 2007 *Nuclear Fusion* **47** 1570-1576
[11] Severo J H F, Tsypin V S, Galvão R M O *et al* 2002 *Brazilian J. of Physics* **32** 13-19
[12] Severo J H F, Nascimento I C *et al* 2007 *Review of Scientific Instruments* **78** 043509
[13] Severo J H F, Nascimento I C, Kuznetsov Yu K *et al* 2009 *Nuclear Fusion* **49** 115026
[14] http://www.nist.gov/physlab/data/asd.cfm

Investigation of Non-inductive Plasma Current Start-up by RF on QUEST

Masaki Ishiguro[1], Kazuaki Hanada[2], Hiqing Liu[1], Ryota Ogata[1], Mitsutaka Isobe[3], Saya Tashima[1], Hideki Zushi[2], Khonosuke Sato[2], Akihide Fujisawa[2], Kazuo Nakamura[2], Hiroshi Idei[2], Mizuki Sakamoto[2], Makoto Hasegawa[2], Yuichi Takase[4], Takashi Maekawa[5], Yasuaki Kishimoto[5], Osamu Mitarai[6], Shoji Kawasaki[2], Hisatoshi Nakashima[2] and Aki Higashijima[2]

[1]IGSES Kyushu University, Kasuga 816-8580, Japan
[2]RIAM Kyushu University, Kasuga 816-8580, Japan
[3]NIFS, Toki 509-5292, Japan
[4]Graduate School of Frontier Science, University of Tokyo, Kashiwa 277-8561, Japan
[5]Graduate School of Energy Science, Kyoto University, Kyoto 606-8502, Japan
[6]Liberal Arts Education Center, Tokai University, Kumamoto 862-8652, Japan

ishiguro@triam.kyushu-u.ac.jp

Abstract. Formations of a closed flux surface (CFS) on QUEST are achieved by fully non-inductive current start-up driven by RF, which is 8.2GHz in frequency and more than 40kW in power. It found that appropriate magnetic configuration with positive n-index and reduction of particle recycling was crucial to achieve the non-inductive plasma current start-up (PCS) successfully. Especially the controllability of particle recycling should be improved by wall conditioning based on successive plasma production and wall cleaning with electron cyclotron resonance heating (ECR) plasmas induced by RF in frequency of 2.45GHz.

1. Introduction

Spherical tokamak (ST) is a tokamak whose aspect ratio ($A = R/a$: R is major radius and a is minor radius) is significantly low. Lower aspect ratio provides the possibility to obtain higher β plasma and therefore ST is a promising candidate for future cost-effective fusion reactors [1].

In tokamaks, toroidal plasma current is required to confine plasmas. For plasma current start-up (PCS) in conventional tokamaks, ohmic heating (OH) due to significant loop voltage induced by the time variation of central solenoid coil current is a popular way. But, for forming ST configuration, the aspect ratio should be reduced and tighter central space should be satisfied. This indicates that the magnetic flux supplied from the coil located inside the plasma is significantly restricted. ST with no central solenoid coil is ideally the best. Thus, the way to make a closed flux surface (CFS) and subsequent PCS under lower or fully non induction of loop voltage is quite essential for obtaining confined plasmas in ST. Moreover PCS with the lower loop voltage is likely to be also required in super-conductive tokamaks like ITER because of the difficulty of making fast flux swing [2]. Therefore, the study for non-inductive PCS is one of the most important issues in future tokamaks and the experiments were done in many devices such as, WT-2, 3, JIPPT-IIU, PLT, TRIAM-1M, JT-60U,

Journal of Physics: Conference Series **511** (2014) 012041 doi:10.1088/1742-6596/511/1/012041

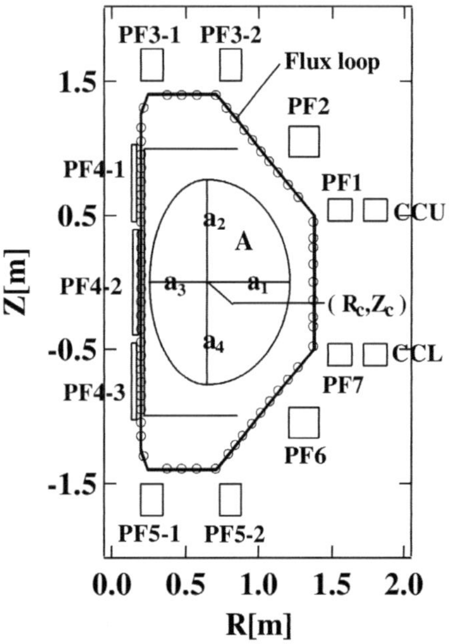

FIG. 1. Cross-sectional side view of QUEST including positions of the PF coils and the flux loops is shown. The ellipse inside the vacuum chamber is the most outer shape of current distribution model as introduced in ref.10.

CDX-U[3], LATE[4], TST-2[5], MAST[6], NSTX[7] and so on. Especially, some devices have made remarkable experimental results in PCS with RF injection [4, 5, 6].

Initial current should be induced by RF injection in an open magnetic field configuration (OMFC) formed by external poloidal field (PF) coils. It was reported that CFS was formed when initial current reached to a certain large value [8]. The formation of CFS is a drastic phenomenon, which contains the topological change of magnetic field, and an essential step for forming tokamak configuration. The formation of CFS on the ST named Q-shu University Experiment with Steady-State Spherical Tokamak (QUEST) is obtained with non-inductive way induced by RF. In this paper, this result will be reported. In section 2, the experimental apparatus of QUEST is shown. The experimental results are shown in section 3 and some discussions are done in section 4. Conclusion and summary is presented in section 5.

2. Experimental Apparatus

QUEST (R=0.68m, a=0.4m, B_T=0.25T at R=0.64m) [9] has two RF systems. One is the 2.45GHz system up to 50kW and the other is 8.2GHz oneup to 400kW. In this study, only 8.2GHz one is used to drive plasma current and 2.45 GHz one is used to make ECR plasmas for wall cleaning.

The schematic view of QUEST is shown in figure 1. QUEST is constructed to study plasma wall interaction (PWI) under an all-metal hot wall in long duration plasma [9]. 11 PF coils and a 16 turns toroidal field coil for confinement of the plasma are set at outside of the vacuum vessel. The PF4-1, 2 and 3 coils, which located at center of the torus, are used as an OH coil. Besides, a pair of cancel coils (CCs), which will be normally used to connect with PF4-1, 2 and 3 in series, are set to make a null point in the vacuum chamber at the start-up phase. It should be noted that the PF4-1, 2, 3 coils and CCs were not used in the experiments presented in this paper.

The diameter and the height of a cylindrical inner limiter are 0.44m, 2.5m, respectively and those of outer wall are 2.74m and 2.79m. A pair of diverter plates is set at $Z = \pm 1$m, where the mid-plane is Z = 0m. The design value of plasma size in QUEST is major radius R_0 = 0.64m, minor radius a = 0.36m and aspect ratio A = 1.78 [9].

In QUEST, 61 flux loops and one Rogowski coil are utilized to reconstruct magnetic surfaces. All flux loops are located along the inside surface of the vacuum chamber wall. Numerical integration is done to obtain the flux values. In this paper, distribution of plasma current is estimated from the measured flux values with the modified method introduced in ref. [10] and then the reconstruction of the magnetic flux surface is done. In the method, the parabolic model of plasma current distribution shown in Fig.1 is assumed. This model has 8 parameter j_0, R_c, Z_c, a_1, a_2, a_3, a_4 and α. The current density in section A shown in Fig. 1 is written as follows,

$$j = j_0 \left\{ 1 - \frac{(R - R_c)^2}{a_1^2} - \frac{(Z - Z_c)^2}{a_2^2} \right\}^{\alpha} \quad (1).$$

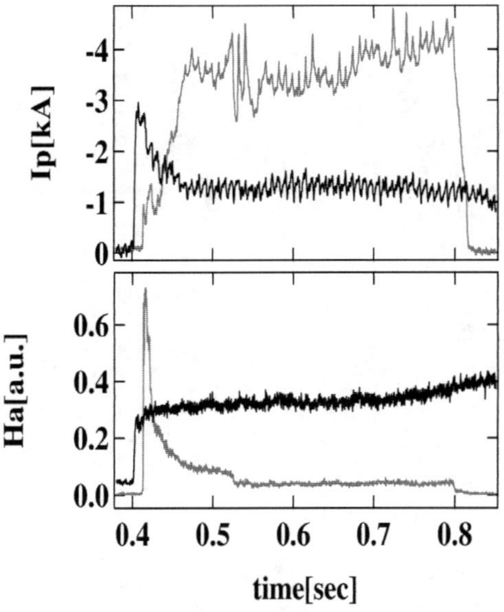

FIG. 2. The comparison of plasma current (top picture) and H_α signal (bottom picture) in the term1 (black line, sn6616) and term2 (gray line, sn6878) are shown. In these two shots, vertical field is 1mT and n-index is 0.35.(Ha should be changed to Hα in the figure.)

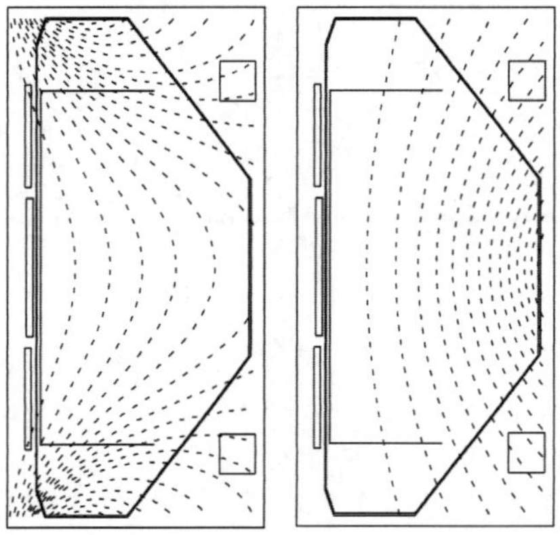

FIG. 3. The open magnetic field configuration with positive n-index (left side) and negative one (right side). Dashed line shows the projection of magnetic line to poloidal cross section. In QUEST layout of coils, positive n-index configuration is made by charging only PF26 coils and negative one is made by charging only PF17 coils.

Where, j_0 is the peak value of current density. In QUEST, j_0 is decided as the total amount of plasma current measured by the Rogowski coil. This is the main modification of the method from that in ref. [10]. The other parameters are scanned to find the best condition by use of the least square methods to the measured flux value.

3. Experimental Results

Formation of CFS can be observed in non-inductive PCS experiment by RF in QUEST. Figure 2 shows the development of plasma current in this experimental campaign. In the experimental term before first observation of formation of CFS (term1), typical plasma current is around 1kA (sn6616: black line in Fig.2) under time-constant external vertical field of 1mT, n-index=0.35 on the mid-plane at R = 0.64m, where n-index is expressed by

$$. n = -\frac{R}{B_z}\frac{\partial B_z}{\partial R} \qquad (2).$$

The projections of magnetic line to poloidal cross section with positive and negative n-indexes are shown in Figure. 3.

In this term, particle recycling indicated by H_α signal is significantly large because of lack of the wall cleaning. After the wall cleaning using 2.45GHz was done, CFS is obtained (term2) in the almost same operational condition with sn6616, and then the plasma current goes up to about 4kA (sn6878: gray line in Fig.2). The distributions of the plasma current and the flux surface obtained by the magnetic measurement in sn6878 are shown in Fig.4. The clear formation of CFS can be observed. The plasma parameters estimated from last closed flux surface are: R_0=0.64m, a=0.41m, A=1.56, elongation κ=1.47 and triangulality δ=0.35. This analysis indicates that a part of current exist out of last closed flux surface. This condition is same as the condition discussed in ref. [4], [8] and [10] and

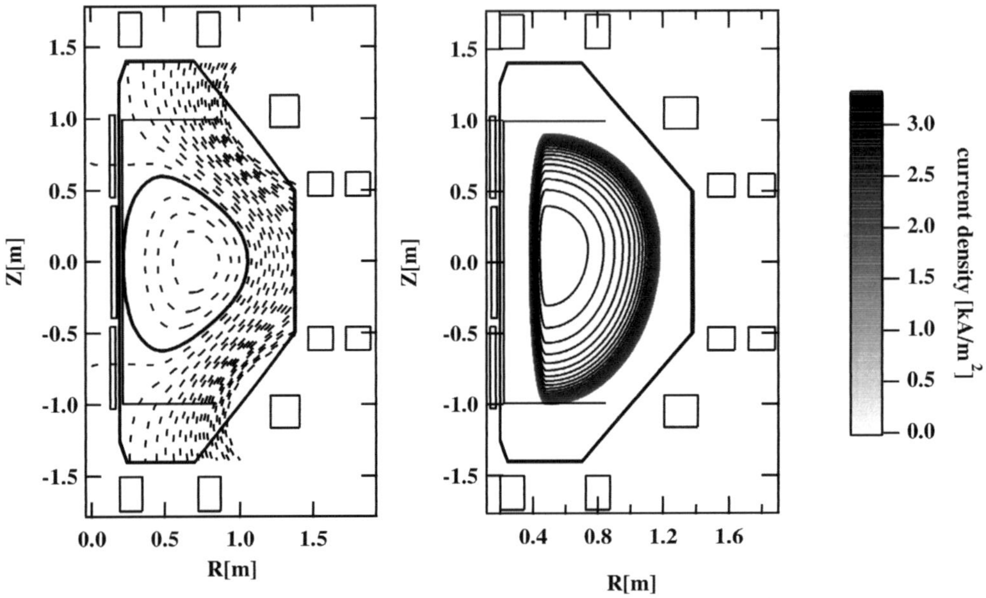

FIG. 4. The plasma current distribution estimated by the magnetic measurement (right picture) and the closed flux surface reconstructed with the current distribution (left picture). Solid line in left picture shows the last closed flux surface.

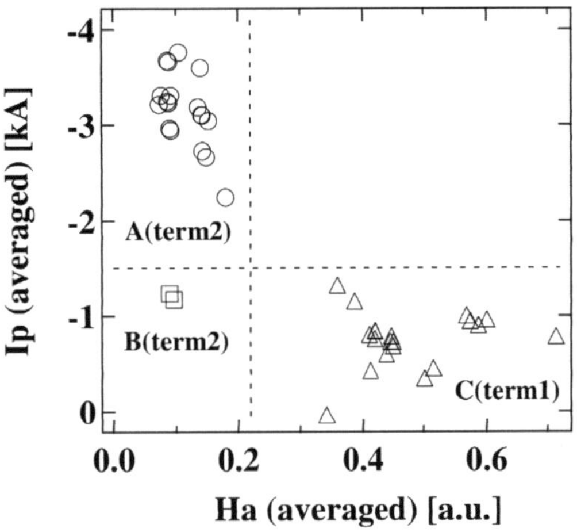

FIG. 5. The relation between plasma current and H_α in several tens of shots is shown. Circles show the shots with positive n-index in term2, square shows the shots with negative n-index in term2 and triangles show the shots with positive n-index in term1. Each shot has different vertical field strength from 0mT to about 3mT. Squares means the shot operated vertical field and n-index is 1mT and -0.067 at R=0.64m.

this is peculiar feature of RF discharge in the STs. Figure 5 shows the plasma current inversely depends on the H_α signal in several tens of shots. This suggests that the recycling of particle should be reduced to achieve CFS by RF. While two shots (square markers) shown in region B of Fig. 5 was not obtained CFS even in term2. In these two shots, external vertical field is 1mT, which is same value with that in sn6878, and n-index is -0.067 at R=0.64m. This indicates that PCS is done successfully only in term 2 with positive n-index configuration (region A in Fig. 5).

4. Discussion

It is shown that positive n-index of magnetic flux induced by external PF coils and low recycling is a key parameter to start-up plasma current successfully as described in the section 3. Lower recycling leads to lower neutral density. These results agree with the idea which plasma current in OMFC is driven by energetic electrons [11]. When the n-index is negative, electrons generated at ECR layer by RF go to low field side and cannot be confined in OMFC. On the other hand, with positive n-index, a part of energetic electrons are well-confined in OMFC according to particle orbit calculation. The confined electrons have asymmetric movement in toroidal direction and this asymmetric movement can generate plasma

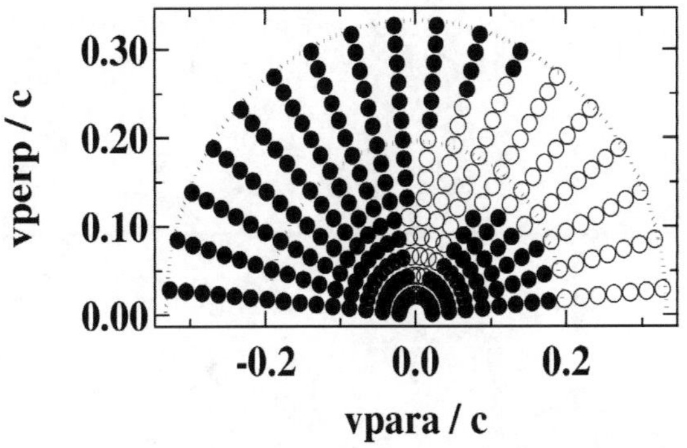

FIG. 6. The relation between particle confinement and initial velocity of electron is shown. The orbit calculation is started from mid-plane on fundamental ECR layer and horizontal and vertical axis shows the velocity of electrons at the start point of the calculation. Open circles show the confined electrons and closed circles show the electrons lost by the direct contact of the orbit to the wall. The dot line in the figure shows the electron energy. Inner, middle and outer dotted circles show the contours of energy at 1keV, 10keV and 30keV, respectively.

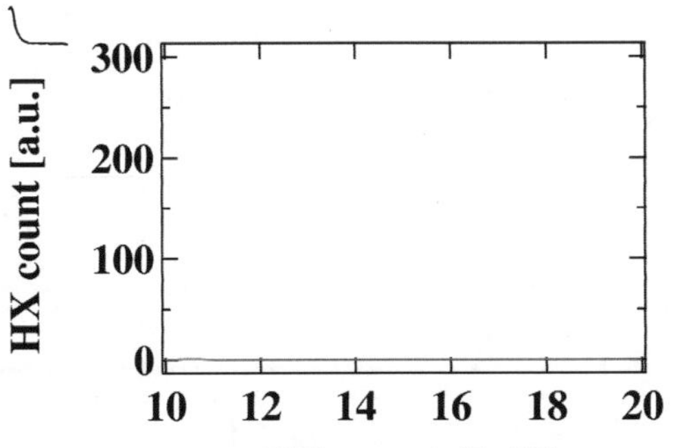

FIG. 7. The results of hard X-ray (HX) measurement are shown at the beginning of plasma dicharges. Black line shows the result in sn10154 (positive n-index case, around 4kA plasma current) and gray line shows the result in sn10159 (negative n-index case around 1 kA plasma current).

current. The contribution of such current to total plasma current in PCS is considered. Larger energy of electrons gives the enhanced asymmetric movement. But, electrons with too high energy cannot be confined in the weak vertical magnetic field. The appropriate range of electron energy in typical experimental condition of QUEST is estimated with the orbit calculation as shown in Fig. 6. At the condition, electrons of energy from a few hundred eV to several tens of keV are confined in OMFC and these electrons have asymmetric movement in toroidal direction. Thus, these electrons can be regarded as candidates of the electrons carrying the plasma current.

Then, the experimental results concerned about confined electrons which may have some contribution to total plasma current are introduced. In figure 7, the results of hard X-ray measurement in sn10159 (negative n-index and around 1kA current) and sn10154 (positive n-index and around 4kA current) are shown. In sn10159 little or no X-ray is detected. On the other hand, X-ray of more than 10keV is detected in sn10154. It should be noted that 10keV is the lower limitation of sensitivity of the X-ray detector. This result indicates that confined energetic electrons exist only in positive n-index configuration. Now, it is difficult to estimate the absolute value of plasma current driven by these electrons, therefore the contribution of these electrons to forming CFS is not clear. It is a future work.

5. Summary and Conclusion

Formation of a closed flux surface can be observed in non-inductive current start-up experiments by RF of 8.2GHz, 40kW in QUEST.

Remarkable reduction of H_α signal is observed after discharge wall cleaning with 2.45GHz RF system and formation of the closed flux surface is obtained. This result indicates that the reduction of recycling is a key for the formation of the closed flux surface. Even in good wall condition, plasma current start-up cannot be done successfully in the magnetic field with negative n-index. These may suggest electrons confined in OMFC carry significant plasma current in the plasma current start-up. The existence of confined energetic electrons only in positive n-index case is confirmed by X-ray measurement.

Acknowledgements

This work was performed under the auspices of the NIFS Collaboration Research Program (NIFS05KUTRO14, NIFS08KUTRO022). This work was partially supported by a Grant-in-aid for Scientific Research from the Ministry of Education, Culture, Sports, Science and Technology of Japan.

References

[1] Y.-K. Peng. et al., Phys. Plasmas, 7, 1681 (2000).
[2] Y. Gribov et al., Nucl. Fusion 47, S385 (2007).
[3] C. B Forest. et al., Phys. Plasmas 1, 1568 (1994).
[4] M. Uchida et al., PRL 104, 065001 (2010).
[5] A. Ejiri. et al, Nucl. Fusion 46, 709 (2006).
[6] V. E. Shevchenko et al., Nucl. Fusion 50, 022004 (2010).
[7] R. Raman. et al., Fusion Sci. Tech., vol. 56, 512 (2009).
[8] T. Yoshinaga. et al, PRL 96, 125005 (2006)
[9] K. Hanada. et al, PFR Volume 5, S1007 (2010).
[10] T. Yoshinaga et al, Nucl. Fusion 47, 210 (2007).
[11] A. Ejiri and Y. Takase, Nucl. Fusion 47, 403 (2007).

Construction of a plasma confinement device TOKASTAR-2 with tokamak-helical hybridization concepts

T Oishi, K Yamazaki, H Arimoto, K Baba, M Hasegawa and T Shoji
Nagoya University, Chikusa-ku, Nagoya 464-8603 Japan

E-mail: t-oishi@mail.nucl.nagoya-u.ac.jp

Abstract. In the research of confinement of torus plasma using magnetic fields, a concept of superposition of the external helical magnetic field to tokamak plasma has been widely proposed. One of the objectives of this concept is to improve the stability of plasmas, for example, to suppress plasma current disruption. A tokamak-helical hybrid plasma confinement device TOKASTAR-2 has recently been constructed to investigate the effect of helical field to tokamak plasma. First plasma was just produced in June, 2009. TOKASTAR-2 machine has Ohmic heating central coils, eight toroidal field coils and two outboard helical field coil segments. The major radius R_p and the averaged minor radius $\langle a_p \rangle$ of the toroidal plasma with nearly circular cross-section is expected to be approximately 0.1 m and 0.04 m, respectively. The toroidal magnetic field B_t is typically 0.1 T. The plasma start-up experiment using the electron cyclotron heating (ECH) by the radio frequency wave with the frequency of 2.45 GHz and the injected power of 2.0 kW is in progress in this machine as the pre-ionized phase of tokamak operation. In this paper, the overview of TOKASTAR-2 device and present states of the study are presented. According to the preliminary experiment, we confirmed the effect of outer vertical or helical magnetic field application to suppress bursting density fluctuation in the pre-ionized ECH plasmas.

1. Introduction

Many low-aspect-ratio torus systems have been proposed for compact and low-cost designs of fusion reactors [1-7]. Those compact systems sometimes employ the hybridization concept among tokamak, helical, and open field. One of the authors (K. Y.) also proposed the tokamak-stellarator hybrid called TOKASTAR [8]. In TOKASTAR configurations, the flux surface can be formed by simple coils. Moreover, if the plasma current is induced, it will support to increase rotational transform. At first, TOKASTAR having the toroidal period $N = 4$ was proposed, followed by the proposal of $N = 1$ or 2 simple coil system named C-TOKASTAR [9]. An experimental device having $N = 2$ coils of C-TOKASTAR was constructed and the formation of the vacuum magnetic flux surfaces was confirmed experimentally [10]. Plasma equilibrium with toroidal plasma current was also analysed in the $N = 2$ TOKASTAR configuration [11]. Based on these achievements, deeper understanding of the relationship among the plasma current, helical magnetic configurations, and confinement property in TOKASTAR configuration is needed. Especially, the research of superposition of the external helical magnetic field to tokamak plasma is important to improve the stability of tokamak plasmas, for example, to suppress plasma current disruption [12,13]. For this purpose, a tokamak-helical hybrid plasma confinement device "TOKASTAR-2" has recently been constructed. First plasma was just produced in June, 2009. In this paper, we would like to describe the overview of TOKASTAR-2 device and the present status of the research.

2. TOKASTAR-2 device

Figure 1 shows the coil configurations of TOKASTAR-2 device. It has ohmic heating (OH) central coils, eight toroidal field (TF) coils and two outboard helical field (HF) coil segments. Each coil consists of cupper conductor with a diameter of 3.2 mm. There are 50 turns in TF coil, 100 turns in vertical field (VF) coil, 42 turns × 2 layers in the central part of OH coil, 22 turns in upper and bottom part of OH coil, and 98 turns in HF coil. These coils are installed in the vacuum cryostat except external VF coils. At maximum three units of 200μF capacitors are used for the power supply to coils. Helium or Hydrogen is used for the working gas. Base pressure of the vessel is in the order of 10^{-3} Pa. Experiment region of the gas pressure is 0.4 – 7.0 Pa when helium is the working gas. A radiofrequency (RF) wave with the frequency of 2.45 GHz and the injected power up to 2.0 kW is used for the start-up of the plasma.

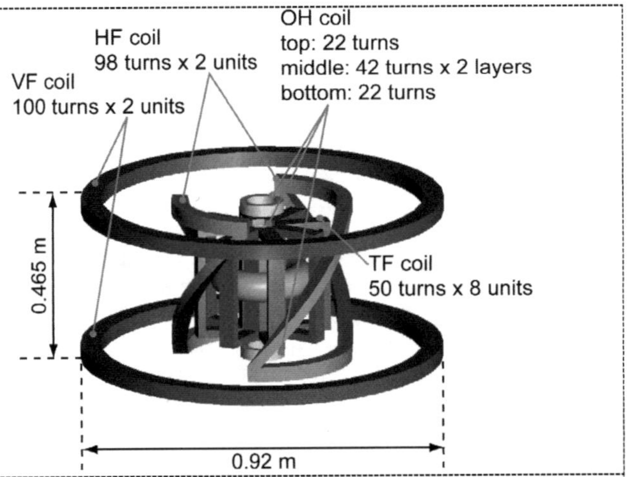

Figure 1. Coils of TOKASTAR-2 device.

3. Production of ECH plasmas as a pre-ionized phase of the tokamak operation

The ECH plasma start-up experiment using the RF wave is in progress in this machine. Figure 2 shows the typical waveform of the electron cyclotron heating (ECH) plasma discharge. The toroidal magnetic fields B_t calculated using the current in the TF coils and the intensity of the visible optical emission measured using an avalanche photodiode detector (APD) are shown together. RF wave with the power of 0.4 kW was injected continuously. As shown in this figure, the onset of the APD signal corresponds to the time that B_t in the innermost region of the volume inside the TF coils (major radius $R = 6$ cm) reaches to 0.0875 T which is the electron cyclotron resonance (ECR) field for 2.45 GHz RF. The APD signal starts to decrease when B_t becomes lower than the ECR field. These facts suggest that plasmas are produced by ECR. In this experiment, the electron temperature and density was in the order of 10 eV and 10^{16} m^{-3}, respectively.

We have a plan to induce plasma current in this ECH plasma. The ECH discharge corresponds to the "pre-ionized phase" of the tokamak operation. The mechanism of initiation and heating of the pre-ionized plasmas should be clarified. In addition, as another objective to study the ECH plasma, if there are radial electric fields in the ECH plasmas, closed surfaces of the particle orbit could be formed by $E\times B$ drift. There is a possibility that outer helical magnetic fields assist to form the closed surface. Based on these motivations, we are conducting the measurement of parameters of this ECH plasma. Figure 3 shows the radial profile of the ion saturation currents I_{is} and the floating potentials V_f of the

ECH plasma. Each profile was measured using Langmuir probes located at three vertical position z = +4.5 cm, 0 cm, and -4.5cm, where the midplane corresponds to z = 0cm. As shown in this figure, both I_{is} and V_f have peak values around R = 10 cm radially and z = 0 cm vertically. This radial position of the peak value is close to the position of the upper hybrid resonance (UHR) layer rather than that of the ECR layer. Therefore, we suspect that the plasma is initiated by the ECR as described above and further heated by the UHR.

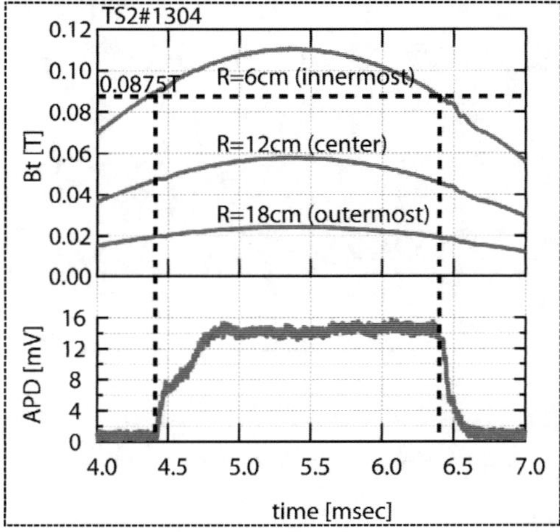

Figure 2. Temporal evolutions of the toroidal field strength at three locations calculated using the TF coil current (upper) and the intensity of visible emission measured using the avalanche photodiode detector (lower).

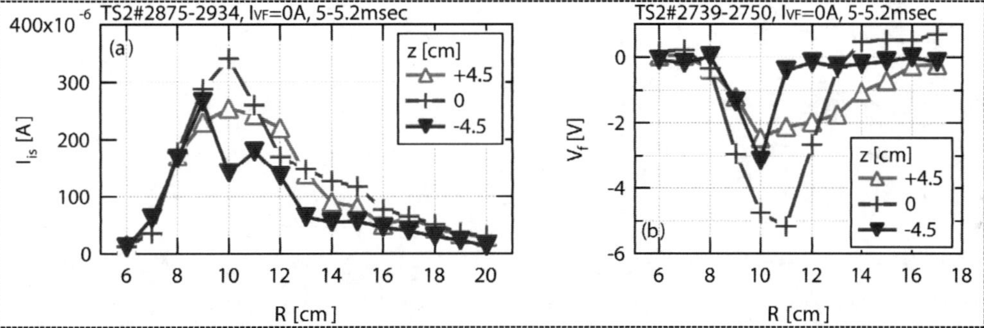

Figure 3. Radial profiles of (a) ion saturation currents and (b) floating potentials measured using Langmuir probes at z = +4.5, 0, -4.5 cm. The midplane corresponds to z = 0 cm.

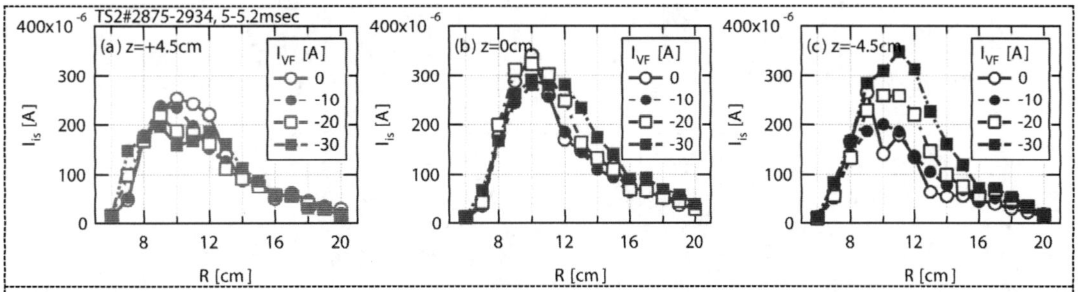

Figure 4. Radial profiles of the ion saturation currents at z = (a) +4.5 cm, (b) 0 cm, and (c) -4.5 cm in the ECH pre-ionized plasma. The current in the vertical field coil was scanned from 0 A to -30 A. Vertical field was in the downward direction.

It was found that the profile of I_{is} is varied by the vertical magnetic field. Figure 4 shows the radial profiles of the I_{is} at z = (a) +4.5 cm, (b) 0 cm, and (c) -4.5 cm. The toroidal magnetic field is in the clockwise direction in the top view. The current in the vertical field coil I_{VF} was scanned from 0 A to -30 A. The negative sign of I_{VF} corresponds to the downward direction of the vertical field. As shown in this figure, I_{is} at z = +4.5 cm decreases while I_{is} at z = -4.5 cm increases with increasing I_{VF}. This suggests that the plasma shifts downward in the case with the larger vertical field. The mechanism of this shift will be discussed later.

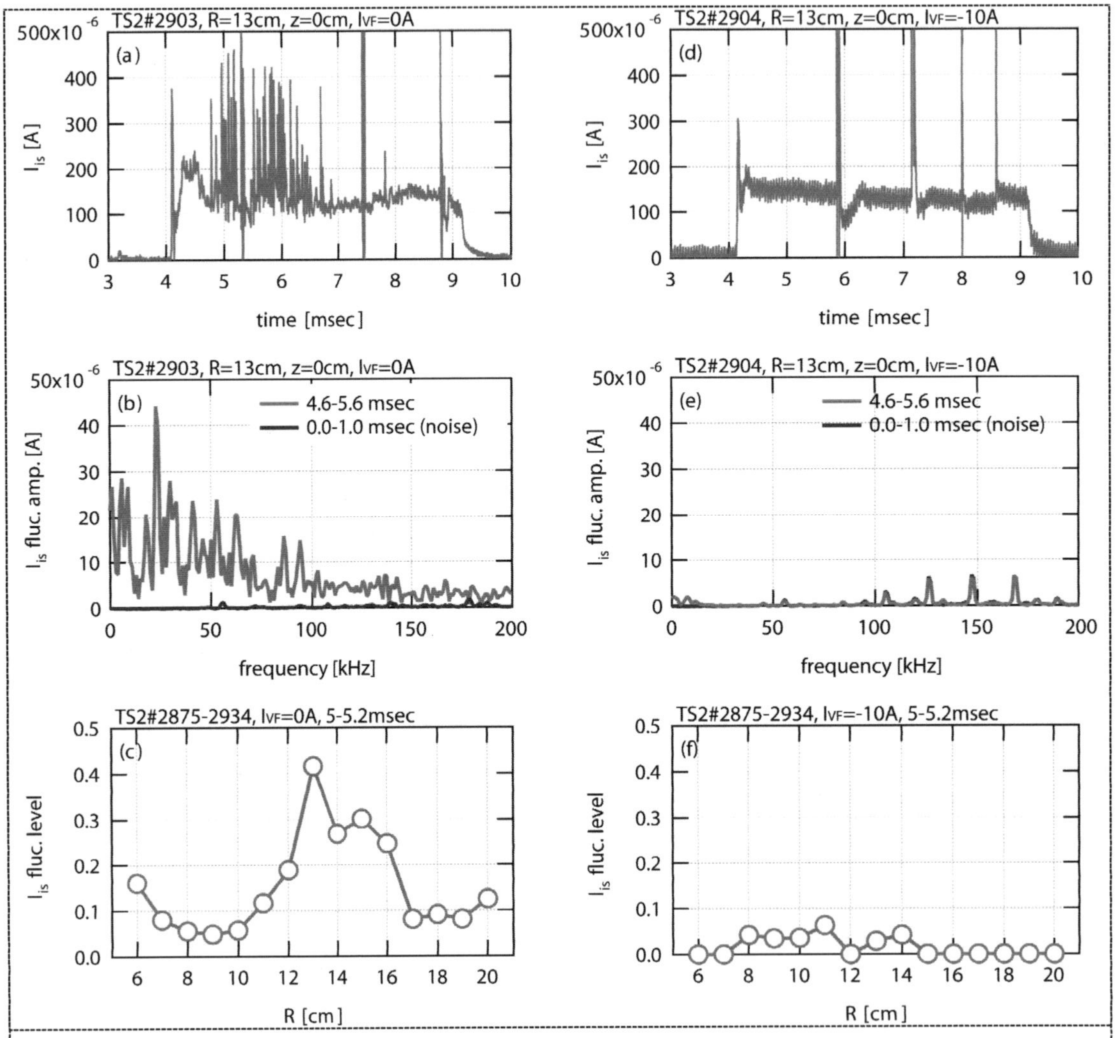

Figure 5. (a) Temporal evolution and (b) frequency spectrum of the ion saturation current at the location where the fluctuation is largest in the ECH pre-ionized plasma. (c) Radial profile of the fluctuation level evaluated using the standard deviations of the signals divided by the DC values. No vertical field was applied in figures (a-c). Same analyses were conducted in the case that the vertical field was applied, which are shown in figures (d-f).

Another important physics is the fluctuation of plasma parameters which are considered to be correlated with the plasma confinement. Figure 5(a) shows that large fluctuation exists in the I_{is} of the

ECH plasma. The frequency spectrum of the fluctuation is shown in figure 5(b). Fluctuation having frequencies of less than 100 kHz is dominant. Figure 5(c) shows the radial profile of the fluctuation level evaluated using the standard deviations of the signals divided by the DC values. The peak value of the fluctuation level is located around R = 13 cm, although the peak location of the profile of I_{is} itself is around R = 10 cm. No vertical field was applied in figures 5(a-c), however, we have found that the fluctuation can be suppressed by imposing the vertical magnetic field. Same analyses as shown in figures 5(a-c) were conducted in the case that the vertical field was applied, which are shown in figures 5(d-f). One can see that the fluctuation is almost completely suppressed. Note that several peaks having frequencies of larger than 100 kHz in the frequency spectrum in figure 5(e) are identical to those in the noise.

Fluctuations in the ECH plasmas were suppressed by applying not only the vertical field but also the helical field. Figure 6 shows the dependence of the fluctuation level of I_{is} on the vertical field or the helical field. Fluctuation of I_{is} decreases with the variation of I_{VF} from 0 to 5 A and 0 to -5 A, and I_{HF} from 0 to 10 A. Suppressed fluctuation level is kept for larger I_{VF} and I_{HF}.

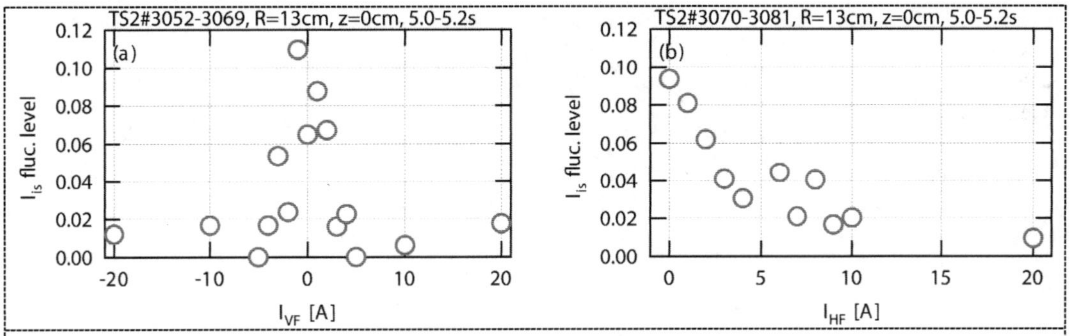

Figure 6. Dependence of the fluctuation level of the ion saturation currents on (a) the vertical field and (b) the helical field.

4. Discussion

In the previous section, we have stated that the profile of I_{is} in the pre-ionized ECH plasma shifts downward vertically by applying the vertical magnetic field in the downward direction. In this experiment, toroidal field is in the clockwise direction; therefore, this shift of I_{is} profile has same direction to that of the ∇B-drift of the ions. However, ∇B-drift of the ions might not be appropriate for the mechanism of the shift of I_{is} profile because I_{is} is proportional to the electron density. We suspect that another candidate of this shift is concerned with the $E \times B$ toroidal flow generated by the radial electric field as shown in figure 3(b) and the vertical magnetic field. In the case that the vertical field is applied to the simple toroidal magnetic field, the shape of magnetic field line becomes helically spiral. If there are toroidal flows in plasmas, the particles will move along the spiral magnetic field line and they will have the vertical component of the velocity. Indeed, we also have experimental results with the counter-clockwise toroidal field which are consistent with this model.

Physics of the growth and suppression of fluctuation has not been clarified yet. We have several hypothesis of the reason to suppress fluctuations as follows. One is that the error magnetic field might be corrected by the outer magnetic fields. The other is that the orbit of magnetic field lines or particle motions might be modified, and it results in the reduction of plasma-material interaction. To investigate these effects, magnetic field line tracing analysis including the vertical field or the helical field is considered as a future plan.

The ECH plasma which is investigated in this study is the pre-ionization phase of a tokamak operation. Therefore, we are trying to induce plasma currents using Ohmic heating. It has been confirmed that the plasma current increases after the onset of the current in the OH coil. However, the current reaches up to only several tens of amperes at maximum. It has to be much improved to generate the sufficient rotational transform of the magnetic field. Plans to increase the plasma current include upgrading power supply circuit to increase the loop voltage, optimization of the vertical field by equilibrium calculation, and evaluation of eddy currents in the vessel.

On the other hand, the calculation of tracing magnetic line has revealed that TOKASTAR-2 device has a characteristic to form the magnetic surface by applying outer helical field, even though the plasma current is not induced [14]. Characteristics of the magnetic surface, such as size, rotational transform or magnetic well depend on the amount and the ratio of currents in each coil, which should be investigated to evaluate the property of plasma confinement.

5. Summary

New torus plasma confinement device having tokamak-helical hybrid magnetic configuration "TOKASTAR-2" was constructed. It can impose outer helical field to tokamak plasma by using a pair of helical coils located outside toroidal coils and along the toroidal direction. It was confirmed that pre-ionized plasma for the tokamak operation is produced by the electron cyclotron resonance. Pre-ionized ECH plasma has profiles of the ion saturation current and the floating potential with a peak at the center in both radial and vertical direction. The profile shifts vertically in cases that the vertical magnetic field is applied. The outer vertical or helical magnetic field application suppresses bursting density fluctuation in the pre-ionized ECH plasmas. Induction of the plasma current by Ohmic heating has to be further improved to realize stable tokamak confinement and to investigate the effect of the outboard helical field on tokamak plasmas.

Acknowledgments

This work is supported in part by the National Institute for Fusion Science collaborative research program (NIFS10KOAF002).

[1] Ikuta K 1968 *J. Phys. Soc. Jpn.* **25** 1484
[2] Furth H P and Hartman C W 1968 *Phys. Fluids* **11** 408
[3] Ohkawa T 1981 *Proc. 3rd Stellarator Workshop (Moscow, 1981)* **1** 111
[4] Sheffield G V, Christensen U R and Furth H P 1981 *Bull. Am. Phys. Soc.* **26** 860
[5] Todd T N 1990 *Plasma Phys. Control. Fusion* **32** 459
[6] Moroz P E 1996 *Phys. Rev. Lett.* **77** 651
[7] Pedersen T S and Boozer A H 2002 *Phys. Rev. Lett.* **88** 205002
[8] Yamazaki K and Abe Y 1985 *Research Report of the Institute of Plasma Physics, Nagoya, Japan* IPPJ-718
[9] Yamazaki K and Kubota Y 2005 *Proc.Plasma Science Symposium 2005 and the 22nd Symposium on Plasma Processing (PSS2005/ SPP-22) (Nagoya, 2005,)* P3-094
[10] Taira Y, Yamazaki K, Arimoto H, Oishi T and Shoji T 2010 *Plasma and Fusion Res.* **5** S1025
[11] Yamazaki K, Taira Y, Oishi T, Arimoto H and Shoji T 2009 *J. Plasma Fusion Res. SERIES* **8** 1044
[12] WVII- A Team 1980 *Nucl. Fusion* **20** 1093
[13] Fujita J, Itoh S, Kadota K, Kaneko O, Kawahata K, Kawasumi Y, Kuroda T, Matsuoka K, Matsuura K, Miyamoto K, Noda N, Ohkubo K, Oka Y, Sakurai K, Sato K, Sato M and Tanahashi S 1981 *IEEE Transaction on Plasma Science* **PS-9** 180
[14] Oishi T, Yamazaki K, Okano K, Arimoto H, Baba K, Hasegawa M and Shoji T 2010 *J. Plasma Fusion Res. SERIES* **9** 69

Thermal stability studies of an experimental nuclear fusion reactor

Julio J. Martinell and Javier E. Vitela[†]

Instituto de Ciencias Nucleares, Universidad Nacional Autónoma de México, A. Postal 70-543, Mexico D.F., Mexico

E-mail: martinel@nucleares.unam.mx

Abstract. The conditions for the operation of a tokamak fusion reactor are studied considering a fiducial operating state and analyze variations about this state. We use a volume averaged 0-D two-temperature model and consider the case of a fusion thermonuclear reactor of the ITER type which will operate with a burning plasma at an energy gain factor of Q=10. The possible operation states are represented in POPCON plots of density n versus electron temperature T_e by varying parameters such as the fusion power, the energy gain and the auxiliary heating power, but keeping the ratio $(T_e - T_i)/T_e$ fixed, in order to determine the optimal operation point. Then, the thermonuclear instability of the burning plasma around this point is studied with a linear analysis. It is found that this instability cannot be excited, due to the fact that the plasma is far from ignition.

1. Introduction

In order to optimize the performance of a fusion power plant it is necessary to minimize the auxiliary heating requirements, which will ideally be zero if it works with an ignited plasma. The optimal operation should maximize the energy gain factor Q for a given fusion energy production and this operation point would be maintained by a control system. A possible control system may be based on an Artificial Neural Network that manages the fuel injection rate and the auxiliary power [1]. In this work we analyze the burn regimes of a tokamak fusion reactor with the design characteristics of ITER for different plasma parameters, using a zero-dimensional, two temperature model in which energy and particle transport are accounted for by global confinement times for energy τ_E, fuel particles τ_{DT} and alpha particles τ_α. Although the energies of ions and electrons evolve independently, the global transport is assumed to be the same. The operating points studied in this work are constrained to keep the ratio $(T_e - T_i)/T_e$ constant.

It is also necessary to make sure that the optimal states that one chooses are stable with respect to the thermonuclear instability which is liable to appear in a burning plasma. This instability arises in certain temperature range due to the fact that the fusion reaction rate there is an increasing function of temperature, which produces more fusion power as the temperature rises, thereby providing a positive feedback. We study the stability of the operational states around the optimal points in order to make sure that the thermonuclear instability is not a threat to a safe operation. We base our analysis on POPCON (Plasma OPeration CONtours) plots which graphically show the relevant range of electron density n_e and temperature T_e.

2. The plasma model

We use a multi-fluid description for the plasma (electrons, main ions and helium ash) which can be reduced to two particle and two energy conservation equations for the total deuterium-tritium density $n_{\rm DT}$, alpha particle density n_α and the electron and ion temperatures T_e, T_i, assuming quasineutrality ($n_e = n_{\rm DT} + 2n_\alpha + Z_{\rm Be}n_{\rm Be} + Z_{\rm Ar}n_{\rm Ar}$). Two impurities (Be and Ar) are included but we assume they are fixed. The equations are,

$$\frac{\partial}{\partial t} n_{\rm DT} = S_f - \frac{1}{2} n_{\rm DT}^2 \langle \sigma v \rangle - \nabla \cdot \vec{\Gamma}_{\rm DT} \ , \quad (1)$$

$$\frac{\partial}{\partial t} n_\alpha = \frac{1}{4}(1 - f_{\rm frac})\ n_{\rm DT}^2 \langle \sigma v \rangle - \nabla \cdot \vec{\Gamma}_\alpha \ , \quad (2)$$

$$\frac{\partial}{\partial t}\left[\frac{3}{2} n_e T_e\right] = \mathcal{P}_{{\rm aux},e} + \frac{1}{4}\ (1 - f_{\rm frac})\ f_e\ Q_\alpha n_{DT}^2 \langle \sigma v \rangle + \mathcal{P}_{\rm OH} - \mathcal{P}_{\rm brem} - \mathcal{P}_{\rm cycl} - \frac{3}{2} n_e (T_e - T_i)/\tau_{\rm ei} - \nabla \cdot \vec{\Gamma}_{{\rm E},e} \quad (3)$$

$$\frac{\partial}{\partial t}\left[\frac{3}{2}(n_{DT} + n_\alpha + n_{\rm Be} + n_{\rm Ar}) T_i\right] = \mathcal{P}_{{\rm aux},i} + \frac{1}{4}(1 - f_{\rm frac})\ f_i\ Q_\alpha n_{DT}^2 \langle \sigma v \rangle + \frac{3}{2} n_e (T_e - T_i)/\tau_{\rm ei} - \nabla \cdot \vec{\Gamma}_{{\rm E},i} \ . \quad (4)$$

where the electron energy losses and sources include bremsstrahlung $\mathcal{P}_{\rm brem}$ and cyclotron radiation $\mathcal{P}_{\rm cycl}$, ohmic heating $\mathcal{P}_{\rm OH}$, fusion power (in terms of the D-T reactivity $\langle \sigma v \rangle$) and auxiliary power $\mathcal{P}_{{\rm aux},e}$. For ions there is only fusion and auxiliary power $\mathcal{P}_{{\rm aux},1}$. There is also the electron-ion energy exchange measured by the relaxation time $\tau_{\rm ei}$. The expression for the reactivity we use is from [2]. Thermalization of the alpha particles produced by fusion is assumed to be instantaneous. The birth energy of the alpha particles is $Q_\alpha = 3.5$ MeV; $f_{\rm frac}$ is the effective fraction of alpha particles which are anomalously lost due to MHD events before they are thermalized; f_e and f_i are the fractions of the alpha particle energy Q_α deposited to the electrons and to the ions, respectively.

The fluxes Γ_k are not considered explicitly because of the poor understanding of particle and energy plasma transport, and instead we take a volume average of the fluid equations, which reduces the transport losses to the consideration confinement times for the electron and ion energy $\tau_{E,e}$ and $\tau_{E,i}$, as well as for the D-T and the helium ash confinement times τ_p and τ_α, respectively. This operation simplifies the model to zero dimensions, since all the radial variation of the plasma quantities is washed out. In doing this we assume constant density profiles $n_k(r) = n_{k0}$ and radial temperature profiles for both ions and electrons of the form,

$$T_k(r, t) = T_{k0}(t)[1 - (r/a)^2]^\gamma \ , \quad (5)$$

with T_{k0} the central temperature of each species and a the tokamak minor radius. We will consider the volume-averaged equations for steady state operation, which are expressed in terms of the central temperatures and density (we drop the 0-subscript for shortness). They take the form,

$$S_f - \frac{1}{2} n_{DT}^2 \langle \sigma v \rangle_{\rm vol} - n_{DT}/\tau_p = 0 \quad , \quad (6)$$

$$\frac{1}{4}(1 - f_{\rm frac}) n_{DT}^2 \langle \sigma v \rangle_{\rm vol} - n_\alpha/\tau_\alpha = 0 \quad , \quad (7)$$

$$\mathcal{P}_{\text{aux,e}} + \frac{1}{4}(1 - f_{\text{frac}})\ f_{\text{e}}\ Q_\alpha n_{DT}^2 \langle\sigma v\rangle_{\text{vol}} - \left(\frac{1}{1+\gamma}\right)\frac{3}{2}n_e\frac{T_e}{\tau_E} - \left(\frac{1+\frac{3}{2}\gamma}{1+\gamma}\right)\frac{3}{2}n_e\frac{T_e - T_i}{\tau_{ei}}$$
$$+A_h Z_{\text{eff}}^{1/2}\left(\frac{1+1.198Z_{\text{eff}}^{1/2}+0.222Z_{\text{eff}}}{1+2.966Z_{\text{eff}}^{1/2}+0.75Z_{\text{eff}}}\right) F_{\text{OH}}\,(1+\frac{3}{2}\gamma)T_e^{-3/2} - A_b Z_{\text{eff}} F_{\text{rad}}\frac{n_e^2}{1+\gamma/2}T_e^{1/2}$$
$$-\frac{A_{\text{cyc}}}{1+\frac{5}{2}\gamma}n_e^{1/2}T_e^{2.5}\frac{1+1.9303\ T_e/511}{1-0.58167\ T_e/511}(1+\frac{18a}{R\sqrt{T_e}})^{1/2} = 0\ , \quad (8)$$

$$\mathcal{P}_{\text{aux,i}} + \frac{1}{4}(1 - f_{\text{frac}})\ f_{\text{i}}\ Q_\alpha n_{DT}^2 \langle\sigma v\rangle_{\text{vol}} + \left(\frac{1+\frac{3}{2}\gamma}{1+\gamma}\right)\frac{3}{2}n_e\frac{T_e - T_i}{\tau_{ei}} - \left(\frac{1}{1+\gamma}\right)\frac{3}{2}n_i\frac{T_i}{\tau_E} = 0; \quad (9)$$

where $n_i = n_{DT}+n_\alpha+n_{\text{Be}}+n_{\text{Ar}}$ and the averaged power densities are given explicitly in MW/m^3, with $A_b = 5.33\times10^{-43}$, $A_h = 0.032$, $A_{\text{cyc}} = 1.23\times10^{-14}$ for n and T in m^{-3}, keV (for ITER parameters). We have included the fudge factor $F_{\text{rad}} = 2$, such that the bremsstrahlung loss term accounts for the combined bremsstrahlung and line radiated power reported at the nominal density and temperatures for ITER[4], which compensates for the lack of a line radiation term in the energy balance equations. Likewise, the factor $F_{\text{OH}} = 7.5$ is used to match the design value. We assumed $\tau_{E,\text{i}} = \tau_{E,\text{e}} = \tau_E$, which is taken from the IPB98(y,2) scaling[3],

$$\tau_{\text{IPB98(y,2)}} = 0.0562\ H\ I^{0.93}R^{1.97}B^{0.15}M^{0.19}\epsilon^{0.58}\kappa^{0.78}n_e^{0.41}P_{\text{s}}^{-0.69}\ , \quad (10)$$

where the factor H expresses the degree of enhancement expected over the predicted value due to improved confinement.

The volume averaged D-T reactivity can be written as

$$\langle\sigma v\rangle_{\text{vol}} = G(\gamma, T_i)\times\langle\sigma v\rangle\ ; \quad (11)$$

where G is a correction factor due to the radial profile in Eq. (5), which is fitted by a polynomial

$$G(\gamma,T_i) = 0.249 + 0.017T_i - 0.00011T_i^2 - 0.13\gamma + 0.023\gamma^2 - 0.0077\gamma T_i + 0.00125T_i\gamma^2$$
$$+0.000067T_i^2\gamma - 0.0000126T_i^2\gamma^2\ , \quad (12)$$

where T_i is the central ion temperature in keV and $\langle\sigma v\rangle$ is the reactivity evaluated at the central ion temperature.

For our analysis we use the energy gain factor Q_G defined as the ratio of the energy generation rate in the plasma due to the fusion reactions to the total external heating power, and measures how close a fusion reactor is to ignition conditions,

$$Q_G = \frac{\langle\mathcal{P}_{\text{fusion}}\rangle_{\text{vol}}}{\langle\mathcal{P}_{aux} + \mathcal{P}_{ohmic}\rangle_{\text{vol}}} \quad (13)$$

Here, $\mathcal{P}_{\text{aux}}$, includes both the auxiliary heating to electrons and to ions and $\mathcal{P}_{\text{fusion}}$ takes into account the total energy produced in the DT fusion reactions, i.e. the energy carried by the alpha particles and by the neutrons. Another quantity to watch is the L-H mode transition power threshold,

$$P_{\text{threshold}} = 4.30\ M_{\text{eff}}^{-1}B^{0.772}n_e^{0.782}R^{0.999}a^{0.975}\ ; \quad (14)$$

and we have to keep the total power avobe this value in order to stay in the H mode.

In the energy range covered during the alpha particles thermalization, the fraction of energy deposited to electrons and to ions is approximately 78% and 22% respectively, so $f_e = 0.78$, $f_i = 0.22$. The reference operating state for the studies presented below is $n_0 = 1.01 \times 10^{20}$ m^{-3}, $T_{e0} = 23.6$ keV and $T_{i0} = 23.0$ keV, for the electron density and the peak temperatures of the electrons and the ions, respectively. The radial profile parameter γ will be taken equal to 1.85 for both electron and ions temperatures. It is assumed that 10% of the alpha particles are anomalously lost before they are thermalized, hence $f_{\rm frac} = 0.1$ and their confinement time is $\tau_\alpha = 6.8\tau_E$. Likewise we take $\tau_{\rm DT} = \tau_\alpha$ for DT ions. The corresponding value of the fixed temperature ratio is $(T_e - T_i)/T_e = 0.025$ for all the operation states considered. In the following analysis we solve Eqs. (6)-(9) for the DT refueling rate, S_f, the fractional density of helium ash, f_α and the auxiliary heating power to electrons and ions, $P_{\rm aux,e}$ and $P_{\rm aux,i}$, varying n_e anf T_e over a range about the reference state.

3. The optimal operating states

For all computations presented we will take the *fraction* of Be impurities, fixed to $f_{\rm Be} = 0.02$, and the *density* of Ar fixed at $n_{\rm Ar} = 1.21 \times 10^{17}$ m^3 for all operating points, and thus the fractional density $f_{\rm Ar}$ varies for the different electron density values. The resulting operating states are shown in POPCON plots (normalized T_e-n_e space) with isolines for some relavant quantities. Fig. 1 shows the contour plots of the auxiliary heating power to electrons and ions. It is seen that for a constant electron density, the auxiliary heating power to the ions decreases when T_e decreases reaching a point where $P_{\rm aux,i} = 0$. Notice that the contour lines with $P_{\rm aux,i} < 0$ are not included because these are not physically plausible steady states. Thus, the line $P_{\rm aux,i} = 0$ represents a boundary for the available operating points of the reactor.

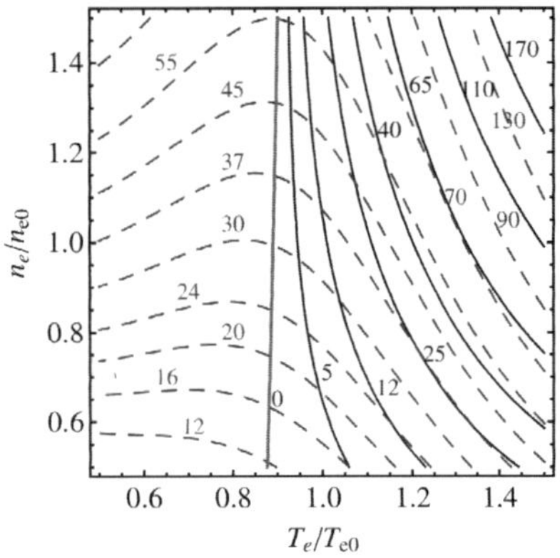

Figure 1. Contour curves of auxiliary power to electrons, $P_{\rm aux,e}$ (dashed line) and to ions, $P_{\rm aux,i}$ labeled in MW. Only positive values are plotted

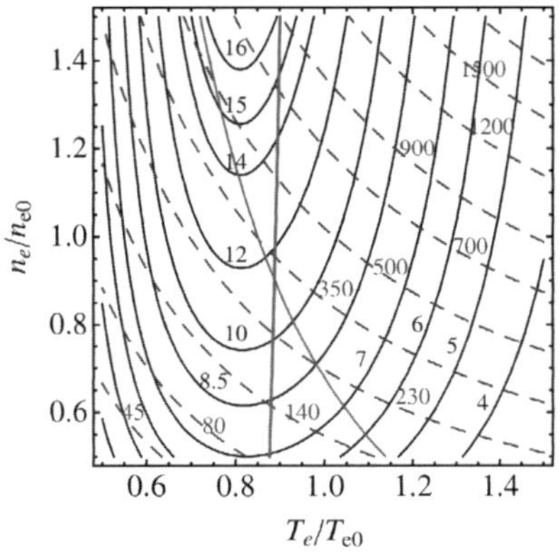

Figure 2. Isolines of total fusion power in MW (dashed line) and the Q-gain factor. Lines of $P_{\rm aux,i} = 0$ (red) and L-H transition threshold (magenta) are shown

Fig. 2 shows the contour lines of the gain factor, Q_G, as obtained from Eq. (13) and for constant total fusion power (including the energy of the neutrons), together with the boundary line $P_{\rm aux,i} = 0$. We observe that for a fixed fusion power value we can decrease the electron temperature increasing simultaneoulsy the electron density, properly moving along the corresponding $P_{\rm fusion}$ = constant line, thus increasing the Q_G monotonically until the boundary

$P_{\text{aux,i}} = 0$ is reached. Thus the optimal states, defined as those having a maximum Q_G for a given P_{fusion}, correspond to the crossings with boundary line $P_{\text{aux,i}} = 0$. Notice that the H-mode operation has to be to the right of the magenta line which represents the L-H transition power threshold. The energy confinement time is presented in Fig. 3 (dashed lines), together with the Q-gain factor. We notice that the line corresponding to the L-H transition power (magenta) is almost identical to the contour line associated with $\tau_E \approx 5.2$ s. Thus, we can assert that the H-mode states have $\tau_E < 5.2$ s. Interestingly, for a given Q_G, in order to increase τ_E one has to reduce the density.

4. Thermal stability

Once the best steady states are determined it is necessary to look a their stability properties. The temperature fluctuations might drive the plasma away from the desired operating states, if the reactor is liable to the thermonuclear instability. This arises when the temperature rise due to the energy released by the fusion reactions increases further the DT reactivity, which provides more heat to the plasma, and has to happen in the energy range with a positive slope of the function $\langle\sigma v\rangle(T)$, i.e. $T < 50$ keV. In order to study this instability we apply a thermal perturbation to the system and analyze its response. We consider two different cases: (a) equal temperatures, $T_i = T_e \equiv T$ and (b) two temperatures $T_i \neq T_e$ as in the analysis of the optimal states. The first case is taken as a simple approximation, for in this case there is a single equation for the total energy $W = W_e + W_i$. Performing a linear analysis, when a perturbation of the form $T = T_0 + \delta T$ on the equilibrium value T_0 is applied, the total energy equation can be expanded in δT and takes the form,

$$\partial \delta T/\partial t = dP(T_0)/dT\delta T \tag{15}$$

where $P(T)$ is the net power to the plasma. Clearly, the perturbation is stable when the term on the right is negative since otherwise the temperature perturbation will increase exponentially. Thus, the stability criterion is $dP(T_0)/dT < 0$. For the second case, since there are two temperatures we have to take separate perturbations $T_e = T_{e0} + \delta T_e$, $T_i = T_{i0} + \delta T_i$ and consider the electron and ion energy equations (3, 4). Linearizing again about the two equilibrium temperatures, we can find an equation for the vector $\delta\mathbf{T} \equiv (\delta T_e, \delta T_i)$, as $\partial\delta\mathbf{T}/\partial t = \mathbf{M}\cdot\delta\mathbf{T}$ with the matrix,

$$\mathbf{M} = \begin{pmatrix} \partial P_e(T_{0e}, T_{0i})/\partial T_e & \partial P_e(T_{0e}, T_{0i})/\partial T_i \\ \partial P_i(T_{0e}, T_{0i})/\partial T_e & \partial P_i(T_{0e}, T_{0i})/\partial T_i \end{pmatrix}, \tag{16}$$

where the powers $P_e(T_e, T_i)$ and $P_i(T_e, T_i)$ are the right hand sides of eqs. (3, 4) respectively. The solution is obtained in terms of the eigenvalues of $\mathbf{M}$, λ_1 and λ_2, as $\delta\mathbf{T} = \delta\mathbf{T}_0 \exp(\lambda_k t)$. The system will be unstable if the real part of at least one of the λ_k is positive. The stability computations are made for all the operating states in the range considered before and the marginal stability curves are presented in a POPCON plot in Fig. 4 for the cases (a) and (b). It is seen that the unstable region is for low T_e and high n_e, as found in other studies [5], for the two cases and there is actually not much difference between them. However, the region of instability falls outside the allowed operating regime since it is to the left of the green line $P_{\text{aux,i}} = 0$, thus it would not be excited for the optimal states found in Section 3, or any other allowed state.

In order to have an idea of the sensitivity of the results to the particular energy confinement scaling used, we considered a case with constant τ_E. In this analysis, τ_E is kept fixed at its equilibrium value when the variations in Eqs. (15, 16) are made, instead of using (10). The results are shown in Fig. 4 with dashed lines for the two cases of equal and different temperatures. It is clear that the difference is quite noticeable, and now the unstable region covers all the range of optimal operating states. Since the constant-τ_E is not compatible with the previous calculations we cannot conclude that the instability will be present, but it shows that the choice of the actual confinement scaling is important.

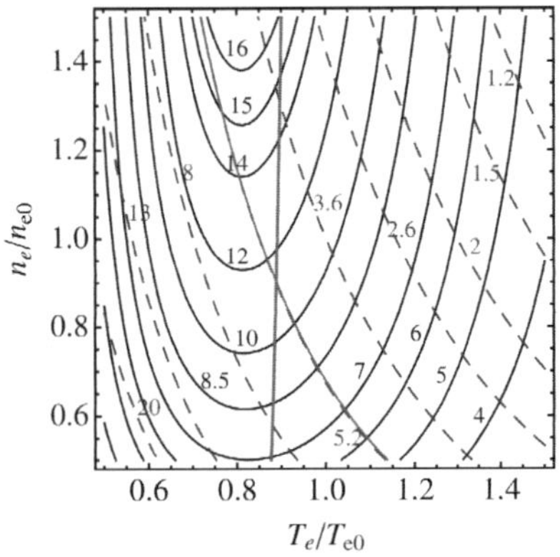

Figure 3. Contour plots for the energy confinement time (dashed) and the gain factor Q_G, showing that for Q_G given, τ_E increases as n_e decreases

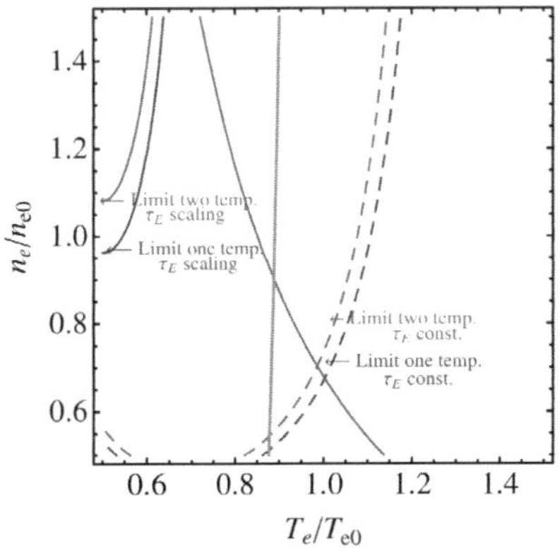

Figure 4. Marginal stability curves for cases of $T_e = T_i$ (blue) and $T_e \neq T_i$ (red). Dashed curves correspond to constant confinement time

5. Conclusions

In the present study we have used a 0D model to describe the operation of a nuclear fusion reactor, obtained by a volume average of the particle and energy equations for electrons, ions and helium ash. The steady states of operation were analyzed by taking variations of the electron density and temperature around the nominal operation point of ITER, maintaining the temperature ratio $(T_e - T_i)/T_e$ fixed. They are presented in POPCON plots showing contours of constant auxiliary power to electrons and ions, where it is seen that there are no states with $P_{\text{aux}} = P_{\text{aux,e}} + P_{\text{aux,i}} = 0$, meaning that ignition is not possible. However, there are states with $P_{\text{aux,i}} = 0$ which represent an operational boundary and they turn out to be also the states with maximum Q_G for a given fusion power. Hence we we identify these operational states as the optimal ones.

The thermonuclear stability is studied using linear analysis by considering temperature perturbations about the equilibrium state. Cases with $T_e = T_i$ and $T_e \neq T_i$ were considered and in both cases we showed that the system will be stable for the possible operating states of ITER parameters. The results are sensitive to the type of τ_E scaling; thus for certain scalings the reactor may become unstable.

Acknowledgments. This work was supported by projects DGAPA-UNAM IN119408, IN109909 and Conacyt 81232-F.

References

[1] Vitela J E and Martinell J J 2001 *Plasma Phys. Control. Fusion* **43**, 99
[2] Bosch H S and Hale G M, 1992 *Nucl. Fusion* **32**, 611
[3] Aymar R, Barabaschi P and Shimomura Y 2002 *Plasma Phys. Control. Fusion* **44**, 519
[4] "ITER Physics Basis" 1999 *Nucl. Fusion* **39** Vol. 12
[5] Mandrekas J and Stacey W M 1991 *Fusion Technol.* **19**, 57

Modelling of the Ignitor scrape-off layer including neutrals

F Subba[1], F Bombarda[2], G Maddaluno[2], G Ramogida[2] and R Zanino[1,3]

[1] Dipartimento di Energetica, Politecnico di Torino, I-10129, Torino, Italy

[2] ENEA, Frascati, Italy

E-mail: roberto.zanino@polito.it

Abstract. Ignitor is a tokamak project aimed at achieving ignition. In the reference scenario, plasma-surface interactions are controlled by a Mo first-wall/limiter, which constitutes a simple engineering solution but, at the same time, a special challenge for edge plasma modelling. Here the ASPOEL plasma fluid code, already applied to Ignitor in the recent past, is coupled with the neutral Monte Carlo code EIRENE. We study the effects of the neutrals on the plasma density and temperature profiles in the Ignitor scrape-off layer, and compute the particle and heat loads onto the Ignitor first-wall limiter.

1. Introduction

The main goal of the Ignitor experiment [1] is that of establishing the "reactor physics" (i.e. the physics of power producing reactors) in regimes close to ignition, where the "thermonuclear instability" can set in with all its associated non linear effects. The machine (see Table 1) is characterized by an optimal combination of high magnetic field ($B_T \lesssim 13$ T), compact dimensions ($R_0 \cong 1.32$ m), relatively low aspect ratio ($R_0/a \cong 2.8$), and considerable elongation and triangularity of the plasma cross section. The optimal central density of the fusing nuclei to achieve ignition is estimated to be about 10^{21} m^{-3}. The corresponding line-averaged density is well below the known density limit (related to the average plasma current density) for magnetically confined plasmas, as the plasma current I_p allowed by the machine design can reach 11 MA. Ignition can be achieved by ohmic heating alone, shortly after the end of the plasma current rise. The peak temperature at ignition is expected to be about $T_{e0} \cong T_{i0} \cong 11$ keV for an energy confinement time $\tau_E \cong 0.6$ sec; the relatively low beta poloidal, i.e. the ratio of the plasma energy density (pressure) to the poloidal magnetic field energy density, is consistent with favourable conditions for macroscopic plasma stability. An ICRH system operating between 80 and 120 MHz is also available, to accelerate the attainment of ignition by applying modest amounts of RF power during the current rise, or to facilitate access to the H-mode regime in double X-point configurations at lower plasma currents.

According to an independent assessment [2], "ignition in Ignitor would be reached under non-steady state conditions, as in fact ignition is intrinsically a time evolving event. The ignited regime would last a few confinement times ($2\text{–}5\times\tau_E$, $\tau_E \approx 0.5$ s) and many alpha particle slowing down times ($50\text{–}100\ \tau_{sd}$), which is adequate from the physics point of view". Here we shall refer to the operating scenario described in figure 1, which was studied in [3].

[3] To whom any correspondence should be addressed.

Table 1. Main design parameters of the Ignitor tokamak

Major radius R (m)	1.32	Plasma current I_P (MA)	11
Minor radius a (m)	0.47	Toroidal field B_T (T)	13
Elongation κ	1.83	Peak plasma density $n(0)$ (m^{-3})	10^{21}
Triangularity δ	0.4	Peak plasma temperature $T(0)$ (keV)	11
Plasma volume V (m^3)	10	RF input power P_{ICRH} (MW)	8-16
First wall area S (m^2)	36	First wall/Limiter material	Mo

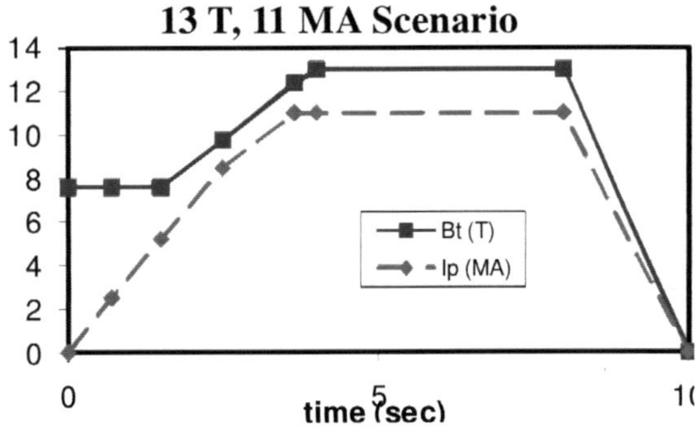

Figure 1. Ignitor operating scenario considered in this paper.

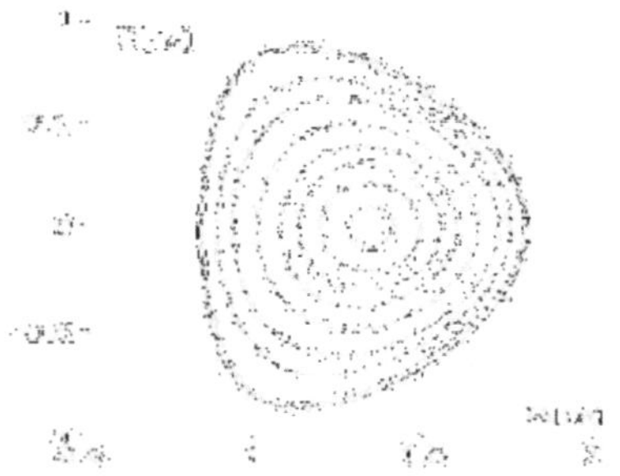

Figure 2. Magnetic equilibrium (dashed) with FWL (solid).

From the point of view of plasma-surface interactions (PSI), Ignitor constitutes a significant and somewhat special challenge to modelers, as it adopts a first wall/limiter (FWL), which is rather different from the now standard divertor configuration of many present and future tokamaks. While 0D-1D models of PSI in Ignitor have been developed long time ago with emphasis on high-Z impurity screening in high-density limiter tokamaks [4], and also validated against FTU data [5], the issues related to 2D modeling accounting for the singularity introduced by the tangency line between FWL and last closed flux surface (LCFS) at the inboard equatorial plane of the machine have been only addressed in recent work: a new code, ASPOEL, was developed for that purpose by our group at Politecnico di Torino [6], [7], and also applied to and validated against divertor far scrape-off layer (SOL) data [8].

Here we extend the work presented in [7] by coupling ASPOEL with the Monte-Carlo code for neutrals EIRENE [9]. The coupling algorithm is described and the main effects introduced by the presence of the neutrals on both plasma profiles and FWL loads will be highlighted. The main purpose of the paper is of computational nature, namely to demonstrate how the coupled tools can work in a realistic FWL geometry, while no claim is made as to the direct physics relevance for Ignitor of the computational test conditions considered here.

2. Approach to edge modeling in FWL configurations

The ASPOEL code adopts a control-volume finite-element (CVFE) hybrid approach [10]; it uses dual CV (polygonal) and FE (triangular) grids, where the FE give the natural interpolation scheme for the approximation of the fluxes on the CV faces. In view of the strong anisotropy of the transport coefficients introduced by the magnetic field, it is well known that it is essential to align the triangles with the local field direction [11]; therefore, the mesh generation in ASPOEL is done such as to guarantee perfect alignment of one side of each triangle. Approaches using less stringent alignment mesh generation criteria [12] are also being pursued by other authors, but will not be discussed here. The basic strategy of mesh generation adopted here starts from an equilibrium computed with the MAXFEA code [13], see figure 2; we then use the CARRE code [14] to build a first grid which ignores the FWL geometry; this grid is finally adapted to the FWL by adding magnetic (poloidal) surfaces at the intersection between the FWL and the given radial surfaces (other approaches to the same problem have been proposed more recently [15]).

ASPOEL solves the standard set of two-fluid equations for electrons and a single fuel ion species, consisting of: continuity, quasi-neutrality, diffusive Ansatz for the (anomalous) fluxes in radial direction, parallel (to the magnetic field) momentum balance, electron and ion energy balance. For the sake of demonstration purposes we assume a pure plasma with deuterium as the only ion species: the model is at this time unable to treat the two fuel ion species (D, T) simultaneously, or any impurity species. Steady state conditions in the edge plasma are assumed, considering that the time scale of relevance in the SOL (~ ms) is much faster than that of the evolution of the main plasma (~ 0.1 – 1 s).

Particle, momentum and energy sources are provided by EIRENE. ASPOEL computes the (background) plasma conditions and the plasma (D^+) flux onto it; EIRENE follows the neutral particles recycling from the FWL (e.g. D, D_2) as a result of the plasma bombardment, until they suffer a collision. The main atomic physics processes included in the model for the case at hand are: electron impact (including ionization), charge-exchange, elastic collisions and volume recombination, for both atomic and molecular neutral hydrogenic particles. About 30000 neutral particles were launched from the FWL and followed to perform the statistics + a comparable number arising from volume recombination. According to now standard practice in the B2-EIRENE coupling [16], EIRENE was called every ASPOEL iteration.

3. Problem definition and boundary conditions

We consider the region between the FWL and the magnetic surface located 2 mm inside the LCFS at the inboard equatorial plane (as customary, see e.g. [16], not only the SOL but also a portion of the main plasma is included in the computational domain, in order to relax the effect of conditions imposed at the inner boundary – indeed, even a larger portion should possibly be profitable in this case, since the LCFS is always very close to the FWL in Ignitor). Up-down symmetry is assumed, so that only the portion of the domain above the equatorial plane will be considered.

For the purpose of the computational test to be presented in this paper, boundary conditions (BC) have been imposed as follows: at the inner boundary, Dirichlet conditions $n = 10^{20}$ m^{-3}, $T_e = T_i = 15$ eV $V_{\parallel} = 0$ m/s; at the FWL, the standard Bohm-Chodura criterion $V_{\parallel} = C_S$, the Teilhaber-Birdsall criterion [17] $V_r = 0.01C_S$, the generalized conditions on energy transmission through the Debye sheath $\Gamma_{En}^{e} = 5.5 * \Gamma_n T_e$ and $\Gamma_{En}^{i} = 3.5 * \Gamma_n T_i$. Concerning these BC it should be noted that somewhat higher n and T at the inner boundary could be more physically relevant for Ignitor; also, as the BC at the FWL are rather uncertain, in view of the continuous transition between tangency and finite incidence of the magnetic field line, it should possibly be useful to perform a sensitivity study of their effect on the solution.

In order to guarantee a minimum of consistency with the main plasma conditions, it is important to consider the main plasma power balance (first principle of thermodynamics)

$$P_\alpha + P_\Omega + P_{ICRH} - P_{rad}^{MAIN} - P_{MAIN \to SOL} = \frac{dW}{dt} \quad (1)$$

For the Ohmic ignition scenario in FWL configuration under consideration, we can neglect P_{ICRH}, while alpha + ohmic heating – variation of the plasma internal energy have been estimated to sum up to a total of ~ 20 MW [3]. As the power advected and conducted from the main plasma to the SOL is computed by ASPOEL starting from the above-mentioned Dirichlet BC, we can obtain from (1) the power radiated in the main plasma, which will contribute the background of the FWL heat load distribution, see below.

Radial (anomalous) diffusivity values of 0.3, 0.3, 4.5 and 0.5 m^2/s were assumed here for particles, momentum, electron and ion energy, respectively.

4. Results

We present here a selection of the results obtained from the coupled ASPOEL-EIRENE simulation. The runs are performed on a fixed grid with ~ 10000 nodes. Reduction of the number of nodes by a factor of 2 in each direction has a weak effect (e.g., a reduction of ~ 10 % of the peak heat flux on the FWL), so that the results below can be considered reasonably grid independent.

The plasma density is significantly affected by the presence of neutrals, with an increase by almost an order of magnitude with respect to the case without neutrals, especially in the region close to the top (and bottom) of the plasma chamber, and peak density more than a factor of 10 above the inner boundary BC, see figure 3. This is due to the strong recycling in that region, related in turn to the relatively high plasma flux, see figure 5 below, reaching the FWL in the only portion of it which significantly departs from tangency to the field lines: angles between poloidal field and the tangent to the wall attain an absolute maximum of almost 10 degrees in this region, see figure 6 below.

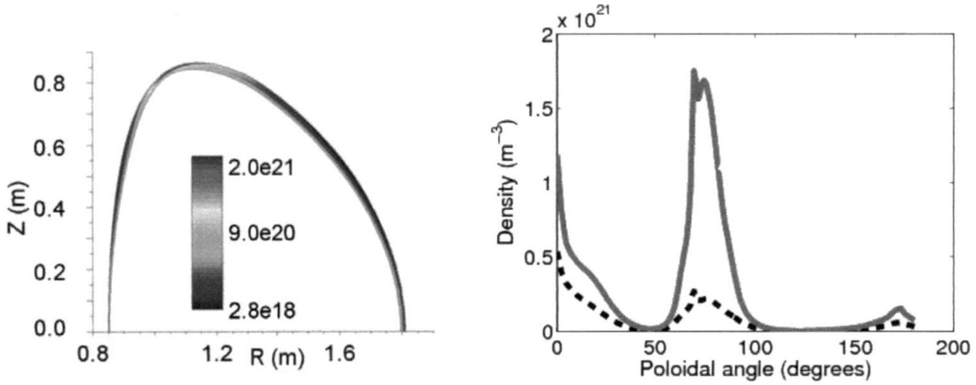

Figure 3. Left: Computed 2D plasma density (m^{-3}) map in the whole computational domain. Right: Poloidal distribution of the computed plasma density in front of the FWL: with neutrals (solid red), without neutrals (dashed black). The poloidal angle is measured from the inboard equatorial plane.

The distribution of the electron temperature in front of the FWL, see figure 4, is also significantly affected by the neutrals, with ionizations leading to strong cooling of the plasma in the top (and bottom) regions of the machine. The ion temperature (not shown) behaves similarly. As, however, the temperature reduces by a factor of four, while the density increases by a factor of almost 10, the particle flux to the FWL, which scales as $nT^{1/2}$, is also peaked near the top (bottom) of the machine, see figure 5.

For the same reasons, impurity production from sputtering at the FWL should be fairly limited in these conditions, as in the region of highest plasma outflux plasma ions reach the FWL with energies close to the Mo sputtering threshold.

The distribution of the heat load on the FWL, due to plasma advection and conduction from the SOL, as well as radiation from the main, is peaked close to the inboard equatorial plane, see figure 7. The load distribution is qualitatively similar to that computed in [7], the main quantitative difference being due to the inclusion here of the neutrals, as well as of the background radiation from the main plasma. About 2/3 of the power reaching the FWL in the plasma channel are due to the electrons.

Since impurities in the SOL (and the associated radiation) are not included in the model, the computed peak load is a rather pessimistic estimate.

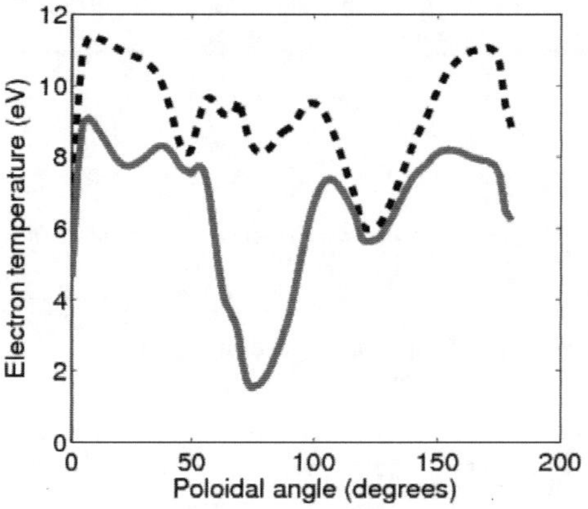

Figure 4. Poloidal distribution of the computed electron temperature in front of the FWL: with neutrals (solid red), without neutrals (dashed

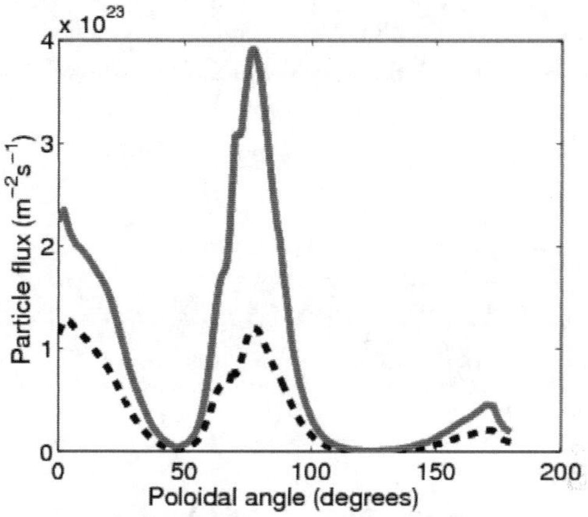

Figure 5. Poloidal distribution of the computed plasma flux onto the FWL: with neutrals (solid red), without neutrals (dashed black).

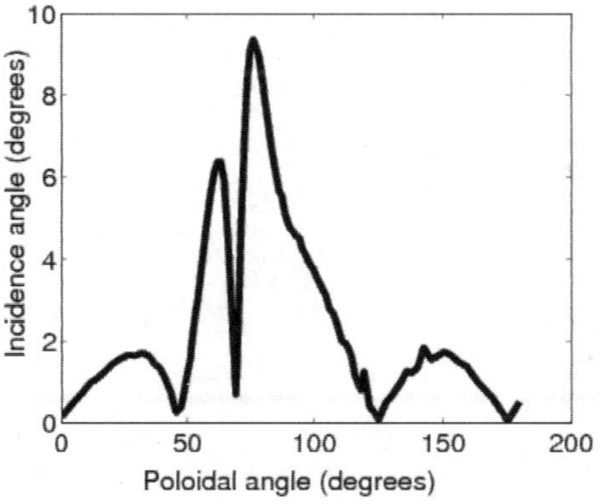

Figure 6. Poloidal distribution of the computed incidence angle of the poloidal magnetic field onto the FWL.

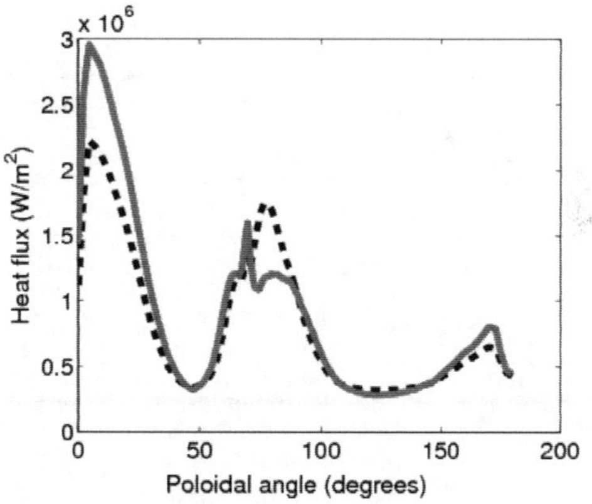

Figure 7. Poloidal distribution of the computed heat flux onto the FWL: with neutrals (solid red), without neutrals (dashed black).

5. Conclusions

Neutrals have been included in the 2D computational model of the Ignitor SOL by coupling the edge plasma fluid code ASPOEL and the neutral Monte Carlo code EIRENE. The case of a first-wall/limiter configuration was considered, which constitutes a special challenge for edge plasma modeling.

The computed peak heat flux on the FWL, close to the inboard equatorial plane of the machine, increases when neutrals are accounted for, but including impurities in the SOL model should reduce the peak value by a non-negligible amount.

A high density / low temperature region is created by ionization of recycling neutrals near the top / bottom of the machine. This temperature reduction could strongly limit the production of impurities by sputtering, but this feature will have to be checked by a self-consistent calculation.

Acknowledgments
We thank P. Boerner and D. Reiter (IPP Juelich, Germany) for helping us with the EIRENE code.

References

[1] Coppi B, Nassi M and Sugiyama L E 1992 Physics basis for compact ignition experiments *Phys. Scripta* **45** 112-32
[2] Horton W, et al. 2002 Ignitor physics assessment and confinement projections *Nucl. Fusion* **42** 169-79
[3] Coppi B, Airoldi A, Bombarda F, et al. 2001 Optimal regimes for ignition and the Ignitor experiment *Nucl. Fusion* **41** 1253-7
[4] Ferro C and Zanino R 1990 Edge parameters and scrape-off layer (SOL) characteristics in IGNITOR *J. Nucl. Mater.* **176-177** 543-7
[5] Zanino R, Ferro C, Frassinesi M and Leigheb M 1999 Self-consistent impurity modeling in the Frascati Tokamak Upgrade *J. Nucl. Mater.* **266-269** 1290-5
[6] Subba F and Zanino R 2004 Modeling plasma-wall interactions in first-wall limiter geometry *Comp. Phys. Comm.* **164** 377-82
[7] Subba F, Airoldi A, Bombarda F, Cenacchi G, Maddaluno G and Zanino R 2007 Development of a computational tool for limiter edge plasma modeling with application to IGNITOR *J. Nucl. Mater.* **363-365** 693-7
[8] Subba F, Bonnin X, Coster D and Zanino R 2008 2D fluid modeling of far SOL divertor plasma including the first wall *Comp. Phys. Comm.* **179** 194-8
[9] Reiter D, Kotov V, Boerner P, et al. 2007 Detailed atomic, molecular and radiation kinetics in current 2D and 3D edge plasma fluid codes *J. Nucl. Mater.* **363-365** 649-57
[10] Patankar S V 1972 A calculation procedure for heat, mass and momentum transfer in three-dimensional parabolic flows *Int. J. Heat Mass Transf.* **15** 1787-806
[11] Zanino R 1997 Advanced finite element modeling of the tokamak plasma edge,*J. Comput. Phys.* **138** 881-906
[12] Adams M F, et al. 2009 Scaling to 150K cores: recent algorithm and performance engineering developments enabling XGC1 to run at scale *JPCS* **180** 012036
[13] Barabaschi P 1993 The MAXFEA code (Presented at the Plasma control technical meeting, Naka, Japan)
[14] Marchand R and Dumberry M 1996 CARRE: a quasi-orthogonal mesh generator for 2D edge plasma modeling *Comp. Phys. Comm.* **96** 232-46
[15] Dekeyser W, et al. 2010 2D edge plasma modeling extended up to the main chamber wall (Presented at the PSI conference, San Diego (CA) USA)
[16] Schneider R, Bonnin X, Borrass K, Coster D, Kastelewicz H, Reiter D, Rozhansky D and Braams B 2006 Plasma edge physics with B2-EIRENE *Contrib. Plasma Phys.* **46** 3-191
[17] Teilhaber K and Birdsall C 1989 Kelvin-Helmholtz vortex formation and particle transport in a cross-field plasma sheath. II: Steady state *Phys. Fluids B* **1** 2260-72

Collisional-Radiative modeling of the transient excitation of a carbon atoms beam crossing a tokamak plasma edge

Arnaud Bultel[1], Ioan F. Schneider[2], Djamel Benredjem[3] and Pascale Monier-Garbet[4]

[1]UMR CNRS 6614 CORIA, Université de Rouen, Avenue de l'Université, BP12, 76801 Saint-Etienne du Rouvray, France
[2]FRE 3102 LOMC, Université du Havre, BP 540, rue Philippe Lebon, 76058 Le Havre, France
[3]LAC, UPR CNRS 3321, University of Paris Sud-Orsay, France
[4]IRFM, CEA Cadarache, Saint-Paul-lez-Durance, France

E-mail: arnaud.bultel@coria.fr

Abstract. A time-dependent collisional-radiative (CR) model is elaborated in the purpose of modeling the penetration of an atomic carbon beam in the edge plasma of a fusion machine in order to probe it. The excited states population densities of the beam are assumed to be modified by electron-induced processes only. All momentum transfer is neglected: the penetration velocity is assumed constant. In typical conditions for the electron density and temperature gradients ($\partial n_e/\partial x \approx 10^{20}$ m^{-4} and $\partial T_e/\partial x \approx 10^7$ K m^{-1}), the results illustrate the electrons efficiency to ionize the beam. With a velocity equal to 1 km s^{-1}, the beam is ionized in 20 μs in these conditions, which corresponds to a penetration depth of 2 cm. During the conversion in ions, the atomic beam is sufficiently excited to emits spectra whose measurement in coupling with the present CR model can provide information on $\partial n_e/\partial x$ and $\partial T_e/\partial x$.

1. Introduction

The high values of the edge electron density and temperature gradients ($\partial n_e/\partial x \approx 10^{20}$ m^{-4} and $\partial T_e/\partial x \approx 10^7$ K m^{-1} in order of magnitude [1]) play a key role in the case of fusion devices [2]. Therefore, these gradients have to be measured. These measurements can be performed using Thomson scattering [3] or Beam Emission Spectroscopy [4]. The latter technique consists in injecting an atomic beam with a sufficiently high center of mass velocity. The resulting penetration through the edge plasma leads to a progressive atomic excitation and to the emission of spectra in part characterized by the values of the gradients [5]. Coupled with a Collisional-Radiative (CR) model, these spectra provide the n_e and T_e profiles. The beam can be produced for instance by ablating plasma facing components made with carbon using a nanosecond laser pulse. In the present paper, we focus our attention on the CR model elaborated to calculate the progressive excitation of such an atomic beam penetrating a typical fusion plasma edge.

2. C excited states considered and assumptions

All excited states belonging to the NIST database have been considered [6]. This set is composed by 268 states. Due to the high gradients values mentioned in Introduction, C atoms in these excited states can undergo ionization: we have therefore taken into account the ground ionic state. Table 1 lists a part of the set used.

Table 1. Some of the states considered in the present CR model on carbon among the 268 considered from the NIST database [6]. Usual notations are used.

No	Configuration	Term	Energy (eV)	J	No	Configuration	Term	Energy (eV)	J
1	[He] $2s^2$ $2p^2$	^{3}P	0.000000	0	70	[He] $2s^2$ 2p 4f	^{2}[7/2]	10.41299	4
2	[He] $2s^2$ $2p^2$	^{3}P	0.002033	1	95	[He] $2s^2$ 2p 5d	^{3}D	10.70759	1
3	[He] $2s^2$ $2p^2$	^{3}P	0.005381	2	120	[He] $2s^2$ 2p 6p	^{1}S	10.82891	0
4	[He] $2s^2$ $2p^2$	^{1}D	1.263725	2	145	[He] $2s^2$ 2p 7p	^{3}D	10.92251	2
5	[He] $2s^2$ $2p^2$	^{1}S	2.684011	0	170	[He] $2s^2$ 2p 7f	^{2}[3/2]	10.99062	2
6	[He] 2s $2p^3$	^{5}S	4.182631	2	195	[He] $2s^2$ 2p 8d	^{1}P*	11.05502	1
7	[He] $2s^2$ 2p 3s	^{3}P	7.480391	0	220	[He] $2s^2$ 2p 10d	^{1}F*	11.13123	3
14	[He] $2s^2$ 2p 3p	^{1}P	8.537096	1	245	[He] $2s^2$ 2p 15d	^{1}P*	11.20760	1
22	[He] $2s^2$ 2p 3p	^{1}D	9.002581	2	268	[He] $2s^2$ 2p 29d	^{1}F*	11.25200	3
44	[He] $2s^2$ 2p 4p	^{3}D	10.01575	1	C^+	[He] $2s^2$ 2p	^{2}P*	11.26030	$^1/_2$-$^3/_2$

The atomic beam injected into the fusion machine interacts with the plasma already formed. The following assumptions have been made:

(i) *the C atoms enter the plasma under weak density and low temperature conditions.* According to its production mode, the atomic beam produced by laser ablation of solid materials can reach a temperature higher than their melting point after few microseconds [7]. However, the beam is produced from a plasma facing component carbon tile whose typical temperature is of $T_0 = 2000$ K. The initial beam temperature has been therefore chosen equal to T_0.

(ii) *the properties of the plasma are not significantly modified by the carbon beam.* The mean density order of magnitude in the edge plasma is $10^{21} - 10^{22}$ m^{-3}. Even if the smallest initial density of the beam obtained by laser ablation is classically of some 10^{22} m^{-3} under weak fluence [8], we can assume that the plasma is very weakly modified by the penetration of the beam owing to the high electron temperature.

(iii) *the beam velocity is uniform.* The density of the plasma is weak. In addition, the plasma is mainly composed by light particles as protons, hydrogen, deuterium, tritium and helium atoms. Their momentum transfer cross section with carbon is therefore weak and the resulting retardation can been neglected. In addition, the atoms velocity distribution function is assumed narrow with respect to the center of mass velocity: the beam is assumed monokinetic and there is no spreading during the propagation.

(iv) *the free electron density and temperature gradients are uniform.* As we will see in the following, the penetration depth is of the order of some centimeters. The characteristic length over which the electron parameters gradients are uniform is higher [9]: these gradients have been therefore considered as uniform.

(v) *the contribution of the protons to the C excitation is negligible.* At same energy, protons are much less efficient than electrons to excite or ionize another particle owing to the mass ratio: proton-induced elementary processes have been therefore disregarded. This assumption leads to neglect the collisions between C atoms as well.

(vi) *the C^+ ions are only produced on the [He]2s^{2}2p ground state.* The first excited electronic state of C^+ is the [He] 2s 2p^2 ^{4}P state whose energy is 16.59203 eV from the C ground state. We have assumed in the present preliminary study that the number of electrons with this energy or higher is negligible.

(vii) *the C emission lines are optically thin.* The divergence of the beam is assumed small [10]: the escape factor is therefore close to unity [11].

3. CR model and balance equation

The previous hypotheses lead to a simple form for the balance equation of a carbon atom on the excited state subscripted j:

$$\frac{D[C_j]}{Dt} = \left(\frac{\partial[X_j]}{\partial t}\right)_{Rad.} + \left(\frac{\partial[X_j]}{\partial t}\right)_{Coll.} \tag{1}$$

where a Lagrangian approach is used. The two terms in the right-hand side are the radiative and collisional terms respectively. The elaboration of the CR model consists in calculating realistically the elementary processes involved in these terms. Using the NIST database for the spontaneous emission probabilities $A_{k\to j}$ and $A_{j\to i}$, the radiative term is:

$$\left(\frac{\partial[X_j]}{\partial t}\right)_{Rad.} = \sum_{k>j} A_{k\to j}[C_k] - \sum_{i<j} A_{j\to i}[C_j] \tag{2}$$

The radiative recombination has been neglected.

The collisional term involves the elementary processes changing the excitation and ionization degrees of the beam and can be written as:

$$\left(\frac{\partial[X_j]}{\partial t}\right)_{Coll.} = \sum_{i<j} k_{i\to j}[C_i]\, n_e - \sum_{k>j} k_{j\to k}[C_j]\, n_e - k_{j\to ion}[C_j]\, n_e - \tag{3}$$

$$\sum_{i<j} k^*_{j\to i}[C_j]\, n_e + \sum_{k>j} k^*_{k\to j}[C_k]\, n_e + k^*_{ion\to j}[C^+]\, n_e^2 \tag{4}$$

where the part (3) of the right-hand side is related to the forward (inelastic) processes and the part (4) is related to the backward (superelastic) processes. The rate coefficients $k_{i\to j}$, $k_{j\to k}$ and $k_{j\to ion}$ are calculated assuming valid the cross sections of Drawin [12] whose main interest is to distinguish correctly the forbidden from the allowed transitions, even if they have been identified in the case of hydrogen-like species as rare gas atoms. These cross sections are integrated over Maxwellian distribution for the electron energy according to the electron temperature T_e and finally written as:

$$\text{Optically allowed transition:} \quad k_{m\to n} = \frac{3.45\times10^{-6}\alpha_A}{[(E_n-E_m)k_B]^{1.6454}} T_e^{0.1454} e^{-\frac{E_n-E_m}{k_B T_e}} \quad [\mathrm{m^3 s^{-1}}] \tag{5}$$

$$\text{Optically forbidden transition:} \quad k_{m\to n} = \frac{5.24\times10^{-17}\alpha_F}{[(E_n-E_m)k_B]^{0.5067}} T_e^{-0.0067} e^{-\frac{E_n-E_m}{k_B T_e}} \quad [\mathrm{m^3 s^{-1}}] \tag{6}$$

$$\text{Ionization:} \quad k_{m\to n} = \frac{3.45\times10^{-6}\alpha_I}{[(E_n-E_m)k_B]^{1.6454}} T_e^{0.1454} e^{-\frac{E_n-E_m}{k_B T_e}} \quad [\mathrm{m^3 s^{-1}}] \tag{7}$$

with $E_n > E_m$. The backward rate coefficients are calculated using the detailed balance principle.

By writing the balance equation (1) for the 269 species here considered, an ordinary differential equations system is obtained and finally solved using the DVODE library [13]. The time evolution of the different $[C_j]$ population densities is derived, which can be easily converted in spatial profiles owing to the constant velocity of the beam.

4. Results

The test-case studied corresponds to usual conditions observed in fusion machine plasma. The gradients values adopted are $\partial n_e/\partial x = 2.75 \times 10^{20}$ m^{-4} and $\partial T_e/\partial x = 3 \times 10^7$ K m^{-1} with $n_e(x = 0) = 0$ and $T_e(x = 0) = T_0 = 2000$ K. The velocity of the beam depends on the laser pulse characteristics. However, the order of magnitude is typically of some km s^{-1} in low fluence conditions [14]: we have therefore adopted $v = 1$ km s^{-1}.

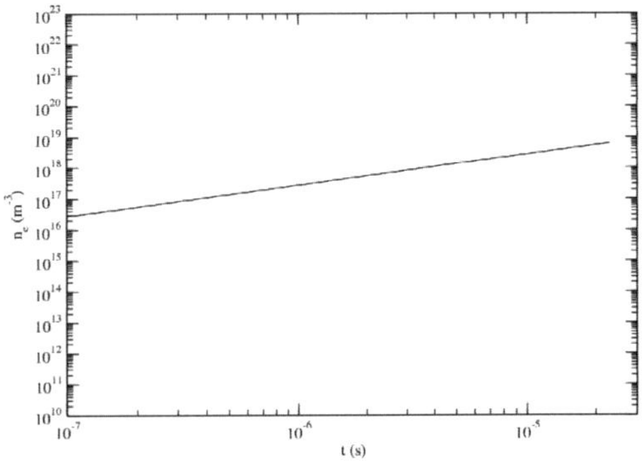

Figure 1. Time-evolution obtained for the electron density n_e in the test-case studied.

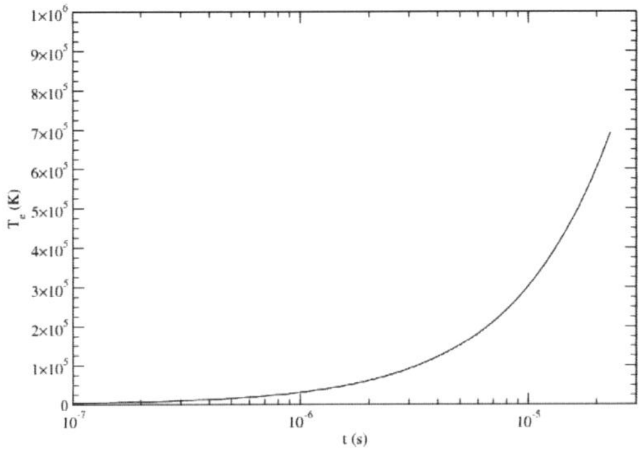

Figure 2. Time-evolution obtained for the electron temperature T_e.

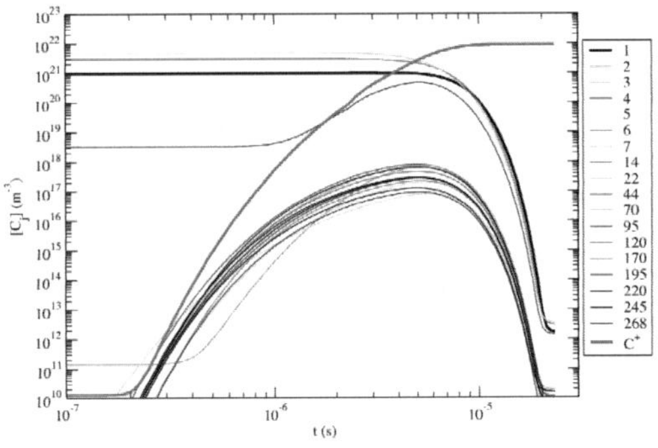

Figure 3. Evolution of the population density of some C excited states listed in table 1. The evolution of $[C^+]$ is also displayed.

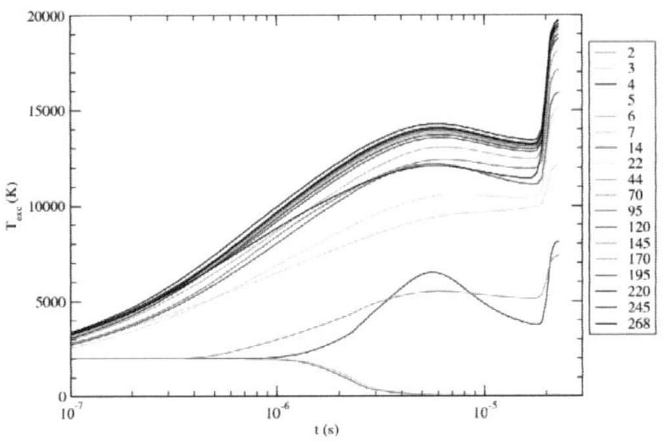

Figure 4. Evolution of the excitation temperature of some C states calculated with respect to the ground state density (see equation 8).

Figures 1 and 2 display the time-evolutions of n_e and T_e obtained in our Lagrangian approach. These values of the electron density and temperature correspond to the fluid particle location at time t. The electron density values are rather weak, but we can expect an important modification of the distribution of the beam particles among the excited states owing to the high electron temperature levels.

Figure 3 illustrates the time-evolution derived for the states of carbon atoms accounted for in table 1 in the test-case conditions. We can notice first that the population densities vary strongly. These excited states can be separated in two groups. The first group corresponds to states weakly excited forming the fine structure of the ground state (levels i such that $1 \leq i \leq 3$) and the two metastable states ^1D and ^1S at 1.26 and 2.68 eV from the ground respectively. These states have population densities well coupled to each other for $t > 8 - 9 \times 10^{-6}$ s. The second group is formed by all other states whose population densities are coupled as early as $t = 10^{-6}$ s. This behavior can be easily explained by the energy distribution of these excited states. The C energy diagram exhibits indeed a cut due to the existence of the two metastable states reinforced by a large empty interval of 3.3 eV between the ^{5}S and ^{3}P states preventing a complete coupling. We have already noted such a characteristics with rare gas atoms whose cut of energy diagram is deeper [15]. Despite this bad coupling, the behavior of both groups is similar: they undergo a brutal depopulation owing to the beam ionization.

Figure 3 displays also the C^+ density evolution. We can note the complete conversion of the beam: firstly only composed by neutrals, the beam is totally ionized under electron impact over almost 1 μs. When a C^+ ion is produced, an electron is of course produced as well. The initial density of the neutral beam is 10^{22} m^{-3}: in these conditions, the electron density rapidly increases and becomes higher than the one resulting from the fusion plasma. We assumed that the beam penetration does not modify significantly the plasma properties: is our approach finally inconsistent? Actually not, because electrons formed during the ionization processes have a low kinetic energy: they cannot penetrate the plasma and are moved perpendicularly to the beam direction owing to the confinement strong magnetic field. Equations 3 and 4 involve therefore only the electron properties of the fusion plasma. The C^+ ions are much more heavy: the magnetic field action is limited and is neglected. Anyway, the C^+ ions play a role only through the three-body recombination. The term $k^*_{ion\to j}[C^+]\,n_e^2$ of equation 4 takes this process into account. A thorough study of the different contribution to $(\partial[X_j]/\partial t)_{Coll.}$ shows that three-body recombination is negligible at any time owing to the increase of T_e.

At time $\tau_p \approx 2 \times 10^{-5}$ s, the beam is thus totally ionized: this corresponds to a position of $x_p \approx 2$ cm from the surface where the beam is produced by the laser pulse. This distance can be therefore considered as the "penetration depth" of the carbon beam. The observation of a rich emission spectrum can be expected over this distance: the final determination of the n_e and T_e gradients can be only performed over x_p.

The calculation of the excitation temperature of each excited state with respect to the ground state gives further insight into the global ionization process. This excitation temperature is calculated by:

$$T_{exc}(j) = \frac{E(j) - E(1)}{k_B} \left\{ \ln \left(\frac{g_e(i)}{g_e(1)} \frac{[C_1]}{[C_i]} \right) \right\}^{-1} \tag{8}$$

Figure 4 displays the resulting time-evolutions. Starting from 2000 K owing to the initial equilibrium, T_{exc} increases for each state according to a very similar way, except for the less excited ones. The different behaviors can be ascribed again to the energy diagram peculiarities discussed above. However, we can notice the moderate increase of the excitation temperature in comparison with the one observed for T_e. This weak increase results from the good coupling previously observed and from the rapid ionization of the ground state. When the conversion of the beam is complete ($t \approx 2 \times 10^{-5}$ s), one can note the quick increase of the excitation temperature followed by a slower one whose interpretation necessitates further investigations.

In the past, laser blow-off (LBO) diagnostics were performed on different fusion machines. They are based on the production of Li-C atoms beam obtained by lighting with a laser pulse the rear side of layers deposited on a glass or a quartz substrate. The beam is then formed by atoms with a center of mass velocity of 10^4 m s $^{-1}$ approximately [16]. In these conditions, a more important penetration depth is expected. The usual value for x_p is indeed around 10 cm [17] which leads to ability of the method to diagnose the edge plasma assessed to $n_e \times x_p = 10^{18} - 10^{19}$ m^{-2} [18, 19]. In our case, this product is five times lower owing to the value of x_p. However the production mode of the plasma is easier: the present method offers an interesting economical alternative to the LBO method to perform edge plasmas n_e and T_e measurements.

5. Conclusion

A Collisional-Radiative model is elaborated in order to provide information on the penetration of a carbon atomic beam in the edge plasma of a fusion machine. Numerous excited states of C are accounted for and their chemistry is assumed governed by the main electron-induced elementary processes (excitation/deexcitation and ionization/recombination). The plasma is assumed unchanged by the beam interaction. The calculations are based on the solution of the ordinary differential equations system obtained by writing the balance equation of each state population density under a Lagrangian approach. The results put forward the rapid conversion of the atomic beam into an ionic beam which lasts approximately 20 μs in the typical conditions adopted. This characteristic delay corresponds to a penetration depth of the beam equal to 2 cm for a beam velocity of 1 km s^{-1}. The conversion occurs through a high correlation between excited states illustrated by the population densities and excitation temperature time-evolutions.

These results are certainly highly depending on the set of cross sections used. Further investigations will be performed with at least one different set. In addition, the interaction with the plasma is assumed reduced to electron-induced processes only. The density and energy of protons are important and the role of momentum transfer can be studied: in future calculations, these processes will be accounted for. Helium atoms exhibit an energy diagram more favorable to the present diagnostic method and can be implemented more easily experimentally: other calculations based on He will also be performed.

Acknowledgments

The authors wish to thank the IEFE Institute (Institut Energie, Fluide & Environnement) in France for its financial support.

References

[1] Kubo H, Takenaga H, Nakano T, Higashijima S, Shimizu K, Sawada K, Kobayashi S and the JT-60 Team 2005 *Divertor spectroscopy with molecular transport* (Nuclear fusion research, Understanding plasma-surface interactions) ed R E H Clark and D H Reiter (Berlin, Heidelberg, New York: Chemical Physics, Springer) chapter 5 pp 121-134
[2] Jenko F, Dorland W and Hammett G W 2001 *Phys. Plasmas* **8** 9 4096-4104
[3] Murmann H, Gtsch S, Rhr H, Salzmann H and Steuer K H 1992 *Rev. Sci. Instrum.* **63** 10 4941-4943
[4] Branas B, Tafalla D, Tabares F L and Ortiz P 2001 *Rev. Sci. Instrum.* **72** 1 602-606
[5] de la Cal E 1998 *Nucl. Instr. and Meth. in Phys. Res. A* **403** 490-498
[6] $http://physics.nist.gov/PhysRefData/ASD/levels_form.html$
[7] Morel V, Bultel A and Chéron B G 2010 *Spectrochimica Acta Part B* doi:10.1016/j.sab.2010.08.002
[8] Amoruso S 1999 *Appl. Phys. A* **69** 323-332
[9] Beigman I L, Kocsis G, Pospieszczyk A and Vainshtein L A 1998 *Plasma Phys. Control. Fusion* **40** 1689-1705
[10] Utterback N G, Tang S P, Friichtenicht J F 1975 *Phys. Fluids* **19** 6 900-905
[11] Holstein T 1947 *Phys. Rev.* **72** 12 1212-1233
[12] Drawin H W 1967 *Euratom-CEA Technical Report* DPh-PFC/SRFC EUR-CEA-FC-383
[13] Brown P N, Byrne G D and Hindmarsh A C 1989 *SIAM J. Scient. Comput.* **10** 1038
[14] Amoruso S, Armenante M, Berardi V, Bruzzese R and Spinelli N 1997 *Appl. Phys. A* **65** 265-271
[15] Bultel A, van Ootegem B, Bourdon A and Vervisch P 2002 *Phys. Rev. E* **65** 046406
[16] Kumar A, Singh R K, Prahlad V and Joshi H C 2008 *J. Appl. Phys.* **104** 093302
[17] Abel G, Stansfield B, Michaud D, Ross G G and Lachambre J L 1997 *Rev. Sci. Instrum.* **68** 1 994-997
[18] Pospieszczyk A and Ross G G 1988 *Rev. Sci. Instrum.* **59** 4 605-609
[19] Pospieszczyk A and Ross G G 1988 *Rev. Sci. Instrum.* **59** 8 1491-1493

Diffusive Transport Analysis in Low Aspect Ratio Reversed Field Pinch

Yasuo Nagamine[1], Atsushi Fukuyama[2], Shoichi Shiina[3] and Masamitsu Aizawa[1]

[1]Institute of Quantum Science, Nihon University, Chiyoda-ku, Tokyo, Japan
[2]Department of Nuclear Engineering, Kyoto University, Sakyo-ku, Kyoto, Japan
[3]National Institute of Advanced Industrial Science and Technology, Tsukuba, Japan

E-mail: nagamine@phys.cst.nihon-u.ac.jp

Abstract. For the evaluation of confinement property of a low aspect ratio reversed field pinch (RFP) plasma, diffusive transport analysis is studied. Numerical analysis based on the neoclassical transport model in a low aspect ratio RFP show that a considerable bootstrap current contibution can be expected in this configuration, when incorporating the effects of non-circular cross-section and low aspect ratio. So-called neclassical RFP equilibrium are also found to attain both high stablity beta limit and high bootstrap current fraction. However, of transport phenomena in an RFP, the energy confinement time τ_E might be determined by anomalous transport due to micro-instabilities. To analyze such transport phenomena in a low aspect ratio RFP, the applicability of the TASK (Transport Analysing System for TokamaK) simulation code are considered.

1. Introduction

In order to evaluate the confinement properties of a low aspect ratio (A) reversed field pinch (RFP) plasma, diffusive transport analysis is studied theoritically. In the conventional RFP plasma with laeger aspect ratio, transport analysis by using the theoritical model or three-dimensional numerical simulations based on the cylinder approximation, has been shown to be nearly consistent with the present experimental results such as the confinement scaling[1,2]. On the other hand, confinement performance in a low aspect ratio RFP is not well understood, but it is considerd to be important to introduce the transport model including the effect of toroidal geometry. Numerical analysis based on the neoclassical transport model in a low aspect ratio RFP show that a considerable bootstrap current contibution can be expected in the RFP configuration, when incorporating the effects of non-circular cross-section and low aspect ratio[3]. In Ref.[3], however, transport coefficients were evaluated by a simple model based on the collisionless approximation. Therefore, in present study we perform more precise analysis of transport coefficients based on the model applicable to all collisinarity regimes. However, of transport phenomena in an RFP, the energy confinement time τ_E might be determined by anomalous transport due to micro-instabilities. Therefore, it must be considered to select an appropriate model for the turbulent transport in the RFP configuration.

For numerical analysis of transport phenomena in a low aspect ratio RFP plasma, the applicability of the TASK (Transport Analysing System for TokamaK) simulation code[4] was considered. In TASK code, the code groups analyzing the physical processes such as equilibrium, transport, wave propagation, etc. are connected to calculate predictably the development as a whole. This code has been developed for the tokamak and helical magnetic configurations. For analysis of an RFP, because of the different nature of the magnetic configuration, validation analysis of the applicability of the physical models implemented in TASK code is required.

2. Low aspect ratio RFP

In this section we consider some characteristic features of low aspect ratio RFP configuration. As for the equilibrium profile, the safety factor (q) value increases on the axis and decreases at the edge, while its profile flattens in the core region and sharpens in the edge as A decreases. So that the number of resonant surfaces decreases and also the toroidal mode number n decreases, that is, resonant tearing mode locations separate more from each other[5]. For example, at A=2.1 we can set the lowest mode m/n=1/4 at r=0.43 (m=1 is dominant in RFP). The elongation has the equivalent effect with the decrease of A. From the above, lowering A could lead to easily attaining Quasi-Single-Helicity(QSH) states[6] with high beta. A helical structure has also been observed experimentally in A=2 RFP device[7].

Another feature is expected to be the advantage of higher bootstrap current. A consideration of neoclassical effect in the RFP equilibrium has led to a significant fraction of bootstrap current in a low aspect ratio, which is quite important to a steady state operation as well as to a high beta stability[8]. The equilibrium attained is characterized by the hollow current profile of plasma self-induced current, which leads to the compatibility of the strong negative magnetic shear and the strong plasma paramagnetism due to the high pitch magnetic line of force toward the plasma boundary region, in addition to the relaxed-equilibrium state with a minimum energy at finite beta[9]. Due to both the large negative magnetic shear and the strong paramagnetism, the transport improvement is also expected in this configuration, which must be investigated by the transport simulation described in present paper.

3. TASK code

TASK code is open source codes, which are mainly developed at Kyoto University[4, 10]. This code consists of some modules listed in Table 1. In TASK code, the code groups analyzing the physical processes such as equilibrium, transport, wave propagation, etc. are connected to calculate predictably the development as a whole.

Table 1. Present Status of TASK code modules.

EQ	**2D Equilibrium**	Fixed boundary, Toroidal rotation
TR	**1D Transport**	Diffusive Transport, Transport models
WR	**3D Geometr. Optics**	EC, LH: Ray tracing, Beam tracing
WM	**3D Full Wave**	IC, AW: Antenna excitation, Eigen mode
FP	**3D Fokker-Planck**	Relativistic, Bounce-averaged
DP	**Wave Dispersion**	Local dielectric tensor, Arbitrary $f(v)$
PL	**Data Interface**	Data conversion, Profile database
LIB	**Libraries**	

EQ module solves the Grad-Shafranov equation which describes MHD equilibrium configurations of the axisymmetric toroidal plasma. As a numerical solution, the poloidal flux $\psi(R,Z)$ in cylindrical coordinates is obtained for given pressure profile $p(\psi)$ and parallel current density profile $j_{\parallel}(\psi)$, or safety factor profile $q(\psi)$. TR module solves diffusive transport equations for the given geometry. Time evolutions of the spatial distribution of plasma density $n(\rho)$, temperature $T(\rho)$, current density $j_{\parallel}(\rho)$ etc are calculated based on a geometric factor $V'(\rho)$, absorption power $P_{abs}(\rho)$, externally driven current $j_{CD}(\rho)$ and so on. Where ρ represents the normalized minor radius.

In TR module, some diffusive transport models based on the neoclassical transport theory and turbulent transport theory are implemented. In present study, applied for the first time to the RFP configuration, the code is exploited for steady state equilibrium calculations with the neoclassical transport model. It is useful to confirm the numerical results by comparing with previously obtained results in Ref.[3].

As the neoclassical transport model, Hinton & Hazeltine[11], Wilson(Hirshman-Sigmar)[12], Sauter[13] and NCLASS model[14] which was developed as a general numerical model for tokamak plasmas are considered in TASK code. NCLASS is a multi-species fluid model which calculates bootstrap current and neoclassical transport in tokamaks of arbitrary collisionality and aspect ratio. This model uses a recent formulation of the neoclassical plasma viscosity for arbitrary shape and aspect ratio (including the unity aspect ratio limit), arbitrary collisionality, and orbit squeezing from strong radial electric fields.

In order to include the description of the RFP geometry, some parts of the code need to be improved. As the most fundamental improvement, there is the taking of normalized minor radius ρ. In TASK code, this ρ is defined by the square root of toroidal flux, i.e. $\rho_t = \sqrt{\psi_t/\psi_{ta}}$, where ψ_{ta} represents the toroidal flux at plasma boundary. For the RFP configuration, however, since the toroidal field reversal occurs near the plasma boundary, the toroidal flux profile is not a monotonic function between the plasma center and edge. Therefor, ρ_t cannot be used as a minor radius. For unified description of the tokamak and RFP configurations, the definition $\rho_p = \sqrt{\psi_p/\psi_{pa}}$ must be used as a normalized minor radius, where ψ_p and ψ_{pa} represent the poloidal flux and its edge value, respectively.

4. Preliminary Results of TASK calculation

As preliminary calculations of TASK code, equilibrium analysis of low A RFP is performed by the modified EQ module. Figure. 1 shows the result of calculations: (a) magnetic surfaces, (b) toroidal and poloidal current density profiles on the midplane, (c) safety factor profile, in the case of parameters A=2.0, ellipticity κ=1.4, triangularity δ=0.4, plasma current I_ϕ=0.15MA and parabolic pressure profile.

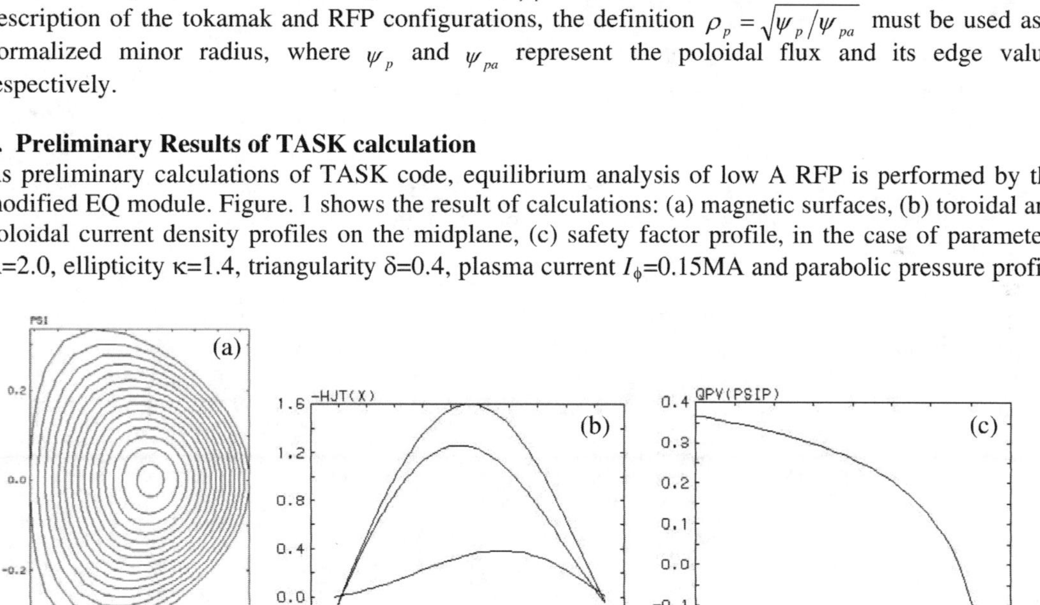

Figure 1. Equilibrium profiles calculated by EQ module: (a) magnetic surfaces, (b) toroidal and poloidal current density profiles on the midplane, (c) safety factor profile.

5. Considerations for turbulent transport analysis

In tokamaks, so many transport analysis have been performd, and developed some turbulent transport models, such as gyro-kinetic or gyro-fluid simulations of ITG(Ion Temperature Gradient) instability and turbulence. Comparisons and physics basis of tokamak transport models and turbulence simulations are summarized in Ref[15].

As the turbulent transport model, CDBM(Current Diffusive Ballooning Mode)[16], GLF23[17], IFS/PPPL[18] and Weiland[19] models are employed in TASK code. The IFS/PPPL model is based on nonlinear gyro-fluid simulations, which predict the fluctuation and thermal transport characteristics of toroidal ion-temperature-gradient driven (ITG) turbulence, along with comprehensive linear gyro-kinetic ballooning calculations, which provide accurate growth rates, critical temperature gradients, and a quasilinear estimate of χ_e/χ_i. The Weiland ITG model is also based on a fluid description in which all moments that are driven by sources (i.e., fueling, heating) are included self-consistently. The fluid moments that are not driven by sources generally decay to zero. The model allows free energy exchange between different transport channels, leading to pinch fluxes. The transport coefficients are derived by using quasilinear theory and a mixinglength rule for saturation. The transport coeffiicients therefore have gyroBohm scaling. However, they have been found to agree well with some nongyro-Bohm L-mode and H-mode experimental data. The Weiland model also includes effects from the impurity profiles, fast ions, and $T_e \neq T_i$, and has been extended to include parallel ion motion and electromagnetic effects.

In order to explain quantitatively the anomalous transport observed for the experimental plasma in a turbulent state, it is necessary to solve the nonlinear gyro-kinetic equation for micro-instability modes including the nonlinear term associated with the convective $\vec{E}\times\vec{B}$ drift, where the particle distribution function and the fluctuation field in saturated turbulent state are calculated. In the gyro-kinetic simulation, however, a large amount of time and memory is required to calculate the distribution function in five dimensional phase space.

In the calculation of the gyro-fluid simulation and the gyro-kinetic simulation due to δf method giving the base of IFS/PPPL model, however, the results for the saturated turbulence level and the anomalous transport coefficient are different each other even in the same condition on the simulation domain, etc. Generally the gyro-fluid simulation gives the larger anomalous transport coefficient than the gyro-kinetic simulation in present stage. The cause of these differences is still unresolved at present.

Recently, as for an RFP, linear gyro-kinetic calculations are applied to the RFP configuration to investigate the occurrence of ITG instabilities[20]. In Ref.[20], linear electrostatic calculations of ITG instability are performed by exploiting the GS2 numerical code[21], which is a widely known gyro-kinetic code based on the electromagnetic nonlinear gyro-kinetic equation. For the description of the RFP geometry, some parts of the code GS2 have been modified. Dependence of ITG instability threshold in different plasma conditions on the most relevant parameters such as the density gradient $-Rd\log n/dr$ and the electron to ion temperature ratio T_e/T_i was studied in the flux tube domain. This analysis shows ITG modes are in general stable in RFP plasmas in the area of experimental parameters, and this type of instability to be only marginally responsible for particle and energy transport. The required gradients could be reached only in correspondance to the temperature slopes arising at the boundary of the helical structure in QSH states. It is also shown the Landau damping effect resulting from the shorter connection length of the RFP is an important vehicle to suppress ITG turbulence. I.e. a quenching of the wave-particle interaction implies a quick increase in the ITG growth rate. It is also reported that the ITG instability threshold remains quite high about five times at midradius as compared with a tokomak plasma.

Numerical calculations in Ref.[20], however, are based on geometry and parameters of the RFX experimental device with aspect ratio $R/a \cong 4$. Also some geometric simplifications such as circular surfaces with no shift and so on are imposed although toroidicity is kept into account. As for an low aspect ratio RFP with shaped magnetic surface and neoclassocal effects, understanding the mechanism of transport is considered to be an interesting subject, for its configuratin has some different characters from the standard high aspect ratio RFP plasma.

6. Summary

In TASK code, several turbulent transport models are implemented as mentioned above. However, this code has been developed for the tokamak and (recently) helical magnetic configurations. For analysis of an low A RFP, because of the different nature of the magnetic configuration, some modifications and validation analysis of the applicability of the physical models in TASK code are required.

References

[1] Bruno A, *et al.*, *Physics of Plasmas* **10**, 2330 (2003)
[2] Scheffel J and Schnack D D, *Nuclear Fusion* **40**, 1885 (2000)
[3] Shiina S, *et al.*, *Physics of Plasmas* **12**, 080702 (2005)
[4] Fukuyama A, *et al.*, *J. Plasma Fusion Res.* **81**, 747 (2005)
[5] Hayase K, *et al.*, *Proceedings of the 19th IAEA Fusion Energy Conference* (Lyon France 2002)
[6] Escande D F, *et al.*, *Phys. Rev. Lett.* **85**, 1662 (1999)
[7] Masamune S, *et.al.*, *Journal of Fusion Energy* **28**, 187 (2009)
[8] Nagamine Y, *et al.*, *Proceedings of the 11th International Congress on Plasma Physics* (Sydney Australia 2002)
[9] Shiina S, *et al.*, *J. Plasma Fusion Res.* **81**, 932 (2005)
[10] http://bpsi.nucleng.kyoto-u.ac.jp/bpsi/en/
[11] Hinton F L and Hazeltine R D, *Reviews of Modern Physics* **48**, 239 (1976)
[12] Wilson H R, *Nuclear Fusion* **32**, 259 (1992)
[13] Sauter O, *et al.*, *Physics of Plasmas* **6**, 2834 (1999)
[14] Houlberg W A, *et al.*, *Physics of Plasmas* **4**, 3230 (1997)
[15] Dimits A, *et al.*, *Physics of Plasmas* **7**, 969 (2000)
[16] Fukuyama A, *et al.*, *Plasma Phys. Contr. Fusion* **37**, 611 (1995)
[17] Waltz R E, *et al.*, *Physics of Plasmas* **4**, 2482 (1997)
[18] Kotschenreuther M, *et al.*, *Physics of Plasmas* **2**, 2381 (1995)
[19] Weiland J and Hirose A, *Nuclear Fusion* **32**, 151 (1992)
[20] Predebon I, *et al.*, *Physics of Plasmas* **17**, 012304 (2010)
[21] Dorland W, *et al.*, *Phys. Rev. Lett.* **85**, 5579 (2000)

Progress towards numerical and experimental simulations of fusion relevant beam instabilities

M King[1], R Bryson[1], K Ronald[1], R A Cairns[2], S L McConville[1], D C Speirs[1], A D R Phelps[1], R Bingham[1,3], K M Gillespie[1], A W Cross[1], I Vorgul[2] and R Trines[3]

[1] SUPA, Department of Physics, University of Strathclyde, Glasgow, G4 0NG, UK
[2] School of Mathematics and Statistics, University of St Andrews, Fife, KY16 9SS, UK
[3] STFC Rutherford Appleton Laboratory, Chilton, Didcot, Oxfordshire, OX11 0QX, UK

Email: m.king@strath.ac.uk

Abstract. In certain plasmas, non-thermal electron distributions can produce instabilities. These instabilities may be useful or potentially disruptive. Therefore the study of these instabilities is of importance in a variety of fields including fusion science and astrophysics. Following on from previous work conducted at the University of Strathclyde on the cyclotron resonance maser instability that was relevant to astrophysical radiowave generation, further instabilities are being investigated. Particular instabilities of interest are the anomalous Doppler instability which can occur in magnetic confinement fusion plasmas and the two-stream instability that is of importance in fast-ignition inertial confinement fusion. To this end, computational simulations have been undertaken to investigate the behaviour of both the anomalous Doppler and two-stream instabilities with the goal of designing an experiment to observe these behaviours in a laboratory.

1. Introduction

Previous work at the University of Strathclyde has focused on the investigation of auroral kilometric radiation, which utilises the cyclotron resonance maser instability[1-4]. This instability involves the transverse bunching of electrons orbiting in an imposed magnetic field which in turn results in the donation of energy to a growing wave within the system. An experiment has been constructed to investigate instabilities driven by particles accelerated along magnetic field lines[5-9]. Following on from this experiment, modifications to the system can be made to study other types of instability. Of primary interest are the anomalous Doppler instability and the two-stream instability, both of relevance to fusion science.

The anomalous Doppler instability can occur in magnetic confinement fusion plasmas. When a plasma has been subject to radio frequency heating, such as in a Tokamak, it is possible to develop a large population of electrons which propagate along an applied magnetic field with a high kinetic energy. These non-thermal electrons can surrender their excess energy by collisions with other particles in the plasma, or by coupling to an electromagnetic wave whose energy may then be dissipated in the bulk plasma. One wave coupling regime which may arise between streams of electrons and transverse slow EM waves in the plasma is the anomalous Doppler effect[10-14]. Here electrons interact with the field components of the wave resulting in an increase of gyrational energy but with a greater loss of translational energy, amplifying the field of the slow wave.

The two-stream instability occurs when two or more charged particle streams flow through one another, resulting in a growing longitudinal wave[15]. This is of importance to fast-ignition inertial confinement fusion where a high power laser pulse is used to accelerate electrons into a compressed deuterium-tritium fuel pellet. As the electrons stream through the fuel pellet, collisions between the particles heat the fuel to induce fusion. However, to improve the efficiency of the process the two-stream instability may be coupled to an ion-acoustic wave that may also heat the fuel[16].

The investigation of the dynamics of both these processes shall be conducted initially by computational simulations followed by controlled laboratory experiments.

2. Physical Principles

2.1. Anomalous Doppler

The electron beam initially has nearly all of its momentum along the axis of the waveguide with little transverse momentum. Electrons then experience a force due to a wave travelling in phase synchronism. This force results in acceleration of the electrons in the radial direction where they do work on the wave's E field. Providing the beam drift velocity exceeds the wave velocity, the pumping of the translational to rotational energy (by the $v_z \times B_\perp$ force) exceeds the dissipation of rotational energy by the E field. These requirements may be satisfied by a Doppler upshifted negative cyclotron harmonic of an electron beam drifting in a fixed magnetic field in the presence of a medium which decreases the wave phase velocity, as illustrated in figure 1. This is known as the anomalous Doppler resonance, where electrons are retarded along the axis of propagation, with energy conserved by the growth of the rotational and wave energy, figure 2. Energy extraction in this way can be efficient as no bunching is required before beam energy is extracted.

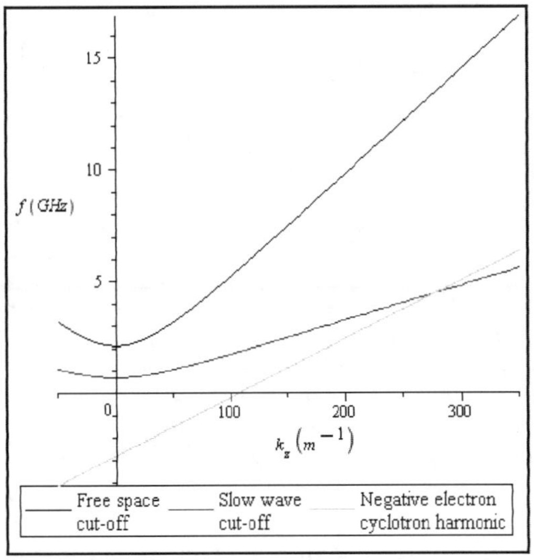

Figure 1. Dispersion showing negative beam harmonic intercepting with a wave in a dielectric loaded waveguide and an unloaded waveguide cutoff

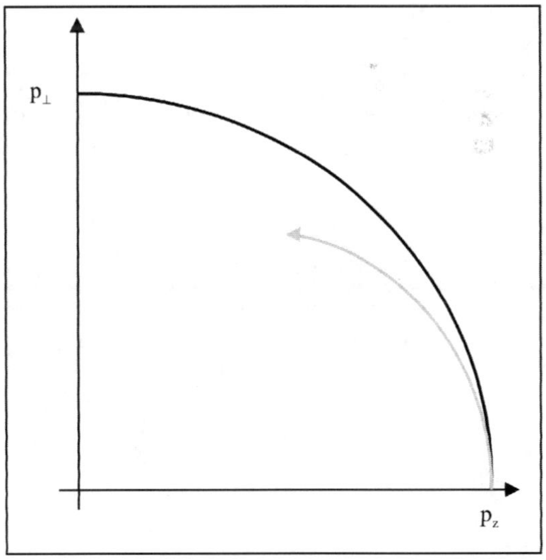

Figure 2. Relaxation of electron energy by anomalous Doppler resonance

2.2. Two-stream

The two-stream instability occurs when there is an interpenetration of two beams, for example an electron beam flowing through an ion beam, or another electron beam. The cause of this instability can be thought of as originating from a point source disturbance within a two-beam plasma. If a density fluctuation arises from this disturbance in one stream of particles, then the electric field will initiate a plasma oscillation at that location. However, these fields can modulate the electron densities of the second stream and the drift of these density modulations through each other can result in energy exchange. This leads to growth of the energy associated with the electric fields feeding from the energy of the initial particle streams. Using a linear theory, the dispersion relation for this mechanism for two co-propagating beams is shown in figure 3.

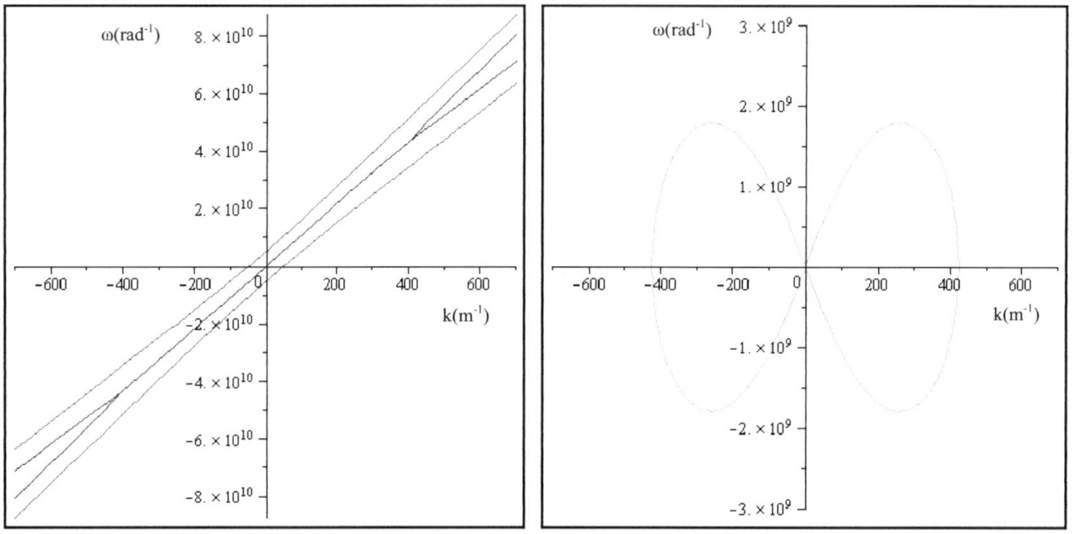

Figure 3. Dispersion relation of the two-stream instability for two co-propagating beams showing (a) the real components (spatial wave-vector) (b) the imaginary components (temporal growth rate)

3. Numerical simulations

3.1. Anomalous Doppler instability simulations

These preliminary simulations (figure 4) make use of a slow wave structure formed of a dielectric loaded waveguide with a small central aperture that allows the passage of an electron beam in place of a full plasma calculation. This configuration allows the negative cyclotron harmonic of an electron beam to interact with the TE_{11} mode in the waveguide, producing a condition where the anomalous Doppler resonance can occur providing $v_z > v_{ph}$. In this simulation, the electron beam with energy 100kV is drifting in a B-field of 0.12T in anomalous Doppler resonance with the TE_{11} mode at 4.6GHz.

Simulations are being conducted with the 3D FDTD Particle-in-Cell code MAGIC. Electrons are confined by a magnetic field to pass through the centre of a waveguide lined with dielectric. In the region of the dielectric the electron beam is in phase synchronism with the TE_{11} mode (as in figure 1). Energy is transferred from the beam to the wave in the interaction region where the magnetic field is uniform and then where the magnetic field decreases the beam is dumped on the waveguide sidewall.

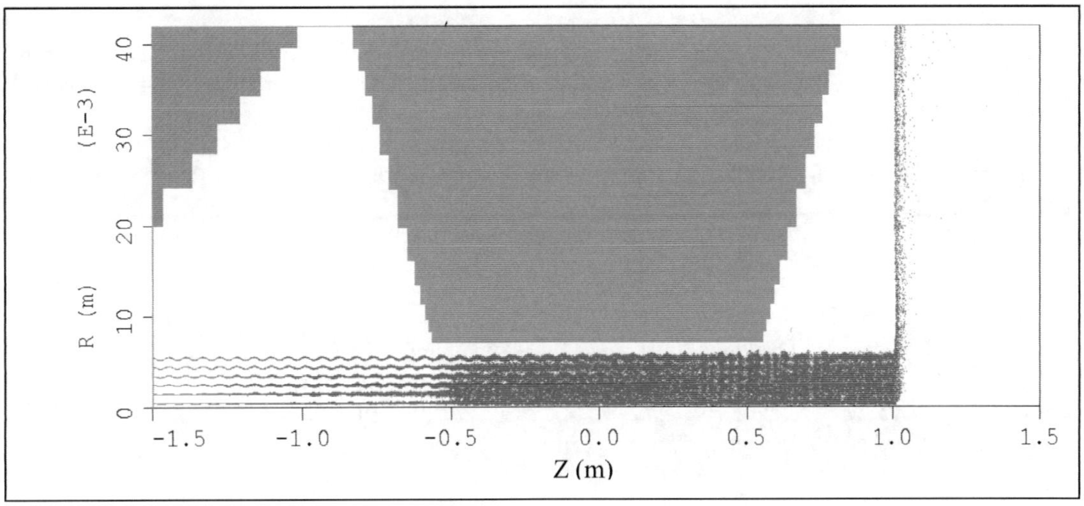

Figure 4. Geometrical illustration of MAGIC simulation showing electrons drifting through a dielectric-lined waveguide

3.2. Two-stream instability simulations

To compare with linear theory, the 2D FDTD Particle-in-Cell code OOPIC Pro was used to create a simple simulation of two co-propagating electron beams at differing velocities. This simplified simulation was undertaken to analyse the features of the two-stream instability without the complication of a background plasma. The beams are of identical radii with one propagating at a velocity of 0.4c and the other at 0.32c. Neither beam has any initial thermal velocity. The beams are contained in a waveguide of circular cross section with a radius of 32mm and are allowed to propagate over a distance of 1.02m. Both beams have the same current of 12A. In order to keep the beams superimposed on each other, a confining magnetic field of 0.1T has been used. From linear theory both the spatial growth length and temporal growth length match the observed behaviour. Further simulations in which one of the beams is replaced with a background plasma are being undertaken. This demonstrates that the presence of a confining B-field does not inhibit the two-stream instability making an experiment to observe this instability easier to design. Therefore it is possible to utilise a Penning trap type of plasma column[17].

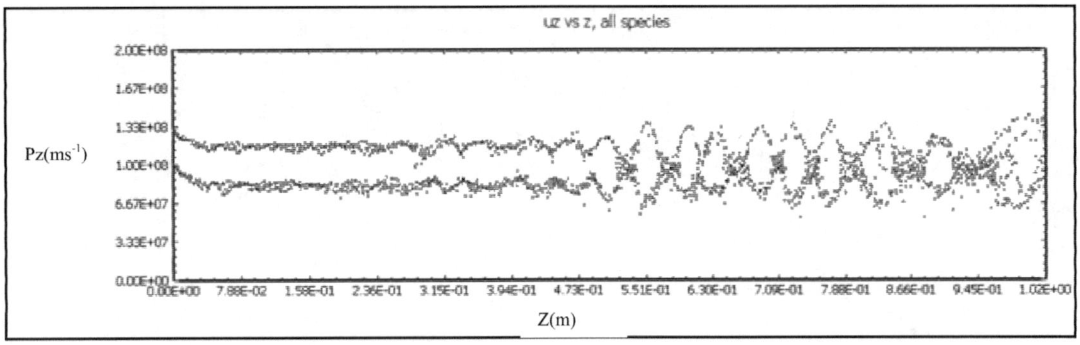

Figure 5. z-momentum of two co-propagating electron beams at 0.4c and 0.32c

4. Summary

Initial calculations for the anomalous Doppler resonance have revealed a rich spectrum of behaviour, including the observation of a radiation signal at the correct frequency for the anomalous Doppler resonance.

The data gathered from the simplified 2D simulation of the two-steam instability for two co-propagating beams will allow for greater understanding of future simulations of the beam-plasma type two-stream instability.

Following from the results of these and ongoing simulations, an experimental setup to investigate both types of instability is being developed.

References

[1] Speirs D C et al 2008 *Plasma Phys. Control. Fusion* **50** 074011
[2] McConville S L et al 2008 *Plasma Phys. Control. Fusion* **50** 074010
[3] Gillespie K M et al 2008 *Plasma Phys. Control. Fusion* **50** 124038
[4] Speirs D C et al 2010 *Phys. Plasma* **17** 056501
[5] Speirs D C, Vorgul I, Ronald K, Bingham R, Cairns R A, Phelps A D R, Kellett B J, Cross A W, Whyte C G and Robertson C W 2005 *J. Plasma Physics* **71** 665-674
[6] Ronald K, Speirs D C, McConville S L, Phelps A D R, Robertson C W, Whyte C G, He W, Gillespie K M, Cross A W and Bingham R 2008 *Phys. Plasmas* **15** 056503
[7] Ronald K, McConville S L, Speirs D C, Phelps A D R, Robertson C W, Whyte C G, He W, Gillespie K M, Cross A W and Bingham R 2008 *Plasma Sources Sci. Tech.* **17** 035011
[8] Cairns R A, Speirs D C, Ronald K, Vorgul I, Kellett B J, Phelps A D R and Bingham R 2005 *Physica Scripta*, **T116** 23-26
[9] Bingham R, Kellett B J, Cairns R A, Vorgul I, Phelps A D R, Ronald K and Speirs D 2004 *Contrib. Plasma Phys.* **44** 382-7
[10] Dendy R O, Lashmore-Davis C N and Montes A 1986 *Phys. Fluids* **29** 4040-4046
[11] Dendy R O 1987 *Phys. Fluids* **30** 2438-2441
[12] Einat M and Jerby E 1997 *Phys. Rev E* **56** 5996-6001
[13] Dendy R O 1991 *Plasma Phys. Control. Fusion* **33** 1069-1076
[14] Ginzburg N S et al 1979 *Radiofiz.* **22** 470-479
[15] Lashmore-Davis C N 2007 *Phys. Plasmas* **14** 092101
[16] Sircombe N J, Bingham R, Sherlock M, Mendonça and Norreys P *Plasma Phys. Control. Fusion* **50** 065005
[17] McConville S L et al. in these proceedings

Formal Proof of Flowers & Ruderman's Instability Mechanism in Magnetic Stars

Pablo Marchant[1] **and Andreas Reisenegger**[2]

Departamento de Astronomía y Astrofísica, Pontificia Universidad Católica de Chile, Casilla 306, Santiago 22, Chile

E-mail: [1] pmarchan@astro.puc.cl, [2] areisene@astro.puc.cl

Abstract. In 1977, Flowers and Ruderman described a perturbation that destabilized a purely dipolar magnetic field in a star. They considered the effect of cutting the star in half along a plane parallel to the symmetry axis and rotating each half 90 degrees in opposite directions, which would cause the energy of the magnetic field in the exterior of the star to be greatly reduced, just as it happens with a pair of aligned magnets. We formally solve for the energy of the external magnetic field and check that it decreases monotonously along the entire rotation.

We also describe the instability using perturbation theory, and see that it happens due to the work done by the interaction of the magnetic field with surface currents.

Finally, we consider the stabilizing effect of adding a toroidal field by studying the internal energy perturbation when the rotation is not done along a sharp cut, but with a continuous displacement field that switches the direction of rotation across a region of small but finite width. Using these results, we estimate the relative strengths of the toroidal and poloidal field needed to make the star stable to this displacement and see that the energy of the toroidal field required for this is much smaller than the energy of the poloidal field.

1. Introduction

Large-scale magnetic fields are present in many stellar objects, and they appear to be long-lived since they do not evolve in a timescale accesible to observations.

These objects are mostly stably stratified so dynamo effects are expected to be irrelevant in keeping the strength of the magnetic field, which should be in a state of stable equilibrium.

Even though these long-lived fields have been known to exist for more than half a century, it has not been possible to find an analytic model for a field that has been shown to be in stable equilibrium. However, stable configurations have been found to exist via numerical calculations , where an initially random field usually evolves into an approximately axisymmetric configuration that is a combination of toroidal and poloidal components of similar energies. In these simulations, once a stable configuration has been achieved, the decay of the field is driven by Ohmic dissipation, and it can be seen to evolve in a timescale comparable to the lifetime of the star.

With regard to the stability of purely poloidal fields, Flowers & Ruderman 1977 argued that any poloidal field that has the shape of a dipole outside of the star should be unstable. In an analogous way to a couple of aligned magnets, one half of the star could turn with respect to the other producing a quadrupole (as shown in figures 1 and 2).

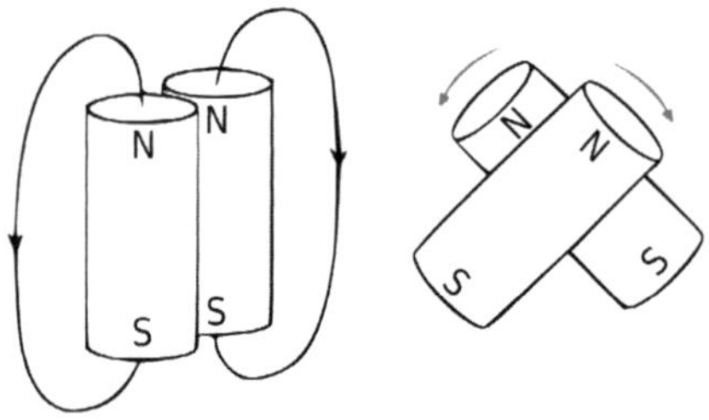

Figure 1. A couple of aligned magnets is in an unstable state of equilibrium.

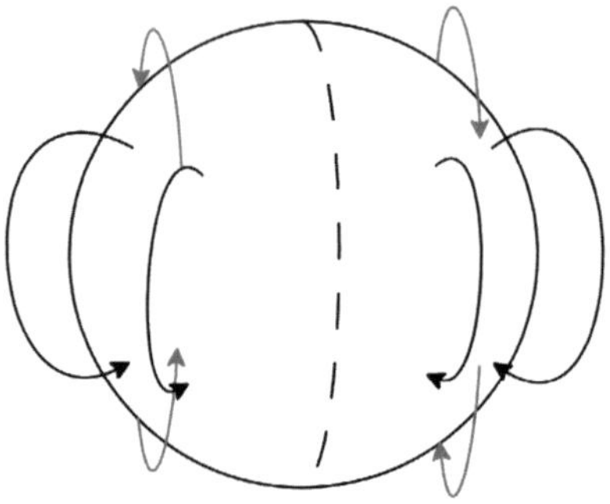

Figure 2. In a similar way to the aligned magnets, each half of a star with a dipole field could rotate in opposite directions, until it reaches a state of stable equilibrium.

We formally prove Flowers and Ruderman's instability for a pure dipole, and study the stabilizing effect of a toroidal field.

2. Formal proof of the instability

To prove Flowers and Ruderman's instability, the following simplifying assumptions are made:

- The star is perfectly spherical, and all hydrodynamic quantities are functions only of r.
- Magnetic flux is completely frozen in the fluid.
- Outside the star there is a perfect vacuum.

Under these conditions, only the energy of the magnetic field outside the star changes, and this was caltulated as

$$E = \frac{1}{8\pi}\int_{r>R} \mathbf{B}^2 \mathrm{d}V = E_0\left[1+\sin^2\Omega(A-1)\right], \tag{1}$$

where E_0 is the initial energy of the dipole outside the star, Ω is the angle of rotation of each half of the star, and A is a numerical constant that results from an infinite series that could not be summed analytically. Also, a quantity Υ defined as

$$\Upsilon = \frac{R^3}{8\pi}\int_{4\pi}\left(B_r^2\right)_{r=R}\mathrm{d}\Omega, \tag{2}$$

was considered. Since the field is only displaced on the surface, Υ is equal to a constant along the entire rotation. It was shown that due to the conservation of Υ, the final energy is smaller than the initial one, which implies that the constant A in (1) is smaller than 1, and the energy decreases monotonically with the angle of rotation.

Using perturbation theory, it was seen that all the net work done was due to surface currents produced by the perturbation.

3. Previous Misconception

Some works, for instance, Roberts 1981 and Braithwaite & Nordlund 2006, mentioned that a second cut in a direction perpendicular to the first one would produce a configuration which resembles a quadrupole. Additionally, they expected that this proccess could be repeated *ad*

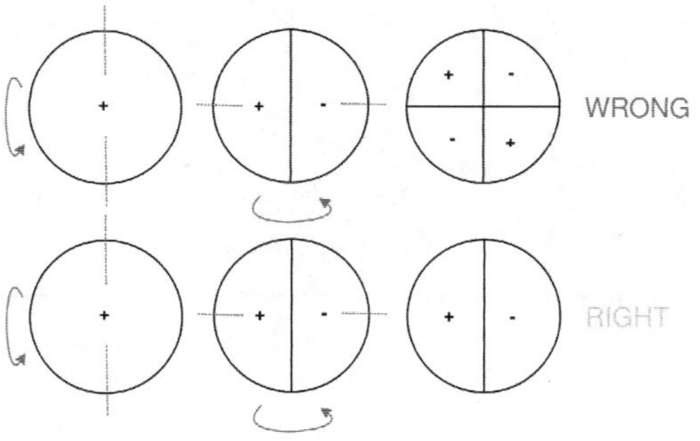

Figure 3. Performing two cuts to the dipole. The top image shows the wrong picture, where an octupole is produced.

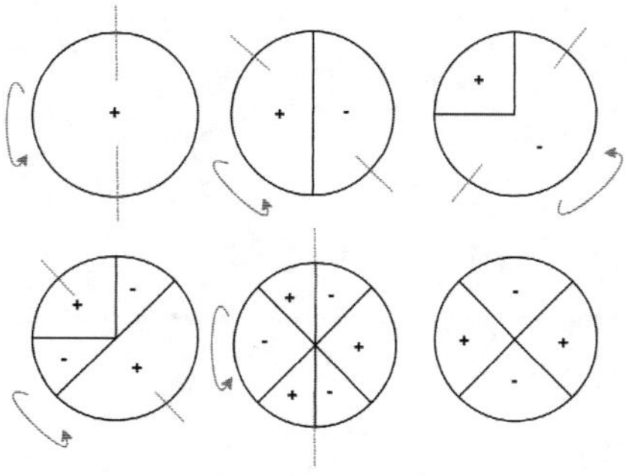

Figure 4. Performing several cuts to the dipole in order to obtain an octupole configuration.

infinitum to produce a configuration that resembles an arbitrary multipole. However, this proccess breaks down just at the second cut.

In this figure, the star is observed from the top of the symmetry axis, and the plus and minus signs indicate field coming out or into the star. The second cut does not produce the octupole, but simply returns the star to the quadrupole configuration.

It is possible to obtain the quadrupole with a significant number of cuts, however, through this proccess the energy is not monotonically reduced, so the star cannot actually follow these displacements.

We therefore expect this mechanism to affect only the initial axisymmetric field.

4. Stabilizing effect of a toroidal field

With a toroidal field, a cut sharp is not possible. However, one can consider each side of the star to rotate in opposite directions, with a thin region in which the direction of rotation switches direction (see figure 5).

In this case, the "bending" of toroidal field lines opposes the displacement, so for a sufficiently strong toroidal field we would expect the field to be stable to Flowers & Ruderman's instability mechanism. We considered particular models for the poloidal and toroidal components with variable strength, which resulted in a dipole field outside the star. The poloidal field was given by

$$\mathbf{B}_P = \nabla\alpha \times \nabla\phi, \quad \alpha(r) = \frac{35B_p}{16}\left[r^2 - \frac{6}{5}\frac{r^4}{R^2} + \frac{3}{7}\frac{r^6}{R^4}\right]\sin^2\theta, \tag{3}$$

and the toroidal field was modeled as contained in a torus, as shown in figure 6, and the strength of the field is given by

$$\mathbf{B}_T = B_T \cos^2\left(\frac{\rho\pi}{2\mu R}\right)\hat{\phi} \tag{4}$$

where the meaning of ρ and μ is illustrated in figure 6.

Under these conditions, we estimated that the system is stable when the ratio of poloidal to total magnetic energy satisfies $E_P/E_T \lesssim 0.99$. This means that a very weak toroidal field is sufficient to stabilize the star.

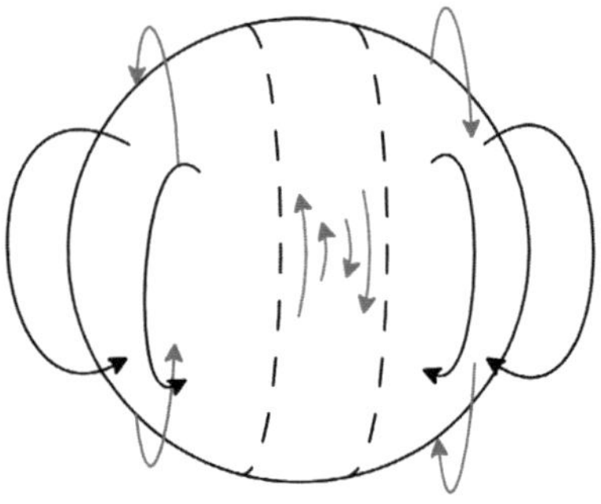

Figure 5. Performing a smooth cut to the star due to the presence of a toroidal field.

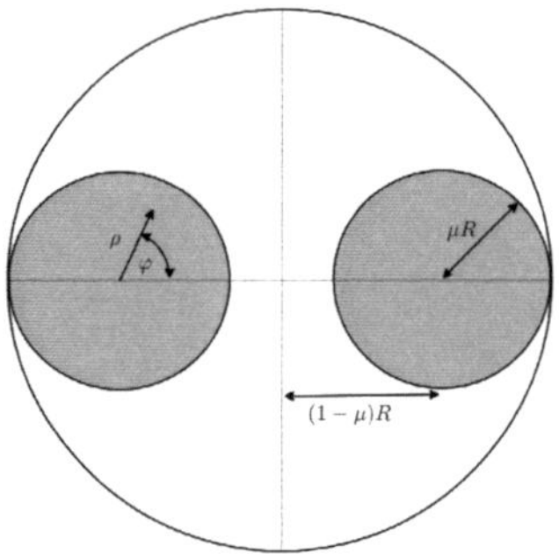

Figure 6. The toroidal field was modeled as contained in a torus of radius μR.

5. Conclusions

We showed that a pure dipole field is unstable to Flowers & Ruderman's instability mechanism by direcly evaluating the external energy of the magnetic field for an arbitrary angle of rotation for each half of the star, and proving that it decreases monotonously along the entire rotation.

It was also demostrated that a second cut could not produce an octupole, as has been said in previous works. To reach an octupole, several cuts must be made, however, in doing so, the external magnetic energy does not decrease monotonically, so this proccess is not physically possible.

To study how a toroidal field could stabilize the star against the perturbation described by Flowers & Ruderman's instability, I considered particular models for the toroidal and poloidal components of the field. In this case, the cut had to be performed smoothly, as shown in figure 5. Under these conditions, the field was stable to Flowers & Ruderman's instability when the ratio of poloidal to total magnetic energy satisfies $E_P/E_T \lesssim 0.99$, so a very weak toroidal field is sufficient to stabilize the star.

Acknowledgments

This project is supported by FONDECYT Regular Project 1060644, FONDAP Center for Astrophysics (15010003), Proyecto Basal PFB-06/2007 and Proyecto Límite VRI 2010-15.

References

[1] Braithwaite, J. and Nordlund, A., 2006, *A&A*, **450**, 1077–1095
[2] Flowers, E. and Ruderman, M.A., 1977, *ApJ*, **215**, 302–310
[3] Marchant, P. *Large-scale instabilities of poloidal magnetic fields in stars*, undergraduate thesis, Pontificia Universidad Católica de Chile, 2010
[4] Roberts, P.H., 1981, *Astron. Nach.*, **302**, 65–70

Dynamics of the Electromagnetic Ion Cyclotron Nonlinear Solitary Structures in the Inner Magnetosphere

T. D. Kaladze,[1] N. L. Tsintsadze,[2] J. W. Van Dam,[3] W. Horton,[3] T. W. Garner[4] and X. R. Fu[3]

[1]I. Vekua Institute of Applied Mathematics of Tbilisi State University, 2 University Str., 0186 Tbilisi, Georgia
[2] E. Andronikashvili Institute of Physics, Academy of Sciences, 0177, Tbilisi, Georgia
[3]Institute for Fusion Studies, University of Texas at Austin, TX 78712 Texas, USA
[4]Space and Geophysics Laboratory, Applied Research Laboratories, University of Texas at Austin, TX 78712 Texas, USA

Abstract. The nonlinear interaction of the electromagnetic ion cyclotron (EMIC) frequency waves with plasma particles in the inner magnetosphere is studied. The emission is considered to be circularly polarized electromagnetic waves propagating along the almost constant dipole geomagnetic field in the equatorial region of the inner magnetosphere. Under the action of the ion cyclotron ponderomotive force excitation of the magnetosonic waves through the amplitude modulation of the EMIC waves is investigated. Two dimensional nonlinear Schrodinger equation for the EMIC waves is derived. In the stationary case two solutions of the nonlinear Schrodinger equation with distinct natures are found. The generation of both vortices and of a quasistatic magnetic field across the geomagnetic field lines is discussed.

1. Introduction

Electromagnetic ion cyclotron waves, commonly called EMICs, appear to play a mul-tifaceted role in the Earth's inner magnetosphere. These waves can be excited by spontaneous plasma instability when plasma is injected via substorms from the geomagnetic tail into the inner magnetosphere (see e.g. [1]). This plasma injection conserves the magnetic momentum of the hot ions in the plasma sheet, producing an energetic ion plasma component with energy up to 30 keV near L-shell values of 4 or 5. Due to its anisotropy in velocity space, this population of energetic ions can drive the EMIC waves through Landau cyclotron resonances. Another common source of electromagnetic ion cyclotron waves is the pulsation of the magnetopause due to variations in the solar wind dynamic pressure [2]. In 2005, the LAZIO-SIRAD detector was installed on the International Space Station with the objective of clarifying the correlation between such electromagnetic waves and earthquakes by measuring changes in the number of charged particles and the strength of magnetic fields at low L-shells.

Electromagnetic ion cyclotron waves have been investigated for a long time with both ground-based instruments and satellite experiments. Initial observations by satellite were made by *Russell et al.* [3], *Gurnett* [4], and *Kintner and Gurnett* [5]. EMIC waves have been seen in association with intense electron fluxes on the auroral field lines with the S3-3 satellite [6, 7], the Freja satellite [8], the Fast satellite [9], and the Polar satellite [10]. EMIC waves have also been observed in the equatorial plane at various L values by the GEOS satellites [11], the Akebono satellite [12, 13, 14], the Equator-S satellite [15], and the CRESS satellite [16, 17]. In this equatorial region, EMIC waves are generated by the ion cyclotron instability driven by the anisotropic distribution of ring-current energetic ions during magnetic storms (see [15,18] and references therein). They are characterized by harmonic structures, with the electric and magnetic field power between the first multiples of the proton cyclotron frequency. EMIC waves are known to take part in the precipitation of electrons (see e.g.,[17-20]). *Parrot et al.* [21] reported the observation of electromagnetic harmonic emissions with the microsatellite DEMETER. These emissions were detected in the ELF range (500 Hz to 2 kHz) in the upper ionosphere during large magnetic storms, and it was shown that they could be related to the electromagnetic ion cyclotron waves.

In the present work, we assume that electromagnetic ion cyclotron waves are present with sufficiently large amplitude so that their nonlinear structure and the conditions for parametric instabilities become important. After displaying the linear dispersion relation with a kinetic-theory response for the ions, we develop a nonlinear two-component fluid description of the electromagnetic ion cyclotron waves and explore their properties with respect to propagation speed versus amplitude.

2. Basic equations for electromagnetic ion cyclotron waves

We consider the nonlinear dynamics of electromagnetic fields in the inner magnetosphere. Our investigation is also applicable to the upper regions of the Earth's ionosphere. We consider the propagation of circularly polarized nonlinear electromagnetic waves propagating along the dipole geomagnetic field in the equatorial region of the inner magnetosphere. We assume that the electromagnetic emission frequency ω_0 is close to the proton gyro-frequency ω_{cp} i.e., $\omega_0 \approx \omega_{cp} =$ $2.98\,L^{-3}$ kHz, where $L = r_{eq} / R_E$ is the shell parameter of the dipole geomagnetic field, r_{eq} is the radial distance to the magnetic field line in the equatorial plane, and R_E is the Earth's radius. Further, we consider regions around *L=2*. The circularly polarized electric field is given by $E_- = E_x - iE_z = E(x,y,t)\exp[i(k_0 y - \omega_0 t)]$. Here k_0 is the y-component of the wave number of the electromagnetic emission. Thus, the pumping electromagnetic waves are assumed to propagate almost along the homogeneous and constant dipole geomagnetic field B_0 oriented along the y-axis. The amplitude $E(x,y,t)$ becomes a slowly varying function of spatial coordinates and time due to the nonlinear interaction of the electromagnetic waves with plasma. The electromagnetic wave fluctuations are in the x and z directions, transverse to the geomagnetic field (y- direction).

Taking the inner magnetosphere plasma to be collisionless and non-degenerate EMIC waves are described by the following dispersion equation [22],

$$N^2 = \frac{k_0^2 c^2}{\omega_0^2} = 1 - \frac{\omega_{pe}^2}{\omega_0 \omega_{ce}} - \frac{\omega_{pp}^2}{\omega_0(\omega_0 - \omega_{cp})} I_+\left(\frac{\omega_0 - \omega_{cp}}{k_0 \mathrm{v}_{tp}}\right), \tag{1}$$

Where ω_{pe} and ω_{pp} are the electron and proton Langmuir frequencies, respectively; v_{tp} is the proton thermal velocity; c is the speed of light in a vacuum and $I_+(z)$ is the appropriate plasma dispersion function. It follows from the dispersion equation (1) that for the range of frequencies $\omega_0 >> |\omega_0 - \omega_{cp}| >> k_0 \mathrm{v}_{tp}$, the refractive index N is complex, i.e., $N = N_0 + i\eta$, where

$$N_0^2 = \frac{k_0^2 c^2}{\omega_0^2} = 1 - \frac{\omega_{pe}^2}{\omega_0 \omega_{ce}} - \frac{\omega_{pp}^2}{\omega_0(\omega_0 - \omega_{cp})}, \quad \eta = \sqrt{\frac{\pi}{8}} \frac{\omega_{pp}^2}{N_0 k_0 \omega_0 \mathrm{v}_{tp}} \exp\left(-\frac{c^2(\omega_{cp} - \omega_0)^2}{2\omega_0^2 N_0^2 \mathrm{v}_{tp}^2}\right), \tag{2}$$

Outside the resonance region $|\omega_0 - \omega_{cp}| >> k_0 \mathrm{v}_{tp}$, the absorption of proton cyclotron waves is exponentially small. From Eq. (2), by maximizing η with respect to k_0, we find that the limit to the resonance occurs at $k_0 = |\omega_{cp} - \omega_0| / \sqrt{2}\mathrm{v}_{tp}$. By using this expression for the frequency difference in the expression for N_0^2 of Eq. (2), which dominates near resonance $(\omega_0 \le \omega_{cp})$, we obtain the maximum value for the refraction index as $N_{0\max} = (c\omega_{pp}^2 / \sqrt{2}\mathrm{v}_{tp}\omega_{cp}^2)^{1/3}$. The EMIC waves are left-hand circularly polarized and resonate with energetic ions. The formation of coherent structures including vortices and solitons, requires the balancing of the dispersion in the $\omega(k)$ waves with nonlinear terms in the wave dynamics.

3. Nonlinear two-component fluid description of EMICs

Here we will discuss the basic equations that describe the nonlinear modulation of electromagnetic ion cyclotron waves propagating along the magnetic field B_{0y}. We assume that the modulation frequency is much smaller than the ion gyro-frequency. Then the slow motion of the plasma can be described by the following single-fluid MHD equations for the plasma density ρ and velocity $\mathbf{v}$:

$$\frac{\partial \rho}{\partial t} + \nabla \cdot (\rho \mathbf{v}) = 0, \quad \frac{\partial}{\partial t}\mathbf{v} + (\mathbf{v}\cdot\nabla)\mathbf{v} = -\frac{v_s^2 \nabla \rho}{\rho} - \frac{1}{4\pi\rho}\mathbf{B}\times\nabla\times\mathbf{B} + \frac{\mathbf{f}}{m_p}, \tag{3}$$

Here, **B** is the total magnetic field, **f** is the ponderomotive force incorporating the motion of the protons, and $v_s = \sqrt{T_e / m_p}$ is the proton sound velocity with T_e the electron temperature.
In our consideration, the ponderomotive force **f** has two components [23]:

$$f_x = -\frac{m_p \omega_0}{\omega_0 - \omega_{cp}} \frac{\partial}{\partial x} \frac{e^2 |E|^2}{m_p^2 \omega_0^2}, \quad f_y = -\frac{m_p \omega_0}{\omega_0 - \omega_{cp}} \frac{\partial}{\partial y} \frac{e^2 |E|^2}{m_p^2 \omega_0^2} + \frac{m_p k_0 \omega_{cp}}{(\omega_0 - \omega_{cp})^2} \frac{\partial}{\partial t} \frac{e^2 |E|^2}{m_p^2 \omega_0^2}, \tag{4}$$

where $|E| = (\omega_0 / c)|A|$, with the electric field **E** being related to the vector potential **A** as **E** = $-\partial \mathbf{A} / c\partial t$.

3.1 Nonlinear Schrodinger equation. Now we will derive the nonlinear Schrodinger equation, using the Maxwell equation for the ion cyclotron waves:

$$\nabla^2 E_- - \frac{1}{c^2}\frac{\partial^2}{\partial t^2} E_- = \frac{4\pi e n_p}{m_p c^2}\frac{\partial P_-}{\partial t}. \tag{5}$$

For our problem, the contribution of the electron current density is smaller than of that of the protons. Here $P_- = P_x - iP_z$ is the ion momentum due to a rapidly varying electromagnetic field. A simple calculation of P_- follows from the equation of motion (3):

$$P_- = \left[\frac{ieA}{\omega - \omega_{cp}} + \frac{e\omega_{cp}}{(\omega - \omega_{cp})^2}\frac{1}{\omega_0}\frac{\partial A}{\partial t}\right] e^{i(k_{0y} - \omega_0 t)}. \tag{6}$$

Recall that according to our assumption all quantities have both fast and slow temporal-spatial scales. Substituting Eq.(6) into Eq.(5), we obtain for the ion cyclotron waves a nonlinear Schrodinger equation of the following form,

$$2i\left(\frac{\partial}{\partial t} + v_g \frac{\partial}{\partial y}\right) A + \frac{c^2}{\omega_0}\left(\frac{\partial^2}{\partial x^2} + \frac{\partial^2}{\partial y^2}\right) A - \frac{\omega_0 \Delta\omega}{\Gamma} A - \frac{\omega_{pp}^2 \omega_0}{\Gamma(\omega_0 - \omega_{cp})}\frac{\delta n_p}{n_{0p}} A = 0, \tag{7}$$

where the group velocity is $v_g = 2(\omega_{cp} - \omega_0)/k_0$ and the expressions for Γ and $\Delta\omega$ are given in [24].

3.2 Parametric Decay instability of large-amplitude plane waves. In order to consider modulational instabilities (leading to the excitation of magnetosonic waves), we linearize Equations (3), (4) and (7) and search for plane wave solutions of the form $\exp\{i(\boldsymbol{qr} - \Omega t)\}$. We obtain the following dispersion equation:

$$\left[(\Omega - q_y v_g)^2 - \left(\frac{c^2 q^2}{2\Gamma}\right)^2\right](\Omega^2 - \Omega_+^2)(\Omega^2 - \Omega_-^2) = B\left(\frac{eA}{m_p c^2}\right)^2, \tag{8}$$

where the expression for B is given in [24]. Here V_A is the Alfven speed, and $q^2 = q_x^2 + q_y^2$. In Eq. (8), $\Omega_\pm$ are the fast and slow magnetosonic frequencies [24]. Equation (8) has several types of complex solutions. For example, when $\Omega \square\, q_y v_g \approx \Omega_+$ or $\Omega \square\, q_y v_g \approx \Omega_-$ we have the excitation of fast or slow magnetosonic waves by the electromagnetic waves. These instabilities also lead, in general, to the existence of two different solutions.

4. Coherent nonlinear EMIC structures

In the previous section, we mentioned that the amplitude modulation of the electromagnetic waves leads to an excitation of magnetosonic waves whose amplitude grows exponentially. Eventually, the wave amplitude stops growing due to the influence of nonlinear terms that were ignored in the linear analysis. Now we take into account nonlinear terms due to the ponderomotive force, which redistributes the protons and changes the density of the plasma. The convective derivative term $(\mathbf{v}_p \cdot \nabla)\mathbf{v}_p$ continues to be ignored, at least as long as the wave does not steepen too much.

4.1 Envelope solitons. Here we show that there are soliton solutions for the nonlinear Schrodinger equation in Equation (7). We look for solutions that propagate as $\xi = y - \mathrm{v}_g t$ and assume that any dependence on this variable is faster than that on the spatial coordinate x or the time t. Hence, for the ponderomotive force in Equation (4), we assume that $|f_y| >> |f_x|$. Thus we can simplify that and the expressions for ponderomotive force f_y and density perturbation may be written as

$$f_y = -m_p c^2 \frac{\omega_{cp}}{\omega_{cp} - \omega_0} \frac{\partial}{\partial \xi} \frac{e^2 |E|^2}{m_p^2 c^2 \omega_0^2}, \qquad \frac{\delta n}{n_0} = \frac{1}{(\mathrm{v}_g^2 - \mathrm{v}_s^2)} \frac{\omega_{cp}}{\omega_{cp} - \omega_0} \frac{e^2 A^2}{m_p^2 \omega_0^2}. \tag{9}$$

Substituting the expressions of Equation (9) into the nonlinear Schrodinger equation (7), we obtain

$$\left(\frac{\partial^2}{\partial \xi^2} + \frac{\partial^2}{\partial x^2}\right) A - \frac{\omega_0 \Delta\omega}{c^2} A + \frac{\omega_{pp}^2 \omega_{cp}^2}{(\mathrm{v}_g^2 - \mathrm{v}_s^2)} \frac{1}{(\omega_{cp} - \omega_0)^2} \frac{e^2}{m_p^2 c^2 \omega_0^2} A^3 = 0. \tag{10}$$

Introducing the new dimensionless variables ξ' and x' and the appropriate dimensionless function U we can obtain the following nonlinear Schrodinger equation describing two-dimensional symmetrical solitons (see [24])

$$\frac{1}{\rho} \frac{\partial}{\partial \rho} \left(\rho \frac{\partial}{\partial \rho} U \right) - U + U^3 = 0, \tag{11}$$

which is written in terms of the cylindrical coordinate $\rho = \sqrt{\xi'^2 + x'^2}$. The solutions of this equation have been described by *Gurevich* and *Shvartsburg* [25], with the boundary conditions $U \to 0$ for $\rho \to \infty$ and either $U = const$ or $dU / d\rho = 0$ at $\rho = 0$. The boundary condition $dU / d\rho = 0$ at $\rho = 0$ leads to U having discrete values U_n at $\rho = 0$. The fundamental mode has $U_1 = 2.2$; this is a soliton solution, peaked at $\rho = 0$ and falling off as $\rho^{1/2} e^{-\rho}$ as $\rho \to \infty$. The higher-order modes for $n > 1$ are damped oscillatory wave functions with $(n-1)$ zero crossings and have initial values $U_2 = 3.3, U_3 = 4.1, U_4 = 4.6$, etc. Hence equation (11) has a soliton solution only for moderate initial values of U. For comparatively large initial values of *U*, this equation describes nonlinear stationary waves propagating with a group velocity v_g. Numerical calculations of Eq. (11) are shown in Fig. 1.

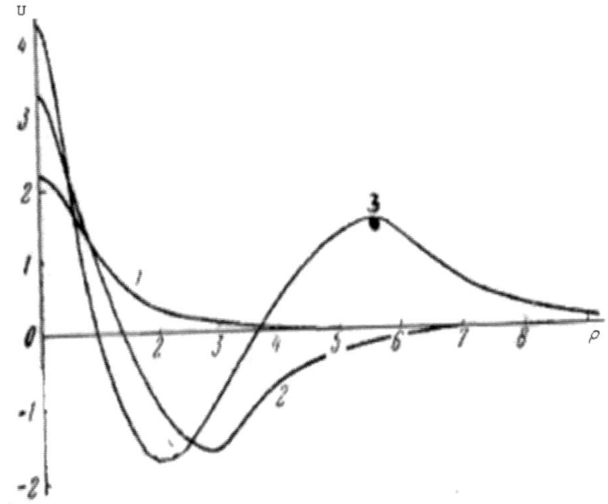

Figure 1. Numerical calculations of Eq. (11)

4.2 Generation of vortices and magnetic field. In this subsection we show that in a homogeneous plasma, electromagnetic ion cyclotron waves can lead to the generation of vortices and a quasi-static magnetic field. Previously, these problems have been considered in laser plasma physics by *Tsintsadze et al.* [26,27].

To derive the equation for vortices, we use the momentum equation for the proton species,

$$\frac{\partial}{\partial t}\left(m_p\mathbf{v}+\frac{e}{c}\mathbf{A}\right)=-e\nabla\phi+\frac{e}{c}\mathbf{v}\times\mathbf{B}+\mathbf{v}\times(\nabla\times m_p\mathbf{v})-\frac{1}{n_p}\nabla p_p-\nabla\left(\frac{m_p v^2}{2}\right)+\mathbf{f}. \tag{12}$$

The expression for f is given in Equation (4), and $\mathbf{B}=B_{0y}\boldsymbol{e}_y+\nabla\times\mathbf{A}$.Taking the curl of both sides of Equation (12), we obtain

$$\frac{\partial\boldsymbol{\Omega}}{\partial t}=\nabla\times(\mathbf{v}\times\boldsymbol{\Omega})+\nabla\times\mathbf{f}\,, \tag{13}$$

where $\boldsymbol{\Omega}=\nabla\times(m_p\mathbf{v}+\frac{e}{c}\mathbf{A})$ is the vorticity of the canonical ion momentum.

If the characteristic spatial scale length and the characteristic time scale satisfy the inequality $l>\mathrm{v}t$, we may then neglect the first term on the right-hand side in Equation (13) compared to the term on the left-hand side and obtain the following simple relationship between the vorticity and the source:

$$\Omega=\frac{\omega_{cp}}{(\omega_{cp}-\omega_0)^2}\mathbf{k}_0\times\nabla\frac{e^2A^2}{m_pc^2}\,. \tag{14}$$

We now introduce the canonical momentum circulation G, defined as

$$\mathbf{G}=\oint_l(\mathbf{P}+\frac{e}{c}\mathbf{A})\cdot dl=\int_S\Omega d\mathbf{S}\,. \tag{15}$$

Due to the existence of the last term in f_y of Equation (4), the flux of $\boldsymbol{\Omega}$ through a surface bounded by any closed ion fluid contour, as well as the canonical momentum circulation, is not conserved, namely,

$$\frac{d\mathbf{G}}{dt}=\frac{d}{dt}\oint(\mathbf{P}+\frac{e}{c}\mathbf{A})\cdot dl=\frac{\omega_{cp}}{(\omega_{cp}-\omega_0)^2}\frac{\partial}{\partial t}\oint_l\frac{e^2A^2}{m_pc^2}\mathbf{k}_0\cdot d\mathbf{r}\,. \tag{16}$$

Hence, if there are initially no vortices, they will be generated by the electromagnetic ion cyclotron waves over time.

Note that since the wave number $\mathbf{k}_0$ is directed along the y-axis and the ponderomotive force is a function of (x,y), then the vorticity has only one component, Ω_z. Thus we can rewrite Equation (13) as

$$\frac{d}{dt}\ln\frac{\Omega_z}{n}=\frac{k_0\omega_{cp}m_p}{(\omega_{cp}-\omega_0)^2}\frac{1}{\Omega_z}\frac{\partial}{\partial t}\frac{\partial}{\partial x}\frac{e^2A^2}{m_p^2c^2}\,. \tag{17}$$

Equation (17) is a generalization of the Hasegawa-Mima equation, and it clearly breaks the frozen-in condition [23], given that $d(\ln(\Omega_z/n))/dt=0$.

A simple expression for the magnetic field that is generated by the non-stationary ponderomotive force can be obtained from Equation (14) if we assume that $|\nabla\times m_p\mathbf{v}|<<eB_z/c$. We then obtain

$$\frac{eB_z}{m_pc}=\frac{\omega_{cp}c^2}{(\omega_{cp}-\omega_0)^2}k_0\frac{\partial}{\partial x}\left(\frac{eA}{m_pc^2}\right)^2\,. \tag{18}$$

Let us estimate this quantity for the following numerical values: ω_{cp} □ 0.37×10^3 Hz, $\omega_0-\omega_{cp}\approx k_0\mathrm{v}_{tp}$, $\partial/\partial x$ □ $1/L_{0x}$□ 1.6×10^{-9} cm^{-1}, k_0□ 10^{-7} cm^{-1}, eA/m_pc^2 □ 10^{-6}, v_{tp} □ 10^5 cm/sec. Substituting these parameters into Equation (22), we obtain the following estimate for the normalized magnetic field perturbation $eB_z / m_p c \cong 3.7\,\text{Hz}$, or, in unnormalized form,

$B_z \cong 0.5 \times 10^{-4} G$.

5. Conclusions

In this paper, we have studied the nonlinear interaction of the electromagnetic ion cyclotron frequency waves with plasma particles in the inner magnetosphere. The emission is considered to be circularly polarized electromagnetic waves propagating along the almost constant dipole geomagnetic field in the equatorial region of the inner magnetosphere. We studied excitation of the magnetosonic waves through the amplitude modulation of the electromagnetic ion cyclotron waves, and obtained a two-dimensional nonlinear Schrodinger equation for the EMIC waves. In the stationary case, we found two solutions of the nonlinear Schrodinger equation with distinct natures. For sufficiently small amplitudes of the EMIC field, there exists a two-dimensional bright soliton, whereas for larger amplitudes the solution is oscillating. The generation of both vortices and of a quasi-static magnetic field across the geomagnetic field lines was discussed.

Acknowledgments. This work was supported in part by award GNSF/ST08/5-442 of the Georgian National Science Foundation, and by National Science Foundation grant ATM-0638480 to The University of Texas at Austin.

References

[1] Cornwall, J. M., *J. Geophys. Res.*, ***70***, 61, doi: 10.1029/JZ070i001p00061, 1965.

[2] Anderson, B. J., and D. C. Hamilton, *J. Geophys. Res.*, ***98***, 11,369, doi: 10.1029/93JA00605, 1993.

[3] Russell, C. T., R. E. Holzer, and E. J. Smith, *J. Geophys. Res.*, ***75***, 755, doi:10.1029/JA075i004p00755, 1970.

[4] Gurnett, D. A., *J. Geophys. Res.*, ***81***, 2765, doi:10.1029/JA081i016p02765, 1976.

[5] Kintner, P. M., and D. A. Gurnett, *J. Geophys. Res.*, ***82***, 2314, doi:10.1029/JA082i016p02314, 1977.

[6] Kintner, P. M., M. C. Kelley, R. D. Sharp, et al., *J. Geophys. Res.*, ***84***, 7201, doi: 10.1029/JA084iA12p07201, 1979.

[7] Temerin, M., and R. L. Lysak, *J. Geophys. Res.*, ***89***, 2849, doi:10.1029/JA089iA05p02849, 1984.

[8] Hamrin, M., P. Norqvist, M. Andr´e, and A. I. Eriksson, *J. Geophys. Res.*, ***107***, 1204, doi:10.1029/2001JA900144, 2002.

[9] Chaston, C. C., J. W. Bonnell, J. P. McFadden, et al., *J. Geophys. Res.*, ***107***, 1351, doi:10.1029/2001JA900141, 2002.

[10] Santolık, O., J. S. Pickett, D. A. Gurnett, and L. R. O. Storey, *J. Geophys. Res.*, ***107***, 1444, doi:10.1029/2001JA000146, 2002.

[11] Young, D. T., S. Perraut, A. Roux, et al., *J. Geophys. Res.*, ***86***, 6755, doi:10.1029/JA086iA08p06755, 1981.

[12] Sawada, A., Y. Kasahara, M. Yamamoto, et al., *Geophys. Res. Lett.*, ***18***, 317, doi:10.1029/91GL00040, 1991.

[13] Kasahara, Y., H. Kenmochi, and I. Kimura, *Radio Science*, ***29***, 751, doi:10.1029/94RS00445, 1994.

[14] Liu, H., S. Kokubun, and K. Hayashi, *Geophys. Res. Lett.*, ***21***, 225, doi:10.1029/93GL02836, 1994.

[15] Mouikis, C. G., L. M. Kistler, W. Baumjohann, et al., *Geophys. Res. Lett.*, ***29***(10), 1432, doi: 10.1029/2001GL013899, 2002.

[16] Fraser, B. J., and T. S. Nguyen, *J. Atmos. Solar-Terr.Phys.*, ***63***, 1225, doi:10.1016/S1364-6826(00)00225-X, 2001.

[17] Meredith, N. P., R. M. Thorne, R. B. Horne, et al., *J. Geophys. Res.*, ***108***, 1250, doi:10.1029/2002JA009700, 2003.

[18] Summers, D., and R. M. Thorne, *J. Geophys. Res.*, ***108***, 1143, doi:10.1029/2002JA009489, 2003.

[19] Lorentzen, K. R., M. P. McCarthy, G. K. Parks, et al., *J. Geophys. Res.*, ***105***, 5381, doi:10.1029/1999JA000283, 2000.

[20] Summers, D., and C. Y. Ma, *J. Geophys. Res.*, ***105***, 2625, doi:10.1029/1999JA900444, 2000.

[21] Parrot, M., A. Buzzi, O. Santolık, J. et al., *J. Geophys. Res.*, ***111***, A08301, doi:10.1029/2005JA011583, 2006.

[22] Alexandrov, A., L. Bogdankevich, and A. Rukhadze, *Principles of Plasmas Electrodynamics*,

Springer, Heidelberg, 1984.
[23] Nishikawa, K., N. L. Tsintsadze, and M. Watanabe, *Soviet J. Plasma Phys.*, ***6***, 1302, 1980.
[24] Tsintsadze, N.L., T.D. Kaladze, J. W. Van Dam, et al., *J. Geophys. Res.*, ***115***, A07204, doi: 10.1029/2009JA014555, 2010.
[25] Gurevich, A.V., and A.B. Shvartsburg, *Nonlinear Theory of Radio Wave Propagation in the Ionosphere,* Science, Moscow, 1973.
[26] Tsintsadze, N. L., K. Mima, L. N. Tsintsadze, and K. Nishikawa, *Phys. Plasmas*, ***9***, 4270, doi:10.1063/1.1506311, 2002.
[27] Tsintsadze, N. L., G. Murtaza, and H. A. Shah, *Phys. Plasmas*, ***13*** (1), 012307, doi:10.1063/1.2158694, 2006.

Laboratory Astrophysics Experiments with Magnetically Driven Plasma Jets

F Suzuki-Vidal[1], S V Lebedev[1], A Ciardi[2], S N Bland[1], G N Hall[1], G Swadling[1], A J Harvey-Thompson[1], G Burdiak[1], P de Grouchy[1], J P Chittenden[1], M Bocchi[1], S C Bott[3], and A Frank[4]

[1] The Blackett Laboratory, Imperial College London, Prince Consort Road, SW7 2BW, London, U.K.
[2] Ecole Normale Superieure, Paris, France
[3] Center for Energy Research, University of California, San Diego, CA, U.S.A.
[4] University of Rochester, Department of Physics and Astronomy, Rochester, NY, U.S.A.

E-mail: `f.suzuki@imperial.ac.uk`

Abstract. We present experimental results on the formation of supersonic, radiatively cooled jets driven by the toroidal magnetic field generated by the 1.5 MA, 250 ns current from the MAGPIE generator. The morphology of the jet produced in the experiments is relevant to astrophysical jet scenarios in which the jet on the axis of a magnetic cavity expanding into an ambient medium is collimated by a toroidal magnetic field. The jets in our experiments have similar Mach number, plasma beta and cooling parameter to those in protostellar jets and additionally the Reynolds, magnetic Reynolds and Peclet numbers are much larger than unity, allowing the experiments to be scaled to astrophysical flows. The experimental configuration generates episodic magnetic cavities, suggesting that periodic formation of jets in astrophysical situations could be responsible for some of the variability observed in astrophysical jets. The dynamics of the formation of laboratory jets are presented, together with new results including preliminary measurements of magnetic, kinetic and Poynting energy of the outflows. In addition first estimates of jet temperature and trapped toroidal magnetic field are presented and discussed.

1. Introduction

Collimated outflows (jets) are observed in different types of astrophysical objects, spanning a very broad range of spatial scales [1]. Some examples include bipolar outflows from galaxies, with typical lengths of thousands of light-years, to jets from young stellar objects, with characteristic lengths of hundreds of astronomical units. Astrophysical jets are typically studied with high-resolution observations and computer simulations. Nevertheless there are still many open questions regarding their dynamics (e.g. the jet-launching mechanism close to the ejection source), and the mechanisms which generate and sustain the high degree of collimation in the outflows far from the launching point.

Laboratory astrophysics, the study of astrophysical phenomena by the means of carefully designed scaled experiments, has seen major advances in the past years due to the development of High-Energy Density Physics (HEDP) facilities [2]. One of the applications of this new research area is the study of astrophysical jets from young stars. Astrophysical jets can be appropriately scaled to laboratory conditions if their dimensionless parameters are sufficiently

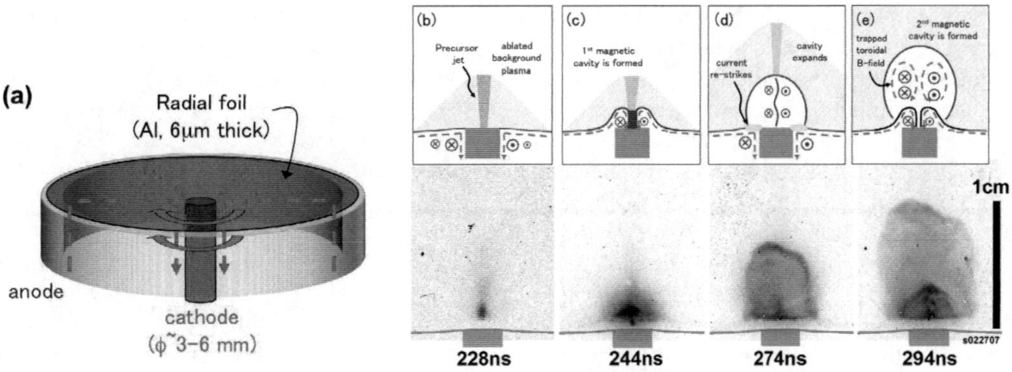

Figure 1. (a) Schematic view of the radial foil setup. (b)-(d) Dynamics of the formation of episodic magnetic cavities from a radial foil, shown schematically (top panel) and from XUV self-emission from the plasma (bottom panel). In both figures the red arrows represent the current path and the blue arrows the toroidal magnetic field.

similar to those in the laboratory system (e.g. Mach, Reynolds, magnetic Reynolds and Peclet numbers [3, 4]).

Astrophysical jet experiments typically use high-power lasers and high-current generators, with both approaches allowing for the generation of supersonic, radiatively cooled plasma jets with similar dimensionless parameters to those from young stars (e.g. [5, 6]). High-current pulsed-power generators, however, allow for the formation of plasma jets in which magnetic fields are dynamically significant in the formation and subsequent collimation of the jet [7, 8]. These experiments are relevant to the study of astrophysical jet launching scenarios, as models propose magnetic fields to be the mechanism behind the formation of jets from young stars. Such models rely on the twisting of an initial poloidal magnetic field due to differential rotation between a star surrounded by an accretion disk. After many rotations, the topology of the magnetic field evolves into that of a predominant toroidal (azimuthal) field. A magnetic cavity is formed which is confined by the ambient medium, confining ejected material as a jet on its axis [9, 10]. Our experiments are characterized by the presence of a predominant toroidal magnetic field which is responsible for the formation of a magnetically driven jet. This configuration makes our experiments relevant to astrophysical jet models that rely on such a topology for driving collimated outflows.

2. Experimental setup

The experiments were conducted in the MAGPIE generator [11], which produces a current pulse of 1.5 MA in 250 ns. The current is introduced into a radial foil [12, 13] (Fig. 1a), an aluminium disc with thicknesses between 6-15 μm. The foil is placed between two concentric electrodes as the load of MAGPIE, with the central electrode having a selection of diameters between $\sim$3-6 mm. The current path along the electrodes and the foil generates a initial toroidal magnetic field, which can reach magnitudes of $B_\phi \sim 100$ T at the radius of the cathode.

The overall dynamics of the radial foil are shown both schematically and from time-resolved XUV emission images, obtained from the same experiment, in Figs. 1b-d. As the current increases, the foil is Ohmically heated and converted into plasma, a process which lasts for the first $\sim$100 ns. The plasma is ablated from the foil following the direction of the $J \times B$ force, which points in a direction normal to the surface of the foil. This forms a low-density background plasma above the foil (Fig. 1b). The Lorentz force is strongest towards the radius of the cathode, and thus a fastest rate of ablation in this region leads to the formation of a radial gap due to the complete ablation of foil material. Plasma is then driven upward by the magnetic pressure, beginning to form a magnetic cavity, with the toroidal magnetic field now moving to a new position above the initial surface of the foil (Fig. 1c). The cavity expands into the ambient plasma due to the magnetic pressure inside it, and the toroidal magnetic field inside the cavity confines a plasma jet on the axis (Fig. 1d).

The formation of the first magnetic cavity is similar to that observed in radial wire array experiments [7, 8, 14]. From a radial foil, however, it is possible to produce subsequent magnetic cavities during the same experiment. This process occurs as the current re-strikes at the base of the first magnetic cavity. This is possible due to the larger mass available in a foil compared to a radial wire array. The generation of multiple cavities is shown schematically in Figs. 1d-e. This process is repeated throughout the duration of the current pulse, typically producing 3-4 magnetically driven jets per experiment.

3. Magnetic and Poynting energy in a magnetic cavity

The formation of a new current inside the magnetic cavities leads to the generation of an additional inductance in the system. This additional inductance was measured using an inductive probe. This diagnostic consisted on an inductance connected to the central electrode, close to the point where the magnetic cavities are formed. The voltage across the radial gap at the base of the cavities was measured (V), which is proportional to the current inside the cavity (I) and the additional, time-dependant inductance due to the new current path (L_{bubb}). Fig. 2a shows the measured voltage as a series of episodic spikes, which are correlated with the formation of the subsequent magnetic cavities in the experiment. From these results it is possible to estimate the magnetic energy ($E_M{=}L_{bubb}I^2/2$) and the Poynting energy, i.e. the total energy delivered to the cavity by the Poynting flux through the radial gap ($E_P{=}\int VIdt$). Fig. 2b shows these estimates, resulting in $E_M{\sim}$100-300 J and $E_P{\sim}$200-600 J.

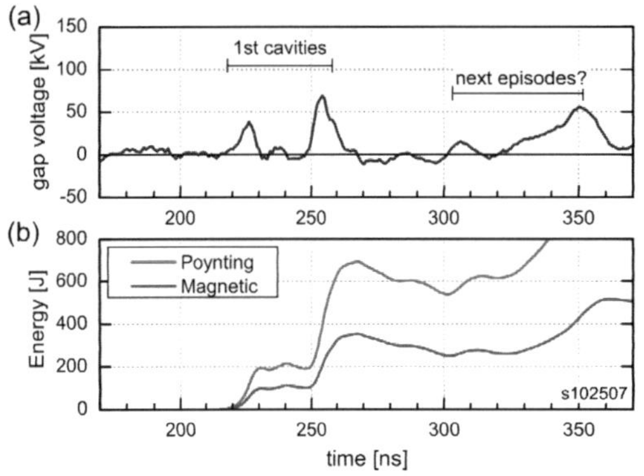

Figure 2. (a) Measured voltage across the radial gap of the episodic magnetic cavities. (b) Measured Poynting and magnetic energy.

4. Kinetic energy estimate

By using optical laser interferometry (λ=532 nm, 0.2 ns exposure) it is possible to obtain 2-D maps of electron density integrated along the probing path ($\int n_e dL{\approx}n_e L$). An example image is shown in Fig. 3. Radial line outs at different heights were obtained from this image, and electron densities were integrated in order to estimate the total number of electrons inside the outflow. The tip velocity of the cavity was measured from time-resolved XUV emission imaging, resulting in $V_Z \sim 130$ km/s. By assuming a level of ionization in the plasma of $Z=5$, obtained from the measured electron temperatures in Sec. 5, the mass inside the cavity m was estimated. The kinetic energy of the outflow was then calculated as $E_K = mV_Z{}^2/2 \sim 100$ J. This estimate is of the right magnitude when compared to the energy delivered to the cavity as Poynting flux, shown in the previous section. Future experiments will allow to measure the distribution of velocities along the jet and thus obtain a more accurate measurement of the kinetic energy inside the cavities.

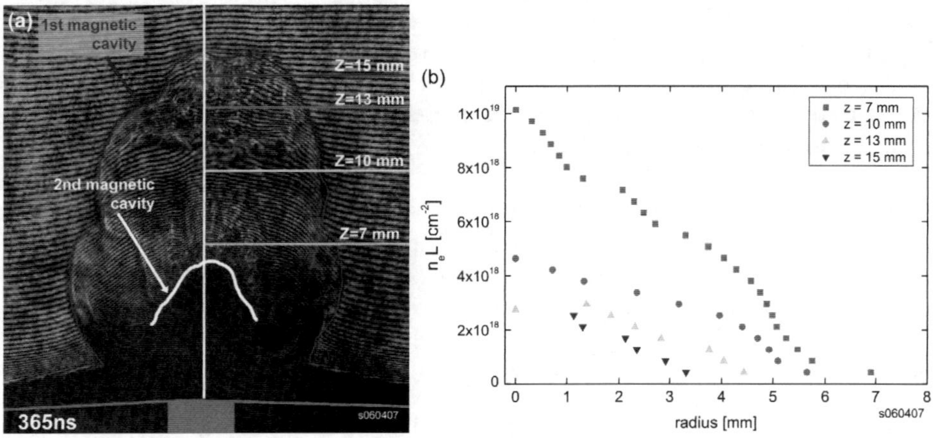

Figure 3. (a) Optical laser interferometry image at 365 ns, showing the formation of two subsequent magnetic cavities. The horizontal lines show the positions for the measured contours of electron density integrated along the probing beam, shown in (b).

5. Trapped toroidal magnetic flux

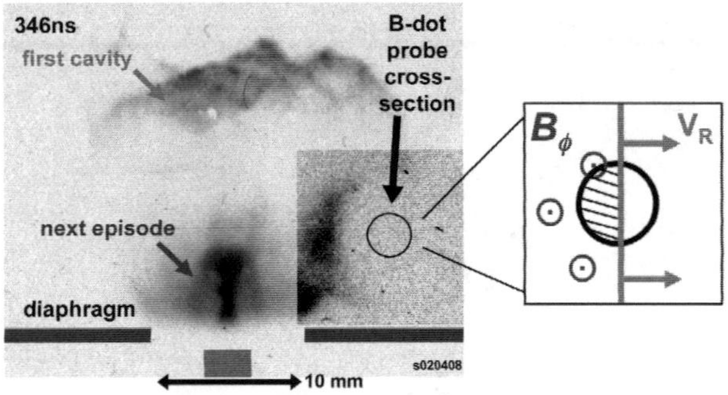

Figure 4. Measurement of toroidal magnetic field inside the expanding magnetic cavities.

The dynamics of the episodic magnetic cavities in radial foils indicate that a toroidal magnetic field is responsible for their formation, as in previous experiments with radial wire arrays ??. The generation of subsequent jet episodes in radial foils could imply that some toroidal magnetic flux will remain in the first cavity at the time when the secondary outflow is formed. The presence of toroidal magnetic field inside the expanding magnetic cavities was measured with a magnetic probe, shown in Fig. 4. The probe was positioned above the foil at a radius of ~13 mm from the axis, and a height of ~10 mm from its surface. A metal diaphragm was placed above the foil in order to shield the probe electrostatically. The probe was positioned in order to measure a toroidal magnetic field. The response of the probe is proportional to the change in the magnetic flux as the cavity reaches the cross-section area of the probe. This changing flux is proportional to the radial expansion velocity of the side-wall of the magnetic cavity V_R. The interaction from the first magnetic cavity with the probe is shown in Fig. 4. The measured radial expansion velocity of the cavity of V_R~90 km/s indicates that the signal from the probe corresponds to a magnetic field of B_ϕ~0.3 T. This estimate can be compared with the expected magnetic field assuming that the initial toroidal magnetic flux in the first cavity is conserved as the cavity expands and the central jet is pinched. The resultant magnetic field is of the right magnitude, though an accurate estimate of the expected field is not possible due to the uncertainty of the current path through the plasma at the time when the cavity reaches the magnetic probe.

6. Temperature of the central jet

The temperature of the central jet was estimated using time-integrated X-ray spectroscopy (photon energies of $h\nu$>1 keV). The spectrum was measured using a spherically-bent mica crystal, which provided spatial resolution along the height of the jet. The electron temperature of the jet was estimated by considering the ratio of the combined intensity of Helium-like aluminium lines. The dependance of these line ratios on temperature has been modelled by [15], assuming similar plasma conditions as those encountered in the jet. Fig. 5 shows the measured temperatures as a function of height along the jet. Electron temperatures of the order of T_e~170-600 eV were measured, and thus the magnetic Reynolds number was estimated. Assuming a value for the ionization of Z=5 (obtained from ionization balance for a temperature of T_e=100 eV), a characteristic spatial scale of the jet of L~1 mm, and a typical flow velocity of V~100 km/s, we calculate a value of Re_M~1000.

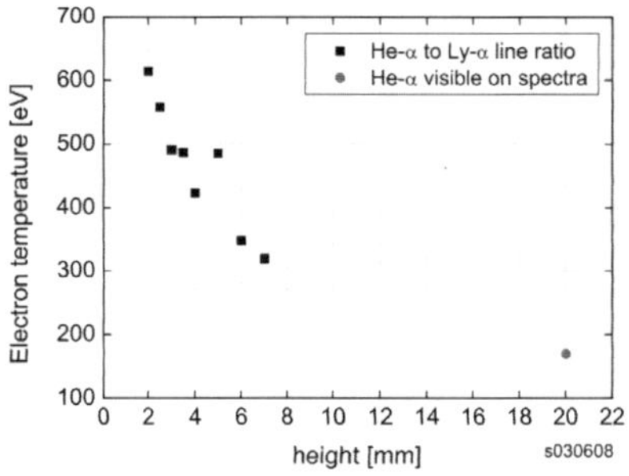

Figure 5. Electron temperature of the jet as a function of height, obtained from time-integrated X-ray spectroscopy.

7. Summary

We have presented new experimental data from radial foil experiments which complement previous results [12] and enhance our understanding on the dynamics of episodic magnetically driven jets. Preliminary measurements of the kinetic energy (E_K~100 J) and magnetic energy (E_M~300 J) inside the magnetic cavities are in good agreement with the total electromagnetic energy injected as Poynting flux (E_P∼600 J), measured with an inductive probe connected at the base of the cavity. The electron temperature of the jet/outflow, determined by time-integrated spectroscopy, resulted in T_e∼170-600 eV, providing a first estimate of this parameter. These high temperatures are consistent with measurements of toroidal magnetic field inside the cavity, indicating that magnetic flux is conserved as the cavity expands and that the plasma has a magnetic Reynolds number larger than unity. Our results indicate values of the order of Re_M∼1000, consistent with previous results from computer simulations [13] thus validating the scaling of these experiments to astrophysical jets.

8. Acknowledgments

This work was supported by the EPSRC Grant No. EP/G001324/1, by the NNSA under DOE Cooperative Agreements No. DE-F03-02NA00057 and No. DE-SC-0001063 and by a Marie Curie RTN fellowship, as part of the JETSET (Jet Simulations, Experiments and Theories) project.

References

[1] Livio M 2002 *Nature* **417** 125–+
[2] Remington B A, Drake R P and Ryutov D D 2006 *Reviews of Modern Physics* **78** 755–807
[3] Ryutov D, Drake R P, Kane J, Liang E, Remington B A and Wood-Vasey W M 1999 *ApJ* **518** 821–832
[4] Ryutov D D, Drake R P and Remington B A 2000 *ApJS* **127** 465–468
[5] Shigemori K, Kodama R, Farley D R, Koase T, Estabrook K G, Remington B A, Ryutov D D, Ochi Y, Azechi H, Stone J and Turner N 2000 *PRE* **62** 8838–8841
[6] Lebedev S V, Chittenden J P, Beg F N, Bland S N, Ciardi A, Ampleford D, Hughes S, Haines M G, Frank A, Blackman E G and Gardiner T 2002 *Astrophysical Journal* **564** 113–119
[7] Lebedev S V, Ciardi A, Ampleford D J, Bland S N, Bott S C, Chittenden J P, Hall G N, Rapley J, Jennings C A, Frank A, Blackman E G and Lery T 2005 *Monthly Notices of the Royal Astronomical Society* **361** 97–108 (*Preprint* arXiv:astro-ph/0505027)
[8] Ciardi A, Lebedev S V, Frank A, Blackman E G, Chittenden J P, Jennings C J, Ampleford D J, Bland S N, Bott S C, Rapley J, Hall G N, Suzuki-Vidal F A, Marocchino A, Lery T and Stehle C 2007 *Physics of Plasmas* **14** 056501–+
[9] Lynden-Bell D 2003 *Monthly Notices of the Royal Astronomical Society* **341** 1360–1372 (*Preprint* arXiv:astro-ph/0208388)
[10] Uzdensky D A and MacFadyen A I 2006 *ApJ* **647** 1192–1212 (*Preprint* arXiv:astro-ph/0602419)
[11] Mitchell I H, Bayley J M, Chittenden J P, Worley J F, Dangor A E, Haines M G and Choi P 1996 *Review of Scientific Instruments* **67** 1533–1541
[12] Suzuki-Vidal F, Lebedev S V, Ciardi A, Bland S N, Chittenden J P, Hall G N, Harvey-Thompson A, Marocchino A, Ning C, Stehle C, Frank A, Blackman E G, Bott S C and Ray T 2009 *Astrophysics and Space Science* **322** 19–23 (*Preprint* 0904.0165)
[13] Ciardi A, Lebedev S V, Frank A, Suzuki-Vidal F, Hall G N, Bland S N, Harvey-Thompson A, Blackman E G and Camenzind M 2009 *ApJL* **691** L147–L150 (*Preprint* 0811.2736)
[14] Suzuki-Vidal F, Lebedev S V, Bland S N, Hall G N, Harvey-Thompson A J, Chittenden J P, Marocchino A, Bott S C, Palmer J B A and Ciardi A 2010 *Plasma Science, IEEE Transactions on* **38** 581–588
[15] Apruzese J P, Whitney K G, Davis J and Kepple P C 1997 *J. Quant. Spectrosc. Radiat. Transfer* **57** 41–61

3D PiC code investigations of Auroral Kilometric Radiation mechanisms

K.M. Gillespie[1], S.L. McConville[1], D.C. Speirs[1] , K. Ronald[1], , A.D.R. Phelps[1], R. Bingham[1,2], A.W. Cross[1], C.W. Robertson[1], C.G. Whyte[1], W. He[1], I. Vorgul[3], R.A. Cairns[3] and B.J. Kellett[2]

[1]SUPA Department of Physics, John Anderson Building, 107 Rottenrow, University of Strathclyde, Glasgow, G4 0NG, Scotland

[2] Space Science and Technology Department, STFC Rutherford Appleton Laboratory, Didcot, OX11 0QX, England

[3] School of Mathematics and Statistics, University of St Andrews, St Andrews, KY16 9SS, Scotland

Email - karen.gillespie@strath.ac.uk

Abstract. Efficient (~1%) electron cyclotron radio emissions are known to originate in the X mode from regions of locally depleted plasma in the Earths polar magnetosphere. These emissions are commonly referred to as the Auroral Kilometric Radiation (AKR). AKR occurs naturally in these polar regions where electrons are accelerated by electric fields into the increasing planetary magnetic dipole. Here conservation of the magnetic moment converts axial to rotational momentum forming a horseshoe distribution in velocity phase space. This distribution is unstable to cyclotron emission with radiation emitted in the X-mode. Initial studies were conducted in the form of 2D PiC code simulations [1] and a scaled laboratory experiment that was constructed to reproduce the mechanism of AKR. As studies progressed, 3D PiC code simulations were conducted to enable complete investigation of the complex interaction dimensions. A maximum efficiency of 1.25% is predicted from these simulations in the same mode and frequency as measured in the experiment. This is also consistent with geophysical observations and the predictions of theory.

1. Introduction

Electrons precipitating into the Earth's magnetosphere, Figure 1, are subject to increasing magnetic field with decreasing altitude. In the absence of collisions and given that the field increases slowly compared to the electron oscillation period, the adiabatic conservation of the magnetic moment comes into effect. Electrons entering the Earth's magnetic dipole structure have a spread in their initial velocity. Those electrons having a small initial component of velocity perpendicular to the magnetic flux lines experience an increase in their rotational component of velocity as they descend towards the atmosphere. The effect of this process is that an initially primarily rectilinear electron beam assumes a horseshoe formation in electron velocity space with a significant number of electrons having high pitch angles $\theta = \arctan(v_{\perp/}v_{//})$ and a region with a positive gradient in number density vs perpendicular velocity, $dn/dv_{\perp,}$ Figure 2. Such horseshoe electron distributions have been measured in the AKR source regions within the polar magnetosphere [2-3].

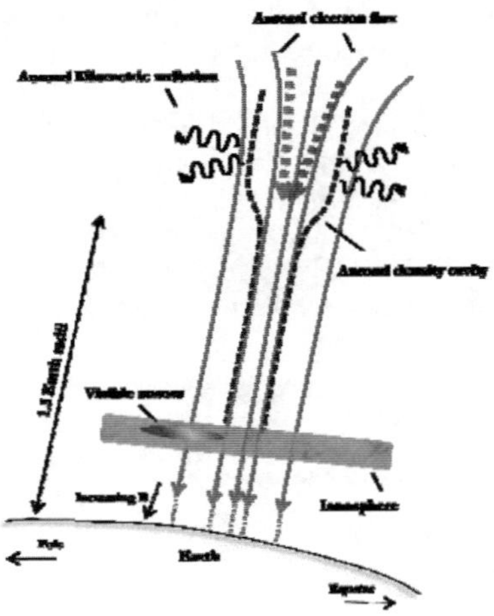

Figure 1: Representation of the AKR source region and terrestrial auroral process.

Part of the electron stream is 'mirrored', reversing its direction of motion along the magnetic field lines once all of its kinetic energy becomes associated with rotational motion providing a second energetic electron population in the auroral magnetosphere, the space bound component. Due to the correlation between the electron cyclotron frequency and the radio frequency, it has been hypothesized for some time that the emissions are due to a cyclotron instability [4-7] in one or both of these populations [8-13].

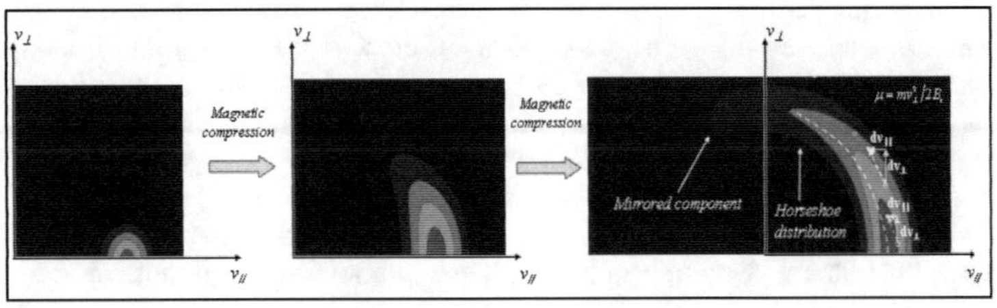

Figure 2 : Formation of the horseshoe in the electron velocity distribution

Over the last few decades, as technology has advanced satellites have been able to obtain quantitative data on the phenomenon of Auroral Kilometric Radiation [14,15]. The magnetospheric measurements have revealed that the radiation emissions are sourced in regions of plasma depletion ($\omega_{pe} << \omega_{ce}$) at an altitude of 1.5-3 Earth radii. The frequency observed at any given altitude extends down to the local electron cyclotron frequency with the strongest emission at around 300kHz. The wave propagation and polarisation directions suggest that the radiation is generated in the X-mode

Journal of Physics: Conference Series **511** (2014) 012051 doi:10.1088/1742-6596/511/1/012051

[16] with a peak power ~10^9 W, corresponding to an estimated radiation efficiency ~1% of the auroral electron precipitation energy [2,3]. The availability of accurate data has resulted in a significant research effort to explain the unusually high efficiency of this natural process. This enhanced data allows for meaningful comparison with theoretical expectations and has resulted in a number of models for the emission mechanisms.

As the equations describing this instability scale with the cyclotron frequency, it became clear that it would be possible for the concept to be tested in a laboratory experiment by increasing the magnetic flux density and scaling the resonance into the microwave regime. In conjunction with the construction of an experimental reproduction of the AKR source region, 2D PiC simulations were conducted. Those simulations have demonstrated the formation of horseshoe distribution in an electron-beam subject to significant magnetic compression and subsequent cyclotron maser emission within an interaction waveguide. [1,17] Although the experimental results [18] were in good agreement with the findings from the 2D PiC simulations, certain aspects of the experimental behavior results require 3D computational approaches for reproduction. Specifically 2D simulations cannot account for an azimuthal index in the modes propagating in the waveguide, aspects of the experimental behavior associated with such modes require 3D simulations. This enhanced numerical simulation work will be used to redesign the laboratory experiment in the future. This paper presents most recent results from simulations conducted in three dimensions.

2. 3D PiC CODE SIMULATIONS

2.1. Simulation objective

2D PiC code simulations provide high execution speed and strong numerical stability. However, 2D simulations although in good agreement with the experiment, can only allow for generation of axisymmetric modes, which means that modes with azimuthal structure cannot be predicted. In conducting 3D simulations a more complete interaction regime may be analysed [19]. The 2D simulations were conducted for electron beam energies of 75-85keV, magnetic compression factors of up to 30 and electron cyclotron frequencies of 4.42GHz and 11.7GHz. At 11.7GHz, beam-wave coupling was observed with the $TE_{0,3}$ mode and an RF output power of 20kW was predicted corresponding to an RF conversion efficiency of 1.3%. At 4.42GHz, excitation of the $TE_{0,1}$ mode was observed with an RF output power of 35kW for a cyclotron-wave detuning of 2%. This corresponds to an RF conversion efficiency of 2.6%. The RF conversion efficiencies obtained are therefore comparable with estimates for the AKR generation efficiency [14]. The 3D simulation geometry, Figure 3, illustrates the main interaction waveguide with a predefined electron horseshoe distribution injected. The gun geometry is not modeled due to the computational requirements of 3D simulations. A uniform axial magnetic field profile was defined from 0-20cm with a linear decrease to 0T from 20-80cm as an initial condition for the PiC code. Recent 3D simulations have been used to study the excitation of modes in the two resonance regimes, which provides data to compare with experimental measurements and illustrate mode competition that 2D simulations were not able to account for.

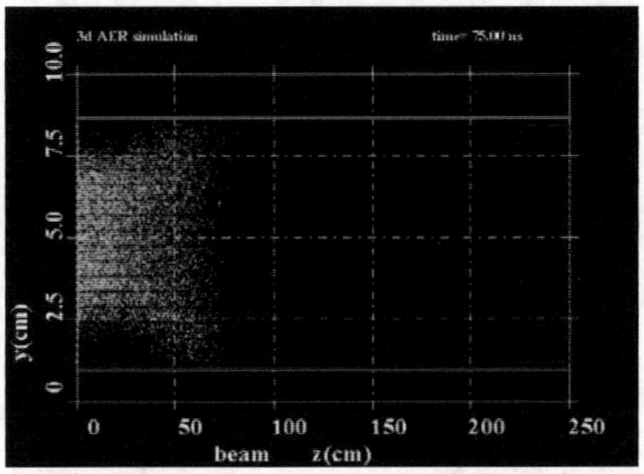

Figure 3: Simulation geometry illustrating waveguide with electron beam PiC particle trajectory, and pre-defined horseshoe distribution.

2.2. Mode analysis

3D simulations plot transverse electric field patterns within the waveguide interaction region. Figure 4(a) illustrates the results of the transverse spatial Fourier analysis output from 3D KARAT for the 11.7GHz resonance, illustrating a $TE_{2,3}$ mode pattern. It is very clear that the simulation also has predicted a strong coupling with the $TE_{0,3}$ mode, Figure 4(b) (a transverse spatial Fourier analysis) and Figure 5(a), along with a few other weaker couplings. These modes were observed in the experimental measurements for this resonance regime. Further analysis of the simulations have been undertaken to study how different factors such as the beam current and the cyclotron detuning affects the modes being excited within the interaction region, and the resultant output power. The detuning is measured as a fractional shift of the cyclotron frequency from the cut-off of the $TE_{0,3}$ mode.

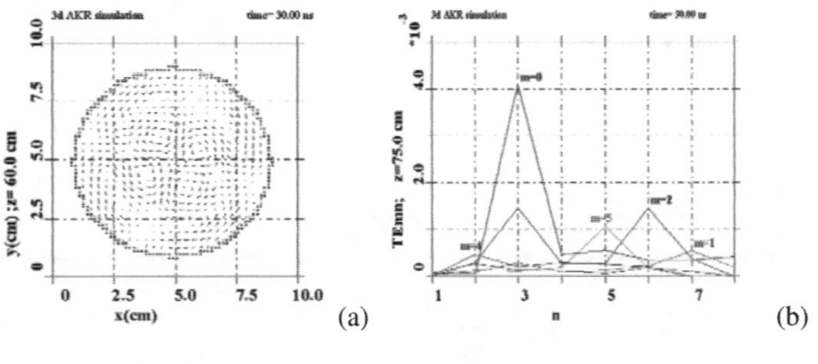

Figure 4: (a) 3D vector predictions of the transverse electric field patterns within the waveguide interaction region, the structure of a $TE_{2,3}$ mode is clearly seen at 11.7GHz. (b) illustrates the results of the transverse spatial Fourier analysis output predicted by KARAT 3D for the 11.7GHz resonance.

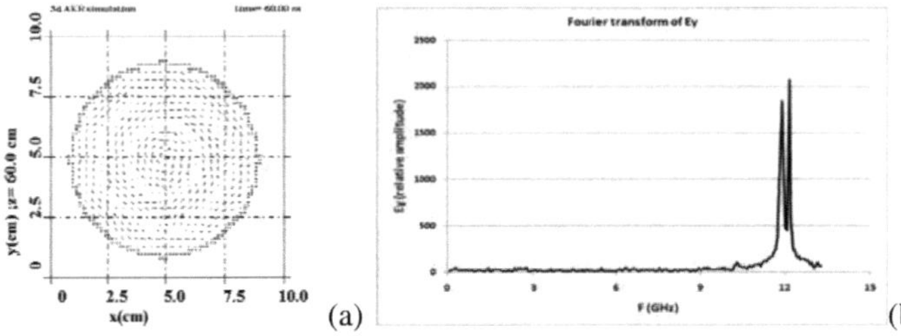

(a) (b)

Figure 5: (a) Illustrates excitation of $TE_{0,3}$ mode. (b) shows a peak at ~11.7GHz along with another at ~12.2GHz, this is illustrative of multiple modes being excited and mode competition.

2.3. Mode competition and Power output

A clear trend was observed throughout the plots, when the beam current becomes lower the average power output decreased. The range of predicted output powers were comparable with the findings from 2D simulations & experimental measurements. The RF output power increased with detuning below resonance but decreased with detuning above resonance. When the detuning was set below resonance the peak power output was high, and the recurring dominant mode was the $TE_{2,3}$. When the detuning increased above resonance the recurring mode excited was the $TE_{0,3}$ whilst the output power decreased. It was noticed that mode competition was more severe as the detuning increased above resonance. Figure 5(b) shows a clear peak at ~11.7GHz, in agreement with the experimental results, along with another at ~12.2GHz, this is illustrative of multiple modes being excited and mode competition. For a beam current of 37A, the predicted RF power output was in the range 40-180kW for a detuning of -3 to 3%. This corresponds to an RF conversion efficiency of ~1-6%, which is comparable with the 2D simulations and experiments. Figure 6(a) illustrates a Poynting flux integral predicted by simulations of the 11.7GHz resonance regime, showing output power rising to 25kW, with a beam current of 13A.

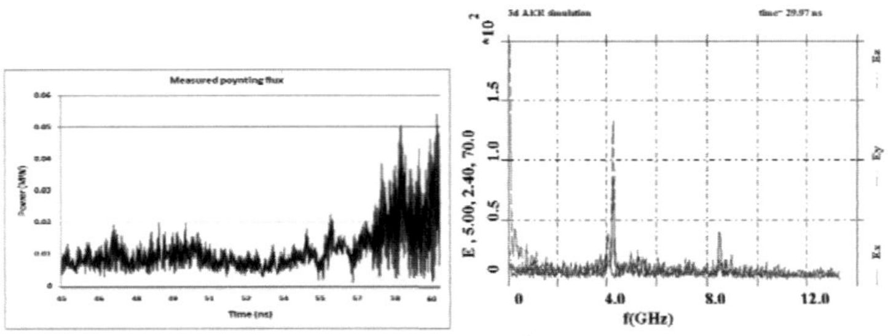

(a) (b)

Figure 6: (a) illustrates a Poynting flux integral for the 11.7GHz resonance regime (b) illustrates some of the initial results taken from the simulations, at 4.42GHz resonance.

These results are comparable with that the experimental measurements, [20-22] and also with the auroral source region, where the radiation spectra has shown detailed structure. Simulations at the resonance at 4.42GHz, are ongoing to study it dynamics in three dimensions. From initial results, the 3D simulations are also proving successful in simulating this resonance regime. Figure 6(b) illustrates some of the initial results taken from the simulations, cleary showing the spectral components measured in experimental data, a large peak at 4.42GHz and a small second harmonic at 8.5GHz.

3. CONCLUSION AND DISCUSSIONS

In summary, it has been shown that the 3D PiC codes are useful in simulating the interaction between a complex electron beam and electromagnetic radiation. These new simulations have proven accurate in identifying the modes and the mode competition that existed in experiment. The peak powers also agree well with those predicted in the experiment and previous simulations.

This enhanced picture of the interaction region will prove valuable for further studies to improve the understanding of AKR and thus the general field of instability and waves in plasmas [23]. Research will continue with simulations being conducted to refine these results investigating the sensitivity to the exact form of the electron distribution function. Lastly these results will aid in design modifications to the apparatus to experimentally measure the sensitivity to detuning in the laboratory experiment for comparison with theoretical predictions.

References

[1] D.C. Speirs et al, *Plasma Phys. Contol. Fusion* **50** (2008)
[2] R.E. Ergun et al, *Astrophyical Journal* **538** pp456-466 (2000
[3] G.T. Delory et al, *Geophys. Res. Lett,* **25** pp2069-2072 (1998)
[4] R.Q. Twiss, *Aus. J. Phys.* **11** pp564-579 (1958)
[5] A.V. Gaponov 1959 *Izv. VUZ. Radiofizika* **2** pp450-462 and pp837
[6] K.R. Chu, *Rev. Mod. Phys.* **76** pp489-540 (2004)
[7] P. Sprangle & A.T. Drobot, *IEEE Trans. on Microwave Theory and Technique* **MTT-25**
[8] J.S. Wagner et al, *Radio Science* **19** pp509-518 (1984)
[9] C.S Wu & L.C. Lee, *The Astrophysical Journal* **230** pp621-626 (1979)
[10] R. Bingham et al, *Contrib. Plasma Phys.* **44** pp382-387 (2004)
[11] R. Bingham et al, *Physica Scripta* **T98** pp160-162 (2002)
[12] R. Bingham et al, *Phys. Plasmas* **7** pp3089-3092 (2000)
[13] I. Vorgul et al, *Phys. Plasmas* **12** Art. 122903 (2005)
[14] D.A. Gurnett 1974 J. Geophys. Research: Space Physics **79** pp4227-4238
[15] D.B. Melrose et al, *J. Geophys. Research: Space Physics* **87** pp5140-5150 (1982)
[16] R.F. Benson *J. Geophys. Res- Space Physics* **90** pp2753-2784 (1985)
[17] D.C. Speirs et al, *J. Plasma Phys.* **71** pp665-674 (2005)
[18] S.L. McConville et al, *Plasma Phys. Control. Fusion* **50** (2008)
[19] K.M. Gillespie et al, *Plasma Physics & Controlled Fusion* **50**, 124038 (2008)
[20] K. Ronald et al, *Plasma Sources, Science and Technology* **17** (2008)
[21] K. Ronald, *Phys. Plamas.* **15** Art. 056503 (2008)
[22] K. Ronald et al, *IEEE Trans. on Plasma Science* **PS-26** pp375-382 (1998)
[23] J.E. Allen & A.D.R. Phelps, *Rep. Prog. Phys.*, **40**, pp1305-1368 (1977)

Numerical simulation of unconstrained cyclotron resonant maser emission

D. C. Speirs[1], K. M. Gillespie[1], K. Ronald[1], S. L. McConville[1], A. D. R. Phelps[1], A. W. Cross[1], R. Bingham[1,3], B. J. Kellett[3], R. A. Cairns[2] and I. Vorgul[2]

[1] SUPA, Department of Physics, University of Strathclyde, Glasgow, G4 0NG, UK
[2] School of Mathematics and Statistics, University of St Andrews, Fife, KY16 9SS, UK
[3] STFC Rutherford Appleton Laboratory, Chilton, Didcot, Oxfordshire, OX11 0QX, UK

Email: david.c.speirs@strath.ac.uk

Abstract. When a mainly rectilinear electron beam is subject to significant magnetic compression, conservation of magnetic moment results in the formation of a horseshoe shaped velocity distribution. It has been shown that such a distribution is unstable to cyclotron emission and may be responsible for the generation of Auroral Kilometric Radiation (AKR) - an intense rf emission sourced at high altitudes in the terrestrial auroral magnetosphere. PiC code simulations have been undertaken to investigate the dynamics of the cyclotron emission process in the absence of cavity boundaries with particular consideration of the spatial growth rate, spectral output and rf conversion efficiency. Computations reveal that a well-defined cyclotron emission process occurs albeit with a low spatial growth rate compared to waveguide bounded simulations. The rf output is near perpendicular to the electron beam with a slight backward-wave character reflected in the spectral output with a well defined peak at 2.68GHz, just below the relativistic electron cyclotron frequency. The corresponding rf conversion efficiency of 1.1% is comparable to waveguide bounded simulations and consistent with the predictions of kinetic theory that suggest efficient, spectrally well defined radiation emission can be obtained from an electron horseshoe distribution in the absence of radiation boundaries.

1. Introduction

Astrophysical radio emissions in association with plasmas having non-uniform magnetic fields have been the subject of particular interest and debate over the last thirty years [1]. Numerous such sources including planetary and stellar auroral radio emissions are spectrally well defined and exhibit a high degree of extraordinary (X-mode) polarization [2,3] with frequencies of emission extending down to the local relativistic electron cyclotron frequency. In particular, for the terrestrial auroral case it is now widely accepted that such emissions are generated by an electron cyclotron-maser instability driven by a component of the precipitating auroral electron flux having a horseshoe shaped velocity distribution [2]. Such a distribution can form in the precipitating electron stream through conservation of magnetic moment in the convergent magnetic field of the polar magnetosphere and theory has shown that such distributions are unstable to cyclotron emission in the X-mode [4-9].

Figure 1 shows a diagrammatic representation of the formation of a horseshoe distribution in electron velocity space as a result of magnetic compression and the associated evolution in the transverse velocity profile. The velocity space is cut off at $v_{//} = 0$ and does not show the magnetically mirrored component. If one considers the electron beam distribution in figure 1a with finite pitch factor and energy spread impingent on an increasing axial magnetic flux density, conservation of magnetic moment μ results in the conversion of axial velocity into perpendicular velocity. As $v_\perp \propto \sqrt{B_0}$, the resultant expansion in velocity space yields a half-horseshoe like profile (figure 1b) with an increasing number of electrons present at high pitch factors across a positive gradient in the transverse velocity profile ($\partial f_e/\partial v_\perp > 0$). With reference to the analytical theory [4-9], electrons across this positive gradient near the tip of the distribution may undergo resonant energy transfer with an electromagnetic wave and as a result this portion of the distribution will appear to spread in $v_\perp$ and predominantly lose energy.

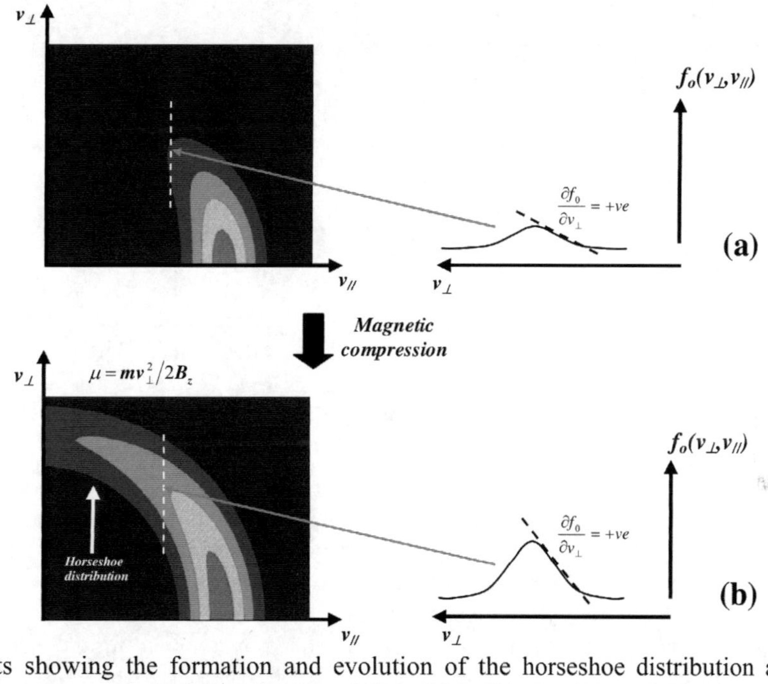

Figure 1. Plots showing the formation and evolution of the horseshoe distribution and transverse velocity profile as the magnetic field is increased from (a) to (b).

2. Numerical Simulations

Previous work at Strathclyde concerned the numerical simulation and laboratory study of the electrodynamics of an electron beam subject to significant magnetic compression within a bounding waveguide structure [10-14]. Cyclotron resonant energy transfer was observed with electromagnetic modes of the waveguide at microwave frequencies [15,16], with RF conversion efficiencies of the order of 1-2%. Here we present the results from PiC (Particle-in-cell) code simulations of the dynamics of the cyclotron emission process in the absence of cavity / waveguide boundaries. Particular consideration was given to the convective growth rate, output spectra and RF conversion efficiencies. In all cases the 2D axisymmetric version of the finite-difference time domain PiC (Particle-in-cell) code KARAT was used. The 2D axisymmetric model allows the distribution of particles in the transverse plane of motion to be monitored via the particle population density in plots of v_ϑ vs v_r. This facilitates the diagnosis of cyclotron-resonant bunching effects in relative electron orbital phase. For the purpose of simulating the unbounded interaction geometry, a 44cm radius region with radially increasing conductivity was defined around the beam propagation path. This represented an idealised absorber of electromagnetic radiation, inhibiting reflections and hence the formation of boundary resonant eigenmodes. An electron beam was injected into this simulation geometry with a predefined horseshoe distribution, comprising a pitch spread $\alpha = v_\perp / v_z$ of $0 \rightarrow 9.5$, beam energy of 20keV±5% and beam current of 14A. Other simulation parameters included a uniform axial magnetic flux density of 0.1T, grid resolution of 0.25cm, PiC particle merging factor of 3×10^6 electrons / PiC particle and a total simulation length (beam propagation path) of 4m.

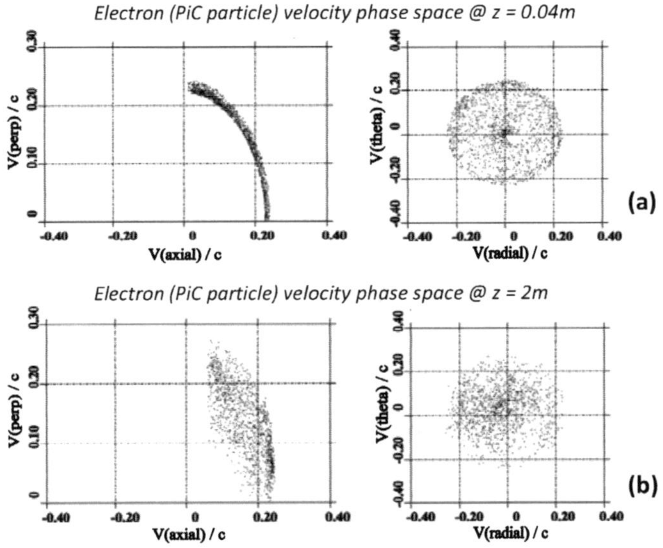

Figure 2. PiC particle velocity distributions measured on transverse planes within the simulation geometry at (a) z = 0.06m and (b) z = 2m

Figure 2 shows the PiC particle velocity distributions at two axial positions within the unbounded simulation geometry after a 200ns run. The injected beam distribution at z = 0.04m contains a well defined pitch spread in $v_\perp$ vs v_z and no evidence of cyclotron bunching effects in the associated v_ϑ vs v_r plot. At an axial position of z = 2m however the picture is very different, with marked smearing in the transverse velocity profile of $v_\perp$ vs v_z over the full pitch range and evidence of electron bunching / phase-trapping in the corresponding v_ϑ vs v_r plot, characterised by a non uniform concentration of PiC particles extending to the origin. Both effects serve as clear evidence for beam-wave energy transfer and the action of a cyclotron maser instability.

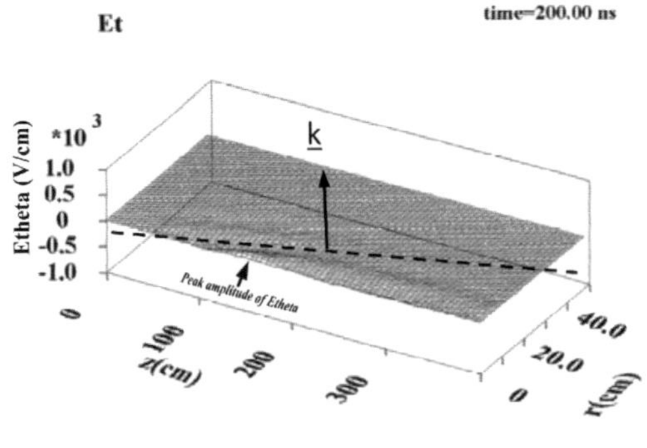

Figure 3. 3D contour plot of E_ϑ within the unbounded simulation geometry.

Figure 3 contains a 3D contour plot of E_ϑ over the simulation geometry after a 200ns run. An electromagnetic wave sourced at z ~ 1.45m is evident propagating near perpendicular to the electron beam with a small backward wave component of the wavevector. A dashed line has been superimposed parallel to the primary wavefront with an arrow indicating the corresponding orientation of the wavevector. A plot of the corresponding radial Poynting flux is presented in figure 4a measured in a plane at r = 3.5cm, over the 4m length of the simulation geometry. A DC offset is present in the measurement due to low frequency electromagnetic field components associated with the electron beam propagation. The rf output power was therefore obtained from the amplitude of the AC signal superimposed on this DC offset. A peak rf output power of ~3kW after 180ns was measured corresponding to an rf conversion efficiency of 1.1%. This is comparable to the ~1% efficiency obtained from waveguide bounded simulations [10-14] and consistent with the estimate of ~1% for the astrophysical phenomena [17,18]. The corresponding output spectrum is presented in figure 4b, showing a well defined signal at 2.68GHz. This represents a 1.1% downshift from the relativistic electron cyclotron frequency of 2.71GHz, consistent with the mild backward wave character observed in figure 3.

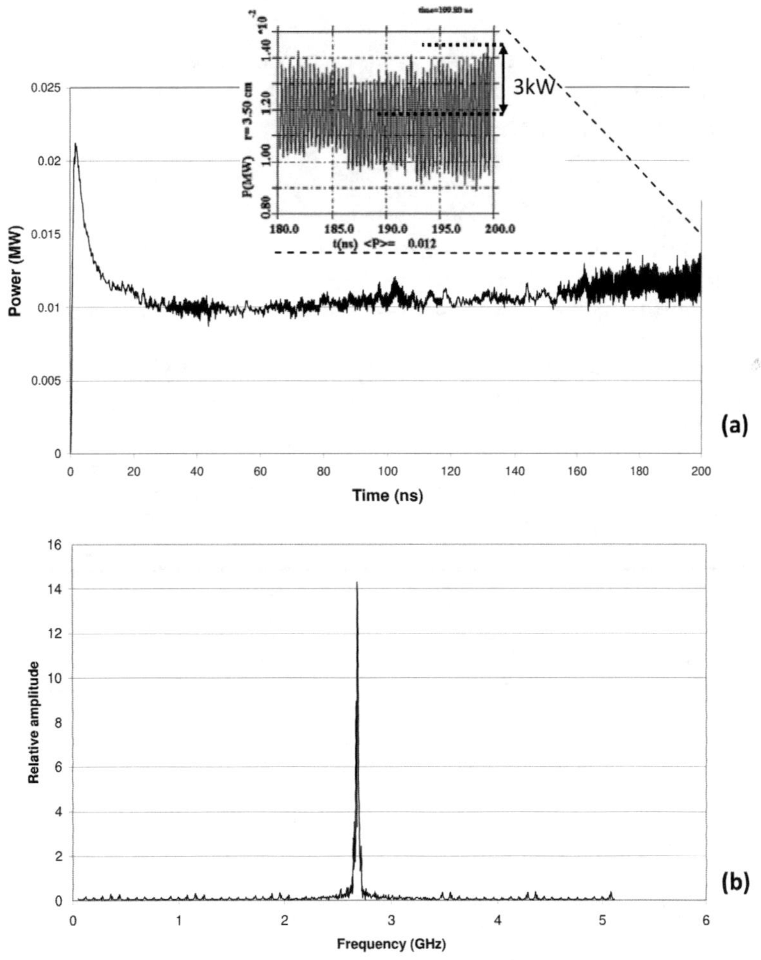

Figure 4. (a) Temporal evolution of the radial Poynting flux measured in a plane at r = 3.5cm spanning the length of the simulation. (b) Fourier transform of E_ϑ from t = 0 →200ns at z = 1.9m.

3. Summary and Conclusions

In conclusion, PiC simulations have been conducted to investigate the dynamics of the cyclotron emission process attributed to numerous astrophysical radio sources [1,17,18] in the absence of cavity boundaries. Computations reveal that a well-defined cyclotron emission process occurs, albeit with a low spatial growth rate compared to waveguide bounded cases [10-14]. RF output is near perpendicular to the electron beam with a slight backward-wave character and well defined spectral output centred just below the relativistic electron cyclotron frequency. The corresponding RF conversion efficiency of 1.1% is comparable to waveguide bounded simulations [10-14] and consistent with the predictions of theory [4-9] and estimates of ~1% for the astrophysical phenomena [17,18].

References

[1] A P. Zarka 1992 *Advances in Space Research*, **12**, 99
[2] D. Gurnett 2008 "Waves in Space Plasmas", *50th Annual APS-DPP Conference*
[3] J.E. Allen & A.D.R. Phelps 1977 *Rep. Prog. Phys.*, **40**, 1305-1368
[4] R. Bingham and R. A. Cairns 2000 *Phys. Plasmas*, **7**, 3089
[5] R. Bingham & R.A. Cairns 2002 *Physica Scripta*, **T98**, 160-162
[6] R.A. Cairns et al 2005 Physica Scripta, **T116**, 23-26
[7] D.B. Melrose et al 1982 *J. Geophys. Research: Space Physics* **87,** 5140-5150
[8] R. Bingham et al 2004 *Contributions to Plasma Physics*, **44**, 382-387
[9] I. Vorgul et al 2005 *Physics of Plasmas*, **12**, 1-8
[10] K.M. Gillespie, D.C. Speirs, K. Ronald et al. 2008 *Plasma Phys. Control. Fusion*, **50**, 124038
[11] K. Ronald, D.C. Speirs, S.L. McConville et al. 2008 *Phys. Plasmas*, **15**, 056503
[12] D.C. Speirs et al 2008 *Plasma Physics & Controlled Fusion* **50**, 074011
[13] D.C. Speirs et al 2005 *Journal of Plasma Physics* **71**, 665-674
[14] D.C. Speirs et al 2010 *Physics of Plasmas* **17**, 056501
[15] S.L. McConville et al 2008 *Plasma Physics & Controlled Fusion* **50**, 074010
[16] K. Ronald et al 2008 *Plasma Sources Science & Technology* **17**, 035011
[17] P.L. Pritchett and R.J. Strangeway 1985 *J. Geophys. Res.*, **90**, 9650
[18] D.A. Gurnett 1974 *J. Geophys. Res.*, **79**, 4227

Detailed kinetic of CO_2 dissociation and C ionization: application to atmospheric Martian entries

Julien Annaloro[1], Arnaud Bultel[1], Pierre Omaly[2]
[1]UMR CNRS 6614 CORIA, Université de Rouen, Avenue de l'Université, BP12, 76801 Saint-Etienne du Rouvray, France
[2] CNES Toulouse, 18 avenue Edouard Belin, 31401 Toulouse
E-mail: julien.annaloro@coria.fr

Abstract. In order to calculate global rate coefficients for atmospheric Martian entries situations, a detailed kinetic model using a state-to-state approach is developped for two important processes: dissociation/recombinaison of CO_2 and ionization/recombinaison of Carbon. The behavior of the species involved is governed by the general balance equation including forward and backward elementary processes whose rate coefficients are in part based on experimental data. Extrapolation methods are used when necessary. During the population densities time-evolution, a quasi steady state can be identified: a global rate coefficient can be derived only during this state. The rates therefore obtained are fitted under a modified Arrhenius law and some of them are finally compared with experimental results.

1. Introduction

The atmospheric entry of a spacecraft or a probe constitutes the more critical step of a spatial flight as illustrated by the dramatic crash of Columbia in 2003. During this phase, the incident flow reaching the fuselage is hypersonic ($v \approx 10$ km s^{-1}). Owing to the high compressibility of the flow, the pressure and temperature undergo a steep increase inside a shock layer near the surface: a complex chemistry is then activated which leads to the formation of excited and ionized species. This gas $\rightarrow$ plasma transition enhances not only the convective flux resulting from the high temperature levels concerned, but also the radiative and chemical parts of the total flux. As a result, the spacecraft or probe surface can be damaged. The amount of reactive species and their inner excitation degree are the key points of the plasma-wall interaction. Accordingly, a clear identification of the underlined chemistry is crucial.

In CORIA, we have already contributed to state-to-state plasma studies devoted to air chemistry [1]. The study of CO_2-N_2 plasma flows in relation with Martian entries is now in progress in the continuity of the previous ones. Two problems to be addressed shortly are identified:

- **dissociation and recombination of CO_2.** CO_2 is the main species of Martian atmosphere: dissociation phenomena occur through state-to-state thermal excitation of vibrational modes. At high temperature, these phenomena are far from being completely understood: one of the objectives of the present study is to give an insight into the dissociation mechanism of CO_2 by developing such a state-to-state approach. The approach adopted lead also to estimates of the recombination rate coefficient.

- **ionization and recombination of C**. At high temperature, atoms like C, N and O can produce ions. The global ionization process takes place through the state-to-state electronic excitation then ionization of atoms. Once again, developing a detailed approach can provide fruitful information not only about ionization, but also for recombination. Such an approach is developed only for C in a first step.

In the following, dissociation/recombination of CO_2 and ionization/recombination of C are studied through the elaboration of relevant state-to-state models.

2. State-to-state models

2.1. *Dissociation/recombination of CO_2*

CO_2 is a molecule with three modes of vibration: a symmetric and an asymmetric non degenerate stretching modes and a doubly degenerate bending mode whose characteristic energies are classically written $h\nu_1$, $h\nu_3$ and $h\nu_2$ respectively. In a first approximation, CO_2 is assumed to behave like a harmonic oscillator. In addition, the energy diagram is assumed totally independent from the Fermi resonance between the first and the second modes. This procedure leads to account for 120 vibrational levels to simulate the dissociation/recombination.

In our conditions, the stepwise vibrational excitation leading to dissociation results from vibration-translation elementary processes. Only mono-quantum transition are taken into account in the same mode in a first approximation. Blauer's data [2] are used: they provide the rate value for the vibrational de-excitation from the first to the ground state in each mode and they are then extrapolated until the dissociation continuum by using the SSH theory [3]. Finally, Boltzmann and Guldberg & Waage equilibrium laws are used to derive rate coefficients for vibrational excitation and dissociation. In the present context, CO_2, CO and O are considered as the main collisional partners.

2.2. *Ionization/recombination of C*

All the excited states of C (225 levels) and the first eight states of C^+ given by the NIST database [4] are taken into account. The temperature ranges being wide, C atoms can be excited or ionized from any excited state according to:

$$C(i) + e^- \leftrightarrow C(j) + e^- \qquad \forall\, i \text{ and } j$$

$$C^+(i) + e^- \leftrightarrow C^+(j) + e^- \qquad \forall\, i \text{ and } j$$

$$C(i) + e^- \leftrightarrow C^+(j) + e^- + e^- \qquad \forall\, i \text{ and } j$$

where electron is assumed to be the main collision partner.

Three different types of transition are possible for the excitation/deexcitation of C or C^+: optically allowed, parity forbidden and spin forbidden transitions. The Drawin's cross sections set is used [5] to calculate the corresponding reaction rates. The backward reaction rates are derived from the detailed balance principle by using the Boltzmann and Saha equilibrium laws.

2.3. *Approach*

Our calculations are systematically performed in constant pressure p and temperature T conditions: in this context, global rate coefficients for either dissociation/recombination or ionization/recombination can be derived. This leads to some difficulties which can be relatively easily addressed. For instance, during a strong ionization phase resulting from an initial strong non equilibrium, electron number increases very fast: accordingly, pressure and temperature cannot be both taken as constant. Indeed, if pressure and temperature are constant, the global density remains unchanged which is incompatible with an ionization condition. This situation requires therefore necessarily a plasma volume which can evolve in time. Even when the ionization is strong, electron number inside the volume considered can freely change under constant pressure and temperature. A global ionization rate can then be derived.

The pressure and temperature are first chosen, then the complete equilibrium composition is calculated. In case the CO_2 dissociation rate is studied, a higher CO_2 density is initially chosen with respect to the density derived from the equilibrium calculation. The same procedure is adopted concerning the initial C density in the case of the C global ionization rate coefficient determination. Recombination rates can be studied by taking the complementary situations.

Time-dependent calculations are performed in our approach. They are based on classical balance equations treatment without neither energy nor particles exchange whose general form is:

$$\frac{d[X_i]}{dt} = \left(\frac{\partial[X_i]}{\partial t}\right)_{Coll.} = G - L \tag{1}$$

where G and L are the gain and loss terms respectively of the source term. These terms take into account the elementary processes discussed in previous sections and the equations set formed owing to the number of species considered is solved using the DVODE library [6].

With this temporal approach, we can therefore follow for instance the time-evolution of the ratios:

$$K_{CO_2}(t) = -\frac{\frac{d[CO_2]}{dt}}{[CO_2]\,\frac{p}{k_B T}} \qquad \text{and} \qquad K_C(t) = -\frac{\frac{d[C]}{dt}}{[C]\,[e^-]} \tag{2}$$

during CO_2 dissociation (left) or C ionization (right). These ratios correspond exactly to rate coefficients when they are not evolving with time: in this case, excited states constitute an intermediate reservoir ensuring the transfer of particles from the CO_2 or C ground states to the continuum of dissociation or ionization respectively. This transient situation is named "quasi steady state" because it lasts while the amount of CO_2 or C particles is high enough. After, the ratios are evolving once again until the new equilibrium is reached.

The previous approach can be easily adapted to the case of recombination situations. Then it is worth to follow the ratios:

$$K'_{CO_2}(t) = \frac{\frac{d[CO_2]}{dt}}{[CO]\,[O]\,\frac{p}{k_B T}} \qquad \text{and} \qquad K'_C(t) = \frac{\frac{d[C]}{dt}}{[C^+]\,[e^-]^2} \tag{3}$$

which correspond to the recombination rate coefficients during the quasi steady-state.

3. Results

3.1. CO_2 dissociation/recombination

Figure 1 displays the population densities evolution in time obtained during the dissociation of CO_2 for p = 10^5 Pa and T = 6000 K. The initial density of CO_2 is too high with respect to the equilibrium density corresponding to these conditions: the dissociation starts approximately when t = 10^{-11} s through the excitation of higher energy levels. Around 10^{-4} s, a steep decrease of the number density of the CO_2 states take place just before a temporary situation where these densities are almost constant. This situation corresponds to the quasi steady-state mentioned in the previous section: a dissociation rate coefficient can be derived from the present calculations only during this state. After, the population number densities evolve again owing to instantaneous departure from equilibrium until it is reached.

Left ratio of 2 is used to derive the global dissociation rate coefficient during the quasi steady-state. The values obtained do not depend on the pressure level, therefore on the total density. The influence of the total density is only observed on time-evolutions owing to its indirect

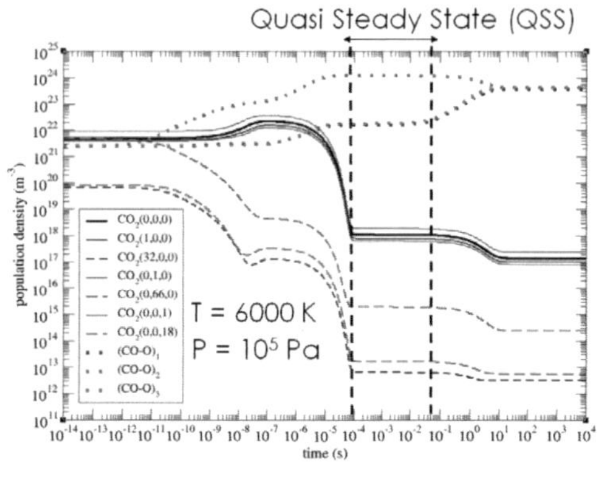

Figure 1. Population densities time-evolution for a typical CO_2 dissociation situation.

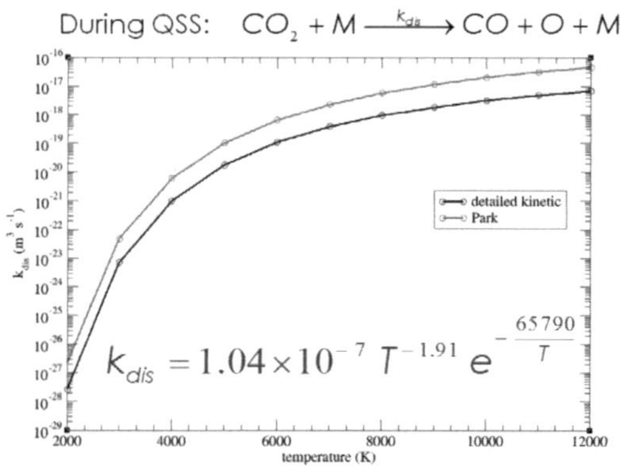

Figure 2. Temperature-dependent CO_2 dissociation global rate coefficient.

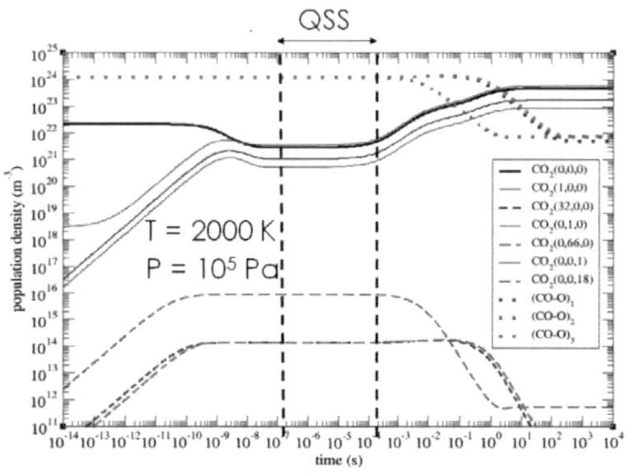

Figure 3. Population densities time-evolution for a typical CO_2 recombination situation.

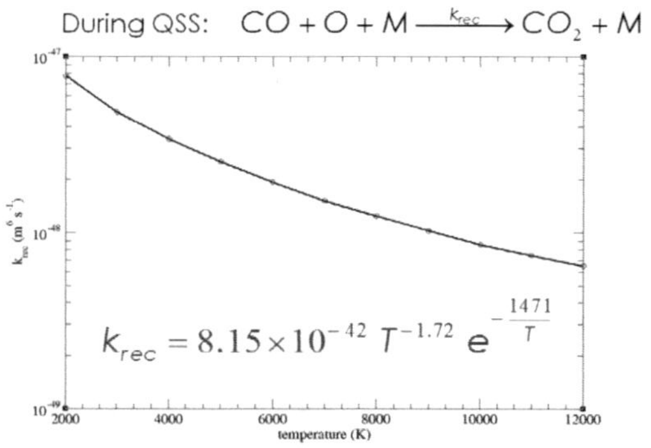

Figure 4. Temperature-dependent CO_2 recombination global rate coefficient.

influence on the collision frequency: a higher density means a higher collision frequency which leads to faster evolution. The quasi steady state is therefore obtained earlier.

A quasi steady-state is observed whatever the conditions adopted for p and T: therefore, a rate coefficient can be systematically derived. Over the range [2000K, 12000K] for T, the results illustrated by figure 4 put forward a logical increase with T of the rate coefficient more important at moderate temperature. These results can be easily fitted by the following modified Arrhenius form:

$$k_{dis} = 1.04 \times 10^{-7}\, T^{-1.91}\, e^{-\frac{65790}{T}} \tag{4}$$

where the characteristic temperature in the exponential term is remarkably close to the dissociation one (64000 K).

This figure displays also the temperature-dependent rate coefficient stated by Park [7] whose chemical model is often implemented in high enthalpy flows calculations related to Martian atmospheric entries. Park's assumptions obviously underestimate the rate coefficient. Our

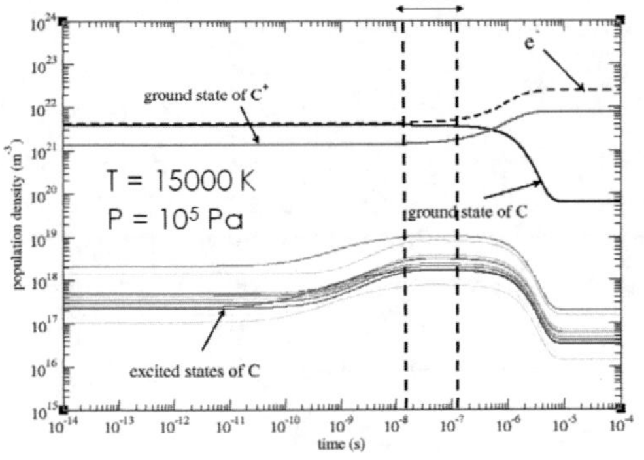

Figure 5. Population densities time-evolution for a ionization situation.

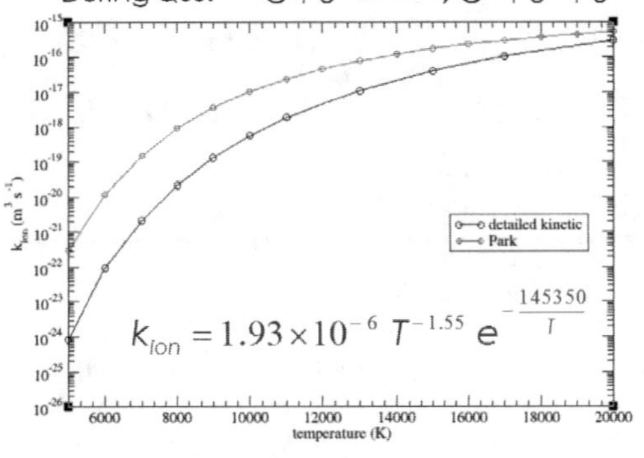

Figure 6. Global rate coefficient for C ionization by electronic impact.

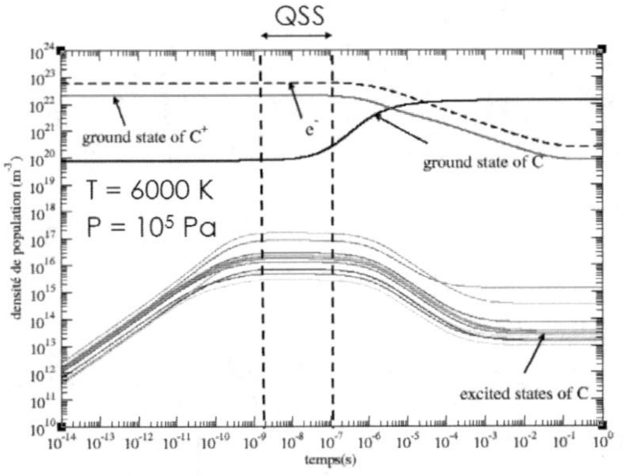

Figure 7. Population densities time-evolution for a recombinaison situation.

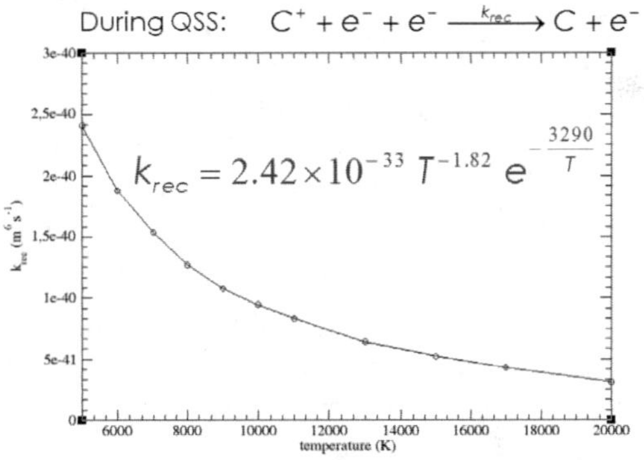

Figure 8. Global rate coefficient for C ionization by electronic impact.

values are indeed very close to the ones obtained experimentally in shock tube experiments [8, 9].

Figure 3 displays the results of the temporal evolutions of densities for the same pressure level, but for T = 2000 K. The recombination is obvious. The three phases put forward previously for the dissociation case are observed once again. In particular, the quasi steady-state is clearly visible. The rate coefficient derived during this transient state is determined by the left equation 3. By generalizing these calculations, one obtains the global rate coefficient as a function of T displayed by figure 4 and very well fitted under the following modified Arrhenius form:

$$k_{rec} = 8.15 \times 10^{-42} T^{-1.72} e^{-\frac{1471}{T}} \quad (5)$$

The characteristic temperature involved in the exponential term is close to 0: this result is logic since the recombination is a process without any threshold.

3.2. C ionization/recombination

The characteristics previously pointed out for the case of CO_2 are also observed for C as a result of the discussion of section 2.3. Figure 5 displays the time-evolution of densities in the case p =

10^5 Pa and T = 6000 K. Figure 6 illustrates the evolution with T of the global rate coefficient for ionization of C. Once again, these results are well fitted by a modified Arrhenius law with:

$$k_{ion} = 1.93 \times 10^{-6}\, T^{-1.55}\, e^{-\frac{145350}{T}} \tag{6}$$

Figure 6 displays also the values given by Park. We can observe an identical evolution, but with higher discrepancies, except at high temperature where both series converge.

The typical evolution in time of population densities in the case of a recombination situation is displayed on figure 7. The quasi steady-state is more difficult to be observed: nevertheless, its existence cannot be disputed. Using right equation 3, the global recombination rate can be deduced during the quasi steady-state. As illustrated by figure 8, they are once again very well fitted by a modified Arrhenius law such that:

$$k_{rec} = 2.42 \times 10^{-33}\, T^{-1.82}\, e^{-\frac{3290}{T}} \tag{7}$$

4. Conclusion

This work addresses a state-to-state approach dedicated to the determination of global rate coefficients for the dissociation or the recombination of CO_2 and the ionization or recombination of C involved in the Martian atmospheric entries chemistry. This state-to-state approach put forward the existence of a quasi steady-state where global rate coefficients can be derived with relevance. Their values follow closely modified Arrhenius forms which can be easily implemented in fluid mechanics numerical codes.

The results displayed in the present communication are preliminary. Some improvements have indeed to be performed and are already in progress:

- **CO_2 dissociation/recombination**. Collisions with CO_2 molecules can induce cross modifications of the vibrational excitation between the three modes. They have to be taken into account. Multiquanta transitions occur also during collisions and will be implemented.
- **C ionization/recombination**. The results obtained depend on the cross sections set chosen to model the chemistry. Other sets will be shortly tested, in particular the one derived from the HULLAC package [10].

Finally, it is worth to notice that a backward rate coefficient is systematically derived from the forward one using directly the related equilibrium law. This approach is not necessarily relevant owing to the departure of equilibrium for the excited states during quasi steady-state. The present approach will give interesting point of view about this question.

Acknowledgments

The authors wish to thank the financial support of CNES (Centre National des Etudes Spatiales) in France.

References

[1] Bultel A, Chéron B G, Bourdon A, Motapon O and Schneider I F 2006 *Phys. Plasmas* **13** 043502
[2] Blauer J A and Nickerson G R 1974 *AIAA Paper* 74-536
[3] Schwartz R N, Slawsky Z I, Herzfeld K F 1952 *J. Chem. Phys.* **20** 10
[4] *http://physics.nist.gov/PhysRefData/ASD/levels_form.html*
[5] Drawin H W 1967 *Euratom-CEA Technical Report* DPh-PFC/SRFC EUR-CEA-FC-383
[6] Brown P N, Byrne G D and Hindmarsh A C 1989 *SIAM J. Scient. Comput.* **10** 1038
[7] Park C, Howe J T, Jaffe R L, Candler G V 1994 *J. Thermophys. Heat Transfer* **8** 1 9-23
[8] Brabbs T A, Belles F E, Zlatarich S A 1963 *J. Chem. Phys.* **38** 8 1939-1945
[9] Ebrahim N A, Sanderman R J 1976 *J. Chem. Phys.* **65** 9 3446-3454
[10] Kuba J, Smith R F, Benredjem D, Moller C, Upcraft L, King R, Klisnick A, Drska L, Pert G J, Gauthier J-C 2003 *JOSA B* **20** 1 208-214

Features of coronal heating by drift waves

J. Vranjes and S. Poedts
Center for Plasma Astrophysics,
and Leuven Mathematical Modeling and Computational Science Research Centre (LMCC),
Celestijnenlaan 200B, 3001 Leuven, Belgium.
E-mail: jvranjes@yahoo.com; Stefaan.Poedts@wis.kuleuven.be

Abstract. The stochastic heating, well known from laboratory plasma physics where it has been confirmed in numerous experiments, has been completely ignored in past studies of coronal heating. In the present study, and in our very recent works, it has been shown that the inhomogeneous coronal plasma is, in fact, a perfect environment for fast growing drift waves and the consequent stochastic heating. As a matter of fact, the large growth rates are typically of the same order as the plasma frequency. The heating rates may exceed the required values for a sustained coronal heating by several orders of magnitude. Some aspects of these phenomena are investigated here. In particular the analysis of the particle dynamics within the growing wave is compared with the corresponding fluid analysis.

1. Introduction

The drift wave is driven by gradients of the background plasma parameters (e.g., gradients of the plasma density, temperature, and magnetic field) that may act either separately or all together. In the literature, the drift wave is frequently called the *universally unstable mode* because it is growing both within the (two- or multi-component) fluid theory and the kinetic theory. In the former (fluid) description, it is either a collisional instability (discussed in [1] in the context of the solar atmosphere) that appears as a combined effect of the electron collisions, the ion inertia, and the density gradient, or a reactive one [2] in the presence of all three gradients mentioned above where now the ions play a major role. The latter (kinetic) theory, on the other hand, can also describe these two instabilities. However, the kinetic description yields an additional (purely kinetic) instability that is absent within the fluid description. In application to the solar corona, this additional kinetic instability is studied in [3]-[5]. This purely kinetic growth of the drift wave is entirely due to electron dynamics and it develops provided the wave frequencies are below the electron diamagnetic drift frequency.

2. Results

For the purpose of the present study, in order to quantitatively describe some details of the heating and acceleration of plasma particles, we shall use the kinetic, density gradient driven instability from [3, 4]. The frequency and growth rate of the kinetic drift wave are given by

$$\omega_r = -\frac{\omega_{*i}\Lambda_0(b_i)}{1-\Lambda_0(b_i)+T_i/T_e+k_y^2\lambda_{di}^2}, \qquad (1)$$

$$\gamma \simeq -\left(\frac{\pi}{2}\right)^{1/2} \frac{\omega_r^2}{|\omega_{*i}|\Lambda_0(b_i)} \left[\frac{T_i}{T_e}\frac{\omega_r - \omega_{*e}}{|k_z|v_{Te}} \exp[-\omega_r^2/(k_z^2 v_{Te}^2)] + \frac{\omega_r - \omega_{*i}}{|k_z|v_{Ti}} \exp[-\omega_r^2/(k_z^2 v_{Ti}^2)]\right]. \quad (2)$$

Here, the geometry is chosen as follows. The constant magnetic field is given by $\vec{B}_0 = B_0\vec{e}_z$, the equilibrium density gradient in the electrically neutral plasma with $n_{i0} = n_{e0} = n_0$ is given by $\nabla_\perp n_0 = -\vec{e}_x n_0' \equiv -\vec{e}_x dn_0/dx$, and the perturbations are of the form $f(x)\exp(-i\omega t + ik_y y + ik_z z)$, with $|k_z| \ll |k_y|$ and an x-dependent wave amplitude f. The local approximation will be used. Hence, it will be assumed that $|d/dx| \ll |k_y|$. The notation in Eqs. (1) and (2) is standard, i.e., $\omega_{*e} = -k_y(v_{Te}^2/\Omega_e)(n_{e0}'/n_{e0})$, $\omega_{*i} = -(T_i/T_e)\omega_{*e}$, $v_{Tj}^2 = \kappa T_j/m_j$, $\Lambda_0(b_i) = I_0(b_i)\exp(-b_i)$, $b_i = k_y^2\rho_i^2$, $\lambda_{di} = v_{Ti}/\omega_{pi}$, I_0 is the modified Bessel function of the first kind and of the order 0, and c_s is the ion sound speed. From Eq. (2) it is seen that within the present instability, the sign of γ can be changed only by the electron term if $|\omega_r| < |\omega_{*e}|$. This is, however, only a necessary condition. A sufficient condition is obtained when the contribution of the rest of the expression (the ion damping terms) is also taken into account.

To provide some details of the particle acceleration and heating by the growing drift wave in the context of the solar corona, we need specific data. Hence, as an example we here use a set of parameters from [3, 4] that are known to yield the instability: $B_0 = 10^{-2}$ T, $n_0 = 10^{15}$ m^{-3}, $L_n = [(dn_0/dx)/n_0]^{-1} = s \cdot 100$ m, $\lambda_y = 0.5$ m, and $\lambda_z = s \cdot 10^4$ m. These are just representative parameters with no particular significance. As discussed in [4], the instability will in fact develop even when the density is varied by several orders of magnitude around the given value, and the same holds for the magnetic field and the temperature. The parameter s is introduced in [3, 4] for convenience only, because it was realized that the ratio γ/ω_r remains exactly the same in case the ratio λ_z/L_n is fixed. Here, L_n is the characteristic scale-length for the inhomogeneous equilibrium density. This further implies that we may go to very different scales, e.g., by taking s in the range $0.1 - 10^4$. As long as λ_y is kept constant, the ratio γ/ω_r will remain the same, although the actual values for these two quantities will certainly change. Hence, for the given parameters, from Eqs. (1) and (2) one finds $\gamma/\omega_r = 0.26$. In the case $s = 1$, this in fact implies $\omega_r = 254$ Hz and $\gamma = 66$ Hz. While in the case $s = 10^3$, this would give a mode with $\omega_r = 0.254$ Hz and $\gamma = 0.66$ Hz. Going to such small values of L_n (tens of meters) in fact makes sense because the perpendicular diffusion, that naturally develops in such a plasma with density gradients, is indeed very small, with the diffusion coefficient $D_\perp$ of the order of 0.01 m^2/s and the corresponding diffusion velocity $D_\perp\nabla_\perp n/n$ of just a few millimeters per second. Short values of L_n in the same time imply very high frequency drift waves and fast growing instabilities, so that the expected heating will develop at time scales that are orders of magnitude shorter than the plasma diffusion. Note that the parameters λ_z and L_n can of course also be varied independently of each other.

A direct numerical integration for plasma particles is required for a full description of the heating mechanism. The numerical simulation has to take into account the actual values of the electric field in the position of a specific gyrating plasma particle. For that purpose, taking the drift-wave electric field with a time-varying amplitude, the appropriate set of equations for a single particle reads

$$y''(t) + y(t) - a(t)\sin\left[y(t) + \frac{k_z}{k_y}z(t) - bt\right] = 0, \quad (3)$$

$$z''(t) - a(t)\frac{k_z}{k_y}\sin\left[y(t) + \frac{k_z}{k_y}z(t) - bt\right] = 0, \quad x'(t) - y(t) = 0. \quad (4)$$

Here, the prime denotes a derivative in time, and the spatial and time coordinates are normalized as $x, y, z \to k_y x, k_y y, k_y z$, $t \to \Omega_i t$, $b = \omega_r/\Omega_i$.

The set of ordinary differential equations (3)-(4) is solved numerically. We do this first for a very small wave amplitude in order to demonstrate some basic features of the particle motion due to the wave electric field. The result is shown in Fig. 1, where we give the projection of the proton positions in the perpendicular plane, $x(t), y(t)$, in a time-varying potential $\phi_1(t)$ due to the above demonstrated drift wave instability. Restricting the time interval to relatively short values (below the onset time of the stochastic heating), yields a spiral trajectory in the x, y-plane. The potential is given by $\phi_1(t) = \widehat{\phi}\exp(\gamma t/\Omega_i)$, $\gamma = 0.26\omega_r$, $\omega_r = 254$ Hz, $\widehat{\phi} = \phi_1(0)$. The maximum time in physical units here is 0.04 s, and in this moment the potential has reached the value of 17 V. Here and further in the text we take $\lambda_y = 0.5$ m, $\lambda_z = 2 \cdot 10^4$ m, which consequently from Eq. (1) yields the wave frequency $\omega_r = 254$ Hz. The normalization is the same as above, viz. $(x, y) \equiv (k_y x, k_y y)$.

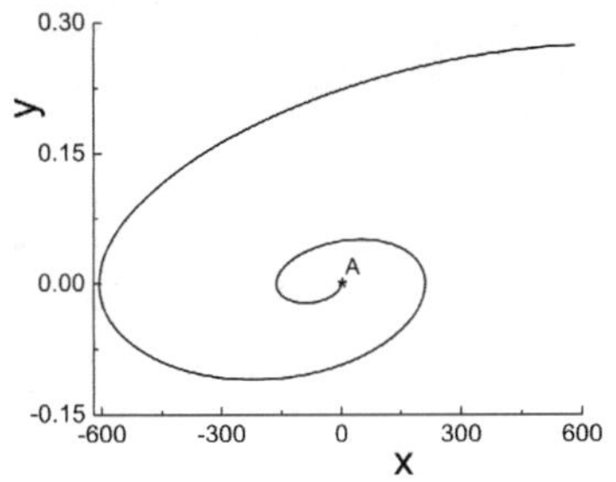

Figure 1. Normalized particle positions in the x, y-plane after it starts from the point A with $(x, y, z) = (0, 0, 0)$, for a growing electric field potential $\phi_1(t) = \widehat{\phi}\exp(\gamma t/\Omega_i)$, and $\widehat{\phi} = 0.86$ V.

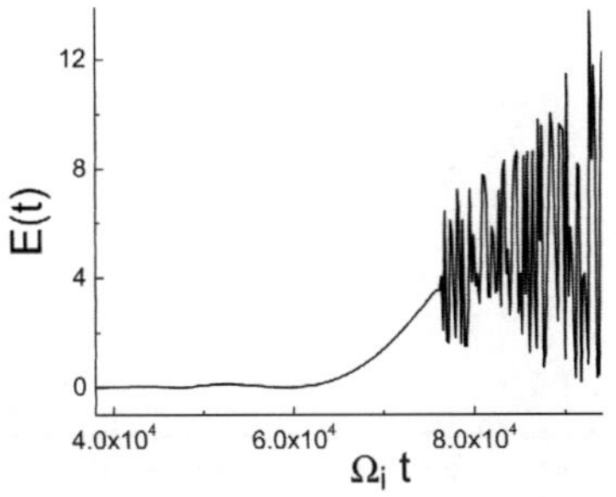

Figure 2. Change in time of the normalized kinetic energy of a particle with a unit mass $E = (v_x^2 + v_y^2 + v_z^2)/2$.

In order to see what happens for larger amplitudes of the electrostatic drift wave when the stochastic heating is supposed to be in action, Eqs. (3)-(4) are solved by allowing a slightly larger time range. One of the results is shown in Fig. 2 for the kinetic energy of a particle with unit mass $E(t)/m = [v_x(t)^2 + v_y(t)^2 + v_z(t)^2]/2$, with normalized velocities as above. The other parameters are the same as in the previous text. Here, the stochastic heating takes place after around 0.078 s, in the moment when the growing wave amplitude reaches the value of around 150 V. Note that this is by about a factor 2.5 larger than the value that can be obtained from the fluid theory [6] and which reads

$$a \equiv k_y^2 \phi_1(t)/(B_0 \Omega_i) = k_y^2 \rho_i^2 e\phi_1(t)/(\kappa T_i) \geq 1. \tag{5}$$

Equation (5) is obtained approximately starting from the guiding center approximation that becomes violated for a large polarization drift, and the analytical expression for the polarization drift in this limit becomes inaccurate. Hence, the difference between the two descriptions is expected. Yet, Eq. (5) describes the general trend for the polarization drift effect; a fast and significant acceleration and heating of the plasma particles will definitely take place, as experimentally verified in [7, 8].

The corresponding plot of the displacement in the direction of the wave $y(t)$, which is in fact also equal to the velocity $v_x(t)$ is shown in Fig. 3.

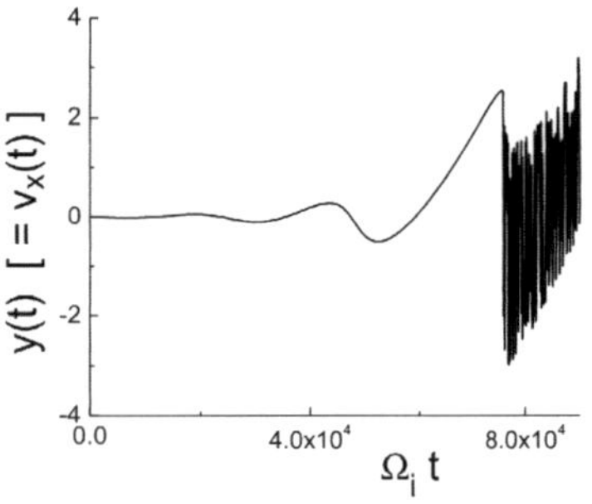

Figure 3. Normalized displacement in the direction of the polarization drift $y(t)$, and the normalized perpendicular velocity component $y(t) \equiv v_x(t)$.

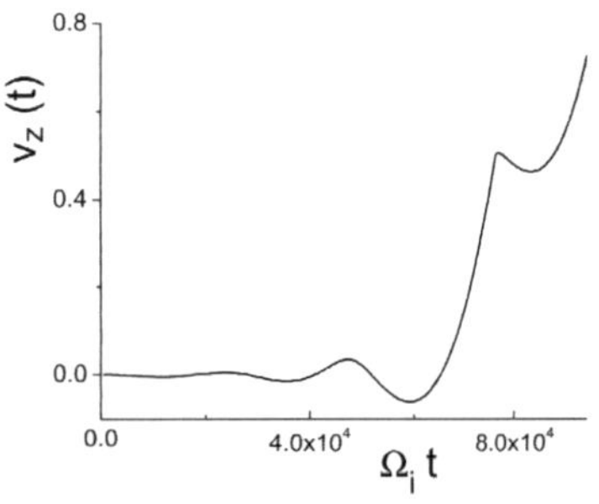

Figure 4. Normalized perturbed velocity in the direction parallel to the magnetic field.

The plot of the parallel velocity $v_z(t)$, presented in Fig. 4, shows that along the magnetic field the particle dynamics remains non-stochastic, as should be expected in view of the analysis presented in the text above. The velocity magnitude is below 1, thus still smaller than the perpendicular velocity which is the consequence of the stochastic heating.

3. Summary

In this work, the trajectories of individual plasma particles have been investigated within the framework of the drift wave instability and the associated stochastic heating, with the aim to describe some fine details of the particle dynamics during the heating process. The calculation of the actual growth rates of the mode is not repeated here because it has been given in detail in our recent studies. From the present study and from our recent works it follows that the general trend of the ion behavior in the wave field with a growing amplitude can be described analytically from the fluid point of view. However, a more detailed picture is obtained by following a single particle which is moving in the wave field. For example, this is seen from the critical threshold for the onset of the heating.

It is also interesting to note that, although the individual particle dynamics becomes stochastic for relatively large wave amplitudes, the overall picture of the macroscopic motion of the plasma, caused by the wave, actually remains well described by the fluid description. This has been directly experimentally verified by [9]. The parameter a [c.f. Eq. (5)] in their experiment was around 2, and thus the stochastic heating was in action, yet the observed electrostatic potential and density profiles nicely matched the theoretical contour plots corresponding to a wave described within the fluid theory and drift approximation. This is in agreement with the nature of the heating; the increased effective temperature T_{max} follows from the stochastic velocity v_{max} which in fact is not random, yet in practical terms it is indistinguishable from standard collisional thermalization and the broadening of the distribution function [6].

The model used here implies a Maxwellian distribution function for plasma species, that is the basis for both the kinetic and the fluid description of the instability. However, the stochastic heating with all its features [2, 4] in fact implies that such a starting distribution function will necessarily evolve. The perpendicular ion temperature will grow and eventually become larger than the parallel one. Hence, the particle distribution may pass through several stages, going through the Oort's distribution with the temperature anisotropy and to the more general loss-cone distribution [10], or to something else. This opens an interesting possibility for eventual numerical simulations because such fine details are beyond the scope of an analytical study.

Acknowledgments

The results presented here are obtained in the framework of the projects G.0304.07 (FWO-Vlaanderen), C 90347 (Prodex), GOA/2009-009 (K.U.Leuven). Financial support by the European Commission through the SOLAIRE Network (MTRN-CT-2006-035484) is gratefully acknowledged.

[1] Vranjes J and Poedts S 2006 *Astron. Astrophys.* **458** 635
[2] Vranjes J and Poedts S 2009 *MNRAS* **400** 2147
[3] Vranjes J and Poedts S 2009 *EPL* **86** 39001
[4] Vranjes J and Poedts S 2009 *MNRAS* **398** 918
[5] Vranjes J and Poedts S 2009 *Phys. Plasmas* **16** 092902
[6] Bellan P M 2006 *Fundamentals of Plasma Physics* (Cambridge: Cambridge Univ. Press) p 106
[7] McChesney J M, Stern R A and Bellan P M 1987 *Phys. Rev. Lett.* **59** 1436
[8] Sanders S J, Bellan P M and Stern R A 1998 *Phys. Plasmas* **5** 716
[9] Bailey A D, Stern R A and Bellan P M 1993 *Phys. Rev. Lett.* **71** 3123
[10] Baronia A and Tiwari M S 1999 *Planet. Space Sci.* 47 1111

Journal of Physics: Conference Series **511** (2014) 012055 doi:10.1088/1742-6596/511/1/012055

Low-pressure plasma discharge of $Ar/N_2/CH_4$ ternary mixture: experimental simulation of Titan's atmosphere

L. Salazar-Flores[1], C. Torres- Segundo[2], P.G. Reyes[2], F. Castillo[3] and H. Martinez[4].

[1]Posgrado en Ciencias Físicas, Universidad Nacional Autónoma de México, Avenida Universidad 1000, 05000, México D.F., México.

[2]Facultad de Ciencias, Universidad Autónoma del Estado de México, Instituto Literario 100, Centro, 50000, Toluca, Estado de México, México.

[3]Instituto de Ciencias Nucleares, Universidad Nacional Autónoma de México, Apartado Postal 70-543, 04510, México DF., México.

[4]Instituto de Ciencias Físicas, Universidad Nacional Autónoma de México, Apartado postal 48-3, 62251, Cuernavaca, Morelos, México.

E-mail: hm@fis.unam.mx

Abstract. A low-pressure DC plasma discharge sustained in a 85.0%N_2-12.0%Ar- 3.0%CH_4, 82.3%N_2- 13.7%Ar- 4.0%CH_4 and 79.6%N_2-15.4%Ar- 5.0%CH_4 ternary mixtures are studied. These plasmas were generated in the total pressure range of 0.5 to 10.0 Torr and power of 7 W. The diagnostic has been made by optical emission spectroscopy (OES) using a spectrometer. The principal species observed were: at 315.93, 337.13, 375.54, 399.84 and 405.94 nm for N_2 $(C^3\Pi_u - B^3\Pi_g)$; at 354.89, 358.21, 376.16, 391.44 and 427.81 nm for $N_2^+(B^2\Sigma_u^+ - X^2\Sigma_g^+)$; at 381.06 nm for $CH^+(A^1\Pi - X^1\Sigma)$; at 392.08 nm for $CN(B^2\Sigma - X^2\Sigma)$, at 428.15 nm for $CN(B^2\Sigma - A^2\Pi)$ and at 708.87 nm for $CN(A^2\Pi - X^2\Sigma)$; at 436.52, 592.34, 653.40, 667.73 and 708.32 nm for $C_2(A^3\Pi_g - X'^3\Pi_u)$; at 392.59 nm for $C_3(^1\Pi_u - {}^1\Sigma_g^+)$; at 486.14 nm for H_β; at 486.13 and 656.27 nm for H_α; at 657.80 nm for C^+; at 739.23,751.47,763.51, 772.42, 794.82 and 811.53 nm for Ar. We present the behaviour of the bands and lines intensities as a function of the pressure and concentration of Ar, N_2 and CH_4 gases. Also, we display more quantitatively the ratios of intensities of N^+_2 (391.44 nm), CN (392.08 nm), and H (486.13 nm) to that of the N_2 (337.13 nm) as function of pressure. The ratios show a slow decreasing behavior as a function of the pressure. Being the CH/N_2 ratio more highest and H/N_2 ratio the lowest one. The variations of excited species at different pressures may change the subsequent chemical reactions in the gas phase significantly. The present results suggest that the ion–molecule and molecule-molecule reactions in the gas phase are likely to play a dominant role in the present pressures.

1. Introduction

In closed systems it is frequently difficult to distinguish between the "pure" volume discharge chemistry and the interaction of the gas with the envelope and electrodes. Our interest here has been concentrated mainly upon the volume effects which by proper choice of the experimental condition are predominant in the initial phase of the discharge (typical times are in the order of 15 min). This in contrast to the long-term effects where the interaction of the gas fills with the materials of the discharge vessel becomes the major effect and which has been studied recently. Qualitative analysis of reaction mechanisms in gas discharge necessitates a detailed knowledge of the ionic and neutral constituents and their variation with time. This is particularly important for closed gas discharges in molecular gases, where complex electrochemical processes take place which may lead to a substantial change in composition before a steady state is reached. Recently, we have been able to see the emergence of papers devoted to the formation of particles in various gas discharges such as hydrocarbon plasmas [1-3], also to the study of methane decomposition in a nitrogen plasma due to a large number of applications in various domains such as the simulation of Titan´s atmosphere [4-5], in deposition of diamond-like carbon layers [6], in optical and electrical characterization of an atmospheric pressure microplasma jet [7], in the conversion of natural gas to C_2 hydrocarbons [8], in the properties and reactions of the CN radical [9] and in metal nitriding [10]. In Titan´s atmosphere the CH_4 abundance is 4.9% at the surface falling to 1.62% at an altitude of 60 Km above which lies Titan´s stratosphere where the methane is uniformly mixed with nitrogen [1, 4]. Moreover, Titan is a test of our understanding of the organic chemistry of planetary atmospheres, since its atmospheric bulk composition is (N_2:CH_4:Ar).

One of the most interesting questions about titan´s atmosphere is the influence of electrical activity on the organic chemistry of its nitrogen and methane rich atmosphere. So, the goal of this work, it is to explore the optical emission spectroscopy (OES) in a low-pressure plasma discharge of Ar/N_2/CH_4 ternary mixture. The dependence of the principal species observed on gas pressure are presented and discussed.

2. Experimental setup

The experimental apparatus and technique to measure the DC plasma emission spectrum is identical to that described in previous work [11, 12]. A brief description is reported here. The system consists of two stainless steel circular plane electrodes, 3 mm thick and 30 mm in diameter. The electrodes are positioned at the center of the reaction chamber with 5 mm gap spacing. The gas was injected into the reaction chamber through the bottom flange. The same gas connection was used for the pressure sensor (MKS, Type 947 convection enhanced Pirani). The DC plasma was produced in three concentration C1 (85.0%N_2-12.0%Ar- 3.0%CH_4), C2 (82.3%N_2-13.7%Ar- 4.0%CH_4) and C3 (79.6%N_2-15.4%Ar- 5.0%CH_4) ternary mixtures at pressures between 0.5 and 10.0 Torr. An ultra pure C_2H_4, N_2 and Ar gases (Praxair 99.99%) were used in the course of the measurements. The discharge power supply (Bertran serie 105, 1 KW) was maintained at an output of 280 Volts and a current of 25 mA (7 W). The base pressure of the plasma chamber of 3.39 X 10^{-3} m^3 volume was 2.0X10^{-2} Torr, achieved using a mechanical pump (SD-301Varian Pump). The vacuum chamber was purged with the working gas several times at a pressure of 1.0 Torr in order to remove the background gas. In order to monitor the plasma discharge, in front of the lateral plane of the electrodes was located an optical fiber (Solarization-resistant UV and fiber diameter size of 400 μm) that was connected to the entrance aperture of a high-resolution Ocean Optics Inc. Spectrometer Model HR2000CG-UV-NIR The inlet and outlet slits were 5 μm wide. The data were obtained in a single accumulation of 9 s integration times. The wavelength scan interval was of 0.47 nm. The fiber optics assembly was positioned to collect the emission light appearing on the CH_4-N_2-Ar plasma.

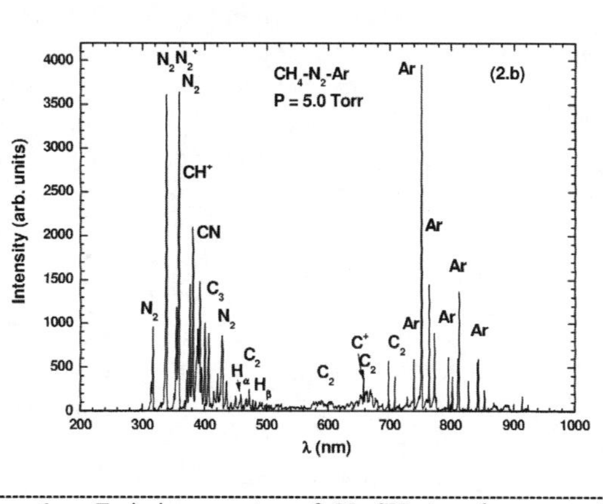

Figure 1. Emission spectra of N_2-CH_4-Ar plasma at 5.0 Torr. For concentration C2.

3. Results and discussion

The chemical and optical properties depend on the deposition pressure. However, the variation in deposition pressure may change not only the collision frequencies of molecules/atoms but also the plasma properties, such as the intensity of UV emission and the charged particles flux. To investigate the plasma properties, the plasma emission was measured with a CCD-array spectrometer (Ocean Optics) with an optical fiber though the Pyrex glass chamber. The plasma-ions flux reaching the cathode can be evaluated from $J=I_D/S$, where I_D is the current discharge (25 mA) and S is the electrode area (7.07 cm^2). So, the plasma- ions flux is ~1.32 x 10^{18} cm^{-2} min^{-1}. A typical OES measurement of mixtures C2 glow discharge plasma at a pressure of 5.0 Torr is displayed in Figure 1, at 7 W of power (only, C2 mixture is shown, due to C1 and C3 display similar behavior). This allowed analysis of the most luminous area, which corresponds to the negative glow near the cathode dark space. Only the most intense spectral lines and bands of the plasma within the 200-1000 range are quoted [13]. The principal species observed were: at 315.93, 337.13, 375.54, 399.84 and 405.94 nm for N_2 ($C^3\Pi_u - B^3\Pi_g$); at 354.89, 358.21, 376.16, 391.44 and 427.81 nm for $N_2^+(B^2\Sigma_u^+ - X^2\Sigma_g^+)$; at 381.06 nm for $CH^+(A^1\Pi - X^1\Sigma)$; at 392.08 nm for $CN(B^2\Sigma - X^2\Sigma)$, at 428.15 nm for $CN(B^2\Sigma - A^2\Pi)$ and at 708.87 nm for $CN(A^2\Pi - X^2\Sigma)$; at 436.52,592.34, 653.40, 667.73 and 708.32 nm for $C_2(A^3\Pi_g - X'^3\Pi_u)$; at 392.59 nm for $C_3(^1\Pi_u - ^1\Sigma_g^+)$; at 486.14 nm for H□; at 486.13 and 656.27 nm for H_α; at 657.80 nm for C^+; at 739.23,751.47,763.51, 772.42, 794.82 and 811.53 nm for Ar.

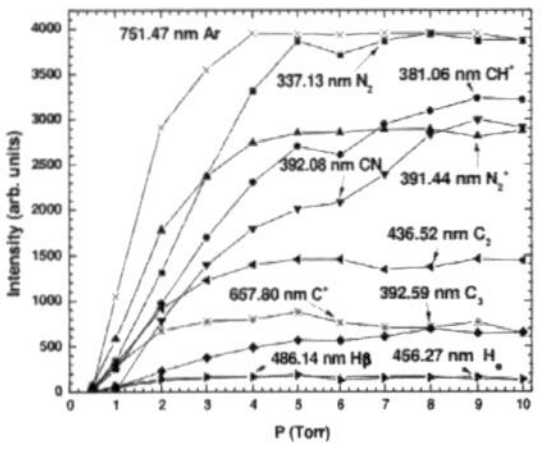

Figure 2. Intensity of the some emission bands and lines as a function of pressure.

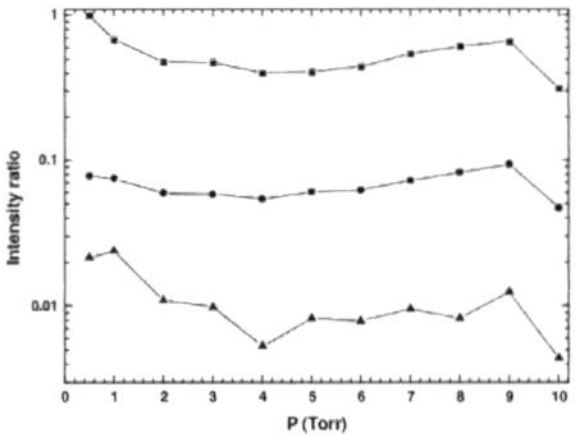

Figure 3. Ratio of intensities as a function of pressure. ■, CN/N_2; ●, N_2^+/N_2; ▲, H/N_2.

At the present plasma condition (V=280 Volts, d=5 mm, P= 1-10 Torr) and assuming T_g= 400 K, the ion gas density (N) is between 0.24x10^{17} and 2.41x10^{17} cm^{-3}, which with n_e □ 10^{10} cm^{-3} (assumed), we obtained an ionization degree of 10^{-7}. As the ionization degree is < 10^{-4}, it may neglect electron-electron interaction [14]. Also, in this case the reduced electric field E/N is between 193.6 and 1936.4 Td (E is the field strength). The measurement conditions, such as the integration time of the spectrometer and the position of the optical fiber, were fixed for every measurement. The limitation factor on long-time measurements was the deposition of yellow-orange layer on the electrodes causing sparks in the discharge. The influence and elimination of this kind of tholins requires further investigations. Although the intensities are not calibrated, the relative intensity is still valid for discussions below. In order to observe more carefully the effect of the pressure on the OES measurements, the pressure dependence of several emission bands and lines are displayed in Figure 2 for the concentration C2 (C1 and C3 have similar behavior). All bands and lines have an increasing behavior between 0.5 and 5 Torr, followed by a saturation behavior within a pressure range of 5.0 to 10.0 Torr. The relative ratios of excited chemical species also depended considerably on pressures. Figure 3 shows more quantitatively the ratios of intensities of N^+_2 (391.44 nm), CN (392.08 nm), and H (486.13 nm) to that of the N_2 (337.13 nm) as function of pressure. The ratios show a slow decreasing behavior as a function of the pressure. Being the CN/N_2 ratio more highest and H/N_2 ratio the lowest one. The presence of the N_2^+ ion in the pressure plasma suggests that the ion–molecule and molecule-molecule reactions in the gas phase are likely to play a dominant role at the present pressures. The variations of excited species at different pressures, such as N_2^+ ions and CN radicals, may change the subsequent chemical reactions in the gas phase significantly. At lower pressures the ionization occurs predominantly by electron impact and it is expectable that the sustaining reduced electric field and the electron density keep their values. On the other hand, in the opposite limit of a higher pressure part of the ionization results from collisions between metaestable species and the ionization rate by electron impact may decrease to maintain the plasma. Thus, the reduced electric field is lower (193.6 Td), as well the excitation of the upper electronic states responsible for the emission of the bands N_2, N_2^+, CH^+, CN, C_2 and C_3.

4. Conclusions

An experimental study of optical emission spectroscopy of a low-pressure DC plasma discharge sustained in a 85.0%N_2-12.0%Ar- 3.0%CH_4, 82.3%N_2-13.7%Ar- 4.0%CH_4 and 79.6%N_2-15.4%Ar-5.0%CH_4 ternary mixtures were studied. These plasmas were generated in the total pressure range of 0.5 to 10.0 Torr and power of 7 W. The principal species observed were from the molecular bands of N_2, N_2^+, CH^+, CN, C_2 and C_3, and atomic lines of $H_{\square}$, H_{α}, C^+ and Ar. All bands and lines have an increasing behavior between 0.5 and 5 Torr, followed by a saturation behavior within a pressure range of 5.0 to 10.0 Torr. The ratios show a slow decreasing behavior as a function of the pressure. Being the CH/N_2 ratio more highest and H/N_2 ratio the lowest one. The variations of excited species at different pressures may change the subsequent chemical reactions in the gas phase significantly. The present results suggest that the ion–molecule and molecule-molecule reactions in the gas phase are likely to play a dominant role in the present pressures.

ACKNOWLEDGMENTS

The authors are grateful to A. González, O. Flores, and José Rangel for technical assistance. This research was partially sponsored by DGAPA IN-105010, CONACyT 28714.

References

[1] Géraud-Grenier I, Massereau-Guilbaud V and Plain A 2001 *European Physics Journal: Appl. Phys.* **14** 187
[2] Berndt J, Hong S, Kovacevic E, Stefanovic I and Winter J 2003 *Vacuum* **71** 377
[3] Chen G Y, Stolojan V, Silva S R P, Herman H and Haq S 2005 *Carbon* **43** 704
[4] Niemann H B, Atreya S K, Bauer S J, Carignan G R, Demick J E, Frost R L, Gautier D, Haberman J A, Harpold D N, Hunten D M, Israel G, Lunine J I , Kasprzak W T, Owen T C, Paulkovich M, Raulin F, Raaen E and Way S H 2005, *Nature* **438** 779
[5] Horvath G, Skalny J D, Mason N J, Klas M, Zahoran M, Vladoiu R and Manole M 2009 *Plasma Source Science and Tech.* **18** 034016
[6] Herrebout D, Bogaerts A, Yan M and Gijbels R 2001 *J. of Appl. Phys.* **90** 570
[7] Yanguas-Gil A, Focke K, Benedikt J and von Keudell A 2007 *J. of Appl. Phys.* **101** 103307
[8] Oumghar A, Legrand J C, Diamy A M and Turillon N 1995 *Plasma Chem. Plasma Process.* **15** 87
[9] Janča J, Tálsk A, Krčma F, Hochard L and Ricard A 1994 *Plasma Chem. Plasma Process.* **14** 197
[10] Ricard A, Oseguera-Pena J E, Falk L, Michel H and Gantois M 1990 *IEEE Trans. Plasma Sci.* **18** 940
[11] Martínez H and Yousif F B 2008 *European Physics Journal D* **46** 493
[12] Villa M, Calixto-Rodriguez M, Martínez H, Poveda J C, Reyes P G, Altuzar P 2010, *Plasma Science and Technology* **12**
[13] Pearse R W B, Gaydon A G, *in The identification of molecular spectra,* University Printing House Cambridge, 1976.
[14] Janda M., Martišovitš V., Morová M., Machala Z., Hensel K. 2007 *Eur. Phys. J. D* 45 309

Coronal Mass Ejections and Disturbances in Solar Wind Plasma Parameters in Relation With Short Term Asymmetric Cosmic Ray Intensity Decreases

P.L.Verma[1], Shiva Soni [2] Anil Kumar Pimpalker[3]

[1]Department of Physics, Govt. Vivekanand P.G. College Maihar Satna M.P.India.
[2]Research Scholar Govt. P.G. College Satna M.P.India.
[3]Research Scholar A.P.S.University Rewa M.P.India.

E-mail: pl_verma2003@yahoo.com

Abstract. Coronal mass ejections(CMEs} are a key aspect of coronal and interplanetary dynamics. They can eject large amounts of mass and magnetic field into the heliosphere ,a key source of interplanetary shocks, disturbances in solar wind plasma parameters and asymmetric cosmic ray intensity decreases (Fds). We have studied asymmetric cosmic ray intensity decreases (Fds) (magnitude$\geq$ 3.0%)observed at Oulu super neutron monitor, during the period of 1986-2006 with coronal mass ejections, interplanetary shocks and disturbances in solar wind plasma parameters (solar wind temperature, velocity, density). We have found that 88.46% are associated with halo and partial halo coronal mass ejections (CMEs). The association rate between halo and partial halo coronal mass ejections are found 89.13% and 10.87% respectively. Most of the asymmetric cosmic ray intensity decreases (Fds) are related to interplanetary shocks (76.00%) and the related shocks are forward shocks. We have also found positive co-relation with co-relation co-efficient .46 between magnitude of asymmetric cosmic ray intensity decreases (Fds) and speed of associated coronal mass ejections. From the further study it is concluded that asymmetric cosmic ray variations are strongly related to the disturbances in solar wind plasma parameters. Positive co-relation has been found between magnitudes of asymmetric cosmic ray intensity decreases (Fds) and magnitude of jump in temperature of associated JSWT events with co-relation co-efficient is .25 between these two events. Positive co-relation has also been found between magnitude of asymmetric cosmic ray intensity decreases (Fds) and maximum velocity associated JSWV events with co-relation co-efficient .37 between these two events.

1. Introduction

The galactic cosmic ray intensity remains constant in time and space outside the heliosphere, within the heliosphere, due to various dynamical processes occurring on the sun and extending into interplanetary space, the long term and short term variation in cosmic ray intensity takes place. There are mainly two specific types of cosmic ray intensity depressions, namely corotating and asymmetric short term (Fds) decreases [7, 10, 13,] Forbush decreases, characterized by a fast decrease within ~1 day followed by a more gradual nearly exponential recovery over a few days, have been observed continuously with neutron monitors since the 1950's. Recurrent modulations of galactic cosmic rays, characterized by a slow decrease and a gradual recovery within a period of ~27 days, comparatively, are less impressive changes in cosmic ray intensity.

It has now been proved by the recent studies that short term cosmic ray variations are strongly related to the coronal mass ejections. Cane H.V.[7]have studied cosmic ray intensity variations with coronal mass ejections and concluded that CMEs are large-scale phenomena that change the configuration of the interplanetary magnetic field (IMF) and clearly modulate the cosmic-ray intensity on short-term (few day) timescales .Forbush decreases which falls under the category of asymmetric short term cosmic ray variations are strongly associated with coronal mass ejections and the interplanetary shocks, Badruddin [4] has reported that abrupt onset of decrease in intensity starts upon the arrival of certain shocks and decreases continue till the passage of post shock turbulent sheath. He has further determined that turbulent shocks are much more effective in producing Asymmetric cosmic ray intensity deceases (Fds) than non-turbulent shocks. Cane et al [5,6] have inferred that the short-term cosmic ray decreases are strongly associated with ejecta and shocks. They have reported that 88% short term cosmic ray decreases are associated with ejecta and 70% of these are associated with shocks. Robert F. Penna et a l [11] have investigate the relation between Forbush cosmic ray decrease recovery time and coronal mass ejection transit time between the Sun and EarthIfedili S.O [9] has studied two-step asymmetric cosmic ray intensity decreases (Fds) with coronal mass ejections magnetic clouds, interplanetary shocks and interplanetary disturbances, interplanetary magnetic field (magnitude and direction). Interplanetary coronal mass ejection (ICME) impacting on slow solar wind, there is a sheath upstream of the ICME led by a fast forward shock and the large IMF variations in this sheath, which sustained the Forbush decreases (FDs) in the cosmic ray intensity. P Subhrmanayam et al [12] have studied asymmetric short term cosmic ray intensity decreases (Fds) with coronal mass ejections and inferred that theses decreases (Fds) are associated with front side coronal mass ejections.

E. A. Chuchkov et al [8] have analyzed the modulation structures of quasi-symmetric ("bays") short-term Forbush decreases. It is concluded that these Forbush decreases were recorded due to the stations flying through coronal mass-ejection regions. The short term variation of cosmic ray intensity recorded with ground based monitors have been studied by several scientists [1,2,3,9] and have inferred that the measure cause of asymmetric cosmic ray decreases (Fds) are coronal mass ejections and their interplanetary manifestations.

2. Experimental data

In this work hourly data of oulu super neutron monitor are used to determine asymmetric cosmic ray intensity decreases in cosmic ray intensity .The data of different types of coronal mass ejections have been taken from SOHO – large angle spectrometric, coronagraph (SOHO / LASCO) and extreme ultraviolet imaging telescope (SOHO/EIT) data. The data of interplanetary shocks are taken from shocks arrival derived by WIND group from WIND observations, ACE list of transient and disturbances.). To determine disturbances in solar wind plasma parameters, hourly data of solar wind plasma velocity density temperature has been used and these data has also been taken from omni web data also.

Table -1 Association of Asymmetric Cosmic Ray intensity Decreases with Coronal Mass Ejections and Disturbances in Solar Wind Plasma Parameters

Asymmetric cosmic ray intensity decreases (Fds)				Temperature		Velocity		CMEs		
date	Onset set timedd (hh)	mag%	shocks	Start time dd(hh)	Magnitude of Jump in Deg k	Start time dd(hh)	Magnitude of Jumpkm /s	Date time dd(hh)	Type H/ P	Speed K/s
06.02.86	06(12)	12	06(13)	06(03)	138008	06(04)	84	nd	nd	nd
04.11.86	04(00)	5.5	03(24)	nd	nd	nd	nd	nd	nd	nd
10.04.87	10(12)	5	na	nj	nj	nj	nj	nd	nd	nd
25.08.88	25(06)	6	25(10)	nd	nd	nj	nj	nd	nd	nd
10.10.88	10(00)	5	10(03)	09(02)	521861	09(01)	58	nd	nd	nd
04.01.89	04(18)	5	4(23)	04(14)	18666	04(11)	32	nd	nd	nd
13.02.89	13(00)	4	na	nj	nj	nd	nd	nd	nn	nd
16.02.89	16(08)	3.5	na	nd	nd	nd	nd	nd	nd	nd
13.03.89	18(00)	15	13(01)	nd	nd	nd	nd	nd	nd	nd
27.03.89	27(00)	6	27(14)	nd	nd	nd	nd	nd	nd	nd
01.07.89	01(06)	3	01(07)	nd	nd	nd	nd	nd	nd	nd
08.11.89	08(13)	6.5	08(11)	nd	nd	nj	nj	nd	nd	nd
28.11.89	28(06)	14	27(22)	27(20)	401301	27(19)	195	nd	nd	nd
30.03.90	30(06)	8	30(07)	30(00)	13004	nj	nj	nd	nd	nd
08.04.90	08(13)	10.5	09(09)	nd	nd	nd	nd	nd	nd	nd
27.07.90	27(10)	5	28(01)	nd	nd	nd	nd	nd	nd	nd
01.08.90	01(13)	4.5	01(08)	01(07)	526211	01(06)	169	nd	nd	nd
09.10.90	09(13)	4	09(13)	08(19)	58997	08(21)	114	nd	nd	nd
02.01.91	02(16)	3	na	nd	nd	nd	nd	dn	nd	nd
24.01.91	24(12)	4	23(21)	23(09)	31788	nj	nj	nd	nd	nd
24.03.91	24(02)	16	24(02)	nd	nd	nd	nd	nd	nd	nd
24.04.91	24(20)	8.5	24(21)	24(17)	326958	24(17)	158	nd	nd	nd
23.05.91	23(18)	6	22(00)	23(13)	154262	23(06)	116	nd	nd	nd
12.06.91	12(12)	18	12(10)	11(19)	140534	11(12)	32	nd	nd	nd
08.07.91	08(12)	9	08(17)	08(01)	156329	07(15)	37	nd	nd	nd
12.07.91	12(09)	7	12(09)	12(03)	513598	12(02)	112	nd	nd	nd
18.08.91	18(20)	9.5	18(19)	18(10)	524912	18(11)	204	nd	nd	nd
27.08.91	27(13)	4.5	27(15)	27(13)	319599	27(13)	129	nd	nd	nd
10.09.91	10(16)	3.5	na	10(10)	226419	10(11)	85	nd	nd	nd
28.10.91	28(08)	17	28(11)	28(05)	1176686	10(06)	347	nd	nd	nd
08.11.91	8(04)	8	08(13)	nd		nd	nd	nd	nd	nd
03.02.92	03(02)	3	na	nd	nd	nd	nd	nd	nd	nd
07.02.92	07(12)	5.5	na	06(22)	27029	07(00)	38	nd	nd	nd
26.02.92	26(18)	8	26(17)	nd	nd	nd	nd	nd	nd	nd
01.03.92	01(06)	4	na	nd		nd		nd	nd	nd
17.03.92	17(10)	5	17(10)	nj	nj	17(07)	201	nd	nd	nd
09.05.92	09(13)	8.5	09(20)	08(19)	264466	nj	nj	nd	nd	nd
10.06.92	10(06)	4	10(04)	09(11)	171411	09(08)	88	nd	nd	nd
20.08.92	20(10)	8	19(22)	19(12)	21895	nj	nj	nd	nd	nd
08.09.92	08(10)	7	09(02)	08(22)	67832	09(00)	163	nd	nd	nd
08.10.92	08(06)	5	08(19)	nd	nd	nd	nd	nd	nd	nd
03.11.92	3(12)	4	na	02(17)	291417	nj	nj	nd	nd	nd
09.11.92	09(13)	4	na	09(04)	41619	09(04)	38	nd	nd	nd
02.01.93	02(03)	3.5	na	nd	nd	nd	nd	nd	nd	nd

20.02.93	20(08)	6.5	20(00)	nj		20(00)	163	nd	nd	nd
09.03.93	09(00)	4.5	08(22)	08(18)	995651	08(20)	263	nd	nd	nd
23.03.93	23(06)	4.5	23(22)	nd	nd	nd	nd	nd	nd	nd
01.07.93	01(03)	3.5	na	nd	nd	nd	nd	nd	nd	nd
16.08.93	16(12)	3	15(16)	15(13)	64951	15(13)	45	nd	nd	nd
22.10.93	22(18)		22(12)	nd	nd	nd	nd	nd	nd	nd
03.11.93	03(06)	3	03(17)	nd	nd	nd	nd	nd	nd	nd
18.11.93	18(06)	3.5	18(12)	nd	nd	nd	nd	nd	nd	nd
07.12.93	07(10)	4	07(12)	nd	nd	nd	nd	nd	nd	nd
05.02.94	05(19)	6	06(09)	nd	nd	nd	nd	nd	nd	nd
21.02.94	21(09)	6.5	21(09)	21(07)	2012229	21(08)	414	nd	nd	nd
16.04.94	16(16)	5	na	nd	nd	nd	nd	nd	nd	nd
24.04.94	24(10)	2	na	23(12)	73812	nj	nj	nd	nd	nd
19.06.94	19(10)	3	na	nd	nd	nd	nd	nd	nd	nd
14.07.94	14(10)	3.5	na	nd	nd	nd	nd	nd	nd	nd
22.03.95	22(03)	3.5	na	nj	nj	nj	nj	nd	nd	nd
10.04.97	10(18)	5	na	09(20)	23450	nj	nj	07(14)	H	905
01.05.98	01(20)	6	01(22)	01(19)	326112	01(10)	255	29(16)	H	1602
04.07.98	04(16)	3	na	03(17)	256121	nj	nj	nd	nd	nd
25.08.98	25(12)	8	26(07)	25(00)	86675	25(10)	77	nd	nd	nd
24.09.98	24(12)	10	24(24)	23(17)	209323	23(17)	152	nd	nd	nd
08.11.98	08(04)	7	08(05)	07(00)	250723	07(07)	42	05(21)		1118
22.01.99	22(20)	7	na	22(03)	522927	22(12)	211	nd	nd	nd
12.12.99	12(16)	8	12(16)	12(15)	660214	12(15)	332	na	na	na
11.01.00	11(12)	6	11(14)	11(12)	285691	11(12)	193	na	na	na
07.04.00	07(00)	3	na	06(15)	325941	06(15)	221	04(17)	H	1199
08.06.00	08(08)	8	08(09)	08(03)	781686	08(08)	253	06(16)	H	1130
15.07.00	15(12)	12	15(15)	14(07)	583648	14(12)	209	14(11)	H	1674
14.09.00	14(20)	3	15(05)	15(14)	92477	15(01)	90	12(12)	H	1839
17.09.00	17(12)	8	na	17(13)	803820	17(15)	298	15(22)	H	537
28.10.00	28(00)	7	28(10)	nj	nj	nj	nj	25(08)	H	948
06.11.00	06(16)	7	06(10)	06(09)	382785	06(07)	41	03(18)	H	643
26.11.00	26(12)	8	26(08)	26(08)	552597	26(03)	252	24(05)	H	1298
03.03.01	3(18)	3	03(11)	03(10)	152228	03(10)	69	01(18.26)	P	947
19.03.01	19(03)	4	19(11)	18(04)	58806	19(08)	169	18(02.26)	H	785
26.03.01	26(06)	6	na	nj	nj	nj	nj	24(21)	H	1185
04.04.01	04(16)	8	04(15)	04(12)	1044352	04(14)	282	01(11)	H	1683
07.04.01	07(12)	6	08(11)	06(12)	42862	nj	nj	05(17)	H	1390
11.04.01	11(16)	8.5	11(14)	11(10)	729729	11(12)	233	09(16)	H	1198
28.04.01	28(04)	6	28(05)	27(19)	787156	27(22)	297	26(12)	H	1171
27.05.01	27(12)	4	27(15)	26(13)	531253	26(18)	194	25(17)	P	962
17.08.01	17(16)	7	17(11)	17(08)	299070	17(09)	264	15(24)	H	1575
27.08.01	27(18)	7	27(20)	27(07)	160985	27(07)	34	25(17)	H	1433
25.09.01	25(20)	8	25(20)	25(20)	974243	25(20)	274	23(20)	H	478
11.10.01	11(16)	6	11(17)	11(14)	544030	11(16)	198	09(11)	H	973
21.10.01	21(16)	5	21(17)	21(15)	422395	21(10)	360	19(170	H	901
06.11.01	06(00)	12	06(02)	5(10)	105551	05(14)	141	04(17)	H	1810
24.11.01	24(12)	10	24(06)	24(03)	2292789	24(03)	504	22(20)	H	1443
15.12.01	15(00)	5	na	14(18)	105398	14(19)	69	13(15)	H	864
30.12.01	30(16)	5.5	30(20)	30(02)	61712	nj	nj	28(20)	H	2239
10.01.02	10(16)	4.5	na	10(04)	396719	10(01)	289	08(18)	H	2210
23.05.02	23(12)	5	23(11)	23(03)	1887414	23(03)	468	21(22)	P	900

10.11.02	10(02)	7	9(18)	09(15)	68657	01(15)	50	09(13.3 1)	H	1838
17.11.02	17(00)	8	na	16(23)	82834	16(23)	52	16(07)	H	1250
22.12.02	22(12)	4	22(13)	22(16)	231230	22(10)	59	19(22.0 6)	H	1318
01.02.03	01(16)	5	na	01(13)	612954	01(06)	393	30(10)	P	869
29.05.03	29(16)	7	29(12)	29(14)	489103	29(14)	103	27(24)	H	964
29.10.03	29(00)	25	29(06)	28(08)	979536	nj	nj	28(11)	H	2686
07.01.04	07(00)	8	06(20)	06(18)	555367	06(18)	155	06(06)	P	1469
21.01.04	21(16)	8	22(01)	21(12)	509255	21(16)	163	20(00)	H	1074
26.07.04	26(16)	10	26(23)	26(15)	1532645	26(15)	417	23(16)	H	824
07.11.04	07(08)	12	07(03)	07(10)	786124	07(09)	386	07(17)	H	1759
08.05.05	08(06)	6	07(19)	07(18)	866682	07(17)	215	06(17)	H	1128
15.05.05	15(00)	7	15(02)	14(18)	117968	14(20)	557	13(17)	H	1689
28.05.05	28(20)	10	28(04)	28(07)	187910	28(04)	188	26(15)	H	586
23.08.05	23(20)	7	24(06)	23(19)	2718176	24(00)	308	22(01)	H	1194
11.09.05	11(00)	12	11(01)	11(00)	1876759	10(20)	346	09(20)	H	2257
14.12.06	14(18)	10	14(14)	13(12)	240834	13(06)	312	13(02.5 4)	H	1931

3. Data Analysis and Results

The number of asymmetric cosmic ray intensity decreases (Fds), associated coronal mass ejections and disturbances in solar wind plasma parameters observed during the period of period of 22nd and 23rd solar cycle are listed in table. The total number of asymmetric cosmic ray intensity decreases (Fds are 112.From the data analysis of asymmetric cosmic ray intensity decreases(Fds) and coronal mass ejections it is observed that out of 112 asymmetric cosmic ray intensity decreases (Fds) we have no data of CMEs for 47 asymmetric cosmic ray intensity decreases (Fds) for association. We have CMEs data for 52 asymmetric cosmic ray intensity decreases (Fds) in which 46 (88.46%) are found to be associated with coronal mass ejection (CMEs). From the further analysis these CMEs associated asymmetric cosmic ray intensity decreases (Fds) 05 (10.87%) are found to be associated with partial halo CMEs and 41 (89.13%) are found to be associated with full halo coronal mass ejections. To know the relationship between magnitude of asymmetric cosmic ray intensity.

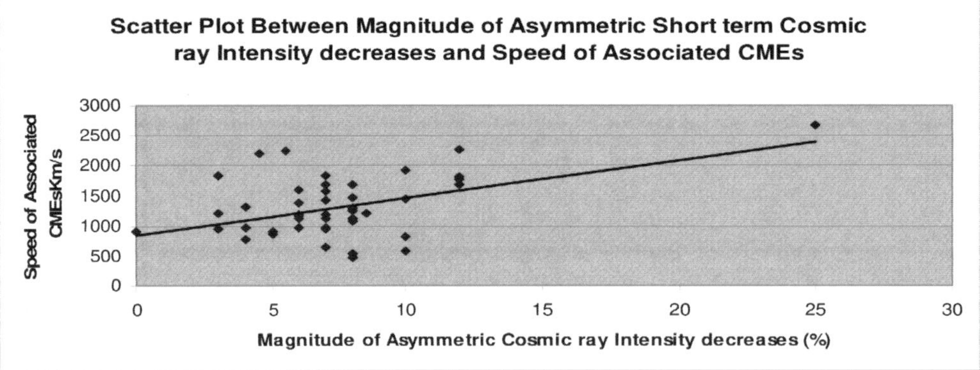

Figure 1 Shows scatter plot between magnitude of asymmetric cosmic ray intensity decreases and speed of associated CMEs showing positive correlation with correlation coefficient 0.46

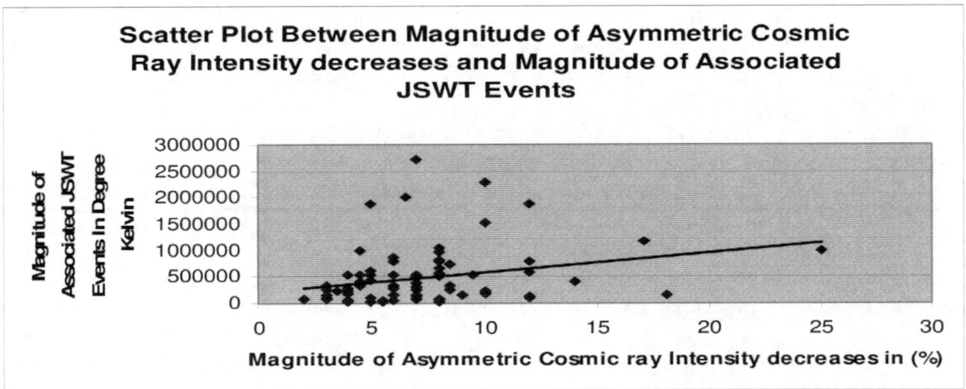

Figure 2 Shows Scatter plot between magnitude of geomagnetic storms and magnitude of Jump in solar wind plasma temperature (JSWT) showing positive correlation with correlation coefficient 0.25

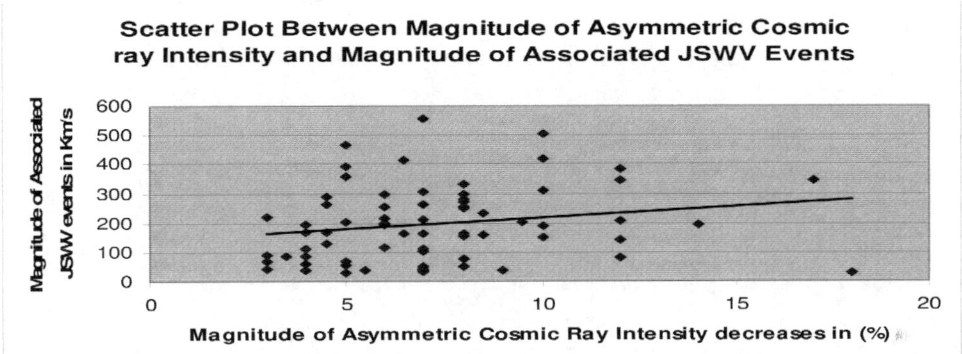

Figure 3 Shows Scatter plot between magnitudes of geomagnetic storms and magnitude of Jump in solar wind plasma velocity (JSWV) showing positive correlation with correlation coefficient 0.37

decreases and speed of associated CMEs we have plotted scatter diagram between magnitude of asymmetric cosmic ray intensity decreases and speed of associated CMEs .The resulting diagram is shown in fig 1 and figure shows positive correlation between these two events .Statistically calculated co-relation co-efficient .46 have bee found between magnitude of asymmetric cosmic ray intensity decreases (Fds) and speed of associated coronal mass ejections.

From the data analysis of asymmetric short term cosmic ray decreases (Fds)and jump in solar wind plasma temperature we have found that the most majority of Asymmetric (Fds) short term cosmic ray decreases are found to be associated with JSWT events. We have 112 Asymmetric (Fds) short term cosmic ray decreases data for which is available for association with JSWT events are 85 .Out of these 85 events, 78 asymmetric (Fds) short term cosmic ray decreases are (91.76%) found to be associated with jump in solar wind plasma temperature.

To sees how the magnitude of asymmetric cosmic ray intensity decreases (Fds) is correlated with JSWT events. We have plotted a scatter diagram between the magnitude of asymmetric cosmic ray intensity decreases (Fds) and magnitude JSWT events in fig.2 It is clear from the fig that most of the asymmetric cosmic ray intensity decreases (Fds) which have large magnitude are associated with such JSWT events which have large magnitude, but the magnitude of these two events do not have any fixed proportion, we have found some asymmetric cosmic ray intensity decreases (Fds) which have large magnitude but they are associated with such JSWT events which have small magnitude. Positive co-relation has been found between magnitudes of asymmetric cosmic ray intensity decreases (Fds) and magnitude of jump in temperature of associated JSWT events. Statistically calculated co-relation co-efficient is .25 between these two events.

The asymmetric cosmic ray intensity decreases are also associated with jump in Solar wind plasma velocity (JSWV). From the analysis we have112 asymmetric cosmic ray intensity decrease (Fds) but data of JSWV events for association are available for 86 asymmetric cosmic ray intensity decrease (Fds). From the data analysis we have obtained that asymmetric cosmic ray intensity decrease (Fds) are closely associated with JSWV events. .Out of these, 86, 69 (80.23%) asymmetric cosmic ray intensity decrease (Fds) are associated with JSWV events.

To see how the magnitude of asymmetric cosmic ray intensity decreases (Fds) are correlated with the magnitude of JSWV events, we have plotted a scatter diagram between the magnitude of asymmetric cosmic ray intensity decreases (Fds) and JSWV events in fig.3 From the figure it is clear that maximum asymmetric cosmic ray intensity decreases (Fds) which have large magnitude are associated with such JSWV events which have relatively large magnitude but magnitude of these two events do not have any quantitative relation their amplitude do not have any fixed proportion. We have found some asymmetric cosmic ray intensity decreases (Fds) which have large magnitude but they are associated with such JSWV events which have small magnitude. Positive co-relation has been found between magnitude of asymmetric cosmic ray intensity decreases (Fds) and magnitude of associated JSWV events. Statistically calculated co-relation co-efficient is. 37 between these two events.

4. Conclusion

From our study 46 out of 52 asymmetric cosmic ray intensity decreases (Fds) (88.46%) are found to be associated with coronal mass ejection (CMEs). The association rates of halo and partial halo coronal mass ejections have been found 89.13% and 10.87 % respectively. Positive co-relation with co-relation co-efficient .46 have bee found between magnitude of asymmetric cosmic ray intensity decreases (Fds) and speed of associated coronal mass ejections. Most of the asymmetric cosmic ray intensity decreases have been found associated with interplanetary shocks. Positive correlation have also been found between magnitude of asymmetric cosmic ray intensity decreases and associated jump in solar wind temperature and solar wind velocity .From theses results it is concluded that majority of the asymmetric cosmic ray intensity decreases(Fds) are caused by halo coronal mass ejections and interplanetary shocks, disturbances in solar wind plasma parameters that they generate .

References

[1] Badruddin, Astrophys. Space Sci., Vol. 246, pp. 171, 1997.
[2] Badruddin, Nuvo Cimento, Vol. 23C, pp. 217, 2000.
[3] [Badruddin, Phys., Vol. 209, pp. 195, 2002.
[4] Badruddin and Singh, Y. P Ind. J. Phys., Vol. 77, pp. 497, 2003
[5] Cane H.V.I.G. Richardson and T.T. Von Rosenving J. Geophys. Res. 10(A10) 21-561, 21572, 1996.
[6] Cane H.V.I.G. Richardson, and G.wibberenze J. Geophys.Res. 102(A4) 7086,1997.
[7] Cane, H. V. Space Sci. Rev., Vol. 93, pp. 55-77, 2000
[8] Chuchkov E. A. Tulupov V. I. . Okhlopkovand V. P Lyubimov G. P. Moscow University Physics Bulletin Volume 64, Number 3 / June, 2009
[9] Ifedili, S. O. J. Geophys. Res., Vol. 109, A02117, 2004.
[10] Richardson, I. G. Space Sci. Rev., Vol. 111, pp. 267, 2004.
[11] Robert F. Penna , Alice C. Quillen, J. Geophys. Res. Vol.110, A09S05, 2005.
[12] Subramanium P Antia H.M.Dugad S.R.GoswamiU.D.,Gupta S.K.Mohanty Nonaka P.K.,Nayak T.,.Tanaka P.K.Tanwar S. C Proc. 29thInt. Cosmic Ray Conf., Pune, Vol. 2, pp. 73, 2005.
[13] Venkatesan D. and Badruddin, Space Sci. Rev., Vol. 52, pp. 121, 1990.

ICPP2010 & LAWPP2010 IOP Publishing
Journal of Physics: Conference Series **511** (2014) 012057 doi:10.1088/1742-6596/511/1/012057

Coronal Mass Ejections, Interplanetary Shocks In Relation With Forbush Decreases Associated With Intense Geomagnetic Storms

P.L.Verma[1], Nand Kumar Patel [2] Mateswari Prajapati[3]
[1]Department of Physics, Govt. Vivekanand P.G. College Maihar Satna M.P.India
[2]Research Scholar, A.P.S. University Rewa M.P.India.
[3]Research Scholar Govt.P.G.College Satna M.P. India.

E-mail: pl_verma2003@yahoo.com

Abstract. Coronal mass ejections (CMEs} are the most energetic solar events in which large amount of solar plasma materials are ejected from the sun into heliosphere, causing major disturbances in solar wind plasma ,Interplanetary shocks , Forbush decrease(Fds) in cosmic ray intensity and geomagnetic storms. We have studied Forbush decreases associated with intense geomagnetic storms observed at Oulu super neutron monitor, during the period of May 1998-Dec 2006 with coronal mass ejections (CMEs), X-ray solar flares and interplanetary shocks. We have found that all the (100%) Forbush decreases associated with intense geomagnetic storms are associated with halo and partial halo coronal mass ejections (CMEs). The association rate between halo and partial halo coronal mass ejections are found 96.00%and 04.00% respectively. Most of the Forbush decreases associated with intense geomagnetic storms (96.29%) are associated with X-ray solar flares of different categories . The association rates for X-Class, M-Class, and C- Class X –ray solar flares are found 34.62%, 50.00% and 15.38% respectively .Further we have concluded that majority of the Forbush decrease associated with intense geomagnetic storms are related to interplanetary shocks (92.30 %) and the related shocks are forward shocks. We have found positive co-relation with co-relation co-efficient .7025 between magnitudes of Forbush decreases associated with intense geomagnetic storms and speed of associated coronal mass ejections. Positive co-relation with co-relation co-efficient 0.48 has also been found between magnitudes of intense geomagnetic storms and speed of associated coronal mass ejections.

1. Introduction

Coronal mass ejections (CMEs) are magnetized structures, which can affect the heliospheric conditions, producing large fluctuations in the heliospheric magnetic field. CMEs traveling at different speeds tend to merge into what are known as complex ejecta, which are seen often in the interplanetary medium during times of high solar activity. The increase of the magnetic field during the passage of ejecta at 1 AU is related to the GCR intensity decrease [6,]. Cane H.V.[5]have studied cosmic ray intensity variations with coronal mass ejections and concluded that CMEs are large-scale phenomena that change the configuration of the interplanetary magnetic field (IMF) and clearly modulate the cosmic-ray intensity on short-term (few day) time scales .

Forbush decreases which falls under the category of asymmetric short term cosmic ray variations have been studied by several scientists are strongly associated with coronal mass ejections and the interplanetary shocks, magnetic clouds, ejecta which are interplanetary manifestations of coronal mass ejections decreases [17,1,3 , ,4 ,10, , 14,7 ,] . Zhang and Burlaga [17] concluded that relatively large decreases in cosmic ray intensity is associated with magnetic clouds that are preceded by a shock, whereas only a small decreases in cosmic ray intensity is associated with magnetic clouds that are not preceded by shock. Cane et al [3,4.] have inferred that the short-term cosmic ray decreases are strongly associated with ejecta and shocks. Robert F. Penna et a l [12] have investigate the relation between Forbush cosmic ray decrease recovery time and coronal mass ejection transit time between the Sun and EarthIfedili S.O [10] has studied two-step asymmetric cosmic ray intensity decreases (Fds) with coronal mass ejections magnetic clouds, interplanetary shocks and interplanetary disturbances, interplanetary magnetic field (magnitude and direction). Interplanetary coronal mass ejection (ICME) impacting on slow solar wind, there is a sheath upstream of the ICME led by a fast forward shock and the large IMF variations in this sheath, which sustained the Forbush decreases (FDs) in the cosmic ray intensity. P Subhrmanayam et al [14] have studied asymmetric short term cosmic ray intensity decreases (Fds) with coronal mass ejections and inferred that theses decreases (Fds) are associated with front side coronal mass ejections.

Several investigators have studied geomagnetic storms with various solar features, solar wind parameters and inferred that CMEs are responsible for the most geoeffective solar wind disturbances and, therefore the largest geomagnetic storms.R - Landi and G - Moreno et al [13] have investigated the role of the coronal mass ejections in producing non recurrent geomagnetic storms in period 1969-1974. They have concluded that coronal mass ejections associated with chromospheric flares, accompanied by type IV radio emission, are the most effective in perturbing the geomagnetic field.N. Gopalswamy et al [9] have studied geoeffectiveness, speed, solar source, and flare association of a set of 378 halo coronal mass ejections (CMEs) of solar cycle 23 (1996-2005). They have compiled the minimum Dst values occurring within 1 - 5 days after the CME onset. They have compared the distribution of such Dst values for the subset of halo CMEs: — disk halos, limb halos, and back side halos CMEs. Defining that a halo CME is geoeffective if it is followed Dst $\leq$ -50 nT, moderately geoeffective if -50nT < Dst < -100nT, and strongly geoeffective if Dst $\leq$ -100nT, they have found that the disk halos are followed by strong geomagnetic storms, limb halos are followed by moderate storms, and back side halos are not followed by significant storms. Badruddin; Singh, Y. P. [2] have studied the subset of CMEs, called interplanetary magnetic clouds (MCs), observed in the heliosphere and their associated features (shock/sheath, IR and HSS). Best-fit equations representing the relation between level of the geomagnetic activity and interplanetary plasma/field parameter have been obtained..Zhao, et al [16]) studied the source locations of 130 solar flares associated with type II radio burst events and interplanetary shocks observed by L1 spacecraft during February 1997-August 2002. They have investigated the relative positions between the flare sources, the heliospheric current sheet (HCS), and the Earth. They have found that (1) Solar flares are usually distributed within [S30°, N30°] in heliographic latitude and [S30°, N30°] × [E10°, W30°] are the predominant source region on the solar disk that includes the majority of geoeffective solar flares.Michalek, G.et al [11] have concluded that halo coronal mass ejections (HCMEs) originating from regions close to the center of the sun are likely to be geoeffective. They have showed that, only fast halo CMEs (with space velocities higher than ~1000 km/s) and originating from the Western Hemisphere close to the solar center could cause intense geomagnetic storms. Echer, E et al [8] have analyzed plasma and magnetic field parameter variations across fast forward interplanetary shocks during the last solar cycle minimum (1995 1996, 15 shocks), and maximum year 2000 .They have observed that the solar wind velocity and magnetic field strength variation across the shocks are the parameters better Verma P.L. et al [15] have studied geomagnetic storms Dst < - 50nT observed during the period of 1997-2006, with halo and partial halo coronal mass ejections associated with X-ray solar flares of different categories and concluded that they have concluded that majority of the observed geomagnetic storms are found that halo and partial halo CMEs associated with X ray solar flares are most potential candidates for production of geomagnetic storms.

2. Experimental Data

In this work hourly data of oulu super neutron monitor have been used to determine Forbush decreases in cosmic ray intensity.The data of different types of coronal mass ejections have been taken from SOHO – large angle spectrometric, coronagraph (SOHO / LASCO) and extreme ultraviolet imaging telescope (SOHO/EIT) data. The data of interplanetary shocks are taken from shocks arrival derived by WIND group from WIND observations, ACE list of transient and disturbances. The data of X ray solar geophysical data report U.S.Department of commerce, NOAA monthly issue and ftp://ftp.ngdc.noaa.gov/STP/SOLAR_DATA/ftp://ftp.ngdc.noaa.gov/STP/SOLAR_DATA/SOLAR_FLARES/FLARES_XRAY/data have been used.

3. Data Analysis and Results

Forbush decreases associated with intense geomagnetic storms, coronal mass ejections and interplanetary shocks during the period of 1998 to 2006 are listed in table.

Table 1 Association of Forbush decreases with intense geomagnetic storms, coronal mass ejections and interplanetary shocks for the perid of 1998 to 2006

(Fds)					Shocks	CMEs			SOLAR FLARES	
date	Onset set timedd (hh)	mag %	Onset set timedd (hh)	Magnitude in nT	Shock	Date time dd(hh)	Type H/P	Speed K/s	Date time dd(hh)	Class
01.05.98	01(20)	6	02(04)	-214	01(22)	29(16)	H	1602	29(29)	M-68
25.08.98	25(12)	8	26(09)	-171	26(07)	nd	nd	nd	24(22)	X10
24.09.98	24(12)	10	25(00)	-203	24(24)	nd	nd	nd	23(07)	M-71
08.11.98	08(04)	7	08(20)	-126	08(05)	05(21)	H	1118	06(15)	M-17
07.04.00	07(00)	3	06(16)	-282	na	04(17)	H	1199	06(02)	M-18
15.07.00	15(12)	12	15(15)	-308	15(15)	14(11)	H	1674	14(11)	X-57
17.09.00	17(12)	8	17(20)	-197	na	15(22)	H	537	16(04)	M-59
28.10.00	28(00)	7	28(21)	-126	28(10)	25(08)	H	948	25(09)	C-40
06.11.00	05(12)	7	06(10)	-134	06(10)	03(18)	H	643	03(20)	C-53
26.11.00	26(12)	8	26(22)	-127	26(08)	24(05)	H	1298	24(05)	X-20
19.03.01	19(03)	4	19(11)	-150	19(11)	18(02.26)	H	785	16(10)	C-54
11.04.01	11(12)	8.5	11(15)	-269	11(14)	09(16)	H	1198	09(16)	M-79
17.08.01	17(16)	7	17(17)	-102	17(11)	15(24)	H	1575	14(10)	C-97
25.09.01	25(20)	8	25(22)	-102	25(20)	23(20)	H	478	23(11)	M-11
21.10.01	21(16)	5	21(16)	-178	21(17)	19(170	H	901	19(16)	X-16
06.11.01	05(12)	12	05(19)	-297	06(02)	04(17)	H	1810	04(16)	X-10
24.11.01	24(12)	10	24(06)	-223	24(06)	22(20)	H	1443	22(20)	M-38
29.10.03	29(00)	25	29(06)	-384	29(06)	28(11)	H	2686	28(10)	X-172
21.01.04	21(16)	8	22(05)	-144	22(01)	20(00)	H	1074	20(07)	M-61
26.07.04	26(16)	10	26(23)	-150	26(23)	23(16)	H	824	23(17)	M-22
07.11.04	07(08)	12	07(10)	-150	07(03)	07(17)	H	1759	07(16)	X-20
08.05.05	08(06)	6	07(20)	-126	07(19)	06(17)	H	1128	06(11)	M-13
15.05.05	15(00)	7	15(05)	-293	15(02)	13(17)	H	1689	13(16)	M-80
23.08.05	23(20)	7	24(08)	-219	24(06)	22(01)	H	1194	22(01)	M-26
11.09.05	11(00)	12	11(02)	-127	11(01)	09(20)	H	2257	09(19)	X-62
14.12.06	14(18)	10	14(21)	-143	14(14)	13(02.54)	H	1931	13(02)	X-34

From the data analysis given in table it is observed that the total number of Forbush decreases associated with intense geomagnetic storms is 26.The onset time of all the associated intense geomagnetic storms are found after the onset time of Forbush decreases. We have found positive correlation with correlation coefficient .48 between magnitude of Forbush decreases and magnitude of associated geomagnetic storms. From the data analysis of Forbush decreases associated with interplanetary shocks it is concluded that majority of the Forbush decreases associated with intense geomagnetic storms are also related to interplanetary shocks .We have 26 Forbush decreases in our list out of which 24 (92.30%) are found to be associated with interplanetary shocks .It is also observed that the related shocks are forward shocks .The onset time of majority of the Forbush decreases are found at ± 10 hours time lag between onset time of Forbush decreases and arrival time of interplanetary shocks. From the data analysis of Coronal mass ejections and selected Forbush decreases, we have determined that all the Forbush decreases are found to be associated with coronal mass ejections. We have 26 Forbush decreases in our list in which, we have no data of CMEs for association for two events.24 out of 24 Forbush decreases are found to be associated with coronal mass ejections .Out of these associated Forbush decreases 23 out of 24 are found to be associated with halo coronal mass ejections and one is related to partial halo coronal mass ejections .The association rates of halo and partial halo CMEs are found 96.00%and 04.00% respectively.Positive co-relation with correlation coefficient .7025 have been found between magnitudes of Forbush decreases associated with intense geomagnetic storms and speed of associated coronal mass ejections .From the data analysis of Forbush decrease and X ray solar flares we have observed that the most of the Forbush decreases associated with intense geomagnetic storms (96.29%) are associated with X- ray solar flares of different categories .The association rates for X-Class, M-Class, and C- Class X –ray solar flares are found 34.62%, 50.00% and 15.38% respectively.

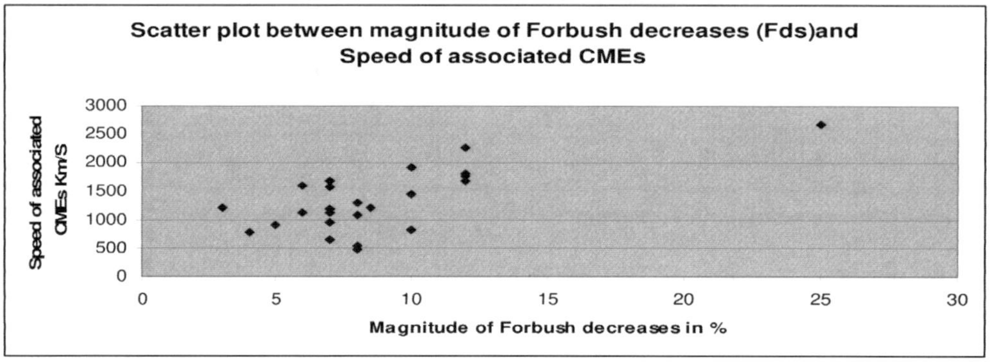

Figure 1 Shows Scatter plot between Magnitude of Forbush decrease associated with intensegeomagnetic storms and speed of associated CMEs

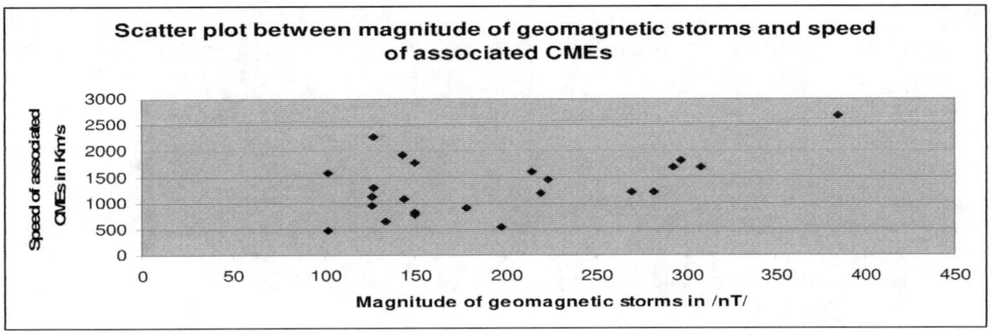

Figure 2 Shows Scatter plot between Magnitudes of intense geomagnetic storms and speed of associated CMEs

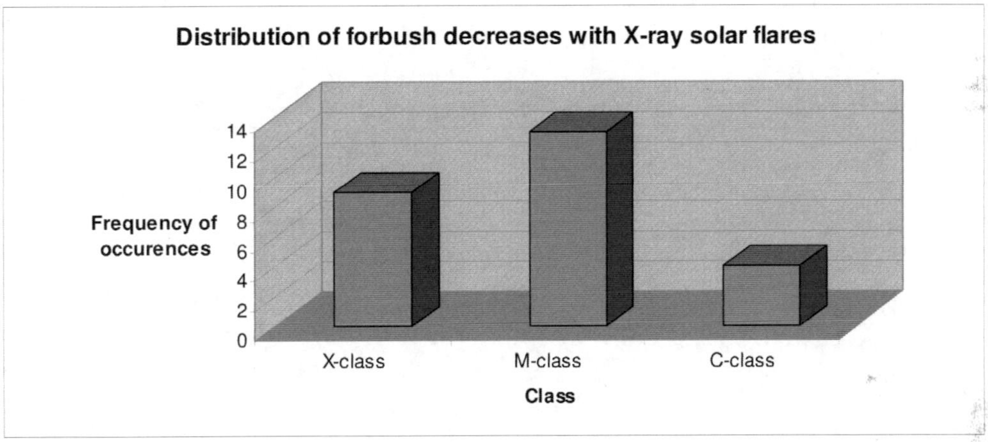

Figure 3 Shows distribution of Forbush decreases associated with intense geomagnetic storms and associated X ray solar flares

4. Conclusion

From our study we have found that Forbush decreases associated with intense geomagnetic storms are associated with halo and partial halo coronal mass ejections (CMEs). The association rate between halo and partial halo coronal mass ejections are found 96.00%and 04.00% respectively. Most of the Forbush decreases associated with intense geomagnetic storms (96.29%) are associated with X- ray solar flares of different categories. The association rates for X-Class, M-Class, and C- Class X –ray solar flares are found 34.62%, 50.00% and 15.38% respectively .Further we have concluded that majority of the Forbush decrease associated with intense geomagnetic storms are related to interplanetary shocks (92.30 %) and the related shocks are forward shocks. We have found positive co-relation with co-relation co-efficient .7025 between magnitudes of Forbush decreases associated with intense geomagnetic storms and speed of associated coronal mass ejections. Positive co-relation with co-relation co-efficient 0.48 has also been found between magnitudes of intense geomagnetic storms and speed of associated coronal mass ejections. From the above results it is concluded that Forbush decreases associated with intense geomagnetic storms are closely related with halo coronal mass ejections associated with X ray solar flares of different categories and interplanetary shocks .Hence These Fds are caused by halo coronal mass ejections associated with hard X ray solar flares and interplanetary shocks that they generate.

References

[1] Badruddin and Singh, Y. P Ind. J. Phys., Vol. 77, pp. 497, 2003.

[2] Badruddin; Singh, Y. P. Planetary and Space Science, Vol. 57, Issue 3, 318, 2009.

[3] Cane H.V.I.G. Richardson and T.T. Von Rosenving J. Geophys. Res. 10(A10) 21-561, 21572, 1996.

[4] Cane H.V.I.G. Richardson, and G.wibberenze J. Geophys.Res. 102(A4) 7086, 1997.

[5] Cane, H. V. Space Sci. Rev., Vol. 93, pp. 55-77, 2000

[6] Cliver, E. W., Ling, A. G., & Richardson, I. G. ApJ, 592, 574,2003.

[7] Chuchkov E. A. Tulupov V. I. . Okhlopkovand V. P Lyubimov G. P. Moscow University Physics Bulletin Volume 64, Number 3 / June, 2009

[8] Echer, E M V Alves and W D Gonzalez, Solar Phys. 221, 361,2004.

[9] Gopalswamy, N S Yashiro and S Akiyama, *J.* Geophys. Res. 112, A06112, 2007.

[10] Ifedili, S. O. J. Geophys. Res., Vol. 109, A02117, 2004.

[11] Michalek, G. Gopalswamy, N. Lara, A Yashiro, S Space. Weather, Volume 4, Issue 10th, 2006.

[12] Robert F. Penna , Alice C. Quillen, J. Geophys. Res. Vol.110, A09S05, 2005

[13] R. Landi and G. moreno J. Geophys. Res Vol. 103 No. A_{20}, 553 – 20 –559 – 1998.

[14] Subramanium P Antia H.M.Dugad S.R.GoswamiU.D.,Gupta S.K.Mohanty Nonaka P.K.,Nayak T.,.Tanaka P.K.Tanwar S. C Proc. 29thInt. Cosmic Ray Conf., Pune, Vol. 2, pp. 73, 2005.

[15] Verma P.L. Tripathi A.K. & Sharma ,Sushil J. Plasma Fusion Res. SERIES, Vol. 8 Page 221-225, 2009.

[16] Zhao, Xinhua; Feng, Xueshang; Wu, Chin-Chun J. Geophys. Res .Vol.112, Issue A6, 2007.

[17] Zhang G. and Burlaga, L. F. J. Geophys. Res., Vol. 93, pp. 2511, 1988.

Symmetric Cosmic Ray Intensity Decreases in Relation with Coronal Mass Ejections and Disturbances In Solar Wind Plasma Parameters

P.L.Verma[1], Omkar Tripathi [2], Ashok Kumar Vishwkarma[3]
[1]Department of Physics, Govt.Vivekanand P.G. College Maihar Satna M.P.India
[2, 3]Research Scholar, Govt. P.G.College Satna M.P.India.

E-mail: pl_verma2003@yahoo.com

Abstract. We have studied symmetric cosmic ray intensity decreases (V Shape, and U Shape) magnitude ≥2.0%, observed at Oulu super neutron monitor, during the period of 1986-2006 with coronal mass ejections CMEs), interplanetary shocks and disturbances in solar wind plasma parameters (solar wind temperature, velocity, density). We have found that 76.92% V shape symmetric cosmic ray intensity decreases are associated with halo and partial halo coronal mass ejections (CMEs). The association rate between halo and partial halo coronal mass ejections are found 40.00%and 60.00% respectively. Most of the V Shape symmetric cosmic ray intensity decreases are related to interplanetary shocks (61.53%) and the related shocks are forward shocks. We have also found positive co-relation with co-relation co-efficient 0.54 between magnitude of V shape symmetric cosmic ray intensity decreases and speed of associated coronal mass ejections. From the further study it is concluded that V shape symmetric cosmic ray decreases are strongly related to the disturbances in solar wind plasma parameters. Negative co-relation has been found between magnitudes of V Shape symmetric cosmic ray intensity decreases and magnitude of jump in solar wind plasma density of associated JSWD events with co-relation co-efficient -0.152 between these two events. Negative co-relation has also been found between magnitude of V Shape asymmetric cosmic ray intensity decreases and magnitude of jump in solar wind plasma velocity of the associated JSWV events with co-relation co-efficient -0.48 between these two events. From the study of U Shape symmetric cosmic ray intensity decreases and disturbances in solar wind plasma parameters, we found negative co-relation with correlation coefficient, -0.473 between magnitude of U shape symmetric cosmic ray intensity decreases and magnitude of jump in solar wind plasma temperature of associated JSWT events, -0.468 between magnitude of U shape symmetric cosmic ray intensity decreases and magnitude of jump in solar wind plasma density of associated JSWD events and -0.26 between U shape symmetric cosmic ray intensity decreases and magnitude of jump in solar wind plasma velocity of associated JSWV events.

1. Introduction

The solar modulation of Galactic cosmic-ray (GCR) intensity has been known since the work of Forbush (1954) and these variations have been classified according to the duration of variation, long term and short term variations. The short-term modulation in galactic cosmic ray intensity, are two types namely corotating and asymmetric short term (Fds) decreases [9,19,21,].Forbush decreases [FDs] are transient and rapid decreases in galactic cosmic ray intensity followed by a more gradual recovery phase typically lasting several days. These decreases are not only observed by ground based detector's but are also observed by space born detector's and so are present in interplanetary medium [22,23,]. These decreases are closely related to coronal mass ejections and interplanetary shocks that they generate. [7,8,20].Recurrent modulations of galactic cosmic rays, characterized by a slow decrease and a gradual recovery within a period of ~27 days, comparatively, are less impressive changes in cosmic ray intensity. Since their discovery through world-wide distribution of ion-chambers, these decreases have been extensively studied to search for their solar source [5,10,] interplanetary structure responsible [3,4,6,9,12,14,16] and the mechanism(s) playing major role in this phenomenon [1, 7,19,].The recurrent or corotating decreases in cosmic ray intensity have been analyzed by several scientists [13,19] it has been found that these decreases are closely related to coronal holes, corotating interaction regions. Some other scientists have also studied these variations with several interplanetary parameters [Richardson et al. [15,17,18 22] to identify potential sources and relative role of corotating interaction regions, high speed streams and heliospheric current sheet in corotating modulation. It is concluded polar coronal holes have been identified as potential solar source of co rotating decreases. Although solar polar coronal holes have been identified as potential solar source of corotating decreases, the relative role of corotating interaction regions, high speed streams and heliospheric current sheet in corotating modulation is yet to be decided.

2. Experimental Data

In this work monthly and hourly data of oulu super neutron monitor have been used to determine V shape and U shape symmetric cosmic ray decreases The data of coronal mass ejections (CMEs) will be taken from SOHO – large angle spectrometric, coronagraph (SOHO / LASCO) and extreme ultraviolet imaging telescope (SOHO/EIT) data. The data of interplanetary shocks have been taken from shocks arrival derived by WIND group from WIND observations, ACE list of transient and disturbances. To determine disturbances in solar wind plasma parameters, hourly data of solar wind plasma velocity density temperature average interplanetary magnetic field has been used these data has also been taken from omni web data.

3. Data Analysis and Results

The total number of V shape and U shape symmetric cosmic ray intensity decreases and associated coronal mass ejections ,interplanetary shocks ,disturbances in solar wind plasma parameters are listed in table no 1 and table no 2.

From the data analysis of V shape symmetric cosmic ray intensity decreases and associated coronal mass ejections it is observed that the number of V shape symmetric cosmic ray intensity decreases during the period of 1986 to 2006 have been found 26.We have no data of CMEs for 13 V-shape symmetric short term cosmic ray decreases for association. The CMEs data available for association is for 13 V shape symmetric cosmic ray intensity decreases, in which 10 (76.92%) V shape symmetric cosmic ray intensity decreases are found to be associated with coronal mass ejections .Out of theses associated CMEs 06 (60%) V-shape symmetric short term cosmic ray decreases are found to be associated with partial halo CMEs where as 04 (40%) V-shape symmetric short term cosmic ray decreases are found to be associated with full halo coronal mass ejections. We have also found positive co-relation with co-relation co-efficient 0.54 between magnitude of V shape symmetric cosmic ray intensity decreases and speed of associated coronal mass ejectionsFrom the data analysis of V shape cosmic ray intensity decreases and interplanetary shocks,16 out of 26 (61.54) V shape symmetric cosmic ray decreases found to be associated with interplanetary shocks. The onset time of 12.5 % V shape symmetric cosmic ray decreases are found after the arrival time of interplanetary shocks where as the onset time of 87.5% V shape symmetric cosmic ray decreases are found before the arrival time of interplanetary shocks. The onset time of half of the V shape symmetric cosmic ray decreases are

Journal of Physics: Conference Series **511** (2014) 012058 doi:10.1088/1742-6596/511/1/012058

found at ± 10 hours time lag between onset time of V shape symmetric cosmic ray decreases and arrival time of interplanetary shocks We have obtained 16 V shape symmetric cosmic ray decreases, which are associated with interplanetary shocks out of which 08(50%) are found at ± 10 hours time lag between onset time of V symmetric cosmic ray decreases and arrival time of interplanetary shocks.From the data analysis of V shape symmetric cosmic ray decreases and JSWD events. We have obtained that the maximum V-shape Symmetric short term cosmic ray decreases events are associated with JSWD events. We have 26 V-shape Symmetric short term cosmic ray decreases in which available data for association with JSWD events are 23, out of which 19 V-shape Symmetric short term cosmic ray decreases (82.61%) are found to be associated with JSWD events. To see how the magnitude of V-shape Symmetric short term cosmic ray decreases is correlated with magnitude of JSWD events. We have plotted a scatter diagram between the magnitude of V-shape Symmetric short term cosmic ray decreases and magnitude JSWD events and resulting figure is shown in fig1. From the fig it is clear that, most of the V-shape Symmetric short term cosmic ray decreases which have large magnitude are associated with such JSWD events which have relatively small magnitude.Negative co-relation has been found between magnitudes of V Shape symmetric cosmic ray intensity decreases and magnitude of jump in solar wind plasma density of associated JSWD events with co-relation co-efficient -0.152 between these two events.

Table No 1 Association of Symmetric V shape Cosmic Ray Intensity with Coronal Mass Ejections and Disturbances in Solar Wind Plasma Parameter

Date	Onset set timedd (hh)	ma g%	Type of decre ases	SSC	Start time dd(hh)	Magni tude of Jump n/cc	Start time dd(hh)	Magni tude of Jump km/s	Date time dd(hh)	TypeH /P	S eed K/s
05.03.86	05(16)	3.5	V	06(11)	04(17)	17.2	04(18)	145	nd	nd	nd
07.10.87	07(00)	3	V	na	06(21)	2.3	nj	nj	nd	nd	nd
09.12.87	09(12)	4	V	9(20)	08(23)	4.2	nj	nj	nd	nd	nd
09.02.88	09(12)	6	V	na	08(22)	17.2	09(05)	47	nd	nd	nd
05.05.88	05(08)	4	V	6(04)	05(02)	4.2	05(05)	210	nd	nd	nd
23.06.88	23(12)	3	V	24(00)	nj	nj	22(19)	77	nd	nd	nd
11.01.89	11(13)	4.5	V	11(12)	nd	nd	nd	nd	nd	nd	nd
15.01.89	15(10)	4	V	na	14(12)	7.7	14(11)	67	nd	nd	nd
27.12.89	27(20)	6	V	na	nd	nd	nd	nd	nd	nd	nd
12.03.90	12(10)	6	V	12(15)	nd	nd	nd	nd	nd	nd	nd
27.04.91	27(14)	3	V	na	27(18)	4.1	nj	nj	nd	nd	nd
01.01.92	01(10)	4.5	V	01(17)	31(11)	6.1	nj	nj	nd	nd	nd
26.11.94	26(04)	3	V	26(07)	25(13)	37.4	25(13)	58	nd	nd	nd
06.10.97	06(08)	3	V	na	10(05)	4.8	nj	nj	na	na	na
17.11.97	17(12)	6	V	na	16(04)	8	16(11)	46	14(13.36)	P	967
09.12.97	9(12)	3	V	10(04)	nj	nj	nj	nj	06(10.27)	P	567
06.01.98	6(00)	3	V	6(14)	06(13)	19.1	06(12)	114	02(23.28)	H	548
05.06.98	05(18)	5	V	5(10)	04(22)	13.4	05(12)	41	04(02.04)	H	1821
23.10.98	23(12)	4	V	na	23(01)	7.5	23(04)	107	na	na	na
12.09.99	12(02)	3	V	12(04)	12(03)	31.4	12(02)	204	10(17.30)	P	723
12.10.00	12(04)	4	V	12(22)	11(05)	1.7	11(08)	40	09(23.50)	H	881
09.01.03	9(18)	3	V	na	09(12)	13.1	09(08)	135	07(08.30)	P	513
07.04.03	07(12)	4	V	08(01)	07(04)	33.2	nj	nj	04(21.50)	P	257
02.04.04	2(18)	3	V	3(10)	nj	nj	nj	nj	29(00.40)	P	489

05.08.05	5(12)	3	V	na	nj	nj	nj	nj	na	na	na
09.07.06	9(18)	3	V	9(21)	09(15)	13.8	09(16)	99	06(08.54)	H	995

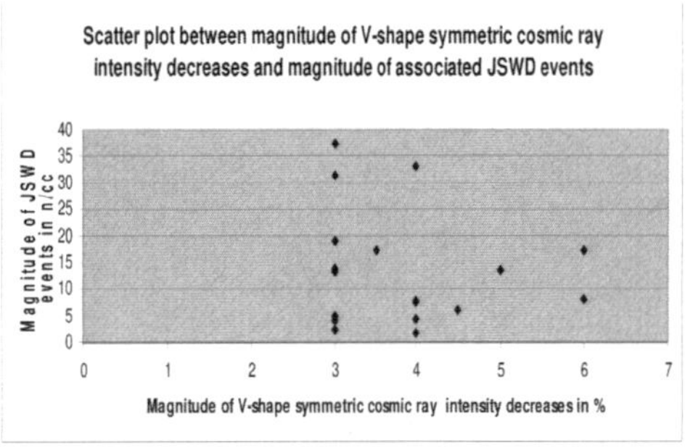

Figure 1 shows Scatter plot between magnitude of V shape symmetric cosmic ray decreases and magnitude of JSWD events showing negative correlation with correlation coefficient -0.152

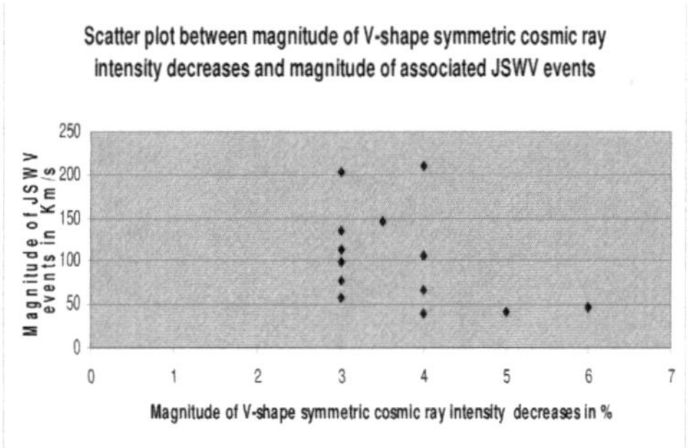

Figure 2 shows Scatter plot between magnitude of V shape symmetric cosmic ray decreases and magnitude of JSWV events showing negative correlation with correlation coefficient -0.48

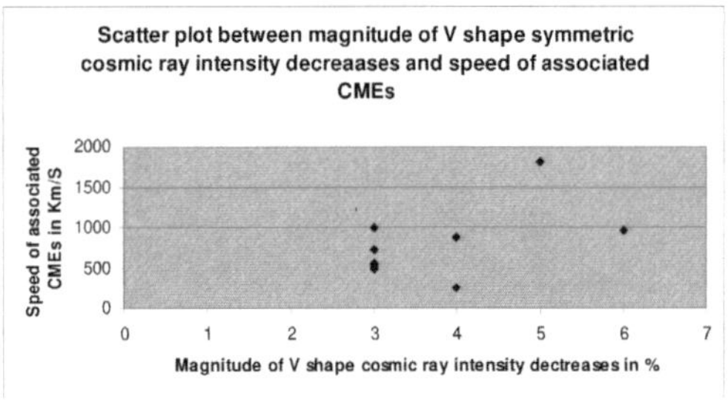

Figure 3 shows Scatter plot between magnitude of V shape symmetric cosmic ray decreases and speed of associated CMEs with correlation coefficient 0.54

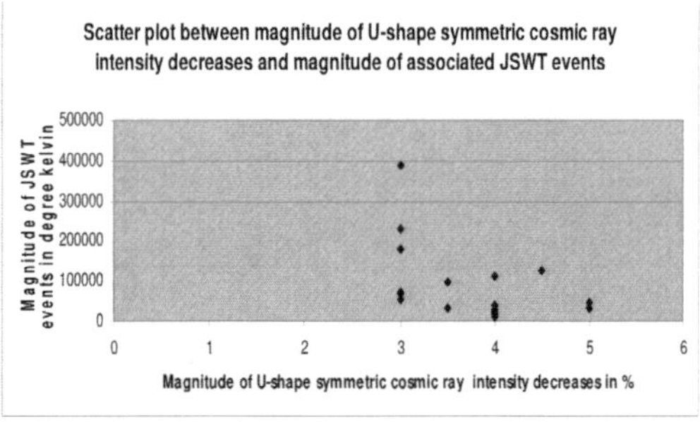

Figure 4 Shows Scatter plot between magnitude of U shape symmetric cosmic ray decreases and magnitude of JSWT events showing -0.473

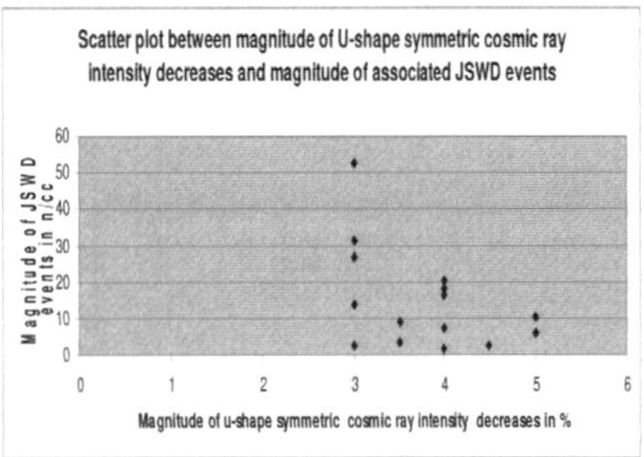

Figure 5 shows Scatter plot between magnitude of U shape symmetric cosmic ray decreases and magnitude of JSWD events showing negative correlation with correlation coefficient 0.468

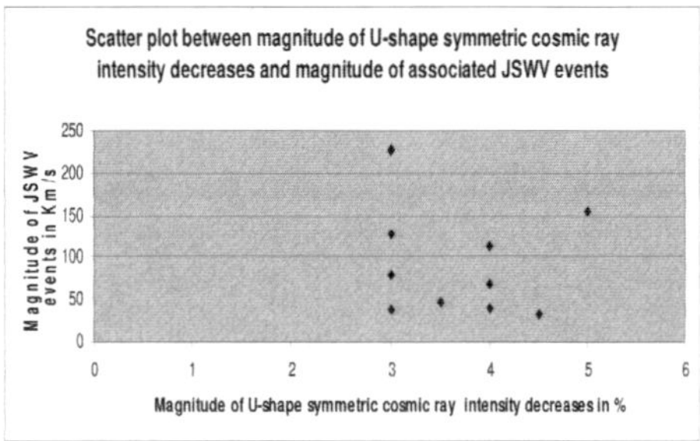

Figure 6 shows Scatter plot between magnitude of U shape symmetric cosmic ray decreases and magnitude of JSWV events showing negative correlation with correlation coefficient -0.26

From the data analysis of V-shape Symmetric short term cosmic ray decreases and JSWV events it is concluded that V shape symmetric cosmic ray decreases are closely associated with JSWV events. We have found 14 out of 23 V shape symmetric cosmic ray decreases are associated with JSWV events (60.87%).

To see how the magnitude of V-shape Symmetric short term cosmic ray decrease are correlated with the magnitude of JSWV events, we have plotted a scatter diagram between the magnitude of V-shape Symmetric short term cosmic ray decrease and JSWV events and the resulting figure is shown in figure 2 .From the it is clear that maximum V-shape Symmetric short term cosmic ray decrease which have large magnitude are associated with such JSWV events which have relatively small magnitude but magnitude of these two events do not have any quantitative relation their amplitude do not have any fixed proportion. These results indicate that although these events have not any quantitative relation but the V-shape Symmetric short term cosmic ray decrease of higher magnitude are generally associated with such JSWV events, which have relatively low magnitude. Negative co-relation has also been found between magnitude of V Shape asymmetric cosmic ray intensity decreases and magnitude of jump in solar wind plasma velocity of the associated JSWV events with co-relation co-efficient -0.48 between these two events.

From the data analysis of U shape symmetric short term cosmic ray intensity decreases and jump in solar wind plasma temperature ,density and velocity we have found 86.96% of U shape Symmetric short term cosmic ray decrease events are found to be associated with JSWT events. To sees how the magnitude of U-shape Symmetric short term cosmic ray decrease is correlated with JSWT events. We have plotted a scatter diagram between the magnitude of U-shape Symmetric short term cosmic ray decrease events and magnitude JSWT events in figure 4 .From the it is clear that, most of the U-shape Symmetric short term cosmic ray decrease which have large magnitude are associated with such JSWT events which have small magnitude. We found negative co-relation with correlation coefficient, -0.473 between magnitude of U shape symmetric cosmic ray intensity decreases and magnitude of jump in solar wind plasma temperature of associated JSWT events. From the analysis of U-shape Symmetric short term cosmic ray decrease with jump in solar wind plasma density (JSWD), 83.33%U-shape cosmic ray intensity decreases are found to be associated with JSWD events. To sees how the magnitude of U-shape Symmetric short term cosmic ray decrease is correlated with magnitude of JSWD events. We have plotted a scatter diagram between the magnitude of U-shape Symmetric short term cosmic ray decrease and magnitude JSWD events and the resulting figure is shown in figure 5. From the fig. it is clear that, most of the U-shape Symmetric short term cosmic ray decrease which have large magnitude are associated with such JSWD events which have small magnitude. We found negative co-relation with correlation coefficient, -0.468 between magnitude of U shape symmetric cosmic ray intensity decreases and magnitude of jump in solar wind plasma density of associated JSWD events. From the analysis of U-shape Symmetric short term cosmic ray decrease with that of jump in solar wind plasma velocity JSWV 61.11% U-shape Symmetric short term cosmic ray decrease is associated with JSWV events. We found negative co-relation with correlation coefficient and -0.26 between U shape symmetric cosmic ray intensity decreases and magnitude of jump in solar wind plasma velocity of associated JSWV events.

Table No 2 Association of Symmetric U shape Cosmic Ray Intensity with Coronal Mass Ejections and Disturbances in Solar Wind Plasma Parameters

	Symmetric U Shape cosmic ray intensity decreases				Tempara ture		Density		Velosity	
Date	Onset set timedd (hh)	mag %	Type of decre ases	Shocks	Start time dd(hh)	Magnitu de of Jump in Deg k	Start time dd(hh)	Magnitu de of Jumpn/c c	Start time dd(hh)	Magnitu de of Jumpm/ s
18.09.88	18(12)	4	U	17(19)	nd	nd	nd	nd	nd	nd
23.02.89	23(09)	4.5	U	na	22(13)	126963	22(22)	2.5	22(16)	33
05.05.89	5(00)	5	U	04(24)	04(22)	31385	04(22)	10.3	04(20)	155
12.06.90	12(06)	6	U	12(08)	nd	nd	nd	nd	nd	nd
20.12.90	20(04)	3	U	na	19(21)	72941	nj	nj	nj	nj
20.02.91	20(12)	4	U	na	20(02)	11653	20(01)	1.9	nj	nj
11.07.92	11(13)	3.5	U	na	10(15)	31654	11(08)	8.9	nj	nj
04.08.92	4(18)	5	U	4(14)	04(05)	45467	04(02)	5.9	nj	nj
03.06.93	03(06)	3	U	na	02(18)	54712	03(03)	26.6	03(02)	127
23.06.93	23(06)	3	U	na	nd	nd	nd	nd	nd	nd
11.12.98	11(12)	4	U	na	10(14)	24850	10(15)	18.1	10(14)	39
05.05.99	5(12)	4	U	5(16)	05(11)	109721	15(02)	16.6	15(14)	113
22.05.99	22(18)	4	U	na	22(09)	29329	nj	nj	nj	nj
22.03.00	22(06)	3	U	na	21(23)	69823	22(01)	52.8	22(01)	227
23.01.01	23(06)	3	U	23(11)	23(02)	230408	23(05)	13.8	22(07)	80
22.07.01	22(18)	3	U	na	22(04)	178323	21(23)	2.5	22(12)	37
03.12.01	03(20)	3.5	U	na	03(06)	96635	03(05)	3.4	03(05)	46
27.01.02	27(18)	4	U	na	nj	nj	nj	nj	nj	nj
09.04.02	9(12)	4	U	na	08(23)	18921	09(08)	20.1	nj	nj
01.11.02	01(18)	4	U	na	01(14)	39058	01(09)	7.3	01(10)	68
09.11.06	9(12)	3	U	9(17)	09(11)	389111	09(07)	31.3	09(11)	229

4. Conclusion

From our study we have found that 76.92% V shape symmetric cosmic ray intensity decreases are associated with halo and partial halo coronal mass ejections (CMEs). The association rate between halo and partial halo coronal mass ejections are found 40.00%and 60.00% respectively. Most of the V Shape symmetric cosmic ray intensity decreases are related to interplanetary shocks (61.53%) and the related shocks are forward shocks. We have also found positive co-relation with co-relation co-efficient 0.54 between magnitude of V shape symmetric cosmic ray intensity decreases and speed of associated coronal mass ejections. Negative co-relation has been found between magnitudes of V Shape symmetric cosmic ray intensity decreases and magnitude of jump in solar wind plasma density of associated JSWD events with co-relation co-efficient -0.152 between these two events. Negative co-relation has also been found between magnitude of V Shape asymmetric cosmic ray intensity decreases and magnitude of jump in solar wind plasma velocity of the associated JSWV events with co-relation co-efficient -0.48 between these two events. From the study of U Shape symmetric cosmic ray intensity decreases and disturbances in solar wind plasma parameters, we found negative co-relation with correlation coefficient, -0.473 between magnitude of U shape symmetric cosmic ray intensity decreases and magnitude of jump in solar wind plasma temperature of associated JSWT events, -0.468 between magnitude of U shape symmetric cosmic ray intensity decreases and magnitude of jump in solar wind plasma density of associated JSWD events and -0.26 between U shape symmetric cosmic ray intensity decreases and magnitude of jump in solar wind plasma velocity of associated JSWV events. These results shows that coronal mass ejections and disturbances in solar wind plasma parameters are the solar and interplanetary phenomena, characteristics of which are closely related to the characteristics of symmetric cosmic ray intensity decreases and these parameters will help in better understanding for theses decreases.

References

[1] Barouch E. and Burlaga, L. F. J. Geophys. Res., vol. 80, pp. 449-455, 1975.
[2] Badruddin, Astrophys. Space Sci., vol. 246, pp. 171-191, 1997.
[3] Badruddin, Nuvo Cimento, vol. 23C, pp. 217-226, 2000.
[4] Badruddin, Phys., vol. 209, pp. 195-206, 2002.
[5] Badruddin and Singh, Y. P Ind. J. Phys., vol. 77, pp. 497-502, 2003.
[6] Cane, H. VJ. Geophys. Res., vol. 98, pp. 3509-3512, 1993
[7] Cane H.V.I.G. Richardson and T.T. Von Rosenving J. Geophys. Res. 10(A10) 21-561, 21572, 1996 .
[7] Cane H.V.I.G. Richardson and T.T. Von Rosenving J. Geophys. Res. 10(A10) 21-561, 21572, 1996.
[8] Cane H.V.I.G. Richardson, and G.wibberenze J. Geophys.Res. 102(A4) 7086,1997.
[9] Cane, H. V. Space Sci. Rev., vol. 93, pp. 55-77, 2000
[10] Duggal S. P. and Pomerantz, M. A J. Geophys. Res., vol. 82, pp. 2170-2174, 1977.
[11] Forbush, S. E. 1954, J. Geophys. Res., 59, 525
[12] Ifedili, S. O. J. Geophys. Res., vol. 109, A02117, 2004.
[13] Kota J. and. Jokipii, J. R Proc. 27th Int. Cosmic Ray Conf., Hamburg, pp. 3577-3580, 2001
[14] Lockwood, J. A. Webber W. R. and Debrunner, H. J. Geophys. Res., vol. 96, pp. 11587-11604, 1991.
[15] Morfill, G. Richter A. K. and. Scholar, M J. Geophys. Res., vol. 84, pp. 1505-1513, 1979
[16] Nagashima, K. Sakakibara, S. Fujimoto, K. Tatsuoka R. and I. Morishita, Nuovo Cimento, vol. 13C, pp. 551-587, 1990.
[17] Richardson, I. G. Wibberenz G. and Cane, H.V J. Geophys. Res., vol. 101, pp. 13483-13496, 1996.
[18] Richardson, I. G. Cane H.V. and Wibberenz, G J. Geophys. Res., vol. 104, pp. 12549-12561, 1999.
[19] Richardson, I. G. Space Sci. Rev., vol. 111, pp. 267-376, 2004.
[20] Subramanium P Antia H.M.Dugad S.R.GoswamiU.D.,Gupta S.K.Mohanty Nonaka P.K.,Nayak T.,.Tanaka P.K.Tanwar S. C Proc. 29th Int. Cosmic Ray Conf., Pune, Vol. 2, pp. 73, 2005.
[21] Venkatesan D. and Badruddin, Space Sci. Rev., vol. 52, pp. 121-194, 1990.
[22] Webber, W.R., J.A. lockwood and S.R. Jokipii (1986). J. Geophys. Res. 91, 4103-4110.
[23] Webber, W.R., FB Mc Donald, J.A. Lockwood and B, Heikkilo (2002); Geophys. Res. Lett. 29 (10), 1377, doi : 10, 1029/2007 GLO14729.

CO and C_2 excited states relaxation in CO_2 plasmas derived from a Collisional-Radiative model

Arnaud Bultel[1], Ioan F. Schneider[2] and Yacine Babou[3]
[1]UMR CNRS 6614 CORIA, Université de Rouen, Avenue de l'Université, BP12, 76801 Saint-Etienne du Rouvray, France
[2]FRE 3102 LOMC, Université du Havre, BP 540, rue Philippe Lebon, 76058 Le Havre, France
[3]Von Karman Institute, Chaussée de Waterloo, 72 B-1640 Rhode Saint-Genèse, Belgium

E-mail: arnaud.bultel@coria.fr

Abstract. In order to explain the behavior of the main excited electronic states of CO and C_2 molecules formed in CO_2 entry or ground test facilities plasmas, a time-dependent Collisional-Radiative (CR) model is elaborated. This CR model is then included in the general balance equation of the species to give an insight on the influence of the transport and elementary processes. The related flows are generally observed under relatively low temperature owing to the numerous internal degrees of freedom to be excited, leading to a rather weak ionization degree. The electron-induced processes are even so taken into account: their influence is studied and commented. The conditions to reach non-equilibrium steady states and the related characteristic time scales are discussed as well.

1. Introduction

The important increase of the translation temperature during the atmospheric entry of a vehicle leads to the formation of a plasma characterized by numerous electronic excited species [1]. Conversely to the case of Earth's re-entries where nitrogen and oxygen lead to the formation of weakly radiative species, the case of Martian entries can be dramatic since the species formed are CO, C_2 and CN which radiate strongly [2]. The net heat flux density to the vehicle's wall can be strongly increased and damage it in part. One of the challenges of ground studies is to identify the kinetic mechanism leading to the formation of the states involved and to estimate their population density [3]. In this communication, we present a Collisional-Radiative (CR) model devoted to the theoretical study of the relaxation of the CO and C_2 main electronic states in the case of the pure CO_2 plasmas obtained in the ground test facilities (ICP torches) of VKI and CORIA institutes. This model takes into account the main processes leading to population densities changes by collisions (excitation/deexcitation, dissociation/recombination, exchange processes under heavies or electron impact) and radiation (spontaneous emission).

2. Species and CO and C_2 states involved

Owing to the possible dissociation of CO_2 under high temperature levels, CO and O can be formed. These species can re-associate to form O_2 and C_2 whose dissociation can produce C. The mixture is therefore composed by the following six species CO_2, CO, O_2, C_2, C and O. The collisions between them can lead to the CO and C_2 electronic states excitation whose less excited ones can modify the behavior of the flow since they may be sufficiently populated. These states are listed in table 1.

Table 1. CO and C_2 states considered and their spectroscopic constants with usual notations [4].

No	State	g_e	T_e (eV)	ω_e (eV)	$\omega_e x_e$ (10^{-3}eV)	B_e (10^{-4}eV)	E_{diss} (eV)	Dissociation state
CO								
1	$X^1\Sigma^+$	1	0.000	0.2690	1.648	2.395	11.092	C(^{3}P)+O(^{3}P)
2	$a^3\Pi$	6	6.036	0.2162	1.780	2.097	5.082	C(^{3}P)+O(^{3}P)
3	$a'^3\Sigma^+$	3	6.921	0.1523	1.298	1.667	4.229	C(^{3}P)+O(^{3}P)
4	$d^3\Delta_i$	6	7.578	0.1453	1.319	1.625	3.576	C(^{3}P)+O(^{3}P)
5	$e^3\Sigma^-$	3	7.964	0.1386	1.325	1.591	3.193	C(^{3}P)+O(^{3}P)
6	$A^1\Pi$	2	8.068	0.1882	2.405	1.998	3.065	C(^{3}P)+O(^{3}P)
7	$D^1\Delta^-$	2	8.174	0.1356	1.265	1.558	2.985	C(^{3}P)+O(^{3}P)
8	$b^3\Sigma^+$	3	10.392	0.2727	1.860	2.462	0.698	C(^{3}P)+O(^{3}P)
9	$B^1\Sigma^+$	1	10.780	0.2619	1.887	2.432	0.316	C(^{3}P)+O(^{3}P)
C_2								
No	State	g_e	T_e (eV)	ω_e (eV)	$\omega_e x_e$ (10^{-3}eV)	B_e (10^{-4}eV)	E_{diss} (eV)	Dissociation state
10	$X^1\Sigma_g^+$	1	0.000	0.2300	1.654	2.256	6.210	C(^{3}P)+C(^{3}P)
11	$a^3\Pi_u$	6	0.089	0.2035	1.447	2.024	6.134	C(^{3}P)+C(^{3}P)
12	$b^3\Sigma_g^-$	3	0.798	0.1823	1.387	1.858	5.436	C(^{3}P)+C(^{3}P)
13	$A^1\Pi_u$	2	1.040	0.1994	1.497	2.004	5.185	C(^{3}P)+C(^{3}P)
14	$c^3\Sigma_u^+$	3	1.650	0.2432	1.699	2.319	4.553	C(^{3}P)+C(^{3}P)
15	$d^3\Pi_g$	6	2.432	0.2217	2.038	2.173	3.732	C(^{3}P)+C(^{3}P)
16	$C^1\Pi_g$	2	4.248	0.2243	1.960	2.211	4.493	C(^{1}D)+C(^{1}D)
17	$e^3\Pi_g$	6	5.058	0.1372	4.868	1.478	2.462	C(^{3}P)+C(^{1}P)
18	$D^1\Sigma_u^+$	1	5.361	0.2268	1.728	2.273	4.799	C(^{1}D)+C(^{1}S)

3. CR model and balance equation

The behavior of these excited species is governed by the general balance equation, which is given under the following form:

$$\frac{\partial [X_i]}{\partial t} = -\vec{\nabla}\left([X_i]\,\vec{v}\right) - \vec{\nabla}\vec{J}_{X_i} + \left(\frac{\partial [X_i]}{\partial t}\right)_{Rad.} + \left(\frac{\partial [X_i]}{\partial t}\right)_{Coll.} \quad (1)$$

with in the right-hand side the convective, diffusive, radiative and collisional terms respectively. The elaboration of the CR model consists in calculating realistically the elementary processes involved in the collisional and radiative terms. (1) can be modified to derive a first order differential equation. The diffusion is mainly due to the fundamental diffusion mode depending on the diffusion coefficient $\bar{D}_{X_i}$ and the characteristic radius R of the flow. In addition, the convection term can be crudely estimated by R and v. We have stated:

$$\vec{\nabla}\left([X_i]\,\vec{v}\right) \approx v\,\frac{[X_i]}{R} \qquad \text{and} \qquad \vec{\nabla}\vec{J}_{X_i} = \bar{D}_{X_i}\,\frac{[X_i]}{L_0^2} \qquad \text{with} \qquad L_0 = \frac{R}{2.405} \tag{2}$$

The radiative term of (1) takes into account the main radiative systems listed in table 2. Radiation produced during spontaneous emission can be trapped owing to the opacity of the flow. An escape factor denoted $\Lambda_{j\to i}$ depending both on the upper j and lower i transition level densities is calculated assuming a global Doppler line broadening under the form:

$$\Lambda_{j\to i} = e^{-\frac{k_0^{j\to i}\,R}{(k_0\,R)_{ref}}} \qquad \text{with} \qquad (k_0\,R)_{ref} = 1.587 \tag{3}$$

where $k_0^{j\to i}$ is the emission coefficient by unit length at the line center for the transition $j \to i$.

With this escape factor, the $j \to i$ system contribution to $(\partial[X_i]/\partial t)_{Rad.}$ and $(\partial[X_j]/\partial t)_{Rad.}$ is:

$$\left(\frac{\partial[X_i]}{\partial t}\right)_{Coll.}^{(j\to i)} = +\Lambda_{j\to i}\,\frac{[X_j]}{\tau_{j\to i}} \qquad \text{and} \qquad \left(\frac{\partial[X_j]}{\partial t}\right)_{Coll.}^{(j\to i)} = -\Lambda_{j\to i}\,\frac{[X_j]}{\tau_{j\to i}} \tag{4}$$

The collisional term is classical and accounts for the usual (excitation/deexcitation, dissociation/recombination, exchange) processes under both heavies and electron impact. The rate coefficients used can be found in [8, 9, 10, 11, 12, 13]. The review of data derived from literature can be found in [14]. The forward processes are not only accounted for: the backward rate coefficients are also included and their values are derived from the forward ones using the detailed balance principle.

Finally the heavy particles are assumed under Maxwellian equilibrium according to a temperature denoted T_A and the electrons with their own temperature denoted T_e. The ordinary differential equations system obtained is solved by using the DVODE library [15].

Table 2. CO and C_2 main radiative systems taken into account in (1) and the related lifetimes τ [5, 6, 7].

CO transition (j→ i)	System	$\tau_{j\to i}$ (ns)	C_2 transition (j→ i)	System	$\tau_{j\to i}$ (ns)
$a'^3\Sigma^+ \to a^3\Pi$	Asundi	10^5	$A^1\Pi_u \to X^1\Sigma_g^+$	Philips	13×10^3
$d^3\Delta_i \to a^3\Pi$	Triplet	250	$D^1\Sigma_u^+ \to X^1\Sigma_g^+$	Mulliken	13
$A^1\Pi \to X^1\Sigma^+$	4th positive	10	$C^1\Pi_g \to A^1\Pi_u$	Deslandres-d'Azambuja	28
$b^3\Sigma^+ \to a^3\Pi$	3rd positive	54	$b^3\Sigma_g^- \to a^3\Pi_u$	Ballik and Ramsay	17×10^3
$B^1\Sigma^+ \to X^1\Sigma^+$	Hopfield-Birge	94	$d^3\Pi_g \to a^3\Pi_u$	Swan	102
$B^1\Sigma^+ \to A^1\Pi$	Angstrøm	50	$e^3\Pi_g \to a^3\Pi_u$	Fox-Herzberg	200

4. Results

The test-case studied is related to typical conditions obtained in plasma torches of the VKI and CORIA institutes: $T_A = 8000$ K, $T_e = 10000$ K, $p = 1500$ Pa, $v = 200$ m s^{-1} and $R = 4$ cm. Successive assumptions are stated in order to identify the cause of departure from equilibrium and to assess the influence of electrons.

Figure 1 illustrates the time evolution of the population densities starting from particular initial state when radiation and electron-induced processes are disregarded. We can notice the relative rapid evolution of these densities until a steady state obtained for $\tau = 0.4\times10^{-6}$ s. The calculation of the excitation temperature of a level i based on the ground state by:

$$T_{exc}(i) = \frac{E(i) - E(X)}{k_B} \left\{ \ln \left(\frac{g_e(i)}{g_e(X)} \frac{Z_{rv}(i)}{Z_{rv}(X)} \frac{[X]}{[i]} \right) \right\}^{-1} \tag{5}$$

provides information on additional characteristics of the final steady state. As we can see on figure 2, all excitation temperatures are equal to T_A at the final stage of the evolution. The diffusion and convective exchanges have no influence in this case. This means that the mixture is in chemical equilibrium: the population densities only depend on the local conditions. We have performed other calculations with different initial conditions. In each case, the final steady state is the same, proving without any doubt that this final state is indeed an equilibrium state.

The time needed to reach the final equilibrium state depends mainly on T_A and the pressure level p such that the following law can be derived:

$$\tau_{eq} = \frac{9.44}{p T_A^{1.0177}} \tag{6}$$

We can therefore assume in a first approximation that the product $\tau_{eq}\, p T_A$ is constant.

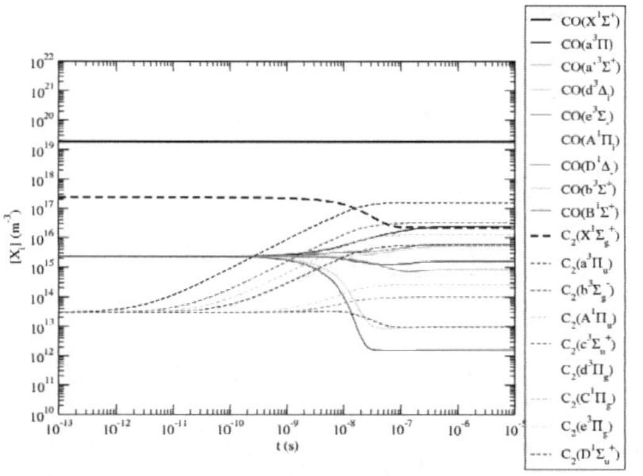

Figure 1. Population densities time-evolution starting from specific conditions without radiation.

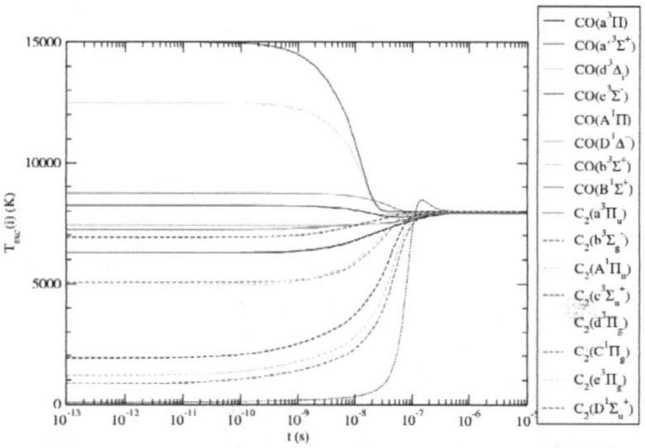

Figure 2. Excitation temperatures calculated by equation 5 in the conditions of figure 1.

These behaviours are no longer observed when the radiative processes are included in $(\partial[X_i]/\partial t)_{Rad.}$. Figure 3 displays the time-evolution of the population densities and figure 4 the time-evolution of the $\Lambda_{j\rightarrow i}$ terms: in this case, the mixture is almost optically thin.

A steady state is clearly observed, but the final excitation temperatures do not correspond to T_A. We can conclude that this steady state is not an equilibrium state. In figure 5 are also displayed the results obtained by changing the initial conditions. In both cases, the time needed to reach the steady state is approximately the same: 0.5×10^{-6} s. Its order of magnitude is remarkably close to the one obtained without considering radiative processes. This similar order of magnitude added to the near optically thin character of the flow explain why a final non equilibrium state is observed. If the pressure is higher, (6) indicates that the time needed to obtain equilibrium is shorter. In this case, the lower states population density is higher which leads to a higher optical thickness of the flow: the mixture is therefore quickly in equilibrium. In

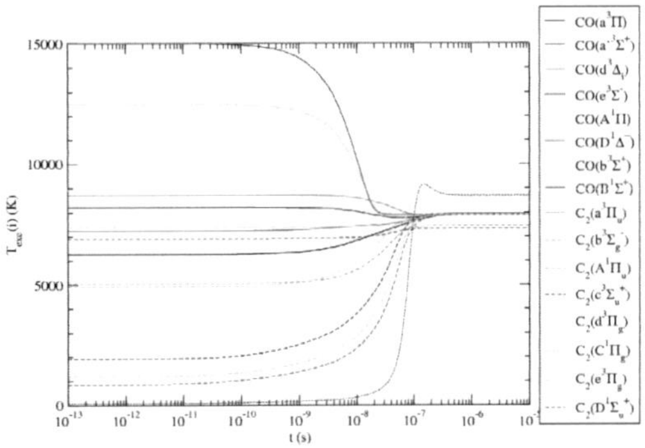

Figure 3. Excitation temperatures time-evolution including radiation (same conditions and initial state as for figures 1 and 2).

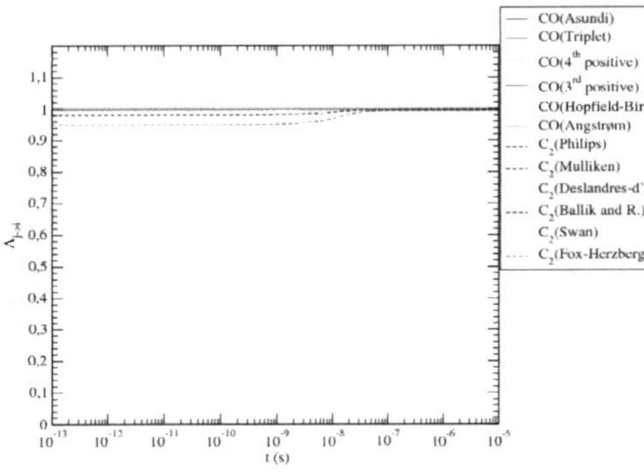

Figure 4. Time-evolution of the radiative systems escape factor. We can notice that $\lim_{t\to\infty} \Lambda_{j\to i} \approx 1$: the mixture is globally optically thin.

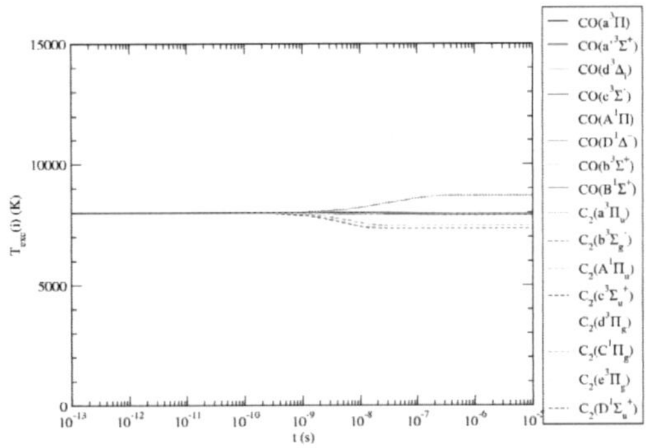

Figure 5. Same as figure 3, but with different initial conditions corresponding to an excitation equilibrium at $T_{exc} = 8000$ K.

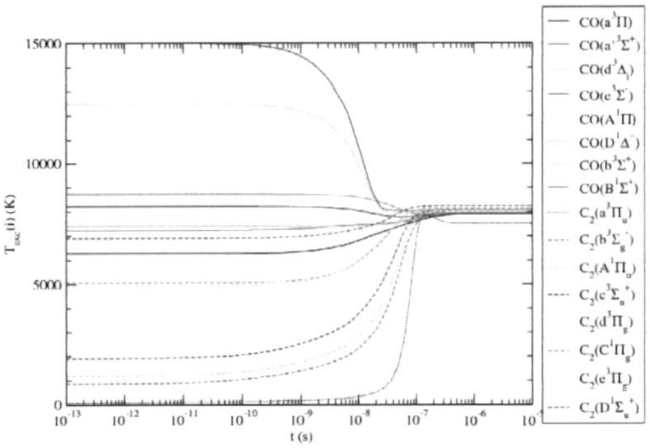

Figure 6. Excitation temperatures time-evolution with radiation and electron-induced processes (same conditions and initial state as for figures 1 and 2).

the contrary case, the mixture is closer to an optically thin medium whose radiation can easily escape according to a time scale logically given by the mean radiative lifetimes reported in table 2. The mixture cannot be longer considered as isolated: the final steady state is therefore a non equilibrium state (case of figures 3, 4 and 5).

CO and C_2 seem to present different behaviors. Carbon monoxide is much more close to equilibrium, contrary to the case of C_2. It is difficult to ascribe the difference to the radiative lifetimes values. These differences remain an open question so far and necessitate further studies.

In ICP torches, the production of the plasma results from an alternative orthoradial electromagnetic field which accelerates the statistically existing electrons in the injected gas and induces ionization phenomena under electron impact: in this area, n_e and T_e are therefore rather strong. Flowing from this zone and recombining owing to energy losses, the plasma is characterized by decreasing electron parameters. Despite this recombination, n_e and T_e can be high enough downstream to influence the chemistry. We have therefore also included electrons in the elementary processes of (1).

Assuming $T_e > T_A$ owing to the creation mode described above ($T_e = 10000$ K), we have searched for the minimum n_e value to obtain departures from the previous behaviors. Figure 6 illustrates the results when $n_e = 10^{19}$ m^{-3} which corresponds to an ionization degree of 0.1. We can notice that C_2 presents a different behavior from CO again: we have to further analyze the situation to explain why this difference is observed. Nevertheless, this threshold is interesting. This electron density is rather high and is presumed difficult to be reached in the case of CO_2 flows: numerous degrees of freedom have indeed to be excited to dissociate the molecule, which leads certainly to a lower n_e value. We can therefore state that the electron-induced processes play a negligible role in the present conditions.

In the case of the plasmas obtained during a Martian entry phase, the pressure level behind the shock front is largely higher than the one considered previously owing to the high compression of the hypersonic flow [16]. We can reasonably assume valid (6) if the influence of N_2 can be considered as negligible[1]. In this case, the time required to reach equilibrium is very short. Departure from equilibrium can therefore only be expected in the boundary layer surrounding the object's surface.

5. Conclusion

A Collisional-Radiative model is elaborated in order to give an insight of the behavior of the CO and C_2 excited states in the conditions of the ground test facilities of VKI and CORIA institutes. Departures from equilibrium can be ascribed to radiative losses mainly. Electron-induced processes can be considered obviously as negligible. CO and C_2 do not have the same behavior: the reasons for such discrepancies have not been yet identified and necessitate further developments.

Acknowledgments

The authors wish to thank the IEFE French Institute (Institut Energie, Fluide & Environnement) for its financial support.

References

[1] Zel'dovich Ya B and Raizer Yu P 2002 *Physics of shock waves and high-temperature hydrodynamic phenomena* Hayes W.D. and Probstein R.F. Mineola, N.Y. (Dover Publications)
[2] Bultel A, Rond C, Boubert P and Chéron B G 2008 *Proc. Int. Work. Radiation of High Temperature Gases in Atmospheric Entry* ESA p32
[3] Bultel A, Chéron B G, Bourdon A, Motapon O and Schneider I F 2006 *Phys. Plasmas* **13** 043502
[4] Capitelli M, Colonna G, Giordano D, Marraffa L, Casavola A, Minelli P, Pagano D, Pietanza L D and Taccogna F 2005 *Tables of internal partition functions and thermodynamic properties of high-temperature properties of high-temperature Mars-atmosphere species from 50K to 50000K* ed D Giordano and B Warmstein (ESA STR-246)
[5] Babou Y 2007 *Ph.D Thesis* University Paris-Sud Orsay, France
[6] Da Silva M L 2004 *Ph.D Thesis* University of Orleans, France
[7] Kirby K and Cooper D L 1989 *J. Chem. Phys.* **90** 4895-4902
[8] Bauer W, Becker K H, Bielefeld M and Meuser R 1986 *Chem. Phys. Lett.* **123** 33-36
[9] Becker K H, Donner B, Freitas Dinis C M, Geiger H, Schmidt F and Wiesen P 2000 *Zeitschrift fur Physikalische Chemie* **214** 503-510
[10] Gorse C, Cacciatore M and Capitelli M 1984 *Chem. Phys.* **85** 165-176
[11] Mangir M S, Reisler H and Wittig C 1980 *J. Chem. Phys.* **73** 829-835
[12] Surzhikov S 2008 *Proc. VKI Lecture Series* RTO-EN-AVT-162
[13] Zalogin G N, Kozlov P V, Kuznetsova L A, Losev S A, Makarov V N and Romanenko Yu V 2001 *Technical Physics* **46** 10-16
[14] Bultel A 2009 *Elaboration de codes modèles collisionnels-radiatifs électroniques spécifiques portant sur CO et C_2* CORIA Technical Note
[15] Brown P N, Byrne G D and Hindmarsh A C 1989 *SIAM J. Scient. Comput.* **10** 1038
[16] Park C, Howe J T, Jaffe R L, Candler G V 1994 *J. Thermophys. Heat Transfer* **8** 1, 9-23

[1] The relative CO_2-N_2-Ar composition of the Martian atmosphere is 95%-3%-1.6%.

Coronal Mass Ejections And Disturbances In Solar Wind Plasma Parameters In Relation With Geomagnetic Storms

P.L.Verma[1], Puspraj Singh[2] and Preetam Singh[3]

[1]Department of Physics, Govt.Vivekanand P.G. College Maihar Satna M.P.India, [2]Research Scholar, A.P.S.University Rewa M.P.India. [3]Research Scholar, Govt. P.G.College Satna M.P.India.

e-mail: pl_verma2003@yahoo.com

Abstract. Coronal Mass Ejections (CMEs) are the drastic solar events in which huge amount of solar plasma materials are ejected into the heliosphere from the sun and are mainly responsible to generate large disturbances in solar wind plasma parameters and geomagnetic storms in geomagnetic field. We have studied geomagnetic storms, (Dst ≤-75nT) observed during the period of 1997-2007 with Coronal Mass Ejections and disturbances in solar wind plasma parameters (solar wind temperature, velocity, density and interplanetary magnetic field) .We have inferred that most of the geomagnetic storms are associated with halo and partial halo Coronal Mass Ejections (CMEs).The association rate of halo and partial halo coronal mass ejections are found 72.37 % and 27.63 % respectively. Further we have concluded that geomagnetic storms are closely associated with the disturbances in solar wind plasma parameters. We have determined positive co-relation between magnitudes of geomagnetic storms and magnitude of jump in solar wind plasma temperature, jump in solar wind plasma density, jump in solar wind plasma velocity and jump in average interplanetary magnetic field with co-relation co-efficient 0 .35 between magnitude of geomagnetic storms and magnitude of jump in solar wind plasma temperature, 0.19 between magnitude of geomagnetic storms and magnitude of jump in solar wind density, 0.34 between magnitude of geomagnetic storms and magnitude of jump in solar wind plasma velocity, 0.66 between magnitude of geomagnetic storms and magnitude of jump in average interplanetary magnetic field respectively. We have concluded that geomagnetic storms are mainly caused by Coronal Mass Ejections and disturbances in solar wind plasma parameters that they generate.

1. Introduction

Coronal Mass Ejections (CMEs) are the drastic solar events in witch huge amount of solar plasma materials are ejected into the heliosphere from the sun and are mainly responsible to generate large disturbances in solar wind plasma parameters and geomagnetic storms in geomagnetic field. CMEs from the Sun drive solar wind (SW) disturbances in terms of magnetic field, speed and density, which in turn cause geomagnetic disturbance in Earth [14].It has been established by now the main cause of geomagnetic storms is believed to be the large IMF structure which has an intense and long duration southward magnetic field component, Bz [3,10 ,]. They interact with the Earth's magnetic field and facilitate the transport of energy into the Earth's atmosphere through the reconnection process.

Several investigators have studied geomagnetic storms with various solar features, solar wind parameters and inferred that CMEs which are the energetic solar features and associated with active regions are responsible for the most geoeffective solar wind disturbances and, therefore, the largest storms and are well associated with geomagnetic storms [1,9,12,13] .Enhanced solar wind speeds and southward magnetic fields associated with interplanetary shocks and ejecta are known to be important causes of storms [4,10]. Mcallister and Crooker [7] have studied effects of solar and heliospheric phenomena on geomagnetic field. They have concluded that major geomagnetic storms, both recurrent and non recurrent, are the result of the combined effects of CMEs and CIRs.Correiaa and De Souza [2] have presented the identification of solar coronal mass ejection (CME) sources for selected major geomagnetic storms. They have inferred that full halo CMEs originating from active regions associated with X-ray solar flares and propagating in the western hemisphere, cause strong geomagnetic storms. Gopalswamy et al. [5] have studied magnetic clouds ,coronal mass ejections and geomagnetic storms, they have found that 86% magnetic clouds are associated with full and partial halos coronal mass ejections .The remaining 14% of magnetic clouds are associated with non-halo CMEs originating from close to the disk center. They have concluded that magnetic clouds associated with partial halo and halo coronal mass ejections are most potential candidates for production of geomagnetic storms .they have further concluded that magnetic clouds associated with non halo CMEs may also cause geomagnetic storms . Verma et al [11] have studied geomagnetic storms Dst < - 50nT observed during the period of 1997-2006, with halo and partial halo coronal mass ejections associated with X-ray solar flares of different categories and concluded that they have concluded that majority of the observed geomagnetic storms are found that halo and partial halo CMEs associated with X ray solar flares are most potential candidates for production of geomagnetic storms.Michalek, G.et al [8] have concluded that halo coronal mass ejections (HCMEs) originating from regions close to the center of the sun are likely to be geoeffective. They have showed that, only fast halo CMEs (with space velocities higher than ~1000 km/s) and originating from the Western Hemisphere close to the solar center could cause intense geomagnetic storms. Gopalswamy [6] has studied the role of halo and partial halo CMEs in producing geomagnetic storms. He has reviewed the results obtained by previous investigators and concluded that the generation of geomagnetic storms rates can be readily explained by the different definition of halo CMEs used by different authors. Partial halos are less energetic and generally originate far from the disk center, so most of them behave similar to the nongeoeffective CMEs and hence most of the partial halo CMEs may not produce geomagnetic storms. He has inferred that halo CMEs originating close to the disk center or very much effective in producing geomagnetic storms.

2. Experimental Data

In this investigation, hourly Dst indices of geomagnetic field have been used over the period 1997 through 2007 to determine onset time, maximum depression time, magnitude of geomagnetic storms. This data has been taken from the NSSDC omni web data system which has been created in late 1994 for enhanced access to the near earth solar wind, magnetic field and plasma data of omni data set, which consists of one hour resolution near earth, solar wind magnetic field and plasma data, energetic proton fluxes and geomagnetic and solar activity indices. The data of coronal mass ejections (CMEs) have been taken from SOHO – large angle spectrometric, coronagraph (SOHO / LASCO) and extreme ultraviolet imaging telescope (SOHO/EIT) data. To determine disturbances in solar wind plasma parameters, hourly data of solar wind plasma velocity density temperature average interplanetary magnetic field has been used and these data has also been taken from omni web data also.

3. Data Analysis and Results

The association between geomagnetic storms < –75nT and coronal mass ejections (CMEs), and disturbances in solar wind plasma parameters for the period 1997 to 2007 are given in Table No.1. We have 124 geomagnetic storms in our list in which we have no data of CMEs for association. From the data analysis it is observed that 76out of 120 (84.93%) geomagnetic storms < –75nT are found to be associated with coronal mass ejections. We have 124 geomagnetic storms in our list in which the CME data for association is available for 120 geomagnetic storms. We have further observed that the majority of related CMEs are halo CMEs.We have 76 geomagnetic storms, which are associated with coronal mass ejections out of which 55 geomagnetic storms (72.37%) are related to the halo coronal mass ejections. Only 21 out of 76 (25.81%) geomagnetic storms are found to be associated with partial halo coronal mass ejections. From the study of geomagnetic storms with disturbances in solar wind plasma parameters i.e. jump in solar wind plasma temperature (JSWT), jump in solar wind density (JSWD) jump in solar wind plasma velocity and jump in average interplanetary magnetic field, it is inferred that geomagnetic storms of higher magnitudes are found to be associated with such JSWT, JSWD, JSWV and jump in average interplanetary magnetic field events, which have relatively higher jump magnitude of temperature, density velocity, and average interplanetary magnetic field .We have determined positive co-relation between magnitudes of geomagnetic storms and magnitude of jump in solar wind plasma temperature, jump in solar wind plasma density, jump in solar wind plasma velocity and jump in average interplanetary magnetic field with co-relation co-efficient 0 .35 between magnitude of geomagnetic storms and magnitude of jump in solar wind plasma temperature, 0.19 between magnitude of geomagnetic storms and magnitude of jump in solar wind density, 0.34 between magnitude of geomagnetic storms and magnitude of jump in solar wind plasma velocity, 0.66 between magnitude of geomagnetic storms and magnitude of jump in average interplanetary magnetic field respectively. We have concluded that geomagnetic storms are mainly caused by Coronal Mass Ejections and disturbances in solar wind plasma parameters that they generat.

Table – Association of Geomagnetic Storms with Coronal Mass Ejections and Disturbances in Solar Wind Plasma Parameters

	Geomagnetic Storm			Temparature		Density		Velocity		IMFB (nT)		CMEs		
S. NO.	Date	Onset time in dd(hh)	Magnitude in nT	Start time in dd(hh)	Jump Magnitude in Degree kelvin	Start time in dd(hh)	Jump Magnitude in N/cc	Start time in dd(hh)	Magnitude of Jump In Km/s	Start time in dd(hh)	Magnitude in nT	Date(hh.mm)	types H/P	Speeds km/sec.
1	10.01.97	10(04)	-85	10(00)	83898	10(00)	6.4	09(23)	69	09(21)	13.5	na	na	na
2	26.02.97	26(23)	-85	26(21)	136171	26(12)	11.6	26(20)	97	26(14)	10.1	na	na	na
3	10.04.97	10(21)	-88	10(16)	169574	10(16)	20.3	10(10)	153	10(17)	7.8	07(14.27)	H	878
4	16.04.97	16(07)	-79	16(11)	326450	16(08)	33.2	16(03)	200	16(12)	9.8	na	na	na
5	21.04.97	21(12)	-101	21(03)	18607	20(21)	26.7	21(07)	57	21(11)	7.4	na	na	na
6	15.05.97	15(04)	-113	15(01)	308796	14(22)	15.7	15(01)	120	15(01)	19.4	12(05.30)	H	464
7	07.06.97	07(06)	-81	06(09)	120542	06(20)	9.2	06(19)	47	07(23)	10.6	na	na	na
8	03.09.97	03(18)	-103	02(21)	97676	03(12)	12.6	02(21)	107	02(18)	16.2	na	na	na
9	01.10.97	01(02)	-100	01(00)	144646	30(20)	15.2	30(20)	122	02(00)	10.9	28(01.08)	H	369
10	10.10.97	10(17)	-109	10(02)	36728	10(09)	25.3	10(13)	40	10(01)	9.3	na	na	na
11	06.11.97	06(23)	-105	06(02)	150720	06(17)	14.3	06(21)	135	06(19)	13	04(06.10)	H	785
12	22.11.97	22(10)	-106	22(04)	395681	22(06)	21.3	22(08)	177	21(19)	24.1	19(12.27)	H	150
13	30.12.97	30(04)	-80	30(03)	106297	30(00)	27.4	29(22)	58	30(08)	6.5	na	na	na
14	06.01.98	06(15)	-83	06(02)	126983	06(13)	19.1	06(12)	114	06(09)	16.3	na	na	na
15	17.02.98	17(14)	-105	17(02)	38149	17(02)	9.1	17(04)	48	17(04)	16.2	na	na	na
16	10.03.98	10(13)	-110	na	na	na	na	10(04)	228	10(01)	18.3	na	na	na
17	21.03.98	21(09)	-76	21(02)	291106	21(06)	16.3	21(03)	295	21(07)	6.5	18(07.33)	P	636
18	02.05.98	02(09)	-203	01(19)	326112	01(19)	9.6	01(10)	255	01(21)	14.2	29(16.58)	H	1374
19	26.06.98	26(00)	-92	25(15)	70984	25(12)	5.4	25(15)	97	25(15)	6.6	na	na	na
20	06.08.98	06(02)	-139	05(22)	21176	06(01)	32.2	06(01)	66	06(02)	13	nd	nd	nd
21	26.08.98	26(11)	-143	26(01)	2034319	25(22)	6	26(01)	449	26(04)	14	nd	nd	nd

22	25.09.98	25(00)	-203	24(23)	731981	24 (14)	11.5	24(23)	403	24(21)	18.8	nd	nd	nd
23	19.10.98	19(02)	-111	19(07)	173366	18(08)	55.9	18(18)	113	18(17)	23.1	na	na	na
24	07.11.98	07(11)	-139	07(00)	250723	07(05)	8.8	07(07)	42	07(19)	27.1	05(02.02)	H	380
25	08.11.98	08(20)	-126	08(14)	56923	08(02)	21.3	08(01)	173	07(20)	26.9	05(20.24)	H	1118
26	13.11.98	13(00)	-129	12(14)	39633	12(13)	28.5	12(14)	96	12(22)	11.3	10(06.18)	P	286
27	13.01.99	13(11)	-105	13(06)	55728	13(09)	22.8	13(09)	96	13(09)	14.1	nd	nd	nd
28	18.02.99	18(03)	-125	17(23)	450164	18(00)	10.2	17(22)	292	18(00)	20.8	na	na	na
29	28.02.99	28(17)	-94	28(12)	55926	28(02)	17.8	28(04)	30	27(15)	8.8	na	na	na
30	16.04.99	16(21)	-90	16(07)	68434	16(08)	45.7	16(10)	86	16(15)	20.1	na	na	na
31	12.09.99	12(09)	-78	12(01)	363153	12(03)	31.4	12(02)	204	12(12)	4.2	10(07.54)	P	1467
32	22.09.99	22(18)	-182	22(07)	295048	22(02)	37	22(08)	248	22(10)	1.7	20(06.06)	H	604
33	22.10.99	22(00)	-214	22(01)	511615	21(22)	29.7	21(09)	104	21(00)	27.7	19(05.50)	P	753
34	10.11.99	10(22)	-89	10(15)	98797	10(20)	1.1	10(16)	58	10(04)	3.2	na	na	na
35	12.12.99	12(18)	-77	12(15)	660214	12(14)	10.2	12(15)	332	13(04)	5.2	na	na	na
36	11.01.00	11(15)	-77	11(12)	285691	11(05)	8.3	11(12)	193	11(10)	7.5	na	na	na
37	22.01.00	22(16)	-96	21(23)	112347	22(00)	20.2	22(00)	72	22(04)	10.6	na	na	na
38	11.02.00	11(07)	-132	11(01)	157720	11(01)	6.3	11(02)	89	11(23)	14.3	08(09.30)	H	1079
39	06.04.00	06(16)	-282	06(15)	325941	06(12)	26.6	06(15)	221	06(10)	26.5	04(16.32)	H	1188
40	15.04.00	15(19)	-81	na	na	na	na	na	na	16(11)	10.8	13(21.30)	P	613
41	17.05.00	17(00)	-86	16(10)	134851	16(23)	11.1	16(21)	128	17(21)	4.8	na	na	na
42	24.05.00	24(00)	-151	23(20)	413126	23(16)	26.6	23(16)	113	23(15)	24.8	22(01.50)	H	649
43	08.06.00	08(15)	-89	08(03)	781686	08(08)	15.8	08(08)	253	08(05)	19.1	06(15.54)	H	1119
44	26.06.00	26(02)	-80	25(21)	172586	25(18)	4.1	25(19)	93	25(23)	7.8	25(07.54)	P	1617
45	15.07.00	15(15)	-308	15(08)	1452345	15(08)	16.9	15(13)	400	15(08)	45.2	14(10.54)	H	1674
46	19.07.00	19(19)	-95	19(14)	244724	19(11)	6.4	19(11)	113	19(11)	11	na	na	na
47	12.08.00	12(01)	-214	11(18)	651965	11(20)	8.3	11(16)	266	11(17)	24.1	09(16.30)	H	702
48	17.09.00	17(20)	-197	17(13)	803820	17(13)	28.9	17(15)	298	17(14)	34.1	16(05.18)	H	1215
49	03.10.00	03(23)	-156	03(14)	15645	03(16)	5.2	02(23)	82	03(00)	12.9	30(18.06)	P	703
50	13.10.00	13(14)	-100	13(02)	173385	12(23)	18.7	12(21)	139	12(19)	12	09(23.50)	H	798
51	28.10.00	28(21)	-126	28(13)	102142	28(07)	32.6	28(05)	101	28(18)	10.1	25(08.26)	H	770
52	05.11.00	05(10)	-150	04(17)	303563	04(22)	1.8	04(11)	193	04(23)	4.8	03(18.26)	H	291
53	10.11.00	10(07)	-102	10(05)	1925089	10(05)	9.8	09(23)	341	10(05)	8.2	08(04.50)	H	474
54	26.11.00	26(22)	-127	26(08)	552597	26(11)	29.3	26(03)	252	26(08)	24.2	24(15.30)	H	1254
55	19.03.01	19(11)	-150	na	na	na	na	19(08)	196	19(09)	15.2	18(02.26)	H	752
56	22.03.01	22(14)	-81	22(14)	101413	22(15)	9.2	22(03)	33	22(10)	15.1	20(03.26)	P	478
57	27.03.01	27(21)	-86	27(06)	290035	27(17)	18	27(10)	216	27(15)	16	24(20.50)	H	906
58	31.03.01	31(04)	-379	30(22)	666839	30(19)	31.2	30(19)	315	30(21)	43.8	28(01.27)	H	427
59	11.04.01	11(15)	-269	11(10)	729729	11(10)	23	11(12)	233	11(09)	30.1	09(15.54)	H	1192
60	18.04.01	18(01)	-106	17(22)	272566	17(22)	24.9	17(23)	162	17(23)	18.9	15(14.06)	P	1199
61	22.04.01	22(00)	-106	21(13)	54930	21(15)	23.1	21(15)	37	21(21)	9.6	20(10.06)	P	1160
62	17.08.01	17(17)	-102	17(08)	299070	17(10)	23.4	17(09)	264	17(07)	28	15(23.54)	H	1575
63	11.10.01	11(17)	-76	11(14)	544030	11(11)	21.8	11(16)	198	11(08)	21.3	09(11.30)	H	973
64	21.10.01	21(16)	-178	21(15)	422395	21(13)	16.4	21(10)	360	21(13)	21.7	19(01.27)	H	558
65	28.10.01	28(01)	-142	28(00)	178222	27(20)	8.9	28(01)	145	27(22)	12.8	25(15.26)	H	1092
66	31.10.01	31(14)	-104	31(08)	63658	31(10)	19.5	31(12)	62	31(18)	7.8	30(03.30)	P	1005
67	05.11.01	05(19)	-297	05(10)	105551	05(09)	31.4	05(14)	141	05(12)	51.2	04(16.35)	H	181

														0
68	24.11.01	24(06)	-223	24(03)	2292789	24(02)	40	24(03)	504	24(04)	51.2	22(20.30)	H	1443
69	02.02.02	02(02)	-77	01(03)	82946	01(03)	15	01(20)	104	01(20)	5.4	na	na	na
70	05.02.02	05(10)	-79	04(23)	135988	04(14)	26.7	05(06)	339	04(21)	10.7	02(14.54)	P	362
71	23.03.02	23(16)	-91	23(09)	82810	23(06)	8.7	23(06)	86	23(11)	2.2	20(17.54)	H	603
72	17.04.02	17(11)	-149	17(08)	353020	17(09)	25	27(06)	286	17(07)	23.2	15(03.50)	H	720
73	11.05.02	11(13)	-103	11(06)	193268	11(07)	48.2	11(08)	100	11(05)	13.3	08(13.50)	H	614
74	23.05.02	23(12)	-89	23(03)	1187414	22(15)	13.7	23(03)	468	22(23)	33.4	22(03.50)	H	1557
75	01.08.02	01(23)	-105	01(03)	292141	01(22)	15.7	01(22)	88	01(01)	7.4	29(23.30)	P	360
76	18.08.02	18(22)	-110	18(11)	484987	18(15)	11.2	18(16)	171	18(18)	9.1	16(12.30)	H	1585
77	04.09.02	04(02)	-102	04(00)	226235	na	na	04(00)	77	03(07)	9.1	na	na	na
78	06.09.02	06(09)	-159	06(03)	26868	05(12)	1.4	05(12)	26	na	na	04(13.31)	P	513
79	01.10.02	01(04)	-156	30(09)	71897	30(09)	29.3	01(00)	53	01(06)	4	na	na	na
80	24.10.02	24(00)	-88	23(22)	436899	23(19)	13.4	24(07)	308	24(00)	5.1	na	na	na
81	20.11.02	20(16)	-79	19(23)	138265	20(11)	4.5	20(07)	9	20(22)	22.2	na	na	na
82	21.11.02	21(02)	-122	21(01)	401981	20(16)	40.7	20(20)	318	20(22)	22.2	19(03.06)	P	938
83	09.05.03	09(09)	-78	09(00)	488764	09(01)	5.3	08(20)	153	09(04)	6.6	06(15.50)	P	345
84	27.05.03	27(23)	-118	27(00)	277056	27(11)	4.6	27(17)	207	27(04)	3.5	na	na	na
85	02.06.03	02(02)	-85	01(14)	365794	01(13)	5.9	01(14)	158	01(22)	3.7	31(02.30)	H	1835
86	16.06.03	16(10)	-136	15(13)	100632	16(02)	8.2	15(12)	100	15(17)	5	14(05.30)	P	1215
87	11.07.03	11(00)	-109	11(03)	516059	10(03)	9.9	10(16)	28	11(03)	3.5	na	na	na
88	26.07.03	26(17)	-75	26(19)	413471	26(14)	28.5	26(10)	446	26(12)	23.4	na	na	na
89	17.08.03	17(17)	-171	17(09)	224467	17(12)	11.1	17(13)	105	17(00)	16.5	14(20.06)	H	378
90	28.10.03	28(06)	-384	28(08)	979536	28(01)	6.2	27(22)	373	28(01)	9.6	27(08.30)	P	1322
91	04.11.03	04(08)	-90	04(04)	947434	04(02)	11.5	04(04)	266	04(01)	13.4	02(09.30)	H	2036
92	20.11.03	20(02)	-461	20(05)	487107	19(20)	14.4	20(02)	262	20(05)	48.1	18(08.05)	H	1660
93	22.01.04	22(05)	-144	22(00)	501487	21(23)	13.1	22(00)	210	21(21)	19.7	20(00.06)	H	965
94	11.02.04	11(10)	-107	11(06)	51108	11(04)	19.8	11(13)	320	11(09)	13.6	na	na	na
95	09.03.04	09(13)	-76	09(11)	556729	09(08)	11.5	09(09)	368	09(13)	10.8	07(10.34)	P	395
96	03.04.04	03(14)	-113	03(07)	80343	03(08)	18.2	03(08)	125	03(23)	10.1	na	na	na
97	05.04.04	05(12)	-77	05(09)	421082	05(04)	9.6	05(10)	215	05(00)	2.3	na	na	na
98	16.07.04	16(22)	-84	16(20)	300586	16(21)	6.4	16(21)	165	16(20)	10.8	na	na	na
99	22.07.04	22(00)	-106	22(08)	507727	22(06)	13.7	22(09)	312	22(14)	10.4	20(13.31)	H	710
100	24.07.04	24(11)	-198	24(00)	472606	23(22)	14.6	24(04)	113	24(05)	16.3	na	na	na
101	30.08.04	30(02)	-119	29(14)	47028	29(06)	10.5	29(06)	50	29(23)	5.9	na	na	na
102	07.11.04	07(20)	-376	07(10)	786124	07(03)	54.1	07(09)	386	07(12)	38.8	04(09.54)	H	653
103	07.01.05	07(12)	-94	07(05)	104824	07(08)	25.1	07(03)	85	07(07)	15.3	05(15.30)	H	735
104	16.01.05	16(20)	-117	16(09)	93358	16(09)	15.9	16(09)	8.4	17(07)	31.1	15(06.30)	H	2049
105	21.01.05	21(19)	-103	21(04)	640871	21(11)	28.4	21(14)	403	21(15)	24.8	19(08.29)	H	2020
106	04.04.05	04(11)	-86	04(04)	429669	03(16)	17.2	04(02)	297	04(21)	5.9	na	na	na
107	11.04.05	11(17)	-77	11(13)	281023	na	na	11(09)	198	11(16)	13.1	09(13.50)	P	542
108	07.05.05	07(20)	-126	07(18)	866682	07(10)	39.2	07(17)	215	07(12)	10.7	05(20.30)	H	1180
109	15.05.05	15(05)	-293	15(02)	883346	14(22)	15	14(23)	545	15(01)	48.4	13(17.12)	H	1689
110	20.05.05	20(04)	-101	19(18)	26401	19(18)	16.6	19(16)	33	20(06)	5.7	16(13.50)	P	405
111	29.05.05	29(22)	-150	29(12)	247526	29(12)	11.3	29(08)	172	29(01)	9.1	26(15.06)	H	586
112	12.06.05	12(17)	-109	12(09)	228070	na	na	12(02)	202	12(07)	18.6	na	na	na
113	22.06.05	22(23)	-92	na	na	na	na	na	na	23(04)	14.2	na	na	na
114	10.07.05	10(11)	-100	10(00)	290138	09(20)	16.1	09(19)	140	10(02)	15.7	07(17.06)	H	683
115	17.07.05	17(06)	-77	17(00)	108021	16(15)	7.4	16(16)	89	17(12)	6.8	14(10.54)	H	211

														5
116	24.08.05	24(08)	-219	24(03)	408410	23(20)	23.1	24(00)	295	24(04)	43.1	22(01.31)	H	119 4
117	31.08.05	31(12)	-138	31(18)	145193	31(08)	278	31(07)	68	31(03)	10.8	29(10.54)	H	160 0
118	11.09.05	11(02)	-127	na	na	na	Na	na	na	10(21)	13	09(19.48)	H	225 7
119	27.12.05	27(14)	-77	27(11)	470574	27(06)	26.9	27(09)	336	27(05)	15.9	na	na	na
120	04.04.06	04(10)	-90	na	na	na	Na	03(13)	180	05(06)	6.6	na	na	na
121	14.04.06	14(10)	-111	13(17)	356943	13(10)	11.4	14(01)	165	13(14)	10.9	na	na	na
122	29.11.06	29(22)	-77	29(06)	73161	29(17)	10.1	30(01)	17	29(04)	11	na	na	na
123	14.12.06	14(21)	-143	14(03)	779800	14(05)	7.7	14(06)	575	14(11)	13.9	13(02.54)	H	177 4
124	20.11.07	20(07)	-77	20(08)	234743	20(06)	8.7	20(20)	222	19(21)	14	na	na	na

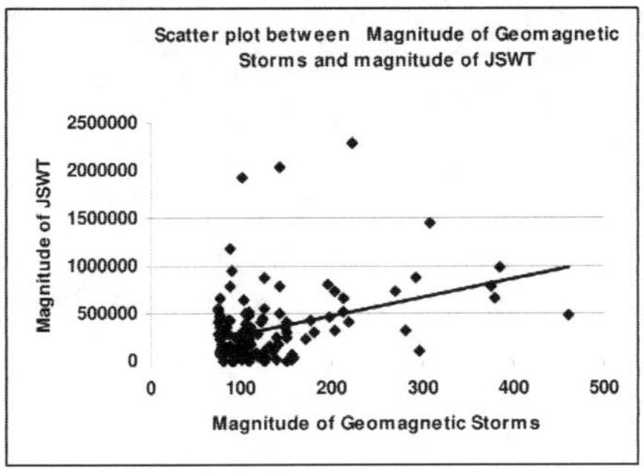

Figure-1 Shows Scatter plot between magnitude of geomagnetic storms and magnitude of Jump in solar wind plasma temperature (JSWT) showing positive correlation with correlation coefficient 0.35

Figure-2 Shows Scatter plot between magnitude of geomagnetic storms and magnitude of Jump in solar wind plasma density (JSWD) showing positive orrelation with correlation coefficient 0.19

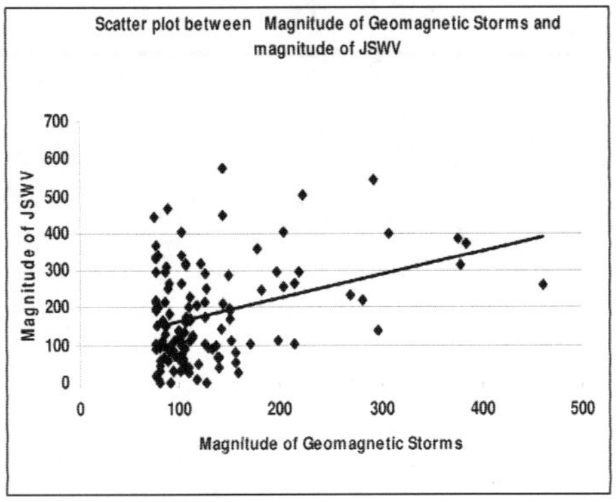

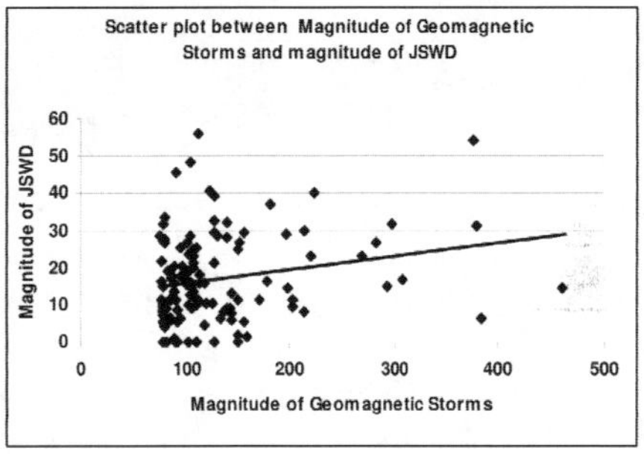

Figure 3-Shows Scatter plot between magnitude of geomagnetic storms and magnitude of Jump in solar wind plasma velocity (JSWV) showing positive correlation with correlation coefficient 0.34

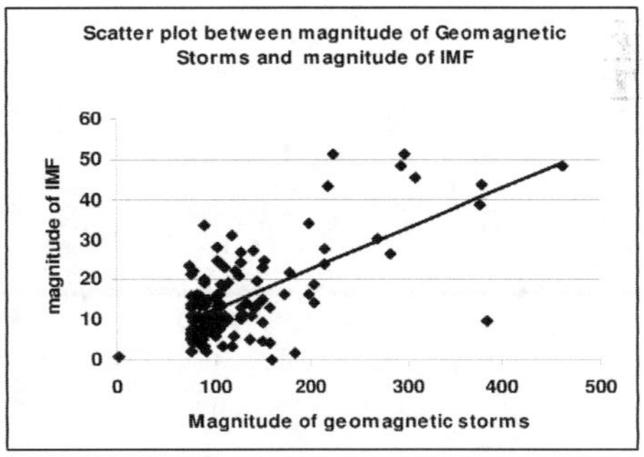

Figure 4-Shows Scatter plot between magnitude of geomagnetic storms and magnitude of Jump in average interplanetary magnetic field (IMF) showing positive correlation with correlation coefficient 0.66

4. Conclusion

From our study 76 out of 120 geomagnetic storms < -75nT have been identified as being associated with coronal mass ejections (63.33%), 55out of 76 have been identified as being associated with halo coronal mass ejections and 21out of 76 as being associated with partial halo coronal mass ejections The association rates for halo and partial halo CMEs are found 72.37% and 27.63% respectively. The positive correlation between magnitude of geomagnetic storms and magnitude of jump in solar wind plasma temperature, density velocity, and average interplanetary magnetic fields suggest that coronal mass ejections and disturbances in solar wind plasma parameters play crucial role in producing geomagnetic storms.

References

[1] Cane, H. V., Richardson, I. G., & St. Cyr, O. C., Geophys. Res. Lett., 27, 3591, 2000.
[2] Correiaa, E. R.V. de Souzaa Journal of Atmospheric and Solar-Terrestrial Physics 67, 1705, 2005
[3] Echer, E M V Alves and W D Gonzalez, Solar Phys. 221, 361,2004.
[4] Gosling, J. T., McComas, D. J., Phillips, J. L., and Bame, J. J. Geophys. Res.,96, 7831, 1991
[5] Gopalswamy, N., S. Akiyama, S. Yashiro, G. Michalek, and R. P. Lepping, J. Atm. Sol. Terr. Phys., 70, 245, 2008.
[6] Gopalswamy N.Letter Earth Planets Space, 61, 1–3, 2009
[7] Mcallister.A.H.and Krooker.N.U Geophysical monograph Vol 99 (303 279-289, 1997
[8] Michalek, G. Gopalswamy, N. Lara, A Yashiro, S Space. Weather, Volume 4, Issue 10th, 2006.
[9] St. Cyr, O.C. et al. J. Geophys. Res. **105**, 18,169–18, 185, 2000
[10] Tsurutani, B T Gonzalez, W D F Tang, S I Akasofu and E J Smith, J. Geophys. Res. 93, 8519, 1988.
[11] Verma P.L. Tripathi A.K. & Sharma ,Sushil J. Plasma Fusion Res. SERIES, Vol. 8 Page 221-225, 2009.
[12] Webb, D. F., Cliver, E. W., Crooker, N. U., St. Cyr, O. C., & Thompson, B. J., J. Geophys. Res., 105, 7491, 2000.
[13] Zhao, X. P., & Webb, D. F., J. Geophys. Res., 108, 1234, DOI: 10.1029/ 2002JA009606, 2003.
[14] Zhang .G.Dere.K.P.Astrophysical Journal,Vol 582,520,2003

Journal of Physics: Conference Series **511** (2014) 012061 doi:10.1088/1742-6596/511/1/012061

Scaled Experiment to Investigate Auroral Kilometric Radiation Mechanisms in the Presence of Background Electrons

S.L. McConville[1], K. Ronald[1], D.C. Speirs[1], K.M. Gillespie[1], A.D.R. Phelps[1], A.W. Cross[1], R. Bingham[1,3], C.W. Robertson[1], C.G.Whyte[1], W. He[1], M. King[1], R. Bryson[1], I. Vorgul[2], R.A. Cairns[2] and B.J. Kellett[3]

[1]SUPA Department of Physics, University of Strathclyde, Glasgow, G4 0NG, Scotland

[2]School of Mathematics and Statistics, University of St. Andrews, St. Andrews, KY16 9SS, Scotland

[3]Space Physics Division, STFC, Rutherford Appleton Laboratory Didcot, OX11 0QX, England

sandra.l.mcconville@strath.ac.uk

Abstract. Auroral Kilometric Radiation (AKR) emissions occur at frequencies ~300kHz polarised in the X-mode with efficiencies ~1-2% [1,2] in the auroral density cavity in the polar regions of the Earth's magnetosphere, a region of low density plasma ~3200km above the Earth's surface, where electrons are accelerated down towards the Earth whilst undergoing magnetic compression. As a result of this magnetic compression the electrons acquire a horseshoe distribution function in velocity space. Previous theoretical studies have predicted that this distribution is capable of driving the cyclotron maser instability. To test this theory a scaled laboratory experiment was constructed to replicate this phenomenon in a controlled environment, [3-5] whilst 2D and 3D simulations are also being conducted to predict the experimental radiation power and mode, [6-9]. The experiment operates in the microwave frequency regime and incorporates a region of increasing magnetic field as found at the Earth's pole using magnet solenoids to encase the cylindrical interaction waveguide through which an initially rectilinear electron beam (12A) was accelerated by a 75keV pulse. Experimental results showed evidence of the formation of the horseshoe distribution function. The radiation was produced in the near cut-off TE_{01} mode, comparable with X-mode characteristics, at 4.42GHz. Peak microwave output power was measured ~35kW and peak efficiency of emission ~2%, [3]. A Penning trap was constructed and inserted into the interaction waveguide to enable generation of a background plasma which would lead to closer comparisons with the magnetospheric conditions. Initial design and measurements are presented showing the principle features of the new geometry.

1. Introduction

Some magnetized planets and stars such as Earth and UV Ceti, exhibit radiation generation at their magnetic poles. For the case of the Earth this radiation emission is known as Auroral Kilometric

Radiation. Approximately 3200km above the Earth's surface there exists a region of plasma depletion known as the Auroral Density Cavity. Particles descend through this region towards the ionosphere following the path of the magnetic field lines. Along their descent the particles undergo magnetic compression. Due to conservation of the magnetic moment there is a transfer of axial to perpendicular velocity leading to a horseshoe shaped distribution function forming in velocity space.

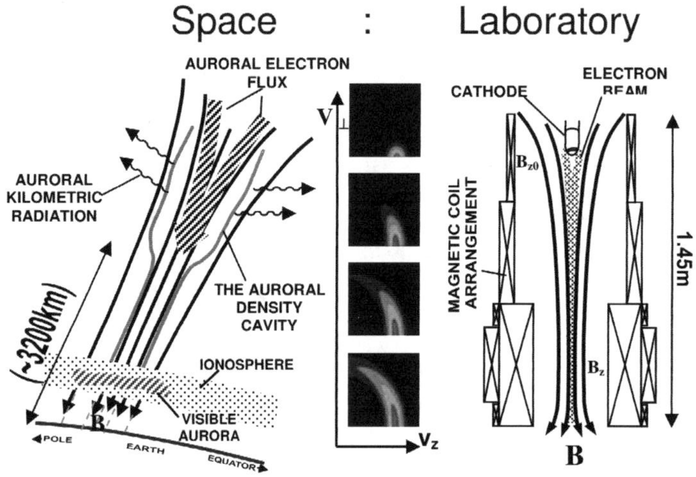

Figure 1: Comparisons in the geometry between magnetospheric and laboratory environment

Figure 1 shows this distribution formation alongside a schematic of the magnetospheric and experimental geometry. The horseshoe distributions have been shown to be unstable to an electron cyclotron resonance maser type of instability [10-12], and are now proposed as the source of free energy responsible for the production of AKR in these regions [13-16]. The emitted radiation has been observed by satellites to be in the X-mode. To provide a controlled environment to study this naturally occurring phenomenon, a scaled laboratory experiment was constructed at Strathclyde University, Figure 1, to replicate the process.

Theoretical studies in this field have been carried out at the University of St Andrews [17,18] where the problem is approached from the kinetic plasma theory point of view. These results correlate well with the experimental and numerical measurements at Strathclyde.

2. Experimental Apparatus

2.1. Magnet Coils

A solenoid system was constructed, Figure 2, which would encompass the electron beam as it traversed through the apparatus, left to right from the cathode region to the anode. The coils are wound individually allowing the magnetic field profile to be varied with a fine degree of precision. Altering the supply current of coil 3 which is the main cavity coil controlled the peak magnetic plateau whilst the ratio of coils 1-3 controls the magnetic compression of the electron beam as it passes through the interaction waveguide where resonance with the waveguide modes will take place. The blue lines in Figure 2 (left to right) show the position of the cathode face and the mid-point of the interaction waveguide respectively.

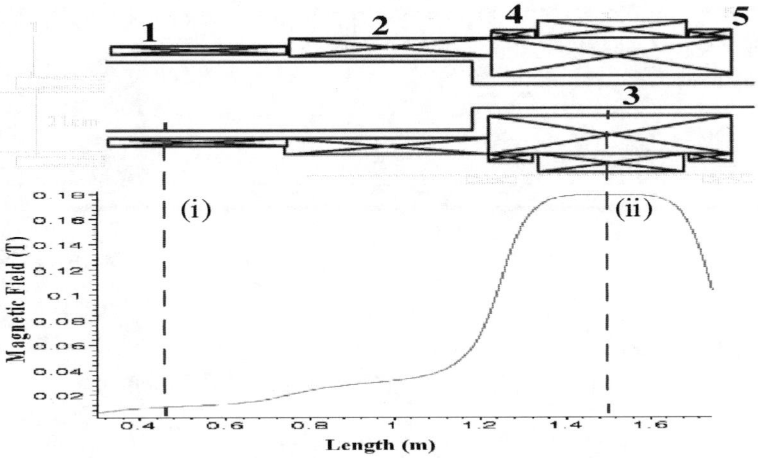

Figure 2: Layout of magnet coils with corresponding magnetic field profile

2.2. Electron Injector

For electron beam generation, an explosive emission cathode was used, Figure 3. As can be seen from the picture there is a velvet ring surrounding the dome of the cathode face. The dome is to provide an initial spread in pitch of the electrons as they leave the cathode, whilst the velvet provides for field emission that occurs at the fibre tips.

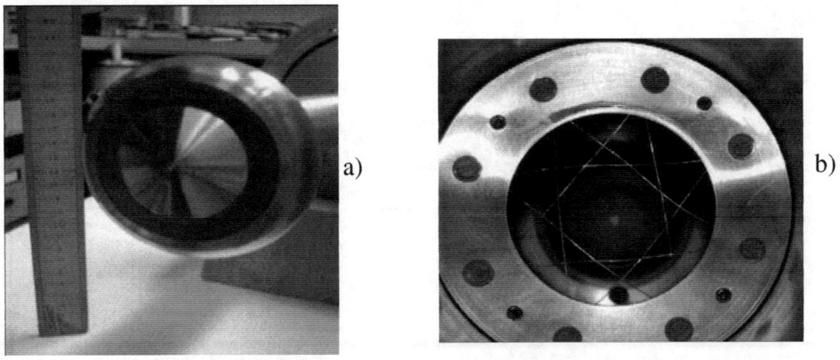

Figure 3: a) Cathode face with velvet ring, b) Anode mesh.

The anode mesh which can be seen in Figure 3b) is situated at ~ 2cm away from the cathode face and is designed in a way that will have minimal disruption to the beam as it passes through the copper wires, whilst helping to concentrate the field back at the cathode face thus allowing for lower accelerating voltages to be used during operation.

2.3. Penning Trap

The design for the Penning trap used in the experiment, Figure 4a), was based on the conventional design of two solid cathodes and cylindrical anode creating a trap in which to contain electrons as they collide with gas particles. To allow an electron beam to pass through the trap a 40mm hole was cut in one cathode plate whilst the second cathode plate was replaced with a mesh to allow egress of radiation. The cathode and anode walls of the trap were constructed using copper and were encased in a nylon jacket to act as an insulator and to prevent the copper walls from touching the walls of the interaction waveguide where it was situated, Figure 4b).

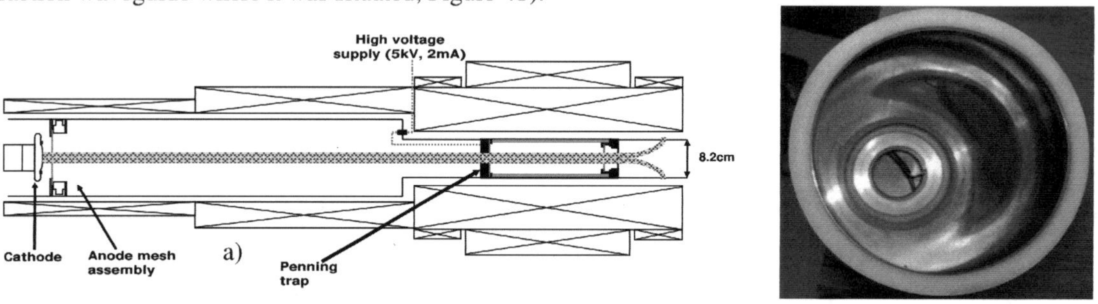

Figure 4: a) Schematic of apparatus showing position of Penning trap inside the interaction region,

b) End view of the Penning trap

A specially designed vacuum feed-through delivered a high voltage (up to +5kV) DC signal to the anode, shown in red in Figure 4a). With the plateau magnetic field set at 0.21T for the TE_{01} resonance in the new dimensions of the interaction region, a voltage was applied to the trap and the anode current and voltage measured. The plateau magnetic field was altered to investigate the effect this had upon the ignition, optical intensity, stability and sustained current and voltage of the discharge.

3. Experimental Results

3.1. Beam Measurements without Penning trap

The initial experiments do not include the Penning trap and so the cyclotron frequency is 4.42GHz, for the interaction waveguide dimension of 4.14cm radius, with an operating plateau magnetic field of 0.18T. With increasing cathode flux density it was seen that the beam current increased as is shown in Figure 5a).

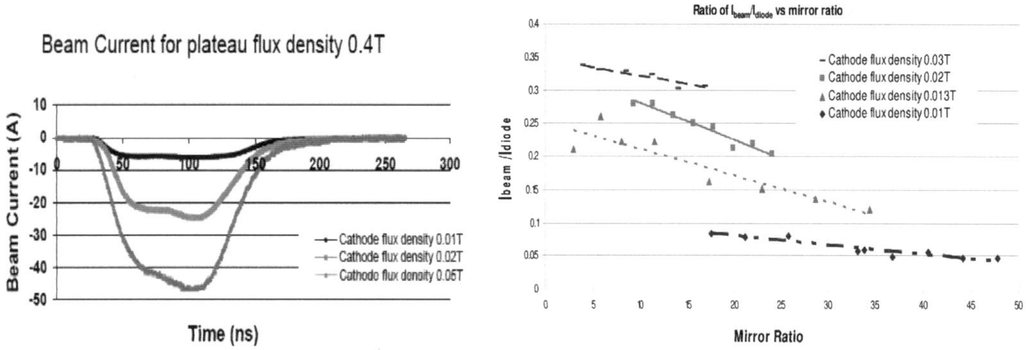

Figure 5: a) Variation of beam current with cathode flux density, b) Effect of increasing plateau magnetic field on beam current measurements

For different cathode flux settings it was seen that with increasing mirror ratio, the measured beam current decreased steadily, showing signs that mirroring was taking place within the experiment and therefore that a horseshoe shape distribution function was forming. This can be seen in Figure 5b).

3.2. Microwave Measurements

Using a fast 12GHz oscilloscope connected to microwave antenna, the operating frequency of the emitted radiation was measured. This was done by capturing the AC waveform signal and performing an FFT on the data giving a value of 4.42GHz, as predicted, and is shown in Figure 6a). The mode scan experiment allowed the mode pattern of the radiation to be plotted and was seen to be that of the TE_{01} mode, as expected, shown in Figure 6b). This mode of operation correlates well with the magnetospheric data where the radiation is polarised in the X-mode. The X-mode shares properties with close to cut-off TE_{01} modes in that they propagate and are polarised perpendicular to the static magnetic field.

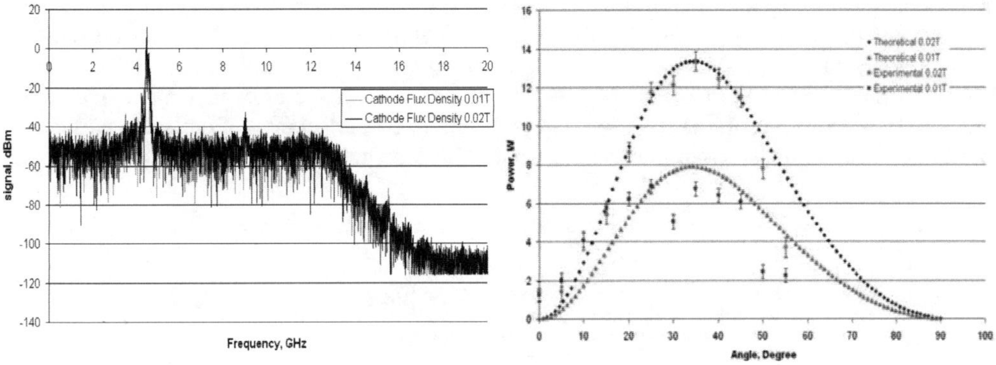

Figure 6: a) Operating frequency of emitted radiation, b) TE_{01} mode pattern obtained through antenna scans

3.3. Penning Trap Characterisation

To establish a stable discharge regime in the Penning trap a range of gasses (Helium, Nitrogen and Argon), pressures ~10^{-6}mBar to 10^{-4}mBar and magnetic field settings were investigated. Figure 7a) shows a graph of the voltage (blue trace) and current (green trace) measured against time. In this particular case the magnetic field is at a setting of 0.27T. Figure 7b) shows an alternative method to plotting the IV characteristics as a straight IV plot with no time axis.

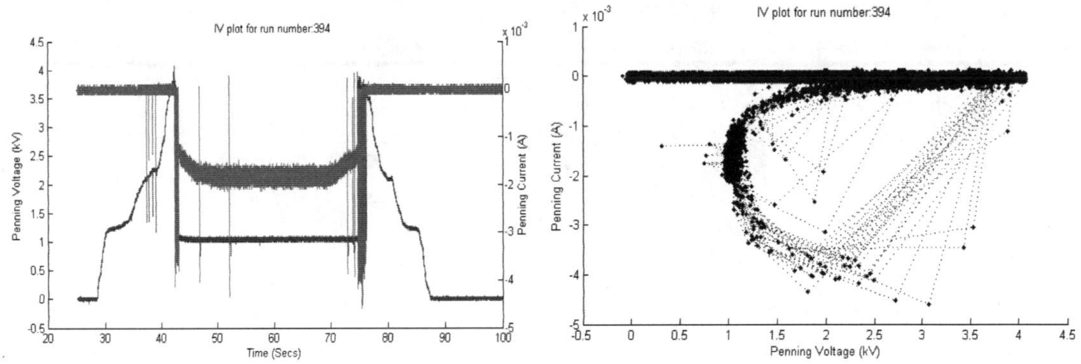

Figure 7: a) IV characteristics of the Penning trap, b) Alternative plot method for IV characteristics

It was found, through an extensive experimental investigation, that the optimum gas and pressure setting for the IV characteristics of the Penning trap was Argon at 10^{-4}mBar, this is plotted in Figure 7a). In general the higher the pressure the less voltage was required to ignite and sustain the discharge.

What can be seen in the above Figure 7 is that as the drive power to the trap was increased, the voltage measured across the trap decreased whilst measured current increased. Both voltage and current attain a steady sustained value once the discharge ignites, before tailing back to zero as the drive power is switched off.

These measurements which characterize the new geometry of the Penning trap will aid in the analysis of results that will be obtained in the future when the Penning chamber is used to create a background plasma to study plasma wave resonance conditions [19].

4. Conclusion/Discussion

The results presented in this paper form the base for future plasma-cyclotron resonance experiments to be carried out. The initial AKR experiments prior to the addition of the Penning trap showed proof of mirroring with steadily decreasing beam current as the magnetic field was increased. For the dimensions of the waveguide the frequency was measured at 4.42GHz as expected close to cut off for the TE_{01} mode, with power of emitted radiation ~19-35kW. The efficiency of emission was calculated to be ~1-2%. These measurements compare well with magnetospheric observations but lack the presence of a background plasma.

The initial investigations demonstrating the ability of the Penning geometry to provide a background plasma lay a foundation to run experiments to analyze the instabilities of a beam which is passed through the plasma discharge in the Penning trap and to measure the radiation emitted. Detailed analysis with insertion of a plasma probe will allow plasma frequency, density and temperature to be determined which will enable comparisons to magnetospheric data.

References

[1] D.A. Gurnett, *Journal of Geophysical Research*, **79**, pp4227-4238 (1974)
[2] P.L. Pritchett & R.J. Strangeway, *Journal of Geophysical Research*, **90**, pp9650-9662 (1985)
[3] S.L. McConville et al, *Plasma Physics & Controlled Fusion* **50**, 074010 (2008)
[4] K. Ronald et al, *Plasma Sources Science & Technology* **17**, 035011 (2008)
[5] K. Ronald et al, *Physics of Plasmas* **15**, 056503 (2008)
[6] D.C. Speirs et al, *Plasma Physics & Controlled Fusion* **50**, 074011 (2008)
[7] K.M. Gillespie et al, *Plasma Physics & Controlled Fusion* **50**, 124038 (2008)
[8] D.C. Speirs et al, *Journal of Plasma Physics* **71**, pp665-674 (2005)
[9] D.C. Speirs et al, *Physics of Plasmas* **17**, 056501 (2010)
[10] R.Q. Twiss, *Australian Journal of Physics*, **11**, pp564-579 (1958)
[11] P. Sprangle & A.T. Drobot, *IEEE trans. on Microwave Theory and Techniques*, **MTT-25**(1977)
[12] R. Bingham & R.A. Cairns, *Physics of Plasmas*, **7**, pp3089-3092 (2000)
[13] R. Bingham & R.A. Cairns, *Physica Scripta*, **T98**, pp160-162 (2002)
[14] R. Bingham et al, *Contributions to Plasma Physics*, **44**, pp382-387 (2004)
[15] G.T. Delory et al, *Geophysical Research Letters*, **25**, pp2069-2072 (1998)
[16] R.E. Ergun et al, *The Astrophysical Journal*, **538**, pp456-466 (2000)
[17] R.A. Cairns et al, *Physica Scripta*, **T116**, pp23-26 (2005)
[18] I. Vorgul et al, *Physics of Plasmas*, **12**, pp1-8 (2005)
[19] J.E. Allen & A.D.R. Phelps, *Rep. Prog. Phys.*, **40**, pp1305-1368 (1977)

Temporal description of aluminum laser-induced plasmas by means of a collisional-radiative model

Vincent Morel, Arnaud Bultel and Bruno G. Chéron
UMR CNRS 6614 CORIA, Université de Rouen, Avenue de l'Université, BP12, 76801 Saint-Etienne du Rouvray, France

E-mail: vincent.morel@coria.fr

Abstract. A 0D numerical approach including a Collisional-Radiative model is elaborated in the purpose of describing the behavior of the nascent plasma resulting from the interaction between a laser pulse ($\lambda = 532$ nm, $\tau = 4$ ns and F = 6.5 J cm^{-2}) with an aluminum sample. The species considered are Al, Al^+, Al^{2+} and Al^{3+} on their different excited states and free electrons. Both groups of particles are characterized by their translation temperature in thermal non-equilibrium state. Besides, each population density is assumed to be in chemical non-equilibrium and behaves freely through the seven involved elementary processes (electron impact induced excitation and ionization, elastic collisions, multi-photon ionization, inverse laser Bremsstrahlung, direct electron Bremsstrahlung and spontaneous emission). Atoms passing from sample to gas phase are described by considering classical vaporization phenomena so that the surface temperature is limited to values less than the critical point. The relative role of the elementary processes is discussed and the time-evolution of the excitation of the species is analyzed.

1. Introduction

The emergence of lasers at the beginning of the 1960's provided the ability to diagnose solid samples by the LIBS (Laser-Induced Breakdown Spectroscopy) technique which consists in analyzing the spectra emitted by the plasma induced by interaction between a nanosecond laser pulse and a sample [1]. According to the level of the laser fluence F, two thresholds can be observed. The first one corresponds to the minimum value to observe the breakdown ($F_0 \approx 1$ J cm^{-2} for aluminum [2]) and the second one corresponds to the observation of an explosion phase where the surface temperature reaches at least the critical point ($F_1 \approx 7$ J cm^{-2} [3]). This explosion phase is also characterized by formation of droplets [4]. Between F_0 and F_1, the interface melts first, then evaporates and the vapor produced is ionized. The plasma formed radiates and its spectra can be analyzed.

The underlined laser-matter interaction is far from being completely understood even when $F \in [F_0, F_1]$. In particular, the mechanism leading to an excitation equilibrium is not well known. This question is crucial because the treatment of LIBS signals is generally based on equilibrium assumptions [5]. The present contribution reports the elaboration of a two-temperatures Collisional-Radiative (CR) model devoted to the verification of these assumptions in the case of a laser pulse ($\lambda = 532$ nm, $\tau = 4$ ns and F = 6.5 J cm^{-2}) impinging an aluminum sample [6]. This CR model takes into account the seven main elementary processes (electron impact induced excitation and ionization, elastic collisions, multi-photon ionization, inverse laser Bremsstrahlung, direct electron Bremsstrahlung and spontaneous emission) influencing the number density of the 106 states of Al, Al^+, Al^{2+} and Al^{3+} considered.

2. Al excited states, assumptions and processes

The test-case studied corresponds to the classical interaction between a nanosecond Nd:YAG laser pulse assumed of the Gaussian type on an aluminum sample. Equations modeling the heating (melting, then vaporization) of the sample are not solved. The surface temperature is assumed to follow such a Gaussian evolution in time with a maximum T_{max} less than the critical temperature $T_c = 6700$ K [7]. Moreover, the atom flux density φ_A emerging from the interface is assumed to follow the Hertz-Knudsen law with a saturated vapor pressure given at T_s [8]. These atoms form the Knudsen layer whose characteristics allow a free expansion of the vapor according to a velocity close to the thermal heavies mean velocity [9].

In order to avoid wasteful calculations, the number of aluminum excited states is reduced with respect to the actual states. Levels of close energy are grouped according to their core configuration. The resulting energy spectrum contains finally 106 fictitious levels which are listed in part in table 1.

Table 1. Some of the aluminum states considered in the present CR model. No "0" figures the species ground state. E is the energy, g is the statistical weight and ℓ is the angular momentum quantum number.

Species	i	E (eV)	Core config.	g	ℓ	Species	i	E (eV)	Core config.	g	ℓ
Al(Z=0)	0	00.000	$3s^2$	2	1	Al^+(Z=1)	0	05.986	3s	1	0
	2	03.143	$3s^2$	2	0		13	21.578	3s	12	1
	6	04.827	$3s^2$	10	2		35	24.669	3s	14	3
	12	05.600	$3s^2$	38	4		36	16.584	3p	5	1
	18	05.963	$3s^2$	150	2		41	24.649	3p	15	2
	19	03.608	3s 3p ^{3}P	12	1	Al^{2+}(Z=2)	0	24.815	$2p^6$	2	0
	26	08.772	3s 3p ^{3}P	6	1		7	45.971	$2p^6$	2	0
	32	09.636	3s 3p ^{3}P	16	1		13	49.782	$2p^6$	10	2
	36	10.019	3s 3p ^{3}P	16	1		17	50.757	$2p^6$	64	5
	41	10.481	3s 3p ^{3}P	24	1		20	51.750	$2p^6$	54	5
	42	10.369	3s 3p ^{1}P	16	1	Al^{3+}(Z=3)	0	53.263	$2p^5$	1	1

The dynamics of these states results from the following elementary processes:

(i) *multi-photon ionization (MPI).* The vapor produced by the sample heating can be ionized by multi-photon ionization. This process is taken into account under the form proposed by Müsing *et al.* [10].

(ii) *inverse Bremsstrahlung (IB).* Inverse Bremsstrahlung phenomenon can heat electron gas during the laser pulse: its contribution is accounted for [11].

(iii) *electron-induced excitation (EE) and ionization (EI).* The increase of electron temperature T_e and density n_e promotes electron-induced excitation and ionization. The forms proposed by Drawin [12] for the cross sections (depending on the angular momentum quantum number ℓ, see table 1) are used.

(iv) *elastic collisions (EC)*. Owing to the increase of T_e by MPI and IB processes, thermal non equilibrium between heavies (with the temperature T_A) and electrons occurs. The resulting elastic collisions play a role and the form given by Decoster *et al.* [13] is adopted for their contribution.

(v) *spontaneous emission (SE)*. The increase of the excited states population density promotes spontaneous emission. The related probabilities obtained from the NIST database [14] are used and corrected by an escape factor resulting from the optical thickness of the plasma. Escape factors are calculated assuming a Stark line broadening.

(vi) *electron thermal Bremsstrahlung (BE)*. The direct Bremsstrahlung [15] is also accounted for owing to its experimental observation at early time [16].

3. CR model and balance equations

The previous elementary processes are involved in the collisional-radiative term of the classical balance equation [17] where the atom flux resulting from the vaporization is taken into account. The electrons and heavies energy balance equations are coupled with the previous ones. We do not reproduce here any details (cf. [6]), but we obtain finally the following equations:

$$\frac{dT_e}{dt} = \frac{2}{3\,k_B\,n_e}\left(P_e + \frac{\varphi_e(t)}{\Delta(t)}\right) - \frac{T_e}{n_e}\frac{dn_e}{dt} \tag{1}$$

$$\frac{dT_A}{dt} = \frac{2}{3\,k_B\,\sum_{Z,i}[\mathrm{Al}_i^{Z+}]}\left[P_A + \frac{\varphi_A(t)}{\Delta(t)} - \sum_{Z,i}\left(\frac{3}{2}\,k_B\,T_A + E_{Z,i}\right)\frac{d[\mathrm{Al}_i^{Z+}]}{dt}\right] \tag{2}$$

In these equations, $\varphi_e(t)$ is the electron flux density emerging from the surface, $\Delta(t)$ is the instantaneous thickness of the plasma layer, and P_A and P_e are the source terms involving the elementary processes listed in section 2.

4. Results

Two typical values for T_{max} (3000 and 4000 K) are adopted. The resulting time-evolutions for T_A and T_e are displayed on figure 1. We observe two consecutive phases: the first one which corresponds to the plasma formation is followed by a second phase where the plasma recombines. The transition between both phases takes place when the laser flux is maximum. At long time, the thermal equilibrium is reached: T_A and T_e are equal when $t = \tau_{eq} \approx 1$ μs with $T_{max} = 4000$ K and when $t = \tau_{eq} \approx 20$ μs with $T_{max} = 3000$ K.

Figures 2a and 2b display the evolution in time of the elementary processes contributions to dT_e/dt (figure 2a) and dT_A/dt (figure 2b) in equations 1 and 2 when $T_{max} = 3000$ K. We can notice the preponderant influence of the multi-photon ionization phenomena on T_e during the plasma formation. The increase of T_e is often ascribed to inverse Bremsstrahlung: we can see that IB phenomena play a minor role. Except MPI phenomena, the radiative processes can be considered as almost negligible: in particular, the contribution of electron thermal Bremsstrahlung is rather weak. Among the collisional contributions listed above, only superelastic electron-induced collisions and elastic collisions play a role: the first type of collisions promotes energy gain to the electron gas whereas the second type induces electron energy losses. Both types are almost counterbalanced: the T_e decrease is slow.

The evolution of T_A follows a different dynamics. Figure 2b shows the preponderance of the energy flux from the sample surface until the recombination starts. MPI phenomena do not play a noticeable role. This is common to all radiative phenomena: in particular, spontaneous emission is negligible. At long time, the elementary processes play on T_A a reverse role with respect to the case of T_e. When the laser pulse ends, the interface is no longer acting on the

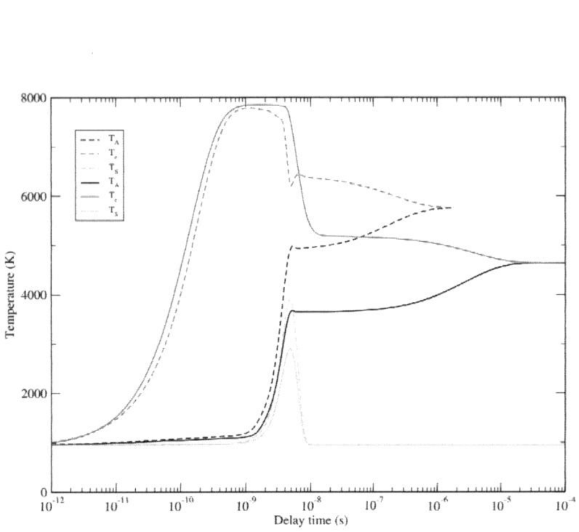

Figure 1. Time-evolution obtained for T_A and T_e when T_s evolves with T_{max} = 3000 K (continuous lines) or 4000 K (dashed lines).

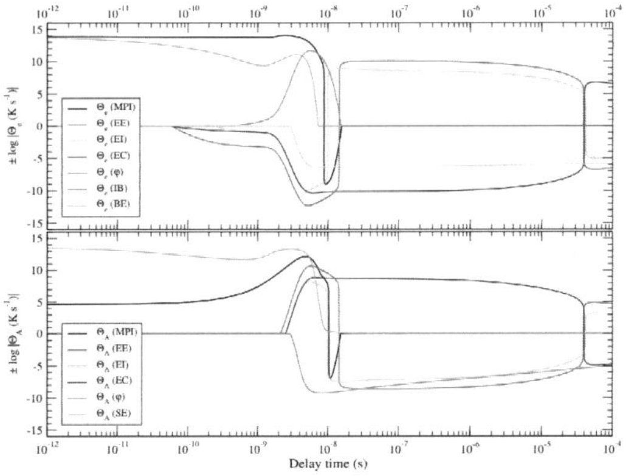

Figure 2. (a) evolution in time of the contributions to dT_e/dt (equation 1). (b) evolution of the contributions to dT_A/dt (equation 2). T_{max} = 3000 K.

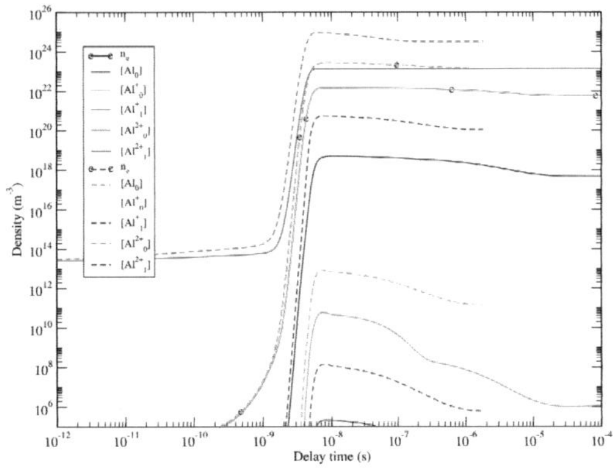

Figure 3. Time-evolution of the number density of some states. The laser pulse maximum corresponds to a delay time of 5 ns (continuous lines: T_{max} = 3000 K, dashed lines: T_{max} = 4000 K).

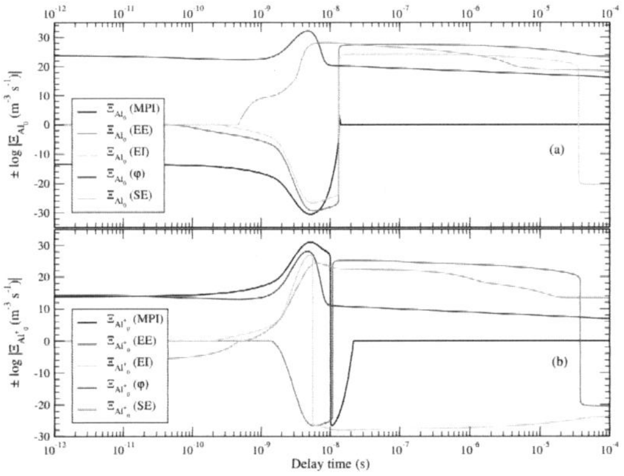

Figure 4. Evolution in time of the elementary processes contributions to the Al (a) and Al^+ (b) ground states population density when T_{max} = 3000 K.

plasma. Electrons and heavies are alone: this situation leads to their mutual interaction and to thermal equilibrium owing to elastic collisions.

The characteristic time scale τ_{eq} required to reach thermal equilibrium is shortened when T_{max} is higher. This is due to the higher levels of population obtained under these conditions. Figure 3 illustrates the influence of T_{max} on these levels. It is interesting to note that the maximum number density obtained for Al ground state with T_{max} = 3000 K is of 10^{23} m^{-3} approximately whereas this density reaches 10^{25} m^{-3} with T_{max} = 4000 K. This strong increase of the plasma density when T_{max} is higher leads to the elastic collisions frequency increase and to an important reduction of τ_{eq}.

Moreover, figure 3 shows the absence of temporal symmetry between the production of the plasma and its recombination. The plasma is indeed formed during the increase of the laser flux

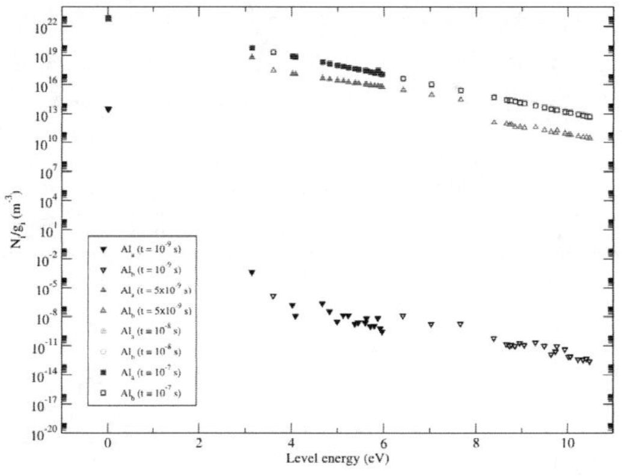

Figure 5. Al Boltzmann plots for t = 1, 5, 10 and 100 ns according to the core configuration with T_{max} = 3000 K.

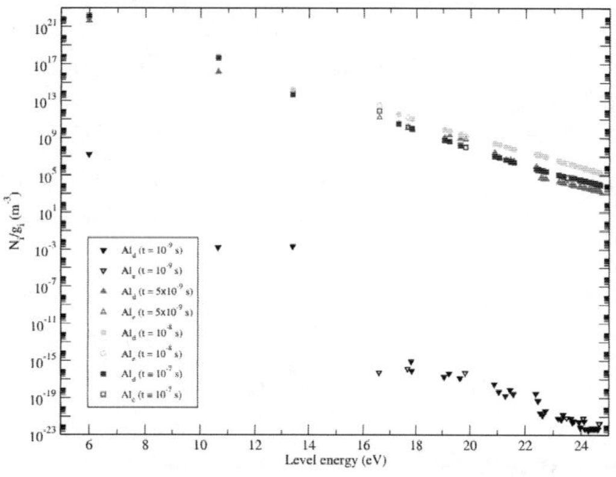

Figure 6. Same as figure 5, but for Al^+.

in few nanoseconds, whereas its recombination lasts some microseconds.

Figure 3 gives also an insight into the behavior of n_e when T_{max} increases. The maximum electron density is concomitant with the maximum laser flux. When T_{max} = 3000 K, this maximum corresponds to $n_e \approx 10^{22}$ m^{-3} whereas $n_e \approx 10^{23}$ m^{-3} with T_{max} = 4000 K. In both cases, the plasma ionization degree remains smaller than 10%. However, additional calculations performed with higher values of T_{max} put forward the rapid increase of the ionization degree with the interface maximum temperature: the plasma can even be totally ionized.

It is worth to further analyze the elementary processes contribution. Figures 4a and 4b display their time-dependent influence on the Al and Al^+ ground states number density respectively. The time scales non symmetry between the plasma creation and relaxation is again observed. Besides, figure 4a illustrates the fundamental role played by vaporization from the sample. The atoms reaching the gas phase are then ionized through MPI phenomena. The end of the laser pulse induces the relaxation of the excited states: this relaxation plays a preponderant role in the Al ground state dynamics. Three-body recombination does not play a significant role and spontaneous emission remains a minor process owing to the escape factors weak value.

The Al^+ ground state behaves of course in a different way. The vaporization from the surface is not the main source of Al^+ ions: this role is in this case played by multi-photon ionization from the Al ground state. The electron-induced processes do not play a significant role, except after the end of the laser pulse. When $t \gg \tau$, the electron excitation of Al^+ decreases and the contribution of electron-induced ionization is negative: the plasma is recombining. All the time, spontaneous emission is totally negligible in the present case.

We have previously mentioned that LIBS signals analysis is in part based on equilibrium hypotheses: in particular, the excitation is often assumed in equilibrium. The present CR model is a particularly suitable tool to verify the validity of this assumption. We have indeed derive Boltzmann plots from our results at different delay times. Figures 5 and 6 illustrate the results obtained for Al and Al^+ respectively. The situation remains far from excitation equilibrium from early time until the laser pulse end. This is the result of the steep ionization of the gas under laser flux and of the interface vaporization. After, electron-induced collisions become sufficiently efficient for ensuring excitation equilibrium: the Boltzmann plot displays a characteristic linear distribution. Long after the pulse, we can note a weak decrease of the number densities well coupled to each other. In the same time, the depopulation of the Al^+ excited states is stronger owing to the plasma recombination (see figure 6). Even for Al^+, the linear distribution is obvious. However, we have to comment cautiously these results owing to the large ranges over which they are shown: a seemingly linear form does not mean that equilibrium is necessarily reached. The excitation temperature T_{exc} defined over a set Σ of well coupled excited states [17] such that:

$$T_{exc}(\mathrm{Al}^{Z+}) = -\frac{1}{k_B\left[\dfrac{d\ln([\mathrm{Al}_i^{Z+}]/g_i)}{dE_i}\right]_\Sigma} \tag{3}$$

allows a largely better estimate of the excitation state. Calculations show that T_{exc} is approximately equal to T_e for $t \geq 0.1$ μs: as a result, the excited states can be then considered in equilibrium with electrons.

5. Conclusion

The behavior of a plasma produced by a nanosecond laser pulse on an aluminum sample is modeled by using a 0D approach involving a Collisional-Radiative (CR) model. Free electrons and heavy species temperatures are assumed different. About hundred excited states of Al, Al^+, Al^{2+} and Al^{3+} are considered. These species are assumed to behave freely through seven different collisional and radiative elementary processes. A simplified approach is adopted: the interface temperature evolves in time independently of the laser flux.

The results put forward the fundamental role played by vaporization from the interface and multi-photon ionization in the formation of the plasma. Even if the electron parameters reach high levels, the influence of free electrons on this formation is rather weak in our conditions. The electron temperature depends weakly on inverse Bremsstrahlung phenomena. When the laser pulse ends, the relaxation of the plasma takes place until equilibrium mainly through electron-induced processes.

Acknowledgments

The authors wish to thank the "Région Haute-Normandie" and the IEFE French Institute (Institut Energie, Fluide & Environnement) in France for their financial support.

References

[1] Noll R, Bette H, Brysch A, Kraushaar M, Mnch I, Peter L, Sturm V 2001 *Spectrochimica Acta Part B* **56** 637-649
[2] Cabalin L M, Laserna J J 1998 *Spectrochimica Acta Part B* **53** 723-730
[3] Fishburn J M, Withford M J, Coutts D W, Piper J A 2006 *Applied Surface Science* **252** 5182-5188
[4] Lu Q, Mao S S, Mao X, Russo R E 2002 *Appl. Phys. Lett.* **80** 17 3072-3074
[5] Cristoforetti G, De Giacomo A, Dell'Aglio M, Legnaioli S, Tognoni E, Palleschi V, Omenetto N 2010 *Spectrochimica Acta Part B* **65** 86-95
[6] Morel V, Bultel A, Chéron B G 2010 *Spectrochimica Acta Part B* doi:10.1016/j.sab.2010.08.002
[7] Morel V, Bultel A, Chéron B G 2009 *Int. J. Thermophys.* **30** 1853-1863
[8] Akhatov I, Lindau O, Topolnikov A, Mettin R, Vakhitova N, Lauterborn W 2001 *Phys. Fluids* **13** 10 2805-2819
[9] Gusarov A V, Smurov I 2002 *Phys. Fluids* **14** 12, 4242-4255
[10] Müsing A, Riedel U, Warnatz J, Herden W, Ridderbusch H 2007 *Proceedings of the combustion institute***31** 3007-3014
[11] Amoruso S, Armenante M, Berardi V, Brizzese R, Spinelli N 1997 *Appl. Phys. A* **65** 265-271
[12] Drawin H W 1966 *Collision and Transport Cross-Sections* Association-Euratom-C.E.A. Technical report (revised) EUR-CEA-FC 383, 656
[13] Decoster A, Markowich P A, Perthame B 1998 *Modeling of Collisions* Gauthier-Villars, Paris
[14] $http://physics.nist.gov/PhysRefData/ASD/levels_form.html$
[15] Koch H W, Motz J W 1959 *Rev. Mod. Phys.* **31** 4 920-956
[16] Morel V, Bultel A, Chéron B G 2010 *Anal. Bioanal. Chem.* in reviewing
[17] Bultel A, van Ootegem B, Bourdon A and Vervisch P 2002 *Phys. Rev. E* **65** 046406

Journal of Physics: Conference Series **511** (2014) 012063 doi:10.1088/1742-6596/511/1/012063

Spectroscopic study of emission coal mineral plasma produced by laser ablation

L P Vera[1,2], J A Pérez[1,2] and H Riascos[1,3]

[1] Universidad Tecnológica de Pereira, Grupo Plasma Láser y Aplicaciones, Colombia
[2] Ingeniería Física, Universidad Tecnológica de Pereira, Colombia
[3] Departamento de Física, A.A 097, Universidad Tecnológica de Pereira, Colombia

E-mail: lilianavera07@hotmail.com

Abstract. Spectroscopic analysis of plasma produced by laser ablation of coal samples using 1064 nm radiation pulses from a Q-switched Nd:YAG on different target under air ambient, was performed. The emission of molecular band systems such as C_2 Swan System ($d^3\Pi_g \rightarrow a^3\Pi_u$), the First Negative System N_2 (Band head at 501,53 nm) and emission lines of the C I, C II, were investigated using the optical emission spectroscopy technique. The C_2 molecular spectra (Swan band) were analyzed to determine vibrational temperature (0,62 eV); the density and electron temperature of the plasma have been evaluated using Stark broadening and the intensity of the nitrogen emission lines N II, the found values of 1,2 eV and 2,2 $\times 10^{18}$ cm^{-3} respectively.

1. Introduction

Laser-produced plasmas are a topic of interest for different fields of science with applications in medicine, astrophysics and synthesis of new materials. For plasmas produced by laser ablation, light energy to remove a portion of a sample by melting, fusion, sublimation, ionization, erosion and/or explosion. Several laser ablation-based methods have been implemented such as laser induced plasma spectroscopy (LIPS), pulsed laser deposition (PLD) [1]. In the laser ablation most of these molecules are formed in their excited states so that spectroscopic measurements offer a excellent means to investigate their evolution and dynamics [2].

Optical emission spectroscopic (OES) is a technique where light emitted from the plasma by electronically excited species, is collected and transmitted to a grating spectrometer; allow species identification, information about the elemental composition of the sample. In this case the carbon molecules are interesting for their unique and fascinating structural and spectroscopic properties, their importance in astrophysical processes and too role in combustion and soot formation [3].

Investigations recent in coals minerals for determine elemental composition, using technologies based in laser [4, 5], shows than the diagnosis of the plasma of coal can give us valuable information about the reaction mechanism involved in combustion and gasification process for production of synthesis gas under plasmas conditions [6], studies of the influence of sample morphology on laser ablation properties of coal, comparing temperature and electron density of the coal plasma with the different particle size [7, 8].

In this paper we used samples of Colombian coal for produced plasma at environmental conditions, and the spectra are analysed for determined temperature and electron density and vibrational temperature.

2. Experimental setup

We used the Nd: YAG laser that provides pulses in the wavelength of 1064 nm with pulse duration of 9 ns and repetition rate of 10 Hz. The laser beam is focused with an f = 23 cm glass lens on the target. The samples were target coal from mines located in Santander, Norte de Santander and Cesar, Colombia. The pressure was atmosphere air environment with a temperature of 26 °C and 55% humidity relative. The laser energy fluence was 7 J/cm2 for 10 minutes. The plume is generated by the laser impact of coal samples (figure 1), detecting radiation emission from plasma was performed by means of an optical fiber, which carries the radiation to the spectrometer. Characterization of plasma was performed by optical emission spectroscopy (OES) using a spectrometer model Jobin Yvon Triax 550 of 0.55m, f = 6.4 equipped with two gratings of 1200 l / mm and 150 l / mm, coupled to a model 3000 CCD camera cooled air multi-channel and 512 × 512 pixels.

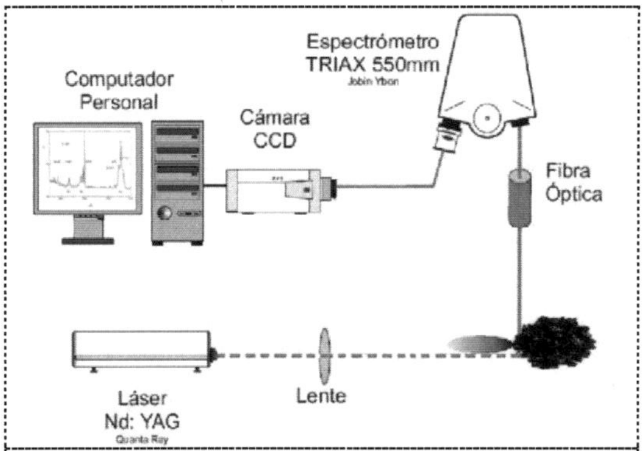

Figure 1. Experimental Setup [9]

3. Results and discussion

Spectral measurements were performed in coal plasma, the electron temperature have been evaluated using the relative intensities of the N II, and electron density of plasma with Stark broadening method, the parameters are available in the literature [10, 11]. In figure 2, carbon plasma produced by laser ablation.

3.1. Temperature and electron density

We report on spectroscopic measurements of plasmas produced by ionization of air at atmospheric pressure using nanosecond laser pulses, obtaining measures consistent with the features observed in nanosecond lasers. The value of electron temperature was 1,2 eV assuming the local thermal equilibrium (LTE).

$$ln\left(\frac{I_{mn}\lambda_{mn}}{A_{mn}g_{mn}}\right) = ln\left(\frac{N}{Z}\right) - \left(\frac{E_m}{kT_e}\right) \quad (1)$$

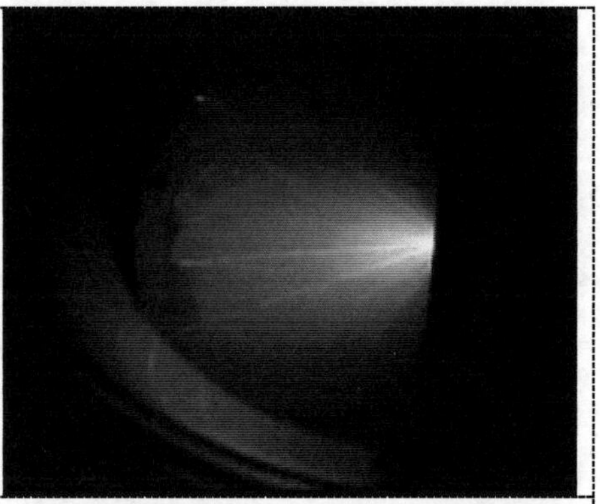

Figure 2. Plasma produced carbon

LTE is valid if collisional processes are more important than radiative decay; particles will have Maxwellian velocity distribution, in dense plasmas Ne $\geq 10^{16}$ cm^{-3} and temperature kT $\leq$ 5 eV, the velocity distribution of free electrons is almost always Maxwellian [12].((archivo 3)) We estimate the number of electron density of plasma by measuring the line width of the individual line of nitrogen. Electron density is related to the line broadening, and for highly ionized and high density plasma, the collisions with charged particles presents a strong electric field that produced a broadening of the transitions between the split atomic levels, these micro electric fields are associated with Stark broadening [13]. The found density was of the 2,2 $\times$ 10^{18} cm^{-3}. In the figure 3, a Lorentz function can be used to fit the line spectra.

Electron density must be sufficiently high, a criterion for LTE necessary but no sufficient is [14].

$$n_e \geq 1.4 \times 10^{14} T_e^{12} (\Delta E_{mn})^3 cm^{-3} \tag{2}$$

Substituting values for T_e and ΔE in (2) the value obtain is 9,57 $\times$ 10^{15} cm^{-3} implying that LTE approximation for these analyses is valid.

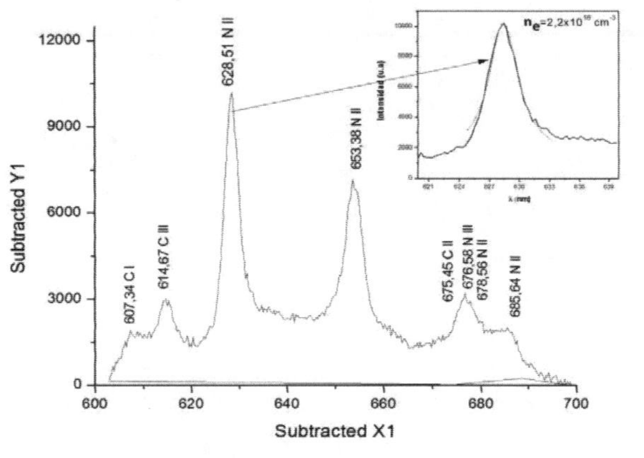

Figure 3. Electron density by Stark Broadening

3.2. Vibrational temperature

The Swan band of C_2 emission intensities were used for calculated vibrational temperature. The strongest system is well know Swan system in the green, is of special interest for provide an estimation of the plume temperature, and very important in astrophysics. They have been observed in the emission spectra of comets and also in the absortion spectra of stellar atmospheres [15]. The Swan systems arise from transitions between the $d^3\Pi_g \rightarrow a^3\Pi_u$ electronic states of the C_2 molecules. Band heads $\Delta v = 0$ (figure 4). The band emission at 510 nm, 515 nm and 520 nm can be utilized to determine the vibrational temperature. These emissions correspond to (0,0), (1,1), (2,2) bands [16].

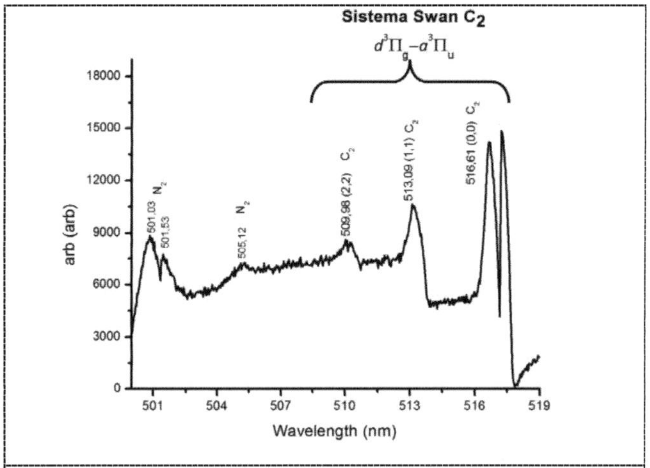

Figure 4. Swan Band of C_2, band heads $\Delta v = 0$

Using the Boltzman distribution (ecuacion), for plasmas in LTE, the value for vibrational temperature for C_2 of figure 5 is 0,62 eV. We can see the emission of the nitrogen molecule (figure 4), which emit in the ultraviolet, visible and infrared, in this case the line emission of first negative system attributed to $B^2\Sigma_u^+ \rightarrow X^2\Sigma_g^+$, occurs readily in a plasma at moderate pressure [17], band head at 501,53 nm.

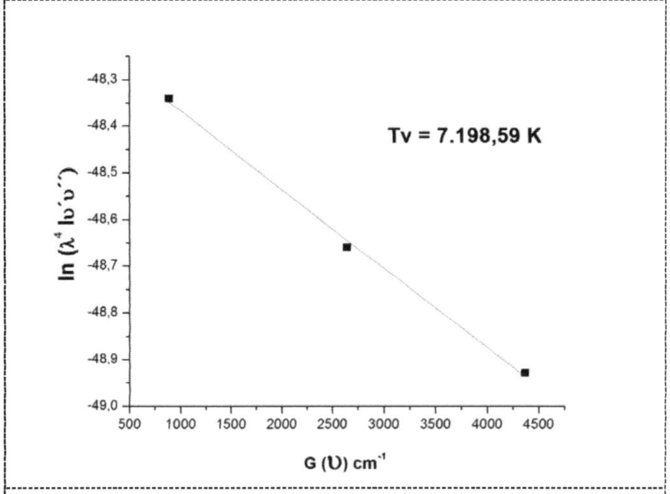

Figure 5. Vibrational temperature with Swan System of C_2

4. Conclusion

The diagnostic of the coal plasmas is very important for analysis parameters such as temperature in the plasma, electron density and the bands system that presents, and this provides information about of the transition and the excited species, incandescence of the carbon particles, and recombination processes and fragmentation. We can observe the formation of the Swan system of C_2 in the visible spectrum as well as the first negative system of nitrogen.

Acknowledgments

The authors (L. P. Vera and J.A. Pérez) wish to express their gratitude for the warm hospitality offered by Dr. Leopoldo Soto, Dr. Felipe Veloso and all the Local Organizing Committee of the ICPP-LAWPP 2010.

5. References

[1] Kleiber L, Fink H, Niessner R and Panne U 2002 *Anal Chem. Sci.* **374** pp 109-114
[2] Harilal S, Riju C, Bindhu C, Nampoori P and Vallabhan C 1997 *J. Appl. Phys.* **30** pp 1703-1709
[3] Harilal S, Riju C, Bindhu C, Nampoori P and Vallabhan C 1997 *J. Appl. Phys.* **81** 8979
[4] Romero C, Saro R, Craparo J, Weisberg A, Moreno R and Yao Z 2010 *Energy Fuels* **24** p 510
[5] Kohse K, Barlow R, Alden M and Wolfrum J 2005 *Proc. Combust. Inst.* **30** pp 89-123
[6] Qiu J, He X, Sun T, Zhao Z, Zhou Y, Guo S, Zhang J and Ma T 2004 *Fuel Proc. Technol.* **85** pp 969-982
[7] Yao S, Lu J, Lu Z, Xie C, Li J, Pan S, Jiang M and Li P 2009 *Acta Optica Sinica* **29** p 1126
[8] Xie C, Lu J, Li J, Yu L and Cui Z 2007 *J. of Eng. Thermoph.* **28** pp 133-136
[9] Franco L, Pérez J A M and Riascos H 2008 *Rev. Col. Física* **176** 2008
[10] Griem H R1964 *Plasma Spectroscopy* (New York: McGraw-Hill)
[11] Griem H R1974 *Spectral Line Broadening by Plasmas* (New York: Academic Pres)
[12] Griem H R1963 *Phys. Rev.* **131** 1170
[13] Man B, Dong Q, Liu A, Wei X, Zhang Q, He J and Wang X 2004 *J. Opt. A:Pure Appl. Opt* **6** pp 17-21
[14] Bekefi G 1976 *Principles of laser plasmas* (New York: Wiley-Intersscience)
[15] Hollas J M 1992 *Modern Spectroscopy.* (New York: John Wiley & Sons)
[16] Pearse R and Gaydon A1976 *The identification of Molecular Spectra* (London: Chapman and Hall)
[17] Karim M, Cameron D and Hashmi M 1994 *Diamond Rel. Mater.* **3** pp 551-554

Characterization of a laser plasma produced from a graphite target

M Ruiz, F Guzmán, M Favre, S Hevia, N Correa, H Bhuyan, E S Wynham and H Chuaqui

Departamento de Física, Pontificia Universidad Católica de Chile, Casilla 306, Santiago 22, Chile

E-mail: marceloruizaraya@gmail.com

Abstract. In order to improve the understanding of pulsed laser deposition (PLD) of diamond-like carbon (DLC) films, we have initiated a detailed study of the plasma dynamics of laser produced carbon plasmas. The carbon plasma is produced by focusing a Nd:YAG laser pulse, 380 mJ, 4 ns at 1.06 μm, onto a graphite target, at a background pressure of 0.3 mTorr. Time resolved optical emission spectroscopic (OES) observations of the carbon plasma plume are obtained, with time and space resolution, using a SpectraPro 275 spectrograph, with a 15 ns MCP gated OMA. Line emission from CII to CIV carbon ions is identified at different stages of the plasma evolution. Line intensity ratios of successive ionization stages, CIII/CIV, was used to estimate the electron temperature throughout the Saha-Boltzmann equation, under the assumption of local thermodynamic equilibrium (LTE), and Stark broadening of CII lines was used to obtain measurements of the electron density. Characteristic plasma parameters, short after plasma formation, are 3.0 eV and $2{\cdot}10^{17}$ cm^{-3}, which after 60 ns of plasma expansion decay to 2.7 eV and $5{\cdot}10^{16}$ cm^{-3}, respectively.

1. Introduction

Pulsed laser deposition (PLD) has become a well established technique for thin film deposition [1, 2]. A short laser pulse is focused onto a solid target, resulting in a laser produced plasma plume that expands away form the target surface. The expansion can take place in vacuum or in low pressure gas background. The effect of several global process parameters on the resulting thin film properties have been studied in detail. These include laser wavelength, pulse energy and pulse duration, background media and substrate conditions, amongst others. Less attention has been given to investigate the laser plasma properties, with time and space resolution, in correlation with the resulting thin film properties. Carbon plasmas produced by pulsed laser ablation of a graphite target in an ambient gas are been used to grow diamond-like carbon (DLC) films and to synthesize carbon nanocompounds [3, 4, 5]. We have recently started investigations of DLC film growth using PLD. In order to improve on the understanding of PLD deposition of DLC films, we have studied in detail the plasma dynamics of laser produced carbon plasmas. By using time and space resolved OES, we have characterized the properties of laser produced carbon plasma plumes,including measurements of the electron density and temperature.

2. Experimental Details

The carbon plasma is produced by focusing a Nd:YAG laser pulse, 380 mJ, 4 ns at 1.06 μm, at 10 Hz repetition rate, onto a rotating graphite target (POCO graphite), using a 12.5 cm focal length lens. Base pressure is kept at 0.3 mTorr. Time resolved OES observations of the carbon plasma plume are obtained with a SpectraPro 275 spectrograph (1200 grooves/mm, maximum resolution 0.05 nm), with a 15 ns MCP gated OMA. Light emission from the plasma plume is collected transversal to the plasma plume, with 0.6 mm space resolution, in the range from 1.6 to 3.4 mm from the graphite target surface.

3. Results and Discussion

Figures 1 and 2 show time resolved spectra of the laser plasma plume, at 2.2 mm from the target surface. Emission lines corresponding to CII to CIV transitions are identified. In order to highlight dynamic features in the plasma evolution, different time windows have been selected for both spectral bands. A particular feature in figure 1 is the broadened line profile of a singly charged ion (CII at 426.7 nm), and a general rapid decrease in the intensities of the carbon lines. Figure 2 shows the temporal evolution of the CIV (580.1 nm) line. After the 60 ns the CIV line declines completely, whereas the CII (589.1 nm) growths over the same time interval. This growth can be ascribed to recombination of higher ionization stages, as inferred from overall line intensity evolution.

Stark broadening of spectral lines is a well established technique to obtain measurements of the electron density in the range of $10^{14} - 10^{18}$ cm^{-3} [6]. The FWHM of the Stark broadened lines $\Delta\lambda_{1/2}$ depends on the electron density n_e through the relation

$$\Delta\lambda_{1/2} = 2W\left(\frac{n_e}{10^{16}}\right)[\mathring{A}] \tag{1}$$

where ion correction factors have been neglected, n_e is in cm^{-3} and W is the electron impact parameter [6]. We have used Stark broadening of the CII line at 426.7 nm to measure, with time and space resolution, the electron density of the carbon plasma. A Lorentzian profile has been fitted to the spectral line to obtain the FWHM. Time and space resolved values of the plasma density are shown in figure 3, for distances between 1.6 and 3.4 mm from the target. The axial displacement of the plasma plume can be appreciated in the time evolution of the axial electron density distribution. As the plasma emission collecting point moves away from the target surface, the time to reach maximum electron density is delayed, reaching lower characteristic values. These features are consistent with and plasma plume that expands in volume, thus decreasing the characteristic electron density as the plume front moves away from the target surface. Relative line intensities of subsequent ionization stages of the same element are very sensitive to temperature changes. In local thermodynamic equilibrium (LTE), the ratio of line intensities is given by [6]

$$\frac{I'}{I} = \frac{f'g'\lambda^3}{fg\lambda'^3}\left(4\pi^{3/2}a_0^3 n_e\right)^{-1}\left(\frac{k_B T_e}{E_H}\right)^{3/2} \times \exp\left(-\frac{E' + E_\infty - E - \Delta E_\infty}{k_B T_e}\right) \tag{2}$$

where the primed symbols represent the line of the atom with higher ionization stage, f is the oscillator strength, g is the statistical weight, λ is the associated wavelength, a_0 is the Bohr radius, E_H is the ionization energy of the hydrogen atom, E the excitation energy, ΔE_∞ is the correction to the ionization energy E_∞ of the lower ionization stage due to plasma interactions, and n_e is the electron density. Here we take advantage from the fact that the electron density has been determined independently for the same plasma conditions. In this particular case we have chosen as higher ionization stage the CIV line at 580.1 nm, and as lower ionization stage the CII line at 569.6 nm. The required date to be fitted in equation 2 for both transitions has been obtained from NIST Atomic Spectra devolution of the electron temperature is again consistent with cooling expanding plasma. The characteristic time span for plasma decay is of the order of 120 ns.

Our results are similar to those reported previously [8] for a laser produced carbon plasma, in which, under equivalent laser irradiation conditions, the plasma temperature was found to be in the 1.5–2.5 eV range, and the electron temperature in the $1.1–2.1 \cdot 10^{17}$ cm^{-3} range. The main difference here is the fact that single shot data allowed both, the electron density and temperature evolution, to be followed in space and time, thus providing valuable data to extrapolate to thin film deposition conditions.

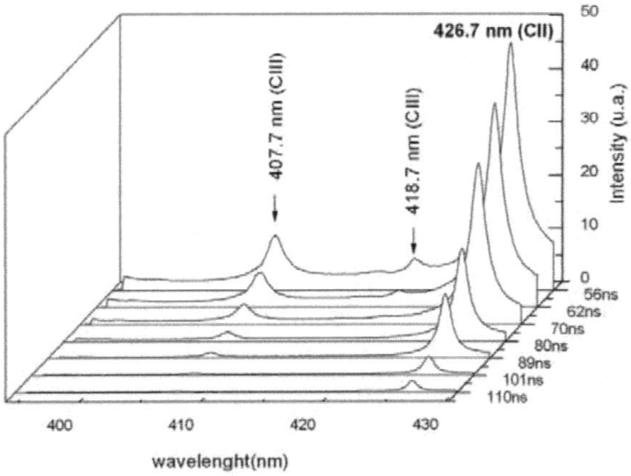

Figure 1. Time resolved spectra in the 395-430nm band.

4. Conclusions

We have characterized the time and space evolution of a laser produced carbon plasma, close to the target surface. This laser carbon plasma characterization will be useful in understanding

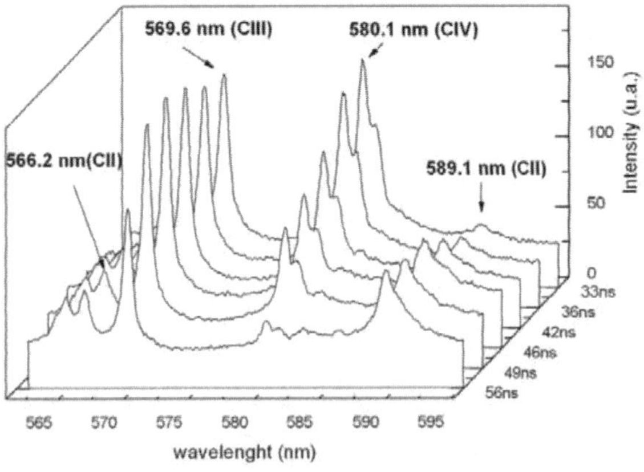

Figure 2. Time resolved spectra in the 465-495nm band.

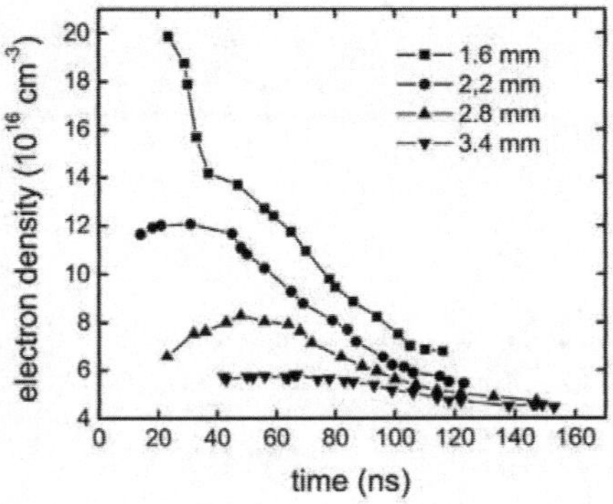

Figure 3. Time and space evolution of the electron density.

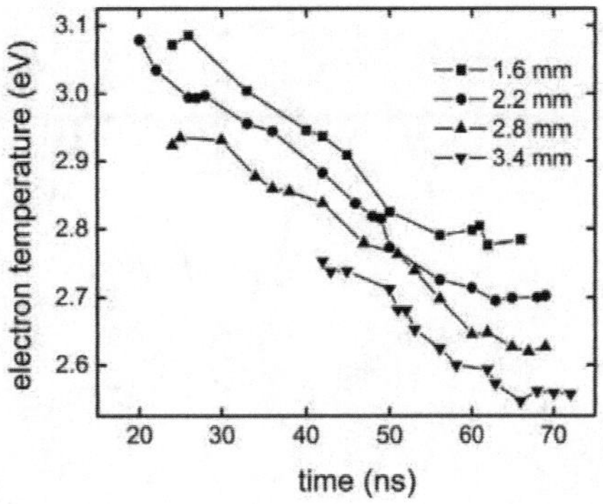

Figure 4. Time and space evolution of the electron temperature.

the properties of nanostructured diamond like carbon (DLC) films produced under identical experimental conditions, which will be reported elsewhere. Additional experiments to characterize a carbon laser plasma produced in an Argon background are also in progress.

Acknowledgments
This work has been funded partially by bridge grant 5/2010 from VRI–PUC and the Physics Department. M Ruiz acknowledges financial support from MECESUP. F Guzmán acknowledges financial support from CONICYT.

5. References

[1] Chrisey D B and Hubler G K 1994 *Pulser Laser Deposition of Thin Films* 1st Ed. (Wiley-Interscience)
[2] Eason R 2006 *Pulsed Laser Deposition of Thin Films: Applications-Led Growth of Functional Materials* (Wiley-Interscience)
[3] Voevodin A A,Jones J G and J. S. Zabinski J S 2002 *J. Appl. Phys.* **92** 724
[4] Yang L, May P W, Yinb L, Smith J A and K.N. Rosser K N 2007 *Diam. Rel. Mat.* **16** 725
[5] Wang H–P and Lin J 2010 *Surf. Coat. Tech.* **204** 2246
[6] Griem H R 1974 *Spectral Line Broadening by Plasmas* (Academic Press, New York)
[7] Ralchenko Y, Kramida A E, Reader J and NIST ASD Team (2010). NIST Atomic Spectra Database (ver. 4.0.0), [Online]. Available: *http://www.nist.gov/physlab/data/asd.cfm* [2010]. National Institute of Standards and Technology, Gaithersburg, MD.
[8] Harilal S S, Bindhu C V, Issac R C, Nampoori V P N and Vallabhan C P G 1997 *J. Appl. Phys.* **82** 2140

Diagnostics of cutting arc plasmas

L. Prevosto[1] and H. Kelly[1,2,*]

[1]Grupo de Descargas Eléctricas, Departamento Ing. Electromecánica, Universidad Tecnológica Nacional, Regional Venado Tuerto, Laprida 651, Venado Tuerto (2600), Santa Fe, Argentina

[2]Instituto de Física del Plasma (CONICET), Departamento de Física, Facultad de Ciencias Exactas y Naturales (UBA) Ciudad Universitaria Pab. I, (1428) Buenos Aires, Argentina

[*]Member of the CONICET

E-mail: prevosto@waycom.com.ar

Abstract. An over-view of several remote and invasive diagnostics to characterize cutting arcs at the nozzle exit-anode gap as well as inside the nozzle is reported. A briefly description of the experimental set-ups, together with the main results obtained in a 30 A high-energy density cutting torch (including the calculation assumptions) are given.

1. Introduction

Cutting arcs are created by a narrow nozzle ($\approx$ 1 mm) inside a torch, where a gas is tangentially injected at a high pressure. The intense gas-vortex convective cooling at the arc fringes enhances the power dissipation in the arc column ($\approx 10^{12}$ W m^{-3}) resulting in extremely sharp plasma quantities gradients, very high heat fluxes (up to 10^{10} W m^{-2} in the jet core), high plasma enthalpies ($\approx 10^8$ J kg^{-1}) and high plasma velocities ($M > 1$, being M the Mach number). As a consequence, the experimental data of such arcs are hard to obtain. In practice, most of the available experimental data are related to spectroscopic measurements in the external plasma region, giving information of only part of the variables involved, mostly temperatures and species concentrations [1]. On the other hand, several remote and invasive diagnostics of cutting arcs have been developed and applied in our Laboratory, such as rotating Langmuir probes, nozzle charge collection, time-of-flight plasma velocimetry and optical refractive techniques. These diagnostics were developed and used with the purpose of basic experimental research and theoretical model validation.

In this work, an over-view of such developed plasma diagnostics techniques is given. The following results correspond to an oxygen high-energy density cutting torch. To avoid plasma contamination by metal vapors from the anode, a rotating steel disk (rotating frequency of the disk equal to 23 s^{-1}) was used as the anode. During the experiment the arc current and the anode –nozzle exit gap were kept constant at 30 A and 8 mm; respectively.

2. Experimental data inside the nozzle

2.1 *Nozzle charge collection*

To obtain the *current – voltage* nozzle characteristic, it was necessary to bias the nozzle that under normal arc operation remains floating. Different nozzle bias voltages (registered with respect to the grounded anode) were obtained using a resistive voltage divider. The nozzle current was calculated from the voltage drop through a small resistance. Alternatively, the nozzle was disconnected to perform nozzle floating voltage measurements. To build the ion branch of the nozzle *current – voltage* curve, the rheostat was connected to the cathode; while it was connected to the anode to obtain the electron branch. The set of sweeping electrostatic probes (see section 3.1) was here employed to determine the plasma floating potential at the nozzle exit. More detail on the nozzle biasing circuit can be found elsewhere [2].

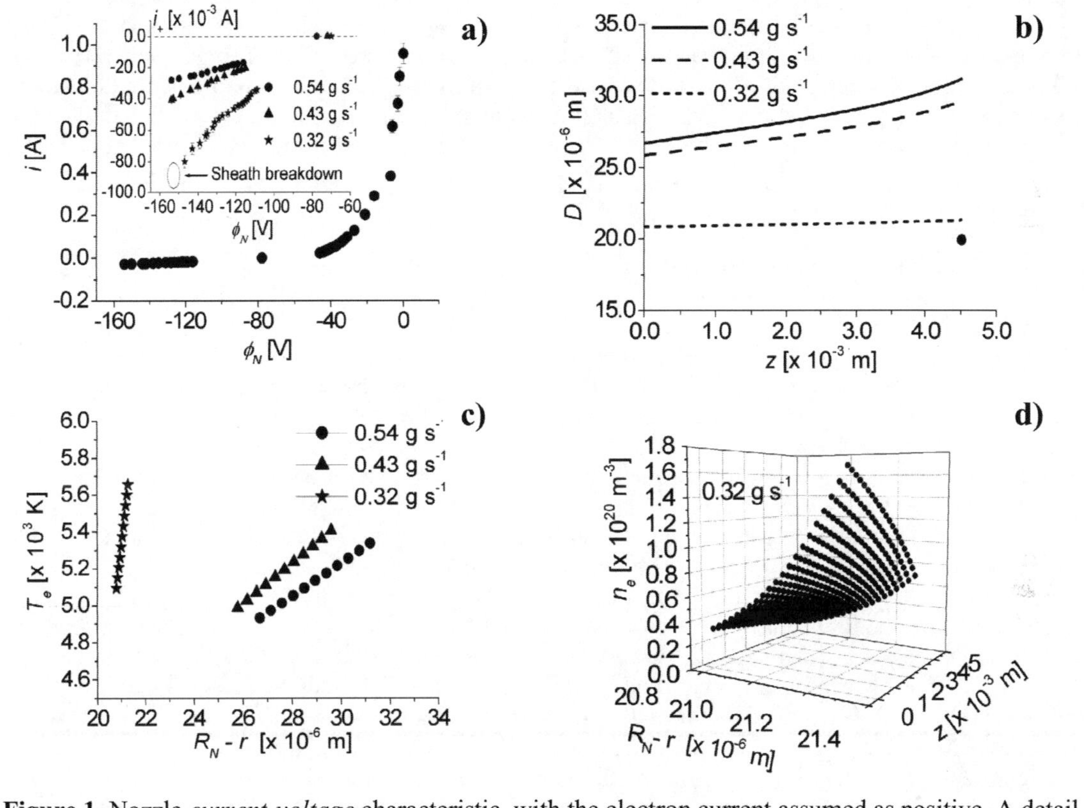

Figure 1. Nozzle *current-voltage* characteristic, with the electron current assumed as positive. A detail of the ion branches for several gas mass flow values with an enlarged current scale is shown at the inserted plot. Note that for the lowest gas mass flow value the sheath breaks-down at a nozzle biasing voltage $\Phi_N \approx$ -155 V. The whole nozzle characteristic curve corresponds to the largest gas mass flow (a). Sheath thickness along the nozzle wall for the largest nozzle voltage value experimentally registered for each gas mass flow value. For comparative purposes, the black circle indicates the sheath thickness value inferred at the nozzle exit using the voltage breakdown measurement and the Paschen's curves (b). Radial profiles of the average (along the nozzle wall) plasma electron temperature as a function of the radial position measured from the nozzle wall, for the different gas mass flow values (c). Spatial distribution of the plasma density close to the nozzle wall for the lowest gas mass flow experimentally registered (d).

Figure 1 a) shows the nozzle *current – voltage* curve for a cutting arc. The plenum pressure and the oxygen gas mass flow rate were varied in the range 0.55 ÷ 0.65 MPa and 0.32 ÷ 0.54 g s^{-1}, respectively. The results show a very low electron density and the lack of electron attachment at the plasma boundary layer. No ion saturation current was found. For the smallest mass flow rate value gas breakdown was found for a biasing nozzle potential close to that of the cathode, but no evidence of such breakdown was found for the larger mass flow rate values. Using an expression for the ion speed at the entry of the colisional sheath formed between the non-equilibrium plasma and the negatively biased nozzle wall together with a generalized (two-temperature) Saha equation coupled to the ion branch of the characteristic, the axial profile of the sheath thickness (figure 1 b)), the radial profile of the electron temperature (figure 1 c)) and the spatial distribution of the plasma density at the plasma boundary (figure 1 d)), were obtained. An electron temperature of about 4700 ÷ 5700 K and a corresponding plasma density of the order of $10^{19} \div 10^{20}$ m^{-3} were found close to the nozzle wall. Very high temperature gradients (of the order of $10^8 \div 10^9$ K m^{-1}, depending on the mass gas flow ratevalue) over the last few electron Debye lengths from the nozzle wall are predicted. In particular, note that the obtained thickness value at the breakdown condition is in good agreement with that obtained from the oxygen Paschen's curve [3] (black circle in figure 1 b)).The obtained results indicates that once the sheath thickness is smaller than the critical value (predicted by Paschen's law), the cold envelope breaks-down leading to double-arcing.

3. Experimental data in the nozzle exit – anode gap

3.1 *Langmuir probe*

The multi-probe system employed for arc mapping consists in a rotating aluminium disk carrying up to 6 identical probes mounted in the radial outward direction at different heights. The probes were made of tungsten or copper wires with radius ranging from 63 to 250 μm crimped on copper sleeves inserted into holes at different heights. This arrangement allowed the axial mapping of the arc in one single measurement with a step of approximately 1 mm, starting at a height of 0.5 mm below the nozzle exit. The probes were swept trough the arc at constant velocity in unbiased (floating) or negatively biased (probe potential below the plasma floating potential) conditions, and the probe voltage were registered with respect to the grounded anode. The disc was rotated using a variable speed motor (with frequencies *f* in the range 8.7 – 43.7 Hz). On the upper disc surface a pair of carbon brushes collected the probe current. Two different circuits were employed for biasing the probes; one of them used an independent biasing power source, while the other took advantage of the voltage distribution occurring in the main arc discharge and employed only resistors with appropriate values to bias the probe. A detailed description of the multi-probe system employed for arc mapping can be found elsewhere [4].

Figure 2 a) shows the ion branch of the probe characteristic obtained at 3.5 mm from the nozzle exit for a gas mass flow of 0.71 g s^{-1}. Figure 2 b) shows several ion current signals corresponding to different probe diameters. As it can bee seen, the ion current collected by negatively biased probes shows no plateau (for both materials), and the signal amplitude is almost independent of the probe radius. To interpret the quoted data, a semi-empirical Langmuir probe model was introduced specially adapted to high energy density cutting arcs [5]. According to the model, the ion drag due to the high velocity plasma flow around the probe limits the effectively collecting area to a small fraction of the probe surface. If, according to the experimental evidence, this fraction is made independent of the probe radius, its value results proportional to the probe bias, and so no plateau is found.

A typical probe floating voltage signal (that shows a characteristic central hump) obtained at 3.5 mm from the nozzle exit is shown in figure 2 c). It was assumed that the hump in these probe voltage signals results from the presence of an electrostatic field directed in the radial direction outward the arc axis that is caused by thermoelectric effects [6]. The probe floating voltage signal was inverted using the generalized Ohm's law together with the Saha equation, thus obtaining the radial profiles of the radial electric field and potential of the plasma at the studied section of the arc. For an arc section located at 3.5 mm from the nozzle exit, the calculated radial profiles of the plasma temperature, electron density (under the local thermodynamic equilibrium – LTE –) and radial electric field together with the plasma potential are given in figure 2 d) to f) respectively. As it can be seen, the temperature profile shows a peak value of about 15000 K while the plasma is almost concentrated within a radius of about 0.3 mm. Note from figure 2 f), that the radial electric field value (≈ 20 V mm^{-1}) is very high, even higher than the axial electric field of about 7 V mm^{-1} [4].

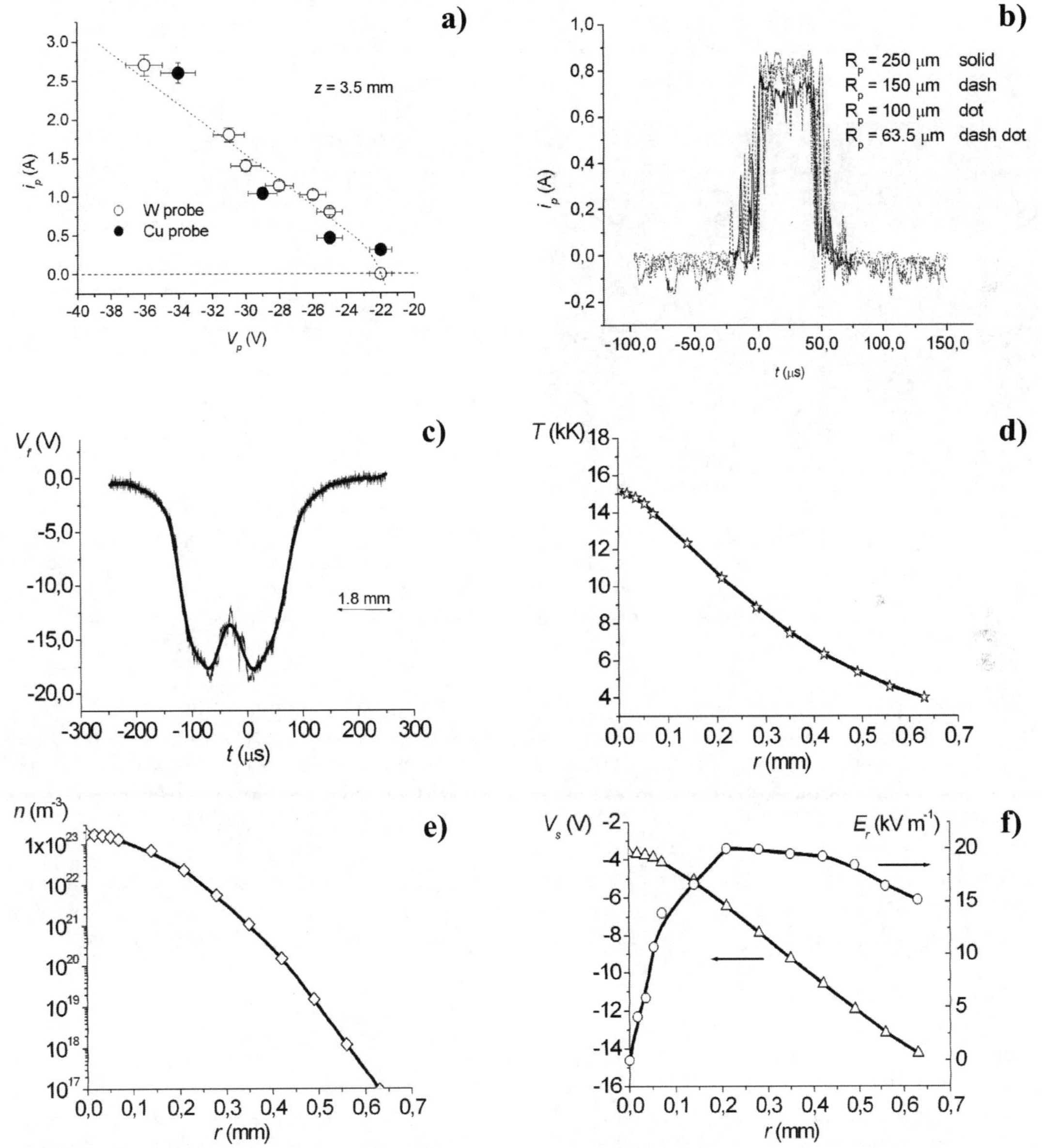

Figure 2. Probe characteristic ion branch (a). Ion current signal for several probe radii (b). Typical probe floating signal at 3.5 mm from the nozzle exit (c). Radial distributions of the plasma temperature (d), electron density (e) and the spatial plasma potential together with the radial electric field (f).

3.2 *Schlieren images*

A Z-type two-mirror schlieren system was used in this research. The system included a 20 mW continuous laser with a main wavelength of 632.8 nm as the light source. The laser beam was expanded to a diameter of 50 mm at the first mirror surface by using an optical expansion system. There were two parabolic mirrors (100 mm diameter) with focal distances of f_1 = 900 and f_2 = 1100 mm for the first and second mirror respectively, and a knife-edge that could be accurately controlled by a 2-D micrometer. The first mirror serves to collimate the incoming light, while the second mirror forms an image of the observation region onto a white screen. A band-pass filter centered at the laser wavelength was used to block the plasma light emitted from the arc. The schlieren images were acquired with a CCD camera at a spatial resolution of 640 × 480 pixels. More detail on the schlieren arrangement can be found elsewhere [7].

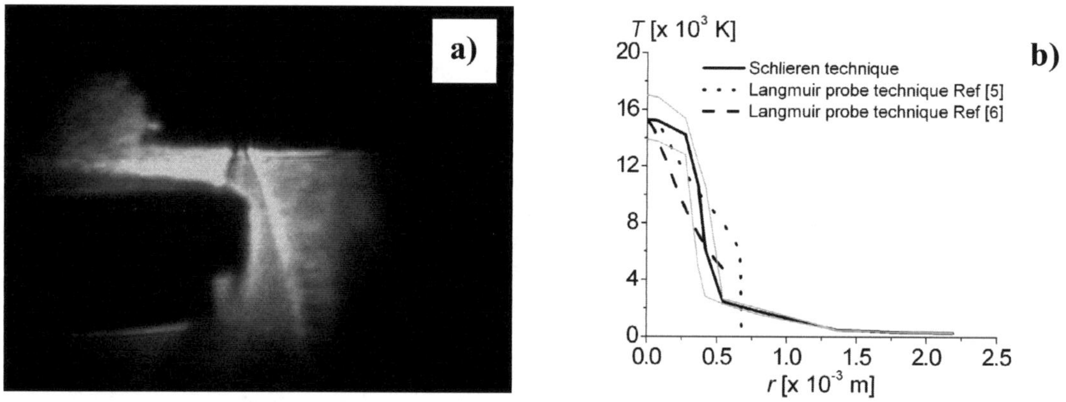

Figure 3. Typical gray-scale schlieren image. The percentage of the knife-edge cut-off was ≈ 50 % (blocking on the left side of the laser beam) (a). Plasma temperature radial profile obtained at 3.5 mm from the nozzle exit. The gray thin lines indicate the limits of the uncertainty range in the temperature values. Plasma temperature profiles obtained by the authors using Langmuir probes are also shown (b).

Due to its great sensibility such technique allows measuring plasma composition and temperature from the arc axis to the surrounding medium by processing the gray-level contrast values of digital schlieren images recorded at the observation plane for a given position of a transverse knife located at the exit focal plane of the system. The technique has provided a good visualization of the plasma flow emerging from the nozzle and its interactions with the surrounding medium and the anode (figure 3 a)). The calculated plasma temperature (under the LTE assumption) is given in figure 3 b). For comparative purposes the temperature profiles previously obtained by the authors using Langmuir probes is also shown. As it can be seen, in spite of that the used techniques are based in quiet different physical principles, a relatively good agreement among the inferred temperature profiles is found.

3.3 *Time of Flight velocimetry*

The optical system consisted simply in a converging lens (with 100 mm of focal distance) which produced a magnified (7X) image of the arc on a plane were two fast photodiodes (separated 7 mm in the axial direction) were mounted. Since the sensitive area of the photodiodes was a 1 × 1 mm square, these sensors were collecting light from two 0.14 × 0.14 mm square arc regions separated 1 mm in the axial direction. Due to its large magnification, the optical system resolution was of the order of the arc radius. More detail on the light detecting circuit can be found elsewhere [8].

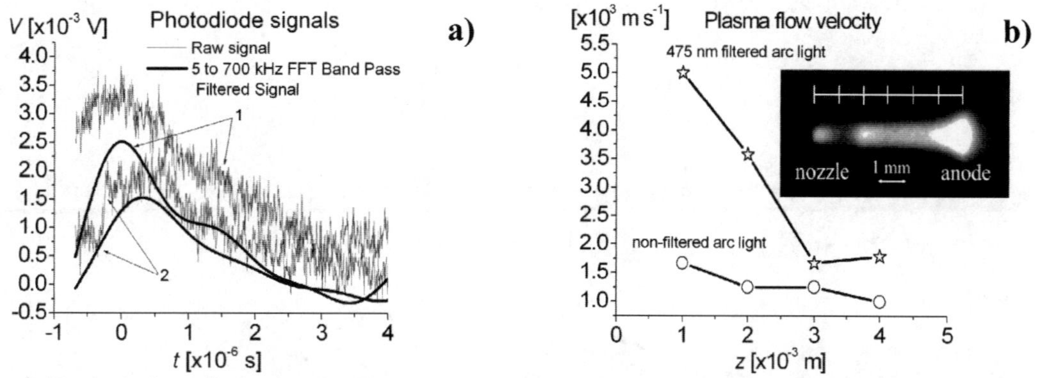

Figure 4. Typical photodiode signals. The signal labeled 1 corresponds to the photodiode focusing at 0.5 mm while the signal 2 corresponds to the photodiode focusing at 1.5 mm from the nozzle exit (a). Flow velocity obtained for different axial positions with and without band-pass optical filter. A visible photograph of the arc is shown in the insert (b).

The velocity diagnostic is based on the analysis of the light fluctuations emitted by the arc that are assumed to propagate with the flow velocity (the Péclet number in this kind of arc is high, $\approx$ 100). These light fluctuations originate from plasma temperature and plasma density fluctuations mainly due to hydrodynamic instabilities. Figure 4 a) show typical photodiode signals. To improve the technique efficiencies, only frequencies higher than 5 kHz and lower than 700 kHz was considered [8]. The arc core velocity was obtained from spectrally filtered light fluctuations measurements using a 475 $\pm$ 50 nm band-pass filter. Figure 4 b) shows the axial profile of the plasma velocity with and without band-pass optical filter. Maximum plasma jet velocities of 5000 m s^{-1} close to the nozzle exit and about 2000 m s^{-1} close to the anode were found.

4. Conclusions

An over-view of several remote and invasive diagnostics to characterize cutting arcs in the nozzle exit-anode gap as well as inside the nozzle has been reported. Diagnostics such as rotating Langmuir probes, nozzle charge collection, time-of-flight plasma velocimetry and optical refractive techniques have been described presenting the basic assumptions used in the data interpretation. The results in terms of the flow velocity, temperature and composition of the plasma jet emerging from the nozzle of a 30 A high-energy density cutting torch have been also reported.

Acknowledgements

This work was supported by grants from the Universidad de Buenos Aires (PID X108), CONICET (PIP 5378) and Universidad Tecnológica Nacional (PID Z 012). H. K. is member of the CONICET.

References

[1] Nemchinsky V A and Severance W S 2006 J. Phys. D: Appl. Phys. **39** R423.
[2] Prevosto L, Kelly H and Mancinelli B 2009 J. Appl. Phys. **105** 013309.
[3] Raizer Y P 1991 *Gas Discharge Physics* (Berlin, Germany: Springer).
[4] Prevosto L, Kelly H and Mancinelli B 2008 IEEE Trans. Plasma Sci. **36** 263.
[5] Prevosto L, Kelly H and Minotti F O 2008 IEEE Trans. Plasma Sci. **36** 271.
[6] Prevosto L, Kelly H and Mancinelli B 2009 IEEE Trans. Plasma Sci. **37** 1092.
[7] Prevosto L, Kelly H and Mancinelli B 2010 J. Appl. Phys. **107** 023304.
[8] Prevosto L, Kelly H and Mancinelli B 2009 J. Appl. Phys. **106** 053308.

C_2 Swan spectrum used as a molecular pyrometer in transferred arc and the influence noise to signal ratio on the temperature values.

H NASSAR
Lebanese University, FSP, section IV, Karak Zahlé, Lebanon.
nassar_hossein@hotmail.com

Abstract. The C_2 Swan system molecular emission spectrum is frequently observed in arc plasmas containing hydrocarbons. The spectra emitted from 5 kw in the transferred arc reactor at atmospheric pressure by $CH_4/CO_2/Ar$ mixture are recorded with the help of an optical system consisting of a linear CCD array coupled with 2m spectrometer. The rotational temperature of 4300±300K is found from the experimental Abel inverted spectra in the arc center after a point–to–point comparison of the spectrum with a computer simulated one. The influence of the noise to signal ratio has been studied, if the noise to signal ratio is about 10% we found an error of 7% at temperature 3000K and 10% at 6000K..

1. Introduction

The Swan band of C_2 molecule is a good tool for diagnosing plasmas containing this molecule, the molecular constants and the lines structure are well-known, and it is present in most of the plasma sources containing hydrocarbon for a temperature between 3000K and 6000K.

Pellerin in 1997 use Swan band spectrum for temperature measurements and to answer some questions on thermodynamic equilibria in a plasma produced from the CH_4 +CO_2 mixture in a transferred arc[1]. Bontemps-Faure in 1997, uses this system to study thermodynamic state in thermal plasmas[2]. In (1998) Laffon [3] use the C_2 (1, 1) and (0, 0) bands to measure the temperature in Comet Hyakutake C/1996 B2. It is shown that the temperature increases with cometo-centric distance. Babou, in 2008, have determined the profile of temperature in CO_2 and CO_2/N_2 plasmas produced by a microwave torch at atmospheric pressure using this band [4].

In the previous publications, we have proposed a numerical method based on the simulation of molecular spectrum for studying the distribution temperature in modified arc Maeker[5] and[6], and in transferred arc[7]. To determine N_2^+/N_2 ratio and the temperature distribution in Glidarc[8], we have made the simulation of both systems first negative and second positive one simultaneously.

In this paper, we have used this method based on the simulation of the (0,0) band of C_2 molecular Swan system. The method proposed permit to evaluate, by comparing point to point a real spectrum with the simulated one, temperature and apparatus function, approximated by the gauss function (the half-width at 1/e height). The real spectrum is emitting from plasma produced in a transferred arc in CH_4+CO_2 mixture. Moreover, it is shown, by noised spectra simulation, the influence of noise to signal ratio at the calculated temperature values.

2. Simulation of the $d^3\Pi_g - a^3\Pi_u$ spectrum of C_2

2.1.The Swan system of C_2 molecule

A complete description of the spectral terms of the diatomic molecules my be found, e.g. in [9] and [10], or in our previous publications [6] and [8].

The Swan system of homo-nuclear molecule C_2 corresponds to radiative transitions between the $d^3\Pi_g$ and $a^3\Pi_u$ electronic state ($(\Lambda = 1)$ at 2.59 ev and 0.19 ev respectively. We are interested in the visible (5138Å ,5170Å) spectral range covering the (0,0) band. The coupling is type '(a) Hund case for lower J values and type '(b) hund case' for J-values higher than 15 [10].

The absolute rotational emission line intensity of the molecule is given by

$$\varepsilon_{ul} = C\sigma^4.S_{J'J''}.\exp\left[-\frac{hcF(J')}{KT}\right] \qquad (1)$$

Where C is a normalization constant, σ is the wave number, Sj'j'' is the Hönl-London factors, h and c are the conventional physical constants, F(J') is the spectral terms and T is the temperature. (marks ' concerne the lower spectral term and the marks " are for the upper one). Formulae for Sj'j'' factors for the intermediate a-b Hund case are taken from Kovacs [10] and the spectral terms are those of Budo [11].

2.3.Numerical simulation of band (0,0)

Line intensities are calculated when a state of thermodynamic equilibrium is fully reached and the corresponding temperature of the gas, considered from a macroscopic and thermodynamic point of view, are the same. In reality, within laboratory plasmas, we usually have to deal with a local thermal equilibrium state instead of thermodynamic equilibrium state.

It is possible, when the population of discrete energy levels follow Boltzmann's Law, to attribute a corresponding temperature for each corresponding energy distribution; electronic, T_e, vibrational T_v and rotational T_r. In a non-equilibrium plasma T_r, T_v and T_e may differ from the kinetic temperature T_g of heavy species. Taking into account the strong coupling between translational and rotational energy states, the temperature derived from experiments is, in a general case, close to the gas kinetic temperature [12] and [13].Thus, *Tg* can be deduced from *Tr*, measured by using the rotational structure of molecular spectra.

All the spectral constants are taken from Herzberg [9]. Computed spectrum of individual lines for a given temperature is then convoluted with our apparatus function approximated by the gauss function whose DX (the half-width at 1/e height):

$$I_0 = \int_{\nu_0-\varepsilon}^{\nu_0+\varepsilon} I_\nu.\exp\left[-\frac{(\nu-\nu_0)^2}{DX^2}\right].d\nu \qquad (2)$$

Where τ =2.5 DX and ν_0 is the wave number of the line.

An integrated cell (pixel) of simulated spectrum is chosen according to the cell of experimental spectrum under consideration. We can, following the situation, group several cells in one.

Practically, for each given set of values (DX, *Tr,*), the line intensities of this band are normalized to its maximum. Examples of simulated spectra, normalized to 100, are shown in figures 1 and 2 for different sets of (DX, *Tr,*) values.

A quick analysis of these simulated spectra shows that the apparatus function has essentially influenced the resolution of the rotational line structure. Particularly, if DX is too high, only the shape of the band

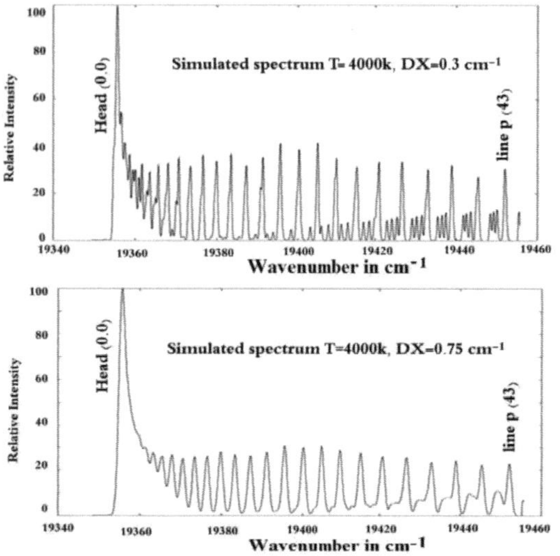

Figure 1. Effect of the apparatus function DX (for rotational temperature T=4000K)

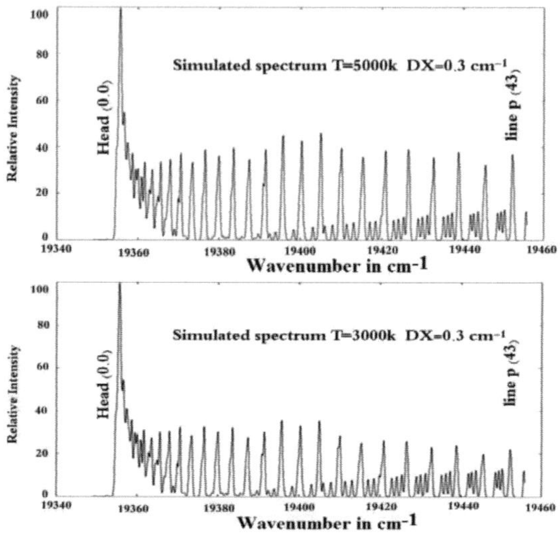

Figure 2. Effect of rotational temperature (the apparatus function DX =0.3 cm^{-1})

can be used to determine the rotational temperature.The rotational temperature has a significant influence on the rotational line structure.

3. Numerical method for rotational temperature evaluation

To evaluate the *Tr* from a given real spectrum, we have chosen to compare this spectrum point-by-point with spectra simulated at different values of *DX Tr.* In our experiments the following procedure was applied. (i) The measured spectrum was corrected for the continuous background, which was assumed to be linear in the studied spectral range. (ii) The experimental spectral intensity was normalized to the maximum value, which was set to 100. (iii) The experimental data were compared with that of the simulated spectrum. The minimization procedure was used to minimize $E(T_r,DX)$) as a function of *Tr, and DX* defined as:

$$E(T_r, DX) = \frac{1}{N}\left[\sum_{i=1}^{N}\left(I_i^r - I_i^s\right)^2\right]^{1/2} \qquad (3)$$

I_i^r et I_i^s are respectively the intensities of the real and simulated i-th pixel of the spectrum containing N points. If the choice of the spectrum reference origin (background) is a bad one, the result will be affected. To avoid this error, we fix several origins, and our computer program searches the best origin to find the E(T) at minimum.

4. Application to simulated noisy spectra

In order to test the sensitivity of the temperature estimator, we applied this method when a noisy spectrum was used instead of the experimental one. The synthetic spectra were calculated for known apparatus functions, temperatures, and noise was introduced using a random number generator.
Examples of simulated spectra noised, normalized to 100, are shown in figure 3 for different sets of N/S ratio. We can see that the rotational spectra are very deformed for the high N/S values.

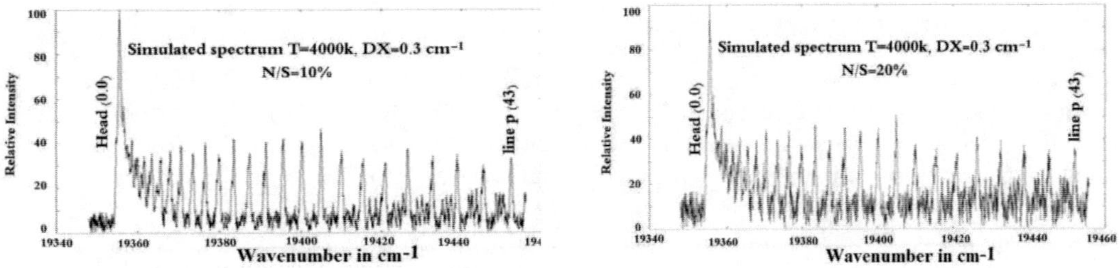

Figure 3. Effect of noise to signal ratio (N/S) (T=4000K and DX =0.3 cm^{-1})

Calculations were performed for noise-to-signal ratios N/S equal to 1%, 5%, 10% and 20% and for temperatures of 3000 and 6000 K. The results obtained are given in table 1.

Table 1. Evolution of the relative error on the temperature with noise/signal ratios.

	ΔT/T%	
Noise/signal %	3000k	6000k
1%	1.5%	3%
5%	4%	7%
10%	7,0%	10.0%
20%	11.5%	18.5%

The results are satisfactory since the relative error in the temperatures is of the same order of magnitude as the noise-to-signal ratio, even if the last one was relatively high.

5. Experimental set-up

5.1 Reactor

The reactor consists of an arc chamber to which an Argon plasma jet of 4 mm diameter emerges through a 2 kw plasma torch nozzle considered as a first anode. This jet makes a first ionization stage. Another tubular transfer anode (5 mm of internal diameter) put at 20 mm from the nozzle aperture allows to obtain a DC transferred arc in which methane and carbon dioxide gases are introduced(see figure 4).
Both copper anodes as well as the torch tungsten cathode are water cooled. The transferred arc chamber is fed at atmospheric pressure by CH_4/CO_2 = 1 (molar ratio) gas mixture at 10 sml flow rate. The electrical power supplied to the arc is 5kw.

5.2 Spectral recording system

The C_2 emission spectra are observed with the help of a 2-meter monochromator equipped with a 1200 groves/ mm grating couples with an optical analyzer based on Thomson linear CCD array (1728 pixels of 13.39 m) The transferred arc column at the half of its lengths is scanned through a quartz window by a step-motor desk computer moved optical system at 12 points at 0.45 mm step following Y axis (figure 5).

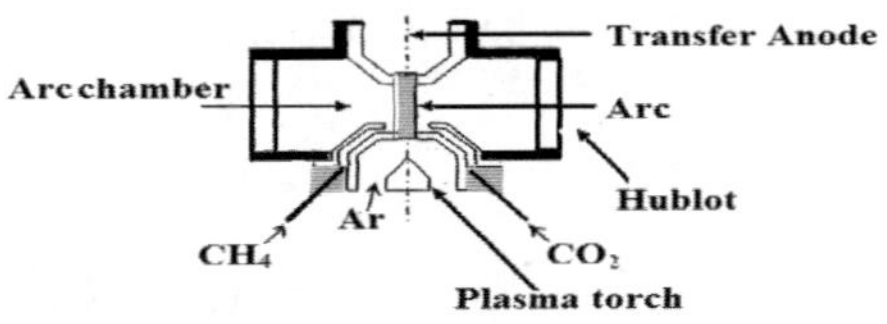

Figure 4. Transferred-arc reactor

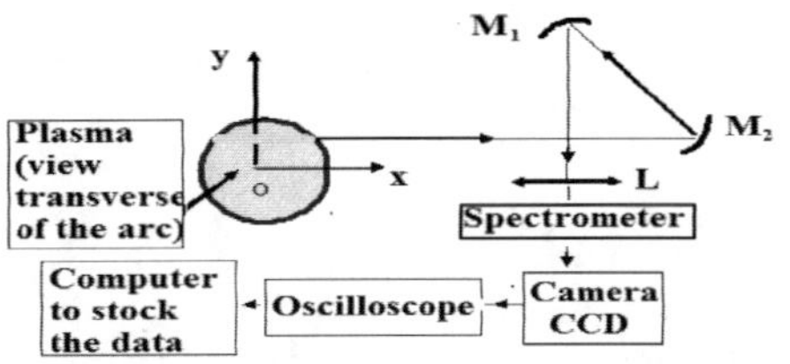

Figure 5. Experimental set. (M_1 and M_2 are Spherical Mirror, L is a lens, O: Center of arc)

6. Results

Twelve spectra records (655 wave lengths points each) were Abel inverted in order to present a local intensity distribution as a function of the radial distance from arc column center. Such distributions are then compared to spectra simulated at large temperature ranges. Three examples of real (experimental) spectra and corresponding simulated one are shown in figure 6,figure 7 and figure 8. The resulting rotational temperature profile in the transferred arc column is shown on figure 9.

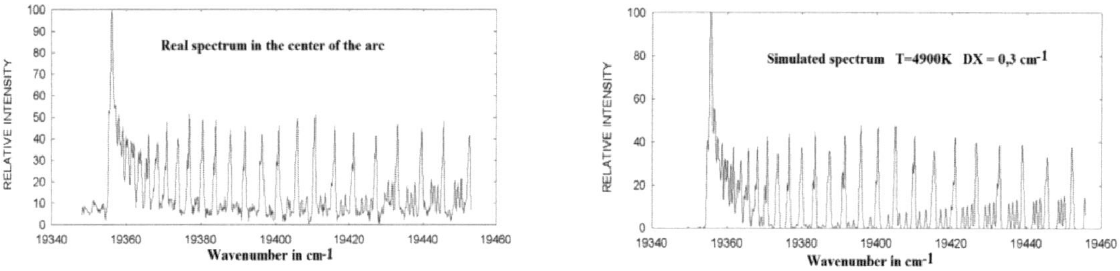

Figure 6.Example of an experimental spectrum in the centre of the arc (after the Abel inversion), and corresponding simulated spectrum.

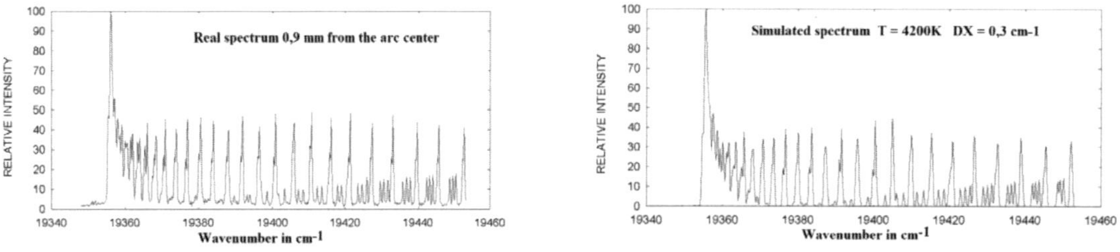

Figure7. Example of an experimental spectrum at 0,9 mm from the arc column center (after the Abel inversion), and corresponding simulated spectrum.

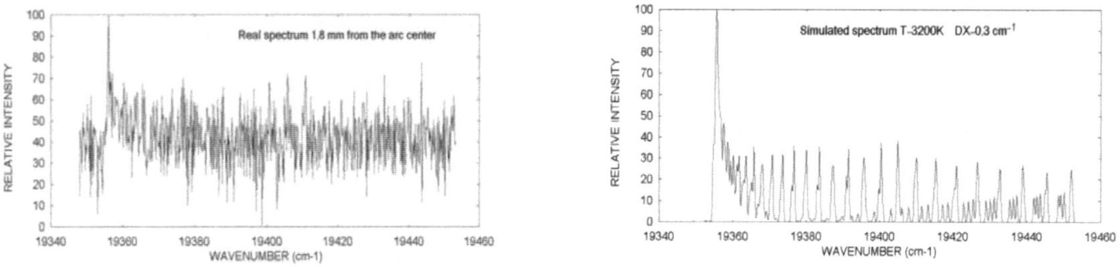

Figure 8. Example of an experimental spectrum at 1.8 mm from the arc column center (after the Abel inversion), and corresponding simulated spectrum.

7. Discussion and conclusion

The rotational temperature of about 4900 ±300 K has been found from the experimental spectrum in the arc column center while at the column fringes the temperature drops to 3200±350 K at 1.8 mm from the axis. This last point was obtained from a quite weak and nosily spectrum (N/S=20%)**.** Looking at different spectra in real points, we observe that the emission is maximum (the signal is very strongest N/S<1%) at 0.9 mm distance from the arc center where the temperature is around 4200 K which means that the concentration of carbon molecule is maximal in this point.The concentration of carbon molecule decreases slightly, approaching the center of the arc where the temperature is about 4900K and N / S = 5%. Instead, this concentration is greatly reduced in periphery zone of the arc, where the temperature is around 3200K and N / S> 20%.

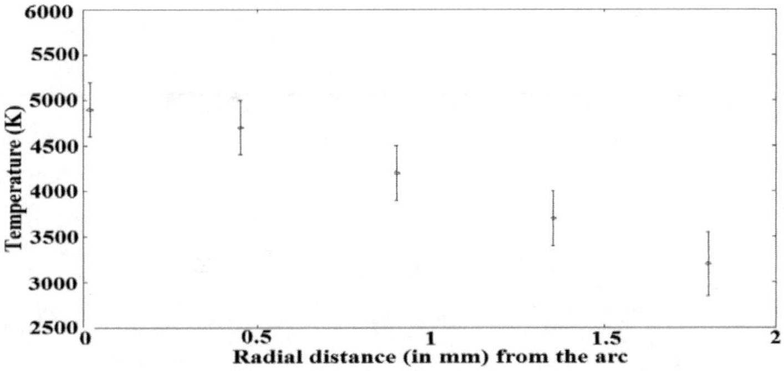

Figure 9.Radial distribution of the rotational temperature in transferred electric arc column in Ar/CO2/CH4 Mixture.

A calculation based on Limoges's computer program [14] of the complete thermodynamic equilibrium for the same $CH_4/CO_2/Ar$ gas mixture (H/C/O/Ar =1/0.485/0.463/0.168 in moles) shows that the C_2 molecule concentration in plasma reaches its maximum at about 4200 K and then drops by two orders of magnitude at about 3200 K. The temperatures for which this molecule can be used as a "thermometer" cover therefore a region from about 3000 to 5000 K.

References

[1] Pellerin S, Musiol K, Motret O, Pokrzywka B and Chapelle J 1996 Application of the (0,0) Swan band spectrum for temperature measurements *J. Phys. D: Appl. Phys.* **29** 2850–65
[2] G.Bontemps-Faure 1997 Calcul de spectres moléculaires, mesures des températures de rotation et de vibration dans les plasmas thermiques *thèse de l'université de Blaise Pascal*
[3] Laffont C, Rousselot P, Clairemidi J, Moreels G 1998 *Planitary and Space Science* **46** 585
[4] Yacine Babou *et al* 2008 *Plasma Sources Sci. Technol.* **17** 045010
[5] Nassar H, Czernichowski A 1990 *J.de phys.* **51**, C5-289
[6] Nassar H 1991 Simulation du premier système négatif de l'azote et évaluation de la température dans les différents sources de plasma *Thèse Université d'Orléans*
[7] Nassar H, Meguernes K, Czernichowski A and Chapelle J 1995 Proc. 12th Int. Symp. on Plasma Chemistry ISPC12 vol 2 (Minnesota, USA) p 651
[8] Nassar H, Pellerin S, Musiol K, Martinie O, Pellerin N, Cormier J-M 2004 J. Phys. D: Appl. Phys. 37 1904–1916
[9] Herzberg G 1962 Spectra of diatomic molecules Molecular Spectra and Molecular Structure vol 1 (New York:Van Nostrand Reinhold)
[10] Kovacs I 1969 Rotational structure in the spectra of Diatomic Molecules (London: Adam hilger)
[11] Budo A 1963 *Z. Phys.*105, 582
[12] Motret O, Hibert C, Pellerin S, Pouvesle J-M 2000 Rotational temperature measurements in atmospheric pulsed DBD—Gas temperature and molecular fraction effects *J.Phys. D: Appl. Phys.* **33** 1493–8
[13] Neumann W 1975 *Progress in Plasmas and Gas Electronics* vol 1, ed R Rompe and M Steenbeck (Berlin: Akademie-Verlag) pp 54–65
[14] Delluc G,Elchinger M F, Pateyron B *ADEP Software* Université de Limoges, France.

Experimental investigation of energy balance in plasma arc cutting process

T. Kavka[1], S. Tossen[2], A. Maslani[1], M. Konrad[1], H. Pauser[3], T. Stehrer[3]

[1]Institute of Plasma Physics AS CR, v.v.i., Za Slovankou 3, Prague 8, 18200, Czech Republic
[2]HELMo-Gramme, Quai du Condroz 28, Angleur, B-4031, Belgium
[3]Fronius International GmbH, A-4600 Wels-Thalheim, Austria

kavka@ipp.cas.cz

Abstract. The present paper describes the power balance of the arc cutting process provided by a plasma torch with steam working medium. The work was concentrated on definition of different power terms including power input as well as effective power utilization and losses as a function of plasma gas flow rate. The work was mostly experimental. The results have shown around 20% of total available power is utilized for material cutting and removing for the studied conditions.

1. Introduction

In the plasma arc cutting a source of heat is an electric arc generated by a plasma torch in the transfer regime. The torch is positioned above the material to be cut. The arc is established between the cathode inside the torch body and the material, which acts as an anode. A strong flow of the plasma gas, which is usually air or oxygen, blows out the nozzle through a narrow orifice, forming a plasma jet. Heat is transported first by convection to the plasma gas and then to material surface to melt it and to superheat the molten material inside a kerf followed by its removing [1].

Successful utilization of this technology requires advanced understanding of the processes, which take place on each step of cutting from plasma generation mechanisms, heat transfer between the flow and material, kerf formation and material removing. Both numerical and experimental investigations of different stages of the cut process have been done and reported in literature. These works report visualization of arc cutting [2], properties of the electric arc [3, 4], arc-material interaction and dross formation [5-7] during cutting with gas plasma. The present work represents experimental study of the cutting of mild steel carried out by the plasma torch working on steam. The main aim of the present work is to describe the energy balance of the steam cutting process and to estimate the process efficiency. The scope of this paper is limited to one type of the material and one process parameters. The only variable was the plasma gas flow rate. The results show distribution of the available electrical power of the torch among different energy terms such as power conduction and convection losses and energy required for cutting.

2. Theoretical description of the energy balance

The energy generated by the arc during cutting is transported to the plasma gas and then to material. A part of this energy is used for melting of the material and removing molten material from the cut and represents effective energy consumption. Some part of this energy represents losses. The main energy terms are represented in Fig. 1. In fact, they can be divided on two different types – the energy sources (P_{tot} and P_{oxy}) and the energy consumption terms (Q_{cool}, Q_{lost_up}, Q_{HCL}, Q_{cut}, Q_{lost_down}).

The total energy balance of the arc cutting process can be described by the following equation:

$$P_{tot} + P_{oxy} = Q_{cool} + Q_{lost\_up} + Q_{HCL} + Q_{cut} + Q_{lost\_down}$$

P_{tot} is the total electric energy brought to the system $P_{tot} = UI$.

P_{oxy} takes into account the eventual exothermic reaction of iron oxidation occurring on the surface of the molten metal in the presence of oxygen. Indeed, liquid iron is very sensitive to oxygen and reacts with it very easily. Not only oxygen from air is brought towards the liquid iron but also reactive species present in the plasma.

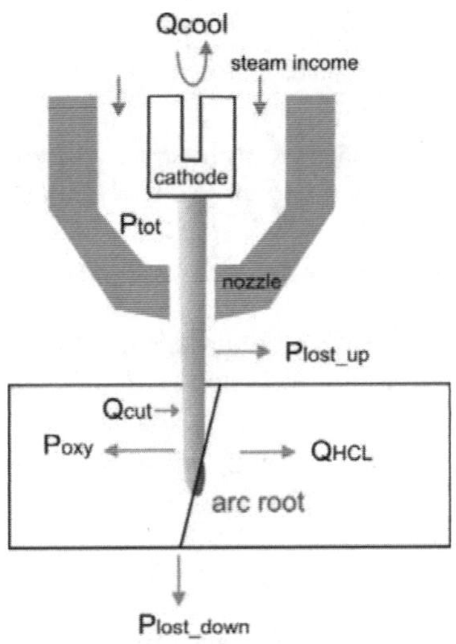

Fig. 1 Energy balance terms in plasma arc cutting

Q_{cool} represents the power losses inside the torch body due to cooling of its components.

Q_{lost_up} is the energy lost by convection and radiation above the material.

Q_{HCL} are the heat conduction losses into material.

Q_{cut} is the energy required to heat material up to its melting point and to melt it. In principle, the material is not heated only to its melting point, but the molten material is being heated up further. The energy utilized for the molten material overheating is also taken into account in this term.

During cutting the plasma jet passes through the kerf taking away residual plasma gas enthalpy downwards the material. This energy term is considered to represent energy losses under the plate Q_{lost_down}.

All these power terms should be estimated to describe the entire power balance.

3. Experimental setup and operating conditions

The present study consisted of several steps. Each step required specific experimental conditions, which will be specified in the following sections.

A modified Fronius TransCut 300 system has been used in the present experiments. The principle diagram of the plasma torch is shown in Fig. 2. The main feature of the torch is the plasma medium. The torch is operated on steam originated from a water-ethanol mixture provided by the producer. The solution is supplied through the channels inside the torch body at definite pressure, where it is heated up to the temperature of vaporization. The formed vapor is heated and ionized utilizing the heat of arc. Thus, the plasma gas consists of products of dissociation and ionization of ethanol solution. The system has been slightly modified to enable higher arc currents up to 90 A. The negative terminal of the power supplies was connected to the cathode, while the positive one was fixed to the working piece. All measurements have been done at arc current I = 60 A. The arc is starting by a contact method between the cathode and torch nozzle in a low current regime (I = 13 A). The plasma gas blows the arc out the nozzle and the arc path is then reconnected to the work piece.

In the present experiments only an effect of pressure has been studied. Three series of experiments have been performed. The pressure of water was set to 4.5, 5.5 and 6.5 bar. The plasma medium flow rate has been measured during all cutting tests. The studied plasma torch parameters are summarized in Table 1. Increase of the water pressure resulted in corresponding increase of the torch chamber pressure and plasma gas flow rate. The plasma jet was always in the supersonic regime. The arc voltage and current has been recorded during cutting procedure using the oscilloscope LeCroy. The average total electric power provided by the arc P_{tot} was then determined. Increase of the gas flow rate resulted in increase of the arc voltage and consequently of the total arc power. Detailed analyses of voltage drop between the cathode, torch nozzle and material have shown that voltage drop increases significantly on the nontransfer part of the arc inside the torch chamber. The voltage drop outside the torch stayed almost unchanged.

Table 1 – Plasma torch parameters for the performed cutting tests

	Pwater	gas flow rate	arc voltage, V	arc power
	bar	g/min	total	W
conditions 1	4.5	9	140	8400
conditions 2	5.5	11	148	8880
conditions 3	6.5	15	150	9000

The same material has been used in all experiments - 15 mm thick plates of mild steel S235JR. The surface of the plates was slightly rusty. The composition of the material was verified using EDX analyses. The result composition is shown in Table 2 together with corresponding values in norm EN 10025. Carbon content appeared to be slightly higher than the norm demands, but it may be caused by pollutions of the analyzer.

The torch was fixed at the same position, while the material was fasten to the moving system. The following cutting parameters have been applied: cutting speed – 30 cm/min, stand-off distance – 2 mm. Each cut was started from the edge of the plate. First, the torch was operated in a pilot arc regime and then the moving system was switched on to provide the cut. The cutting was carried out for around 150 mm always in the same direction to eliminate an effect of the flow rotation. After each cutting test the metal piece was cut some centimeters from the edge by a saw in such a way to enable analysis of the cut surface and kerf geometry and dimensions. An average kerf width was then evaluated to estimate amount of removed material for the energy balance evaluation.

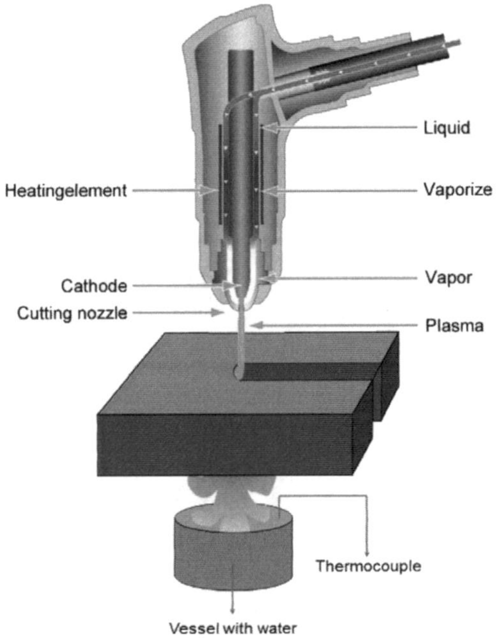

Fig. 2 Plasma torch TransCut and experimental arrangement for measurement of Q_{lost_down}

4. Experimental determination of energy terms

The present chapter describes experimental determination of the energy terms for evaluation of the energy balance of the cutting process. The parameters of the kerf after

Table 2 – Composition of the mild steel S235JR evaluated through EDX analyses and corresponding normalized values (EN 10025)

element	vol. %	
	measured	norm
C	*0.635*	<0,19
Mn	1.5	<1,5
Si	0.43	
S	0.01	<0,045
Ni	0.11	
Cr	0.09	
Cu	0.3	<0,6
Al	0.18	
Fe	97	

cutting tests are shown in Table 3 together with material melting rate. It is obvious that increase of the gas flow rate resulted in reduction of the amount of melted material. Indeed, both material melting rate as well as amount of swarf was smaller for higher plasma gas flow rate.

Table 3 - Kerf geometry and material melting rate for the studied conditions

	swarf	cut width	material melting rate	
	g/cm	mm	g/min	g/cm
conditions 1	1.695	2.45	86	2.87
conditions 2	1.58	2.175	76.3	2.54
conditions 3	1.4	2.025	71.1	2.37

Main energy terms have been evaluated for all three reported conditions and are shown in the following. It should be noted that the energy losses above the plate Q_{lost_up} are very difficult to estimate. They are considered to be small and were not taken into account. Energy losses inside the torch Q_{cool} is only the energy taken away by the cathode cooling water. Experiments have shown that they are insignificant in comparison to other terms (less than 2% of total power) and they were not taken into account.

4.1. Energy required for cutting

The following quantities have to be considered to evaluate this term: the volume of melted material, the melting point of material, and the temperature of the molten overheated material. Volume of the melted material was considered to be equal to volume of the removed material and was determined based on the kerf geometry. It is rather difficult to define material properties and melting point of the mild steel as they are strongly affected by the material composition even in the limits of the norm. The literature reports the following ranges for S235JR steel: melting point - 1450÷1500°C, specific heat - 450÷620 J/kgK, thermal conductivity – 40-70 W/mK. Moreover, material properties are changing with temperature. In the present calculations the following values were chosen (Table 4).

The molten material keeps absorbing energy from the arc, which results in its overheating. A series of tests have been done to estimate temperature of the overheated molten material. It is rather difficult to provide measurement of the temperature inside the kerf during cutting. That is why so-called 'edge-cutting' experiments have been carried out. The plasma torch was moving along the edge of the material and plasma was melting and removing some part of it. Infrared pyrometer Minolta Land Cyclops 152A was focused onto the place of the leading edge continuously measuring its temperature. First, the pyrometer measured temperature on the cut surface when plasma torch was under operation. Then, the plasma torch and the material moving system were stopped simultaneously and the temperature reduction was measured as well. Such a procedure allowed estimation of temperature in the kerf during cutting and it was supposed to be around 1800°C. It appeared that the emission coming from the jet is negligible compared to the emission of material and the values measured during the torch run are valid and were used in the calculations. The estimated temperature values are rather rough as the field of view of the pyrometer was much higher than the size of the leading edge during the cutting.

Knowing the material melting rate and its temperature, it was possible to estimate amount of energy required for cutting Q_{cut} (Fig. 4).

Table 4 – Material properties of the used mild steel S235JR

Melting point	1475 °C
Specific heat	520 J/kgK
Heat of fusion	289 J/g
Density	7,86 g/cm3
Thermal conductivity	52,4 W/mK

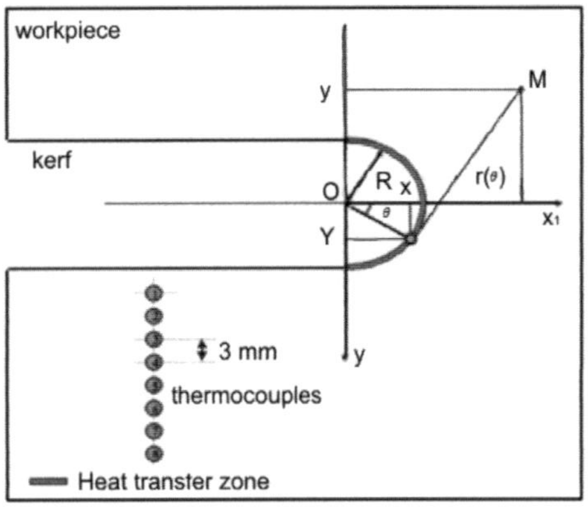

Fig. 3 Experimental setup and system of coordinate for the heat conduction model

4.2. Conventional energy losses to cut material

The term Q_{HCL} is difficult or even impossible to measure directly. That is why both theoretical as well as experimental evaluation of heat propagation inside the metal plate should be considered. To do so a well-known model of heat propagation during welding developed by Rosenthal [8] and successfully applied by Nemchinsky [5] for the arc cutting process has been used. The model solves the equation of heat propagation for the case of 'quasi-stationary' state meaning that an observer positioned at the torch will notice no change in the temperature distribution around the heat source. The heat source is considered to be a half-width of the cylinder representing a kerf (red line in Fig. 3). The solution of the heat propagation equation describes the temperature distribution inside the work-piece by the following equation:

$$T(M) = T_0 + \frac{Q_{HCL}}{2\pi\kappa_t h} * \frac{2}{\pi} * \int_0^{\pi/2} \exp\left(-\frac{\rho C v_t x_1(\theta)}{2\kappa_t}\right) * K_0\left(\frac{\rho C v_t r(\theta)}{2\kappa_t}\right) d\theta$$,

where k_t – thermal conductivity of the material, ρ – density, C – heat capacity, V_t – cutting speed, $x_1 = x - V_t t$, R is the average half-width of the kerf measured after each cutting, K_0 is the zero order and the second kind of modified Bessel function and Q_{HCL} is a term of conduction losses inside the material. This model has been used to determine heat conduction losses to the work-piece. To do so a set of thermocouples was installed inside the work-piece perpendicular to the cutting direction at the distance 5 mm from the cut and 3 mm from each other (Fig. 3). Then the temperature inside the work-piece was measured during cutting. Data from the thermocouples were collected by the data logger Omega OM-DAQPRO-5300 and analyzed in the Matlab environment. The term Q_{HCL} was then determined by fitting the analytical data and experimental curves from thermocouple measurements. The results have shown that Q_{HCL} =3,2 kW for the used material for conditions 2, which corresponded to 36% of the total energy available. The effect of the plasma gas flow rate on this term has not been studied yet and it was assumed to be 36% of the total power for all considered conditions.

4.3. Energy losses below material

Energy losses below material Q_{lost_down} have been determined from calorimetrical measurement in the vessel filled with water and positioned below the plate at the place of cutting (Fig. 2). Water temperature increase has been measured while cutting. The result calculations took into account the following factors:

- Energy generated by the pilot arc before the cut was absorbed by the calorimeter as well and therefore had to be deduced from the energy balance;
- After each cut swarf was collected at the bottom of the vessel. These particles also bring additional heat, which is already taken into account in Q_{cut} term. The swarf was weighted and the heat bringing by them was also deduced.

These losses depend on the plasma gas flow rate and represent a big part of the total available energy (Fig. 4).

5. Energy balance of the cutting process

The result terms of the energy balance for studied conditions is shown in Table 5. The most important terms are also plotted in Fig. 4. It is evident, that energy utilized for cutting is decreasing with the plasma gas flow rate. However, the energy leaving the kerf is also decreasing. The reason for such behavior is not clear and more tests should be done to get better correspondence between power input and output terms. Moreover, effect of the plasma gas flow rate on losses to material Q_{HCL} should be done as well.

Moreover, one should not forget that steam can partially dissociate into hydrogen and oxygen atoms, which may also react with molten material. However, it is still not clear if oxidation is taking place during the steam cutting.

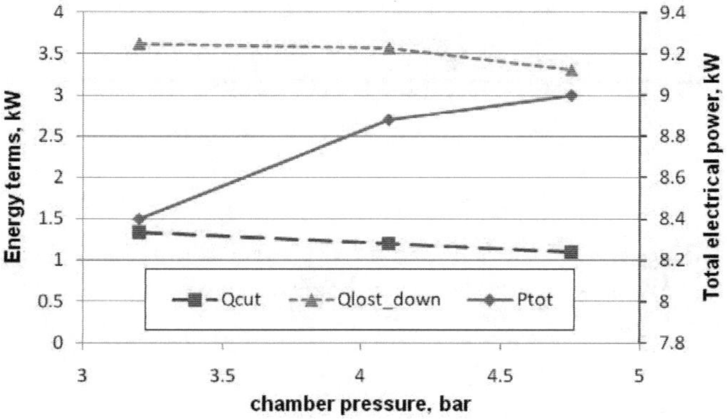

Fig. 4 Total available electric arc power and main energy terms as a function of chamber pressure

Table 5 – Energy balance of the steam arc cutting process as a function of the plasma gas flow rate

flow rate	Ptot	Qlost_down		Q_{HCL}		Qcut		Δ energy
g/min	kW	kW	%	kW	%	kW	%	kW
9	8.4	3.621	43.10714	3.02	36	1.337345	22	0.42
11	8.88	3.566	40.15766	3.20	36	1.202437	19	0.91
15	9	3.302	36.68889	3.24	36	1.105357	17	1.35

6. Conclusions

Energy balance of the arc cutting process by a plasma torch operated on steam was evaluated experimentally. It was shown that the biggest part of the available energy is either consumed by the material itself representing conductive losses or leaves material flowing through the kerf in the form of heated gas. About 20% of available energy is sufficiently utilized for the cutting itself. The efficiency of the energy utilization is reducing with plasma gas flow rate.

Acknowledgement. The research has been supported by Fronius Int. GmbH.

References

1. Nemchinsky VA, Severance WS: J. Phys. D: Appl. Phys. 39(22) 2006:R423-R38.
2. Colombo V, Concetti A, Ghedini E, Dallavalle S, Vancini M.; Plasma Sources Science & Technology. 18(2) 2009.
3. Ramakrishnan S, Shrinet V, Polivka FB, Kearney TN, Koltun P.; J. Phys. D: Appl. Phys. 33(18) 2000, p.2288-99.
4. Ramakrishnan S, Rogozinski MW. ; J. Phys. D: Appl. Phys. 30(4) 1997, p. 636-44.
5. Nemchinsky VA.; J. Phys. D: Appl. Phys. 30(18) 1997, p.2566-72.
6. Nemchinsky VA, Severance WS, Showalter MS; J. Phys. D: Appl. Phys.. 32(12) 1999, p. 1364-9.
7. Teulet P, Girard L, Razafinimanana M, Gleizes A, Bertrand P, Camy-Peyret F, et al; J. Phys. D: Appl. Phys.. 39(8) 2006, p. 1557-73.
8. Rosenthal D.; Welding Journal; 1941. p. 220-34.

Carbon arc plasma: characterization and synthesis of nanosized SiC

O Łąbędź[1], A Huczko[1], J Gawraczyński[1], H Lange[1], C Czosnek[2] and J F Janik[2]
[1]Department of Chemistry, Warsaw University, 02-093 Warsaw, Poland
[2]AGH University of Science and Technology, 30-059 Cracow, Poland

E-mail: olabedz@gmail.com; ahuczko@chem.uw.edu.pl

Abstract. Silicon carbide SiC is an important ceramic material with many applications. The nanomaterials can posess novel electronic and mechanical properties. Thus, we attempted to produce SiC nanopowder via a fast, one-step direct plasma synthesis. Two experimental systems were tested regarding the efficiency of SiC formation (for comparison): (i) arc plasma and (ii) high-temperature aerosol route. The products were characterized by wet chemical analysis, X-ray diffraction (XRD) and scanning electron microscopy (SEM). Almost total plasma conversion of starting Si-bearing reactants (Si, SiO and SiO_2) was achieved with SiC nanopowder (well below 100 nm) as a main product. From the emission spectroscopy measurements the arc plasma temperatures were evaluated to be within 3500-6000 K.

1. Introduction

Silicon carbide (SiC) and its composites have found wide applications because of its high Young's modulus and hardness, excellent oxidation and corrosion resistance, high strength at elevated temperatures, good thermal shock durability and electronic characteristics [1]. In addition, nano-crystalline SiC attracts considerable attention because of the wide band gap and quantum size effects [2]. Thus, the SiC nano-crystallites can be used to fabricate electronic devices which are durable at high temperature, strong radiation and fierce erosion. Several methods to produce nano-SiC have been proposed, such as combustion synthesis [3], chemical vapour deposition (CVD) [4], ion implantation [5], heat synthesis [6] and electrochemical etching [7]. Those routes usually suffer, however, from low productivity and long reaction duration. In this paper, we compare two different novel approaches to prepare SiC nanocrystallites. Not only we study the overall process efficiency, but also the oxygen presence in the reaction zone has been tested since in our earlier work we found [8] its profound influence on the formation of β-SiC nanowires synthesized via a combustion route (SHS). The on-line emission spectroscopy was carried out for a carbon plasma diagnostics, too.

2. Experimental

The plasma runs were carried out in the experimental system shown in figure 1 and described in details elsewhere [9]. The arc gap and its position against the optical axis was controlled automatically. The system allowed for holding the electrode gap between 1 and 2 mm for at least 5 minutes of arc operation, during which the radiation emission was measured. The home-made sintered anodes were prepared using the powdered mixtures of carbon black with either Si elemental, SiO or with SiO_2. Arc sublimation/reduction of reactants were performed under He-O_2 low pressure (300 hPa) at the arc current within 10-50 A. The gas-phase condensed products were collected for the following analyses. For comparison, an attempt was also made to obtain SiC nanopowder via a 2-step high-temperature pyrolysis of a hexamethyldisiloxane and polydimethylsiloxane in the system and applying the procedure outlined elsewhere [10]. The reaction products were further examined using SEM, XRD and wet chemistry techniques. Carbon arc set-up with spectroscopic arrangement is shown in figure 1. The details of the experimental and computational protocols of carbon arc plasma diagnostics were presented earlier [11].

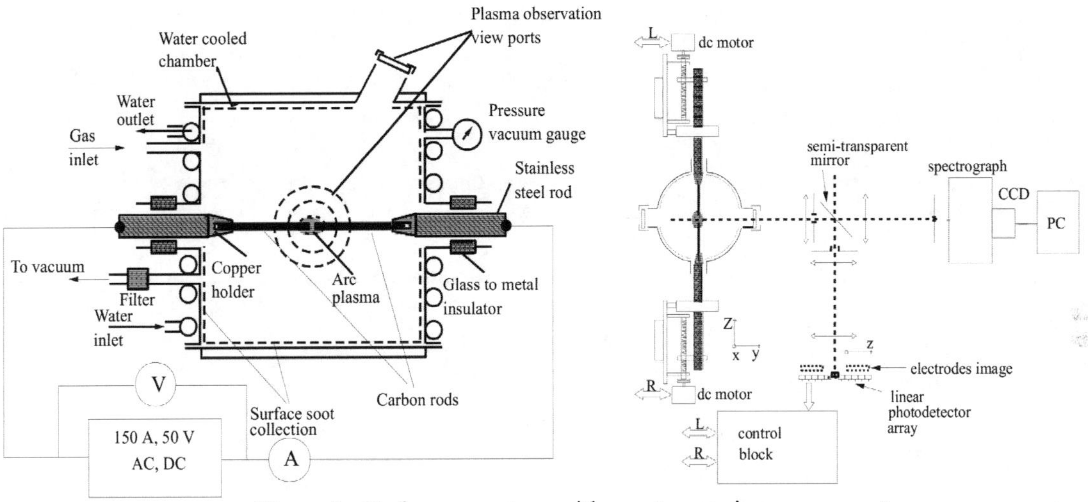

Figure1. Carbon arc set-up with spectroscopic arrangements.

3. Results and discussion

3.1. *Plasma spectroscopy*

The aim of the carbon arc plasma diagnostics was to determine the temperature distribution in the coalescent zone, i.e. in the vicinity of the arc gap. For the diagnostics of the plasma containing carbon gas the most suitable is the Swan band system (d $^3\Pi_g$ – a $^3\Pi_u$), particularly the mostly exploited 0-0 band at 516.5 nm, emitted by C_2 radicals. The same molecular spectrum can also be used when the gas phase contains additionally silicone vapor and oxygen. Indeed, in such a case C_2 bands are the major constituent of the plasma radiation in the optical range. The carbon vapor pressure in the arc under conditions favoring formation of nanocarbons, e.g., fullerenes , CNTs and silicon carbides can be quite high. Therefore, depending the electrode erosion rate, this resonant band, and mainly its 0-0 head, can be disturbed by the self-absorption phenomenon. Thus, it has to be taken into account when gas temperature is determined from the rotational structure of the band. Moreover, if significant, this effect can be used for the simultaneous temperature and C_2 radical concentration determination. We applied here a method based on the effect of self-absorption within the band, which has been routinely used by us for years [11]. Generally, the method is based on the calculation of the spectrum assuming different temperatures and optical densities associated with the transition a $^3\Pi_u$,v"=0 – d $^3\Pi_g$,v'=0. Next, the calculated spectra are fitted into experimental ones. The best fit is equivalent to the particular values of C_2(d $^3\Pi_g$,v'=0) column density (product of density and plasma column length), as well as average temperature along a plasma column. The details of the calculation procedure and its practical application have been described elsewhere [12]. When calculating spectra, a Voigt profile was taken into account. In contrast to pure carbon arc under similar graphite electrode erosion rate, the presence of Si and O in arc plasmas lowers significantly C_2 concentration resulting in much lower optical density of C_2 radical radiation. Two examples of C_2 (0-0) Swan spectra for low (10 A) and high (50 A) arc currents, taken at plasma coordinate 8 mm from the arc centre, are shown in figure 2, left and right respectively. Indeed, in the first case the C_2 column density was too low for any self-absorption detection. At the higher arc current the self-absorption was higher, but still on the verge of its detection.

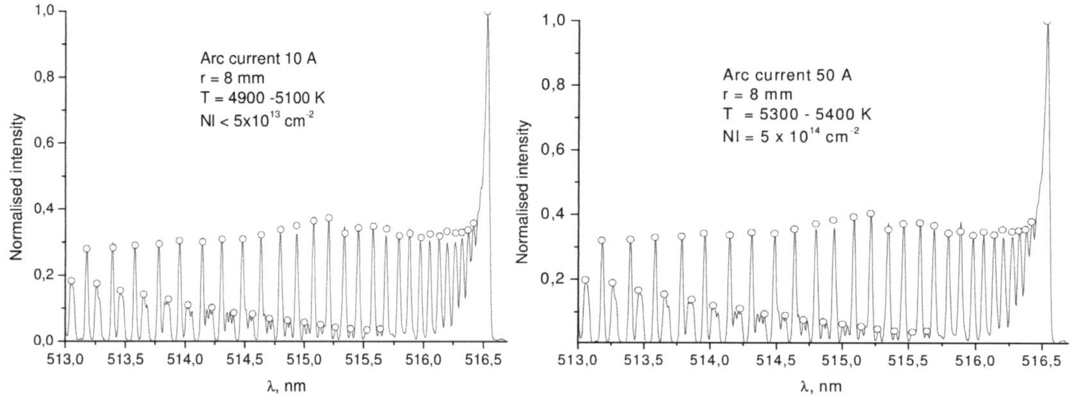

Figure 2. Examples of experimental spectra C_2 (d $^3\Pi_g$ – a $^3\Pi_u$, 0-0). Markers are characteristic points taken from simulated spectra.

On the basis of the temperature and C_2(d $^3\Pi_g$,v'=0) density values one could estimate the global carbon vapor pressure assuming thermodynamic equilibrium. The results point to pressure values about 1 and 7 kPa, respectively, for the arc conditions in figure 2.
Radial temperature distributions are shown in figure 3. Different S-bearing reactants (Si, SiO, SiO_2) used to fabricate the anodes do not alter these results.

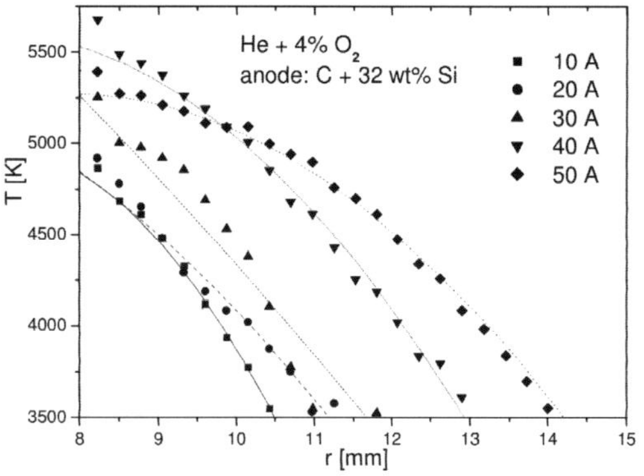

Figure 3. Radial temperature distributions in the arc vicinity.

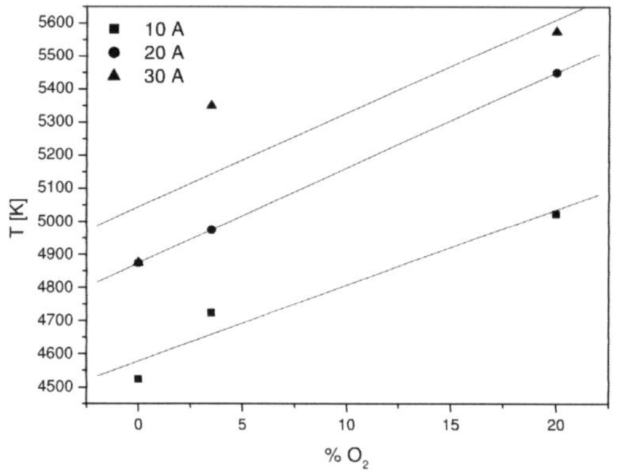

Figure 4. Influence of O_2 on temperature, r = 8 mm.

The only parameters influencing the temperature distributions are the arc current (figure 3), which causes mainly plasma zone broadening, and also O_2 content in He. With the increasing O_2 partial pressure the temperature increases, this due to possible exothermal effect of partial carbon oxidation, accompanying the SiC formation. Figure 4 confirms this observation.

3.2. Carbon arc plasma processing

From the presented SEM images (figure 5) a nanosized dimension of collected powders is evident. Spherical objects well below 100 nm dominate (soot and SiC nanoparticles). Some well-crystallized SiC 3-D microcrystallites can be also spotted. Some images reveal a web-like morphology of products. The study is under way (via controlled oxidation) to verify whether the obtained 1-D nanostructures are either carbon nanotubes or β-SiC nanowires.

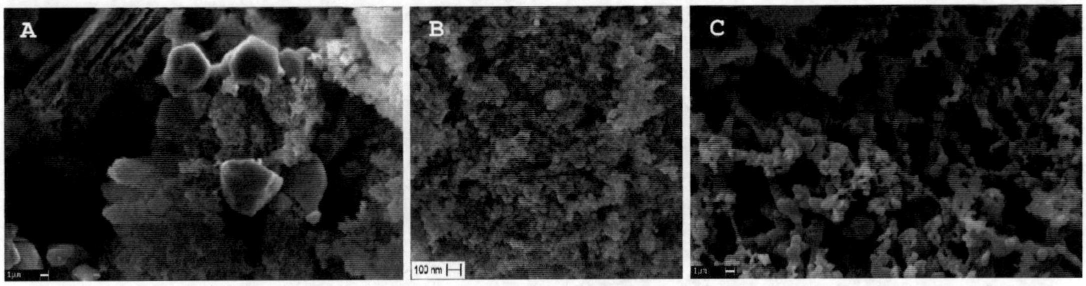

Figure 5. SEM images of arc plasma products (A – SiO_2 in He, B - Si in He + 20% O_2 and C – SiO_2 in He witth 20% O_2).

Figure 6 shows XRD spectra of produced powders. Relatively low intensity of characteristic patterns (almost entirely assigned to β-SiC) proves low crystallization and nano-dimensionality of a product.

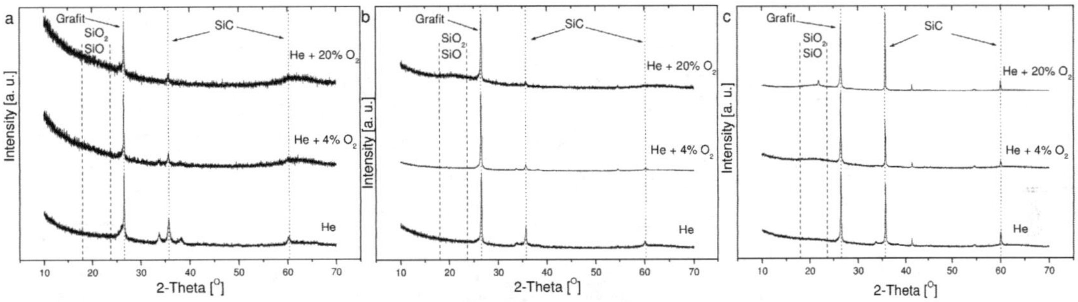

Figure 6. XRD spectra of products obtained from electrodes containing a – Si, b – SiO and c - SiO_2.

The products resulting from carbon arc-sublimated Si-containing graphite anodes (A: 68 wt% C – 32 wt% Si; B: 58 wt% C – 42 wt% SiO and C: 49 wt% C – 51 wt% SiO_2) in He-O_2 atmosphere were chemically analysed (via reaction with boiling 30 wt% KOH) for the determination of free Si. The results are shown in table 1. Very low content of un-reacted Si in the products proves high degree of reduction/conversion of the starting Si-bearing reactant into SiC.

Table 1. Results of chemical analysis of products.

Atmosphere (300 hPa)	Starting electrode A	Starting electrode B	Starting electrode C
He pure	2,8 wt% Si	0 wt% SiO	2,9 wt% Si
He 96% - O_2 4%	2,1 wt% Si	0 wt% SiO	1,8 wt% Si
He 80% - O_2 20%	1,6 wt% Si	~ 25 wt% SiO	0 wt% Si

3.3. *Aerosol-route pyrolysis*

An attempt was made to use various Si-bearing organic precursors to obtain nano-SiC powder of relatively large specific surface (determined by BET) by a pyrolytic route. Figure 7 shows SEM images of nanoproducts obtained using (A) DC Ar plasma, for comparison, and (B) aerosol routes . The products from the aerosol route have a spherical, homogeneous morphology.

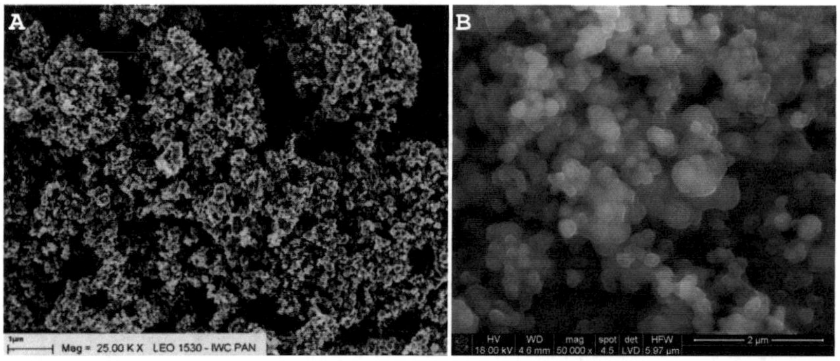

Figure 7. SEM images of SiC products resulting from DC plasma, flow system (1A) and aerosol pyrolysis (1B).

XRD spectra of products are shown in figure 8 and confirm the presence of regular β-SiC phase. Plasma product is a very fine, semi-amorphous SiC powder, while SiC from aerosol route is much better crystallized. FT-IR spectra for plasma (A) and aerosol-route (B) products have (figure 9) a wide absorption at ca 820 cm^{-1} (Si-C binding oscillations) while 1046 cm^{-1} absorption can be assigned to Si-O oscillation. BET measurements also showed relatively high specific surface (up to 50 m^2/g) for aerosol route-synthesized nano-SiC.

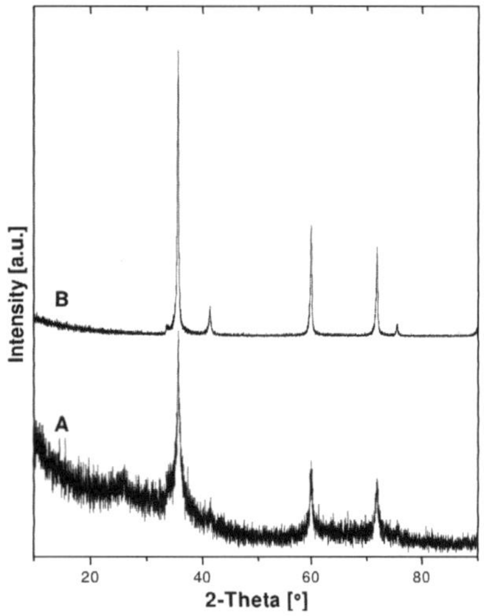

Figure 8. XRD spectra of products resulting from plasma and 2-stage aerosol route.

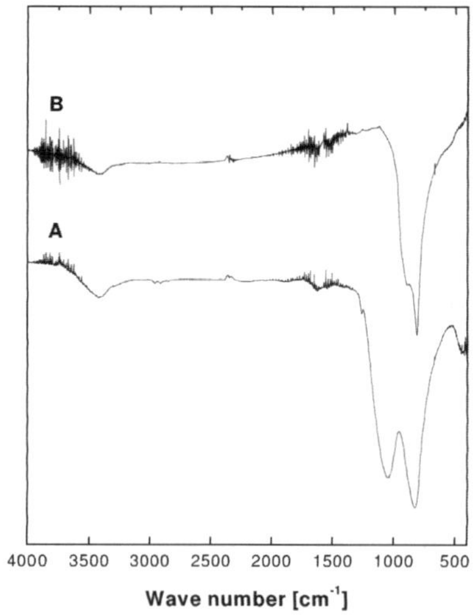

Figure 9. FT-IR absorption of products resulting from plasma and 2-stage aerosol route.

4. Conclusions

Si-bearing reactants efficiently yield SiC nanopowder, regardless of their nature and the experimental system used. The total conversion does not rely directly either on oxygen content in the starting compounds or an oxygen content in the arcing environment. It proves that under plasma conditions silicon carbonization towards SiC is faster comparing to simple oxidation. We envisage the silicon mass transfer via gasoeus SiO which is a relatively stable molecule in the gas phase. Thus, the oxygen-containing atmosphere obviously will favour the efficient formation of SiO. In fact, SiO involvement and Si catalytic role in the process of SiC growth have been already suggested [13] Surprisingly, the plasma temperature (within 3500-6000 K) does not seem to depend on process operating variables. Comparing to the plasma processing, a pyrolytic aerosol route yields more homogeneous and better crystallized nano-SiC, but the productivity is lower and the total reaction time is much longer.

5. References

[1] Liu G, Yang K, Li J, Yang K, Du J and Hou X, 2008 *J. Phys. Chem. C* **112** 6285
[2] Du K, Yang H, Wei R, Li M, Yu Q, Fu W, Yang N, Zhu H and Zeng Y 2008 *Mat. Res. Bull.* **43** 120
[3] Huczko A, Lange H, Chojecki G, Cudziło S, Zhu Y Q, Kroto H W and Walton D R M 2003 *J. Phys. Chem. B* **107** 2519
[4] Mo Y H, Shajahan M, Lee K S, Kim K C, Cha O H, Suh E K and Nahm K S 2002 *Diamond Relat. Mater.* **11** 1701
[5] Liao L S, Bao X M, Yang Z F and Min N B 1995 *Appl. Phys. Lett.* **66** 2382
[6] Shen G Z, Chen D, Tang KB, Qian Y T and Zhang S Y 2003 *Chem. Phys. Lett.* **375** 177
[7] Petrova-Koch V, Sreseli O, Polisski G, Kovalev D, Muschik T and Koch F 1995 *Thin Solid Films* **255** 107
[8] Huczko A, Osica M, Rutkowska A, Bystrzejewski M, Lange H and Cudziło S 2007 *J. of Physics: Cond. Matt.* **19** 395022
[9] Lange H, Baranowski P, Byszewski P and Huczko A, 1997 *Rev. Sci. Instrum.* **68** 3723
[10] Czosnek C and Janik J F, 2008 *J. Nanosci. and Nanotechn.* **8** 907
[11] Lange H and Huczko A, 2001 *Diamonds and Frontier Carbon Technology* **11** 399
[12] Lange H, Saidane K, Razafinimanana M and Gleizes A, 1999 *J.Phys. D: Appl. Phys.* **32** 1924
[13] Zhang J, Wang Q, Wang F, Chen X, Lei W, Cui Q and Zou G, 2009 *J. Phys. D: Appl. Phys.* **42** 035108

6. Acknowledgments

This research was co-financed by the European Regional Development Fund within the Innovative Economy Operational Program 2007-2013 under the project "Development of technology for a new generation of the hydrogen and hydrogen compounds sensor for applications in above normative conditions" (grant No. UDA-POIG.01.03.01-14-071/08-03). CC and JFJ acknowledge a financial support from AGH University of Science and Technology under the grant No. 11.11.210.120.

Optimization of a PIII&D System Using a Cathodic Arc with Titanium

M. Fazio[1], A. Kleiman[1*], D. G. Lamas[2†], D. Grondona[1†], A. Márquez[1†‡]

[1] Instituto de Física del Plasma, FCEN, Universidad de Buenos Aires – CONICET, Ciudad Universitaria C1428EGA, Argentina

[2] Centro de Investigaciones en Sólidos (CINSO), CONICET-CITEDEF, San Juan Bautista de La Salle 4397, Villa Martelli B1603ALO, Buenos Aires, Argentina

E-mail: adriana@tinfip.lfp.uba.ar

Abstract. A plasma immersion ion implantation and deposition (PIII&D) system was recently built at INFIP. A dc cathodic arc with a Ti cathode of 5 cm in diameter and an annular anode of 8cm in diameter was employed as the plasma source. The substrate chamber was electrically insulated and connected with the main discharge chamber through a straight magnetic duct. The discharge current was run at 100 A. The substrate was biased with a pulsed generator (30 kV, 30 A, 0.05 - 3 kHz) based on a pulse transformer controlled by IGBT switches. In this work the optimization of the process as function of the pulse parameters is presented. The characteristics of Ti coatings on steel substrates obtained varying the pulse amplitude from 2 to 12 kV and the pulse frequency from 200 Hz to 400 Hz were analyzed and compared with films grown without biasing the substrate. The thickness was determined weighting the samples before and after the treatment. The morphology was observed with an atomic force microscope. The film structure was studied by x-ray diffraction.

1. Introduction

Plasma immersion ion implantation combined with deposition (PIII&D) using cathodic arcs as plasma sources is a technique that has demonstrated to be useful for the fabrication of nanostructure coatings. Applying PIII&D, the substrate is immersed into the plasma plume produced with a cathodic arc and is biased with high-voltage pulses. When the high negative potential (typically many kilovolts) is on, a plasma sheath forms around the workpiece and ions are accelerated towards the surface and implanted. While the bias is off ions condense on the surface resulting in the deposition of a coating or a film. The use of high voltage pulses to bias the substrate allows to obtain denser films and to relax internal stress improving the adhesion, tribological and mechanical properties [1].

In this work the optimization of the process as function of the pulse parameters is presented. The characteristics of Ti coatings on steel substrates obtained varying the pulse amplitude from 2 to 12 kV and the pulse frequency from 200 to 400 Hz were analyzed and compared with films grown without biasing the substrate.

[*] Fellow of the CONICET
[†] Member of the CONICET
[‡] Corresponding author

2. Experimental setup

A plasma immersion ion implantation and deposition (PIII&D) system was recently built at INFIP, which is shown schematically in figure 1. A dc cathodic arc with a Ti cathode of 5 cm in diameter and an annular copper anode of 8 cm in diameter was employed as the plasma source. A tungsten striker was brought into contact with the cathode surface and later removed to trigger the discharge which was run continuously at 100 A. The discharge circuit consisted in a current supply (18 kW, 150 A) in parallel with a capacitor bank (165 mF) connected to the electrodes through a series inductor (2.8 mH) in order to provide arc stability. Both cathode and anode were refrigerated during the arc discharge by a water cooling system by means of a centrifuge pump with a nominal discharge of 50 l/min. The substrate chamber was electrically insulated and connected with the main discharge chamber through a straight duct. One vacuum system, composed of mechanical and diffusion pumps, pumps separately the plasma generation and deposition chambers to a base pressure of less than 10-4mbar. The pressure was measured by means of an Active Pirani Gauge (Edwards APG-L-NW16).

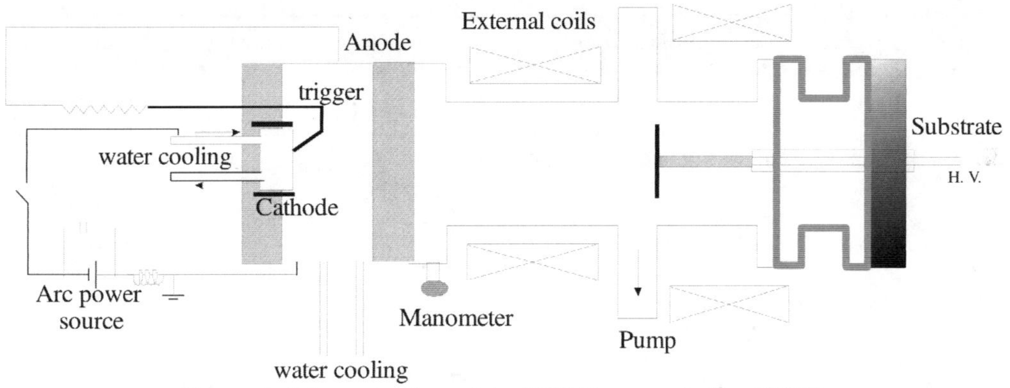

Figure 1. Schematic view of the PIII&D system built at INFIP.

The substrate was biased with a pulsed generator (30 kV, 30 A, 0.05 - 3 kHz) based on a pulse transformer controlled by IGBT switches, shown in figure 2. The pulse amplitude varied from 2 to 15 kV and the frequency from 100 to 300 Hz. A high voltage probe (Tektronic P6015A, 1000X, 3.0 pF, 100 MΩ) was used to measure the pulse parameters previous to and during the arc discharge.

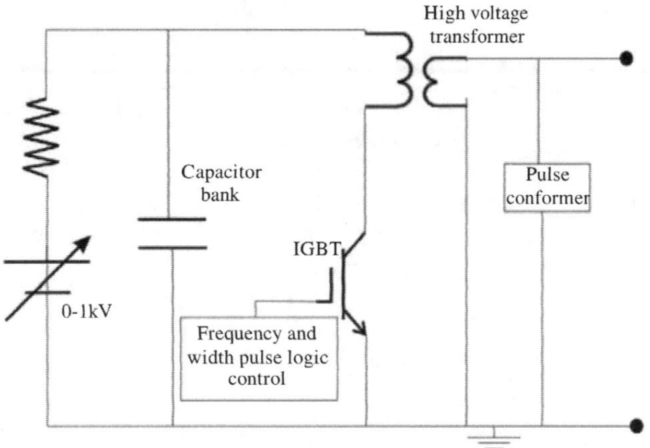

Figure 2. Schematic view of the pulse generator.

Most of the samples were made of AISI 316 steel and placed 25 cm and 40 cm from the cathode. The exposure time to the deposition process was 4 minutes. Some films were grown on silicon boron doped substrates with one side polished, in order to use a smooth surface for morphology characterization of the film. The samples were weighted prior to and after the coating. The crystalline structure of the films was identified by X-ray diffraction (XRD) using a Philips PW 3710 diffractometer with a CuK_{α} source. It was operated with glancing angle geometry by using a Philips thin film attachment, with an angle of incidence of 1 deg. The film morphology was studied by tapping mode atomic force microscopy (AFM) with a Nanoscope III Digital-VEECO.

3. Results

The deposition rate (*r*) is defined as the thickness of the coating over the time of exposure to the arc discharge. The thickness was determined from the deposited mass on the substrate taking the density value as the Ti bulk density (4.507 g/cm3). The deposition rate studied as a function of the pulse amplitude, varying the distance between the cathode and the substrate, is shown in figure 3. Changing the axial position from 25 to 40 cm *r* diminished to a 40 %, leading to a loss of accuracy in its determination for the samples placed further from the cathode. On the other hand, the deposition rate did not show a significant dependence on the pulse amplitude and presented a tendency to increase with the frequency of the pulse. This fact suggests that the contribution of sputtering to the deposition rate is not significant for the employed voltages and the increasing with the frequency could be attributed to an improvement on the ion sticking coefficient when the high voltage pulse is on.

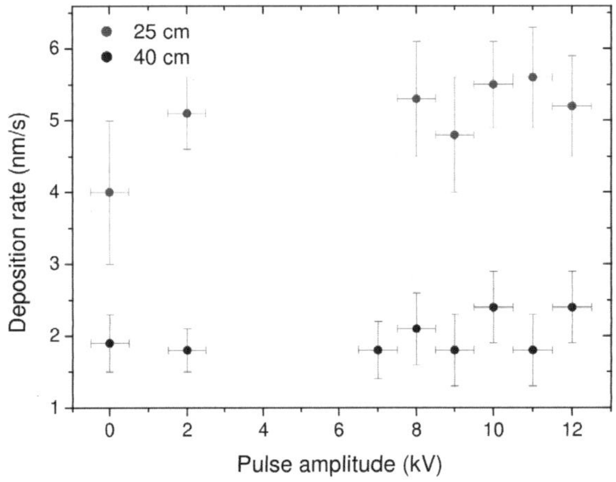

Figure 3. Deposition rate as a function of the pulse amplitude for different distances between the substrate and the cathode.

The dependence of the rate deposition with the pulse frequency was also investigated. The substrates were placed 25 cm from the cathode with pulse amplitude of 8 kV. The deposition rate versus the frequency of the pulse is shown in figure 4. The deposition rate presented a slight increase with the frequency of the pulse.

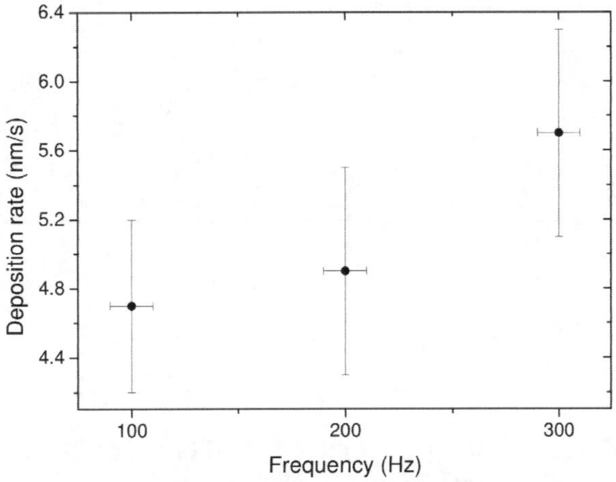

Figure 4. Deposition rate as a function of the frequency of the pulse for a pulse amplitude of 8 kV at a distance of 25 cm between the substrate and the cathode.

AFM images of films grown on silicon substrate are presented in Figure 5. The morphology of films obtained by the PIII&D technique (figure 5 a)) was compared with the surface of titanium coatings deposited without implantation (figure 5 b)). The surfaces of the Ti coatings with and without implantation did not present significant differences; both films were very compact and presented a columnar growth. The surface profiles of the films obtained from AFM images are shown in figure 6. From these profiles, the column size was found in the range 20 – 40 nm; the roughness was estimated in 0.23 nm.

a) **b)**

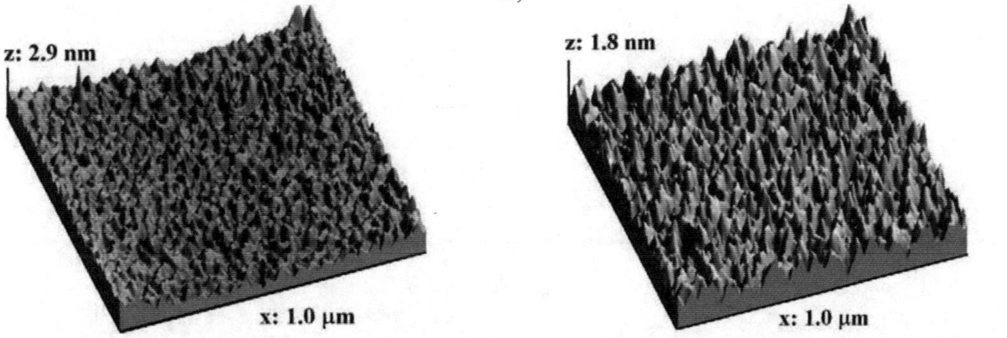

Figure 5. AFM picture of Ti coatings: **a)** deposited by PIII&D and **b)** deposited without implantation.

XRD registers obtained from films deposited at 25 cm from the cathode with a pulse frequency of 200 Hz and with different voltages are shown in figure 7. The main peaks of Ti were observed in all films, but in difractograms corresponding to PIII&D a broadening of Ti peaks was also observed. The changes in the difractograms can be attributed to the Ti implantation, this process could introduce titanium in the crystal lattice modifying the interplanar distances in the film.

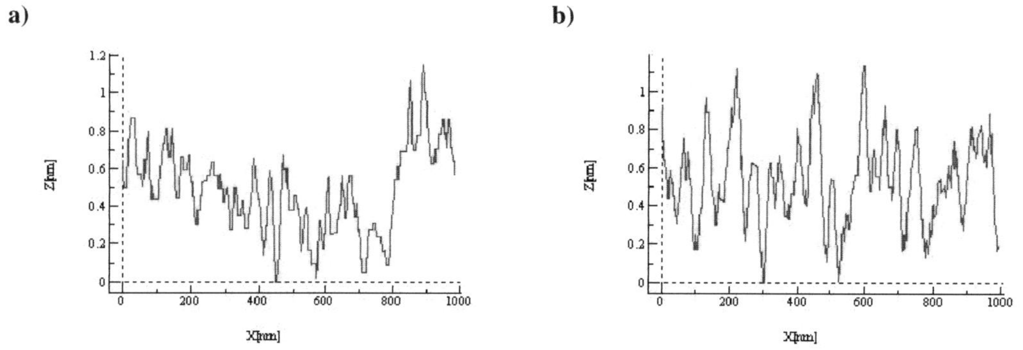

Figure 6. Surface profiles obtained from AFM images of films: **a)** deposited by PIII&D and **b)** deposited without implantation.

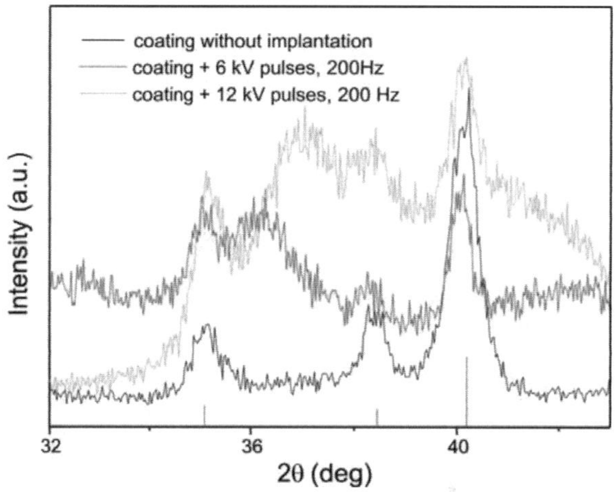

Figure 7. Diffractograms of Ti coating without implantation and Ti coating by PIII&D technique for different pulse amplitudes.

4. Conclusions

The deposition rate with implantation did not present significant differences from the deposition rate without implantation, then sputtering was not affecting the treatment. The increase of the deposition rate with the frequency suggested an improvement on the ion sticking coefficient with the pulsed bias for implantation. The PIIID technique did not introduce significant changes in the morphology of the films, obtaining very dense structures, with a columnar growth and a low roughness. However, PIII&D affected the film structure introducing changes in the interplanar distances of the crystal lattice.

Acknowledgments

This work was supported by grants from Universidad de Buenos Aires and CONICET.

References

[1] Attenberger W, Thorwarth G, Manova D, Mändl S, Stritzker B and Rauschenbach B 2001 *Surf. Coat. Technol.* **142-144** 412

Characterization of Titanium Films Deposited with a Cathodic Arc Using a Straight Magnetic Duct

F. Bermeo [1,2], H. Della Torre[3], A. Kleiman[3*], F. Minotti[3†], A. Márquez[3†‡]

[1]Universidad Santiago de Cali, Calle 5 # 62-00, Cali, Colombia

[2]Universidad del Valle, Calle 13 # 100-00, Cali, Colombia

[3]Instituto de Física del Plasma, FCEN, Universidad de Buenos Aires – CONICET, Ciudad Universitaria, C1428EGA Buenos Aires, Argentina

E-mail: adriana@tinfip.lfp.uba.ar

Abstract. Nanostructured Ti films were obtained employing a cathodic arc with a straight magnetic filter. The films were characterized using X-ray diffraction, scanning electron and atomic force microscopy. The films were found to be dense and with columnar grains, whose size increased with the exposure time. The number of macroparticles, the film roughness and the deposition rate were also analyzed, and the latter compared with the results of a fluid plasma model. Number of macroparticles and film roughness in samples located ahead of the magnetic duct inlet were higher than those determined from samples placed inside the magnetic duct. The deposition rate depended on the axial and radial position inside the duct. The thickness along the radial position was more uniform for samples located at axial positions near the filter extremes, but the mean deposition rate was lower at these positions. Measured and modeled deposition rates agreed reasonably well.

1. Introduction

The application of cathodic arc devices to thin film growth has been widely investigated due to the plasma characteristics, mainly a high charge density and ions with high kinetic energy, which promote the formation of dense nanostructured films with good adhesion and a very high deposition rate [1].

In cathodic arc discharges a metallic plasma jet is ejected from the cathode surface [2]. If a substrate is located facing the plasma jet its surface is coated with a metallic film. The substrate conditions, temperature and bias with respect to the plasma potential used during the deposition process influence the structure, morphology and adhesion of the film. The main disadvantage of vacuum arc deposition is the emission of macroparticles (MPs) from the cathode. MPs in the coatings produce protuberances and depressions on the substrate surface that not only increase its roughness, but also degrade its mechanical properties. Employing a magnetic field parallel to the plasma jet, the ions are concentrated in this direction and the plasma density increases, so that the deposition rate increases and the fraction of MPs decreases. However, the use of magnetic fields leads to a loss of

[*] Fellow of the CONICET
[†] Member of the CONICET
[‡] Corresponding author

uniformity in the film thickness. Counting with a fluid model for the plasma, which allows to calculate the ion density as a function of the position inside the straight duct for different magnetic field values, helps to determine the best magnetic field value and axial position to obtain thickness uniformity adequate to the employed sample.

In this work titanium films deposited by a cathodic arc on stainless steel and silicon substrates with a straight magnetic duct were study. A two dimensional fluid model for the plasma flow through the magnetized duct was employed to fit the experimental behavior of the deposition rate as a function of the axial and radial positions. Values of the floating potential at the entrance of the filter were used as input for the model.

2. Experiment

The experiment was carried out in a d-c vacuum arc system, which is shown schematically in Figure 1. The discharge was produced between a cylindrical Ti cathode (60 mm in diameter) and a grounded annular Cu anode (80 mm in diameter). Both electrodes were mounted on an insulating piece that set an electrode separation of about 15 mm. The cathode was surrounded by a floating shield. The discharge circuit consisted in a current supply (18 kW, 150 A) in parallel with a capacitor bank (165 mF) connected to the electrodes through a series inductor (2.8 mH) in order to provide arc stability. A tungsten striker brought into contact with the cathode surface and later removed was employed to trigger the discharge. A stainless steel duct 25 cm long and 10 cm in diameter was connected electrically isolated behind the anode. A magnetic field was established by an external coil wrapped around the stainless steel tube. The solenoid beginning was located at 20 cm from the cathode surface. The arc current was 120 A. The magnetic field in the center of the solenoid was 120 G. The vacuum system, composed of a mechanical and two diffusion pumps, kept the pressure chamber at 10^{-5} mbar.

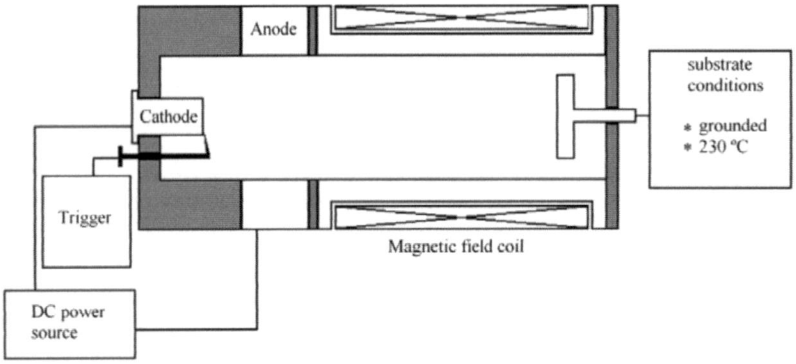

Figure 1. Schematic diagram of the d-c vacuum arc system.

The substrates were disks (19 mm in diameter and 2 mm in thickness) of stainless steel 316 and pieces of silicon (100) doped with Boron. They were placed at different axial distances from the cathode surface and mounted on a heater that holds the temperature at (210 ± 10) ºC. The exposure time to the discharge for growing the films was varied from 10 to 30 s.

The samples were weighted prior to and after the coating. From the measured deposited mass, the mean deposition rates were evaluated. The film thicknesses at different radial positions for the steel samples were determined by employing a Calotester. In this technique a hemispherical crater across the coating was eroded with a steel ball (30 mm in diameter) which was coated with a standard metallurgical diamond lapping paste and rotated during 2 min. The local deposition rates were estimated from the thickness. The film structure was analyzed by x-ray diffraction (XRD) with a

Philips PW 3710 diffractometer using $Cu\mathrm{K}_\alpha$ radiation. It was operated with glancing angle geometry by using a Philips thin film attachment, with an angle of incidence of 1 deg.

The film surface was characterized. The morphology was studied by scanning electron microscopy (SEM) with a Philips ElectroScan 2010 microscope and by atomic force microscopy in tapping mode with a Nanoscope III Digital-VEECO.

Three probes located at different radial positions were employed to measure the floating potential along the filter. The probes were spherical Cu tips of diameter D = 3 mm with holders insulated from the plasma by double glass tubes. A high-impedance resistor was employed to register the floating potential of the collecting probes. The electrical signals were registered in a four channel digitizing oscilloscope (Tektronix 2014, sampling rate of 250 Ms/s, analogical bandwidth of 500 MHz).

3. Plasma model

We consider time independent, axially symmetric fluid equations for the plasma electrons and ions. At the low densities considered collisions between electron and ions will be neglected, and temperatures of each species taken as constant [3]. In this way, the momentum equation for the electrons, neglecting their inertia, is

$$T_e \nabla \ln n_e + e\left(-\nabla\varphi + \mathbf{u}_e \times \mathbf{B}\right) = 0, \tag{1}$$

with e the elementary charge (absolute value), T_e the electron temperature, n_e the electron number density, $\mathbf{u}_e$ the electron fluid velocity, and φ the electrostatic potential. This equation is projected along a magnetic field line to obtain $T_e \ln n_e - e\varphi = const$ for each magnetic line. The magnetic line which at $z = 0$ has coordinate $r = r_0$ has the representation $r_0 = r_0(r, z)$ so that one can write

$$T_e \ln n_e(r,z) - e\varphi(r,z) = T_e \ln n_e\left[r_0(r,z),0\right] - e\varphi\left[r_0(r,z),0\right]. \tag{2}$$

One can thus relate the electron density to the electrostatic potential. Also, by quasi-neutrality, the ion number density is $n_i = Zn_e$, with Z the average ion charge state. In this way, the equation for the ion fluid velocity $\mathbf{u}$ is

$$m_i\left(\mathbf{u}\cdot\nabla\right)\mathbf{u} = -T_i \nabla \ln n_i + Ze\left(-\nabla\varphi + \mathbf{u}\times\mathbf{B}\right) - \lambda\mathbf{u}, \tag{3}$$

for ions of mass m_i and temperature T_i. Here, $\lambda \approx m_i V_0^2 / (c_s a)$ models ion-sound turbulent viscous effects (a is the filter radius, c_s is the ion sound velocity, and V_0 a characteristic ion velocity). The equation for φ, given the relation (2) and quasi-neutrality, can be obtained from the ion continuity equation,

$$\frac{1}{r}\frac{\partial}{\partial r}\left(r j_r\right) + \frac{\partial j_z}{\partial z} = 0, \tag{4}$$

where $j_z = n_i u_z$, and

$$j_r = n_i u_r - D_A \frac{\partial n_i}{\partial r}. \tag{5}$$

The last term in Eq. (5) accounts for radial ambipolar diffusion of ions, limited by electron Bohm diffusion across magnetic lines [4], so that

$$D_A = \left(Z + T_i / T_e\right)\frac{T_e}{16eB}. \tag{6}$$

In order to obtain a system of ordinary differential equations the following explicit r dependence is considered:

$$u_r(r,z)=K(z)r\,;\; u_\varphi(r,z)=\Omega(z)r\,;\; u_z(r,z)=V(z)\,;\; \varphi(r,z)=A(z)\;\; r^2+C(z)\,. \tag{7}$$

Replacement of expressions (7) in Eqs. (3)-(4), together with quasi-neutrality, after equating equal powers of r, result in a system of ordinary first order equations for $A(z)$, $C(z)$, $K(z)$, $\Omega(z)$, and $V(z)$ that can be solved by standard methods once the values of these functions are given at $z = 0$.

4. Results and discussion

Typical XRD spectra obtained from a coated sample is depicted in Figure 2. In the same picture, as reference, a XRD spectrum of the steel substrate and the 2θ-positions of the Bragg peaks corresponding to Ti are included.

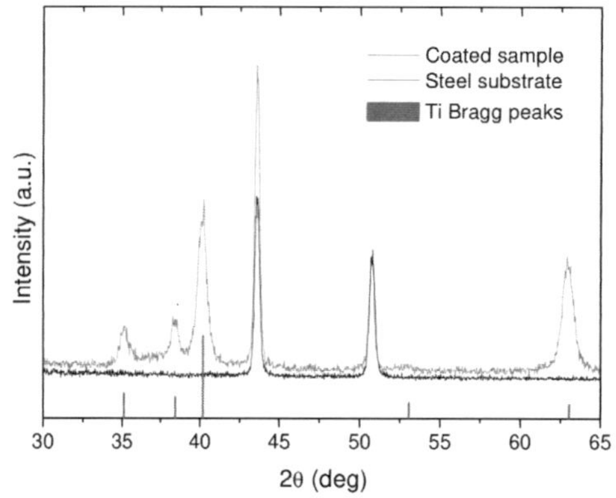

Figure 2. XRD patterns obtained from Ti deposited on stainless steel substrates.

Figure 3 shows typical SEM micrographs (magnification of 500x) of films obtained at different axial position and at an exposure time of 30s. Fig. 3 a) corresponds to a sample placed outside the magnetic duct, 5 cm before the solenoid beginning (z = -5 cm), being the position nearest the cathode. Fig. 3 b), c) and d) were taken from samples situated inside the duct at z = 0, 5 and 7.5 cm, respectively. Position z = 0 indicates the solenoid entrance and at z = 7.5 cm the magnetic field already reached the maxim value (120 G). It can be noted from the figures that the number and size of macroparticles decreased as the axial distance was increased. At z = -5 cm the largest macroparticle diameter was ~ 15 nm whereas at z = 7.5 cm was ~ 7 nm.

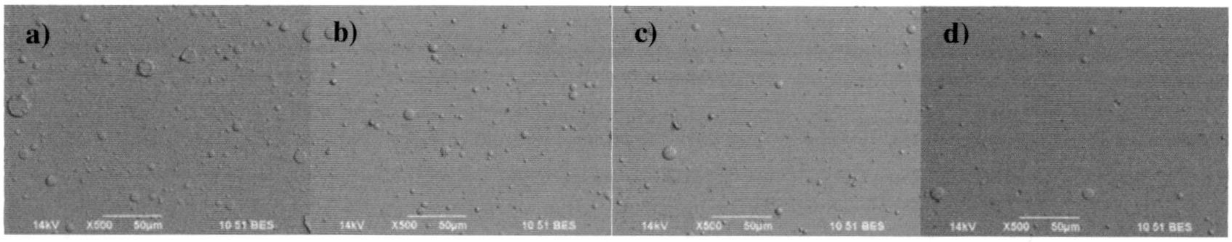

Figure 3. SEM micrographs of films obtained at different axial positions: a) z = -5 cm, b) z = 0, c) z = 5 cm, and d) z = 7.5 cm. Magnification of 500x.

The number of macroparticles also depended on the exposure time, the number increased as the time increased. AFM images acquired from films exposed to the discharge at 2 s and 20 s, on regions without the presence of macroparticles (z = 7.5 cm), display the surface morphologies shown in Figure

4. In both cases, the film surfaces are composed of columnar grains, pretty dense and without the presence of voids. From height profiles and roughness measurements registered on AFM images of films with different exposure time, the grain size and the roughness mean square were plotted as function of the exposure time in Figure 4 c). The columnar grain size increased as the time increased enlarging the surface roughness. The grain size varied from 2 nm to 55 nm for exposure time from 2 to 30 s, respectively.

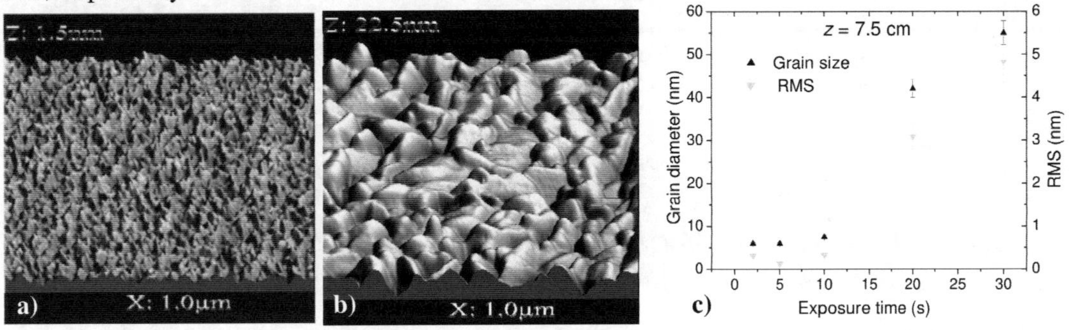

Figure 4. AFM images from films grown at z = 7.5 cm with a) 2 s and b) 20 s exposure time. c) Film roughness and grain size as functions of the exposure time.

Mean deposition rate determined from the deposited mass weighting and the local deposition rates at the center, at 0.5 cm and at 1 cm off the center found from the thickness measurements are plotted in Figure 5. These results were obtained from samples exposed to the discharge during 120 s.

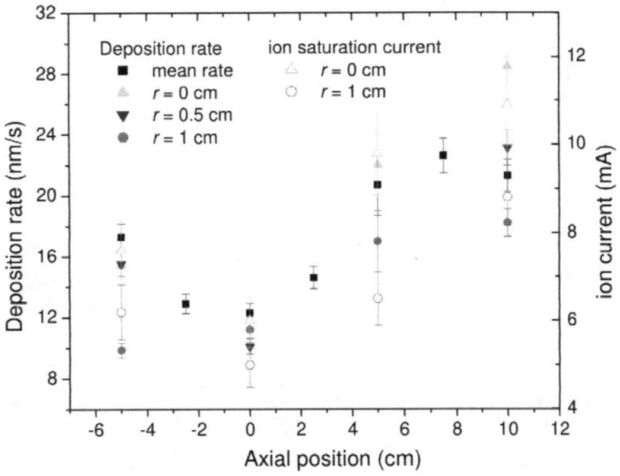

Figure 5. Deposition rate and ion current at different axial positions.

Since the ion energy loss can be neglected at the working pressure, the deposition can be taken proportional to the ion density. The deposition rate and the floating potential as function of z for the different radial positions were compared with the output obtained from the 2D fluid model for the plasma in Figure 6. The floating potential measured outside the solenoid at z = -5 cm was employed as input for the model. The deposition rate position inside the duct agreed reasonably well with the ion density profiles determined from the 2D fluid model for the plasma jet, and the potential profiles were reasonably fitted.

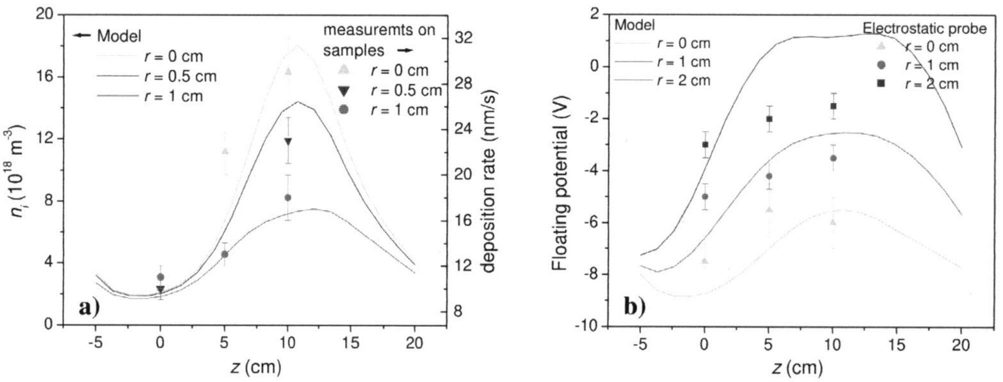

Figure 6. a) Modelled ion density (lines) and measured deposition rates (symbols). b) Modelled (lines) and measured (symbols) plasma potential.

5. Conclusions

Nanostructured Ti films were obtained employing a cathodic arc with a magnetic straight filter. The films were dense and with columnar grains, whose size increased with the exposure time from 5 to 50 nm for time ranging from 2 to 20 s, respectively. The number of macroparticles and the surface roughness in samples located ahead of the magnetic duct inlet were higher than those determined from samples placed inside the magnetic duct. The deposition rate depended on the axial and radial positions inside the duct. The thickness along the radial position was more uniform for samples located at axial positions near the filter extremes, but the mean deposition rate was lower at these positions.

The deposition rate as a function of the position inside the duct agreed reasonably with the ion density profiles determined from the 2D fluid model for the plasma jet, turning this model a useful tool to predict the radial profile of the coating.

Acknowledgments

This work was supported by grants from Universidad de Buenos Aires, CONICET and MinCyT-Colciencias (CO/08/11). The authors would like to thank Diego G. Lamas for his collaboration in the XRD measurements at CINSO (CONICET-CITEDEF)

References

[1] Vyskocil J and Musil J 1992 *J. Vac. Sci. Technol.* A **10** 1740
[2] Boxman RL, Sanders DM and Martin PJ (Eds.) 1995 *Handbook of Vacuum Arc Science and Technology, Fundamentals and Applications* (Park Ridge, NJ: Noyes)
[3] Giuliani L, Minotti F O, Grondona D and Kelly H 2007 *IEEE Trans. Plasma Sci.* **35** 1710
[4] Lieberman M and Lichtenberg A 2005 *Principles of Plasma Discharges and Material Processing* (Hoboken, NJ: Wiley-2nd Edition) pp 152-154

Journal of Physics: Conference Series **511** (2014) 012071 doi:10.1088/1742-6596/511/1/012071

Numerical modeling of a cutting torch

B R Mancinelli[1], **F O Minotti**[1,2] **and H Kelly**[1,2]
[1] Grupo de Descargas Eléctricas, Departamento Ing. Electromecánica, Universidad Tecnológica Nacional, Regional Venado Tuerto, Laprida 651, Venado Tuerto (2600), Santa Fe, Argentina
[2] Instituto de Física del Plasma (CONICET), Departamento de Física, Facultad de Ciencias Exactas y Naturales (UBA) Ciudad Universitaria Pab. I, (1428) Buenos Aires, Argentina

E-mail: bmancinelli@arnet.com.ar

Abstract. A two-dimensional turbulent model of a low-current intensity (30 A) cutting plasma torch was developed using the local thermodynamic equilibrium approximation. A good agreement was found between measured and modelled results of plasma temperature and velocity, the latter of which has not been previously reported in the literature for high energy density torches. The cutting performance was also studied in terms of the heat flux to the work-piece and the value of the force exerted by the impinging plasma jet.

1. Introduction

Plasma hydrodynamic modeling by numerical simulation in cutting torches is a common tool to predict the values of the fundamental physical quantities, namely the plasma temperature, the particles concentration and the fluid velocity. These numerical codes are employed to understand the relevant physical processes ruling the plasma behavior in order to interpret the experimental results of several plasma diagnostics, and ultimately to obtain optimized designs of such devices [1].

The purpose of this work is to present a turbulent two-dimensional local thermodynamic equilibrium (LTE) [2] model of a 30 A oxygen high-energy density cutting torch. Two configurations were used: the experimental one in which the arc is transferred to the rim of a rotating anode 6 mm away, and the cutting one in which a solid work-piece, acting as the anode, is 3 mm away from the nozzle exit. In the following section the plasma torch, model assumptions, governing equations, boundary conditions and the physical details of the model are presented. The calculated distributions of temperature and velocity and its comparison with the experimental data together with the calculated values of the heat flux to the work-piece and the force exerted by the impinging plasma jet are shown in section 3. The conclusion is summarized in section 4.

2. Mathematical model

2.1 Computational domain

The schematic of the modeled domain for the experimental configuration simulation is presented in figure 1. A mass flow rate of 0.71 g s^{-1} with a vortex injection that leads to a ratio of the azimuthal to the axial inlet velocity of $\tan(13^{\circ}) \approx 0.23$ was used at the torch inlet CD. More details on the torch characteristics can be found elsewhere [3].

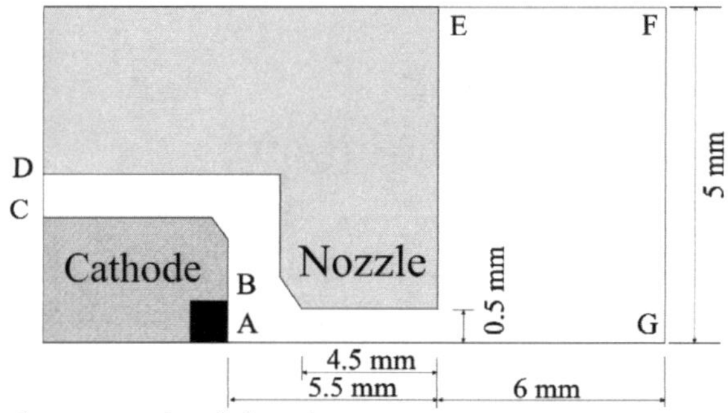

Figure 1. Cutting torch computational domain.

2.2 Model assumptions

(a) The plasma flow is two-dimensional and axisymmetric.
(b) The plasma is considered as a Newtonian fluid following the Navier-Stokes equation.
(c) The plasma gas is assumed to be pure oxygen in LTE, whose thermodynamic and transports coefficients were calculated by Murphy [4].
(d) Hall currents and gravitational effects are considered negligible.
(e) In the energy equation the viscous dissipation term is considered negligible.
(f) The anode was considered as a porous free boundary characterized by its electrostatic potential in the experimental configuration, while was considered as an extensive solid plate in the cutting one.

2.3 Governing equations

The fluid part of the thermal plasma model can be expressed as:

Total mass conservation

$$\frac{\partial \rho}{\partial t} + \nabla \cdot (\rho \bar{u}) = 0. \tag{1}$$

Momentum conservation

$$\frac{\partial (\rho \bar{u})}{\partial t} + \nabla \cdot \left(\rho \bar{u} \otimes \bar{u} + p \bar{\bar{\delta}} - \bar{\bar{\tau}}\right) - \bar{J} \times \bar{B} = 0. \tag{2}$$

Internal energy conservation

$$\frac{\partial (\rho e)}{\partial t} + \nabla \cdot (\rho \bar{u} e + \bar{q}) - \bar{J} \cdot \bar{E} + p \nabla \cdot \bar{u} + 4\pi \varepsilon_N \frac{p}{p_{ATM}} = 0, \tag{3}$$

where ρ represents the total mass density, $\bar{u}$ the fluid velocity (having, axial $-u_x-$, radial $-u_y-$ and azimuthal $-u_z-$ components), p the pressure, $\bar{\bar{\delta}}$ the identity tensor, $\bar{\bar{\tau}}$ the stress tensor, $\bar{J}$ the current density, $\bar{B}$ the magnetic field (only the azimuthal component was considered) , e the internal energy, $\bar{q}$ the total heat flux, $\bar{E}$ the electric field and ε_N the plasma radiation net emission coefficient (NEC) [5].

Two further equations are required to describe the electromagnetic part of the plasma model. The first is the current continuity equation

$$\nabla \cdot \bar{J} = 0, \tag{4}$$

where

$$\bar{J} = -\sigma \nabla \phi, \tag{5}$$

and the second is one of Maxwell's equations

$$\nabla \times \bar{B} = \mu_0 \bar{J}, \tag{6}$$

where σ is the electric conductivity, ϕ is the electrostatic potential and μ_0 the magnetic permeability of free space.

The total heat flux in (3) describes the heat transported by conduction and the enthalpy transport by mass diffusion, and is defined as

$$\bar{q} \equiv -\kappa_e \nabla T + \bar{\Gamma}_e h_e, \tag{7}$$

where κ_e is the effective thermal conductivity and $\bar{\Gamma}_e$ is the electron mass diffusion that can be approximated by

$$\bar{\Gamma}_e \approx -\frac{m}{e'} \bar{J}, \tag{8}$$

where e' is the elementary electric charge and m is the electron mass. Equation (8) neglects the charge transported by ions. In (7) $h_e = 5 k_B T / (2m)$ represents the specific electron enthalpy (k_B is the Boltzmann's constant).

The effective viscosity is

$$\mu_e = \mu_l + \mu_t, \tag{9}$$

and the effective thermal conductivity is

$$\kappa_e = \kappa + \frac{\mu_t C_p}{P_r}, \tag{10}$$

where C_p, μ_l and κ are the plasma specific heat at constant pressure, viscosity and thermal conductivity, respectively. The turbulent Prandtl number was adopted as [6]

$$P_r = 1. \tag{11}$$

The turbulent viscosity μ_t was calculated taking into account the effect of the vortex injection [1]

$$\mu_t = \rho l_m^2 \left(\left(\frac{\partial u_x}{\partial y} \right)^2 + \left(y \frac{\partial}{\partial y} \left(\frac{u_z}{y} \right) \right)^2 \right)^{1/2}, \tag{12}$$

where the Prandtl mixing length was chosen as [6]

$$l_m \equiv c \lambda, \tag{13}$$

here c is a parameter and λ is a local thermal radius which characterizes the boundary of the high velocity arc core, defined as the radial distance from the axis to the point at 2000 K.

The NEC in (4) was taken for one atmosphere and a plasma radius of 0.5 mm [5].

2.4 Boundary conditions

Table 1 summarizes the prescribed values of the physical quantities (or its spatial derivatives) on the boundaries shown in figure 1. At the hafnium insert AB the maximum value of the axial current density on the axis of the geometry was limited to $\leq 170\ \mathrm{A\,mm^{-2}}$ [1]. Besides, at the interface between the plasma and the anode, in order to maintain the conservation of the energy flux and current intensity at this boundary, the following relations (neglecting radiation [1]) were used to calculate the local thermal and electric conductivities

$$\left[-\kappa \left(\frac{\partial T}{\partial x} \right) \right]_{anode} = \left[-\kappa \left(\frac{\partial T}{\partial x} \right) \right]_{plasma} + J_x \left(\frac{5}{2} \frac{k_B}{e} \left(T - T_{anode} \right) + \varphi_A + \varphi_W \right), \tag{13}$$

$$\left[-\sigma\left(\frac{\partial\phi}{\partial x}\right)\right]_{anode}=\left[-\sigma\left(\frac{\partial\phi}{\partial x}\right)\right]_{plasma}, \qquad (14)$$

here φ_A and φ_w are the anode voltage drop and the anode work function, respectively.

Table 1. Boundary conditions.

	p	u	T	ϕ
AB	–	0	3500 K	–
BC	–	0	500 K	–
CD	–	mass flow 0.71 g s^{-1}	300 K	$\frac{\partial\phi}{\partial x}=0$
DE	–	0	500 K	–
EF	1 atm	$\frac{\partial u_x}{\partial y}=\frac{\partial u_y}{\partial y}=\frac{\partial u_z}{\partial y}=0$	300 K	$\frac{\partial\phi}{\partial y}=0$
FG (exp. conf.)	–	$\frac{\partial u_x}{\partial x}=\frac{\partial u_y}{\partial x}=\frac{\partial u_z}{\partial x}=0$	$\frac{\partial T}{\partial x}=0$	$\phi=0$
FG (cutting conf.)	$\frac{\partial p}{\partial x}=0$	$u_x=u_y=u_z=0$	1500 K	$\phi=0$
GA	–	$\frac{\partial u_x}{\partial y}=0,\ u_y=u_z=0$	$\frac{\partial T}{\partial y}=0$	$\frac{\partial\phi}{\partial y}=0$

2.5 Numerical aspects

The unsteady form of the model equations was solved using a time-marching method [7,8]. The specific values used for the initial guesses did not impact the final converged results. The set of governing equation was discretized in time using a Taylor series first-order accuracy and in space using the finite volume method and solved with the given boundary conditions on a 81 × 15 non uniform internal grid and 39 × 47 non uniform external grid, by using the predictor-corrector algorithm [7,8]. The time-step used in the time-marching algorithm was chosen so that the CFL criterion was fulfilled [7,8]. The calculation was stopped when the relative variation of the plasma variables between two consecutive time iterations was < 10^{-3}. This convergence criterion was found to be sufficient. The accuracy of the calculations was tested by repeating them with a 38 × 15 internal grid and 19 × 47 external grid. The change in the plasma temperature was everywhere less than 15 %, while the changes in the axial velocity were less than 20 %. The finer grid was then used for generating the results to be presented in the following Section.

3. Model Results

Figures 1 a) and b) show the comparison among the theoretical profiles (corresponding to c = 0.08) and the experimentally derived temperature and velocity profiles, respectively. Calculations correspond to the same experimental configuration that was used in the experiments [9-12]. As it can be seen, the model results are in good agreement with both, temperature and velocity experimental data. Model validations for the high velocity values characteristic of the high energy density torches have not been reported previously.

The theoretical distribution of the heat flux to the work-piece (located at 3 mm from the nozzle exit) is given in figure 2. As it can be seen, the energy carried by the electron current is the dominant term at the centre of the distribution reaching about 10^9 W m^{-2}, while the conduction term prevailing at larger radial values. The calculated power delivered by the plasma flow at the work-piece was about 1800 W. The calculated pressure force due to the impinging high-velocity plasma jet was around 1 N.

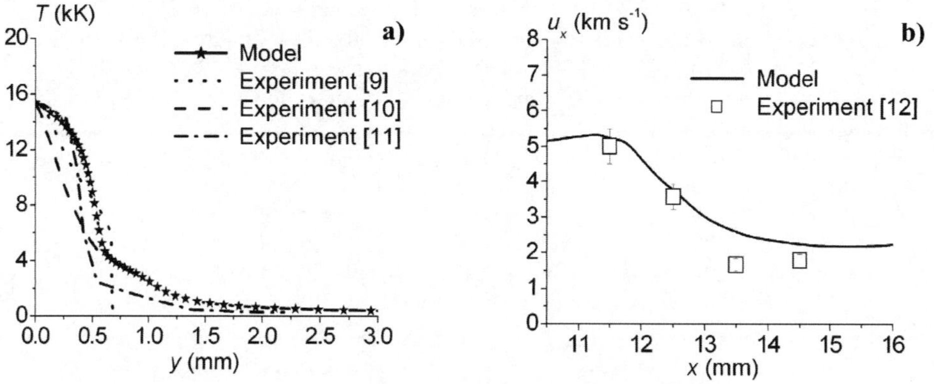

Figure 1. Radial profiles of the plasma temperature (a) and velocity (b) predicted by the model at 3.5 mm from the nozzle exit together with experimental data.

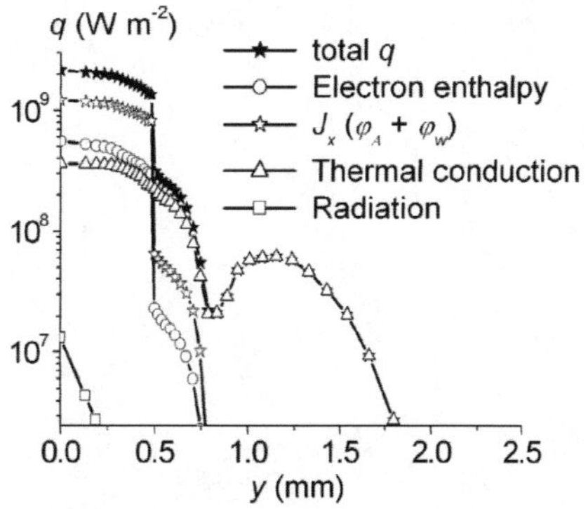

Figure 2. Radial profile of the heat flux to the anode located at 3 mm from the nozzle exit.

4. Conclusions

A turbulent two-dimensional LTE model of a 30 A oxygen high-energy density cutting torch has been reported. The calculated distributions of temperature and velocity in the nozzle exit-anode gap have been validated using experimental data. Within the experimental uncertainties, it was found that the model reproduces both the experimental data of velocity and temperature. The cutting process performance was studied in terms of the calculated values of the heat flux to the work-piece and the force exerted by the impinging plasma jet.

Acknowledgements

This work was supported by grants from the Universidad de Buenos Aires (PID X108), CONICET (PIP 5378) and Universidad Tecnológica Nacional (PID Z 012). F. O. M. and H. K. are members of the CONICET.

References

[1] Gleizes A, Gonzalez J J and Freton P 2005 J. Phys. D: Appl. Phys. **38** R153.

[2] Boulos M, Fauchais P and Pfender E 1994 *Thermal plasmas fundamentals and applications* vol.1 (Plenum Press).
[3] Prevosto L, Kelly H and Mancinelli B 2008 IEEE Trans. Plasma Sci. **36** 263.
[4] Murphy A B and Arundell C J 1994 Plasma Chem. Plasma Process. **14** 451
[5] Naghizadeh-Kashani Y, Cressault Y and Gleizes A 2002 Phys. D: Appl. Phys. **35** 2925.
[6] Yan J D, Nuttall K I and Fang M T C 1999 J. Phys. D: Appl. Phys. **32** 1401.
[7] Ferziger J H and Períc M 2002 *Computational Methods for Fluid Dynamics*, (Springer-Verlag).
[8] Fletcher C A J 1991 *Computational Techniques for Fluid Dynamics* vol 1, (Springer-Verlag).
[9] Prevosto L, Kelly H and Minotti F O 2008 IEEE Trans. Plasma Sci. **36** 271.
[10] Prevosto L, Kelly H and Mancinelli B 2009 IEEE Trans. Plasma Sci. **37** 1092.
[11] Prevosto L, Kelly H and Mancinelli B 2010 J. Appl. Phys. **107** 023304.
[12] Prevosto L, Kelly H and Mancinelli B 2009 J. Appl. Phys. **106** 053308.

AFM, XRD and Optical Studies of Silver Nanostructures Fabricated under Extreme Plasma Conditions

Yashi Malhotra[1] and M.P. Srivastava[1,2]
Department of Physics and Astrophysics, University of Delhi, Delhi-110007, INDIA

[2]E-mail: mps02@rediffmail.com

Abstract. This paper reports the deposition of silver on silicon and glass substrates using ions generated in a 3.3 KJ dense plasma focus device. The hot and dense argon plasma formed during the focus phase ionizes the silver disc placed above the top of the anode. Glass and silicon substrates are placed at an axial position of 4.0 cm and 5.0 cm from the top of the anode and are exposed by two focus shots. The deposited materials obtained were characterized using X-ray diffraction, Atomic Force Microscopy (AFM) and UV-vis Spectroscopy. The X-ray diffraction shows the [111] and [200] reflections of silver deposited on silicon substrates whereas on glass only [111] plane is observed. The AFM of silver deposited on silicon substrates shows nanostructures which have triangular like shape. The UV spectra for silver nanostructures placed at 4.0 cm and 5.0 cm on glass substrate shows the absorption peak which originates from the surface plasmon absorption of nanosized silver particles. A red shift of ~13 nm is observed for silver deposited on glass substrate placed at the distance of 5.0 cm from the top of anode.

1. Introduction

The development of nanoscience and nanotechnology has allowed us to create new nano-sized materials having unique electronic and optical properties quite different from those of their bulk material [1]. These size-dependent properties of nanomaterials have been studied and utilized in various electronic and optical devices [2]. Simultaneously, intensive research efforts have been made to understand and control the collective properties of nanoparticles [3–5]. Metal nanoparticles have been the subject of extensive research for many years because of their anomalous electromagnetic properties originating from the resonant interaction between electromagnet waves and collective conduction electron oscillations, so-called surface plasmons [6, 7] It is well known that the plasmon resonant peak is sensitive to the size [8] and shape of the nanoparticle [9-11] and metallic species [12]. Hybrid approach for the nanofabrication can be achieved by plasma methods in laboratory. Recently, it has been established that plasma aided nanofabrication [13-14] is among the best methods of deposition. It has been established recently [13] that ions generated by high density, high temperature and extremely non-equilibrium plasma of modified Dense Plasma Focus (DPF) [15] for nanofabrication of materials without dc biasing, heating of substrate or annealing. However, it was pointed out by some researchers that the ions produced in high temperature, high density and extremely non equilibrium plasma such as prevailing under fusion conditions are not useful for nanofabrication.

DPF device produces high density (10^{26} m^{-3}) and high temperature (1-2 KeV) pulsed plasma for a duration of ~ 100 ns which is a source of highly energetic ions, X-rays etc. The plasma produced under fusion conditions of high density, high temperature and strongly non equilibrium generates ions of material to be deposited in addition to ions of gas. High temperature and high density and strongly non-equilibrium plasma generating energetic ions, have been established by our group to have many applications in material processing such as phase change of materials [16-17], thin films deposition [18-19] and more recently fabrication of nanomaterials and nanostructures [20-22]. In this paper, we report the fabrication of silver using ions generated in modified device for two focused shots on silicon and glass substrate placed at a distance of 4.0 cm and 5.0 cm from the top of anode.

2. Experimental Details

A schematic arrangement of the experimental set up is shown in Figure 1. A modified DPF device consists of (i) rotary vacuum system and gas inlet (ii) plasma chamber and (iii) viewing port. It also consists of high voltage, low inductance capacitor (30 μF, 15 KV) with swinging cascade spark gap arrangement, triggering system and high voltage charger. For fabricating thins films and nanostructures an arrangement is made for inserting substrates from the top of chamber and the anode of the electrode assembly has been modified with a detachable arrangement on the top to host the cylindrical disc of the material to be deposited. A shutter is located above the anode and slightly below the substrates that protects the substrates from the impact of unfocused energetic ion. The axial movement of substrate holder, shutter and its distance from anode is controlled by a brass rod from the outside. A pure silver disc is made from silver rod and is fixed at top of anode. The silicon and glass substrates are cleaned and mounted on a substrate holder. The focus chamber is evacuated by a rotary pump and the chamber is admitted with argon gas through gas inlet. It was found earlier that if a device is operated at pressure of 80 Pa we obtain good focusing as can be seen on digital storage oscilloscope. The formation of plasma can be understood in a following manner: The capacitor is charged by a power supply and triggering circuit. The gas breakdown occurs between anode and cathode near the insulating sleeve. Image charges on the insulator sleeve initiate the discharge between the anode and the cathode forming weak current filaments leading to current density having an axially downward component.

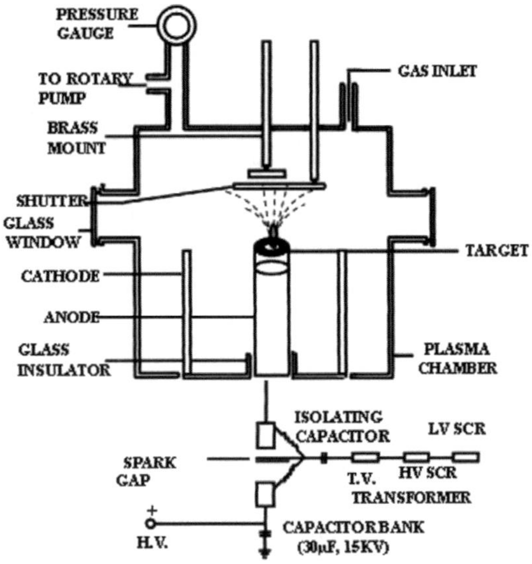

Figure. 1 Dense Plasma Focus device with its modifications.

The current in the anode creates an azimuthal magnetic field. The current filament then moves outwards due to radially outward Lorentz force arising due to axially downward component of current density and azimuthal magnetic field resulting in an inverse pinch phase. The current filament reaches the cathode due to this Lorentz force. As soon as the current sheath reaches the cathode the current filament has dominant radial component of current density. The radial component of current density and azimuthal magnetic field gives rise to axially upward component of Lorentz force which is responsible for axial phase of current sheath. This takes the current sheath towards the top of the electrode assembly. At the top of the anode, axially upward component of current density becomes dominant and this along with the azimuthal magnetic field give rise to a Lorentz force whose direction is radially inward causing the pinching of the plasma. The current sheath is accelerated towards the axis of the anode and plasma collapses to form a thin column of hot, dense plasma at the top of the anode. This is the focused phase having electron density of the order of 10^{26} m^{-3} and temperature 1-2 KeV. In this work we report the studies of silver nanostructures fabricated using two focused shots.

3. Results and Discussion

3.1 Structural Studies

The XRD spectrum is carried out using D8 Discover (with GADDS) with the CuKα radiation (λ = 1.54056A°) in 2θ ranging from 20° to 50°. The XRD pattern obtained for the silver deposited on any substrate placed either at 4.0 cm and 5.0 cm show identical behavior. Figure 2 shows the XRD pattern of silver deposited on silicon and glass substrate. Figure 2(a) show peaks at 2θ value of 38.1° and 44.5° corresponding to [hkl] plane of [111] and [200] respectively of fcc silver (JCPDS No. 4-783). The peak shown in figure 2(b) at 38.3°corresponds to [111] plane of silver on glass substrate.

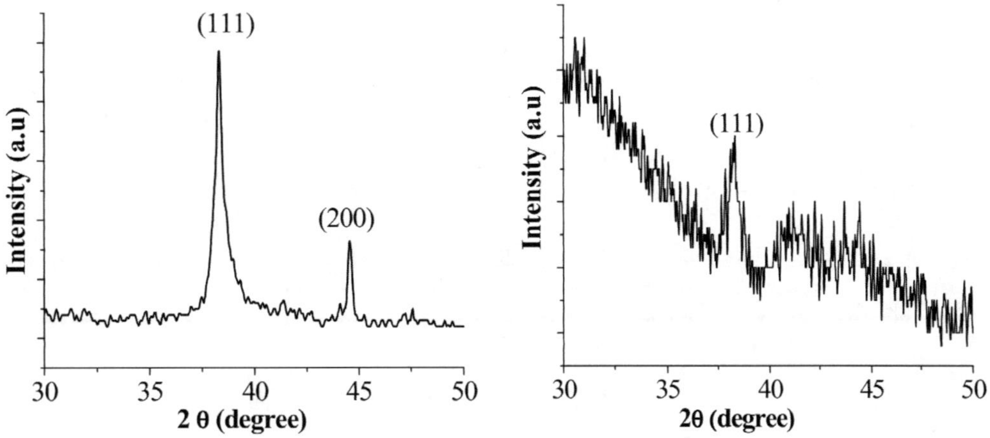

Figure 2. X-ray diffraction of grown silver nanostructures on (a) silicon and (b) glass substrate.

3.2 Surface Morphology

The surface morphology has been studied with the help of a CP-II atomic force microscope (AFM) in a non-contact mode. The AFM images were examined in terms of the surface roughness and the estimated height of silver nanostructures. AFM images of deposited silver nanostructures on silicon substrate placed at a distance of 4.0 cm and 5.0 cm from the top of the anode with two focused shot are shown in (a) and (b) of figure 3 respectively. The variation of mean height ($\overline{Z}$), peak to valley height (Rp-v), rms roughness (Rq), average roughness (Ra) and average height of deposited nanostructures with distance are given in table 1. The nanostructures formed are triangular in shape. This may be due to the agglomeration of the particles forming different structures or this may be due to the condensation of ions on the substrates. The formation of nanostructures can be understood in the following manner. The highly energetic ions produced in the post focus phase reach the substrates, causing heating of the substrates prior to deposition. The high temperature argon plasma produces the silver ions. The silver ions are then mixed with argon ions moves vertically upwards in a fountain like structure and lose its energy to ultimately deposit on the substrate placed above the top of anode. The energetic ions hitting substrate surface may create nucleating centres and help in enhancing the adhesion of the deposited material to the substrate surface. The AFM of samples deposited on glass substrate shows cracks and pits. This may be due to the impact of highly energetic and high fluence ions that hit the substrate.

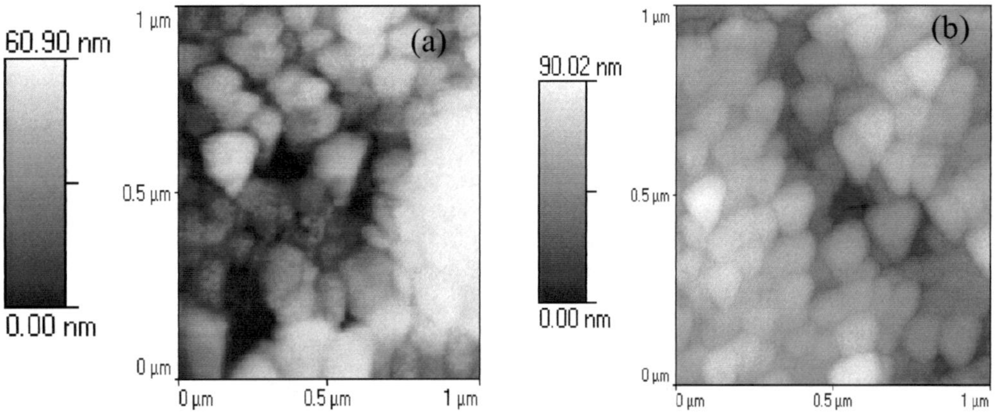

Figure 3. AFM image of silver nanostructures deposited on silicon placed at a distance of (a) 4.0 cm and (b) 5.0 cm from top of anode. The vertical bar on the left side of each image indicates height.

Table 1 Variation of $\overline{Z}$,Rp-v, Rq, Ra and Avg height with substrate placed from top of anode :

Substrates placed at (cm)	$\overline{Z}$ (nm)	Rp-v (nm)	Rq (nm)	Ra (nm)	Avg height (nm)
4.0	33	62	7	6	30
5.0	50	90	14	12	50

3.3 UV-Vis Spectra

In metal nanoparticles such as in silver, the conduction band and valence band overlap each other in which electrons move freely. These free electrons give rise to a surface plasmon resonance (SPR) absorption band, occurring due to the collective oscillation of electrons of silver nanoparticles in resonance with the light wave. The absorption spectrum of silver deposited on glass substrate is investigated. (a) and (b) of figure 4 shows the absorption spectra of silver nanostructures deposited on glass substrate placed at a distance of 4.0 cm and 5.0 cm from op of anode with two focused DPF shots respectively. The absorption peak at ~ 400 nm is observed for silver nanostructures deposited on glass substrates placed at a distance of 4 cm from top of anode is shown in (a) of figure 4. (b) of figure 4 corresponds to absorbance of silver deposited on glass substrate placed at 5 cm from top of anode and it shows a peak at 413 nm for silver nanostructures. This peak which is observed in the spectra is attributed due to surface plasmon resonance (SPR). Silver nanostructures deposited at a distance of 5.0 cm from anode top on glass substrate display absorption peak at 413 nm, which is red-shifted about 13 nm as compared to the silver nanostructures deposited at 4.0 cm from anode top. Mulvaney [12] reports that the surface plasmon resonance (SPR) spectrum depends upon the refractive index of the surrounding medium, particle size, shape of particle and absorption substance of their surface. The red shifted in SPR is assigned due to bigger aspect ratio [23] or it can be due to increase in edge length and tip sharpness of silver nanoparticles.

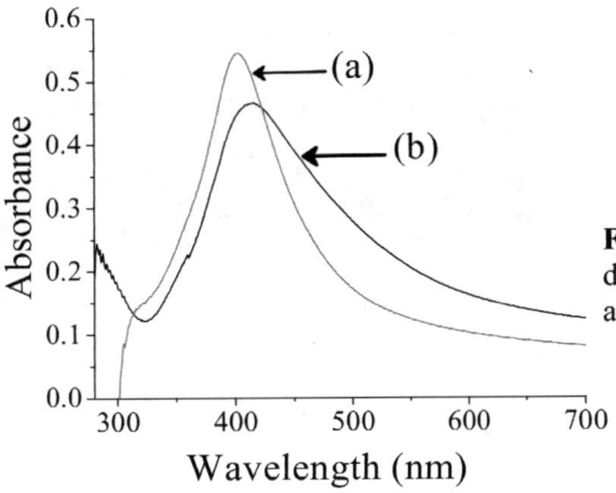

Figure 4. UV spectra of silver nanostructures deposited on glass placed at a distance of (a) 4.0 cm and (b) 5.0 cm from top of anode.

4. Conclusion
Highly energetic, high-fluence ion originated from the fully ionized gas and evaporated silver disc are deposited onto silicon and glass substrates without using any additional substrate biasing or heating which is most common in other devices. XRD shows the fcc planes corresponding to silver. AFM studies show the triangular shape like nanostructures. UV spectra show the peak which correspond to surface resonance plasmon.

5. Acknowledgement
One of us (YM) is thankful to University Grant Commission and Centre of Advanced studies for Physics and Astrophysics, Delhi University for the award of JRF.

6. References
[1] Alivisatos A P 1996 *Science* **271** 933
[2] Ellingson R J, Beard M C, Johnson J C, Yu P, Micic O I, Nozik A J, Shabaev A and Efros A L 2005 *Nano Lett.* **5** 865
[3] Markovich G, Collier C P, Henrichs S E, Remacle F, Levine R D and Heath J R 1999 *Acc. Chem. Res.* **32** 415
[4] Lu Y, Liu G L and Lee L P 2005 *Nano Lett.* **5** 5
[5] Reinhard B M, Siu M, Agarwal H, Alivisatos A P and Liphardt J, 2005 *Nano Lett.* **5** 2246
[6] Kreibig U and Vollmer M, 1995 *Optical Properties of Metal Clusters* (Springer, Berlin)
[7] Bohren C F and Huffman D R, 1983 *Absorption and Scattering of Light by Small Particles* (Wiley, New York)
[8] Berciaud S, Cognet L, Tamarat P and Lounis B 2005 *Nano Lett.* **5** 515
[9] El-Sayed M A 2001 *Acc. Chem. Res.* **34** 257
[10] Kelly K L, Coronado E, Zhao L L and Schatz G C 2003 *J. Phys. Chem.* B **107** 668
[11] Jin R C, Cao Y W, Mirkin C A, Kelly K L, Schatz G C and Zheng J G 2001 *Science* **294**

1901
[12] Mulvaney P 1996 *Langmuir* **12** 788
[13] M.P. Srivastava M P (2003), First Cairo Conference on Plasma Physics and Applications, Ed. Kunze H-J, El-Khalafawy T, Hegazy H, Schriften Des Forschungszentrum Julich GmbH, **34**, 40
[14] Srivastava M P 2005 *Advanced in Applied Plasma Science* **6**,227.
[15] Mather J W 1964 *Phys. Fluids Vol.* **7**, 5
[16] Rawat R S, Srivastava M P, Tandon S and Mansingh A 1993, *Phys. Rev. B*, **47,** 4858
[17] Sagar R and Srivastava M P 1996, *Phys. Lett. A*, **183**, 209
[18] Kant C R, Srivastava M P and Rawat R S 1997, *Phys. Lett. A*, **226,** 212
[19] Kant C R, Srivastava M P and Rawat R S1998, *Phys. Lett. A*, **239**, 109
[20] Gupta R, Srivastava M P, Balakrishnan V R, Kodama R and Peterson M C 2004, *J.Phys.D: Appl. Phys*., 37, 1091
[21] Srivastava M P 2007 *Adv. in Appl. Plasma Sci*., **6**, 227
[22] Malhotra Y, Roy S, Srivastava M P, Kant C R, Ostrikov K 2009, *Journal of Phy D: Appl Phys*, **42**,155202
[23] Jana N R, Gearheart L, Murphy C J, 2005 *J. Phys. Chem. B* **105** 4065.

ICPP2010 & LAWPP2010 IOP Publishing
Journal of Physics: Conference Series **511** (2014) 012073 doi:10.1088/1742-6596/511/1/012073

Investigation of the ion beam emission from a pulsed power plasma device

A. Henríquez, H. Bhuyan, M. Favre, M. J. Retamal, U. Volkmann, E. Wyndham and H. Chuaqui

Pontificia Universidad Católica de Chile, Departamento de Física, Casilla 306, Santiago 22, Chile.

E-mail: aphenriq@uc.cl, hbhuyan@fis.puc.cl

Abstract. Plasma Focus (PF) devices are well known as ion beam sources with characteristic energy among the hundreds of keV to tens of MeV. The information on ion beam energy, ion distribution and composition is essential from the viewpoint of understanding fundamental physics behind their production and acceleration and also their applications in various fields, such as surface properties modification, ion implantation, thin film deposition, semiconductor doping and ion assisted coating. An investigation from a low energy, 1.8 kJ 160 kA, Mather type plasma focus device operating with nitrogen using CR-39 detectors was conducted to study the emission of ions at different angular positions. Tracks on CR-39 detectors at different angular positions reveal the existence of angular ion anisotropy. The results obtained are comparable with the time integrated measurements using FC. Preliminary results of this work are presented.

1. Introduction

In plasma focus (PF), a high density, high temperature, short duration plasma column is formed, following the radial compression phase that characterizes PF discharges [1]. In addition to x-ray and neutron pulses, the PF is a well-known source of energetic ion beams, of characteristic energy from hundreds of keV to tens of MeV [2]. This characteristic energy range exceeds up to several hundred times the maximum energy to be expected from the capacitor bank voltage. Most of the earlier studies on ion beam emission in PF were carried out in deuterium medium, to investigate the correlation between neutron production and deuteron acceleration. Several properties of PF emitted ion beams have been measured, including their angular distributions and energy spectra [3]. A comprehensive characterization of the PF ion beams, in terms of energy, flux, spatial distribution and ion compositions is desirable in order to advance the understanding of the fundamental physical processes involved in their generation. Several theoretical and computational models have been developed on the ion production and acceleration mechanism for PF discharges [4], however a complete explanation of the ion emission mechanism of the different ranges of PF bank energies is still not available. Low energy PF operating in the few kilojoules range are characterized by a modest operational cost and allow easy modification. These particular features of PF devices, combined with the wide energy range and high characteristic flux of the emitted ion beams, together with the possibility of high frequency operation, have attracted the interest of several researchers, in view of the potential applications in different technological fields, such as surface properties modification, ion implantation, thin film deposition, semiconductor doping and ion-assisted coating [5, 6, 7]. In the past, studies have been carried out on the general characteristics of ion beam emission in PF [3, 8, 9]. Sadowski et al [10] observed two to three separate ion pulses each having 20-60 ns full width at half maximum (FWHM), using Thomson parabola techniques. They reported that narrow bunches (with divergence less than 8°) of very high-energy ions are emitted among numerous low energy ion beams pulses. Visualizing the importance of energetic ions in the field of materials processing, we have performed an experimental analysis on the ion emission from a low energy PF device operating in nitrogen.

2. Experimental Apparatus

The experiment was conducted in a low energy, 1.8 kJ, Mather type [1] PF device. A schematic diagram of the experimental arrangements along with the discharge circuit is shown in figure 1. The electrode assembly of the PF chamber consists of a hollow central anode of length 10 cm and outer diameter 2.4 cm encircled by six cathode rods of length 9.7 cm and diameter 0.8 cm uniformly spaced coaxially at a diameter of 6.3 cm. The electrode material is copper. The two electrodes are separated by a 3 cm long Pyrex glass insulator sleeve fixed around the bottom of the anode as shown in figure 1. A hollow anode configuration was used in order to decrease plasma contamination with heavy ions resulting from anode ablation. The PF device is powered by a capacitor bank of 9 μF, at 30 kV maximum charging voltage. At 20 kV operating voltage the capacitor bank delivers a maximum current of 160 kA with 1.8 μs quarter period. The system inductance at the maximum current is around 146 nH. A resistive divider and a single groove Rogowski coil are used as voltage and current probes, respectively. Further details of the PF device were reported elsewhere [11]. In this case, high energy photon illumination of the collectors results in a positive polarity signal. A BPX65 PIN diode masked with a 3 μm thick aluminium filter is placed at the side port of the PF chamber, to record the x-ray signal from the side-on plasma emission. In PF operation, intense soft x-ray pulses are mainly emitted at the time of maximum radial compression, when a hot and dense plasma column is formed. Because of the negative biasing, the ion probes also act as x-ray detector (XRD). Thus, the x-ray emission recorded by the PIN diode and the x-ray signal in the ion probes can be used to establish a time correlation between plasma evolution and ion beam emission during the pinch formation process. The electrical signals from the voltage probe, current probe, PIN diode and ion probes are recorded on a four-channel, 500 MHz, 2 Gs s^{-1} sampling rate oscilloscope. Ion beam measurements were performed in pure nitrogen, in the pressure range 0.25-0.55 Torr.

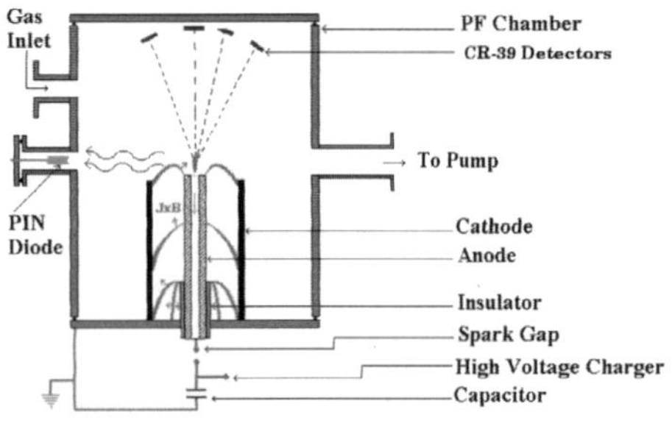

Figure 1. PF configuration using CR-39 track detector at angles of 0, 10, 15 and 20 degrees from the anode axis.

The measurements were done using solid state nuclear track detector (CR-39) at a distance of 15 cm from the top of the anode at different angles (0, 10, 15 and 20 degrees) respect to the anode axis. The shutter was removed after the achievement of optimum operational condition, without disturbing the in vacuum conditions and thereby exposing the track detectors to the energetic ions of a single shot. The samples were etched for 2.5 hours in 6.25 N NaOH solutions at (70±1) ^oC. The etched samples were investigated by Optical Microscope (OM) and Atomic Force Microscope (AFM).

3. Experimental Results and Discusion

Figure 2 shows simultaneous oscillograms of voltage, current derivative and PIN diode. The singularities seen in the voltage and current derivative signals around 1.6 μs indicate strong PF action, which means that a tight, hot and dense plasma column has formed. Earlier we have investigated the anisotroppy of ion emission with Faraday cups (FC), in which case the emission instant of ions was considered at the moment of the appearance of x-ray peak in the x-ray PIN diode. In this measurements, the CR-39 detectors were irradiated by the incident ion beam for single discharge and its depth profile are investigated.

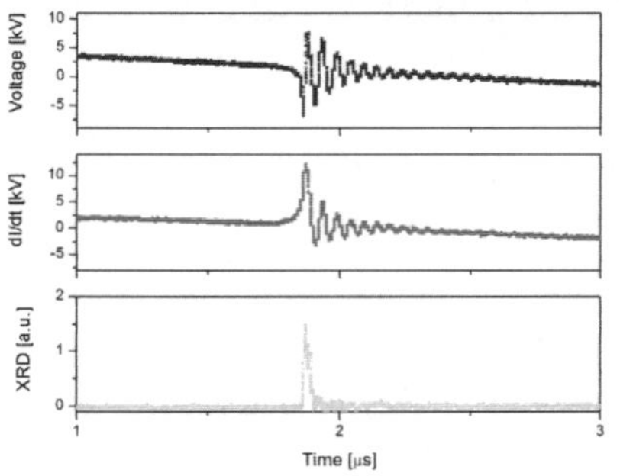

Figure 2. Characteristic electrical signals. From top to bottom: voltage, time derivative of the discharge current and x-ray pin diode.

Figures 3-6 shows AFM images and depth profile of ion exposed CR-39 detectors at different angular position with respect to the anode axis. The grayscale on the AFM images indicates the depth profile of ions. Depth profile for some selected tracks of each AFM images are shown in figures 3b, 4b, 5b and 6b. The variations in the ion tracks are clearly visible on the AFM images, particularly track intensities and track diameters. AFM analysis of the CR-39 detectors at different angular positions shows the variation in track's number density, heights and diameters, indicating the different ion flux and ion energy. The penetration depth of the ions is related to the incoming ion beam energy. The track diameter and track height are the crucial parameters for the ion energy calculation. The different depth and diameter of the tracks are the evidence of the existence of different group of ions in the PF ion beams with different ion spectrum. This result is in agreement with earlier published works [8] that shows an angular anisotropy of the ion beam emission which can be explained in terms of ion Larmor radius effects during the characteristic z-pinch like plasma formation phase.

Measurements of the ion energy distribution in a 4.75 kJ PF operating with nitrogen, performed with a Thomson analyser, shows that the main ion species are N^{+1} , N^{+2} and N^{+3} , with acharacteristic energy in the range 0.17-4.0 MeV, and total (time integrated) average ion flux $8 \cdot 10^{12}$ ions/stereorad [12]. In our case, the estimated track density on the detector is $\sim 10^{12}$ ions/m^2 and considering an ion irradiation time over the samples of about 1 μs, it can be estimated an ion flux of $I_F \sim 10^{18}$ $ions/m^2s$.

The figure 7 shows the ion flux as a function of the penetration depth of the ion beam. The histograms indicate a Gaussian distribution, irrespective of the angular positions respect to the PF axis. The estimated maximum ion penetration depth is of the order of 1800 nm, whereas ion fluxes are maximum for the energy range 200-800 nm. At 0° the maximum flux of ions is observed at around 600 nm depth, whereas at 10° it is around 376 nm. This indicates that

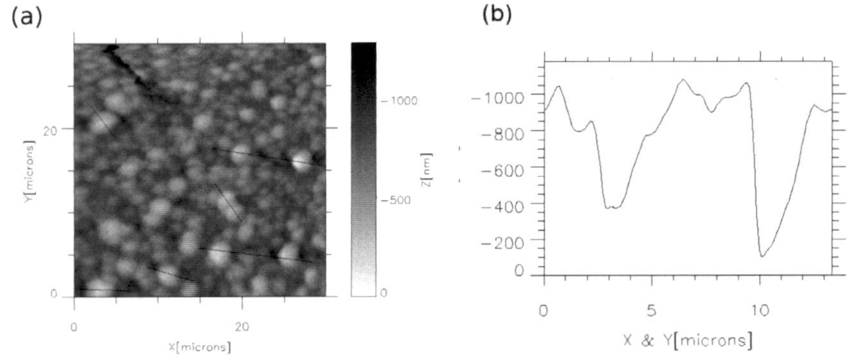

Figure 3. (a)AFM image and (b) depth profile for the CR-39 detectors at 0 degrees.

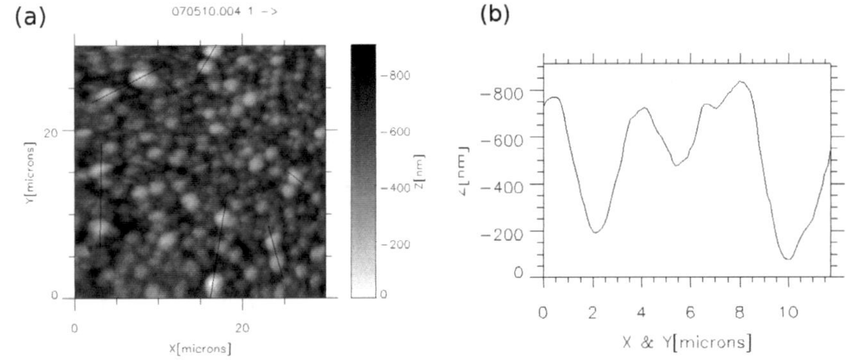

Figure 4. (a)AFM image and (b) depth profile for the CR-39 detectors at 10 degrees.

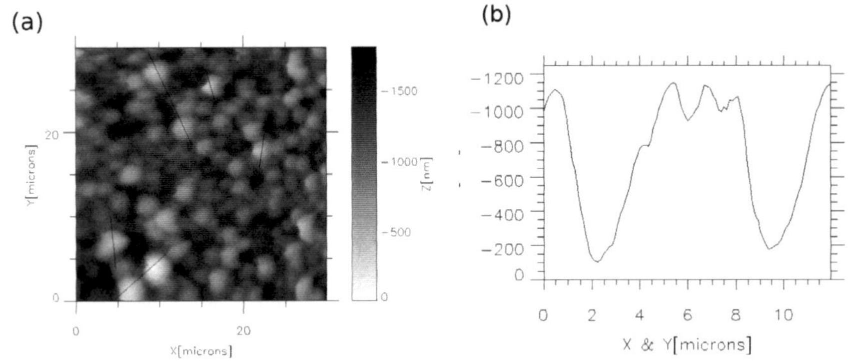

Figure 5. (a)AFM image and (b) depth profile for the CR-39 detectors at 15 degrees.

maximum ion energy is on the axis on 0° whereas maximum ion flux is observed at 10°. This result is in well agreement with the previous results obtained with FC in the same PF, where

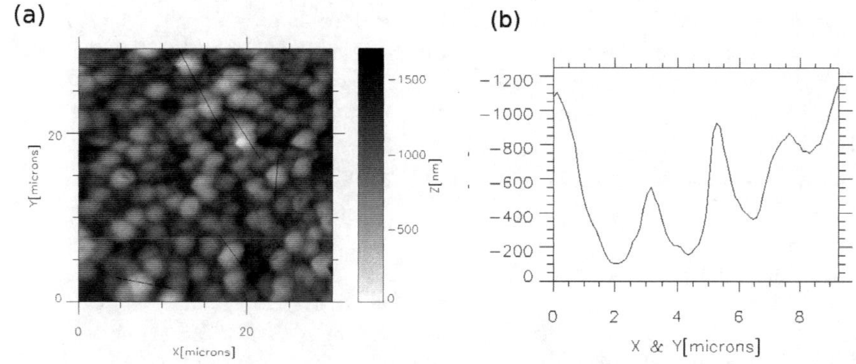

Figure 6. (a)AFM image and (b) depth profile for the CR-39 detectors at 20 degrees.

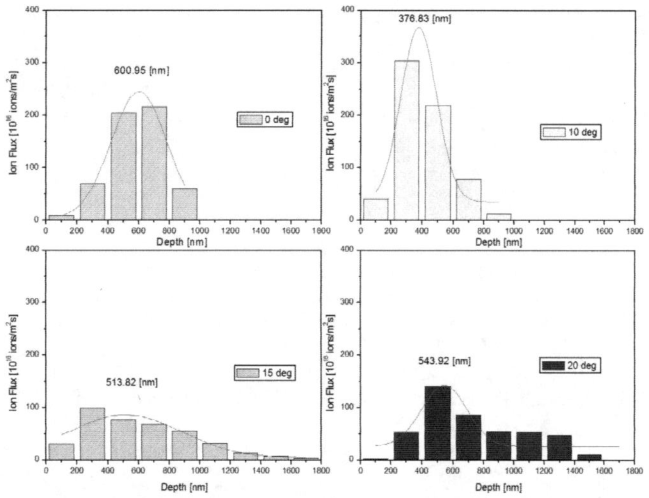

Figure 7. Depth histogram at diferent angles from the PF axis. From top to bottom, and from left to right: 0°, 10°, 15° and 20°.

the angular distribution of hydrogen and carbon ions shows a highly anisotropic character[3]. The flux of hydrogen ions is found to be maximum at the PF axis, whereas in the case of carbon species it is seen to maximize off-axis. Maximum ion emission is confined within a solid angle of 15°. The nitrogen ions detected within a solid angle of 15° emitted from a 2.2 kJ PF, using time of flight (TOF) technique, revealed that the ion flux is maximum at an angle of 5° and minimum at an angle of 0° with respect to the PF axis [13].

Despite the available experimental data, the exact mechanism for ion acceleration is still not clear.Basically, four different possibilities have been proposed to explain the ion beam formation. The first one is a re-distribution of current at the end of the radial compression phase, changing from an annular distribution to a centre peaked one, giving rise to intense, high gradient, axial electric fields [4], the second is an axial voltage increase due to a fast inductance change in the radial collapse phase [14], the third is ion runaway during the radial compression phase [15] and the fourth is the formation of m = 0 MHD instabilities in the plasma column, which result in magnetic field energy being transferred to the kinetic energy of the ions[16]. Bhuyan et al [9] have measured the basic properties, charge states, energy spectrum and flux, of ion beams generated in a low energy PF operating with methane using FC in time of flight technique. The dominant charge states of the carbon ions appear to be consistent with a model in which ion acceleration takes place before the high density, high temperature phase of the PF discharge.

4. Conclusions

We have investigated the ion emission of a 1.8 kJ PF device operating in nitrogen at different angular positions, using CR-39 track detectors. From the depth histogram at different angular positions it is observed that the ion beam emission crearly depends on the angular position. The maximum ion emission is confined within a solid angle of $\sim 20^o$. We observe an ion flux of $\sim 10^{18}$ ions/m^2s. The mass dependence of the ion Larmor radius, together with scattering with neutrals in the background gas, might cause the observed angular anisotropy. In addition, various complex phenomenon like heterogeneity of local electro-magnetic field, energy exchange between different species of ions and electrons etc, which might also affect the ion emission processes. Further works are in progress.

Acknowledgments

This work has been funded by Pontificia Universidad Católica de Chile, Proyecto Puente number 46. The technical assistance of Dirk De Jong and David Alvarez is highly appreciated.

References

[1] Mather J W 1965 *Phys. Fluids* **8** 366
[2] Sadowski M, Zebrowski J, Rydygier and Kucinski J 1988 *Plasma Phys. Control. Fusion* **30** 763
[3] Bhuyan H et. al. 2007 *Journal of Physics D: Applied Physics* **42**, 205207
[4] Bernstein M J 1970 *Phys. Fluids* **13** 2858
[5] H Bhuyan et al *J. Phys. D: Appl. Phys.* **42** 205207
[6] Bhuyan H, Favre M, Valderrama E, Avaria G , Chuaqui H, Mitchell I, Wyndham E, Saavedra R and Paulraj M 2007 *J. Phys. D: Appl. Phys.* **40** 127-31
[7] Bhuyan H, Favre M, Henriquez A, Vogel G, Valderrama E, Wyndham E, Chuaqui 2010 *Surf.Coat. Technol.* **204** 2950-53
[8] Bhuyan H et. al. 2006 *Journal of Physics D: Applied Physics* **39** 3596
[9] Bhuyan H, Chuaqui H, Favre M , Mitchell I and Wyndham E 2005 *J. Phys. D: Appl. Phys.* **38** 1164-69
[10] M J Sadowski 2006 *Czechoslovak Journal of Physics* **56** Suppl. B.
[11] Favre M, Silva P, Choi P, Chuaqui H, Dumitrescu C and Wyndham E 1998 *IEEE Trans. Plasma Sci.* **26** 1154
[12] Kelly H, Lepone A and Márquez A 1997 *IEEE Trans. Plasma Sci.* **25** 455
[13] Mohanty S R, Bhuyan H, Neog N, Rout R K and Hotta E 2005 *Japan. J. Appl. Phys.* **44** 5199
[14] Browne P F 1988 *J. Phys. D: Appl. Phys.* **21** 596
[15] Deutsch R and Kies W 1988 *Plasma Phys. Control. Fusion* **30** 263
[16] Vikhrev V V 2004 *Proc. 15th Int. Conf. on High-Power Particle Beams (Saint Petersburg, Russia)* submited

A Permanent Magnet Hall Thruster for Pulsed Orbit Control of Lunar Polar Satellites.

Brunno Silva Moraes[1], **José Leonardo Ferreira**[1], **Ivan Soares Ferreira**[1]
Othon Cabo Winter[2], **Décio Cardozo Mourão**[2]

1 - Institute of Physics. University of Brasília – UnB, CEP 70919-970, Brasília – DF – Brazil
2 - São Paulo State University – UNESP, Av. Ariberto Pereira da Cunha, 333, CP 205, CEP 12516-410, Guaratinguetá - SP - Brazil

E-mail: brunno.moraes@gmail.com

Abstract.
Future Moon missions devoted to Lunar surface remote sensing, for example, will require very fine and accurate orbit control. It is well known that Lunar satellites in polar orbits will suffer a high increase on the eccentricity due to the gravitational perturbation of the Earth. Without proper orbit correction the satellite lifetime will decrease and end up in a collision with the Moon surface. It is pointed out by many authors that this effect is a natural consequence of the Lidov-Kozai resonance. We studied different arcs of active lunar satellite propulsion, centered on the orbit apoapsis or periapsis, in order to be able to introduce a correction of the eccentricity at each cycle. The proposed method is based on an approach intended to keep the orbital eccentricity of the satellite at low values.

1. Introduction

Recently, the ESA's Small Mission for Advanced Research in Technology (SMART-1) completes the mission of scanning the Lunar surface from its own orbit. It was the first spacecraft which achieved the Moon's orbit employing plasma rockets (for details, see e.g., [1]), and therefore become a very successful start of an epoch of Moon exploration using plasma technology.

Certainly, in few years it will be build several Lunar bases, mainly in the near poles region, where the probability to found water reservoirs is larger. Because of that, the need of reconnaissance satellites in polar orbits is increasing. However, high inclination orbits around the Moon are susceptible to the Lidov-Kozai resonance [2], which will limit the satellite lifetime by increasing the orbit eccentricity, until the collision between Moon's surface and the satellite.

At the Lidov-Kozai resonance scenario, eccentricity and inclination of a small object's orbit are coupled according third Delaunay momentum, therefore a perturbation would have no influence on semi-major axis. Those quantities could undergo changes and, for high inclined orbits, the eccentricity will grow without control[9].

Winter et al (2009) [3] proposed, in a previous work, a method to increase the lifetime of polar lunar satellites in near circular orbit using low thrust propulsion and the locally optimal thrust. This approach is suitable considering a costly attitude control and missions in which large fluctuations of radial distance are acceptable.

In the present paper, we studied the employment of permanent magnets Hall-effect thrust to introduce a correction of the eccentricity at each cycle of lunar orbit, and therefore sustain very low values to the eccentricity of the orbit. In the next section we describe these thruster, and after that we describe the simulation methods and we show and comment our results.

2. Hall-effect plasma thrusters

Hall-effect plasma thrusters are now a very good option for space probes primary propulsion and also for station keeping of medium and large satellites. This is because of their high specific impulse and thrust efficiencies, which is related to an economy in terms of propellant mass utilization and satellite lifetime. First Hall thrusters were developed in mid 1950's, and they were also called Closed Drift Thrusters, as become clear that the closed electron drift current inside the source channel, promoted by the crossed electric (axial) and magnetic (radial) fields configurations, is responsible for the acceleration of the plasma beam (for a historical review, see [4]). Instead of a good comprehension of the acceleration mechanism and the steady state plasma dynamics, today a lot of problems on the role of electrons transport and plasma fluctuations and instabilities still open (e.g. [5]).

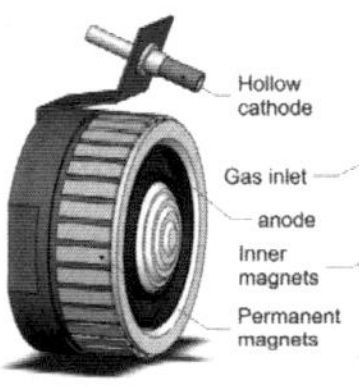

Figure 1. Permanent Magnet Hall-effect Plasma thruster design showing its main components. The external shields are omitted for clarity.

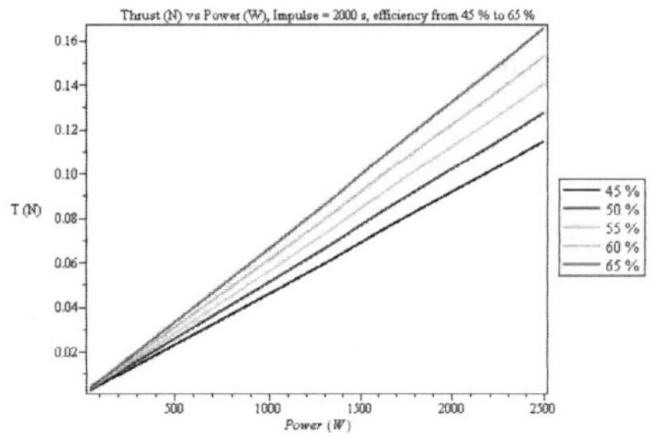

Figure 2. Calculated thrust as function of input power for some values of thruster efficiency.

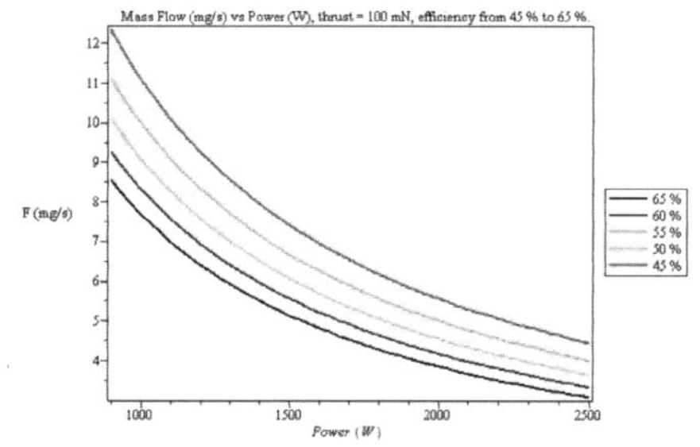

Figure 3. Calculated mass flux, as function of power, to a constant thrust of 100 mN, for several values of thruster efficiency.

Typically, well developed Hall thrusters shows a total efficiency between 50 and 60 %, for a maximum specific impulse of 2500 s ([6]). One limiting factor in the account for a larger total efficiency is electric efficiency, which is the amount of input power that is actually used to accelerate the beam. For an ordinary Hall thruster, this efficiency also includes the power needed to generate the magnetic field. Looking on this aspect, the Permanent Magnets Hall Thruster (PMHT) was originally proposed [7]. In PMHT, an array of permanent magnets, carefully designed and placed both beside the inner and outer walls, generates the magnetic field. This solution should be responsible for a increase of the total efficiency of the next generation of thrusters. We can see in Figure 1 the permanent magnets Hall-effect thruster schematics. And in Figures 2 and 3 is plotted the calculated thrust and the calculated mass flux, respectively, both as function of input power and for different efficiencies. Those values will be used as input for the orbit control simulations.

Currently, there are several ongoing studies in order to understand the dynamics of the plasma in the PMHT interior and exterior (plume regions). The main issue is the presence of all three magnetic field components in relevant intensity inside the channel and near the external wall of the plasma source [7]. This is very different from what is regularly found in Hall thrusters, i.e., a mainly radial magnetic field, present only in the plasma acceleration channel exit. Despite of that, the use of the PMHT in space missions, including deep space maneuvering, is already underway, as for an example in the work by Santos et. al. (2008)[8].

3. Method

To perform the numerical simulations, has been considered the 3-Body problem: Moon, Earth and satellite. In that case, the coordinate system was chosen to be centered in the Earth-Moon barycenter and the equations of motion of the satellite are given by:

$$\ddot{\vec{x}} = \sum_{i=1}^{2} G\frac{m_i}{|\vec{x_i} - \vec{x}|^3}(\vec{x_i} - \vec{x}) \tag{1}$$

$$\ddot{\vec{y}} = \sum_{i=1}^{2} G\frac{m_i}{|\vec{y_i} - \vec{y}|^3}(\vec{y_i} - \vec{y}) \tag{2}$$

$$\ddot{\vec{z}} = \sum_{i=1}^{2} G\frac{m_i}{|\vec{z_i} - \vec{z}|^3}(\vec{z_i} - \vec{z}) \tag{3}$$

In this system, the Earth's motion is described by:

$$\ddot{\vec{x_i}} = \sum_{j=1,j\neq i}^{2} G\frac{m_i}{|\vec{x_j} - \vec{x_i}|^3}(\vec{x_j} - \vec{x_i}) \tag{4}$$

$$\ddot{\vec{y_i}} = \sum_{j=1,j\neq i}^{2} G\frac{m_i}{|\vec{y_j} - \vec{y_i}|^3}(\vec{y_j} - \vec{y_i}) \tag{5}$$

$$\ddot{\vec{z_i}} = \sum_{j=1,j\neq i}^{2} G\frac{m_i}{|\vec{z_j} - \vec{z_i}|^3}(\vec{z_j} - \vec{z_i}) \tag{6}$$

Where m is the mass and the index $i = 1$ refer to the Earth and $i = 2$ refer to the Moon.

At each integration cycle, the classical keplerian osculating elements were computed and monitored.

During the orbit propagation, when the value of true anomaly stay between certain values of angular position, it was satisfied one condition for turning on the propulsion system. Those angular regions were defined as *propulsion arcs* and centered on the orbital apoapsis or periapsis.

The approach proposed is a very simple one and the idea is to introduce a correction on eccentricity by means of constant vector thrust. This thrust is activated only inside the propulsion arcs centered on apocentre or pericentre.

The procedure consists in fixing a limit value to the eccentricity e_{max} and then activating the thrust every time the spacecraft is inside any of propulsion arcs if that limit value is exceeded.

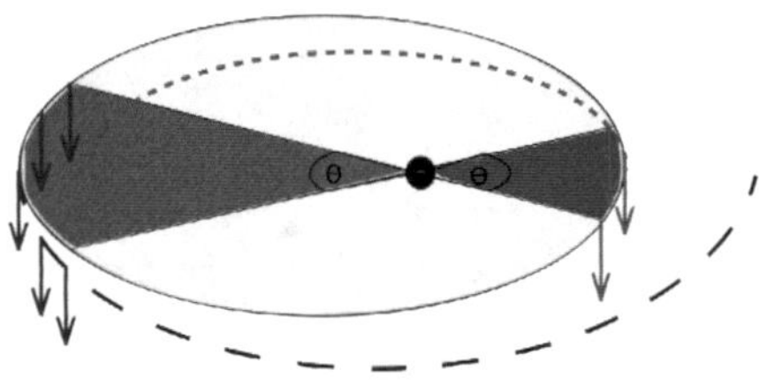

Figure 4. Schematic view of maneuver plan where the angular apertures centered on the orbital apoapsis or periapsis has been shown with their proper directions for the thrust vector in order to introduce the due corrections on orbital eccentricity.

As the value of orbital eccentricity was limited, the values of distances of apocentre or pericentre needed to be kept inside certain intervals. If the spacecraft was inside the propulsion arcs and if those distances stay beyond these intervals, it was satisfied all the conditions for turning on the propulsion system. As one can see in the Figure 4, if the spacecraft was inside the pericentre propulsion arc and its orbital apocentre distance was greater than $(1 + e_{max})\,a$ (where a is the orbital semi-major axis) or the spacecraft was inside the apocentre propulsion arc and its orbital pericentre distance was less than $(1 - e_{max})\,a$, the propulsion system was activated in order to maintain the value of orbital eccentricity below e_{max}.

The set of initial conditions for the simulations were listed on Table 1. We used typical values for the thrust and spacecraft mass (see Figures 2 and 3) as well we input the expected value for altitude of a polar orbit reconnaissance satellite.

Table 1. Initial conditions used on the simulations.

	Initial value
Spacecraft mass (Kg)	1,000
Thrust magnitude (mN)	100
Altitude (Km)	500
Inclination (deg)	90
Angular aperture of propulsion arcs ϑ (deg)	6 to 122
Limit eccentricity (e_{max})	0.05
Initial eccentricity (e_0)	0.04

4. Results

Following, the results of our simulations. In each run has been measured the running time (R_t) needed to keep the excentricity below e_{max} during mission times between 0 and 10,000 days. From this data, was obtained an empirical expression for the uptime as function of duration of mission.

One can see in Figure 5 that the growth of eccentricity was controled during all the mission time. In that case the colision with Moon surface was properly avoided.

Figure 6 shows the uptime of PMHT as function of duration of mission. Samples of the results for ϑ equal to 12°, 32°, 52° and 72° are shown in both figures.

It was performed many linear fit ($R_t = A(\vartheta)[t_k - 339.933]$) for every tested angular apertures. The values of $A(\vartheta)$ were plotted in Figure 7 where can be seen that there is a minimum when ϑ is around 12 degrees. Figures 8 and 9 presents basically the same results as previous ones. The main diference is the set of maximum eccentricities maintained.

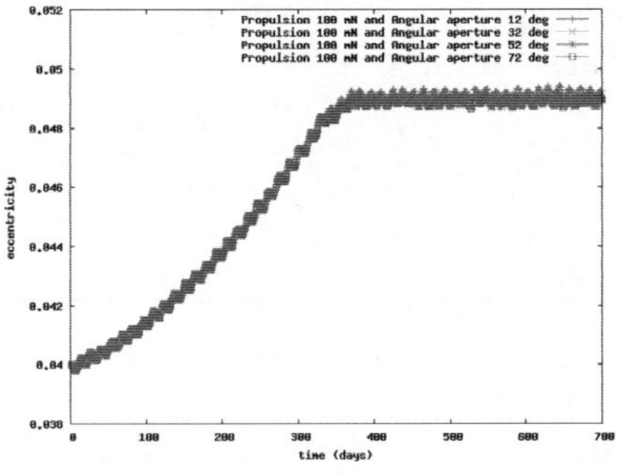

Figure 5. Stabilization of eccentricity values with the correction provided by PMHT. Many sizes of propulsion arcs were considered.

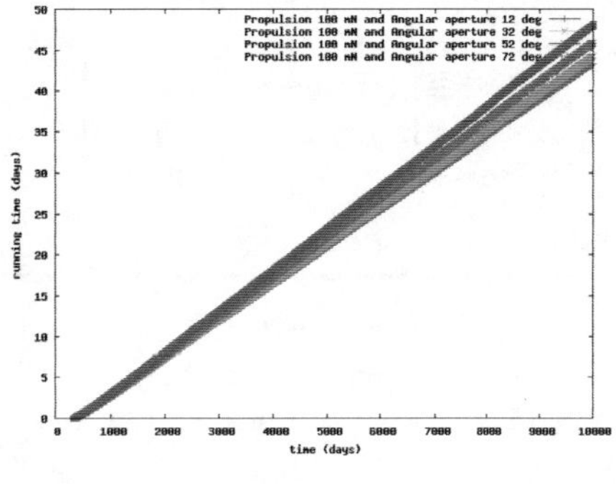

Figure 6. The up time of thrusters as function of the total time used to the simulations.

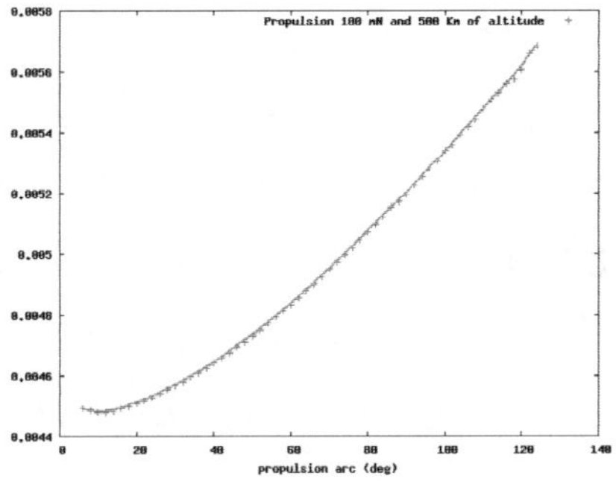

Figure 7. Angular coeficient $A(\vartheta)$, resulted from linear fitting of data shown in Figure 6 versus lenght of propulsion arcs.

One can see (from the Figures 8 and 9) that the sooner the thrusters are activated the lower will be the Angular coeficients $A(\vartheta)$ or the running time.

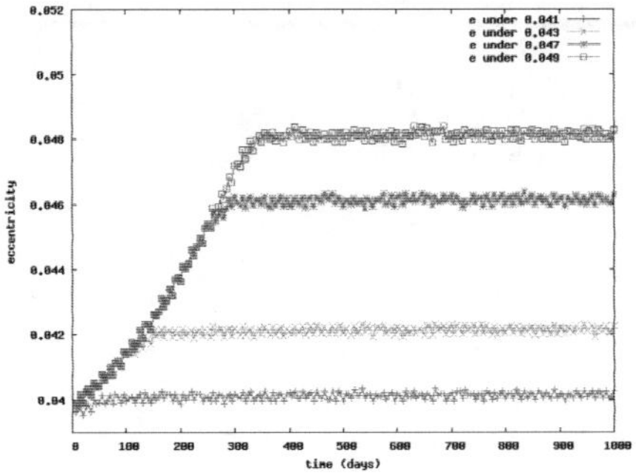

Figure 8. Samples of eccentricties as functions of time.

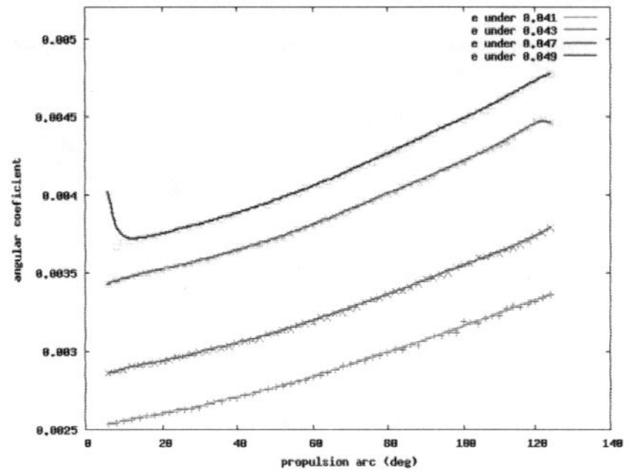

Figure 9. Samples of Angular coeficients as functions of *propulsion arcs* ϑ.

5. Conclusions

It is confirmed that the growth of eccentricity of a lunar polar satellite orbit can be controlled employing a PMHT in a pulsed manner.

There is the possibility to keep the satellite's eccentricity *stable* under 0.05, during a time t_k, with a running time of system given by:

$$R_t = A(t_k - 339.933) \tag{7}$$

The angular coeficient $A(\vartheta)$ presents a minimum value when the propulsion arc is around 12 degrees. Therefore we can assume there are an optimal lenght of these angular apertures in order to minimize the fuel consumption. The next step will be the seeking for empirical expressions of optimal ϑ as function of eccentricity maintained, altitude and mission time.

Acknowledgments

We are in debt to Brazilian Space Agency (AEB), FAPDF, IFD/UnB, DPP/UnB and the CAPES. Moraes thanks particularly the support of The Brazilian School of Public Administration - ENAP.

References

[1] Estublier, D. , et. al., Electric propulsion on SMART-1 - A technology milestone, 2007, *ESA Bulletin*, 129, pp. 40 - 46.

[2] Kozai, Y., Secular perturbations of asteroids with high inclination and eccentricity, *The Astronomical Journal*,vol.67,pp.591-598,1962.

[3] Winter, O. C. et al, "Controlling the Eccentricity of Polar Lunar Orbits with Low Thrust Propulsion" *Mathematical Problems in Engineering*, vol. on Space Dynamics (2009).

[4] Choueiri, E., A critical history of electric propulsion: the first 50 years (1906-1956), *J. Propulsion and Power*, Vol. 20, n 2, 2004.

[5] Smith, A. W., Cappelli, M.A., On the role of fluctuations, cathode placement, and collisions on the transport of electrons in the near-field of Hall thrusters, *Physics of plasmas*, 17, 093501(2010).

[6] Goebel, D., Katz, I., Fundamentals of electric propulsion: ion and hall thrusters, *JPL*, Pasadena, 2008.

[7] Ferreira, J. L. et. al., Plasma Diagnostic and Performance of a Permanent Magnet Hall Thruster, arXiv:physics/0410170

[8] Santos, D. et. al., Optimal trajectories towards Near-Earth-Objects using solar electric propulsion (SEP) and gravity assisted maneuver, 2008, *J. Aerospace Engineering*, Vol. 1 - 2, pp. 51 - 64.

[9] Murray, C.D. and Dermott, S.F., Solar system dynamics, *Cambridge University Press*,1999.

Treatment of polycarbonate by dielectric barrier discharge (DBD) at atmospheric pressure

K G Kostov[1], Y A A Hamia[1], R P Mota[1], A L R dos Santos[1] and P A P Nascente[2]

[1] Universidade Estadual Paulista – UNESP, 12516-410, Guaratinguetá – SP – Brazil

[2] Federal University of São Carlos – UFSCar, 13565-905, São Carlos – SP – Brazil

E-mail: kostov@feg.unesp.br

Abstract. Generally most plastic materials are intrinsically hydrophobic, low surface energy materials, and thus do not adhere well to other substances. Surface treatment of polymers by discharge plasmas is of great and increasing industrial application because it can uniformly modify the surface of sample without changing the material bulk properties and is environmentally friendly. The plasma processes that can be conducted under ambient pressure and temperature conditions have attracted special attention because of their easy implementation in industrial processing. Present work deals with surface modification of polycarbonate (PC) by a dielectric barrier discharge (DBD) at atmospheric pressure. The treatment was performed in a parallel plate reactor driven by a 60Hz power supply. The DBD plasmas at atmospheric pressure were generated in air and nitrogen. Material characterization was carried out by contact angle measurements, and X-ray photoelectron spectroscopy (XPS). The surface energy of the polymer surface was calculated from contact angle data by Owens-Wendt method using distilled water and diiodomethane as test liquids. The plasma-induced chemical modifications are associated with incorporation of polar oxygen and nitrogen containing groups on the polymer surface. Due to these surface modifications the DBD-treated polymers become more hydrophilic. Aging behavior of the treated samples revealed that the polymer surfaces were prone to hydrophobic recovery although they did not completely recover their original wetting properties.

1. Introduction

Polymers are frequently applied in modern industry owing to their high performance, relatively low cost and easy recycling. Among these materials, polycarbonate (PC) has high creep resistance, high toughness and excellent optical transparency, making it an attractive polymer material. PC is used in variety of applications as viable substitute for glass e.g. automobile headlamps, corrective lenses, plastic vessels, and compact discs. However, the low hardness, low scratch resistance and degradation by ultraviolet (UV) radiation of PC have restricted his application. To remedy these limitations, attempts of producing hard protective coatings on PC have been tested [1]. However the adhesive strength between the deposited film and the PC substrate is poor due to the low surface energy of the polymer. As a matter of fact, many polymers commonly used in industry have disadvantageous surface properties with regards to gluing, printing, painting, inking or coating because of their low surface energy [2]. Therefore, to enhance the surface energy of polymers, their surface should be activated by some kind of chemical or plasma treatment [3].

Plasma treatment is one of the most versatile techniques in surface modification, which can alter the outmost surface layers of polymer without changing its advantageous bulk properties [4]. The plasma is usually generated by ionization of a feed gas that produces positive and negative ions, energetic electrons, UV photons, along with active free radicals. The effect of plasma treatment of polymeric surfaces can range from branching and crosslinking to etching and fictionalization of surface groups. The extent of these effects depends mostly on the processing parameters such as the treatment time, gas pressure, kind of feed gas, discharge power, frequency and so on. The chemical effect of plasma treatment consists of reactions between the plasma species and polymer surface can form different polar groups, such as, -OH, C-O, C-OH, C=O etc. These polar groups cause increasing of the surface energy and an enhancement of the material wetting properties [4, 5]. Atmospheric plasma treatments are especially attractive due to the elimination of expensive vacuum equipment, easier handling of the samples and simple scalability for industrial in-line processing. Therefore, in resent years, a lot of effort has been directed into the development of non-thermal plasma reactors working at (or near) atmospheric pressure. One promising technology for producing atmospheric plasmas is based on the use of dielectric barrier discharge (DBD) [6]. Here, a dielectric barrier covers one or both of the reactor electrodes leaving a small gas gap between them. The DBD discharge can produce large volumes of non-thermal plasma at atmospheric pressure. The mean electron energy is in the range of 1–10 eV, while the chemical binding energy of polymers is less than 10 eV [7]. Therefore, energetic particles in DBD can easily break the chemical bonds of polymers and active radicals abundant in the plasma can react with the surface, modifying its properties. Since plasmas modify the chemical and physical nature of the surfaces, the wettability of polymers is therefore normally altered by the plasma treatment [4,5, 8]. Some authors have investigated the enhancement of PC scratching resistance as well as improvement of polymer wetting properties by low-pressure plasma treatment [9, 10]. More recently atmospheric pressure plasma treatments, such as an air plasma torch [11], a DBD in helium [12] and a homogeneous DBD in nitrogen [8] were used for modification of PC surface properties.

This work describes the surface modification of PC polymer by atmospheric pressure filamentary DBD and compares the efficiency of the treatment in atmosphere of air and nitrogen. The samples utilized in this study were cut from commercially available PC used for protective glasses. Material characterization was carried out by contact angle measurements, surface energy assessments and X-ray photoelectron spectroscopy (XPS).

2. Experimental

The experimental arrangement used to generate DBD discharge at atmospheric pressure is sketched in figure 1.

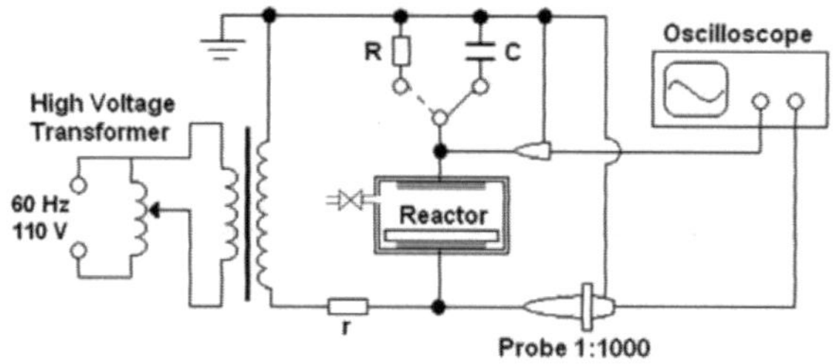

Figure 1: Experimental set-up

The DBD discharge is generated between two 9.5-cm-diam parallel aluminum electrodes enclosed within a 150-mm-diam cylindrical chamber made from Delrin®. A glass layer with 2.0 mm thickness completely covered the bottom reactor electrode. The upper electrode was grounded and the lower one was employed as high-voltage electrode. The distance between the two electrodes was kept constant during the DBD treatment at 3 mm. The cylindrical chamber was connected to a flow regulator for introducing gas precursor into the plasma reactor. The reactor outlet was connected to an exhaust. Before starting the plasma processing the samples (10x15 mm^2), cut from 0.5-mm-thick commercial polycarbonate film, were placed on the lower electrode. After introduction of the samples into the reactor it is flushed with the discharge gas at a rate of 5.0 slm for 5 min. This purging step is performed in attempt to remove the residual air from the reactor. Subsequently the feed gas flux into reactor chamber is maintained at 2.4 slm using a rotameter. In this work two different feed gases (Air Liquid, Brazil) are used: dry air and nitrogen.

Prior the DBD treatment, the samples were ultrasonically cleaned in distilled water and detergent for half an hour. The samples were washed by rinsing in isopropyl alcohol for 10 min to remove organic contaminants from the surface and after that dried at room temperature. The high-voltage power supply is consisted of a step-up high-voltage transformer (110/20000V of V_{rms}, 60Hz) driven by an autotransformer Variac. The AC voltage applied to the reactor lower plate is measured by using a 1000:1 high-voltage probe (Tektronix P6015A, 75MHz) and displayed on a digital oscilloscope (Tektronix TDS 2024B, 200MHz). For each feeding gas all plasma treatments were carried out at a fixed voltage of $30kV_{p\text{-}p}$ and gas flux of 2.4 slm while the treatment time was varying. Following the plasma treatment, the samples were subjected to contact angle measurements and surface chemistry analysis.

The contact angle (CA) was obtained by the sessile drop method on a standard Rame-Hart goniometer, model 300 using the DROPImage software. The measurements were carried out at room temperature in controlled environment kept at relative humidity of 60%. The volume of each liquid drop was 2 μℓ and two different test liquids: de-ionized water and diiodomethane were used. At least five different drops were deposited on the polymer surface to obtain the average value of the contact angles. The maximum error in the contact angle assessments did not exceed 3%. Surface energies were calculated by using the Owens-Wendt method on which polar and non-polar (dispersive) contributions are considered to explain the interaction between the liquid and the solid phases. The geometric-mean method was used to determine the surface energy and its polar and dispersive components using the equation system:

$$\gamma_L(1-\cos\theta_i) = 2\sqrt{\gamma^P\gamma_L^P} + 2\sqrt{\gamma^D\gamma_L^D}$$

where θ_i stands for the contact angle of each test liquid.

Surface chemical characterization was carried out by X-ray photoelectron spectroscopy using a Kratos XSAM HS system. The base pressure in the analyzing chamber was kept below 10^{-7} mbar. As excitation source the Mg K_α line (hν = 1253.6eV) was employed with the emission voltage and current of the source set to 6.0 kV and 5.0 mA, respectively. The obtained spectra were processed by the code provided by the apparatus manufacturer using the Shirley method for background subtraction. All peaks were fitted using Gaussian curves. The value of 284.8 eV of the hydrocarbon C1s core level was used for calibration of the energy scale.

3. Results and discussion

The DBD discharge used for PC treatment operates in filamentary mode i.e. it is constituted by many tiny streamers distributed over entire area of the dielectric barrier. The discharge current consists of large number of short micro-pulses typical for the DBD discharges. To calculate the discharge power of the DBD, the current measuring resistor was replaced by a capacitor C = 0.91 μF. The charge $Q(t)$ stored on the electrodes is represented as a function of $V_p(t)$ for one period to form so-called Lissajous figure. In case of filamentary DBD driven by a sinusoidal power source the typical shape of its *Q-V* curve is a parallelogram [6]. The area enclosed by the Lissajous figure gives the electrical energy consumed for voltage cycle. The discharge power P is then calculated by multiplying the electrical energy by the frequency of the applied voltage (60 Hz). Figure 2 represents the *Q-V* figures of DBD in different gases. As can be seen in figure 2 the shape of the Lissajous figures depends on the kind of feeding gas however the discharge energy per cycle, represented by Lissajous's figure area in both cases is almost the same. The mean discharge power measured for DBD in nitrogen and air gas precursors was 3.3 W, and 2.8 W, respectively.

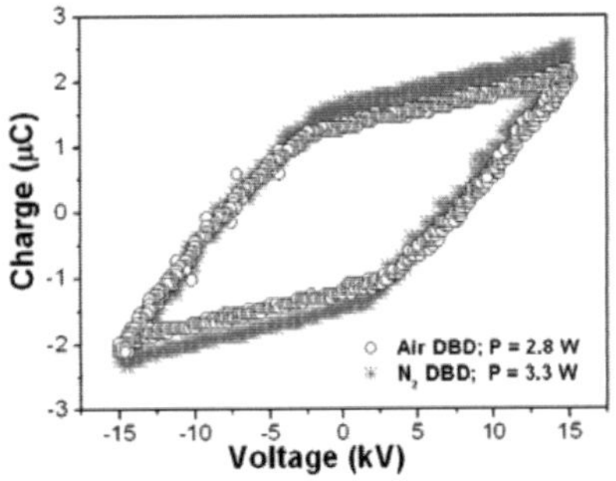

Figure 2: Q-V diagrams for DBD in air and nitrogen.

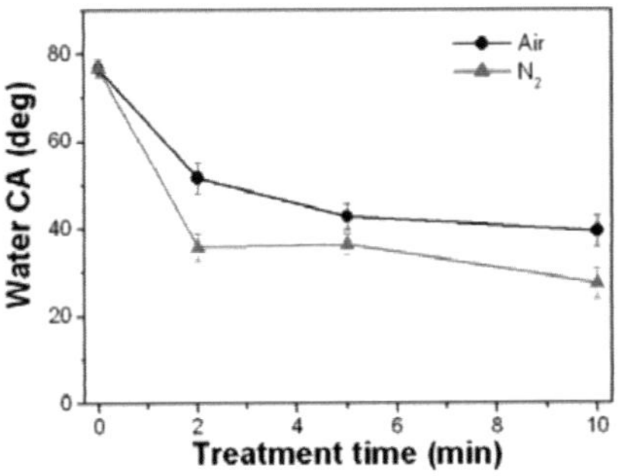

Figure 3: Water CA of PC samples measured soon after the treatment.

To investigate how surface wetting properties of PC are affected by the plasma exposure PC samples were treated for 2, 5 and 10 min using different feed gases. The PC water contact angle measured soon after the DBD is shown in figure 3. As expected all plasma treatments lower the water CA with respect to the untreated PC sample. The initial CA of pristine PC (76°) was reduced to about 40° after 10 min air plasma treatment. No significant changes in the contact angle with further increase of the plasma exposure time were detected. The DBD treatment in N_2 seems to be more efficient in reducing water CA of PC, reaching saturation value of about 30° after 10-min N_2 plasma exposure.

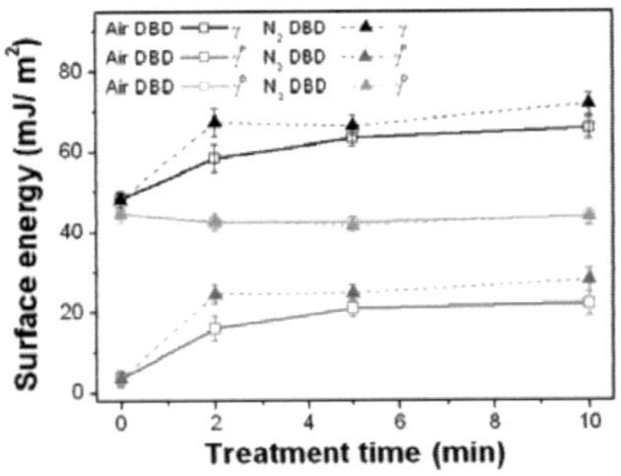

Figure 4: Variation of PC surface energy and its polar and dispersive components.

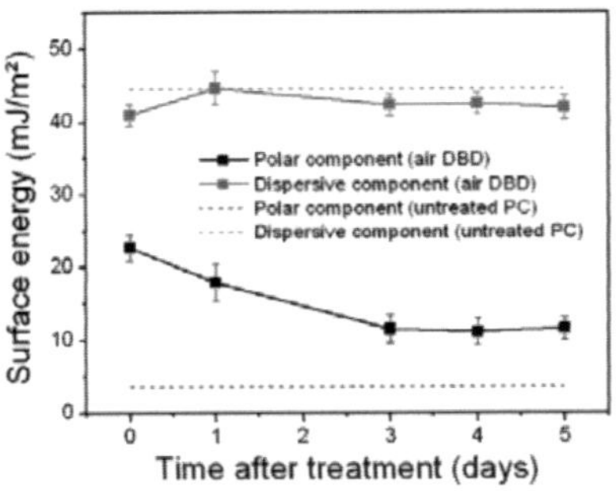

Figure 5: Evolution of SE polar and dispersive components.

Figure 4 shows the variation of surface energy as well as its polar and dispersive components of PC samples treated with air and N_2 DBD. The surface energy of pristine PC was 48.2 mJ/m^2 with dispersive and polar components of 44.6 mJ/m^2 and 3.6 mJ/m^2, respectively. After 10 min of plasma treatment the surface energy of PC samples increased up to 71.8 mJ/m^2 and 66.0 mJ/m^2 for N_2 and air plasma, respectively. As can be seen in figure 4 the increase of polymer surface energy is mainly associated with the increase of its polar component. They increase with the treatment time reaching saturation values beyond 5 min of treatment. Generally, for most polymeric materials the DBD treatment results in no or little variation of the dispersive component of surface energy [5]. Based on this it can be concluded that the polar component plays an important role in the improvement of surface energy of DBD treated polymers. This enhancement of the surface energy polar component is due to the oxygenation of the polymer surface by the plasma. Reactions between the plasma species and the polymer chain introduce polar oxygen groups (C-O, C-OH, C=O, O-C-O, O-C=O) on the surface, which results in enhancement of polar component of the surface energy [8]. Nitrogen plasma turns out to be more efficient in promoting enhancement of the surface energy polar component, which results in better wetting properties of PC surface DBD. This finding will be further discussed when presenting results from XPS analysis.

It is known that the polymer surfaces treated by plasma and left in ambient conditions tend to partially recover their original wetting properties [13].This process is referred to as hydrophobic recovery Contact angle measurements with both test liquids were made immediately following the plasma treatment and after 1, 2, 3, 4 and 5 days of storage at controlled ambient (temperature of 21°C and 60% relative humidity).

The evolution of both components of PC surface energy can be observed in figure 5. This surface aging effect is attributed to the orientation of polar groups located on the surface into the bulk of the material and diffusion of nonpolar groups from the volume to the surface [13]. Both processes result in a partial reduction of the surface energy polar component while the dispersive component remains practically unchanged (see figure 5). The process of hydrophobic recovery predominantly occurs during the first couple of days after the treatment and further aging seems to be less important.

Table 1. Atomic composition of PC samples

	C (%)	O (%)	N (%)	O/C	N/C
Untreated PC	80.51	19.49	-	0.24	-
Air plasma	70.68	28.03	1.29	0.40	0.02
N_2 plasma	69.25	27.18	3.57	0.39	0.05

When polymers are exposed to plasma the plasma energetic species (electrons and UV photon) preferentially break the weaker C-C and C-H bonds on the surface. Atmospheric plasmas are abundant of reactive oxygen atoms which react with the dangling bonds on the polymer surface forming different species such as mono-oxidized C-O, C-OH carbons, bi-oxidized C=O or O-C-O carbons and tri-oxidized O-C=O carbons [5]. These oxygen-related polar groups are responsible for the enhanced hydrophilicity of the DBD treated polymers [8]. Surface atomic composition of the plasma-modified polymers, determined by XPS analysis, is shown in table 1. The measurements were performed at least 2 weeks after the DBD treatment therefore the XPS analysis presents the atomic composition of polymeric surfaces when they had already underwent the process of hydrophobic recovery and the surface chemical composition reached a state-state. As can be seen in table 1 both plasma treatments lead to increase of polymer O/C ratio to about the same extend. Since our experiment is conducted in an open environment it is impossible to ensure DBD discharge in pure N_2 atmosphere. Due to residual air trapped in the reactor chamber there are always some oxygen impurities in the discharge. Even in little quantity as in case of N_2 DBD, the oxygen atoms are very reactive and they are incorporated onto the polymer surface. Also, after the treatment all samples are kept in ambient conditions where unsaturated bonds on the polymer surface can be subjected to reactions with the atmospheric oxygen. That is why the oxygen content on the surface of polymers treated in different discharge gases is about the same. Nevertheless the O/C ratio is almost the same for both treatments the DBD in N_2 gas shows higher efficiency in reducing water CA as shown in figure 3. This finding may be explained by the fact that the DBD in nitrogen plasma added more N atoms on the PC surface (see Table 1).

The characteristic photoelectron C1s peak obtained by XPS for PC before and after the DBD treatments is shown in figure 6. All peaks were resolved into 4 components C1 (C-C/C-H), C2 (C-O), C3 (C=O or C(=O)-N) and C4 (O-C(=O)-O). The peaks binding energies and relative areas are shown in figure 6. For both DBD-treated PC samples the intensity of C1 component decreases while the intensity of the other tree components increases as observed by other authors [8-10]. As can be seen in figure 6 the composition of the oxidized carbons species on the surface of PC samples DBD treated in air and N_2 is practically the same. For the N_2 DBD in the originally empty N 1s region appears a new peak at 399.8 eV which is assigned to the amide group C(=O)-N. Similarly for air DBD a new N 1s peak develops that can be decomposed into two components N1 at 399.7 (amide group, area 72%) and a smaller component at 401.7 eV with 28% relative area assigned to the protonated amine [3].

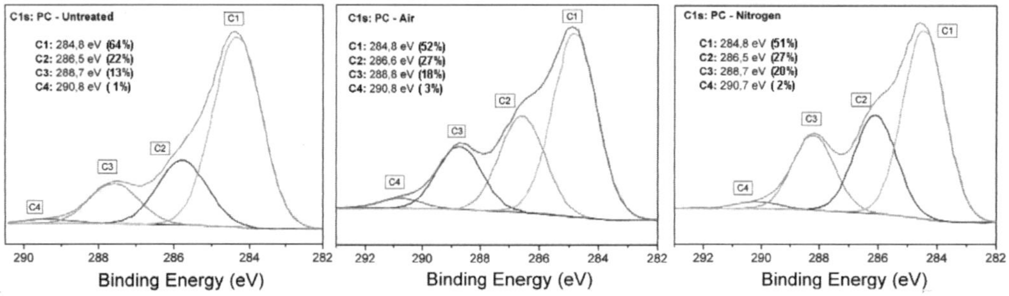

Figure 6: Deconvolution of C1s peak of PC samples.

4. Conclusions

Polycarbonate samples were subjected to atmospheric pressure DBD plasma in air and nitrogen. Both treatments result in incorporation of polar oxygen and nitrogen groups on PC surface which leads to enhancement of material wetting properties. The DBD process in nitrogen turns out to be more efficient probably due to the higher amount of N atoms added on the surface. After the treatment the samples were stored in air at ambient temperature to study the aging behaviour of plasma-modified samples. The polymer surfaces are prone to hydrophobic recovery however after 5 days of storage they only partially restore their original wetting properties.

Acknowledgments

This work was supported by FAPESP, CNPq and FUNDUNESP under grants 2009/05329-5, 471928/2008-9 and 346/2010, respectively.

References

[1]. Zajickova L, Bursikova V, Perina V, Mackva A, Subedi D, Janca J, Smirnov S, 2001 *Surf. Coat. Technol.* **142-144** 449.

[2]. Bismarck A, Brostow W, Chiu R, Lobland H E H, Ho K K C, 2008 *Polymer Engin. Sci.*, 1971.

[3]. Toth A, Kereszturi K, Mohai M, Bertoti I, 2010 *Surf. Coat. Technol.* **204** 2898.

[4]. Lai J, Sunderland B, Xue J, Yan S, Zhao W, Folkard M, Michael B D, Wang Y, 2006 *Appl. Surf. Sci.* **252** 3375.

[5]. Kim H. Jung S J, Han Y H, Lee H Y, Kim J H, Jang D S, Lee J J, 2008 *Thin Solid Films* **516** 3530.

[6]. Wagner H E, Brandenburg R, Kozlov K V, Sonnenfeld A, Michel P, Behnke J F, 2003 *Vacuum* **71** 417.

[7]. Kogelschatz U, 2003 *Plasma Chem. Plasma Process.* **23** 1.

[8]. Šira M, Trunek D, St'ahel P, Bursikova V, Navratil Z, 2008 *J. Appl. Phys. D* **41** 015207.

[9]. Pelagade S, Singh N L, Shah S, Qureshi A, Rane R S, Mukherjee S, Deshpande U P, Ganesan V, Shripathi T, 2010 *J. Phys. Conf. Serie* **208** 012107.

[10]. Qureshi A, Shah S, Pelagade S, Singh N L, Mukherjee S, Tripathi A, Deshpande U P, Shripathi T, 2010 J. *Phys. Conf. Serie* **208** 012108.

[11]. Abenojar J, Torregrosa-Coque R, Martinez M A, Martin-Martinez J M, 2009 *Surf. Coat. Technol.* **203** 2173.

[12]. Sharma R, Holcomb E, Trigwell S, Mazumder M, 2007 *J. Electrostatics*, **65** 269.

[13]. Geyter N D, Morent R, Leys C, Gengembre L, Payen E, 2007 *Surf. Coat. Technol.* **201** 7066.

Effects of the magnetic field over the nanometric growth morphology on the material synthesis in a liquid spray thermal plasma reactor

L. Beiras , G. Torrente [1], J. Puerta , F. Blanco
Departamento de Física, [1] Departamento de Mecánica,
Universidad Simón Bolívar, Apdo: 8900, Caracas - Venezuela
E-mail: laurischiq@gmail.com, gtorrente@usb.ve, jpuerta@usb.ve, fblanco@usb.ve

Abstract. It is possible that the magnetic field affect the growth morphology of the materials at nanometric scale while there are synthesized. In this work is developed a thermal plasma reactor in fluidized bed assisted by magnetic mirror for material synthesis using a liquid spray. An aluminum solution is carrier in the plasma reactor with a without magnetic external magnetic field applied. We found from the characterization of material synthesized that: Aluminum nanowire and alumina nanofiber are generated when the external magnetic mirror is applied to the thermal plasma reactor.

1. Introduction

Recently the interest on the works of investigation about the synthesis under the influence of electric and magnetic fields has increased. Possibly this interest has been generated by the advances on investigations of nanostructured materials. Since 2002 in Lomonosov Moscow State University the professor Khomutov [3] studied the effect of applies electric and magnetic fields in the growths of amorphous colloidal particles. He found that the size and the forms of nanoparticles synthesized changes dramatically aligning itself when a external magnetic field parallel to the plane is applied during the synthesis. He accomplished that to control the morphology of nonstructural materials by synthesis under magnetic fields will be a promising proposal for the engineering of nanofaces and nanotech. In 2005 other authors notified the modification of morphology growths and crystalline direction by the applications of external magnetic fields, among them L. Bârdos [1] from Uppsala University in Sweden, he informed about the modification of the morphology of growth of coating of TiN by the application of magnetic fields, and Takahashi [4] from Yamagata University in Japan reported the synthesis of carbon fibers aligned by the application of magnetic fields. Nowadays plasma spraying is a good established technology used in laboratory and in the nanofacturing. Several interesting uses are in biomedical, petrochemical, etc [2]. In our case this phenomena that can govern the formation of the plasma jet inside the thermal reactor has been simulated with and without magnetic field generated by a mirrors type machine in order to improve [5] the plasma conditions.With the magnetic field we have observed better properties in nanostructured material. Around several nanometers has been obtained for this system through the machine we use. In the reactor its routing the liquid using a peristaltic pumps and in a conventional (DC) plasma torch. This paper show the results of this experiments using Al in solutions.

2. Experimental Model

To perform good predictable capabilities of our plasma torch experiments, 2D simulation has been carried out, but in order to make this job in a better way 3D simulations must be made in order to understand the physics here involved assuming thermodynamic equilibrium in the stationary phase. In our experiment the gas mixture is N_2 with a nozzle of 5mm in diameters and the liquid injection is made transversally into the plasma via a peristaltic pump. Good spray condition was obtained with the parameters given by the table.

The reactor parameters :

V_{EF},I_{EF}	$220Volts$, $54Amp$
$\dot{m}_g$	Nitrogen $13.2lpm$, $40psi$
R_r	$0.7874inch$
$\dot{m}_p$	Aluminium $22\mu m$, $0.3g/s$

The experimental procedure was realized using an stainless steel cylindrical reactor. Inside this one was placed a magnetic coil(see fig. 1) and in backside of the reactor we find the plasma torch, " Plasjet 105/15" by Thermal Dynamics of 15Kw total power. Connected to the system we have a peristaltic pump in order to introduce the neutralized Al-solution. In fig. 2 we show a schematic of the complete experiment. The magnetic field intensity inside of the magnetic coil along the z - axis, was realized with a gauss meter and shown in fig.3. The structure of the field present the minimum B field configuration. This type of field was obtained due to the positive susceptibility of the bobbin(made of steel 1050), in such a way that a counter magnetic field is achieved near of the center and quasi same sense at both sides(front and back) of the coil. We speak also, about of a mirror field configuration.In our point of view, that is a nobel experiment considering the literature in this area of plasma spray. The solution of Al was prepared using HCl and after this neutralized with NaOH in the appropriate Al concentration useful for our experiment. We analyze the products via TEM. Figure 1 - 4 shown some of the preliminary results we obtained.

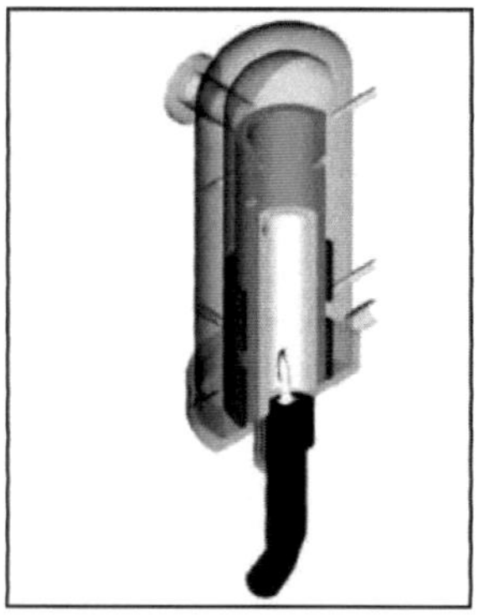

(a) Thermal Plasma Reactor Assisted by a Magnetic Mirror Configuration in Fluidized Bed

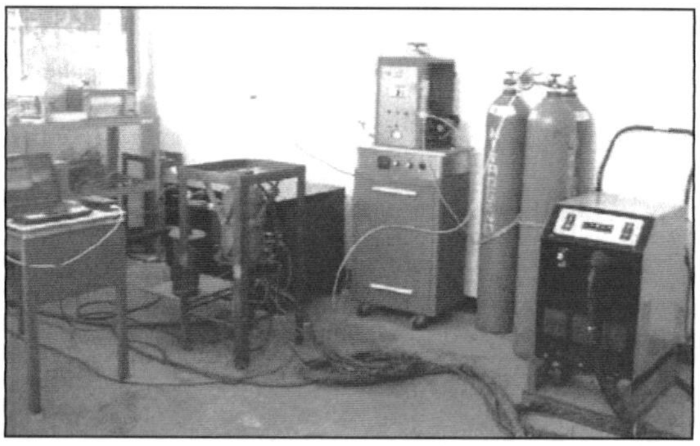

(b) Experimental Setup

Figure 1.

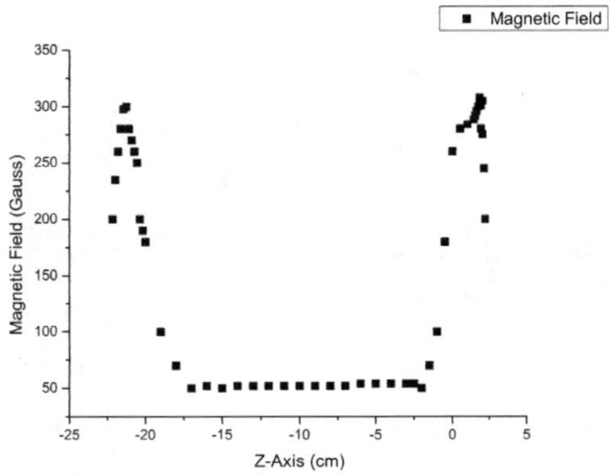

Figure 2. Magnetic field measurement inside the coil using a gauss-meter model YOKOGAWA 325i

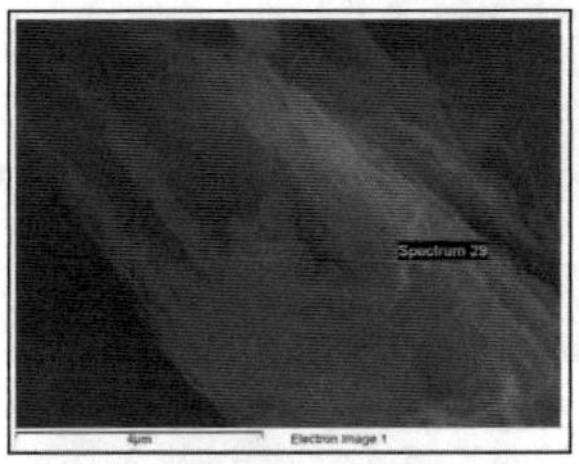

(a) Espectro 1

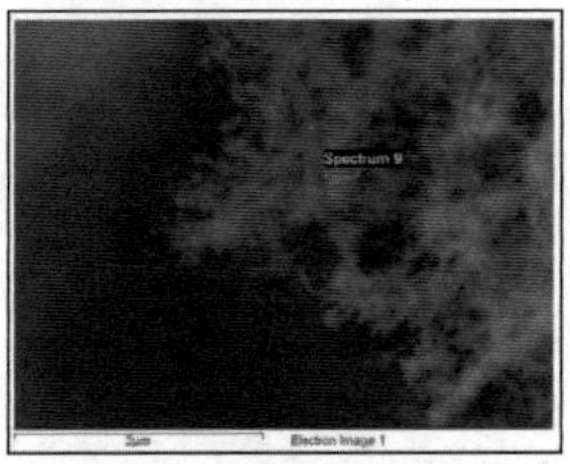

(b) Espectro 2

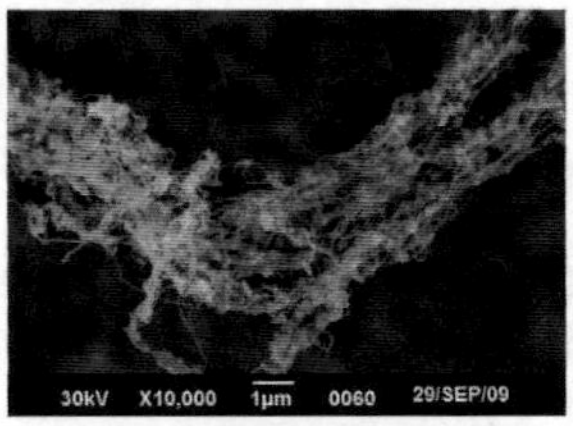

(c) Espectro 3

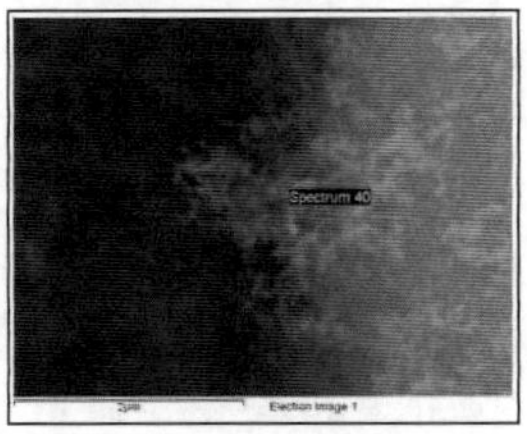

(d) Espectro 4

Figure 3. Espectros

3. Conclusions

In figures 3 we shown some results from our past and present preliminary experiments. The results has been very successfully.

The plasma temperature profile (Fig.4) inside the reactor assisted by a magnetic field shows that the highest temperature are near the plasma torch and decrease along the plasma, far away from the plume. The electron temperature profile inside the reactor assisted by a magnetic field shows that the highest electron temperature are near the plasma torch and better confined along the reactor in comparison with the case without magnetic field.

It is interesting to mention here, that a result of our past simulation(paper Bochum), the heating of the aluminium powder in the center is higher than near the wall of the reactor. That is due to the fact that in the center the plasma temperature of the plume is higher than near the

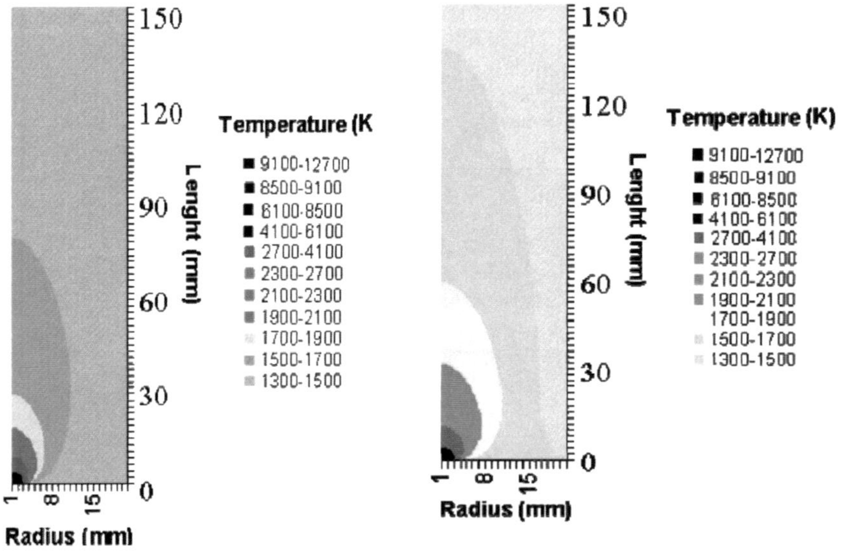

Figure 4. Electron Temperature profile without and with magnetic field

wall and also due to the residence time for the powder in the reactor is longer than for lighter particles. Finally it is to conclude, that the application of the magnetic mirror reduces the plasma flow velocity on radial direction, displacing it to the reactor's walls and confining it partially. These decrease the heat losses in the gas flow and mainly on the electron flow. The diminution of the axial and radial plasma velocity by effect of the magnetic field, directly affects the collision frequency between heavy particles and electrons. Therefore the collision frequency is the main mechanism of energy transfer between these components in the plasma. These two factors shift away the plasma from thermal equilibrium increasing the temperature difference between the heavy particles and electrons, observing a slight increase of heavy particles temperature but a greater increase on electronic temperature.Figure 4 show the effect of the magnetic field in the plume via a computer simulation where we observe that the electron temperature increase alon the z-axis compared with the case with no magnetic field.

4. Acknowledgments
Work supported by DID, Universidad Simón Bolívar

5. References
[1] L. Bârdos, L. E. Gustavsson, and H. Barnkov. Effect of ferromagnetic substrates on the film growth in magnetized plasma systems. *Surface and coatings technology*, 200:1862–1866, 2005. Issues 5-6.

[2] J.R. Davis, editor. *Handbook of thermal spray technology*. ASN International, 2004.

[3] G. B. Khomutov, S. P. Gubin, V.V. Khanin, A. Yu. Obydenov, V. V. Shorokhov, E.S. Soldatov, and A.S. Trifonov. Formation of nanoparticles and one-dimensional nanostructures in floating and deposited langmuir monolayers under applied electric and magnetic fields. *Colloids and Surface A*, 198-200:593–604, 2002.

[4] Tatsuhiro Takahashi, Taichi Murayama, and Ayumu Higuchi. Aligning vapor-grown carbon fibers in polydimethylsiloxane using dc electric or magnetic field. *Carbon*, 44:1180–1188, 2006. Issues 7.

[5] G. Torrente, J. Puerta, and N. Labrador. Two- temperature single-flow model of a thermal plasma reactor in fluidized bed assisted by magnetic mirror. *Physica Scripta*, T131(014010), 2008.

Journal of Physics: Conference Series **511** (2014) 012077 doi:10.1088/1742-6596/511/1/012077

Titanium oxidation by rf inductively coupled plasma

R Valencia-Alvarado[1], A de la Piedad-Beneitez[2], R López-Callejas[1,2], S R Barocio[1], A Mercado-Cabrera[1], R Peña-Eguiluz[1], A E Muñoz-Castro[1], B G Rodríguez-Méndez[1], and J M de la Rosa-Vázquez[3]

[1] Instituto Nacional de Investigaciones Nucleares, Plasma Physics Laboratory, AP 18-1027, 11801, México, D.F., Mexico

[2]Instituto Tecnológico de Toluca, AP 890, Toluca, Mexico, Mexico

[3]ESIME-ZACATENCO-IPN, 07738, Mexico, D.F., Mexico

E-mail: raul.valencia@inin.gob.mx

Abstract. The development of titanium dioxide (TiO_2) films in the rutile and anatase phases is reported. The films have been obtained from an implantation/diffusion and sputtering process of commercially pure titanium targets, carried out in up to 500 W plasmas. The experimental outcome is of particular interest, in the case of anatase, for atmospheric pollution degradation by photocatalysis and, as to the rutile phase, for the production of biomaterials required by prosthesis and implants. The reactor employed consists in a cylindrical pyrex-like glass vessel inductively coupled to a 13.56 MHz RF source. The process takes place at a 5×10^{-2} mbar pressure with the target samples being biased from 0 to -3000 V DC. The anatase phase films were obtained from sputtering the titanium targets over glass and silicon electrically floated substrates placed 2 cm away from the target. The rutile phase was obtained by implantation/diffusion on targets at about 700 ^{o}C. The plasma was developed from a 4:1 argon/oxygen mixture for ~5 hour processing periods. The target temperature was controlled by means of the bias voltage and the plasma source power. The obtained anatase phases did not require annealing after the plasma oxidation process. The characterization of the film samples was conducted by means of x-ray diffraction, scanning electron microscopy, x-ray photoelectron spectroscopy and Raman spectroscopy.

1. Introduction

Titanium dioxide (TiO_2) thin films are of considerable interest because of their outstanding chemical inertness, thermal stability, corrosion resistance, mechanical, optical, electrical, and electronic properties, as well as excellent photocatalytic applications in water purification, the treatment of air and the development of biocompatible coatings in human implants [1-7]. Titanium dioxide structured thin films have been processed using a wide range of deposition and implantation methods such as sputtering [8], chemical vapor deposition [1,5], sol-gel [4], laser ablation [9], plasma immersion ion implantation [10], inductively coupled plasma [2, 11], or atmospheric pressure plasma spraying [6] among other techniques.

The reactive direct current magnetron sputtering may well be the most employed method to obtain TiO_2 structures as it allows the control of the composition and properties of the TiO_2 films through the deposition conditions [12-13]. This control leads to get high quality films with thickness uniformity and controlled stoichiometry. TiO_2 biocompatibility properties, widely used in orthopedic and dentistry surgical implants, depend on the treated layer being thick enough. The photocatalytic property of titanium dioxide takes place when it is illuminated by photons with energies higher than the TiO_2 band gap one. This process generates electron-hole pairs required for the photocatalytic action.

In this paper we describe the simultaneous growth of thin films in the anatase and rutile phases through an implantation/diffusion and sputtering process in an inductively coupled radio frequency plasma of calibrated argon/oxygen gases mixtures without post annealing. The influence of both the deposition time, and bias on the film morphology, composition and structure have been analyzed using different techniques such as x-ray diffraction (XRD), scanning electron microscopy (SEM), x-ray photoelectron spectroscopy (XPS) and Raman spectroscopy. The aim of the experiment is studying the feasibility of obtaining anatase and rutile TiO_2 films as a function of target and substrates temperatures.

2. Experimental procedures

The implantation and sputtering process was conducted within a pyrex-like glass cylindrical vessel 50 cm long and 20 cm in diameter around which a 3 mm diameter copper pipe antennae is wound helicoidally. The plasma is developed from calibrated argon/oxygen (80%/20%) gas mixtures by transferring 500 W to the antennae through an automatic coupling network from a 1.2 kW RF generator where induction works as in a transformer with the plasma playing the role of a secondary coil. Fig. 1 exhibits this experimental set up, with its substrate holder, reported elsewhere [14-15].

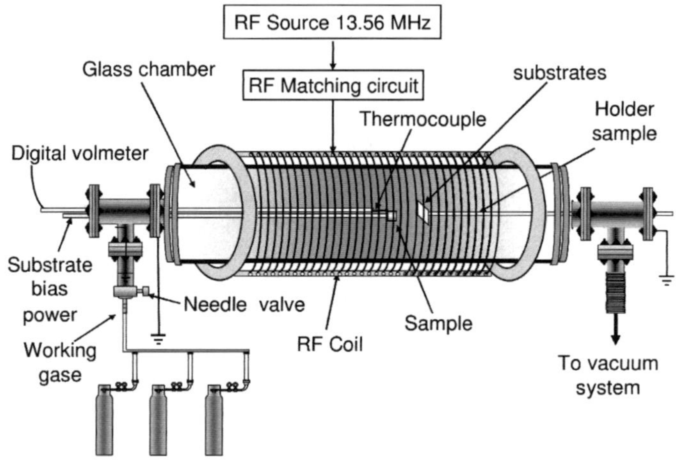

Fig. 1. Diagramme of the plasma reactor

The vessel was evacuated at pressures $<5\times10^{-6}$ mbar by a turbomolecular pump operating at 500 l/s. After attaining the base pressure, the work argon/oxygen mixture was admitted through a precision leak valve up to 5×10^{-2} mbar pressure at which the highest plasma density is obtained. Titanium commercially pure (CpTi) cylindrical samples, 9 mm in diameter and 5 mm long, were introduced to the vessel and oxidized by the plasma during periods up to 5 hours by means of a sputtering and implantation/ diffusion process. The samples were biased up to -3000 V DC. In the unbiased case, the sample temperature was kept about 250 °C by the action of the plasma alone and no significant evidence of either sputtering or oxidation was found. In the opposite -3000 V DC case, the ion bombardment raises the sample temperature around 700 °C, as measured by a commercial K type thermocouple. Insulation between the biased piece and the thermocouple-reading meter is provided by an isolation amplifier. The electrically floated silicon and glass substrates were placed on the holder 2 cm away from the CpTi sample, becoming thermalised between 250 °C (by the plasma alone) and 290 °C, during the oxidation process. Both the titanium samples target and the glass/silicon substrates were finally cut in cross sections in order to measure their TiO_2 film depth. XRD and Raman spectroscopy techniques, used to characterize the oxidised samples, indicate the growth of the anatase and rutile phases. XRD spectra were used to estimate the mean crystallite size from the Scherrer equation. The surface morphologies and titanium/oxygen atomic percentage on sample and silicon sputtering films were characterized by SEM and XPS give ratio for purely TiO_2 stoichiometry.

3. Results

The depth and surface morphology of implantation and deposited films, so much in the titanium target just as on glass and silicon substrates were analyzed by SEM. Figs. 2.(a) and 2.(b) shows images of cross sections of titanium dioxide thin films resulting from 5 hours processing. Fig. 2.a represents the case of implantation/diffusion when the titanium target was biased around -3000 V DC for 5 hour; the layer depth is between 14 and 20 μm with an atomic concentration rate of 73 % at of oxygen and 27 % at titanium. Fig. 2.(b) corresponds to silicon coating by sputtering. The layer is in the order of 2.5 μm and the deposition rate ~6 nm/min. Fig. 2.(c) shows the anatase phase microstructure on the silicon substrate.

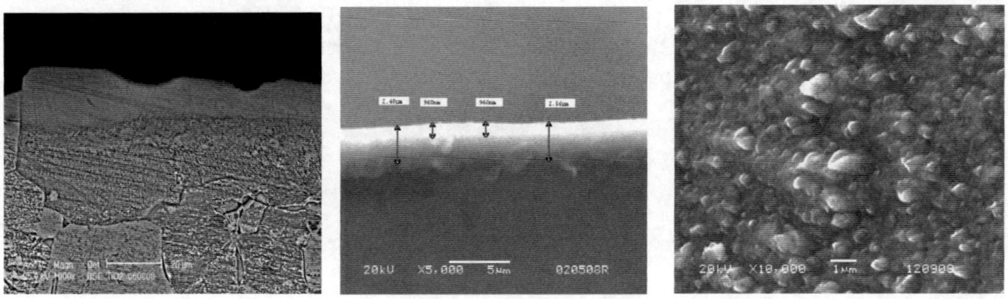

a b c

Fig. 2. Layer depth: (a) implantation/diffusion, (b) silicon coating by sputtering and (c) anatase phase microstructure.

The coating was analysed using the technique electron dispersive spectroscopy (EDS) in conjunction with the SEM. The atomic percentages found in average in silicon were ~55 % at. of oxygen and 35 at. % of titanium. The other hand, different percentages were identified in the glass (SiO_2+CaO+Na_2O) samples (70 at. % oxygen and 23 at. % titanium), indicating the chemical composition of the film albeit not the stoichiometric ratio of two parts of oxygen per one of titanium. The sample surface and substrates change in colour during the oxidation process according to the temperature and the bias voltage of the treatment. It is observed that the colours are directly related to the composition of the oxide formed as well as to the thickness of the oxide layer. As the bias voltage varies along the 0 to -3000 V DC range, the titanium sample surface typically acquires a dark blue shade which turns silvery and, by the end of the process, into opaque lead grey. This evolution indicates, in an empirical way, the achievement of oxidation in the rutile phase.

Figure 3 shows the relative atomic composition of the anatase phase on a silicon substrate which was analysed with EDS. The spectrum point out the presence of stainless steel (Fe, Ni, Cr) attributable to sputtering from the sample holder.

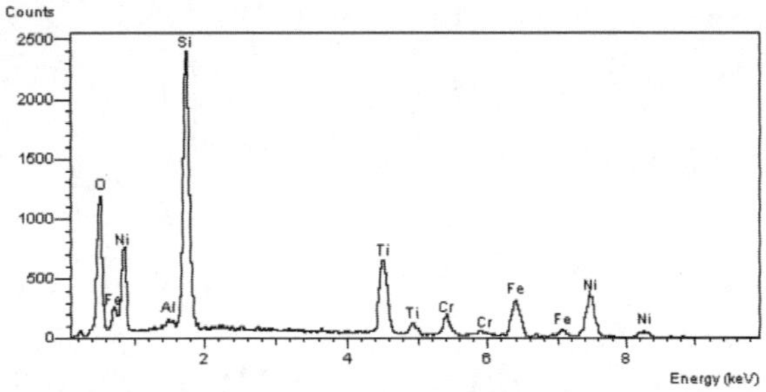

Fig. 3 EDS analysis of film anatase deposited on silicon substrate

The case of the glass and silicon substrates involves the appearance of a multicoloured circular pattern, ranging from purple red to olive green, pointing to the existence of crystalline phases acting as prisms.

Furthermore, the multiplication of such concentric strips suggests an increase of the crystalline deposited film depth as the processing time goes by. Finally, the temperature level of the treatment excludes most other phases than the anatase one.

The new crystalline phases of the titanium sample surface modified by implantation and diffusion have been analysed by XRD using CuK_α radiation within a 2θ=20° to 90° scan range. Fig. 4 displays the XRD patterns of the 5×10^{-2} mbar mixture pressure sample at -3 kV of target bias. This spectrum reveals that the rutile phase is present at R(110)=27.7°, R(101)=36.2°, R(111)=41.5°, R(211)=54.6°, R(301)=70° and, for the titanium target Ti(100)=35.1, Ti(002)=38.4, Ti(101)=40.3, Ti(102)=52.6, Ti(110)=62.8, Ti(200)=74.3, Ti(112)=76.2 and Ti(201)=77.4. Given the Scherrer equation, the grain size of the TiO_2 film in the rutile phase implanted with the preferential crystalline orientation (110) was 44 nm.

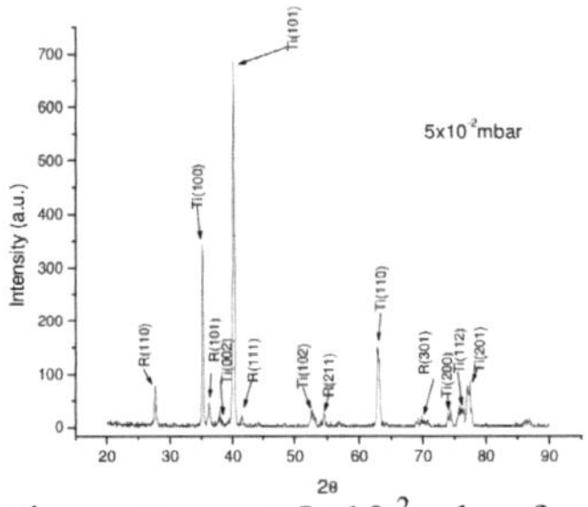

Fig. 4. X-ray diffraction patterns at 5×10^{-2} mbar from de oxidized target.

The deposited films on the glass/silicon substrates were characterized by Raman spectroscopy. The spectrum from glass showed in Fig. 5.(a), confirms the presence of the anatase phase grown at temperatures below 300 °C it is agree with Löbl [16] and Mändl [17]. The Raman amplitudes at wavelengths 150.3, 400.6, 523.5 and 643.2 cm^{-1} correspond to the anatase phase. The resulting ~6 nm/min deposition rate is considered acceptable in this kind of oxidation process. Fig. 5.(b) describes a similar growth on silicon. The differences at wave lengths between Figs 5.(a) and 5.(b), are due to the microstructural from the grain size. Similar results have been reported by de Vicente and Oja [18-19].

A Thermo Scientific* K-Alpha spectrometer was applied to the characterization of TiO_2 films deposited on glass/silicon substrates. The high resolution XPS spectrum (not shown) corresponding to the silicon surface was revealed by a main doublet composed of two symmetric peaks situated at 458 eV and 464 eV corresponding to Ti2p$_{3/2}$ and Ti2p$_{1/2}$ respectively, and 529 eV for O1s.

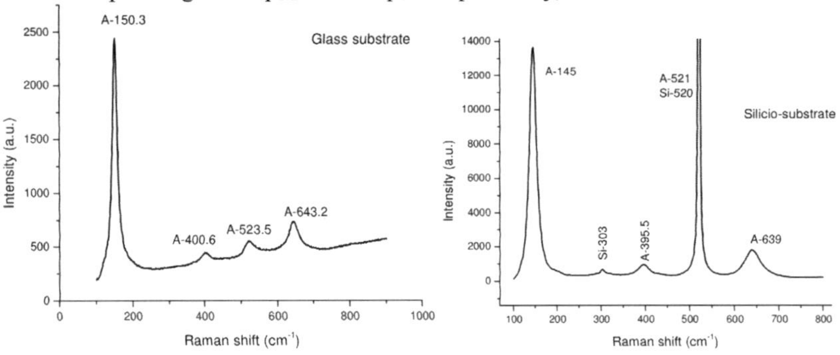

a **b**

Fig. 5. Raman spectra of anatase phase on substrates: a) glass, b) silicon.

4 Conclusions

An inductively coupled plasma source supplied by a specifically designed and built RF generator has allowed obtaining TiO_2 films in the rutile phase by means of the implantation/diffusion of oxygen atoms into a titanium target while, at the same time, films in the anatase phase were developed by titanium sputtering with argon ions. Atomic concentration rates above 2 are due to the presence of TiO-like oxides in addition to the TiO_2 An oxygen atomic percentage of 55 at.%, versus a 35 at.% of titanium leads to a stoichiometry well below to 2. By contrast, percentages on the glass are 70 at.% vs 23 at.%, respectively, which can be attributed to the silicon oxide composition of the glass itself. Colour changes of the samples during their implantation/diffusion processing suggest the evolution of their composition and structure. Thus, a transition from deep blue to silvery white is typically achieved by the end of the process, frequently accompanied by iridescent concentric circular patterns whose number can be related with the deposited film depth. The deposited anatase phase did not require any additional annealing after the oxidation.

Acknowledgements
This project has been partially funded by CONACyT and DGEST, México. The authors are grateful to Dr. Manuel Espinoza and Mr. Jorge Pérez for their technical support in the XRD and EDS diagnostics, respectively. The technical assistance received from Pedro Ángeles Espinoza, María Teresa Torres M. and Isaías Contreras Villa is much appreciated.

References

[1] Baroch P, Musil J, Vlcek J, Nam K H and Han J G 2005 *Surf. & Coat. Tech.* **193** 107-11
[2] Li Z G, Wu Y X and Miyake S 2009 *Surf.& Coat. Tech.* **203** 3661-68
[3] Boukrouh S, Bensaha R, Bourgedis S, Finot E and Marco de Lucas M C 2008 *Thin Solid Films* **516** 6353-58
[4] Jiaguo Y, Xiujian Z, Qingnan Z and Jincheng D 2000 *Chinese J. of Mat. Research* **14-2** 203-
[5] Mc Curdy P R, Sturgess L J, Kohli S and Fisher E R 2004 *Appl. Surf. Sci.* **233** 69-79
[6] Chu P K 2007 *IEEE Trans. on Plasma Sci.* **35-2** 181-7
[7] Lee S H, Kim H W, Lee E J, Hi L H and Kim H E 2006 *J. of Biomaterials Applications* **20** 195-208
[8] Toku H, Pessoa R S, Maciel H S, Massi J and Mengui U A 2008 *Surf. & Coat. Tech.* **202** 2126-31
[9] Matsubara M, Yamaki T, Itho H-S, Abe H and Asai K 2003 *Japanese J. Appl. Phys.* **42** L479-L81
[10] Mandl S, Thorwarth G, Schreck J, Stritzker B and Rauschenbach B, 2000 *Surf. & Coat. Tech.* **125** 84-8
[11] Okimura K and Furumi T 2005 *Japanese J. of Appl. Phys* **44-5A** 3192-95
[12] Yamagishi M, Kuriki S, Song P K and Shigesato Y 2003 *Thin Solid Films* **442** 227-31
[13] Eufinger K, Janssen E, Poelman H, Poelman D, de Gryse R and Marin G 2006 *Thin Solid Films* **515** 425-29
[14] Valencia-Alvarado R, de la Piedad-Beneitez A, López-Callejas R, Barocio S R, Mercado-Cabrera A, Peña-Eguiluz R, Muñoz-Castro A E, de la Rosa-Vázquez J 2009 *Vacuum* **83** S264-7
[15] Valencia-Alvarado R, de la Piedad-Beneitez A, de la Rosa-Vázquez J, López-Callejas R, Barocio S R, Mercado-Cabrera A, Peña-Eguiluz R and Muñoz-Castro A E 2010 *Int. J. Nanomanufacturing* **5-1/2** 62-8
[16] Löbl P, Huppertz M and Margel D 1994 *Thin Solid Films* **251** 72-9
[17] Mändl S, Thorwarth G, Schreck M, Stritzker B and Rauschenbach B 2000 *Surf. & Coat. Tech.* **125** 84-8
[18] de Vicente F S, Bernardi M I B, Leite F L, Herrmann P S P and Hernandes A C 2006 *Rev. Bras. de Aplica. de Vácuo*, **25-1** (2006) 24-30
[19] Oja I, Mere A, Krunks M, Nisumaa R, Solterbeck C-H and Es-Souni M 2006 *Thin Solid Films* **515-2** 674-77

Optical emission spectroscopy of Aluminum Nitride thin films deposited by Pulsed Laser Deposition

J A Pérez[1,2*], **L P Vera**[1,2], **H Riascos**[1,3], **J C Caicedo**[4]

[1] Universidad Tecnológica de Pereira, Grupo Plasma Láser y Aplicaciones, Colombia
[2] Ingeniería Física, Universidad Tecnológica de Pereira, Colombia
[3] Departamento de Física, A.A 097, Universidad Tecnológica de Pereira, Colombia
[4] Grupo películas delgadas, Universidad del Valle, Cali – Colombia

*E-mail: jaimeandres@ingenieros.com

Abstract. In this work we study the Aluminium Nitride plasma produced by Nd:YAG pulsed laser, (λ = 1064 nm, 500 mJ, τ = 9 ns) with repletion rate of 10 Hz. The laser interaction on Al target (99.99%) under nitrogen gas atmosphere generate a plasma which is produced at room temperature; with variation in the pressure work from 0.53 Pa to 0.66 Pa matching with a applied laser fluence of 7 J/cm^2.The films thickness measured by profilometer was 150 nm. The plasma generated was at different pressures was characterized by Optical Emission Spectroscopy (EOS). From emission spectra obtained ionic and atomic species were observed. The plume electronic temperature has been determined by assuming a local thermodynamic equilibrium of the emitting species. Finally the electronic temperature was calculated with Boltzmann plot from relative intensities of spectral lines. .

1. Introduction

Pulsed laser deposition (PLD) using nanosecond pulses is considered to be one of the most promising techniques for the synthesis and deposition of thin films [1–4]. This method has advantages, such as high reproducibility, control of the film growth rate and stoichiometry, and low impurity concentration in the composition of deposited films. On the other hand aluminum nitride (AlN) exhibit attractive properties, such as thermal and chemical stability, high thermal conductivity, high dielectric permittivity, breakdown field, high-speed piezo-acoustic wave and the mechanical hardness [1]. AlN thin films with good morphological quality are essential for applications in surface acoustic wave (SAW) and UV optical devices. That's why the scientific interest is focused on the preparation and study of high-quality films AlN. The aim of this work is to study the optical emission spectroscopy in of aluminum nitride thin films deposited by pulsed laser deposition on Si (100) substrate for use in optical applications.

2. Experimental

The experiments have made in usual PLD configuration consisting of a laser system Fig.1, a multiport stainless steel vacuum chamber equipped with a gas inlet, a rotating target and a heated substrate holder. The Nd:YAG laser that provides pulses at the wavelength of 1064 nm with 9 ns pulse duration and repetition rate 10 Hz was used. The laser beam was focused with an f = 23 cm glass lens on the target at the angle of 45°, with respect the normal. The target rotated to 2.2 rpm to avoid fast drilling. The distance between the target and the substrate was 6.5 cm. Before deposition the vacuum chamber was evacuated down to $1 \cdot 10^{-5}$ mbar by using a turbo-molecular pump backed with a rotary pump.

Figure 1. AlN plume during the deposition process.

The films were deposited in nitrogen atmosphere as working gas, in an atmosphere of nitrogen reactive the nitrogen gas pressure varied between 0.39 Pa and 0.93 Pa and aluminum target (99.99%). The Films were deposited with a laser fluence of 7 J/cm^2 for 10 minutes on silicon (100) substrates.

The Plasma characterization was performed by optical emission spectroscopy (OES) by using a spectrometer model Jobin Yvon Triax 550 of 0.55 m, f = 6.4 equipped with two gratings of 1200 l/ mm and 150 l/mm, coupled to a CCD camera model 3000 air-cooled multi-channel and 512 × 512 pixels Fig.2. For the morphological study used a scanning electron microscope Philips XL 30. It was determined the grain size and roughness by using Atomic Force Microscopy (AFM) measurements in atomic force microscopy Asylum Research MFP-3D-SA. The films thickness was measured with a Dektak 8000 profilometer.

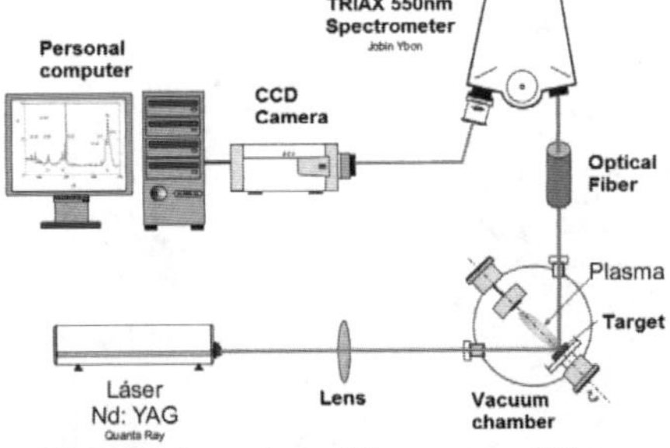

Figure 2. Schematic diagram of the Nd:YAG laser deposition system [5].

Chemical composition analysis of the coatings was done with a Philips XL 30 FEG scanning electron microscopy, an X-ray detector and secondary electrons detector of Lithium Beryllium inside the chamber with the propose of amplify the signal in the EDS analysis.

3. Results and discussion

3.1 Plasma Diagnostics

The study of pulsed laser ablation plumes has increased the attention recently due to its importance in laser deposition. Plasma is a transient phenomenon in nature with characteristic parameters dependent on the component species and rapidly evolving. These parameters are highly dependent on the irradiation conditions, laser intensity, pulse duration, wavelength, composition and atmosphere [6]. Taking in account that the relationship between plasma and morphological quality in the films is very important, in this sense the AlN films are used as substrates for SAW sensors where the surface quality is a decisive factor in the sensors performance. This study focuses on the variation of nitrogen gas pressure between 0.53 Pa and 0.66 Pa in aluminum nitride deposition.

Identified a large number of emission lines attributed to emission bands of aluminum nitride, atomic spectral lines are also indicating the presence of Al and atomic N_2 (Figs 3b and 3c). The oxygen presence it was observed in optical emission for the AlN with 0. 53 Pa and 0.66 Pa, of which is a product of contamination in the vacuum chamber. All atomic emission lines were identified through the database of the National Institute of Standards and Technology-NIST. In Fig. 3 the most intense lines are emission aluminum species. Apparently, the main species emitted in the ablation of aluminum species, being once ionized aluminum (Al II). The strongest lines in the spectrum of XII in plasma are at 631.337 nm, for electron configuration $1s^3p$ - $1s^3d$. Also it was observed emission lines of nitrogen species (neutral and multiply ionized) with most intense peaks at 618.909 nm (NI), 644.902 nm (N III), 740.359 nm (NIII). The emission peak of atomic Nitrogen was dominant compared to the emission peaks of atomic aluminum.

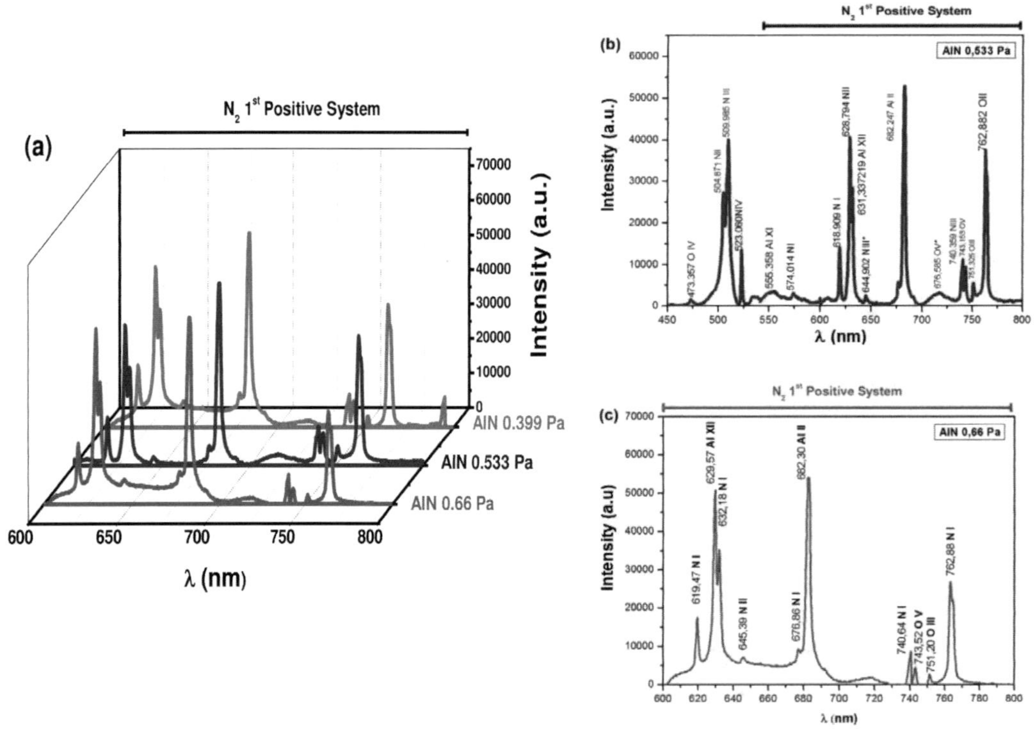

Figure. 3 Optical emission for the AlN plume with different values of nitrogen pressure: (a) emissivity for N_2 1st positive system between 600nm and 800 nm, (b) Optical emission for the AlN with 0. 53 Pa, and (c) optical emission for the AlN with 0. 66 Pa.

The oxygen presence of is also attributed low flow of nitrogen gas during the degassing processes. The oxygen species are observed O II 762.882 nm 751.325 nm O III and OV in 676.585 nm and 743.153 nm. In 509.985 nm observed emission band of AlN (0.0) [7]. A second emission band, weaker, is analyzed in 523.060 nm for AlN (1.0) [8]. In this work the oxygen bands only are evidenced for a working pressure of 0.53 Pa. Moreover, in this work was found participle density in Debye sphere Nd = $2.46x10^{-1}$.

In this case is possible take in account that the plume is in local thermodynamic equilibrium [9], therefore, the emission line intensity (I) in a specific wavelength (λm) may be expressed by:

$$\ln\left(\frac{I_{mn}\lambda_{mn}}{A_{mn}g_{mn}}\right) = \ln\left(\frac{N}{Z}\right) - \left(\frac{E_{mn}}{kt_e}\right) \quad (1)$$

where λmn is the transition wavelength, Imn is the observed intensity transition line, Amn is the transition probability, gmn is the degeneracy of the upper level, Emn is the energy of the emitting level, k is the Boltzmann constant and T_e is the electronic temperature. A typical plot is reported in the Fig. 4 for the emission of the AlN plume. The higher temperature calculated in the presence under 0.53 Pa of N_2 can be associated to recombination phenomena occurring during plume expansion and the thin films deposition, in relation to local thermodynamic equilibrium (LTE) the electron density, such as is showed by the next equation:

$$n_e \geq 1,4x10^{14}T_e^{\ 12}(\Delta E_{mn})^3 cm^{-3} \quad (2)$$

In this paper is reported a value of the $5.90x10^{13}$ cm^{-1} for the LTE approximation, which is agree with the literature [7,9].

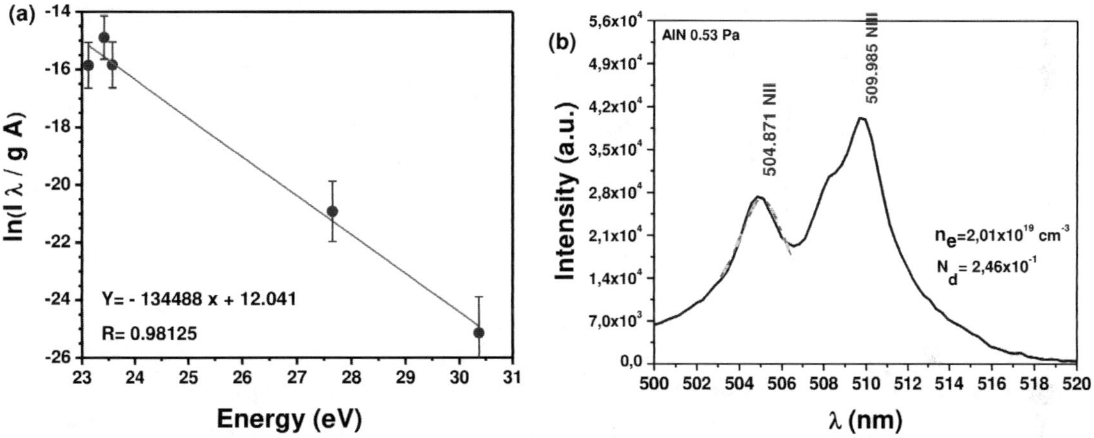

Figure. 4 (a) determination of electronic temperature Te = 8.832.3 K from AlN plume emission in 0.533 Pa N_2 by using the line N II and (b) the plasma electron density n_e calculated from Stark broadening, for NII transition *($2s^2p^2$—4P) 3p* at the line 504.871 NII, and the $n_e = 201x10^{19}$ cm^{-3}

3.2 Film characterization

EDS spectra from the AlN films surface showed the presence of (Al, N, O) element, which is characteristic of those materials. The presence of oxygen has been often found during the production of AlN is normally associated to residual oxygen in the chamber [7]. Therefore, is possible to observe de dependence of work pressure on decreasing of Al content. The areas of the peaks were used to calculate the composition of both coatings; thus, the values from Fig. 5 indicated that our films were substoichiometrics. On the other hand, a careful correction has to be done in all stoichiometric analysis because EDS has low reliability for nitrogen concentration. Therefore, EDS elemental concentrations were obtained using the ZAF correction method; because certain factors related to the sample composition, called matrix effects associated with (atomic number (Z), absorption (A) and fluorescence (F)), can affect the X-ray spectrum produced during the analysis of electron microprobe and therefore, these effects should be corrected to ensure the development of an appropriate analysis.

The correction factors for a standard specimen of known compositions were determined initially by the ZAF routine. The relative intensity of the peak K was determined by dead time corrections and a referent correction for the X-ray measured. So, before each quantitative analysis of an EDS spectrum, a manual background correction and an automated ZAF correction was carried out [10]. Thus, Fig. 5 shows the energy-dispersive X-ray spectroscopy (EDS) values of AlN films deposited with differents work pressures. All samples were observed via SEM and chemical analyses were done with an amplification of 20000x.

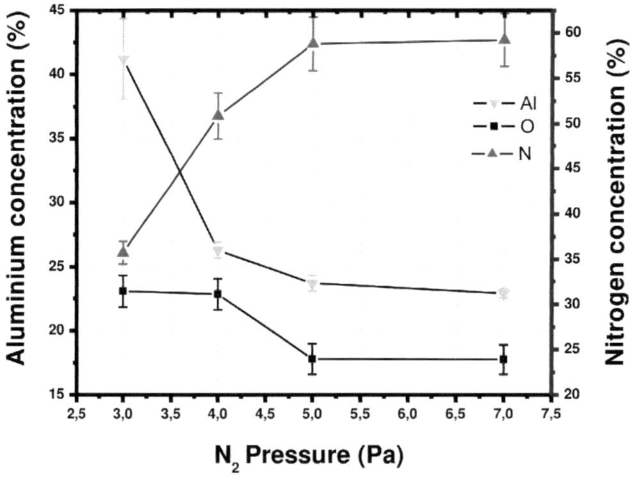

Figure 5 Correlation between work pressure in the Al-N plasma and aluminum, nitrogen and oxygen contents for deposited AlN thin films.

The observed dependence of the film surface morphology under nitrogen pressure during deposition is closely related with the film growth mechanism, associated to the surface diffusion length L which is given by:

$$L \sim (D\tau)^{1/2} \qquad (3)$$

where D is the diffusion coefficient and τ is the residence time of adatoms. Larger values of diffusion length imply more time for the adatoms to find energetically favorable lattice positions, thus, reducing the density of surface defects and improving the crystal quality.

Figure 6 shows the surface morphology AlN films was studied by recording AFM images together SEM micrographs where is observed the random distribution of micro-particles or micro-droplets on these surfaces as function of deposition pressure (0.66 Pa and 0.53 Pa). Therefore, the deposition pressure affects clearly the increase of micro-drops; this can be possible due to low surface mobility when the pressure was varied from 0.39 Pa to 0.93 Pa. This surface mobility reduces the possibility that the micro-drops are anchored on the surface when arrives with high energy on AlN film. Other possible reason can be associated with the mean free path that produces surface diffusion of nano-drops or micro-drops which can decrease the overall number of particles.

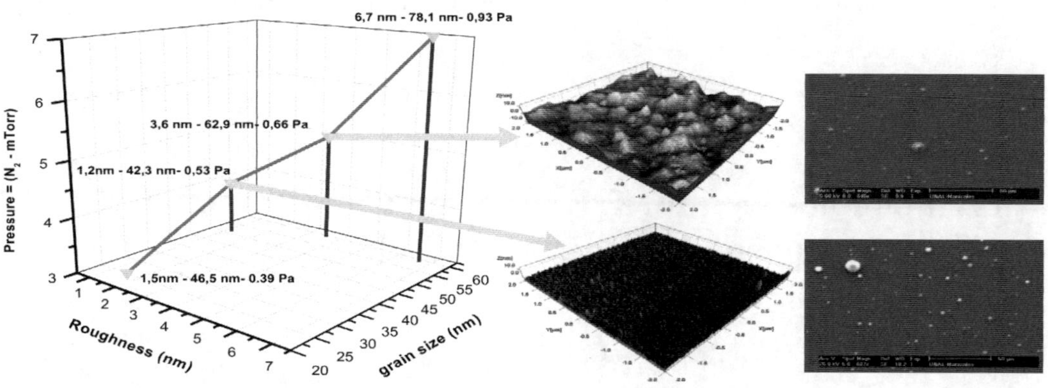

Figure 6. Relationship among surface roughness, grain size and plasma pressure for AlN thin films within vacuum chamber for AlN thin film deposited with a pressure of 0.66 Pa and 0.53 Pa.

4. Conclusion

In this work was found a dependency in relation to nitrogen concentration, roughness and grain size in the AlN films with the nitrogen work pressure, increasing in this sense the nitrogen concentration and the roughness on AlN films. The plasma pressure affects the stoichiometry and the morphological nature in the AlN films. The variation of nitrogen work pressure exhibit low effect on intensity of spectral lines emitted. The electron temperature value (Te = 8832.3 K) presented in our aluminum nitride plasma is similar to previous works.

5. Acknowledgements

The authors (L. P. Vera and J.A. Pérez) wish to express their gratitude for the warm hospitality offered by Dr. Leopoldo Soto, Dr. Felipe Veloso and all the Local Organizing Committee of the ICPP-LAWPP 2010. This work was supported in part by the Universidad Tecnologica de Pereira, Colombia to the project E6-10-6.

6. References

[1]. Murakami, K., (1992), Laser Ablation of Electronic Materials—Basic Mechanisms and Applications, Fogarassy, E. and Lazare, S., Eds. *(Amsterdam: Elsevier)*, p. 125.

[2]. Baek J, Ma J, Becker MF, Keto JW, Kovar D. *Thin Solid Films* (2007) 515:7096.

[3]. Sudhir GS, Fujii H, Wong WS, Kisielowski C, Newman N, Dieker C, et al. *Appl Surf Sci* (1998);127–129:471.

[4]. Kaluza N, Steins R, Cho YS, Sofer Z, Hardtdegen H. In: Proc. *11th European Workshop on MOVPE. Lausanne*, Switzerland; (2005) 179–81.

[5]. L.M. Franco, J.A.Pérez, H.Riascos, Estudio Espectroscópico de Plasmas del Aire, Cobre y Aluminio Producidos por Láser Pulsado, *Rev. Col. Fís*., (1) 176, (2008)

[6]. Bakalova S, Szekeres A, Cziraki A, Huhn G, Havancsak K, Grigorescu S, *Vacuum*. 84 155-157 (2009)

[7]. J. Hermann, C. Dutouquet, *J. Phys. Appl. Phys*. 32 (1999) 2707.

[8]. A. Santangata, V. Marotta, S. Orlando, R. Teghil, M. Zaccagnino, A. Giardini, *Appl. Surf. Sci*. 208–209 (2003)

[9]. H.R. Griem Spectral line broadening by plasmas, *Academic Pres, New York* (1974).

[10]. C.R. Brundle, C.A. Evans Jr., S.Wilson, L.E. Fitzpatrick (Eds.), *Handbook of Materials Characterization:* Surfaces, Interfaces, Thin Films, Butterworth-Heinemann,Woburn, MA, (1992), pp. 120–121.

Correlation between the OES Plasma Composition and ZnMnO films properties during Pulsed Laser

H Riascos[1], D Ramírez[1], LM Franco[2]

[1] Plasma, Láser y Aplicaciones, Universidad Tecnológica de Pereira, A. A. 097, Pereira, Colombia.
[2] Departamento de Física, Universidad del Valle, A. A. 25360, Cali, Colombia.

E-mail: hriascos@utp.edu.co

Abstract. Manganese doped zinc oxide was ablated on (100) silicon using the fundamental harmonics of Nd:YAG laser at various oxygen ambient gas pressures and temperature substrate was kept at room temperature during the deposition. In all experiments we used ZnMnO target. Optical emission spectroscopy (OES) is used to survey the plasma composition during deposition. The intensities of the ZnI and ZnII emission lines are correlated to the oxygen gas pressure and composition and morphology of the films deposited. Strong emission lines of ZnI, Zn II and neutral oxygen are observed. Oxygen and Mn emission lines in its atomic or ionic states have not been detected in this investigation, which is thought ZnMnO as being exclusively produced on the substrate surface and not generated into the plasma, which is in agreement with previous studies of ZnO plasma plume. Atomic Force Microscopy (AFM) is used to study the influence of oxygen gas pressure on the ZnMnO film morphology. An X-ray diffraction was tried to verify a crystal structure of the sample. We have demonstrated that gas pressure is an important parameter for the quality and performances of ZnMnO structures obtained by pulsed laser deposition.

1. Introduction Pulsed Laser Deposition (PLD) of material processing is a technique which has given good results for various applications of thin films of wide range materials [1]. In PLD, a pulsed laser is focused onto a target of the material to be deposited. For sufficiently high laser energy density, each laser pulse vaporizes or ablates a small amount of the material creating a plasma plume. The ablated material is ejected from the target in a highly forward-directed plume. The ablation plume consists primarily of excited atomic, diatomic, ionic and others species. The plasma properties of the plume largely determine the quality of the thin films deposited on the substrate. Most of the work about PLD is now devoted to improve the quality of the deposited films. To this end, it is of great importance to understand the ablation process and the transport phenomena of materials from the target to the collecting substrate. The stream of atoms, ions, molecules, and clusters ejected from the target surface emit radiation that can be analyzed using optical emission spectroscopy to gather information about the composition and dynamics of the plasma plume.

ZnO is a semiconductor possessing a large number of interesting properties that have been the focus of intensive research for many years. Due to recent and exciting developments in several key areas, pure or doped ZnO films appear to be a highly promising new and advanced functional material. Initially interest in oxide based in semiconductors was stimulated by the theoretical prediction of Dietl et al [2] that ZnO would exhibit ferromagnetism at room temperature after doping with Mn. Until now several research groups attempted to obtain experimental evidence for such ferromagnetism in thin films form of ZnO doped with transition metals. Thin Films of ZnO and ZnO-related compounds have been grown by a wide a variety of chemical as well as physical methods: sol-gel synthesis [3], chemical vapour deposition [4], magnetron sputtering [5], molecular beam epitaxy (MBE) [6] and Pulsed Laser Deposition [7].

In this work we report the analysis of plasma generated by pulsed laser ablation using fundamental wavelength (1064 nm) of Nd:YAG (neodymium-doped yttrium aluminum garnet) laser. The effect of O_2 gas pressure on plasma plume and some thin films were studied comparatively.

2. Experimental Methods

The plasma plume was produced by ablating a ZnMnO target using a Nd:YAG laser (wavelength: 1064 nm, pulse width: 9 ns, repetition rate: 10 Hz) at various oxygen ambient pressures from 13 to 95 mTorr. The ZnMnO target was mounted on a holder rotated at a constant rate of 2.2 rpm so that the laser beam continuously irradiated a fresh region on the surface. The laser beam was focused onto the target at an incident angle of 45° with respect to the target surface at power of 1.2 W. The optical emission spectra from plume were measured using a spectrometer (Jobin Yvon Triax550) with a wavelength resolution of 2 nm and 1200 grooves/mm grating attached to a CCD camera. The ablation was done in O_2, atmosphere at pressures ranging from 13–95 mTorr.

The surface morphology and crystallite size of ZnMnO in the films were recorded by atomic force microscopy (AFM). Also, other properties of thin films were observed by X-ray diffraction (XRD).

3. Analysis of the plasma

To investigate the plume dynamics of the ejected material accompanying laser ablation of the ZnMnO target, OES measurements were carried out, obtaining information on the nature of the ejected species. The optical emission spectra of plasma plume were recorded in the wavelength range from 400 to 800 nm at various gas pressures and substrate temperature.

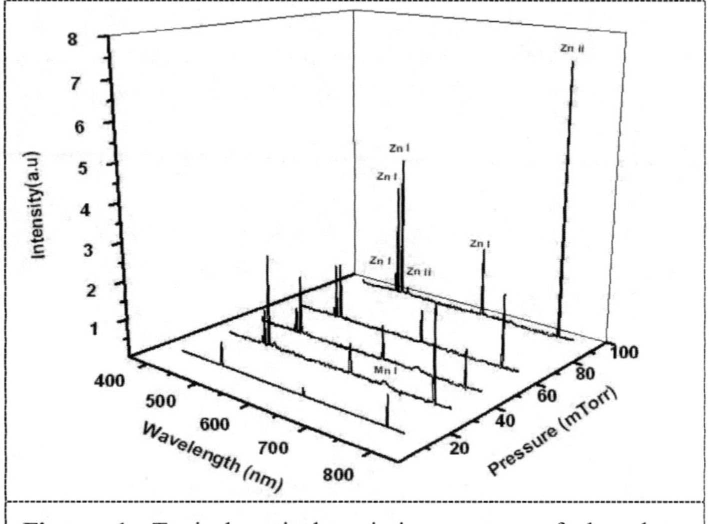

Figure 1. Typical optical emission spectra of the plume accompanying the 1064 nm pulsed laser ablation of the ZnMnO target in oxygen atmosphere at a pulse power of 1.2W.

Table 1. The strongest observed emission lines of the plume accompanying the 1064 nm pulsed laser ablation of the ZnMnO target in oxygen atmosphere at a pulse power of 1.2W

Lines ZnO		
ZnII	492.4 nm	$4d^2D_{5/2}$-$4f^2F_{7/2}$
ZnI	468.01 nm	$4s4p^3P_0$-$4s5s^2S_1$
ZnI	472.21 nm	$4s4p^3P_1$-$4s5s^3S_1$
ZnI	481.05 nm	$4s4p^3P_2$-$5s^3S_1$
ZnI	636.23 nm	$4s4p^1P_1$-$4s4d^1D_2$

Strong blue light emission from the plasma due to the emissions from the neutral Zn atoms could be observed. Typical emission spectra of the plasma plume formed during laser irradiation in an O_2 atmosphere and at a laser power 1.2 W are shown in figure 1. These spectra showed that the plasma emission in mainly due to atomic and singly ionized Zn. These results are similar to those previously reported for pulsed-laser ablation of ZnO [8–10] and, Saji et al [11] who only observed zinc atoms and ions during the ablation of a sintered ZnO target by Q-switched third harmonic Nd:YAG laser (355 nm) at different oxygen partial pressures. Strong emissions were observed around 481.05 and 636.23 nm from the neutral Zn atoms (Zn I) and 773.25 from the singly ionized Zn atoms (ZnII) in the all gas pressure, others neutral spectral lines such as 468.01, 472.1, and the ionic line 492.4 appear in all pressures except at 13 mTorr. Practically at 13 mTorr there are a few presences of Zn emission lines. The peaks of the neutral and ionized Mn atoms and O species could not be observed, possibly due to a lower density of these compared with that of the neutral Zn atoms. On the other hand, during ablation, the ionization potential of atomic Zn is much lower than those of the other atomic species in the plume (9.39 eV for atomic Zn, 13.62 eV for atomic O), [9]. Therefore, Zn species are the dominant in the dense plasma. Significant variations in the intensities of emission lines of various atomic and ionic species of the plasma were observed with changes ambient gas pressures, it is observed that both electronically excited Zn atoms and the singly ionized Zn atoms peaks increase linearly with gas pressure as shown in figure 2. The intensities of the emission lines corresponding to the Zn I and Zn II reach maximum values at 95 mTorr. The intensity increase of the spectral lines is due to the effect of cooling by the background gas resulting in confinement of the plasma plume. The strongest observed emission lines are listed in table 1 with corresponding spectroscopic data. From figure 2, it is seen that the emission intensity of strongest lines increase with Oxygen gas pressure.

4. Deposition of ZnO films

ZnMnO thin films were deposited on silicon (111) substrate at room temperature, various gas pressure from 13 to 95 mTorr by PLD using a Nd:YAG laser with 1.2W per pulse. The focused laser radiation illuminated a rotating target at an incident angle of 45°. The target was 99.9% pure ZnO doped with 0.05wt% Mn. The distance target-substrate was 5.5 cm.

4.1. XRD analysis

The ZnMnO thin films grown at room temperature on various oxygen gas pressures have been characterized with regards to the structure by XRD. Amorphous films were obtained, except the film grown at 95 mTorr as shown in figure 3. This suggests that the ablated species reaching the substrate may not have the minimum energy required for crystallization and thereby getting amorphous films. The X-ray diffraction peak corresponding to ZnO (002) plane is observed. Observation of the (002) plane indicates that the film is c-axis oriented. This indicate that the surface energy of (002) plane is the lowest in ZnO crystal [12].It is known that the number of particulates arriving at the substrate is

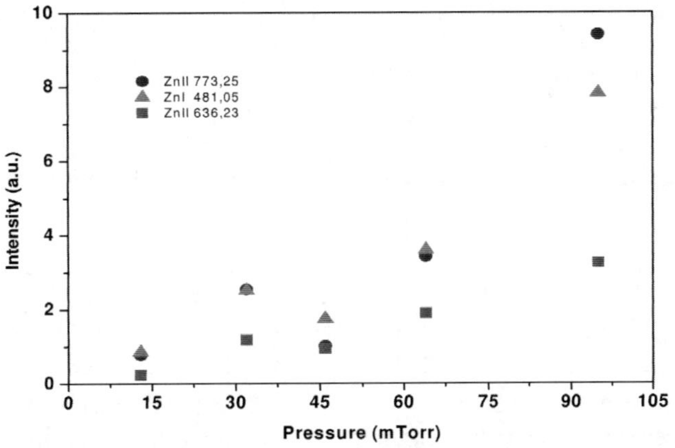

Figure 2. (color online) Variations of Intensity emission lines of the plume corresponding to ZnI (481.05 nm, 636.23 nm) and ZnII (773.25 nm) with increase of oxygen pressure.

Table 2. Structural parameters for the ZnMnO samples obtained from AFM at RT.

Oxigen Pressure (mTorr)	rms surface roughness (nm)	Grain size (nm)	Thickness (nm)
13	0.55	15.89	65.9
32	2.87	19.90	360.0
46	1.20	26.57	97.8
64	2.74	40.16	267.0
95	3.92	44.59	180.0

decreased as the pressure of an ambient gas increases. Deposited particulates in a ZnO film degrade film quality. Therefore, the crystal quality is improved as the ambient gas pressure increase [13].

4.2. Surface morphology by AFM

Films deposited under different oxygen pressures at room temperature were also evaluated by AFM. The scanning area was 1 μm×1 μm.

Figures 4a) to 4e) show the surface morphology of these samples as seen by AFM. The images show the effect of gas pressure on the surface roughness and grain size. It is seen that the grain size increase with oxygen gas pressure, with a grain-size distribution between 15.9 and 44.6 nm for the samples grown at room temperature. In contrast, the roughness behaves very different; there is no a clear trend of increasing roughness with increasing gas pressure.

Table 2 summarizes the results of AFM measurements on ZnMnO thin films grown at different pressures. The results can be understood by considering the ZnMnO plume size. As the ambient pressure increases, the number of particles arriving on the substrate decreases, resulting in deterioration of the deposited film.

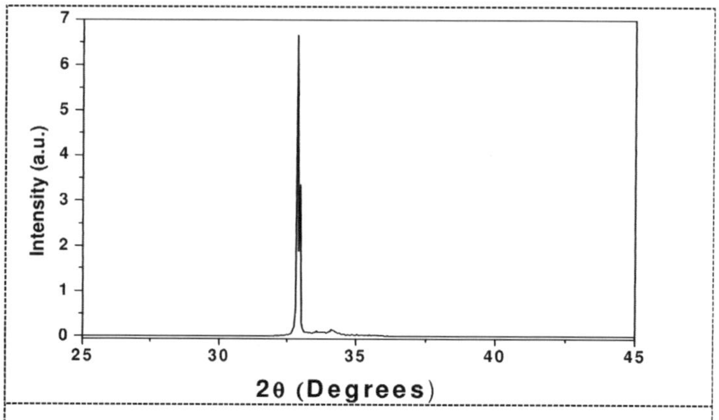

Figure 3. XRD 2θ scan of film deposited on Silicon (100) substrate at room temperature for 95 mTorr oxygen pressure by PLD.

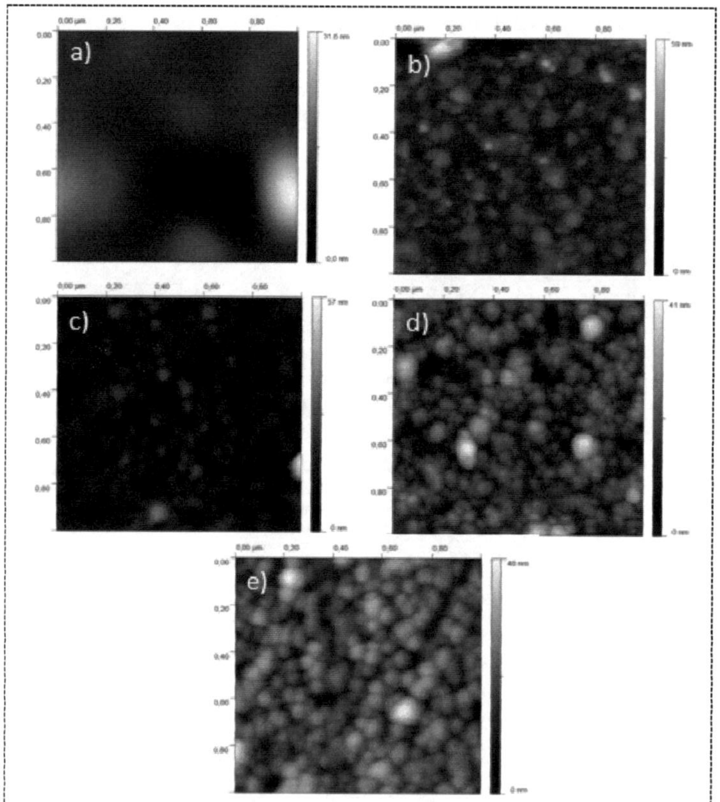

Figure 4. (color online) Images of ZnMnO films prepared in oxygen atmosphere at various pressures and room temperature by Nd:YAG (1064 nm, 9 ns, 1.2W, 10 Hz) on silicon substrate.

5. Conclusions

We observed that only at 95 mTorr ZnMnO crystalline films was grown. Optimization of the deposition conditions makes it possible to produce high-quality thin films. Structure and surface morphology of all obtained samples demonstrated the strong influence of the gas pressure. OES investigations are in agreement with the studies of films, showing plasma richer in Zn species Our studies evidenced the prevalent presence of ZnI in the plasma generated under action Nd:YAG laser operated at 1064 nm.

6. References

[1] Robert Eason, "Pulsed laser deposition of thin films: applications-led growth of functional materials", A Wiley-Interscience publication. (2007)
[2] T. Dietl, H. Ohno, F. Matsukura, J. Cibert, D. Ferrand, *Science* ***287*** (2000) 1019
[3] Y.M. Kim, M. Yoon, I.-W. Park, J.H. Lyou, *Solid State Commun.* **129** (2004) 175.
[4] P. F. Zhang, H. Y. Wei, G. W. Cong, W. G. Hu, H. B. Fan, J. J. Wu, Q. S. Zhu and X. L. Liu. *Thin Solid Films* **516** (2008) 925.
[5] H. Li, X. J. Pan, M. Qiao, Y. Z. Zhang, T. Wang and E.Q. Xie. *Vacuum* **82** (2008) 459.
[6] A. Bakin, A. El-Shaer, A.C. Mofor, M. Kreye, A. Waag, F. Bertram, J. Christen, M. Heukena and J. Stoimenos. *J. Cryst. Growth* **287** (2006) 7.
[7] R. O'Haire_, E. McGlynn, M.O. Henry, J.-P. Mosnier, *Superlattice Microst.* **42** (2007) 468–472
[8] R K. Thareja, H. Saxena and V. Narayanan, *J. Appl. Phys.* **98** (2005)034908
[9] F Claeyssens,. A.Cheesman, S.J. Henley, M.N.R.*J.* Ashfold, *Appl. Phys.* ***92*** (2002) 6886.
[10] S. Namba, R. Nozu, K. Takiyama and T. Oda, *J. Appl. Phys.* **99** (2006) 073302
[11] K. J. Saji, R. Manoj, R. S. Ajimsha, and M.K. Jayaraj, *Proc. of SPIE* Vol. **6286** (2006) 62860D
[12] K. L. Chopra, S. Major, and D. K. Pandaya, *Thin Solid Films* **102** (1983) 1
[13] Y. R. Ryu, S. Zhu, S. W. Han, and H. W. White, *J. Vac. Sci. Technol. A* **16** (1998) 3058.

Behavior of AISI SAE 1020 Steel Implanted by Titanium and Exposed to Bacteria Sulphate Deoxidizer

Ely Dannier V. Niño[1,3], Hernán Garnica[2], Veleriy Dugar-Zhabon[3], Genis Castillo[2]

[1] Grupo de Investigación en Tecnología del Plasma – GINTEP, Universidad Pontificia Bolivariana, Bucaramanga, Santander, Colombia
[2] Corporación para la Investigación de la Corrosión - CIC, Piedecuesta, Santander, Colombia
[3] Grupo de Física y Tecnología del Plasma – FITEK, Universidad Industrial de Santander, Bucaramanga, Colombia

E-mail: ely.valbuena@upb.edu.co

Abstract. A hybrid technology to treat solid surfaces with the pulse high voltage and electric arc discharges of low pressure with a three-dimensional ion implantation technique (3DII) is applied. This technology is used to protect AISI SAE 1020 steel against a microbiological corrosion. The titanium ion implanted steel samples (coupons) are subjected to a medium of bacteria sulphate deoxidizer (BSD) which are very typical of the hydrocarbon industry and are potentially harmful for structures when are in contact with petroleum and some of its derivatives.

The used technology aims to find an effective hybrid procedure to minimize the harmful effects of bacteria on AISI SAE 1020 steel. The hybrid technology efficiency of superficial titanium implantation is estimated through the measurements of the point corrosion characteristics obtained after testing both the treated and non-treated coupons. The three-dimensional surface structures of the samples are reconstructed with help of a confocal microscope.

1. Introduction

The problem of the control over the microbiological corrosion is quite notorious for the hydrocarbon segments of the metal-mechanic industry. The costs to prevent and correct microbiological corrosion damages caused in the process of extraction and transportation of hydrocarbon are very high, for example, in the USA these costs are estimated as much as 2 billion dollars per year (Buck, E., *et al.*, 1996; Koch G. H., *et al.*, 2001). However difficult it may be to evaluate accurately the costs, the estimations show that about 40% of the corrosion damages of the interior walls of the tubes fabricated

[1] ely.valbuena@upb.edu.co
[2] hgarnica@corrosion.uis.edu.co
[3] vdougar@uis.edu.co

for natural gas transportation can be attributed to the microbe activity (Pund, *et al*, 1998). The recovery costs of the damages caused by the hydrogen sulphate produced by the bacteria sulphate reducer (BSR) (Zuo *et al.*, 2004) amount to 4 to 6 billion dollars.
In Colombia, the costs associated with metal corrosion have been estimated as 1300 million dollars per year, which almost equals 1.2% of the Annual Revenue (Arroyave *et. al.*, 1997), with 30 or 40% of which corresponding to the losses caused by biological corrosion (Srivastava, V., 1992). For this reason, the problem of the control over the microbiological corrosion is highly imoportant both for science and economy.
Plasma technologies which can significantly change the composition and structure of metal surfaces through diffusion, deposition and implantation processes can be a good alternative to the chemical technologies which are used to reduce corrosion damages. The plasma processes are realized in the high voltage and arc discharges (Dougar V. et al, 2003).
A set of preliminary experiments to test the efficiency of titanium ions implantation into metals as a method for attenuating of corrosion processes in low carbon metals is being carried out at the Industrial University of Santander and the Corporation for the Investigation of the Corrosion (CIC).

2. Experimental

The dimensions and geometry of the surface defects of metal samples exposed to the bacteria sulphate reducer are determined. The samples are coupons of cylindrical geometry of 25.4 mm in diameter and 5 mm in height; their surface is treated by emery papers of 1200 (the roughness of 0.003 nm). The coupons are exposed to high voltage pulse discharges in the atmosphere of titanium vapour produced by an arc discharge. Then the samples in static conditions are introduced into the 0.03% salt solution (which is used as reference to the water of the petroleum production) enriched by BSR to be kept there for 50 days. The images of the coupons surface topography are reconstructed by a high resolution microscope with the aim to determine the difference between the inoculated coupons and non-inoculated coupons as well as the plasma treated and plasma non-treated ones.

3. Material

The coupons are made out of AISI SAE 1020 steel because this type of steel is the material used for gravimetric and bio tests in the hydrocarbon industrial segment in accordance with the ASTM G1-03, ASTM G4-01, NACE RPO775-05 standards and the Basic Corrosion Course, chapter 7 of the NACE.

4. Set-up

The experimental procedure consists in the solution media treatment of the coupons not inoculated and inoculated without any implantation treatment as well as the coupons, non-inoculated and inoculated, but subjected to titanium ion implantation (see table: 1).

Table 1. Implantation parameters

Process	Time	High Voltage	Frequency	Pressure	Duration Pulse	Element
Sputtering[a]	25 min	5 Kv	30 Hz	3.2 – 2.5 Pa	0.25 ms	N2
Sputtering	10 min	5 Kv	30 Hz	0.28 – 0.25 Pa	0.25 ms	Ti
Implantation	10 min	10 Kv	30 Hz	0.31 – 0.28 Pa	0.25 ms	Ti

[a] These ***must*** be set to 0 cm. In addition, please make sure the *Mirror Margins* option is ***not*** selected.

The experiments evidence the coloration and turbidity of the solution has changed due to the corrosion product, with a greater degree in the case of the non-implanted coupons (see Figure 1, 2 y 3).

A B C

Figure 1. Experiment assembly in translucent bottles of 250 ml with a threaded top and a coupon holder: A) Non-inoculated sample, 1st day, B) Inoculated sample, 50th day, C) Inoculated and implanted sample, 50th day.

5. Results

After the 50 days exposure of the treated samples in the aqueous solution, the local defects appear on the surface samples due to the bacteria activity. Fig. 2 shows the coupons surfaces both on the non-implanted and the titanium ions implanted samples (see Figure 3). The dimensions of the local defects on an analyzed area of 142x106 m^2 are determined as 2 m in diameter, 3 m in depth for the implanted coupons with the help of the microscope of brilliant field image 01M ZIISS. For the non-implanted coupons, an area of 284x212 m^2 is chosen. The geometry of the defects on the coupon surfaces is as follows: the diameters are 15 - 30 m and the depths are 8 – 15 m (as shown in Figure 4).

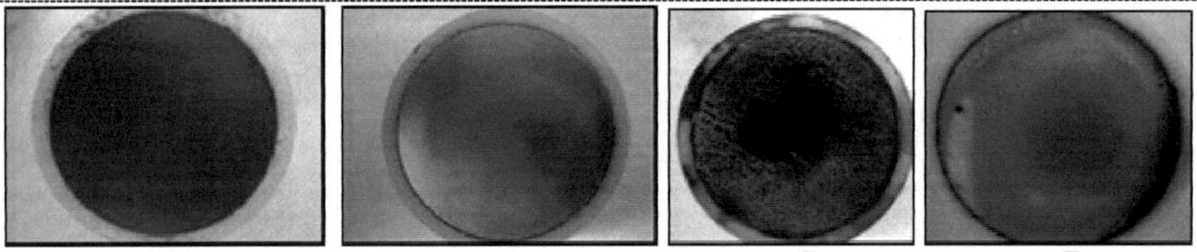

A B C D

Figure 2. Coupons surfaces during the aqueous solution exposure: A) Model sample, B) Implanted surface, C) Inoculated sample, 50th day, D) Implanted and inoculated sample, 50th day.

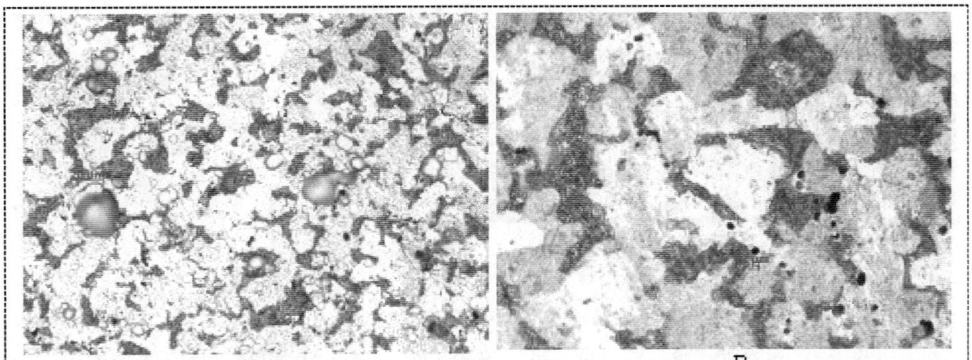

Figure 3. Samples surface topography: A) not implanted, B) implanted.

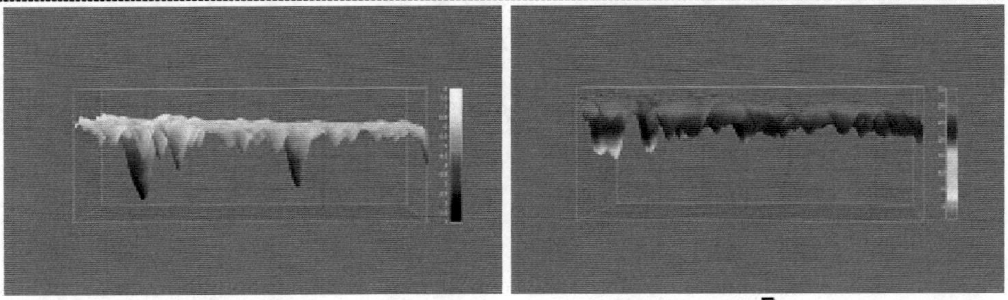

Figure 4. Three-dimensional reconstruction of the coupons surfaces: A) not implanted, B) implanted.

6. Conclusions

The titanium implantation technique is found efficient for protecting the AISI SAE 1020 steel surfaces from the BSR bacteria actions: it reduces the defect dimensions to 0.003 mm. The preliminary results make it possible to use the suggested method to protect some other types of steel from the microorganism aggression.

7. Recommendations

The bacteria behaviour in time should be further studied because of the metabolic processes that can form titanium oxides causing still greater defects.

Acknowledgments

The authors would like to express their gratitude to Biocorrosion, Physic-chemical and Plasma Physics groups, to the workshop of fine mechanics of the CIC, to the group of Optics and Signal Treatment, to Dr. A. Plata and Dr. P. Tsygankov.

8. References

[1] Buck, E., G. C. Maddux, and R. L. Sullivan. 1996. Internal corrosión cost impact study-United States natural gas exploration and production industry. GRI-96/0056 document N. 96-1466. Gas Research Institute.

[2] Pund, B. G. 1998. Gap analysis of the pipeline research committee international (PRCI)/Gas Research Institute (GRI) research program on internal corrosion. Topical report SF26363.000/AOTO/1198/BO02. Gas Research Institute.

[3] Zuo, R.; Ornek, D., Syrett, B.C., Green, R.M., Hsu, C.H., Mansfeld, F.B. & Wood, T.K. (2004). "Inhibiting Mild Steel Corrosión From Sulfate-reducing bacteria Using Antimicrobial-producing Biofilms in Three-Mile-Island Process Water". Appl. Microbiol. Biotechnol. 64: 275-283.

[4] Arroyave, C., et. Al. (1997). "Corrosion Problems and their Incidence on Colombian Economy". Grupo de Corrosión y Protección de la Universidad de Antioquia. Proc. 10th Asia Pacific Corrosion Control Conference. Bali, Indonesia, 27-31 October. Paper 12.1/6.

[5] Srivastava, V. (1992). "Environmental Science & Technology Center-Biotechnology for Gas Operation", GTI.

[6] Dougar-Jabon V., J. Dulcé, P.A. Tsygankov, High voltage pulse discharge for ion treatment of metals. In: Rev. Sci. Instr., Vol. 73, No. 2, (2002); p. 828-830.

Syngas Generation from Organic Waste with Plasma Steam Reforming

G. Diaz, E. Leal-Quiros, R.A. Smith, J. Elliott and D. Unruh
University of California, Merced, 5200 North Lake Road, Merced CA 95343, U.S.A.

E-mails: gdiaz@ucmerced.edu, edbertho@gmail.com

Abstract
A plasma steam reforming system to process waste is in the process of being set up at the University of California, Merced. The proposed concept will use two different plasma regimes, i.e. glow discharge and arc torches to process a percentage of the total liquid waste stream generated at the campus together with shredded local organic solid waste. One of the main advantages of the plasma technology to be utilized is that it uses graphite electrodes that can be fed to the reactor to achieve continuous operation, thus, electrode or nozzle life is not a concern. The waste to energy conversion process consists of two stages, one where a mixture of steam and hydrogen is generated from the liquid in a glow-discharge cell, and a second stage where the mixture of exhaust gases coming out of the first device are mixed with solid waste in a reactor operating in steam reforming mode interacting with a plasma torch to generate high-quality syngas. In this paper, the results of a thermodynamic model developed for the two stages are shown. The syngas composition obtained indicates that the fraction of CO_2 present decreases with increasing temperature and the molar fractions of hydrogen and carbon monoxide become dominant. The fraction of water vapour present in the product gases coming out of the second stage needs to be condensed before the syngas can be utilized in a prime mover.

1. Introduction
Plasma gasification of waste is a concept introduced relatively recently [1-5]. Plasma torches operating over 3000 K break the chemical bonds of waste to generate hydrogen-rich synthesis gas (syngas) from the organic compounds. In most cases, the interaction of this plasma with the organic molecules produces the dissociation of the molecules. Magmafication [5] happens from the interaction of plasma with inorganic compounds. The plasma temperature may be accurately controlled for the production of vitrified materials, which can be used to manufacture architectural tiles, bricks, and aggregates for the construction industry.

2. Status of the technology
Plasma processing of waste to generate energy, liquid fuels, and other products is a process that has not been fully characterized yet. The hydrogen-rich syngas can be used to generate power and heat by means of generator sets, lean-combustion turbines, or solid oxide fuel cells. The current state of the art of the technology includes a plant located in Utashinai, Japan operating since 2003 that processes 200-280 tonnes/day of auto shredder residue to produce electricity [6]. Another plant in Yoshi, Japan uses plasma gasification to produce bricks for construction purposes and heat. About 20 projects around the world are in the planning or construction stages. UC Merced is collaborating with the

company Foret Plasma Labs (FPL) [7] in the installation of a plasma steam-reforming unit that will process liquid and solid waste.

3. Campus waste streams

Being a relatively new campus has allowed facilities managers to obtain accurate data of the available waste streams and energy demand and their projections for the near future. Table 1 shows the waste generation at UC Merced and its estimates until 2014.

Table 1. Waste generation at UC Merced and estimate of their growth until 2014.

Predicted Waste Stream - UC Merced	2008	2009	2010	2011	2012	2013	2014
Potential Fuel Stream (kg/h)							
Single Stream	11.09	13.87	16.64	19.63	21.99	24.19	26.61
Green	14.91	18.64	22.37	26.40	29.57	32.52	35.78
Compost	0.25	0.31	0.37	0.44	0.49	0.54	0.60
OCC	7.58	9.47	11.36	13.41	15.02	16.52	18.17
Pallets	1.47	1.84	2.21	2.61	2.92	3.21	3.54
Paper	7.76	9.71	11.65	13.74	15.39	16.93	18.62
Total solids (kg/h)	30.50	38.13	45.76	53.99	60.47	66.52	73.17
Sewage Avg Volume (m^3/h)	2.36	2.95	3.54	4.18	4.68	5.15	5.67
Sewage Avg Weight (kg/h)	**2721.58**	**3401.98**	**4082.37**	**4817.20**	**5395.26**	**5934.79**	**6528.27**

4. Proposed concept and description of plasma system

The proposed concept will process a percentage of the total sewage stream and solid waste generated on campus. One of the main advantages of the equipment from FPL is that it uses graphite electrodes that can be fed continuously for continuous operation, and thus, electrode life is not a concern as with traditional plasma-torch nozzles. Figure 1 shows a representation of the equipment to perform the gasification of sewage and solid waste.

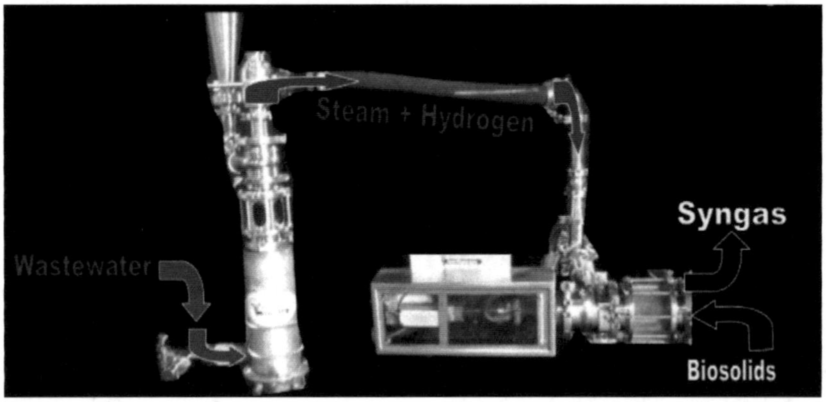

Figure 1. Plasma gasification of seage and solid waste using a dual plasma system (Courtesy of Foret Plasma Labs).

The reactor is comprised mainly of two plasma systems that operate in different plasma regimes. The first stage uses a plasma glow discharge to process the swage and generate steam and hydrogen. The second stage uses a plasma torch and the output gases from the first stage to process bio-solids. Syngas, mainly CO and H_2, is produced in the second stage.

5. Plasma in the DC discharge

The physics of glow discharges and plasma torches have been well described in the plasma literature, for example in E. Leal-Quiros [1-5] and others. However, the objective of this paper is to develop a thermodynamic model for both processing stages described above with the intent of analyzing the composition of the products gases after the first and second stages under a variety of operating conditions.

5.1 Cascade process of ionization

To understand the interaction between the plasma generated from the torch with the organic materials, we can visualise the phenomenon as a collision between the plasma components and the neutral atoms of the material. As a matter of illustration the cascade process can be considered as initiated by an energetic electron from the plasma. One incident electron (e−) collides with a neutral atom to produce a second electron and an ion. Then, there are two electrons and one ion. After these two electrons have independently collided with another neutral atom, four electrons and three ions are produced. This process continues, and after about 20 successive sets of collisions, millions of electrons and ions will be rapidly formed (the mean free path between collisions is very small at atmospheric pressures). Smaller sizes of the organic materials promote a better interaction with the plasma in the cascade process and a more efficient syngas generation is obtained.

6. Thermodynamic model and basic chemical reactions

Although there are a number of chemical reactions occurring during the plasma-arc gasification process, the assumption of equilibrium of the chemical reactions allows us to develop an approximate model [4,8-10]. The basic chemical reactions occurring during the gasification process correspond to:

$C + CO_2 = 2CO$ (Boudouard equilibrium)
$C + 2H_2 = CH_4$ (Hydrogenating gasification)
$C + H2O = CO + H_2$ (Heterogeneous water gas shift reaction)
$CH_4 + H_2O = CO + 3H_2$ (Methane decomposition)
$CO + H_2O = CO_2 + H_2$ (Water gas shift reaction)

Low temperature or low steam content promotes the formation of soot so the first three equations are relevant. At higher fractions of steam addition and high temperature, carbon (soot) is not formed so that the last two equations become dominant [9]. Following the models presented in [9-11], an approximate thermodynamic model was developed. Obtaining the ultimate analysis of the solid waste, such as, cardboard, paper, etc, and sewage, a global gasification reaction becomes:

$$aC+bH+cH2O+dO=n_1H_2+n_2CO+n_3CO_2+n_4H_2O+n_5CH_4+n_6C \tag{1}$$

where the coefficients a, b, c, and d are obtained from the ultimate analysis and the coefficients n_1 to n_6 are unknowns. Performing a molar balance for each species and adding three equilibrium constants given by eqn. (2) for the methane decomposition, primary water gas shift, and heterogeneous water gas shift reaction, the system of equations is solved iteratively with constants K1, K2, and K3 obtained from the eqn (3).

$$K_1 = \frac{y_{CO} y_{H_2}^3}{y_{CH_4} y_{H_2O}}, K_2 = \frac{y_{CO_2} y_{H_2}}{y_{CO} y_{H_2O}}, K_3 = \frac{y_{CO} y_{H_2}}{y_{H_2O}} \tag{2}$$

$$-\ln K_i = \frac{\Delta G_i^0}{RT} \tag{3}$$

7. Numerical results

The results obtained with the thermodynamic model are presented in the following sections.

7.1 *Generation of steam and hydrogen in the first stage*

When the sewage interacts with plasma glow discharge, steam and hydrogen are produced, as shown in Figure 2, where the hydrogen comes from a plasma electrolysis process. Other species, such as CO_2, CO, etc., are also formed but their molar fractions in the product-gas composition are small. The figure shows the molar fraction vs. the temperature of the products generated from the plasma glow interaction with sewage for a flow rate of 250 kg/hr. The sewage and solid waste properties have been obtained from an ultimate and proximate analysis.

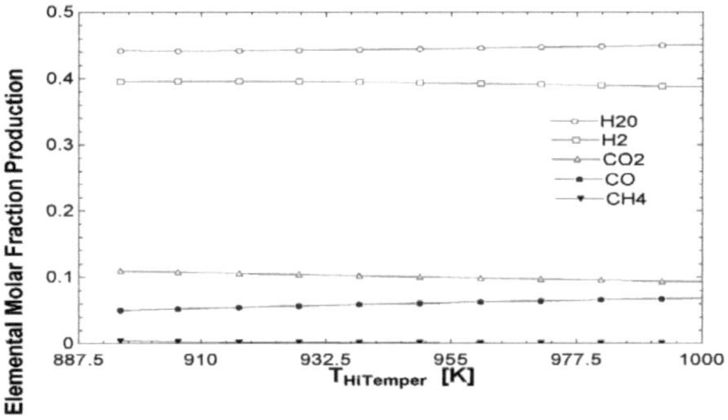

Figure 2. Molar fraction vs. temperature of the products generated in the first stage of the reactor for a flow of sewage of 250 kg/hr.

7.2 *Synthesis gas generation on the second stage*

Figure 3 shows the molar fraction for the composition of syngas produced in the second stage. The syngas is the result of the interaction between the plasma torch, the product gases from the first stage, and the organic solids being fed at a rate of 30 kg/hr. The temperature of the syngas depends on the temperature from the product gases from the first stage and the energy added for the plasma torch in the second stage. The syngas acquires a high temperature but far less than the plasma torch temperature.

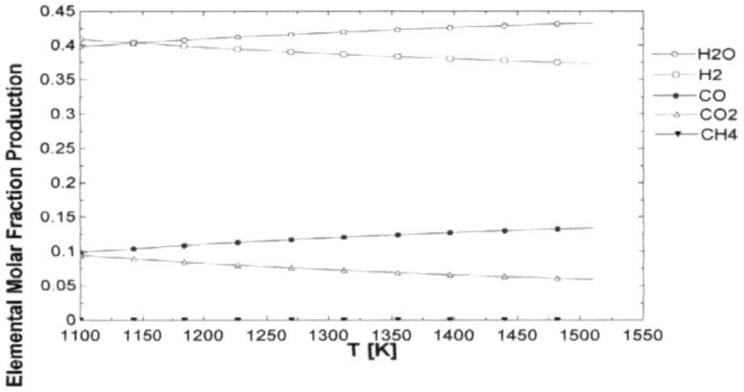

Figure 3. Molar fraction of the syngas vs. temperature acquired by the syngas during the steam reforming process in the second stage. Solid flow rate of 30 kg/hr for a sewage flow rate in the first stage of 250 kg/hr have been used. The temperature of the syngas varies depending on the power utilized for the plasma torch.

In practice, the vapour content in the syngas is condensed before the gas is supplied to a prime mover, such as, a generator set, turbine, etc. Different flow rates of sewage and bio solids produce different syngas compositions. The syngas composition also varies depending on the type of liquid and solid waste utilized.

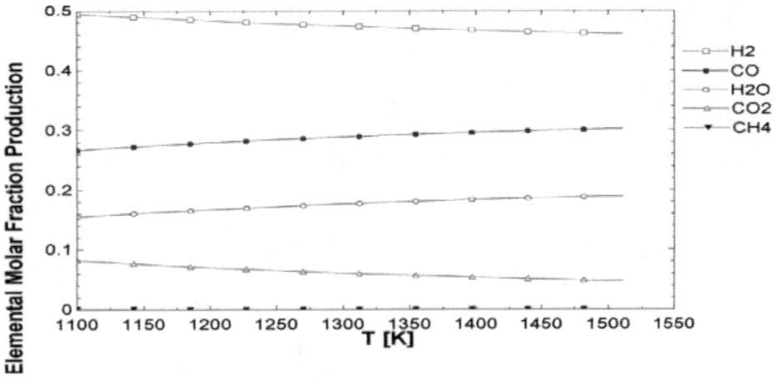

Figure 4. Molar fraction of the syngas generated vs. temperature acquired by the syngas. Solid flow rate of 200 kg/hr for a sewage flow rate in the first stage of 250 kg/hr.

Figure 4 shows the syngas composition produced by the system for a flow rate of bio-solids of 200 kg/hr. The molar fraction of CO_2 decreases as the temperature increases. The molar fractions of H_2 and CO are significantly larger, but there is still a percentage of water-vapour present.

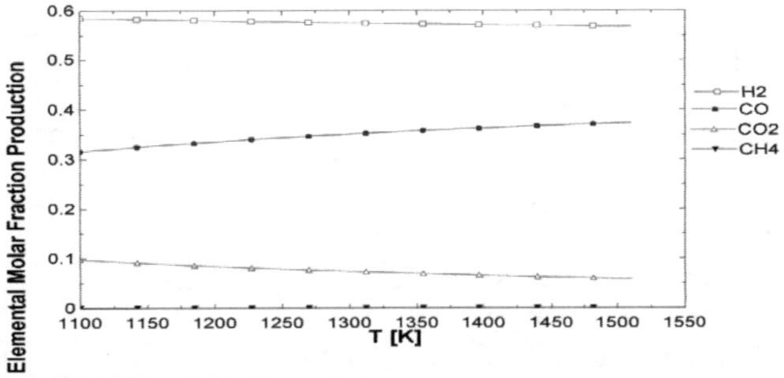

Figure 5. Molar fraction of the syngas generated vs. temperature acquired by the syngas, without water vapour in the products of the second stage.

Figure 5 shows the syngas composition on a dry basis, i.e. after the water vapour has been condensed. The larger fraction of hydrogen obtained is the consequence of adding the hydrogen content generated in the first stage, through glow discharge electrolysis, and the hydrogen generated by the steam reforming of bio-solids in the second stage.

8. Conclusions

A two-stage system for plasma gasification of waste is in the process of being installed at the University of California at Merced. A thermodynamic model has been developed to predict syngas composition based on operating-condition parameters, such as, flow rate of solid and liquid waste, power supplied, and waste composition.

The results of the model indicate that a large molar fraction of hydrogen and carbon monoxide is obtained with the system. The flow rates of solid and liquid waste, as well as, the power required, and the composition of waste utilized affect the final syngas composition which shows a fraction of water vapour present. In practice, this water vapour has to be condensed before the syngas is supplied to a prime mover. The benefits of the proposed concept not only relate to the reduction of solid waste that is sent to landfills or the reduction of the amount of sewage that is sent to the sewer network. This project will also provide an invaluable source of information about the operating parameters and performance of a plasma waste gasification plant.

9. Acknowledgements

This work was partially supported by California Energy Commission RESCO Project: "Piloting an Integrated Renewable Energy Portfolio for the UC Merced Community", award number PIR-08-036. We thank Foret Plasma Labs for providing pictures of their plasma gasification technology, and Robert Avalle and the facilities department at UC Merced for helping find a suitable location for the pilot plasma gasification plant, and for providing campus information concerning energy demand and waste streams.

10. References

[1] Edbertho Leal-Quiros. "Plasma processing of municipal solid waste". Brazilian Journal of Physics, Vol. 34, no. 4B, pp 1587-1593, Dec (2004).

[2] Leal-Quiros, E, Villafane, C.R. "An assessment of the power generated with plasma processing of sludge from wastewater treatment plants". IEEE Transactions on Plasma Science, Vol. 35, no. 6, pp 1662-1627 Dec (2007).

[3] Edbertho Leal-Quiros, Igor Matveed. "Plasma Coal Ignition for Utility Power Plants," 29th ICPIG July 12-27, 2009. Cancun, Mexico.

[4] G. Diaz, E. Leal-Quiros "Sustainable Waste Management Using Plasma Gasification at University of California-Merced." 29th ICPIG, July 12-27, 2009, Cancun, Mexico.

[5] Edbertho Leal-Quiros, "Magmavication, Vitrification, Gasification, and Generation of Energy with High Temperature Plasma Processing of Waste," Current Trends in International Fusion Research – Proceedings of the Fifth Symposium, pages: 341-357. Edited by E. Panarella and R. Raman. NRC Research Press, National Research Council of Canada, Ottawa, ON K1A 0R6 Canada, 2008.

[6] K.C. Bhasin. Plasma arc gasification for waste management. Part 2 of 2. Electronics for you. February (2009).

[7] Foret Plasma Labs, LLC (http://www.plasmawhirl.com). US Patents #7,422,695, #7,578,937, and # 7,622,693.

[8] Shuster, G. et al. Biomass steam gasification – an extensive parametric modeling study. Bioresource Technology, 77 (2001) 71-79.

[9] Zainal, Z.A. et al. Prediction of performance of a downdraft gasifier using equilibrium modeling for different biomass materials. Energy Conversion and Management, 42 (2001) 1499-1515.

[10] Mountouris, A. et al. Plasma gasification of sewage slidge: Process development and energy optimization. Energy Conversion and Management. 49 (2008) 2264-2271.

Extended Numerical and Analytical Model for Heat Transfer with and without Phase Change in non Ferric Powder for Thermal Spray Plasma Conditions

M. De Abreu [1], J.Puerta, F. Blanco, A. Guerrero [1] , J. Lira [1]
Departamento de Física, [1] Centro de Ing. de Superficies,
Universidad Simón Bolívar, Apdo: 8900, Caracas - Venezuela
E-mail: jpuerta@usb.ve, fblanco@usb.ve, aguerrero@usb.ve, jlira@usb.ve

Abstract. Splat generation of metal particles on substrates is a very important process for a variety of depositions techniques like plasma thermal spray. In this article an analytical and a numerical model are presented to describe the heating and melting process(heating with phase change), using adequate boundary conditions, global thermal capacity and the Stefan condition at the interface. An analytical dimensionless expression is determined for the temperature distribution as well the position and velocity of the moving liquid-solid interface in terms of Bessel functions. Comparison with the numerical results are carried out.

1. Introduction
The plasma spray process has demonstrated to be a very important technical process to produce high densities metal thin films, specially super conducting deposited on ceramic substrates. In those process a DC, RF or a Hybrid plasma torch are used in order to melt the solid state particles existing in the previously injected powder and deposited on the substrate, thus it is necessary the control of this process. In this sense, Sobolev [1] , Weir [2] and Mayer [3] have proposed several models to describe solid state particles behavior in a plasma among other contribution. In this attempt, a novel approach has been undertaken in order to achieve the temperature distribution for solid particles as well the dynamic of the melting processes in plasma [4]

2. Theoretical Model 1: Heating without phase change
1- Temperature distribution in solid particles.
In this article a description of the plasma thermal process in order to obtain the temperature distribution during the heating is here presented assuming global thermal capacity and following assumption:

a.- Particles are of a pure homogenous metal.
b.- Particles have spherical form at all.
c.- Heat loses by radiation are neglected.
d.- Heat transfer to the particle is assumed to be uniform.
e.- The Biot number is considered to be: $B_i \leq 0.01$.
f.- The particle flow is in the Stokes regime $R_e \leq 02$.

g.- Evaporation is neglected.

In this part of the problem the heat equation is used in spherical coordinates together with the convection boundary condition in the form

$$\frac{1}{\alpha}\frac{\partial T(r,t)}{\partial t} = \frac{2}{r}\frac{\partial T(r,t)}{\partial r} + \frac{\partial^2 T(r,t)}{\partial r}, \quad 0 \leq r \leq R, \quad t > 0. \tag{1}$$

$$\lambda_p \frac{\partial T(r,t)}{\partial r} = h(T_g - T(r,t)), \quad r = R_p \quad t > 0. \tag{2}$$

Where h is given by

$$h = \frac{N_u R_p}{\kappa}. \tag{3}$$

Here we use the relations obtained by Lee [5] to calculate the *Nusselt number* considering properties of the plasma as well for the particle, where the gas properties are evaluated by the film temperature $T = (T_g + T_p)/2$.

$$Nu = (2 + 0.6 R_e^{0.5} P_r^{0.33})(\frac{\rho_g \mu_g}{\rho_s \mu_s})^{0.6}(\frac{C_{pg}}{C_{ps}})^{0.3} \tag{4}$$

Now, in this section we put the equations in dimensionless form prior we find a solution using following relations

$$\theta = \frac{T - T_0}{Tm - T_0} \quad x = \frac{r}{R_p}, \quad \tau = \frac{\alpha t}{R_p^2} = F_o, \quad B_i = \frac{R_p h}{k_p} \tag{5}$$

In this way Eq(1) and Eq(2) can be written

$$\frac{\partial \theta}{\partial \tau} = \frac{2}{x}\frac{\partial \theta}{\partial x} + \frac{\partial^2 \theta}{\partial x^2} \quad 0 \leq x \leq 1, \quad \tau \geq 0 \tag{6}$$

Now, in an attempt to solve our equations we use the Laplace transform obtaining an expression that can be solved via Bessel functions where we observe $J_0(-is\sqrt{s}x)$ have simple zeroes in $-i\sqrt{s} = \gamma_1, \gamma_2 \cdots \gamma_n$, hence we get the solution for the temperature distribution inside the particle in the form

$$\theta(x,\tau) = 2\theta_r - \frac{2e^{-B_i\tau} J_0(\sqrt{B_i}x)\theta_r}{J_0(\sqrt{Bi})} + \sum_{n=0}^{\infty} \frac{2e^{-\gamma_n^2\tau} J_0(\sqrt{\gamma_n}x)}{\gamma_n(B_i + \gamma_n^2)J_1(\gamma_n)} \tag{7}$$

Here we define $\theta_r = (T_\infty - T_0)/(T_m - T_0)$. Eq(7) is the analytic solution solution of Eq(1). In figure 1 and 2 it is shown the temperature distribution in function of the position for aluminium, zinc and tin. A finite difference code was developed in order to calculate numerically and compare this results with the analytical calculation. Good agreement has been found with the numerical calculation and other authors.

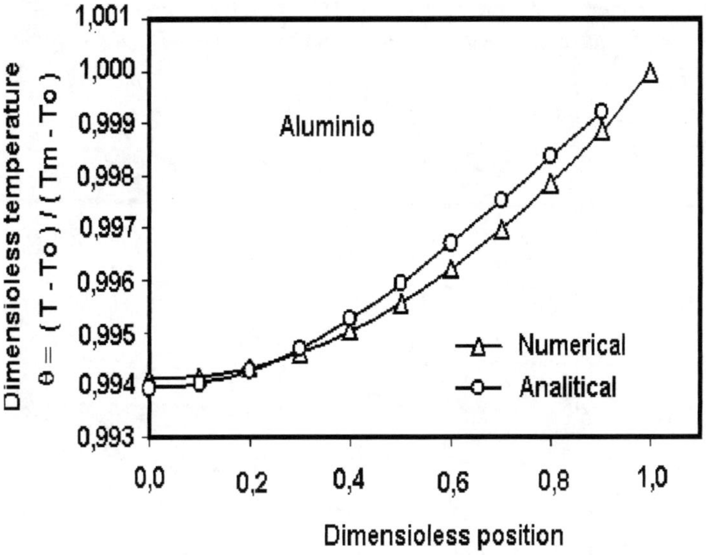

Figure 1: Temperature profile by the heating process in plasma for Aluminium particles with $50\mu m$ *in comparison with the numerical results*

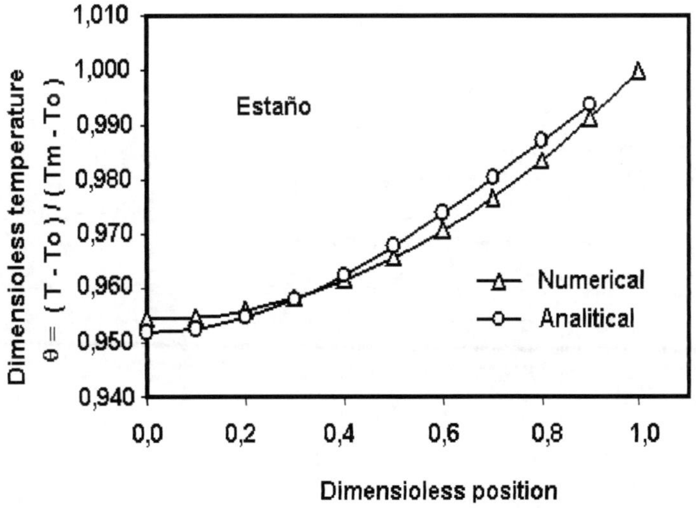

Figure 2: Temperature profile by the heating process in plasma for Tin particles with $50\mu m$ *in comparison with the numerical results*

3. Theoretical Model 2: Heating with phase change, melting problem

We consider a particle at melting temperature T_m has radius r. At time $t = 0$ the temperature of the boundary at $r = R_p$ is raised raised to T_0 in plasma, which is normally higher than T_m and maintained at this temperature for $t > 0$. As a result of that condition melting start at the surface where $r = R_p$ and the solid - liquid interface moves in that directions where r goes down to zero. Also, considering the solid phase being permanently at T_m, therefore the problem can be treated as a one phase problem. No density changes or temperature gradients here considered. Now, using Eq(1) and the boundary energy balance we get. [6]

$$k_s \frac{\partial T_l}{\partial r} - k_l \frac{\partial T_l}{\partial r} = \rho L \frac{ds(t)}{dt} \qquad r = s(t) \quad t > 0, \tag{8}$$

where L is the latent heat. In its dimensionless form we have

$$\frac{\partial \theta_s}{\partial r} - \frac{k_l}{k_s}\frac{\partial \theta_l}{\partial r} = \frac{1}{S_{te}}\frac{ds(\tau)}{d\tau} \qquad r = s(\tau) \quad \tau > 0, \tag{9}$$

In the case we have to discuss , one phase problem

$$-\frac{k_l}{k_s}\frac{\partial \theta_l}{\partial r} = \frac{1}{S_{te}}\frac{ds(\tau)}{d\tau} \qquad r = s(\tau) \quad \tau > 0, \tag{10}$$

where

$$S_{te} = \frac{C_{ps}(T_m - T_0)}{L} \qquad \theta_l = \frac{T_l(x,\tau) - T_m}{T_m - T_0} \tag{11}$$

By derivation of Eq(1) it easy to obtain

$$\frac{\partial \theta(x,\tau)}{\partial x} = \frac{2e^{-B_i \tau} J_1(\sqrt{B_i x})\theta_r}{J_0(\sqrt{B_i})} + \sum_0^{\infty} \frac{2e^{-\tau\gamma_n^2} J_1(\gamma_n x)}{(B_i + \gamma_n^2) J_1(\gamma_n)} \tag{12}$$

Now, at $x = s(\tau)$ and $\tau > 0$ we get

$$\begin{cases} \theta_l(s,\tau) = & \theta_m(s,\tau) = 1 \\ -\frac{k_l}{k_s}\left(\frac{\partial \theta_l}{\partial x}\right) = & \frac{1}{S_{te}}\frac{\partial s(\tau)}{\partial \tau} \end{cases} \tag{13}$$

From Eq(13) we achieve finally the position and velocity of the moving interface. Simultaneously we apply the method of finite differences like other authors to solve our system obtaining good agreement with the numerical results taken several systems like Aluminium , Tin and Zinc for the problem of heating an melting of metallic powders during plasma spray conditions . We shown this in figure 3, 4 and 5.

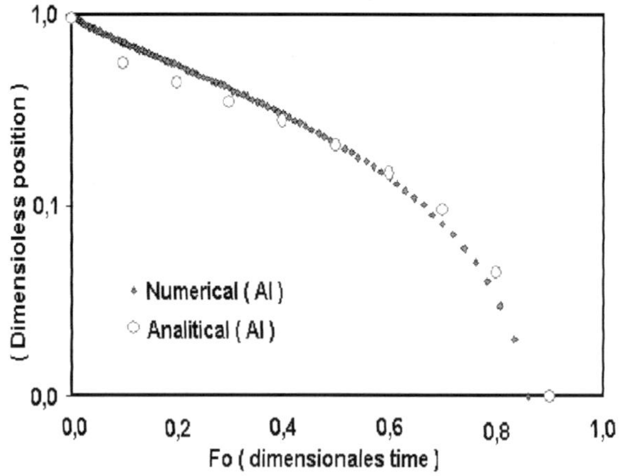

Figure 3: The melting front moves on toward the direction where r decreases from $r = R_p$ up to $r = 0$ for Al particles. It is shown good agreement with the numerical simulation.

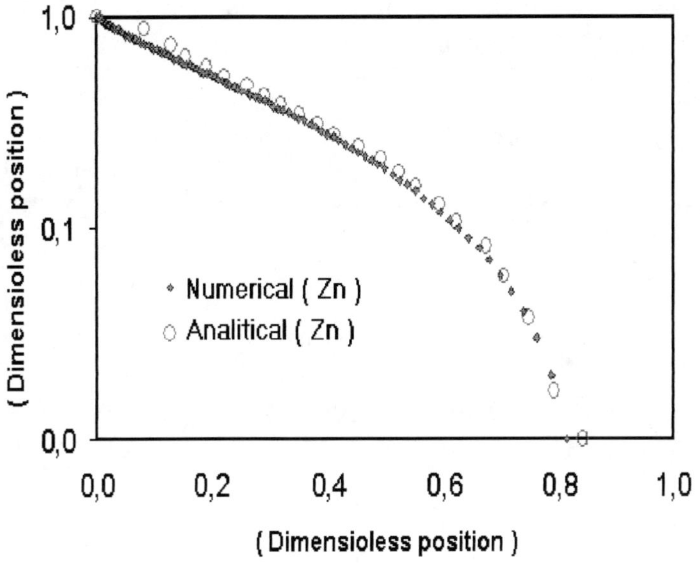

Figure 4: The melting front moves on toward the direction where r decreases from $r = R_p$ up to $r = 0$ for Al particles. It is shown good agreement with the numerical simulation.

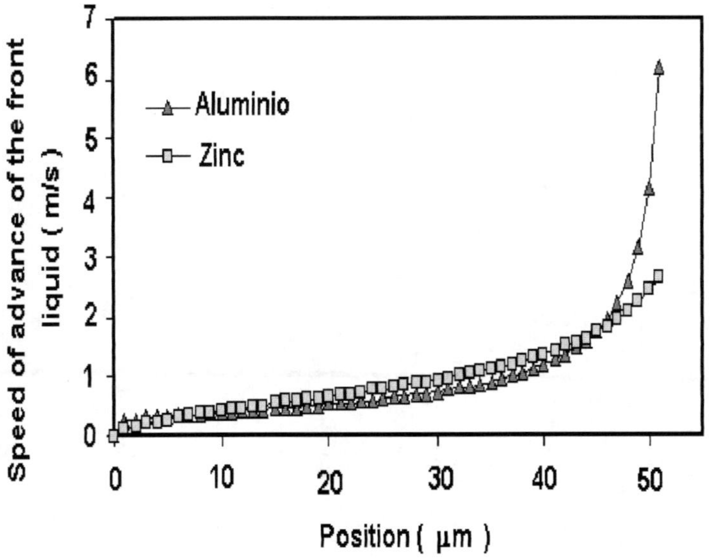

Figure 5: Velocity of the melting front in function of the radius for Al and Zn particles.

4. Conclusions

The heating problem without phase change and melting has been analyzed using a nobel analytical and a numerical approach to determine the temperature distribution inside of solids particles in plasma obtaining good agreement between both procedures. In the case of heating with phase change we use the Stefan condition at interface $s(x,t)$ and solving the system analytically and numerically. In the corresponding figures it is even to sees similar agreement as for the case without phase change. The velocity was obtained in an easy way for the movement of the melting front. It is important to point out, that our approach do not take into account density variation or density gradients.

5. Acknowledgments
Work supported by DID, Universidad Simón Bolívar

6. References
[1] V. Sobolev and M. Guilemany, Int. Material Reviews, **41** , 13 (1996)
[2] D. Weir, D. Apelian and B.Farouk, Journal of Heat and Mass Transfer, **109** , 971(1995)
[3] H. Mayer, Das Verhalten von Pulvern in Plasmastrahl, *Report Deutsche Keramische Gesselschaft*, **39**, 115(1962)
[4] M.N. Oezisik, *Heat Conduction*, Wiley and Sons N.W., pp. 201 - 226(1993)
[5] Y. C. Lee, Y. O. Chyou and Pfender, *Particle Dynamics and Particle Heat and Mass Transfer*, Porc. Plasma Chem. **5**, 391(1985)
[6] D. M. Christopher, Numerical Heat Transfer B, **39** 189(2001)

Journal of Physics: Conference Series **511** (2014) 012084 doi:10.1088/1742-6596/511/1/012084

Study of magnetic field enhanced plasma immersion ion implantation in Silicon

E J D M Pillaca[1], K G Kostov[1], M Ueda[2]

[1]Dept. of Physics and Chemistry, Faculty of Engineering – FEG, State University of São Paulo – UNESP, Av. Dr Ariberto Pereira da Cunha, 333, Guartinguetá, SP, Brazil

[2]Associated Laboratory of Plasma, National Institute for Space Research, S. J. Campos, SP, Brazil

e-mail: elver_jdd@feg.unesp.br

Abstract. A comparison between experimental measurements and numerical calculations of the ion current distribution in plasma immersion ion implantation (PIII) with external magnetic field is presented. Later, Silicon samples were implanted with nitrogen ion to analyze the effect on them. The magnetic field considered is essentially non-uniform and is generated by two magnetic coils installed on vacuum chamber. The presence of both, electric and magnetic field in PIII create a crossed **E**x**B** field system, promoting drift velocity of the plasma around the target. The results found shows that magnetized electrons drifting in **E**x**B** field provide electron-neutral collision. The efficient ionization increases the plasma density around the target where a magnetic confinement is formed. As result, the ion current density increases, promoting significant changes in the samples surface properties, especially in the surface wettability.

1. Introduction

Plasma immersion ion implantation is a widely used technique for modification material surface properties of complex-shaped three-dimensional (3-D) objects [1]. Basically, the sample to be treated is placed on a sample holder and immersed in plasma and is pulsed to negative high-voltage of several kV. The resultant electric field drives electrons away from the substrate in the time scale of the inverse electron plasma frequency ω_{ep}^{-1} (~ 10^{-9} sec). Thus an ion matrix sheath, which is a region depleted of electrons, forms around the sample holder. The negatively biased substrate will accelerate the ions within a time scale of the inverse ion plasma frequency ω_{ip}^{-1} (~ 10^{-6} sec) towards the substrate and subsequently implant them into the surface. For time order of times ω_{ip}^{-1} the ion flux during the voltage pulse is maitained by the continual expansion of the plasma-sheath edge until it reaches the stationary state for a long time ($t >> \omega_{ip}^{-1}$), where the sheath thickness does not change with the time.

In order to apply the PIII technique successfully to complex-shape workpiece is important to understand the plasma sheath dynamics. Since that sheath dynamics is responsible for the energies distribution of implanted ions several studies on plasma sheaths in PIII has been realized. One of the first to describe the sheath plasma was Langmuir [2]. Due to the importance of the sheath in application such as plasma technology and fusion research there are several works on it [3, 4, 5] and at the present time plasma sheath has received a considerable amount of attention [1]-[21]. A simple one-dimensional mathematical model was proposed by Lieberman [6] to analyze the sheath expansion in PIII. More realistic situations in PIII considering complex geometry, collision processes, etc. are usually difficult to treat analytically, so numerical methods are required and approaches as fluid model [7], kinetic model [8] and PIC model [9] have been used. There are some observations with respect to the sheath thickness, which if sheath thickness in PIII is not great enough compared with the workpiece size the conformal implantation is not ensured [10]. On the other hand, the sheath thickness should not be too thin (in high pressure) because it can cause electrical breakdown due to very high electrical field strength at the surface [11] [12].

An alternative approach in controlling the sheath thickness is to use a magnetic field. The presence of plasma anisotropy due to the magnetic field does make the analytic treatment very difficult. Consequently a large number of models consider only cases with planar geometry, where the magnetic field intersects the surface target at oblique angles [13][14]. Those studies show that a sheath is significantly affected by the magnetic field. Therefore, it is expected that a magnetic field will affect the sheath dynamics under PIII conditions. Initially, PIII with magnetic fields have been proposed for the suppression of secondary electrons. At the present time, several studies of PIII with external magnetic field have been realized [15][16][17]. On the other hand, the use of strong magnetic field is not desirable in PIII, since it could affect ion trajectories toward the substrate and influence the dose [18].

In this work we analyze the effects caused by the presence of a no-uniform magnetic field in PIII using experimental and numeric techniques. Silicon samples (100) were submitted to PIII treatment with external magnetic field.

2. Experimental and numerical set up

A schematic diagram of the equipment used is shown in figure 1. Cylindrical vacuum vessel of volume ~ 0.02 m^3 is equipped with coils wound on it in order to produces an axial magnetic field. In order to produce the magnetic field the coils are powered by a DC power supply to produce several coil current (I_{coil}). Due to problem of coil heating, the I_{coil} was limited to operate at current below 10 A. Nitrogen plasma is generated by DC glow discharge at pressure ~ $9.4\text{x}10^{-3}$ mbar. The procedure is described in detail in [19]. A rectangular steel holder (4x4x16 cm) was built to support Silicon samples (100) on it. It was positioned on the camera axis doing coincides with the center between the magnetic coils. The size of the samples was 1.5x1.5 cm. Before treatment, the samples were ultrasonically cleaned in acetone. The typical treatment parameters were as follow: gas working pressure $9.4\text{x}10^{-3}$ mbar, the pulse repetition rate 400 Hz. The pulse voltage was fixed at -3 kV, and the pulse length was 40 μs. The implantation time was 60 minutes.

Before the implantation process the magnetic field inside of the vacuum chamber was mapped using a gaussimeter. The characteristic voltage and current waveforms were recorded using a digital oscilloscope to later investigation. After the implantation process, the Silicon samples were characterized by Atomic Force Microscopy (AFM). A contact angle goniometer (Rame – Hart model 300) was used to measure the water contact angles of the samples in ambient conditions.

To have better insight of the PIII process with magnetic field a numeric simulation was realized. It has been carried out using the computer code KARAT [20]. It employs the particle-in-cell (PIC) algorithm for the calculation of charged particles motion in electromagnetic field. In this simulation we use the 2.5 D computer simulation in cylindrical geometry. Previously, numeric simulation using KARAT code has been used to study the PIII with several configurations of magnetic field [16][17][21] where enhanced of the PIII process was obtained.

KARAT code is implemented with the probabilistic Monte Carlo algorithm to simulate the ionization process. The emission of secondary electrons from the target upon the ion bombardment is also considered in the simulation. In our numerical calculations, rectangular cells with size $\Delta z = 3$ mm and $\Delta r = 1.3$ mm and a time step of $2.0\text{x}10^{-12}$ s were used.

The dimensions of the simulation vacuum chamber as well as the parameters of voltage and gas pressure were close the one experimental parameters described before. The simulation started with an uniform plasma, $n_i = n_e$, with initial density of $n_o = 5x10^{14}$ m^{-3}, electron temperature $T_e = 3$ eV, ion temperature $T_i = 0$ eV, and secondary electron coefficient of $\gamma = 2.4$ for N^+_2 ions in steel.

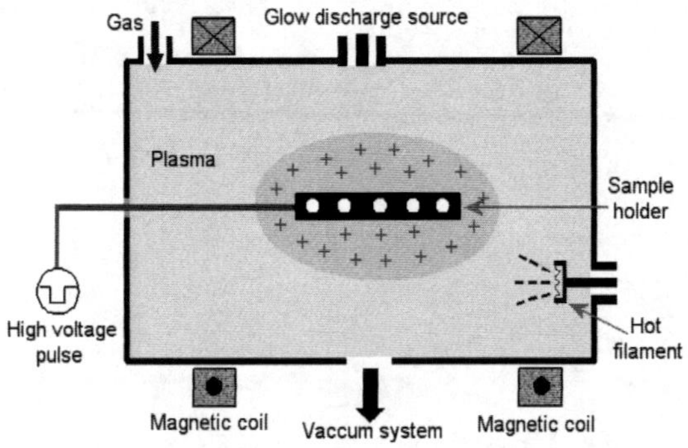

Figure 1. Schematic diagram of PIII vessel.

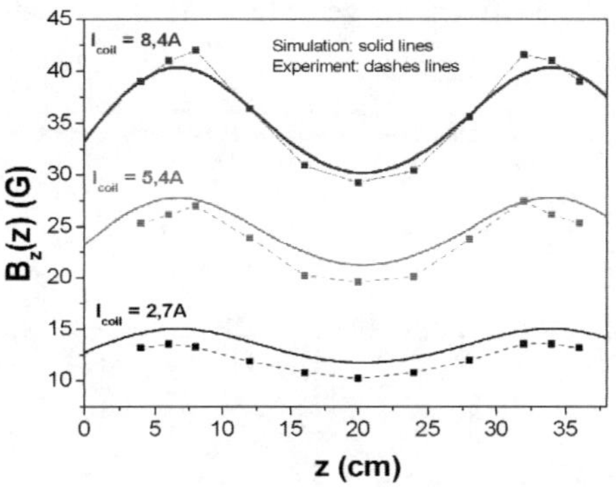

Figure 2. Distribution magnetic field B_z component measured inside of the vacuum chamber.

3. Results and discussions

Figure 2 shows the distribution of the axial magnetic field B_z along the z axis (r = 0 cm) produced by I_{coil} of 2.7, 5.3 and 8.4 A. The distribution is not uniform presenting a minimum at $z \sim 20$ cm. That point of symmetry correspond to the center of the target where B_z is weak. This distribution shows two characteristic regions with different amplitude of magnetic field. A minimum value (B_{min}) at $z \sim 20$ cm corresponds to the center between the coils (see figure 1), the maximum values (B_{max}) at $z \sim 7.0$ cm and $z \sim 35$ cm correspond to the coils locations. This configuration is well known as magnetic bottle.

We will define the ratio $R = B_{min}/B_{max}$. Observing figure 2 we can see that for low I_{coil}, $R \rightarrow 1$, that would roughly correspond to an uniform distribution of the magnetic field. The uniformity is lost when $R << 1$, where I_{coil} is high.

The non-uniform static magnetic field showed in figure 2 together with the radial electric field produced by application of the negative voltage to the target creates the system of **ExB** crossed field. In this system the behavior of magnetized electrons can be explain in terms of the guiding centre approximation. In presence of **ExB** they are accelerated along the magnetic field line [22]:

$$\frac{du_{\|}}{dt} = -\frac{\mu_m}{mB}[(\mathbf{B}\cdot\nabla)\mathbf{B}] \tag{1}$$

and drifts in the perpendicular direction:

$$u_{\perp} = \frac{\mathbf{E}x\mathbf{B}}{|\mathbf{B}|^2} + \frac{\mu_m}{qB^2}[(\nabla B)\mathrm{x}\mathbf{B}] + \frac{mu_{\|}^2}{qB^4}[(\mathbf{B}\cdot\nabla)\mathbf{B}]\mathrm{x}\mathbf{B} \tag{2}$$

where μ_m, m, q, **E** and **B** denote the magnetic moment, mass of the electron, electric charge, electric field and magnetic field, respectively. The terms of the right hand in (2) are the drift velocity due **ExB** fields, the magnetic field gradient drift and magnetic field curvature drift, respectively. Since that the magnitude of the $\mathbf{ExB/B^2}$ velocity is much larger than the gradient and curvature terms it produces the dominant effect in PIII.

 IOP Publishing
doi:10.1088/1742-6596/511/1/012084

Electrons between two cusps on both sides of the target are reflected in the direction of decreasing magnetic field according to (1), so an axial confinement happen since electrons are reflected from the strong magnetic field regions [17]. Supposing the electric field above the target central region is uniform, then the drift velocity at the fixed radial position is $u_\theta=E/B_z(z)$, where $B_z(z)$ has the axial distribution shown in figure 2.

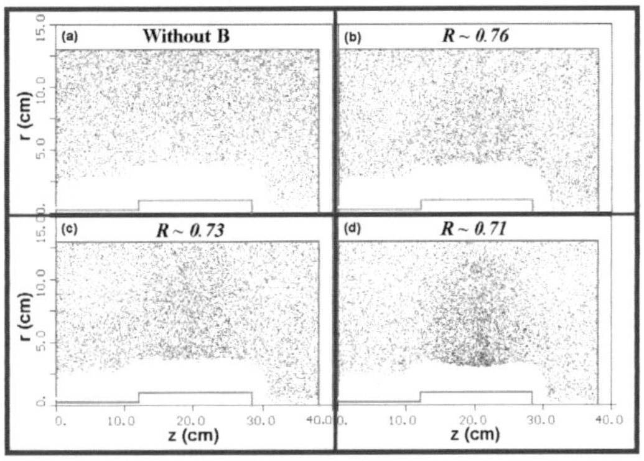

Figure 3. Plasma electron space distribution (a) without magnetic field (c)-(d) with magnetic field at 0.8 μs of simulation time.

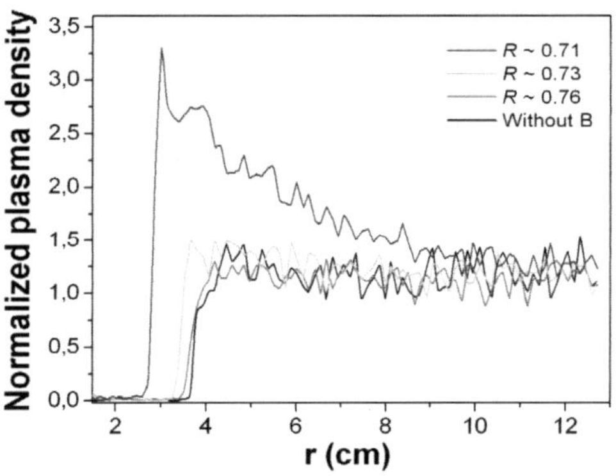

Figure 4. Radial plasma density without and with magnetic field at 0.8 μs of simulation time.

Figure 3 shows the plasma electron space distribution at 0.8 μs of simulation time. The electrons gyrating at azimuthal direction collide with neutrals providing efficient ionization of the background gas. The ionization by magnetization electrons creates a region with high plasma density. As we can deduce from figure 2, the region with high plasma density depends of the B_z distribution, or in other words, depend of ratio R. For low R, the magnetic confinement is more efficient (see figure 3(d)), while at high R the plasma confinement is weaker (see figure 3(b)). The radial plasma density distribution is shown in figure 4. The numeric results show that the plasma density increases by a factor of over two when $R \sim 0.71$. This region is localized between 3 cm to 5 cm radius of at the target middle plane ($z \sim 20$ cm). The growth of the plasma density creates plasma density gradients that affect the dynamics of the sheath (defined as the distance between the target and the plasma edge), reducing it in about 30 % with respects the one without magnetic field. Seeing the figure 2, figure 3 and figure 4 one can perceive that the plasma density growth happens in the region where the magnetic field has minimum value. The plasma sheath edge is also modified in front of target in the case of low ratio R. It can be attributed to the format of the magnetic field distribution over the target. The sheath dynamics as well as the sheath size affect the distribution of implantation current as is shows in figure 5. By decreasing ratio R the dose of implanted ion current onto the target increases, showing a non-uniform distribution at the center.

The relationship between the pulse current waveform and magnetic field was studied. Figure 6 shows the experimental current pulse for different ratios R. The increase of R modifies the current waveform at the characteristic region known as *pulse on*. As we can observe the regions *rise time* and *fall time* remain unaffected. As was previously discussed the increase of the current is attributed to the increase of the plasma density above the target due the efficient ionization of the gas and in presence of magnetic mirror which additionally enhance the process.

Both, experimental and numerical results show an increase of the ion current. It is attributed basically to magnetized electrons that play a key role in the increase of plasma density and consequently in the sheath dynamics. This increase can be predicted by the theory, where the relationship between ion current density and sheath size are related by:

$$j = \frac{4}{9}\varepsilon_0 \sqrt{\frac{2e}{M}} \frac{V^{\frac{3}{2}}}{R_t R_s(t)\beta^2} \quad (3)$$

that is the Child-Langmuir law in cylindrical coordinate. Here ε_0 is the free-space permittivity, e is the electron charge, M is the ion mass, R_s and R_t are the sheath and the target radii, respectively, and β is the tabulated function of R_s/R_t.

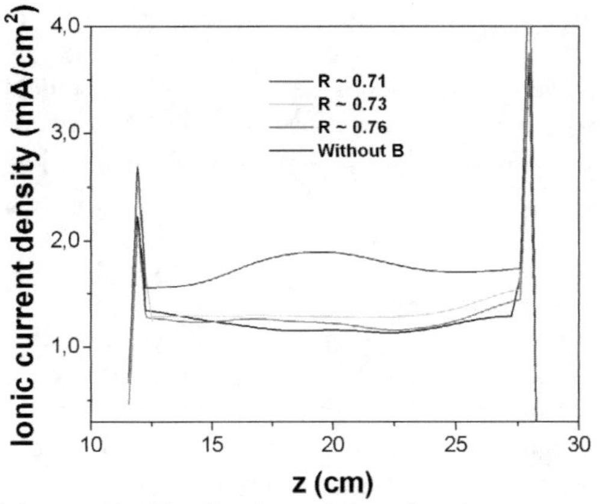

Figure 5. Distribution of ion density current along to target without magnetic field and several ratio R.

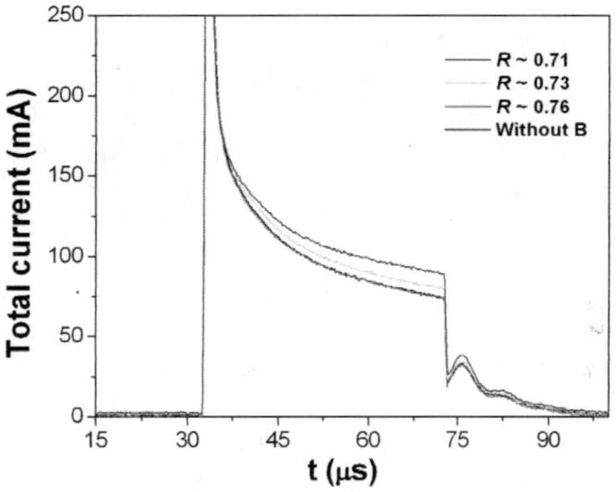

Figure 6. Waveform of the total current without magnetic field and several ratio R.

Silicon samples were implanted with N^+_2 and the results are shown in figure 7, which presents AFM images of a reference sample and samples after implantation process. The scan size of all images was 1x1 μm^2 area. The surface topography is modified when the magnetic field is applied. With the decreases of the ratio R the value of the roughness (R_q) decreases in about 36 % (see table 1) in comparison with the sample treated to case without **B**. This can be attributed to the modification of the ions orbit in presence of **ExB**. Although the ions radius is quite large comparing to the dimension of the vacuum chamber the ion incidence angle onto the surface of the target is influenced by the magnetic field.

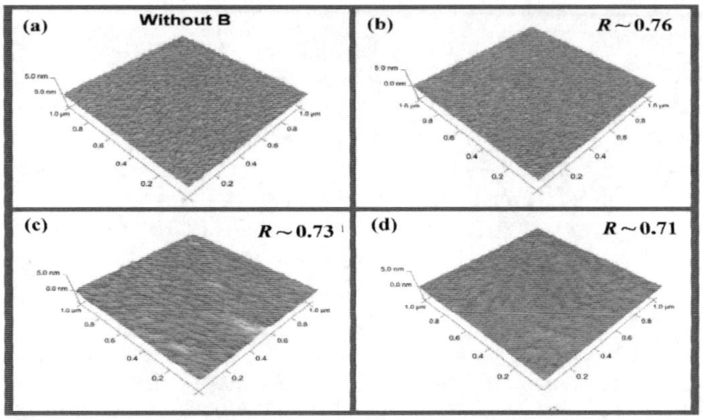

Figure 7. AFM images. (a) Without magnetic field; (b)-(d) with magnetic field.

Table 1. **Values of roughness and water contact angle of Silicon sample**

	Not treated	Without B	R ~ 0.76	R ~ 0.73	R ~ 0.71
Rougghness (R_q)	0.26	0.14	0.13	0.11	0.09
Contact angle (^{o})	63.9	83.2	99.8	99.7	99.9

The increase of the ion incident angle would produce more sputtering of the surface as observed in the figure 7(b)-(d). Modification of the roughness leads to modification of the contact angle as is shown in table 1. As we can observe, samples submitted to treatment without magnetic field shows increase in the value of the angle contact in about 31 % with respect the non-treated ones. In presence of the magnetic field this value increases in about 56 % for all the samples. From the increase of the contact angle can be related the different degree of sputtering in the presence of magnetic field. However, further studies are necessary to better understand this relationship.

4. Conclusions

We have presented experimental and numerical results of PIII in the presence of non uniform external magnetic field. We found that plasma density was increased at several *R*. This finding was attributed to collisions between magnetized electrons drifting around the target and neutrals. Numerical results showed that this happens in the region where the magnetic field has minimum value. As a consequence the increase of the plasma density leads to a growth of ion current density. Finally, Silicon samples were implanted and the effects of the magnetic field on their surface properties were analyzed. We found that surface morphology was modified, resulting in higher in roughness and water contact angle. The resulted obtained in this work show that plasma density modification by an external magnetic field with magnetic mirror configuration can have important consequences in the PIII process, as well as in applications in material treatments.

5. References

[1] Conrad J R 1987 *J. Appl. Phys.* **62** 777
[2] Langmuir I 1929 *Phys. Rev.* **33** 954.
[3] Riemann K U 1991 *J Phys. D* **24** 493
[4] Franklin R N 2003 *J. Phys. D* **36** R309
[5] Scheuer J T and Emmert G A 1990 *Phys. Fluids B* **2B** 445
[6] M. A. Lieberman M A 1989 *J. Appl. Phys.* **86** 2926
[7] Gong Y, Wang X, Duan P, Yu J, and Wang 2005 D *Phys. Plasmas* **12** 043501
[8] Scheuer J T and Emmert G A 1988 *Phys. Fluids* **31** 1748.
[9] Briehl B, Urbassek H M 2002 Surface and Coatings Technology **156** 131
[10] Rej D J, Faehl R J, Matossian J N 1997 *Surf. Coat. Techol.* **96** 45
[11] Anders A 2000 *Appl. Phys. Lett.* **76** 28
[12] Keidar M 2006 *IEEE Transactions on Plasma Science* **34** 804
[13] Ahedo E 1997 *Phys. Plasmas* **4** 4419
[14] Masoudi S F, Esmaeili S S, Jazavandi S 2010 *Vacuum* **84** 382
[15] Ueda M, Tan I H, Dallaqua R S, Rossi J O 2007 Surf. Coat. Technol. **201** 6597
[16] Kostov K G and Barroso J J 2006 IEEE Transactions on Plasma Science **34** 1127
[17] Kostov K G, Algatti M A, Pillaca E J D M, Kayama M E, Mota R P, and Honda R Y 2009 *Eur. Phys. J. D* 205
[18] Tan I H, Ueda M, Oliveira R M, Dallaqua R S, Reuther H 2007 Surf. Coat. Technol. **201** 4826
[19] Ueda M, Berni A, and Gomes G F, Beloto A F and Abramot E 1999 *J. Appl. Phys.* **86** 4821
[20] Tarakanov V P 1994 *User's Manual for Code KARAT* Berkeley VA: Berkeley Res. Assoc. Inc.
[21] Kostov K G, Barroso J J, and Ueda M. 2004 *Brazilian Journal of Physics* **34** 4B
[22] Schmidt G 1979 *Physcs of High Temperatur Plasma* second edition Academic Press INC

ICPP2010 & LAWPP2010 IOP Publishing
Journal of Physics: Conference Series **511** (2014) 012085 doi:10.1088/1742-6596/511/1/012085

On the possibility of measuring the optical spectrum of light sources by means of a rotating mirror

Luis Bernal[1], Luis Bilbao[2]

[1] Department of Physics, University of Mar del Plata, Argentina.

[2] Instituto de Física del Plasma, CONICET-UBA,Ciudad Universitaria, Pab. I, 1428 Buenos Aires, Argentina.

E-mail: bilbao@df.uba.ar

Abstract. We report the construction and operation of an apparatus for measuring optical Doppler shift based on a rotating mirror, and its preliminary application to the measurement of the spectrum of some light sources. We describe preliminary results, and discuss the detection limits and the possibility of using the apparatus for plasma diagnostic

1. Introduction

In a previous paper [1], we described an apparatus for measuring the Doppler shift using a rotating mirror. Optical Doppler shift experiments are not a simple task since the light source cannot usually be moved in a sufficiently smooth and uniform manner to keep the level of noise well below of that of the signal.

Using a rotating mirror it is possible to overcome many of the noise generating effects. The beam from a light source passes a beam splitter that separates the beam, part is reflected from the advancing side of a rotating mirror; part is reflected from the receding part of a rotating mirror. After reflection the beams are recombined using a beam splitter and measured by a fast photo-detector. Beat between the two beams produces fluctuations in the light intensity at the photo-detector.

Since the spectrum of the measured interference signal is proportional to the spectrum of the original source, then it is possible to use this experiment as an online scanner of the light spectrum of the source.

2. Apparatus

Using a rotating mirror in which one beam is reflected from the advancing side and the other beam is reflected from the receding part of a rotating mirror two effective sources with different velocities can be created. The Doppler beat frequency, ν_b, between the two sources is given by [1]

$$\nu_b = \frac{4\pi dF}{\lambda} \tag{1}$$

where F is the frequency of the rotating mirror, d the beam separation and λ the wavelength of the light. In figure 1 a sample beat pattern of a laser source is shown. The FFT (also shown in figure 1) exhibits a peak at the frequency ν_b given by (1).

Note that the beat frequency is a function of the rotational frequency. Therefore, in order to provide a reliable measurement the old motor (used in the previous version) was replaced by an outrunner brushless motor located at the bottom of the mirror. Also an electronic controller was built so the rotational frequency can be varied in a 1 Hz step up to 100 Hz. In figure 2 the new mounting is shown at rest (left) and operating at 50 Hz (right) were the reflected laser beams (separated by 2.4 cm) are also visible. The experimental setup is shown in figure 3, were two sources may be used simultaneously.

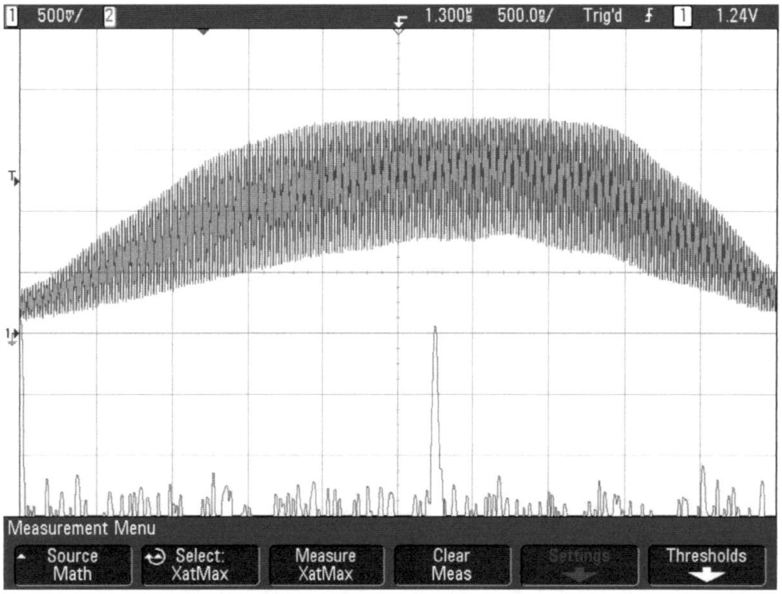

Figure 1. Sample beat pattern for a laser (upper trace) and its correspondent FFT (in dB, lower trace).

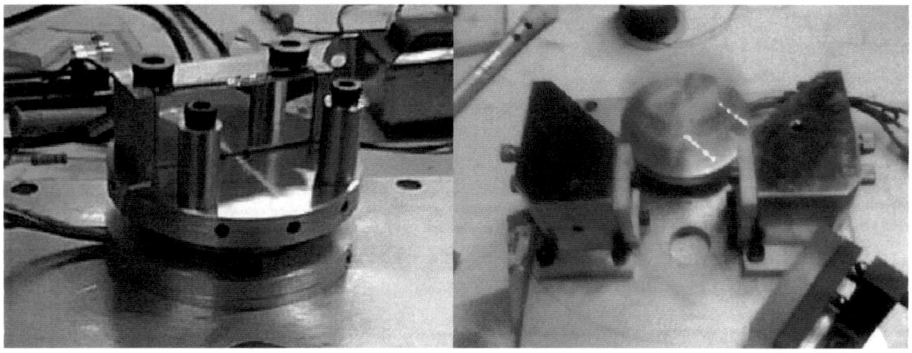

Figure 2. Improved mounting and driver of the rotating mirror (left). The mirror working at 50 Hz (center-right) in between the optical positioners. The green laser beams are also visible.

3. Calibration and preliminary tests

The first idea was to use the apparatus for resolving spectral lines of a given source. Working at 50 Hz and a beam separation of 0.024 m the Doppler beat of the visible spectrum lies between 20 and 40 MHz.

Using a 500,000 record length, 2 GSa/s digitizer the error of the beat frequency due to the FFT resolution is lower than the error due to the rotational stability that is about 0.1%. Using these values the theoretical resolution will be of about 0.5 nm.

The actual resolution will depend also on the optical components. During these preliminary feasibility tests we have used teaching-grade positioners and optics (i.e. from standard teaching kits).

The calibration of the apparatus is performed using monochromatic sources of known wavelength. In figure 4 the simultaneous Doppler beat of a red (658.1 nm) and a green lasers (543.5 nm) is shown. The two peaks of the FFT are clearly resolved and can be used to define the scale for the wavelength.

Finally, a Mercury lamp was used as a source. In figure 5 the Doppler beat is shown. Acquiring at 100 MSa/s the peaks 576.9+579.0 nm and 546.1 nm are resolved with a resolution of about 2 nm.

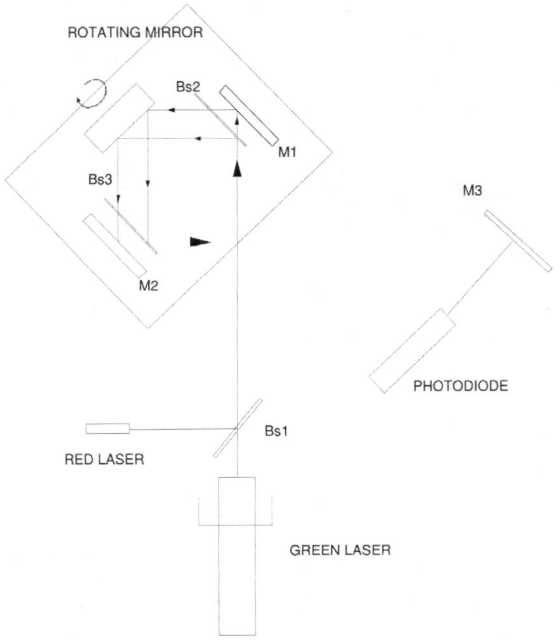

Figure 3. Experimental setup.

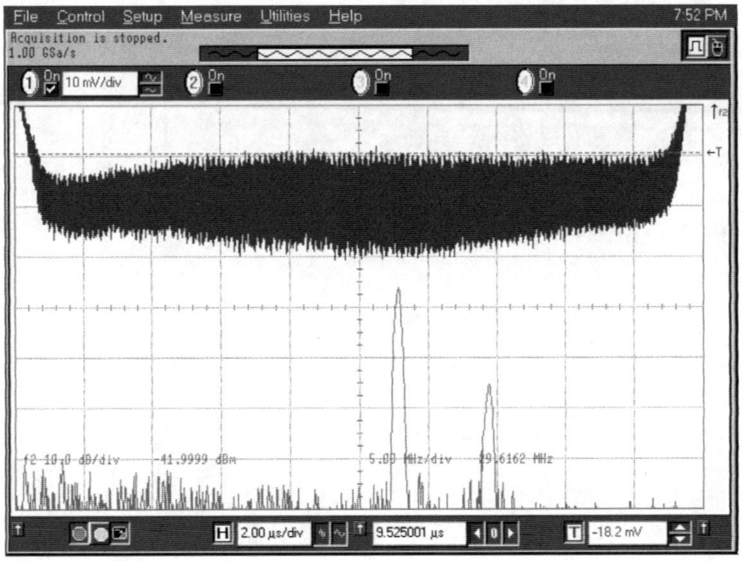

Figure 4. Sample beat pattern at 1 GSa/s for a red and a green lasers (upper trace) and its correspondent FFT (in dB, lower trace). The FFT shows the peaks of the red laser 658.1 nm (left peak) and the green laser 543.5 nm (right peak).

Journal of Physics: Conference Series **511** (2014) 012085 doi:10.1088/1742-6596/511/1/012085

4. Conclusions

Using a simple Doppler technique based on the beat between the two beams that reflects off a rotating mirror, it is possible to obtain the line spectrum (or at list a part of it) of a light source with an inexpensive apparatus.

By varying the rotational speed and the beam separation it is possible to scan a wide part of the electromagnetic spectrum. The method is not limited to visible light. The method can be used in any range of the electromagnetic spectrum as long as beam-splitters and mirrors are available.

Work is still in progress. According to these preliminary results a resolution of 0.1 nm will be possible to achieve by an appropriated choice of parameters.

Plasmas are an interesting source of electromagnetic radiation. The present method can be applied to stationary or quasi-stationary plasmas (the maximum time resolution is given by the rotational frequency) to obtain an online scanner of the light spectrum as a complement to more sophisticated and expensive spectrometers.

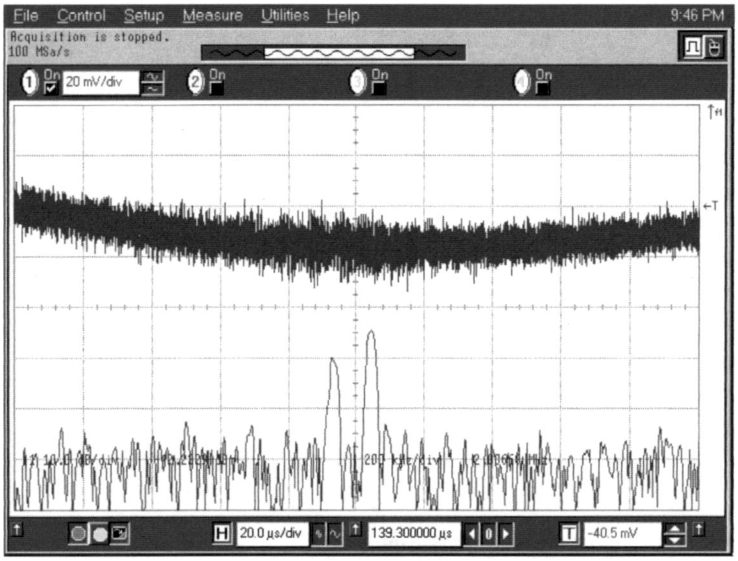

Figure 5. Sample beat pattern at 100 MSa/s for a Hg lamp (upper trace) and its correspondent FFT (in dB, lower trace). The FFT shows the peaks for 576.9+579.0 nm lines (left peak) and for 546.1 nm line (right peak).

5. Acknowledgements

Work partially supported by grant X206 University of Buenos Aires.

References

[1] L. Bernal and L. Bilbao 2007 *Am. J. Phys.* **75** 216.

Adaptive Embedded Digital System for Plasma Diagnostics

Angel González, Othoniel Rodríguez, Osvaldo Mangual, Eduardo Ponce and Xavier Vélez
Polytechnic University of Puerto Rico, San Juan, PR 00918.
E-mail: agonzalez@pupr.edu

Abstract. An Adaptive Embedded Digital System to perform plasma diagnostics using electrostatic probes was developed at the Plasma Engineering Laboratory at Polytechnic University of Puerto Rico. The system will replace the existing instrumentation at the Laboratory, using reconfigurable hardware to minimize the equipment and software needed to perform diagnostics. The adaptability of the design resides on the possibility of replacing the computational algorithm on the fly, allowing to use the same hardware for different probes. The system was prototyped using Very High Speed Integrated Circuits Hardware Description Language (VHDL) into an Field Programmable Gate Array (FPGA) board. The design of the Embedded Digital System includes a Zero Phase Digital Filter, a Derivative Unit, and a Computational Unit designed using the VHDL-2008 Support Library. The prototype is able to compute the Plasma Electron Temperature and Density from a Single Langmuir probe. The system was tested using real data previously acquired from a single Langmuir probe. The plasma parameters obtained from the embedded system were compared with results computed using matlab yielding excellent matching. The new embedded system operates on 4096 samples versus 500 on the previous system, and completes its computations in 26 milliseconds compared with about 15 seconds on the previous system.

Introduction

This work reports the design and implementation of an Adaptive Embedded Digital System (AEDS) to perform plasma diagnostics using electrostatic probes [1]. The Plasma Engineering Laboratory at Polytechnic University of Puerto Rico (PUPR) currently performs plasma diagnostics using a Keithley 2400 source-meter [2] to apply a ramp voltage to the probe and capture the probe's current, and a digital computer running a LabView [3] program that performs the diagnostics computations. The existing system is able to compute plasma parameters for Single and Double Langmuir probes, Emissive probe, and Ion/Electron Energy Analyzers[4]. The system is notoriously slow and difficult to maintain. The new System improves the speed, size and cost, and flexibility of the system. The main components of the AEDS are a 12-bit Digital To Analog Converter (DAC), a 14-bits Analog to Digital Converter (ADC), a Zero Phase Digital Filter, an Exponential Unit (EU), a Computational Unit (CU). A key feature of the design is the possibility of replacing the Computational Unit on-the-fly, allowing to use of the same hardware for probes with different computational algorithms. The system was prototyped using an Xilinx Spartan-3A FPGA Starter Kit board [5] for the Single Langmuir probe, and is able to compute the Plasma Electron Temperature and Density.

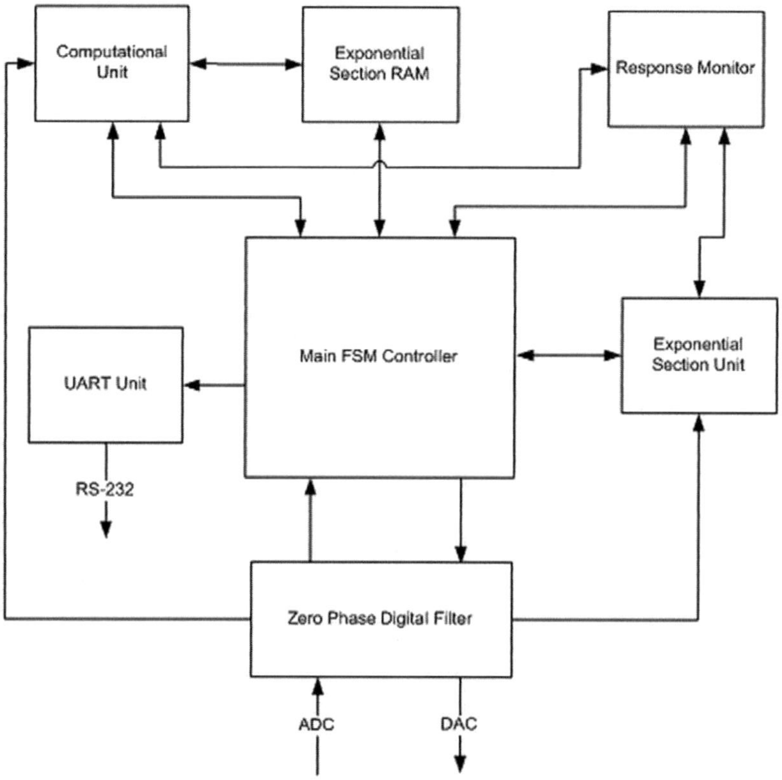

Figure 1. System Block Diagram

Description of the Hardware

This section describes the main components of the system. The hardware was designed using VHDL [6], and using the new VHDL-2008 Support Library [7], which provides for fixed-point and floating-point types required by the design. Figure 1 shows the system block diagram. The system is developed in a modular fashion to enable the substitution of the Computational Unit depending on the type of probe being used.

Zero Phase Digital Filter

Noise degradation of the I-V characteristic causes computational errors. A Zero Phase Digital Low Pass Filter was used in this project to remove the noise from the input signal avoiding phase distortion. This filter performs its function by processing the input data forward and reverse directions, canceling the phase shift produced by the filtering. Figure 2 shows the digital filter block diagram, consisting of a Filter Module, and a Dual-Port RAM module for storing the intermediate and final filtered values.

Exponential Section Unit

The Exponential Section Unit identifies the boundaries of that section, by detecting the zero crossing point of the probe current (leftmost boundary) and the maximum of the derivative of the probe current where the probe voltage equals the space potential (the rightmost boundary).

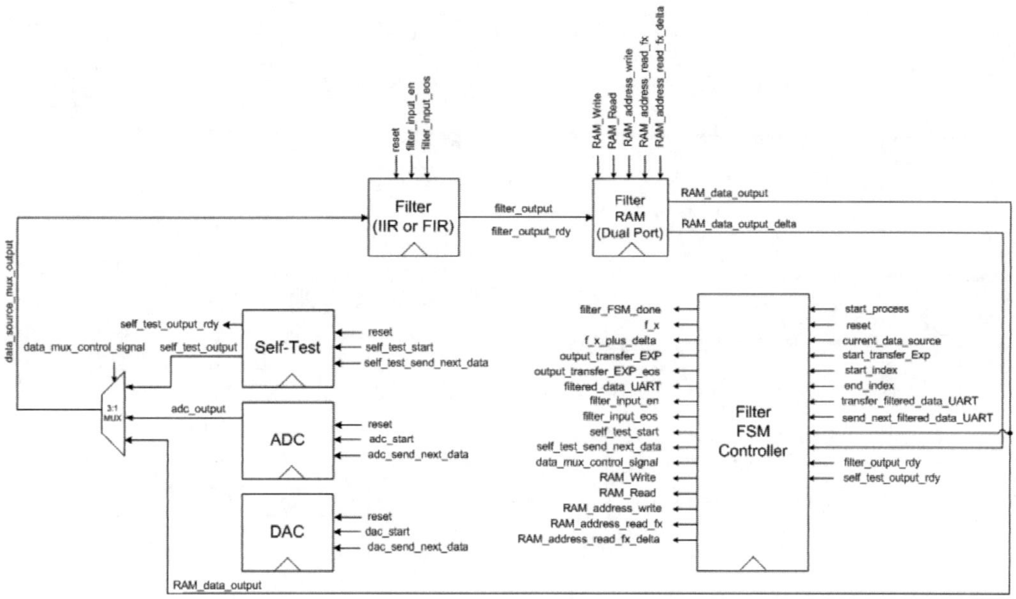

Figure 2. Schematic diagram of Zero Phase Digital Filter

The Exponential Section Unit also identifies three type of errors that could be present in the signals:

(i) No zero crossing;

(ii) Multiple zero crossings;

(iii) Index of the maximum derivative before the index of the zero crossing.

Each error is specifically associated with a control signal to indicate the Response Monitor of the error that has occurred.

Computational Unit

The Computational Unit performs the computations to obtain the parameters, from the I-V characteristic of the probe. The prototype uses equations 1 and 2 for the Electron Temperature KT_e and Density n respectively [8].

$$KT_e = \frac{1}{\max \frac{d\ln|I|}{dV}} \qquad (1) \qquad\qquad n_e = -\frac{I_{es}}{eA\sqrt{\frac{KT_e}{2\pi m_e}}} \qquad (2)$$

where m_e is the electron mass, e is the electron charge, and A is the probe area.

Main Finite State Machine (FSM) Controller

The Main FSM Controller controls and coordinates the system operation.

Response Monitor

The Response Monitor is responsible of receiving any errors signals from the EU and alert the user about them.

Journal of Physics: Conference Series **511** (2014) 012086 doi:10.1088/1742-6596/511/1/012086

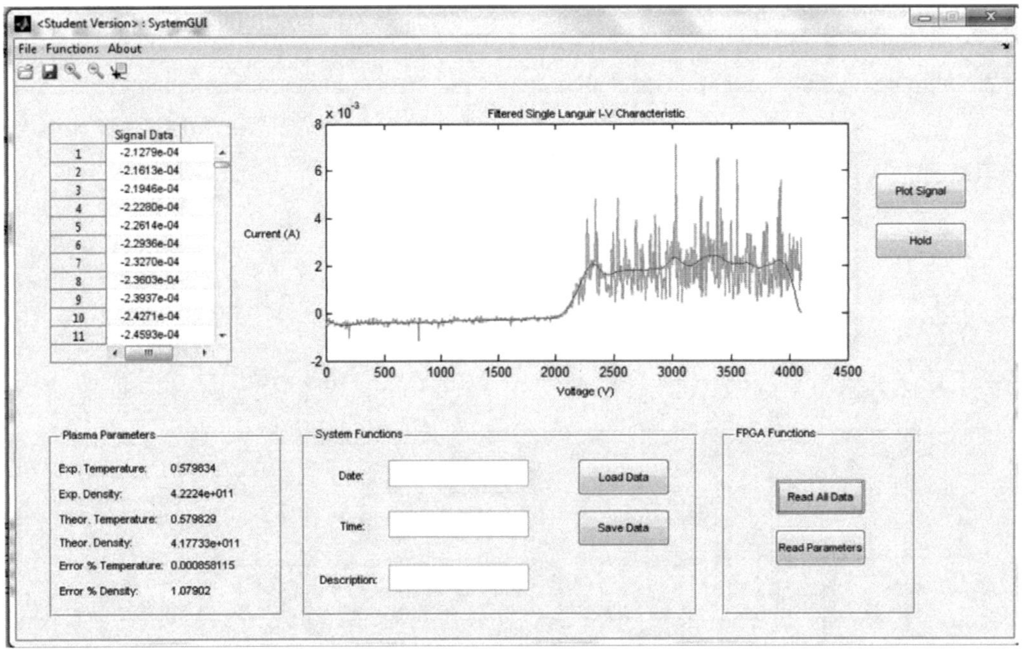

Figure 3. Result using a test signal

Simulations and Results

Simulations were performed for each individual module, as well as for the complete system. The design was loaded into the FPGA and its functionality validated in hardware. Figure 3 shows KT_e and n_e obtained from the system using data previously acquired. These experimental parameters were compared with theoretical results, with and error of 0.000858115% for KT_e and 1.07902% for n_e. Several signals were used as stimulus to the system; the average error percent was 2.644% for KT_e and 2.13% for n_e. The accuracy of the results is a function of the number of bits used for the calculations. This project used 27-bits words for the data because of the size of the FPGA used for prototype. Notice that this is not a design limitation: by increasing the quantity of bits for the number scheme in the system the results will be more accurate. Figure 4 depicts a graph showing the relationship between the number of bits in the fixed point numbers of the system and the error percents of the Plasma Electron Temperature and Density.

The new embedded system operates on 4096 samples versus 500 on the previous system, and completes its computations in 26 milliseconds compared with about 15 seconds on the previous system.

Conclusions

An Adaptive Embedded Digital System was designed, implemented and tested to perform plasma diagnostics with electrostatic probes. Two novel features are included in this design: The adaptability of the computational unit for various probe types, and the use of a zero phase filter to remove the undesired noise from the current signal from the probe. The system was tested with data previously acquired from a Single Langmuir probe showing an acceptable accuracy when compared with computations made with software. The system is able to detect errors in

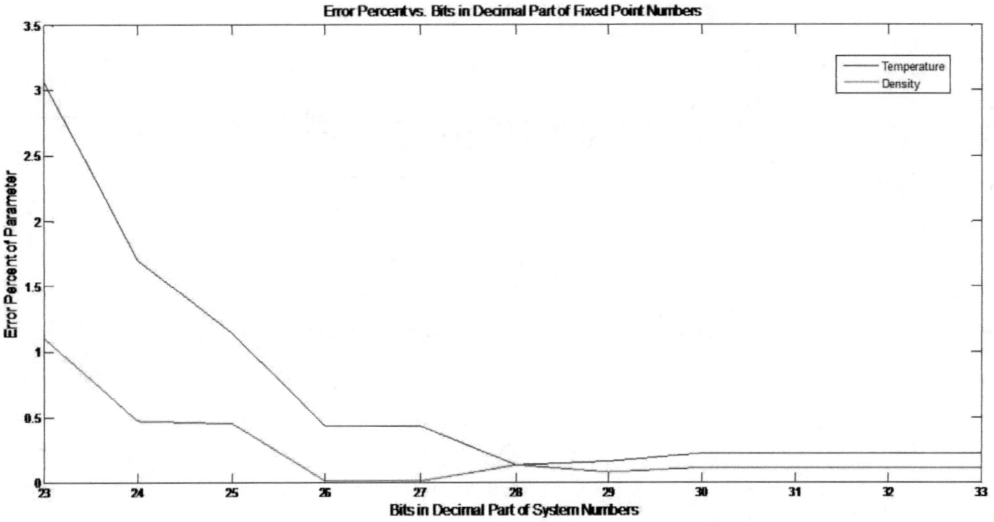

Figure 4. Relation between quantity of bits and the error percent in results

the data acquisition, and interrupt the measurement in those events. Further advantages of the new system are: low hardware cost, minimum size, high scalability, portability, and very high speed computations capabilities.

References

[1] Mercado-Santiago M 2008 Design of a FPGA-based data analysis circuit for the PUPR mirror/cusp plasma machine department of Electrical and Computer Engineering and Computer Science
[2] Keithley Instruments I 2002 *2400 Series Source Meter User Manual* Seventh Printing Cleveland, Ohio, U.S.A.
[3] Communications T 2003 *LabVIEW User Manual* National Instruments Corporation 11500 North Mopac Expressway Austin, Texas 78759-3504
[4] Leal-Quirós E 2005 Basic plasma diagnostic; probes & analyzers URL `http://www.pupr.edu/plasma`
[5] Xilinx I 2007 *Spartan-3A FPGA Starter Kit Board User Guide* Xilinx, Inc 2100 Logic Drive, San Jose, CA 95124, U.S.A. ug330 (v1.3) ed for Revision C Board
[6] Jr Charles H Roth L K J 2007 *Digital Systems Design Using VHDL* 2nd ed (Boston, MA: CL-Engineering)
[7] Bishop D W 2009 Vhdl-2008 support library URL `http://www.vhdl.org/fphdl/`
[8] Leal-Quiros E 1989 *Novel Probes and Analyzers for RF Heated Plasmas and Microwave Heated Plasmas for Controlled Fusion Research: The hyperbolic Energy Analyzer, The Magnetic Moment Analyzer, The Double Energy Analyzer, and The Variable Energy Analyzer* Ph.D. thesis University of Missouri-Columbia

Absolute Calibration of Proportional Counter Based Fast Pulsed Neutron Detectors with Resolution Below 10^5 neutron/pulse

A Tarifeño-Saldivia[5,2], R E Mayer[3,4], C Pavez[1,2] and L Soto[1,2]

[1] Comisión Chilena de Energía Nuclear, Casilla 188-D, Santiago, Chile
[2] Center for Research and Applications in Plasma Physics and Pulsed Power, P4, Chile
[3] Centro Atómico Bariloche, Comisión Nacional de Energía Atómic, Argentina
[4] PLADEMA, Argentina

[5] Universidad de Concepción, Chile.

atarisal@gmail.com

Abstract. A method for absolute calibration of proportional counters for pulsed fast neutrons is presented. The method is based on the use of an isotopic standard source and development of a model for counting detected events from area of a signal compounded by single piled up neutron pulses. Effects of detection counting statistics and electrical background noise are also considered. The method is applied in detectors used for D-D neutron yield measurements in low emission plasma focus devices.

1. Introduction.

The measurement of pulsed fast neutron production from plasma focus devices (pulse width $T << 1\mu s$) is usually done by foil activation techniques (Ag, In, Be). These techniques allow detection limits down to 10^5-10^6 n/pulse. For yields less than 10^6 n/pulse (low emission regime) neutron moderation and proportional counters such as ^3He or BF_3 tubes could be used [1]. When a moderated proportional counter is irradiated by an extremely short pulse of fast neutrons (10-100 ns pulse width), neutrons are moderated to thermal energies by multiple scattering and thus the pulse is spread over tens of microseconds. Depending on the neutron fluence and the preamplifier RC time constant, neutron signals will be piled up making impossible to count single events. The above notwithstanding the output detector signal area will be proportional to the number of detected events. Two schemes have been proposed to tackle the ensuing pulse pile up: I) Use of high counting rate electronics and control of the count rate by detector-source separation. This technique has been reported to be successful in the 10^6 n/pulse range and is not proved in lower emission ranges [2]. II) Cross calibration on the accumulated charge (preamplifier output signal) generated in the tube by the neutron burst using as neutron reference a foil activation detector [3, 4]. As result detection limits are transferred by the calibration process from the reference detector, thus no improvement in the detection resolution is obtained by this method for the low emission regime despite the high efficiency of neutron gaseous proportional counters. At present, there is a lack of an absolute calibration technique for pulsed fast neutron detectors based on proportional counters. In this work, we propose a calibration method to overcome this deficiency, it is based on the use of isotopic standard sources as neutron reference and a counting model on the net signal area which also accounts for the different sources of fluctuations in our detection system (detection counting statistics, pulse pile up statistics and electrical noise).

2. Experimental setup and detection system.

Our detection systems (two units) are based on a paraffin wax moderator ($45x15x15cm^3$) and a ^{3}He tube (model LND 2523) sheathed by a 3-mm thick lead sheet to stop x-rays and prevent their subsequent detection. The moderator is surrounded by a cadmium sheet to absorb thermal neutrons. The tube is connected to a preamplifier (CANBERRA 2006) and the high voltage feed is set to 4kV. In order improve the signal-to-noise ratio, the preamplifier is adjusted with a conversion gain of 235 *mV/M-ion-par* which produces a 5x voltage output factor scale. Its RC time constant is set nominally to 50 μs, this is the nominal fall time for individual pulse when risetime is less than hundred nanoseconds. Signal output from the preamplifier is recorded by an oscilloscope (Tektronix model TDS 684) at 1 MΩ input impedance, AC coupling, and 20MHz band width. Time scale is set to 200 μs/div, and horizontal scale is set in accordance with moderator decay time (governing neutron feeding into the ^{3}He tube). Direct measurements of the total signal area are obtained using the internal integration oscilloscope function on a 1500μs time window, which starts close to the trigger point. The oscilloscope is triggered by the electromagnetic pulse generated from the discharge. In the absence of electrical discharge, there is electrical background noise (EBN), which still contributes to the total signal area. For a specific time window, the EBN depends on oscilloscope horizontal scale; it is affected by changes in laboratory room temperature and should be monitored *in-situ* together with neutron measurements. Its values are well described by a normal distribution. In practical situations, the net signal area ($X_{T,net}$) should be obtained by discounting the EBN mean value from the total signal area. Natural background counting is negligible for the 1500μs time window (r ~ $0.9s^{-1}$). Single neutron signal were characterized using non-linear least squares methods, as results it was found that in the detection system neutron signals have an exponential rise and fall with characteristic constants of $1.2\mu s$ and $75\mu s$ respectively [5]. In general a single neutron area signal is then given by $A=FA\cdot V_0$, where V_0 is the signal pulse height and FA a constant which depends on the detection system. Consequently, the pulse area spectrum for the counter is easily obtained from the pulse height spectrum. For the case of our detectors, it was found that $FA=91.63\mu s$.

3. Counting model and measurement of neutron production.

The expected number of detected events in the counter is given the model

$$m(X_{T,net}, dB) = [a \cdot X_{T,net}] \pm \sqrt{f_0 \cdot [a \cdot X_{T,net}] + a^2 \cdot dB^2}\ , \tag{1}$$

where dB is the standard deviation of the observed distribution for the electrical background noise, and a and f_0 are constants which depend on the detector system. The values of these constants are obtained from numerical calculations. From the experimental probability density function and using the Monte Carlo method is possible to reproduce the random detector system response including fluctuations by counting statistics, piling up statistics and electrical background noise. From these calculations, the value for a and f_0 were carefully selected to ensure the predictability power of the model [6]. The constant a is related with the convergence to the central limit in the piling up statistics, while f_0 includes effects of counting statistics and piling up statistics in the fluctuations of the counting model. The term $a^2 \cdot dB^2$ in eq. (1) includes the effect of the EBN in the uncertainties of the model.

The absolute calibration of our detection system is given by the fact that counting efficiency and thus the detector calibration factor (j_c) is obtained using an isotopic source as neutron reference. The last is valid in the approximation that source and fusion neutron spectra produce identical results after undergoing multiple scattering moderation inside the detector.

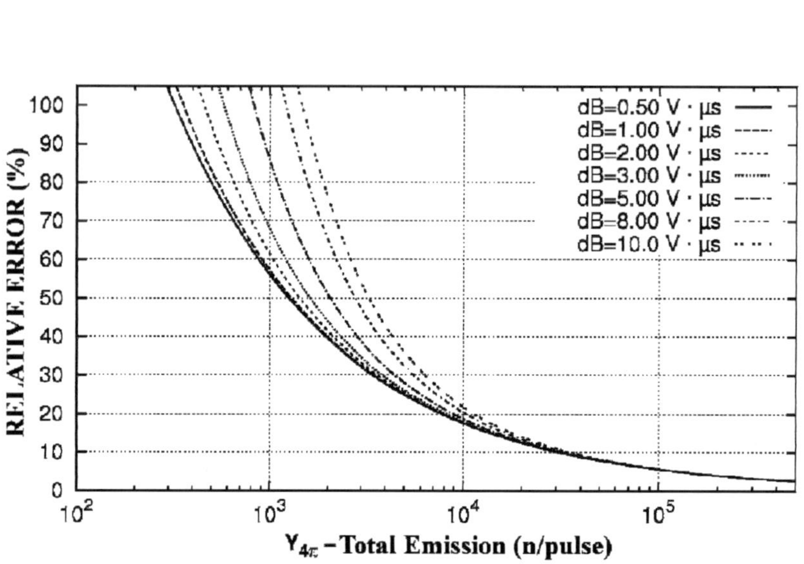

Figure 1. Relative standard error for our detection systems at distances from the source in the range $14\text{cm} \le r \le 40\text{cm}$.

Using the counting model, the neutron yield from the source by solid angle is obtained from

$$Y = j_c \cdot m\left(1 \pm \sqrt{\left(\frac{dj_c}{j_c}\right)^2 + \frac{f_0}{m} + \frac{a^2 \cdot dB^2}{m^2}}\right). \qquad (2)$$

For our detection systems characteristics values $f_0 = 1.0658$ and $a = 0.44617$ were found. When detectors are placed at distances from the source in the range $14\text{cm} \le r \le 40\text{cm}$, a mean value $j_c = 14.2$ is obtained. In figure 1 are shown different plots of the relative standard error for measurements in the low range emission. It is seen that reasonable uncertainties are obtained with the methodology presented in this work for emissions lower of 10^4 n/pulse, which means an improvement in the detection limit of almost two order of magnitude in comparison with state of art measurement techniques.

4. Acknowledgments.

Authors would like to acknowledge financial support from the programs PBCT-ACT26 and PSD-01, Center for Research and Applications in Plasma Physics and Pulsed Power – P^4 Chile.

References

[1] Mayer R E, Milanese M M, Moroso R, Pouzo J and Gillette V H 1998 *Proc. Latin American Workshop on Plasma Phys.* vol. 1.

[2] Waheed A, Aqil-ur-Rahman, Zakaullah M, and Murtaza G 2000 *Journal of Fusion Energy* **19**(1), 91-92.

[3] Moreno J, Birstein L, Mayer R E, Silva P and Soto L 2008 *Meas. Sci. Technol.* **19** 087002.

[4] Verma R, Roshan M V, Malik F, Lee P, Lee S, Springham S V, Tan T L, Krishnanty M, Rawat R S 2008 *Plasma Sources Sci. Technol.* **17** 045020.

[5] Tarifeño-Saldivia A, Mayer R E, Pavez C and Soto L 2010 *AIP Conf. Proc.* **1265** 501-504.

[6] Tarifeño-Saldivia A 2011 *Experimental Study of a Tens of Joules Small Fast Plasma Focus Discharge Emitting Neutrons*. Phd thesis, Universidad de Concepción, Chile (In Spanish).

Journal of Physics: Conference Series **511** (2014) 012088 doi:10.1088/1742-6596/511/1/012088

Time-resolved tuned diode laser absorption spectroscopy of pulsed plasma

P Adámek[1], H T Do[2], M Čada[1*], Z Hubička[1] and R Hippler[2]

[1] Department of low temperature plasma, Institute of Physics of AS CR, v.v.i., Na Slovance 2, 182 21 Prague 8, Czech Republic
[2] Institut für Physik, Ernst-Moritz-Arndt-Universität Greifswald, Felix-Hausdorff-Str. 6, 17489 Greifswald, Germany

E-Mail: cada@fzu.cz

Abstract A novel method for time-resolved tuned diode laser absorption spectroscopy has been developed. In this paper, we describe in detail developed electronic module that controls time-resolution of laser absorption spectroscopy system. The TTL signal triggering plasma pulse is used for generation of two signals: the first one triggers the fine tuning of laser wavelength and second one controls time-defined signal sampling from absorption detector. The described method and electronic system enable us to investigate temporal evolution of sputtered particles in technological low-temperature plasma systems. The pulsed DC planar magnetron sputtering system has been used to verify this method. The 2" in diameter titanium target was sputtered in pure argon atmosphere. The working pressure was held at 2 Pa. All the experiments were carried out for pulse ON time fixed at 100 μs. When changing OFF time the discharge has operated between High Power Impulse Magnetron Sputtering regime and pulsed DC magnetron regime. The effect of duty cycle variation results in decrease of titanium atom density during ON time while length of OFF time elongates. We believe that observed effect is connected with higher degree of ionization of sputtered particles. As previously reported by Bohlmark et al., the measured optical emission spectra in HiPIMS systems were dominated by emission from titanium ions [1].

1 Introduction

The laser absorption spectroscopy (TDLAS) is a versatile method for analysis of particle density in liquids, gases or low-temperature plasma [2-5]. Especially in sputtering systems, the estimation of sputtered particles density in chamber volume is the crucial parameter for optimisation of deposition process. The LAS method is typically used for determination of ground state concentration of sputtered particles and their kinetic temperature from Doppler broadening of absorption line [6]. The laser beam traversing the magnetron plasma volume is absorbed by e.g. metallic atoms in the ground state at relevant resonant frequency. Consequently, the excitation of these atoms cause a drop in intensity of the passing laser beam. The sweeping of laser frequency enables to observe a absorption curve and as a result to calculate the particle density and temperature of studied atoms along the laser path in the plasma volume [6].

*To whom any correspondence should be addressed.

During the last two decades, the pulsed DC magnetron sputtering has been becoming a very promising method for deposition of thin films with improved structure, stress, composition and adhesion. For better understanding how the plasma parameters influence the thin film growth, the extensive studies employing various time-resolved plasma diagnostics have been carried out [7]. The TDLAS method has been used for time-averaged measurements of Al sputtered particles density in pulsed DC magnetron discharge operates in non-reactive mode and in reactive mode[6]. Results revealed that the measured time-averaged Al density in pulsed DC magnetron discharge is almost by one order of magnitude higher than in DC magnetron for the same mean power delivered into the plasma.

The aim of the research presented in this paper is to develop a technique which will be able to provide the time-resolved measurement of the sputtered particles density and their temperature with help of TDLAS method.

2 Experimental set-up

A schematic configuration of low-pressure planar magnetron sputtering system used for our experimental work can be seen in figure 1. The UHV rig was continuously pumped by a turbo-molecular pump backed by combination of Roots and rotary vane pumps. After achieving a base pressure of < 10^{-5} Pa, the chamber was backfilled with argon to an operating pressure of 2 Pa. The gas flow rate was set on 30 sccm of argon. The Lesker Torus planar balanced magnetron was equipped with 2" in diameter titanium target with typical purity 99.7%. The pulsed DC magnetron discharge was driven using a pulsed DC power supply. The pulsed DC supply comprises an Advanced Energy MDX-1K (U_{max}= 1.5 kV and I_{max}= 1 A) and an in-house made power switch unit based on charging of big capacitors during the idle part of the pulses. The power switch unit was capable of drawing up to

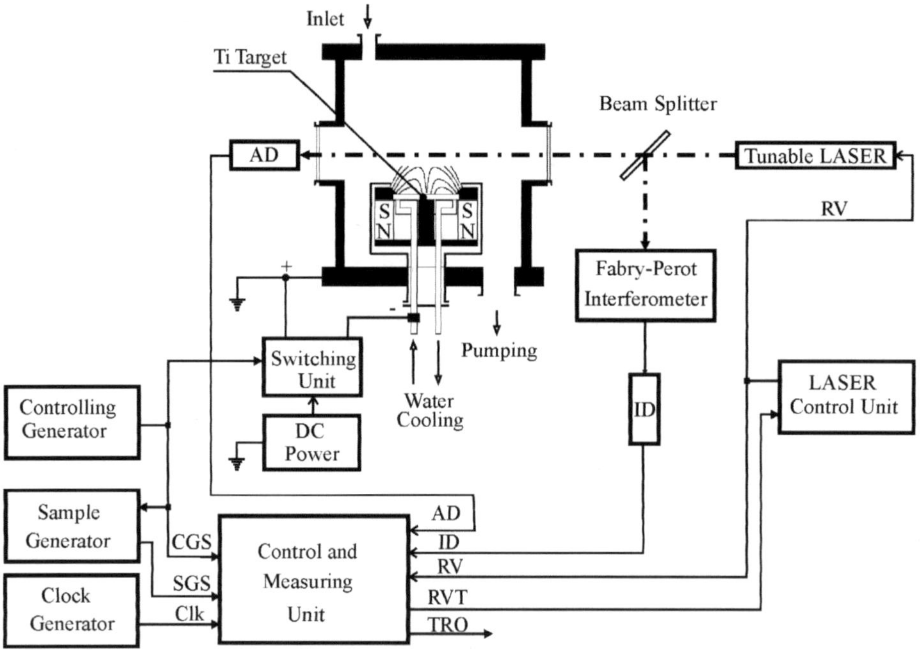

Figure 1. Experimental arrangement.

200 Amps. A ballast resistor of 5.5 Ω incorporated into the power switch unit stabilised the discharge current. Repetition frequency of pulsed discharge was held at 250 Hz (duty cycle 50%) and 500 Hz with the length of the active plasma pulse time 100 μs (duty cycle is then 5%). The mean current was set on I_D = 250 mA.

The TDLAS system used in our experiments is commercially available and consists of: Toptica™ DL100 single mode extended cavity diode laser with optical feedback into the laser diode from the first order of spectrally selective grating, Fabry-Perot etalon with free spectral range 1 GHz for frequency calibration of absorption signal, the laser control unit allowing to adjust the laser wavelength coarsely by temperature changing and finely by changing of current injected into laser diode. The sweeping of the laser wavelength was implemented by means of a piezoelectric element moving with the optical grating in the laser set-up to scan laser frequency over 10 GHz without mode hoping. In this paper, we have studied argon metastable atoms with absorption wavelength 811.53 nm and titanium atoms with absorption wavelength 398.29 nm both excited from ground state. The laser beam was traversed the processing chamber on path length 20 cm and at distance of 70 mm below the target face.

2.1 Control and measuring unit
In figure 1, the block diagram of prototype of control and measuring unit (CMU) is depicted. The CMU is the essential part of the diagnostics system and performs the time-resolved measurements.

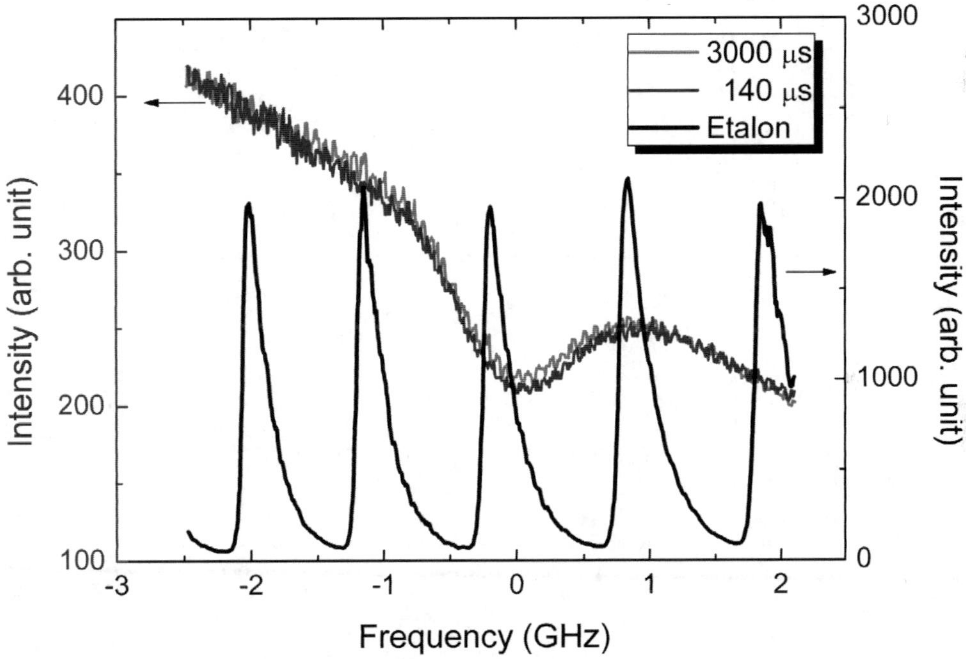

Figure 2. Typical measured time-resolved absorption peaks of Ar^* with Fabry-Perot signal. Processing gas pressure 2 Pa, pulsing frequency 250 Hz, duty cycle 50%

Hardware of the control and measuring unit is driven by Clk signal of clock generator. A triggering of the commercial laser control unit Toptica® DL100 generates the time defined ramp up and ramp down of controlling voltage for tunable diode laser. The triggering signal RVT for laser control unit is derived from the signal CGS of controlling generator and signal SGS of sample generator. A special software running at the Personal Computer that support fast hardware of the unit has been developed. The CMU enables by means of controlling computer to set mode of the fast hardware control circuit, starting and stopping of measuring process and data acquisition from the A/D converters digitising signals from the absorption detector AD, Fabry-Perot interferometer ID and ramp voltage RV. Measured data may be averaged for selected number of repeated measurements and then saved into the text files at the hard disc.

The measured data transferred into the PC are evaluated by special programs for further analyzation and presentation. The detected absorption peak by detector AD, the wavelength calibrating signal from the Fabry-Perot interferometer ID and the sweeping ramp voltage RV of the laser control unit are displayed on-line during a experiment. The on-line displaying of all the signals manages the visual checking of the correct function of the apparatus. The on-line monitoring of AD, ID and RV signals are following: checking of the laser frequency stability and observation of intensity of noise of the absorption detector signal. Laser frequency hop is visualized by distorted interference peaks of calibrating signal from the Fabry-Perot interferometer detector. The intensity of the absorption detector noise is possible to suppress by averaging of several repeated measurements. The number of repeated measurements is also adjustable by controlling software. The signal TRO is optional for triggering of measurement with other processes.

3 Results and discussion

Typical measured time-resolved absorption spectra of Ar metastable atoms for for time 140 µs and 3000 µs, i.e. within the active plasma pulse and during the plasma off-time are depicted in figure 2.

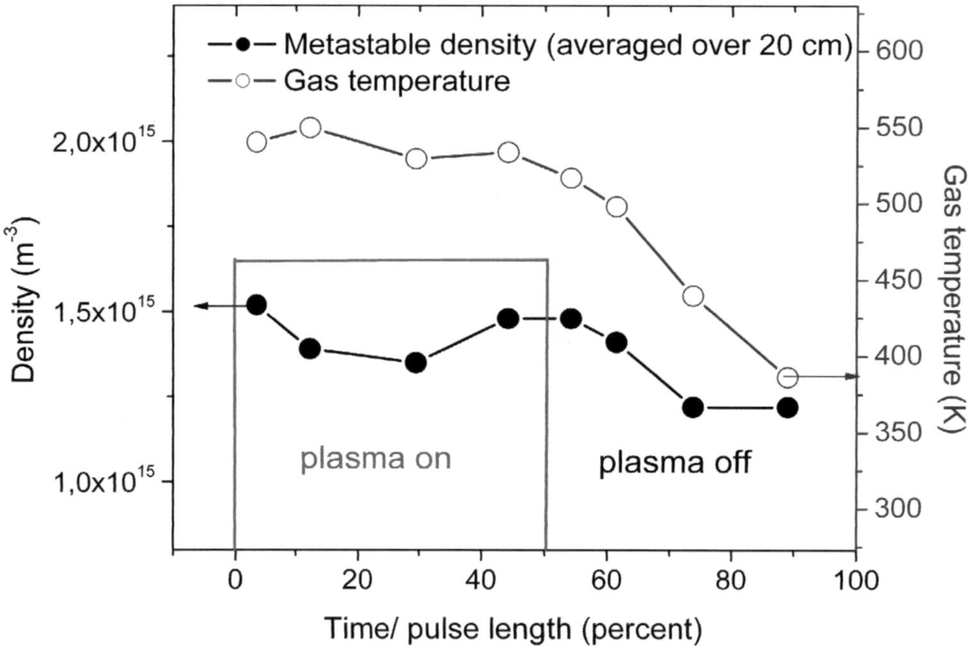

Figure 3. Evaluated time-resolved density and temperature of Ar^* under the same plasma condition as in figure 2.

The magnetron plasma was operated at pulsing frequency 250 Hz and duty cycle was set on 50%. Data demonstrate different absorption peak minimum indicating a change in the Ar^* density. The absorption spectra have been evaluated according to Lambert-Beer law of absorption and atom density and temperature have been numerically calculated according to the method published elsewhere – see [6]. The measured absorption profiles clearly demonstrate the ability of the CMU to acquire time-resolved data of the absorption signal correlated with sweeping frequency of the controlling laser unit.

Evaluated Ar metastable density during the plasma period is plot on a graph in figure 3. It can be seen that Ar^* is roughly 1.5 10^{15} m^{-3} during the active plasma pulse. The time evolution of Ar metastable density demonstrates the decrease in density during the first half of the active plasma pulse followed by the gradual increase to the initial values as at the plasma pulse beginning. During the off-time we can observe a almost linear decrease in the Ar^* density followed by the constant value up to the end of the plasma period. We can assume that at the beginning of the off-time the number of Ar metastable is depleted by collisions in relatively dense plasma and as the plasma density decreases the long-life Ar metastable remain in the gas.

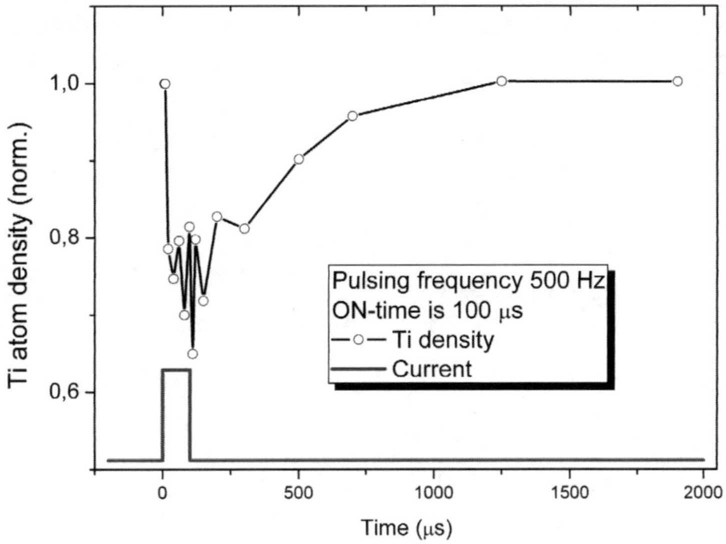

Figure 4. The time-resolved relative Ti atom density for working gas pressure 2 Pa, pulsing frequency 500 Hz and duty cycle 5%.

The Ar metastable temperature reaches 550 K during the active plasma pulse and remains practically constant during the on-time. When plasma decays, we observed the gradual decrease in the temperature almost up to room temperature. The measured time-resolved data indicate enhancement of Ar^* density and temperature when magnetron discharge operates in pulsed dc mode. It can results in higher plasma activity to produce more complex thin film.

The normalised time-resolved Ti atom density is depicted in figure 4. Because of a high level of noise in the absorption detector signal the evaluation of absorption lines was not completely available. For that reason, we plotted only the relative absorption peak height into the figure 4. When pulsing frequency decreased the signal to noise ratio was more and more smaller (not show in this paper) as the Ti atom density during the active plasma pulse decreased due to Ti atom ionisation by very dense HiPIMS plasma [1]. Nevertheless, time-resolved results clearly demonstrate a rapid drop in the Ti atom density as the plasma pulse is developing. While the data measured during the off-time indicate gradual increase in the Ti atom density up to values measured during the early stage of the active plasma pulse. We assume during the plasma afterglow the Ti ions recombine and by diffusion process travel to the chamber walls where they are deposited. We believe that our time-resolved measurements prove the massive ionisation of sputtered metallic particles when high power impulse magnetron sputtering system is used. Furthermore, the Ti ions can dwell in the chamber volume for the relatively long time after the plasma pulse was switched off and can still bombard the substrate. The relative Ti atom density enable us to estimate ion to neutral density ratio as far as we assume the sputtered Ti atom density is measured at the end of plasma period. Then up to 30% of the sputtered Ti atoms is ionised at the distance 70 mm from the target face during the plasma pulse. The previous study demonstrated by means of OES number of ionised sputtered particles up to 90% in vicinity of the target [1]. Therefore our measurements seems to be realistic.

Journal of Physics: Conference Series **511** (2014) 012088 doi:10.1088/1742-6596/511/1/012088

4 Conclusions

Herein presented time-resolved measurement of Ar metastable density, temperature and Ti atom density in pulsed dc magnetron sputtering system confirmed a possibility of using the developed measuring system for time-resolved laser absorption spectroscopy. This method is applicable for diagnostics of discharges working in the pulsed dc mode with ability to observe time evolution of concentration and temperature of neutrals and many other kinds of particles. Actual disadvantage lies in limited frequency of laser sweeping.

Acknowledgements

This work was supported by projects KAN301370701, KJB100100805 of the ASCR, by project 1M06002 of the Ministry of Education, Youth, and Sports of the Czech Republic and by grant 202/09/P159 of the Czech Science Foundation.

References

[1] Bohlmark J, Alami J, Christou Ch, Ehiasarian A P, Helmersson U 2005 *J. Vac. Sci. Technol.* **A23** 18.

[2] Wang W, Fejer M M, Hammond R H, Beasley M R, Ahn C H, Bortz M L, Day T 1996 *App. Phys. Lett.* **68** 729.

[3] Scheibner H, Franke St, Solyman S, Behnke J F, Wilke C, Dinklage A 2002 *Rewiew Sci. Inst.* **73** 378.

[4] Röpcke J, Davies P B, Käning M and Lavrov B F 2001 *Low Temperature Plasma Physics* ed. Hippler et al (Berlin: Viley-VCH) p 173.

[5] Wolter M, Do H T, Steffen H and Hippler R 2005 *J. Phys. D: Appl. Phys.* **38** 2390-2395.

[6] Olejníček J, Do H T, Hubička Z, Hippler R, Jastrabík L 2006 *Jap. J. Appl. Phys.* **45** 8090.

[7] Bradley J W, Welzel T 2009 *J. Phys. D: Appl. Phys.* **42** 093001.

Antecedents and perspectives on the development of nuclear energy in Chile

Gonzalo Gutiérrez

President of the Chilean Nuclear Energy Commission, 2009-2010,
Departamento de Física, Facultad de Ciencias, Universidad de Chile,
Casilla 653, Santiago, Chile

E-mail: gonzalo@fisica.ciencias.uchile.cl

Abstract. A review of activities and future prospects of nuclear energy in Chile is presented. It shows that the development of this sector is closely linked to the history of the Chilean Nuclear Energy Commission, CCHEN, established in 1965. We can distinguish three stages in its development: the installation, culminating at the beginning of 1974 with the La Reina experimental nuclear reactor went critical. Then a phase of consolidation, which reduces the activity of scientific research and stresses the application, between the years 1974 and the end of the eighties, and finally the last twenty years, a stage of continuity marked by a loss of importance and leadership of the institution at the national level, despite having important specific initiatives in the 2000s, such as the successful development of fuel elements, the installation of a cyclotron that allows the execution of PET diagnosis for the first time in the country and the consolidation of a research laboratory on plasma and nuclear fusion. At present there is a consensus that nuclear sector in Chile needs to be placed in accordance with the development experienced by the country, and therefore requires a profound restructuring of the CCHEN, reviewing its vision, mission and work program. Here is a proposal in this regard.

1. Introduction

Chile learned of nuclear energy in a relatively early stage, when the Rector of the University of Chile, Juan Gomez Millas, created the Nuclear Physics Laboratory in 1954. This entity was the prelude to the Chilean Nuclear Energy Commission, (Comisión Chilena de Energía Nuclear, CCHEN, in Spanish) founded in 1965 with the aim to "address the problems associated with the production, acquisition, transfer, transport and pacific use of the atomic energy and fertile, fissile and radioactive materials". Chile, like most Latin American countries did not want to be left out of the tremendous possibilities offered by this new form of energy, which was revealed from the boost given for peaceful uses by various international programs, such as "Atoms for peace" in the mid-1950s, driven by the U.S.

At the foundation of CCHEN there was a clear view of the need for a state agency that was concerned of nuclear issues in the broadest sense. The discussions of the time shows that it was perfectly understood that nuclear issues were strategic for the country's development. This refers not only to have an informed opinion on its geopolitical and defense aspects, but especially on the issue of energy sources and science and technology of nuclear applications in health, mining, agriculture, construction, to name a few. "How can we put this enormous energy released by the atom to serve the country?" was the question asked in 1955 by Dr. Eduardo Cruz Coke, well-known scientist, congressman and the first CCHEN´s President. And he concluded that certainly because the diversity of subjects, the degree of specialization, the urgent need for safe operation and their strategic importance for the country, nuclear issues could not be in the hands of a private organization, nor in a purely academic institute. It had to be a specialized agency of the state who took that responsibility on its shoulders.

In this article we briefly review the development of the nuclear sector in Chile, and then describe the main topics that underpin the strategic plan designed for the near future.

2. The Early Years: 1965-1973

The initial design of CCHEN envisaged a vigorous scientific and technological development plan that included the installation of a nuclear reactor and the study of the application of nuclear techniques in health, agriculture and industry, as well as its possible use as a source for electricity production, that is, nuclear power. Interesting groups were formed for research in close link with the universities. Along with this, the formation of expert researchers, engineers and technicians was encouraged: nuclear physicists and engineers, radio-chemists, physician, lawyers and so on. During these early years, these people began to produce radioisotopes and radiopharmaceuticals (1968), to investigate the applications of irradiation (1969), and the effects of radiation on health and environment, addressing topics such as radiation safety. The first applications to industry and environment started with the development of tracers (1969).

In parallel, it began the research related to the fuel cycle, such as working in geology (1969) and establishing a metallurgical laboratory (1969). A highlight of this stage is the acquisition, installation and the critical set up of the experimental reactor La Reina (1969-1974), located in east side of Santiago. During this time several agreements were also signed with a number of national and international institutions for training specialists, and scientific cooperation. Alongside this, that we could call the "exploitation" of nuclear energy, the safety and security activities were also developed, that is, the tasks of supervising and controlling nuclear activities from the standpoint of the physical security of facilities, processes and staff.

This first phase of installation was interrupted by the Coup d'Etat of September 11th, 1973 that overthrew the constitutional president Salvador Allende. Several scientists and technicians had to emigrate, which caused the dismantling of laboratories and cessation of several activities. The institution failed to reach scientific and technological leadership and could not articulate it to national development [1].

3. Application Stage: 1974-1990

Starting in 1974 there was a shift in the activities of the CCHEN. The military took control of it, resulting in important administrative and operational changes, as the annexation of the Nuclear Studies Center of the Army [2], with its facilities in Lo Aguirre and its research reactor, opened in 1978, but that has never been used on a regular basis. During this stage, the applications were mainly boosted, leaving low scientific research and innovative development. In spite of this fact, it is interesting to note that during this period several of the activities planned in the early years were consolidated: since 1975 the routine production of radiopharmaceuticals and radioisotopes (1975) and tracer for mining (1976) and industry (1984) began; a multipurpose experimental irradiation plant (1978-79) was built, and applications in agriculture (1981) were implemented. On the other hand, there was significant progress on the fuel cycle, including the uranium prospecting of a part of the country (1975-85), and the development of hydrometallurgy, obtaining yellow cake in an experimental basis. In addition to the efforts to produce fuel for research reactors, the operation of the fuel elements plant continued. As before, the tasks of regulation, supervision and control went on and advanced. Also, a Section of production and services began to operate, providing isotopes, instrumentation and chemical analysis to public and private sectors.

The CCHEN administrative structure was also consolidated. A Department for outreach and training, as well as a Section of international relations was established. During this time CCHEN worked significantly in the drafting of laws, decrees and regulations on nuclear activity. This was crowned in 1983, when a Supreme Decree that approved policies and objectives for nuclear development, helped to consolidate the progress in respect of applications, and order the legislation concerning the regulatory aspect was issued. Also, various aspects of nuclear materials, including mining were organized and regulated.

An important aspect developed during these years, which concentrated great efforts, was the studies on the feasibility of having nuclear power in the country: these studies date back to 1969, then 1975-76 and 1983. In all cases, they were dismissed for economic reasons (see [1] for details).

4. The Chilean Nuclear Energy Commission in democracy

Since the beginning of 1990`s, with the return of democracy, the discussion about the role of CCHEN in this new scenario started, but it was not easy at all. During the first years of democracy, there were issues that caused some tension in the Board of Directors, which fortunately, was surpassed later [3]. However, the results of these debates were reflected in the development of a nuclear policy, enacted by Supreme Decree in1995 under the title "National Nuclear Development Plan". However, this plan failed to bring a new CCHEN, according to the challenges of the new era that opened in the country. Somehow, there was a policy of continuity, and in fact, there was a continuation of the same activities of previous years, with few exceptions, such as the installation of a cyclotron (2002) that enabled the development of PET technology in the area of health; the international certification of the fuel elements plant (2007), and the consolidation of the Laboratory of plasma and nuclear fusion (2000), which returns the scientific research on nuclear issues to the CCHEN.

In recent years, the leadership in Latin American and worldwide of the Laboratory of plasma physics and nuclear fusion (which now become in the Thermonuclear Plasma Department) on various subjects, particularly in the production of dense plasmas in small devices (2000-2009) has stood out. There is also progress on the use of nuclear techniques. Among the main milestones we can mention the environmental application like gases and wastewater treatment (1993); chemical metrology (1993); measurement of trace metals in sea water and aerosols; operation of a tissue bank (2000); putting into operation the SPEED 2, donated by the University of Dusseldorf [4]; landmine detector, (2008); remote monitoring, (2009). With respect to fuel cycle and metallurgy, it stands out the development of vacuum sintering technology (1990-93), the conversion of UF6 to metal U (2003-2010) and the production of fuel elements for the RECH-1 (1994 - 2000), an initiative that is crowned by the international certification of the plant gained in Peten, Netherlands (2007).

However, despite these achievements, several years ago there was a feeling that CCHEN was without any clear goal: the country had changed, and there was not a vision or mission of CCHEN in accordance with the needs and demands of our society and state in the perspective of the new millennium. These considerations triggered, some years ago, a deep analysis of the situation not only of CCHEN, but also of the nuclear sector in Chile, in order to rethink it and study the decision that should be taken to bring it up to the current times.

5. The Chilean nuclear sector and its perspective for the future

Status review of the nuclear sector was motivated by several factors, one of which was especially important in the recent past years: the growing energy demand in the country. Chile suffered a severe energetic crisis in 2006-2007, with the lack of natural gas coming from Argentina, what put into sharp focus the discussion on the need to develop a national energy plan which considers diversifying our energy matrix and making it compatible with the protection of the environment. That led to seriously wonder about the possibility of incorporating nuclear power, which in turn led to look how we are in that area. And we found that we were actually in a very wrong foot: nuclear energy specialists are no more than 15 people, several of them close to retirement. The discipline is only taught at one university, the University of Chile, but with one facility, once of first order, now operating at minimum capacity. To make matters worse, CCHEN, the state institution dedicated to these activities, was mired in a morass, with clear signs of stagnation and demoralization.

Indeed, in year 2004 when the energy crisis in Chile due to the problems of gas supply from Argentina began to be glimpsed, the need to diversify the energy matrix immediately raised. This led President Lagos to say that "we also need to think about nuclear energy". After that the crisis became more severe, compounded by drought that hurt hydroelectric plants. Given this scenario, President Bachelet designed a robust energy sector plan that would address the demand the country needed smoothly, without improvisations. The main elements of that policy were embodied in the document "Energy Policy, new guidelines," published in 2008 by the Minister President of the Energy Commission, Mr. Marcelo Tokman [5]. There expressed the urgent need to diversify the energy matrix, begin an intensive energy efficiency program and build a new institutional sector, whose starting point was the creation of the Ministry of Energy.

Consistent with this, in 2007 a Working Group on Nuclear Power, led by Mr. Jorge Zanelli was formed. This group issued a report which set out three main conclusions to evaluate the incorporation of nuclear power in the country: 1) Chile must keep all energy options open. Nuclear power is not an option to discard and could cooperate in the security of electricity supply, 2) nuclear energy is reliable at the levels of security that its industry has reached, but it require permanent concern, discipline and rigor, and 3) nuclear power is a potentially competitive option, especially because of the current fossil fuel prices in international markets [6]. The Zanelli Working Group recommended to advance the discussion of the problem, and to do several studies about institutional issues, economic factibility, environmental, health and territorial aspects, as well as human resources and public opinion, among others.

Then, being prepared to incorporate any energy option, regardless of the final decision, requires to be up to date with the requirements of such an enterprise and therefore, to stress all stakeholders. It is important to point out that the current stage in the country (year 2010) is not to incorporate nuclear power, but to prepare the way for it. The final decision, that is, whether the nuclear-power is incorporated or not to our electrical grid is a decision to be taken after a public debate at a national level. However, actions taken now will be beneficial for the country, regardless of the decision taken. Indeed, part of these actions is the creation of a nuclear regulatory institution (like NRC of the USA), to update all relevant legislation, to modernize and regulate the electricity sector, to allow the incorporation of other generators (not just nuclear but also wind, solar, etc.), to perform geological and seismic studies, and to prepare adequate staff, among other tasks [7].

6. Restructuration of the Nuclear Sector

This, then, is the context and the framework of the analysis and projections of the country's nuclear industry and in particular the future design of the Chilean Nuclear Energy Commission, CCHEN. More than four years ago we became aware of the need to develop a comprehensive strategic planning. In order to do that, we reviewed and made a diagnosis with the aid of specialized external consulting companies, conducting assessments, self assessments, focus groups, etc. Importantly, the strategic planning process included the active collaboration of the vast majority of workers of CCHEN, its head staff and the Trade Union, being directed at various stages by the Directive Board of the institution. In July 2010 this analysis, with the diagnoses and recommendations, was presented by the author, as President of the Directive Board, to the Minister of Energy Mr. Ricardo Raineri. Below we point out the highlights of this plan.

The diagnosis showed several interesting findings. The main problems can be grouped into strategic / organizational and cultural issues. We will present each of them.

A) Strategic / organizational problems: the central point was the lack of clear mission according to the needs and demands of society; it was found that it was a multifaceted and complex institution, but one that has lost focus. For example, CCHEN performs public service and scientific research activities at the same time; it is a regulatory and supervisory agency, and it also works in production activities (radioisotopes for medicine for instance), dissemination and training on nuclear techniques, and also

Journal of Physics: Conference Series **511** (2014) 012089 doi:10.1088/1742-6596/511/1/012089

performs some functions of great social impact, such as health applications, which certainly cannot only be assessed by economic profitability. Regarding its internal structure, it is clear that neither organic structure nor its workers framework nor the wages respond to current needs and have no correlation with the country's progress. On the other hand, there is a disconnection with innovation policy and competitiveness that the country leads. Also, there is limited relationship with other agencies in the country, including universities and institutes.

B) Cultural problems: these account for a culture of individual work, hierarchical, and compartmentalized. There is a lack of incentives for creativity and collaboration, both within the CCHEN as well as with the world outside. Much of the activities are performed routinely; important part of the workers is demotivated and demoralized. There is little human capital management, and renewal of professionals is very difficult. Other aspects are the lack of knowledge management and weak internal CCHEN positioning to the national authority and public opinion.

Having an institution in these conditions could certainly mean not only the loss of opportunities, but also unacceptable hazards that could lead to relaxation in supervisory or production activities, with the possible risk of nuclear or radiological incidents. It also means economic waste when having duplication and inefficiency, loss of investment and knowledge gained in research and development over the past 40 years, and weakening of the link to the IAEA and international relations.

However, the analysis also concluded that if appropriate measures are taken on time, the above situation can be resolved and the institution could become a model for other similar public bodies, not only in Chile but perhaps it could also serve to similar countries in Latin America. These decisions require a profound change, both internal and external, of the institution. Regarding the domestic level, it is clearly a need to consider new targets, which implies an institutional redesign. As for the outer level, new laws and decrees are required, beginning with the early establishment of a nuclear regulatory institution. Currently, this function is fulfilled by CCHEN, but obviously a good practice always advises to separate the regulatory body from the operator body (Here there are several schemes that can be taken and it is instructive to dwell on international experience). Another measure needed is a new law for workers' structure, for remuneration, and a redesign of the public-private relationship. Only a renewed institution can fully meet its obligations in a society which is certainly very different from the one of 1965, when it was founded. All these measures mean a complete restructuring of the nuclear sector.

Thus, the **vision** of CCHEN we need consists of the following points:

-to be the reference body of the state in nuclear issues;

-to be a producer of goods and services needed for the country.

-to be a leading institute in nuclear science, technology and radiological issues, at national and international level, whose results, products and services impact positively on the economy and living conditions of the population.

What can be gathered from the above statements is that the **mission** of the new CCHEN

- collects information and is aware of the world state of art in nuclear issues.
- conducts research and development in the nuclear and radiological service in the country.
- acts as a national laboratory in the country by making its unique facilities, specialized professional and technical support available. Trains researchers, professionals, technicians and students.
- produces goods as well as nuclear and radiological services essential for our country and its people.
- associates with other institutions both from the public and private sector and international organizations.

All these activities performed with high standards of quality and safety.

The previous approaches need to be implemented. We propose to do so on the basis of three main areas: I) the work programs, II) the safety culture, and III) a new institutional design. The development of each of these areas should be in parallel and each at its own pace. We will give a brief explanation of each of them.

I) Work programs: this axis determines the daily work of the institution. It aims to develop different aspects: a) radiation for life, industry and environment, b) material science, c) nuclear energy plasmas, and d) the nuclear regulatory body (its creation must be the subject of a special law and its operation will be outside CCHEN, so we will not detail their function here. Let us just say that its mission is to supervise, control and license nuclear facilities, to ensure their safety in the operation.) These work programs were chosen because they represent the most necessary for the future, and also reflects the best practices carried out in the CCHEN during its 40 years of life. Below we will briefly explain the contents of each of them.

a) Radiation for life, industry and environment: its main focus is on enhancing the development of radioisotopes according to their relevance and usefulness. This is, we plan to develop those products that are necessary for the country and that cannot be provided by other agencies (private, etc.) because the complexity of its production. Similarly, once the techniques are mature, the idea is to transfer them to other institutions for their production, and thus the CCHEN is always in the forefront of knowledge. This area is also raised to develop new projects, one of which is to produce fission Mo-99 at industrial scale. This radioisotope decays into technetium-99, which is the most widely used radionuclide in nuclear medicine. The other new project is to boost the creation of a National Laboratory of Molecular Imaging to develop preclinical research in biomedicine and related areas. This is an ambitious but achievable goal. As in the beginning CCHEN pioneered the development of nuclear medicine in Chile, now it corresponds to step forward and to be pioneers in preclinical research with cutting edge techniques.

b) Material Science: this is a key area for the development of the country, both for mining and industry as well as for construction and defense. The CCHEN can play a fundamental role in research, in conjunction with universities, because it has unique facilities and laboratories, as the reactor itself, which can and must be available for the country to study and characterize materials. This area will be developed in three aspects. One is the fuel cycle, from uranium mining to the production of fuel elements for research reactors. This, somehow, has been done, with certain stages more developed than others. What is needed now is to further systematize and professionalize this activity in all its dimensions. Another aspect is the study of advanced materials. Currently, there are groups working on this issue, at analytical chemistry and diffraction laboratories. The challenge is to interact with universities and institutes, both in Chile and abroad, so that the research carried out is in accordance with the needs of the country and its people. The last aspect to be addressed in this area is the creation of an Institute of Lithium. Indeed, lithium is a material of strategic importance not only for its potential future use as fuel in fusion processes, but for its increasing use in electric batteries (Li-Ion and others). It is anticipated that due to the depletion of fossil fuels, the use of hybrid and electric cars is increasing, and therefore the battery can be considered from this viewpoint, a powerhouse. Chile is the world's largest producer and also has the major reservations, but paradoxically, has almost no studies on this material. However, there is an agreement with CORFO that the money from the royalty of lithium should be used to fund a scientific research institute on lithium technology. CCHEN is the place to start these activities.

c) Nuclear energy and plasma physics: this is the most typical activity of CCHEN. There are three aspects involved in the development of this area. First, to support the nation in the necessary studies to make an informed decision on nuclear-power option and to have the necessary preparation if the

Journal of Physics: Conference Series **511** (2014) 012089 doi:10.1088/1742-6596/511/1/012089

decision is positive. This means, exercising the role of CCHEN as a study group and government adviser on nuclear issues at the highest level. A second aspect is related to foster the development of portable non-radioactive sources for use in applications (exploration and detection of landmines), which until now has been successful in CCHEN, together with military institutions and the ministry of defense. Finally, an area where CCHEN has leadership, not only in Latin America but worldwide, is nuclear fusion research. In fact, devices using small and medium-sized plasmas have been successfully developed at CCHEN. Here the objective is to strengthen this research even further, and study fusion-fission hybrid reactor (plasma focus as seed neutron fission) and pulsed power to generate shock waves, with obvious applications in industry and mining. Of course, we see all this activity in the framework of international collaborations, particularly we want to promote a Latin American program of nuclear fusion.

II) The second axis of development is the safety culture. This is based on the requirement that none of the nuclear industry can be developed without being permeated by a highest standard safety culture, which is well defined by the IAEA. This is reinforced by a disturbing internal report of the Head of the Section on Quality, 2007, which stated that CCHEN has not yet reached a reliable safety culture. Fortunately, in its 45 years of life CCHEN has not had any accident or incident involving security. This theme will work as recommended by the IAEA, and includes joint efforts by all stakeholders: political commitment at the structural level, the authorities of CCHEN, structural commitment of operator behavior, and commitment of each official.

III) Finally, the third axis is related to the need of an institutional design to realize the necessary restructuring of the nuclear sector. First, we must establish policies and criteria of the new public-private relationship of the institution as well as basis to discuss how to address the issue of nuclear and strategic materials that depend on CCHEN, like uranium, thorium and lithium. It is a question of making decisions about granting, operation, taxation and other aspects of these minerals, always having these public goods as a precautionary, and ensure the benefit of the state. A second aspect has to do with the new legislation. We need to pass a new law, of which there is an initial draft, on nuclear and radiation safety. Similarly, a law is needed to create the regulatory body and a new law for workers` structure and remunerations. It is also necessary to conduct a detailed human resources plan and establish a new management model.

So, these are the central ideas and approaches to which the nuclear sector in Chile should go the coming years.

7. Conclusion

Since 1965, CCHEN has made immense contributions to the country. Just to name one: there is virtual unanimity among scholars that the discipline of nuclear medicine, with all the benefits it brings to the population is possible thanks to the CCHEN and its workers. However, despite the many important contributions, it is necessary to restructure the nuclear sector precisely to allow to continue at the forefront of knowledge, as it did in its early years. The situation today is different: we live in a context of depletion of fossil fuels, global warming as result of increased greenhouse gases and the certainty that we are in an environment where natural resources are finite, and therefore we must take care of them. This has brought the issue of energy sources to the center of public debate. In addition, fuels we use in Chile are all imported, except hydropower, and we have a growing demand of energy product of our route to economic development. This is the framework to discuss the incorporation of new sources to our energy matrix. The nuclear power is an option. Whatever decision we make as a society to be informed and prepared for it is a must. That is why I have posed the challenge, which is explained in this article, to restructure the nuclear sector to be according to the XXI century. There is a lot much that CCHEN can give to the country, apart from a possible nuclear power: radioisotopes supplies for medicine, food immunization, research of new materials, sources for material characterization with increasing use in industry and mining, just to name a few. This is, in Chile at least, the use of nuclear energy for peaceful purposes has many pages to be written.

8. Acknowledgements

The ideas raised here are the result of several conversations, discussions and information exchange with many people. In particular, I am very grateful to Jorge Zanelli from CECS, Luis E. Pérez, Gonzalo Torres, Leopoldo Soto and Marcelo Zambra from CCHEN, Juan José Rivas from Ministerio de Energía, and especially to the Directive Board of Chilean Nuclear Energy Commission 2009-2010.

9. References

[1] A. Cubillos, Desarrollo Nuclear en Chile: un analisis comparado con Argentina y Brasil 1964-2008, Ph.D. Thesis in American studies, Universidad de Santiago, (2009).

[2] See the Law decrees DL 1507 (1974) and DL 747 (1976). An interesting account of the military participation in the chilean nuclear development, see A. Cubillos, El Ejército y el desarrollo nuclear nacional, AHM (2004).

[3] Interestingly, CCHEN is managed by a Directive Board and a CEO. The Directive Board consists of seven members: one representative of the President of Chile, acting as President of the Board, and representatives of the Minister of Energy, Minister of Health, the Council of University Rectors, as well representatives of the Commanders in Chief of the Army, Navy and Air Force.

[4] The SPEED 2 is the most powerful generator for experiments in dense transients plasmas in the Southern Hemisphere and the only one in operation in Chile that produces currents of mega-ampere.

[5] M. Tokman, Política energética. Nuevos lineamientos, edited by Comisión Nacional de Energía, Chile (2008).

[6] These findings should certainly be seen now in the perspective of what happened in Japan with the Fukushima nuclear power plant after the earthquake tsunami suffered there in March 2011.

[7] See more details in M. Tokman, Núcleo-electricidad en Chile: posibilidades, brechas y desafíos, edited by Ministerio de Energía, (2010).

AUTHOR INDEX

AUTHOR INDEX

AUTHOR INDEX